Hazardous Materials

Managing the Incident

REVISED FOURTH EDITION

Gregory G. Noll *and* Michael S. Hildebrand
Contributions By: Glen Rudner *and* Rob Schnepp

Jones & Bartlett Learning
World Headquarters
5 Wall Street
Burlington, MA 01803
978-443-5000
info@jblearning.com
www.jblearning.com

Jones & Bartlett Learning books and products are available through most bookstores and online booksellers. To contact Jones & Bartlett Learning directly, call 800-832-0034, fax 978-443-8000, or visit our website, www.jblearning.com.

Substantial discounts on bulk quantities of Jones & Bartlett Learning publications are available to corporations, professional associations, and other qualified organizations. For details and specific discount information, contact the special sales department at Jones & Bartlett Learning via the above contact information or send an email to specialsales@jblearning.com.

Hazardous materials emergency response work is extremely dangerous and many emergency responders have died or sustained serious injury and illness while attempting to mitigate an incident. There is no possible way that this text can cover the full spectrum of problems and contingencies for dealing with every type of emergency incident. The user is warned to exercise all necessary cautions when dealing with hazardous materials. Always use a risk-based decision making process and place personal safety first.

It is the intent of the authors that this text be a part of the user's formal training in the management of hazardous materials emergencies. Even though this book is based on commonly used practices, references, laws, regulations, and consensus standards, it is not meant to set a standard of operations for any emergency response organization. The users are directed to develop their own written Standard Operating Procedures, which follow all system, agency, or employer guidelines for handling hazardous materials. It is the user's sole responsibility to stay up to date with procedures, regulations, and product developments that may improve personal health and safety.

18837-0

Production Credits

General Manager and Executive Publisher: Kimberly Brophy
Executive Editor: Bill Larkin
Senior Development Editor: Janet Morris
Production Manager: Tina Chen
Senior Marketing Manager: Brian Rooney
V.P., Manufacturing and Inventory Control: Therese Connell
Composition: S4Carlisle Publishing Services
Cover Design: Michael O'Donnell
Rights & Media Specialist: Robert Boder
Cover Image: © Courtesy of U.S. Coast Guard (top left) and Rob Schnepp (bottom left and right)
Printing and Binding: LSC Communications
Cover Printing: LSC Communications

To order this product, use ISBN: 978-1-284-18834-9

Library of Congress Cataloging-in-Publication Data
Unavailable at time of printing.

Printed in the United States of America
22 21 20 19 18 10 9 8 7 6 5 4 3 2 1

Brief Contents

Contents

Skill Drills

Resource Preview

Hazardous Materials: Managing the Incident

Chapter Resources

This well-known resource has a successful 30-year history of training fire fighters, law enforcement officers, industrial response specialists, and military special-forces personnel how to safely respond to hazardous materials emergencies. The revised fourth edition complies with all requirements for Hazardous Materials Technicians and Hazardous Materials Incident Commanders as described in the 2018 Edition of NFPA 472, *Standard for Competence of Responders to Hazardous Materials/Weapons of Mass Destruction Incidents* as well as the 2017 Edition of NFPA 1072: *Standard for Hazardous Materials/Weapons of Mass Destruction Emergency Response Personnel Professional Qualifications*. It is supported by a complete suite of student, instructor, and technology resources that will engage students in the learning process and revolutionize how training is delivered.

Hazardous Materials: Managing the Incident, Revised Fourth Edition serves as the core of a highly effective teaching and learning system. Its features reinforce and expand on essential information. These features include:

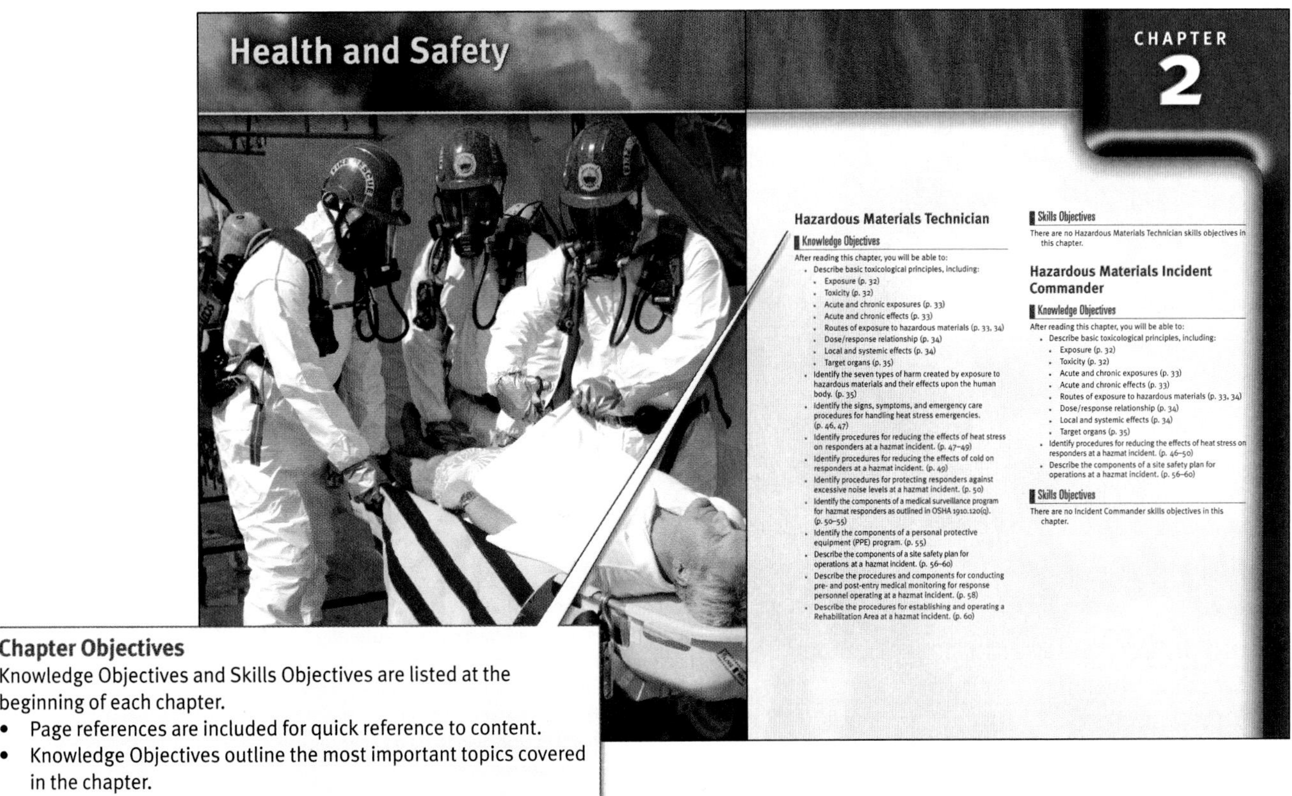

Health and Safety

CHAPTER 2

Hazardous Materials Technician

Knowledge Objectives

After reading this chapter, you will be able to:

- Describe basic toxicological principles, including:
  - Exposure (p. 32)
  - Toxicity (p. 32)
  - Acute and chronic exposures (p. 33)
  - Acute and chronic effects (p. 33)
  - Routes of exposure to hazardous materials (p. 33, 34)
  - Dose/response relationship (p. 34)
  - Local and systemic effects (p. 34)
  - Target organs (p. 35)
- Identify the seven types of harm created by exposure to hazardous materials and their effects upon the human body. (p. 35)
- Identify the signs, symptoms, and emergency care procedures for handling heat stress emergencies. (p. 46, 47)
- Identify procedures for reducing the effects of heat stress on responders at a hazmat incident. (p. 47–49)
- Identify procedures for reducing the effects of cold on responders at a hazmat incident. (p. 49)
- Identify procedures for protecting responders against excessive noise levels at a hazmat incident. (p. 50)
- Identify the components of a medical surveillance program for hazmat responders as outlined in OSHA 1910.120(q). (p. 50–55)
- Identify the components of a personal protective equipment (PPE) program. (p. 55)
- Describe the components of a site safety plan for operations at a hazmat incident. (p. 56–60)
- Describe the procedures and components for conducting pre- and post-entry medical monitoring for response personnel operating at a hazmat incident. (p. 58)
- Describe the procedures for establishing and operating a Rehabilitation Area at a hazmat incident. (p. 60)

Skills Objectives

There are no Hazardous Materials Technician skills objectives in this chapter.

Hazardous Materials Incident Commander

Knowledge Objectives

After reading this chapter, you will be able to:

- Describe basic toxicological principles, including:
  - Exposure (p. 32)
  - Toxicity (p. 32)
  - Acute and chronic exposures (p. 33)
  - Acute and chronic effects (p. 33)
  - Routes of exposure to hazardous materials (p. 33, 34)
  - Dose/response relationship (p. 34)
  - Local and systemic effects (p. 34)
  - Target organs (p. 35)
- Identify procedures for reducing the effects of heat stress on responders at a hazmat incident. (p. 46–50)
- Describe the components of a site safety plan for operations at a hazmat incident. (p. 56–60)

Skills Objectives

There are no Incident Commander skills objectives in this chapter.

Chapter Objectives

Knowledge Objectives and Skills Objectives are listed at the beginning of each chapter.

- Page references are included for quick reference to content.
- Knowledge Objectives outline the most important topics covered in the chapter.
- Skills Objectives map the skill drills provided in the chapter.

You Are the HazMat Responder

Each chapter opens with a case study intended to stimulate classroom discussion, capture students' attention, and provide an overview for the chapter. An additional case study is provided in the end-of-chapter Wrap-Up material.

You Are the HazMat Responder

You are the Health and Safety Coordinator for a chemical manufacturing and distribution facility and have been assigned the task of developing a medical surveillance program for the members of the facility Hazardous Materials Response Team (HMRT). Facility hazards include flammable liquids and gases, corrosive liquids, and poisonous liquids.

The facility HMRT is trained to the "Hazardous Materials Specialist Employee A" level, as defined in NFPA 472, and can provide a Technician-level response capability for the products and risks found within the plant. HMRT members include personnel from the operations, maintenance, safety, and administration departments.

Your management team has asked for your recommendations on several specific issues pertaining to the health and safety of HMRT members.

1. What should be the components of the HMRT medical surveillance program?
2. What criteria should be used for the initial entry medical monitoring?
3. How could the facility coordinate its plant emergency response program with the local hospital?

Introduction

The health and safety of all emergency responders is a critical issue for both management and labor. Regardless of size or nature, every hazmat incident presents responders with a potentially hostile environment. Although preventing exposure to hazardous materials is always a primary concern, command personnel must also evaluate the physical working conditions, work intervals, and the stress of working in personal protective clothing and equipment. Our goal is simple—we want every responder to come home in the same physical, mental, and emotional condition as upon arrival at the incident.

Hazmat incidents are characterized by work environment hazards, which may pose an immediate danger to life and health but which may not be immediately obvious or identifiable. These hazards can vary according to the tasks being performed and a responder's location at the incident site and can change as response activities progress.

Protecting the health and safety of emergency response and support personnel, as well as the general public, must always be the Incident Commander's (IC's) primary concern. In this chapter, we will examine these health and safety concerns in detail.

Toxicology

Toxicology is the study of chemical or physical agents that produce adverse responses in the biological systems with which they interact. Chemical agents include gases, vapors, fumes, dusts, and so on, while physical agents include radiation, hot and cold environments, noise, and so forth.

Toxicity is defined as the ability of a substance to cause injury to a biological tissue. In humans this generally refers to unwanted effects produced when a chemical has reached a sufficient concentration at a particular location within the body.

Responder Safety Tip

To properly protect those personnel operating at a hazmat incident, both Command and Hazmat Group personnel must be able to understand the basic concepts of toxicology, as well as toxicological and exposure terms, and interpret toxicity and exposure data.

Several factors determine the toxicity of a chemical and the potential harm that may result. A simple way of understanding this concept and its potential for harm is the health hazard equation:

Exposure + Toxicity = Health Hazard

1. Exposure means you have had contact with the chemical. Common methods are inhalation, ingestion, skin absorption, or direct contact.

Voices of Experience

As an employee of the Indiana Department of Homeland Security, in the hazmat section, I have responded to many large-scale incidents. One that has particular bearing on the importance of monitoring devices was a large plastics fire in an old heavy-timber building, which produced huge volumes of black, acrid smoke.

Building and Contents Description

The building was of heavy timber construction with a limited number of windows and doors. In the past it had been used by a distillery. As is quite common, the building contents did not conform to occupancy/suppression standards. The building was approximately 100' x 100' and stood six stories tall. Every floor in this building was full of plastic materials and eventually became involved with fire.

Suppression Efforts

With site access severely limited and a lack of suitable windows, there was very little on-scene fire departments could do as far as fire suppression. With the limited access it was quite apparent that this fire was going to burn for a long time. And, with the huge volume of plastics in the building, great amounts of black, acrid smoke was being produced.

Monitoring Requirements

With the wind being very light the smoke carried downwind for about 2 blocks and then started settling, impacting residential, commercial, and high hazard occupancies, including a hospital and a nursing home. Hazard Control Zones had to be determined on a priority basis. The only accurate and defendable way to establish these hazard zones was by using the correct, properly calibrated monitoring device. This monitoring was performed using PIDs, 4-gas, and the appropriate colorimetric tube (formaldehyde, ammonia, and benzene). By using exposure limits referenced in the NIOSH we were able to determine when the hospitals and nursing homes HVAC air intakes had to be shut down. Monitoring was conducted continuously for 2 days. Because the air quality monitoring was performed correctly not one person reported any ill effects from this incident.

Tim Thomas
Indiana Department of Homeland Security
Indianapolis, Indiana

Voices of Experience

In the Voices of Experience essays, veteran hazardous materials specialists share their accounts of memorable incidents while offering advice and encouragement. These essays highlight what it is truly like to be a hazardous materials responder.

Chapter 1 The Hazardous Materials Management System 

Responder Tip

SARA, Title III, is the primary federal legislation that directly affects the local hazardous materials emergency preparedness program. Responders should identify the local agency responsible for the coordination of the LEPC, as well as those organizations that make up the local emergency response community.

- **CAA—The Clean Air Act.** This law establishes requirements for airborne emissions to help protect the environment. The Clean Air Act Amendments of 1990 addressed emergency response and planning issues at certain facilities with processes using highly hazardous chemicals. This included the establishment of a National Chemical Safety and Hazard Investigation Board, EPA's promulgation of 40 CFR Part 68—*Risk Management Programs for Chemical Accidental Release Prevention*, and OSHA's promulgation of 29 CFR 1910.119—*Process Safety Management of Highly Hazardous Chemicals, Explosives and Blasting Agents*. In addition, some facilities are required to make certain information available to the general public regarding the manner in which chemical risks are handled within their facility.

nificant federal regulations that affect hazmat emergency planning and response.

Hazardous Waste Operations and Emergency Response (29 CFR 1910.120). Also known as HAZWOPER, this federal regulation was issued under the authority of SARA, Title I. The regulation was written and is enforced by OSHA in those 23 states and 2 territories with their own OSHA-approved occupational safety and health plans. In the remaining 27 "non-OSHA" states, public sector personnel will be covered by a similar regulation enacted by the EPA (40 CFR Part 311).

The regulation establishes important requirements for both industry and public safety organizations that respond to hazmat or hazardous waste emergencies. This includes firefighters, law enforcement and EMS personnel, hazmat responders, and industrial Emergency Response Team (ERT) members. Requirements cover the following areas:

- Hazmat Emergency Response Plan
- Emergency Response Procedures, including the establishment of an Incident Management System (IMS), the use of a buddy system with back-up personnel, and the establishment of a Safety Officer
- Specific training requirements covering instructors and both initial and refresher training
- Medical Surveillance Programs
- Post-emergency termination procedures

Responder Tips

Responder Tips provide additional advice and information from masters of the trade.

Chapter 2 Health and Safety 39

IDLH oxygen-deficient atmosphere is 19.5% oxygen or lower. IDLH was not originally designed as an exposure level for evaluating protective actions. However, EPA has noted in the *Technical Guidance for Hazard Analysis* that using one-tenth (10%) of the IDLH value may be an acceptable level of concern for evaluating hazmat release concentrations and public protective options.

- Emergency Response Planning Guidelines (ERPG)—ERPGs are air concentration guidelines for single exposures to hazardous materials. Developed by the American Industrial Hygiene Association (AIHA) as an emergency planning tool for public protective action options, ERPGs have been developed for approximately 100+ chemicals, most of which are extremely hazardous substances (EHS) with airborne hazards (e.g., chlorine, anhydrous ammonia, hydrogen sulfide Table 2-7).

Although there are three ERPG tiers, the ERPG-2 level is the most commonly cited guideline when using public protective action decision-making support tools. The ERPG-2 is defined as the maximum airborne concentration below which it is believed that nearly all individuals could be exposed for up to one hour without experiencing or developing irreversible or other serious health effects or symptoms that could impair an individual's ability to take protective action. Additional information can be obtained at the AIHA website at

Responder Safety Tip

Exposure values should be regarded only as guidelines, not absolute boundaries between safe and dangerous conditions. In addition, they should not be used when combinations of materials are involved. To be most effective, exposure values must be combined with monitoring instrument readings and interpreted by response personnel who are familiar with their proper application and limitations.

When evaluating the establishment of hazard control zones at hazmat emergencies, the various TLV and the IDLH values are generally the most informative. When evaluating public protective action options, the ERPG, AEGL or one-tenth of the IDLH value are useful. Remember—the lower the reported concentration, the more toxic the material.

Controlling Personnel Exposures

The primary objective of using these various exposure guidelines is to minimize the potential for both public and responder exposures. One method for understanding these exposure guidelines and applying them at a hazmat emergency is the concept of safe, unsafe, and dangerous. Originally developed

Responder Safety Tips

Responder Safety Tips reinforce safety-related concerns.

Chapter 7 Hazard Assessment and Risk Evaluation 

4. pH paper will detect both acids and caustics (bases). By wetting pH paper with an unknown liquid you can determine whether the product is an acid or caustic (base) or a substance that does not register a pH reading. Initially wetting the pH paper with sterilized water and holding it in the vapor space of an unknown liquid or gas can help to determine whether the unknown is producing a corrosive vapor. This, in turn, can lead responders to the classification of the hazard(s) and potentially identification of the product. **(photo 4)**

By using this basic cadre of instruments, you can address the fundamental tasks of (1) product identification and/or classification, and (2) identify potential hazards present. When the findings from these direct-reading instruments are combined with on-scene indicators, you should be capable of making a timely and effective risk assessment of the incident.

Using Equipment Provided By the AHJ

The AHJ or testing authority must determine which detection and monitoring devices and equipment will be available. The selected equipment should reflect a range of monitoring capabilities and hazards, and should not solely focus upon

screening could be done on an individual container and should be done prior to taking any samples.

To properly use equipment provided by the AHJ in performing field screening, follow the steps in Skill Drill 7-2:

1. A field screen of any scene should include an initial check for explosive devices. This includes not only a thorough visual scan for obvious signs of devices, but also for some of the chemicals that could be used to make explosives. For example, discovering containers of acetone and hydrogen peroxide could be an indicator of the production or attempted production of the explosive tri-acetone tri-peroxide (TATP). In some cases, a Bomb (EOD) Technician could be called to the scene to check for explosive devices. **(photo 1)**
2. Field screening must include a check for radiation. Responders should never assume radioactive sources are not present. Use the instruments provided by the AHJ—it could save your life. **(photo 2)**
3. Flammability risks should be evaluated with a combustible gas indicator. This is commonly performed using a multi-gas instrument with an LEL sensor and oxygen sensor, but it could be any technology that evaluates the atmosphere for the presence of flammable gases or vapors. **(photo 3)**
4. Corrosivity should be checked using either pH paper or a pH meter. Depending on the procedures established by your AHJ, dry or hydrated pH paper may be used. pH paper can be used to swipe across surfaces, wetted and placed gently into the head space of containers after they have been carefully opened, or perhaps inserted into unknown atmosphere at the end of a long pole.

   If you suspect an unknown liquid is a solvent, for example, and a pH check reveals no change on the pH paper—you are one step closer to confirming the presence of a solvent. Conversely, if you suspect an acid, and the result of a pH check confirms your finding, you are closer to confirming your suspicion. In some cases, when a total unknown is being field screened, a pH check can be a quick means to rule out a number of possibilities and help guide you in your effort to identify or classify. Although pH checks are generally only done on liquids, they can be used to evaluate vapors or unknown atmospheres. **(photo 4)**
5. Oxygen levels (deficiency and enrichment) should also be checked, especially when operating in an indoor or confined area. This is commonly accomplished using a multi-gas meter with an oxygen sensor, but single oxygen meters may also be used. **(photo 5)**
6. An unknown container or atmosphere should also be evaluated for the presence of volatile organic compounds (VOCs). This is typically done with a photoionization detector (PID), which may be integrated into a multi-gas instrument. Unknown substances that give off certain readings (e.g., VOC with low or high vapor pressure) can be easily classified and even identified by the reaction time and speed of the PID. **(photo 6)**

Skill Drill 7-2

NFPA 472-7.2.1.3.5

Using Equipment to Perform Field Screening

1 Visually scan the scene for hazards and or dangerous situations. Pay attention to the details!

2 Radiation cannot be detected by sight, sound, or smell—make sure to evaluate the scene for this potential threat.

3 Evaluate flammability risks with a combustible gas indicator.

4 If placing pH paper into the head space of a liquid container has not changed the color of the pH paper, draw a small amount of liquid into a pipette and drop onto the end of the pH paper.

5 Check oxygen levels (deficiency and enrichment) using a multi-gas meter with an oxygen sensor.

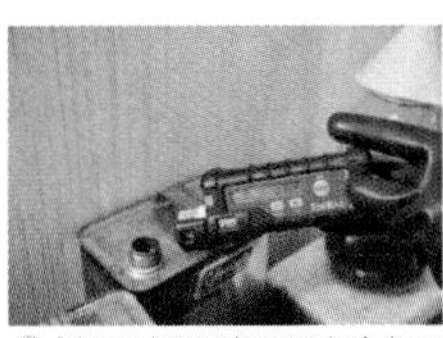

6 Evaluate an unknown container or atmosphere for the presence of volatile organic compounds (VOCs).

Skill Drills

Skill Drills provide written step-by-step explanations and visual summaries of important skills and procedures. This clear, concise format enhances student comprehension of sometimes complex procedures.

Wrap-Up
End-of-chapter activities reinforce important concepts and improve students' comprehension. Additional instructor support and answers for all questions are available in the Instructor's Toolkit CD.

Chief Concepts
Chief Concepts highlight critical information from the chapter in a bulleted format to help students prepare for exams.

Hot Terms
Hot Terms are easily identifiable within the chapter and define key terms that the student must know. A comprehensive glossary of Hot Terms also appears in the chapter Wrap-Up and in the glossary at the end of the text.

HazMat Responder in Action
This activity promotes critical thinking through the use of case studies and provides instructors with discussion points to enhance the classroom presentation.

Wrap-Up

Chief Concepts

- Terrorism is the unlawful use of violence or threats of violence to intimidate or coerce a government, the civilian population, or any segment thereof, to further political or social objectives. This broad definition encompasses a wide range of acts committed by different groups for different purposes.
- The goal of terrorism is to produce feelings of fear in a population or a group.
- Terrorism can occur in any community, so it is essential that responders be aware of all potential targets in their area.
- Terrorists can turn ordinary objects into weapons.
- Secondary devices are intended to explode some time after the initial device detonates.
- Weapons of mass destruction include chemical, biological, and radiological agents, as well as conventional weapons and explosives.
- A part of the response to a potential terrorism incident is important to be able to identify which type of agent is involved.
- When dealing with a potential terrorist-related incident, responders should establish a staging area at a safe distance from the scene and follow the direction of the incident commander.
- Interagency coordination is an important part of responding to a terrorist event.

Hot Terms

Agroterrorism The intentional act of using chemical or biological agents against the agricultural industry or food supply.

Alpha particles A type of radiation that quickly loses energy and can travel only 1 to 2 inches from its source. Clothing or a sheet of paper can stop this type of energy. Alpha particles are not dangerous to plants, animals, or people unless the alpha-emitting substance has entered the body.

Ammonium nitrate fertilizer and fuel oil (ANFO) An explosive made of commonly available materials.

Anthrax An infectious disease spread by the bacterium *Bacillus anthracis*; typically found around farms, infect-

such as metal, plastic, and glass, can stop this type of energy.

Biological agents Disease-causing bacteria, viruses, and other agents that attack the human body.

Blistering agents A chemical that cause the skin to blister. Also known as a vesicant.

Blood agent Chemicals that interfere with the utilization of oxygen by the cells of the body. Cyanide is an example of a blood agent.

Chlorine A yellowish gas that is approximately 2.5 times heavier than air and slightly water soluble. Chlorine has many industrial uses. It damages the lungs when inhaled; it is a choking agent.

Choking agent A chemical designed to inhibit breathing, and typically intended to incapacitate rather than kill.

Color-coded threat-level system The Department of Homeland Security's system for communicating with public officials and the public so that protective measures can be implemented to reduce the likelihood or impact of a terrorist attack.

Cyanide A highly toxic chemical agent that prevents cells from using oxygen.

Cyberterrorism The intentional act of electronically attacking government or private computer systems.

Ecoterrorism Terrorism directed against causes that radical environmentalists think would damage the earth or its creatures.

Explosive ordnance disposal (EOD) personnel Personnel trained to detect, identify, evaluate, render safe, recover, and dispose of unexploded explosive devices.

Forward staging area A strategically placed area, close to the incident site, where personnel and equipment can be held in readiness for rapid response to an emergency event.

Gamma radiation A type of radiation that can travel significant distances, penetrating most materials and passing through the body. Gamma rays are the most destructive type of radiation to the human body.

Homeland Security Information Bulletins Federally issued guidelines that communicate information of interest to U.S. critical infrastructures that does not meet the timeliness, specificity, or significance thresholds of warning messages.

Homeland Security Threat Advisories Federally issued guidelines that contain actionable information about an incident involving, or a threat targeting, critical national networks or infrastructures or key assets.

HazMat Responder
in Action

It is summertime and you have been called by Law Enforcement to a house that has a fatality and an individual who is described as "acting strange." When you arrive you learn that the incident is a suspected illicit laboratory (i.e., drug lab), and drug enforcement agents are on scene processing evidence. The agents tell you that the deceased person is in the back bedroom and the other perpetrator is now in custody, physically in the back of the patrol car behind which your team is parked.

Questions

1. Assuming that both perpetrators were in the same location over the same time frame, why has one died and the other not?
 - A. Routes of exposure are different
 - B. It is a function of the dose/response relationship
 - C. Toxicity of one chemical was greater than another
 - D. All the above

2. The Law Enforcement Commander approaches you and states that one of the SWAT entry team members is starting to act in a fashion similar to the second perpetrator who is in custody. What is your primary line of questions going to be?
 - A. What personal protective equipment was the officer wearing at the time of entry?
 - B. Is the officer taking any medications?
 - C. Is this normal for the officer after a drug raid?
 - D. Does the officer have a history of doing this?

3. The Law Enforcement Commander states that they have found a substance that is labeled Paraoxone. You look it up in your reference library and find that it is toxic at 14mg/kg. What does this information tell you about potential hazards present?
 - A. This chemical is moderately hazardous
 - B. It has an extremely low toxicity and is not to be concerned with
 - C. It is a seriously toxic substance
 - D. It is slightly toxic

4. Which exposure guideline should you use to achieve a safe environment for responders to work in?
 - A. LD50
 - B. IDLH
 - C. PEL
 - D. STEL

Hazardous Materials: Managing the Incident

What Is Unsafe?

A general rule for responders should be—if the material has been released from its container, assume that an unsafe atmosphere may exist and some form of PPE is required. Prolonged exposures at high concentrations can lead to injury; however, acute injuries may not be lethal (e.g., headaches, nausea, irritation to the eyes, nose, or throat). Unsafe atmospheres do not become seriously dangerous unless the exposure continues or the concentration of contaminants increases.

The unsafe atmosphere is an area where some responders may ignore the signs and symptoms of overexposure. The fire service was often guilty of this during the days when "eating smoke" was viewed as an acceptable risk and part of the firefighting culture. In many respects, that culture continues today

Scan Sheet 2-A

Toxicity and Exposure Terminology

The terms used to describe chemical toxicity and exposures can seem complicated, and some have similar meanings, further complicating the issue. All relate to how long an individual can safely work in a chemical or hazardous atmosphere. There is a safety factor incorporated into each of these values, as each person can react differently based upon their age, sensitivity, and pre-existing medical conditions. All of these terms and values are based on a person wearing NO skin or respiratory protection.

Remember: HEALTH HAZARD = EXPOSURE + TOXICITY

"Poison lines" are an excellent tool for visually illustrating the significance and inter-relationship between health exposures and other hazards. Examine the following poison lines for chlorine (poison gas), methanol (flammable liquid with health hazards), and anhydrous ammonia (corrosive gas that is toxic by inhalation and flammable in certain atmospheres). Note that (1) as concentrations rise above the TLV, harm occurs; and (2) as the concentration increases, harm increases. At some point, the IDLH is reached.

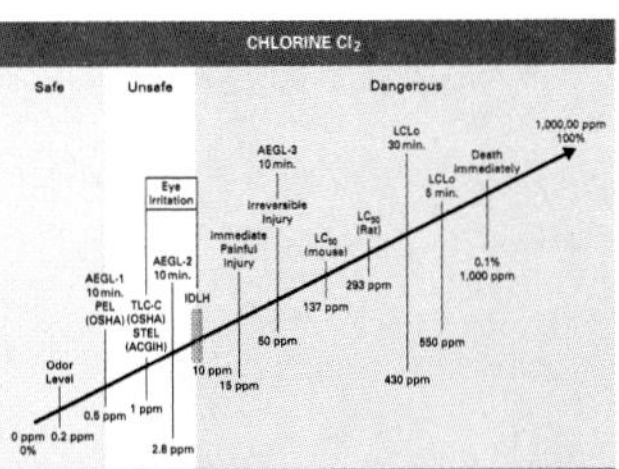

Adapted from original artwork by Michael Callan.

Scan Sheets
Scan sheets provide valuable information relevant to all levels of hazardous materials responders.

Instructor Resources

Instructor's ToolKit
ISBN: 978-1-284-13648-7
This CD includes:

- Adaptable PowerPoint Presentations
- Detailed Lesson Plans
- Electronic Test Bank
- Image and Table Bank
- Skills Sheets

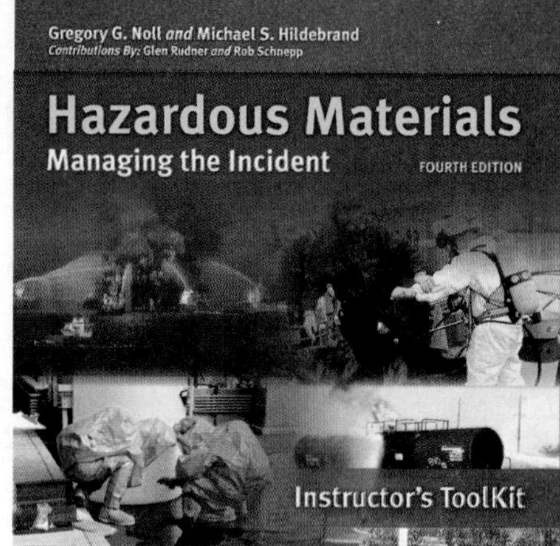

Student Resources

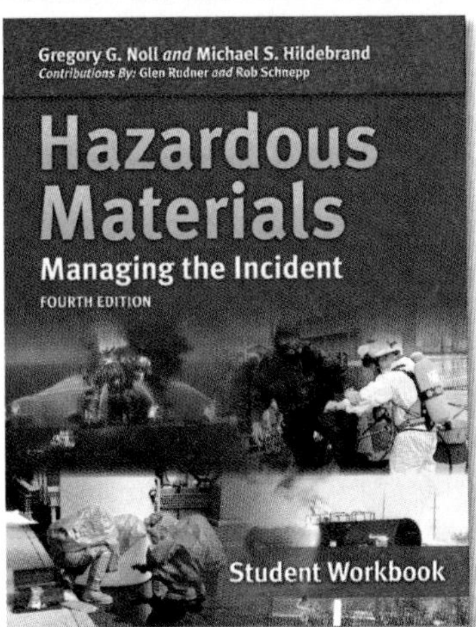

Student Workbook

ISBN: 978-1-4496-8829-5

This resource is designed to encourage critical thinking and aid comprehension of the course materials. The student workbook uses the following activities to enhance mastery of the material:

- Chapter Orientation
- Learning Objectives
- Abbreviations and Acronyms
- Study Session Overview
- Practice
- Important Terminology
- Study Group Activities
- Summary and Review
- Self-Evaluation

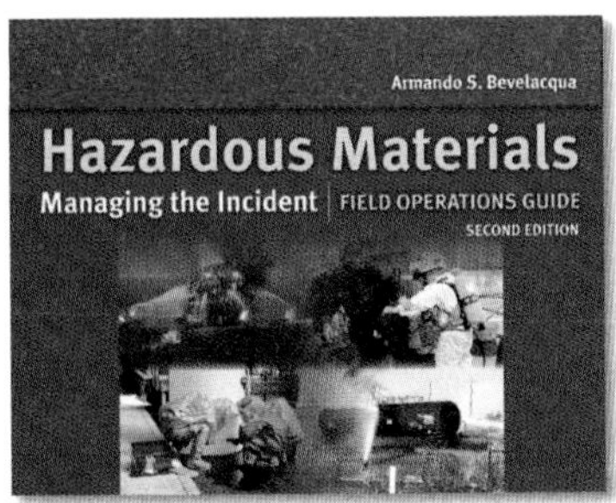

Field Operations Guide (FOG)

ISBN: 978-1-4496-9672-6

The FOG includes detailed tactical checklists that follow the Eight Step Process©, a section on identification and recognition of containers, data cards on the top 50 hazardous materials and CBRNE's, as well as a matrix of WMD and drug lab precursor chemicals. The FOG is designed to be used at the incident scene and as a classroom reference guide to strategic and tactical decision making.

Navigate TestPrep: Hazardous Materials: Managing the Incident

ISBN: 978-1-4496-9670-2

Navigate TestPrep: Hazardous Materials: Managing the Incident is a dynamic program designed to prepare students to sit for local, regional, or state examinations by including the same type of questions they will likely see on the actual examination.

It provides a series of self-study modules, organized by chapter, offering practice examinations and final examinations using multiple-choice questions. All questions are page referenced to *Hazardous Materials: Managing the Incident, Fourth Edition* for remediation to help students hone their knowledge of the subject matter.

Students can begin the task of studying for examinations by concentrating on those subject areas where they need the most help. Upon completion, students will feel confident and prepared to complete the final step in the examination process–passing the examination.

Acknowledgments

About the Authors

Greg Noll and Mike Hildebrand have more than 45 years experience in industry and government. They have served as firefighters, hazardous materials technicians, incident commanders, and instructors. Their experience and expertise includes hazardous materials, weapons of mass destruction, and operations security (OPSEC). They are both Certified Safety Professionals and serve on the NFPA 472 Technical Committee on Hazardous Materials Response. In addition, they are recipients of the International Association of Fire Chiefs' (IAFC) Chief John M. Eversole Lifetime Achievement Award for their leadership and contributions to further enhance the hazardous materials emergency response profession.

About the Contributing Authors

Rob Schnepp is the Division Chief of Special Operations for the Alameda County (California) Fire Department. Chief Schnepp has over 30 years of fire service experience, with a specialty in hazardous materials response. He is the author of Hazardous Materials: Awareness and Operations by Jones & Bartlett Learning, and is on the editorial advisory board for Fire Engineering magazine. Rob serves on the NFPA Technical Committee on Hazardous Materials Response and is a member of the task group charged with revising NFPA 473—Standard for Competencies for EMS Personnel Responding to Hazardous Materials.

Glen D. Rudner is the retired Northern Virginia Regional Hazardous Materials Officer for the Virginia Department of Emergency Management. Over the last 38 years he has been involved in the development, management, and delivery of many local, state, federal, and international training programs. Glen has authored numerous public safety journal articles as well as being a member and secretary for the NFPA Technical Committee on Hazardous Materials Response (NFPA 471, 472, and 473).

Acknowledgments

We have been in the bad day business for 40 years as emergency responders and safety professionals. During that time we have seen emergency responders pay a heavy price to keep the public and our environment safe from hazardous materials, criminals, and terrorists. Some responders received debilitating injuries; others paid with their lives.

Our profession takes a heavy toll on the families of first responders. Special operations work requires years of dedication to training, exercises, and self-improvement. To stay on top of our game we had to spend a lot of time away from home. Anyone who lasts and is good in this business owes a great deal to their family for their steadfast support.

In our travels around the world we have had an opportunity to learn from thousands of firefighters, law enforcement officers, health professionals, industrial response personnel, and members of the military. These people are among the best of the best. Many we have met took the time to share their personal stories, their insights, their successes and failures, all of which added depth to this book and help anchor it in the real world. To all of you who shared and encouraged us to think about better and safer ways to get the job done—Thank You!

One major improvement to the Revised Fourth Edition has been the addition of the NFPA 1072 *Standard for Hazardous Materials/Weapons of Mass Destruction Emergency Response Personnel Professional Qualifications* objectives.

Every book with a long history has a cast of characters who have been behind the scenes helping to produce the various editions. In particular we want to acknowledge the contributions of Don Sellers and Jim Yvorra in creating the First Edition. Without their talent and leadership in the early days, there would never have been a starting point for us as co-authors. We also want to pay homage to George and Katie Dodson who produced the Second and Third Editions. We want to acknowledge the contributions of Toby Bevelacqua, Mike Callan, and JoAnne Hildebrand to the previous editions of the corresponding Field Operations Guide, Instructor's Guide, and the Student Workbook. We have covered a lot of history with all of these wonderful and talented people, and we are better authors—and better people—because of this association.

This Revised Fourth Edition is published by Jones & Bartlett Learning. The Jones & Bartlett production team members have been wonderful folks to work with. We owe a great deal to Executive Acquisitions Editor, Bill Larkin, who helped us make a smooth transition from the Third to the Fourth Edition and for providing sound guidance on how to further integrate our textbook into a larger hazardous materials learning system. We would also like to thank Janet Morris who was our editor. We appreciate the significant effort it took to integrate 52 different reviewers' comments into a single, cohesive manuscript and to help us express complicated, technical concepts using fewer and clearer words.

We would like to offer a heartfelt "thank you" to Thomas Jordan of the Virginia Department of Emergency Management and Chris Sadler, Battalion Chief of Technical Services and Special Operations at York County Fire and Life Safety, for sharing their state-of-the-art training facility and their technical assistance, along with members of the York County Fire and Life Safety Hazardous Materials Team, enabling us to capture in photos all of the various skills applications.

We are strong advocates of third-party reviews when safety and accuracy is concerned. We actively applied this approach to this Fourth Edition. When you consider the reviewers involved in all four editions it includes hundreds of experienced responders. To all of the reviewers who have contributed, past and present, we offer our sincere appreciation.

Contributors and Reviewers

Armando "Toby" Bevelacqua
Orlando Fire Department (retired)
Clermont, Florida

Benjamin Anderson
Loveland Fire Rescue Authority
Loveland, Colorado

Brent Willis
Martinez-Columbia Fire Rescue
Martinez, Georgia

Brian D. Daake
Beatrice Fire and EMS
Beatrice, Nebraska

Carl Raymond
South Crescent Technical College
Thomaston, Georgia

Casey Sobol
South Puget Sound Community College
Olympia, Washington

Christopher Tracy
Fairfield Fire Department
Fairfield, Connecticut

Chuck Thomas
Commercial Airline Pilot
Aurora, Colorado

Dan Roeglin
Minnesota Fire/EMS Safety Center
Saint Paul, Minnesota

Darin Keith
Rock Island Arsenal Fire Department
Rock Island, Illinois

Doug Rohn
Madison Fire Department
Madison, Wisconsin

Dustin Willett
Rapid City Fire Department
Rapid City, South Dakota

Gary Seidel
Los Angeles City Fire Department (retired)
Hillsboro, Oregon

Jason Goodale
Loveland Fire Rescue Authority
Johnstown, Colorado

Jason Loeb
Illinois Fire Service Institute
Oak Forrest, Illinois

Jason Krusen
Columbia Fire Department
Columbia, South Carolina

James Zeigler, Ph.D.
J.P. Zeigler, LLC
Mechanicsburg, Virginia

Jeff King
Flower Mound Fire Department
Flower Mound, Texas

Jeffery L. Combs
Rossburg Fire Department
Rossburg, Ohio

Jeffry J. Harran
Buckskin Fire Department
Parker, Arizona

Jeffrey H. Smith
Pueblo Magnet School
Tuscon, Arizona

Jerry Janick
Illinois Fire Service Institute
LaSalle, Illinois

Jimmy Talton
Reece City Volunteer Fire Department
Atalla, Alabama

John T. Dean
Phoenix Fire Department
Phoenix, Arizona

John F. Ryan
Fort Gordon Fire & Emergency Services
Fort Gordon, Georgia

Judith A. Hoffmann
Hoffmann Consulting
South Bend, Indiana

J. Nathan Kempfer
City of Lawrence Fire Department
Lawrence, Indiana

Kevin Bartoe
Robins Air Force Base Fire and Emergency Services
Robins AFB, Georgia

Kevin Hammons
Community College of Aurora Franktown, Colorado

Larry Crang
Office of the Fire Marshal
Ontario, California

Lori Stoney
Homewood Fire & Rescue Service
Homewood, Alabama

Mark Martin
Perry Township Fire Department
Massillon, Ohio

Mark D. Schuman
North Washington Fire Protection District Denver, Colorado

Michael A. Blann
Lafayette Fire Department
Lafayette, Indiana

Michael L. Farrell
United States Capitol Police –Hazmat Response Team
Washington, D.C.

Mike Pannell
DeKalb County Fire Department
Douglasville, Georgia

Neil R. Fulton
Norwich Fire Department
Norwich, Vermont

Nick Zamiska
SERT Hazmat Brecksville, Ohio

Patrick M. O'Connor
Providence Fire Department
North Providence, Rhode Island

Richard A. Fritz
High Point Fire Department (Retired)
Trinity, North Carolina

Richard J. Kosmoski
Middlesex County Fire Academy
Sayreville, New Jersey

Robert Wayne Green
Culpepper County Office of Emergency Services Culpepper, Virginia

Scott Pascu
University of Akron Uniontown, Ohio

Sharon Sparks
Georgia Fire Academy Forsyth, Georgia

Stephen M. Hardesty
Howard County Fire & Rescue
Columbia, Maryland

Steve Hergenreter
Fort Dodge Fire Department
Fort Dodge, Iowa

Timothy B. Thomas
Indiana Department of Homeland Security
Elwood, Indiana

Todd Mesick
Oneida Co HMRT
Utica, New York

Tyler Bones
Fairbanks – North Star Borough HMRT Fairbanks, Alaska

William Ferguson
Franklin County Public Safety Rocky Mount, Virginia

William Hadley
Louis F. Garland Fire Academy
(U.S. Department of Defense)
San Angelo, Texas

William Hand
HazMat Response Team (retired) Houston Fire Department League City, Texas

Photographic Contributions

We would like thank Glen E. Ellman, the photographer for this project. Glen is a commercial photographer and fire fighter based in Fort Worth, Texas. His expertise and professionalism are unmatched!

We would also like to thank the following members of the York County Department Fire and Life Safety: Richard Burgess (Lieutenant), Brian Williams, Mark Shields, Robert Kudley, Wade Dunlap, and Reggie Rivers, Haz Mat Technicians.

Dedication

This book is dedicated to our wives, Debbie Noll and JoAnne Hildebrand, and our families. Without their unfailing love, dedication, and good counsel we would surely have been like a ship lost at sea.
Love—Greg and Mike

A Friend Remembered–Deputy Fire Chief James G. Yvorra

With the release of the Revised Fourth Edition, it seems hard to believe that the first edition of ***Hazardous Materials: Managing the Incident*** was actually published 30 years ago. One thing for certain, without Deputy Fire Chief James G. Yvorra there would never have been a Hildebrand and Noll partnership and there would be no ***Hazardous Materials: Managing the Incident*** textbook.

Jim Yvorra introduced Hildebrand and Noll to one another in 1979 at the International Association of Fire Chiefs Conference in Kansas City, Missouri—which turned out to be the beginning of a great partnership. We collaborated on many projects including the Eight Step Process©, a nascent idea begun on a cocktail napkin at the National Fire Academy over a couple of beers.

Jim Yvorra was one of the original authors of this textbook, and without his early efforts, you would not be reading this edition. The late Chief Alan Brunacinni once said that Jim Yvorra was the complete package; he was book smart, street smart, and, to our good fortune, he was also an exceptional journalist. Jim eventually convinced us that we could "write a book together." We could not have done it without him.

Jim had this incredible ability to take complex topics and explain them in simple, understandable terms. We learned this and many other significant things from Jim: about writing, organizing content, developing a manuscript, and how the publishing business worked. Without realizing it then, we were the students and he was the teacher.

Jim was usually most happy when he was helping other people. We saw him in action hundreds of times when he cooked chicken dinners for a Berwyn Heights Volunteer Fire Department fundraiser, shared a T-Shirt or patch with a visiting firefighter, or when he was risking his life to save someone else's. In the end, he gave the supreme sacrifice—Jim died in the line of duty in January 1988. Even after three decades we still miss him like it was yesterday.

In honor of Jim, his family and friends established the Yvorra Leadership Development Foundation, now in its 30th year serving America's emergency response community. For more information on leadership development, to make a donation, or to receive a scholarship application, go to http://www.yld.org.

Additional Hazardous Materials Publications

Above Ground Bulk Storage Tank Emergencies, Second Edition by Michael S. Hildebrand, Gregory G. Noll, Bill Hand ISBN: 978-1-284-11277-1

Gasoline Tank Truck Emergencies, Fourth Edition by Michael S. Hildebrand, Gregory G. Noll, Bill Hand ISBN: 978-1-284-11273-3

Hazardous Materials Awareness and Operations, Third Edition by Rob Schnepp ISBN: 978-1-284-14070-5

Hazardous Materials Chemistry, Third Edition by Armando S. Bevelacqua, Laurie Norman ISBN: 978-1-284-04199-6

Intermodal Container Emergencies, Second Edition by Michael S. Hildebrand, Gregory G. Noll, Bill Hand ISBN: 978-1-284-11275-7

The Hazardous Materials Management System

CHAPTER 1

Hazardous Materials Technician

Knowledge Objectives

After reading this chapter, you will be able to:

- Identify the role of the hazardous materials technician during hazardous materials/WMD incidents. (p. 4)
- List the key legislative, regulatory, and voluntary consensus standards that affect hazmat emergency planning and response operations. (p. 5–12)
- Describe the concept of "standard of care" as applied to hazardous materials training, planning and response. (p. 13)
- List and describe the components of the Hazardous Materials Management System for managing the hazardous materials problem within the facility or community. (p. 14–25)

Skills Objectives

There are no Hazardous Materials Technician skills objectives in this chapter.

Hazardous Materials Incident Commander

Knowledge Objectives

After reading this chapter, you will be able to:

- Define and explain the following terms:
  - Hazardous materials (hazmat) (p. 5)
  - Hazardous substances (p. 5)
  - Extremely Hazardous Substances (EHS) (p. 5)
  - Hazardous chemicals (p. 5)
  - Hazardous wastes (p. 5)
  - Dangerous goods (p. 5)
  - Weapons of mass destruction (page 5)
- Identify the steps for implementing the emergency response plans required under Title III Emergency Planning and Community Right-to-Know Act (EPCRA) of the Superfund Amendments and Reauthorization Act (SARA) Section 303, or other state and emergency response planning legislation. (p. 6–9)
- Identify the importance of pre-incident planning relating to safety during responses to specific sites. (p. 16)
- Identify the primary government agencies and identify the scope of their regulatory authority (including the regulations) pertaining to the production, transportation, storage, and use of hazardous materials and the disposal of hazardous wastes. (p. 6–11)
- Identify the reporting requirements of the federal, state, and local agencies. (p. 20)
- List and describe the components of the Hazardous Materials Management System for managing the hazardous materials problem within the facility or community. (p. 14)

Skills Objectives

There are no Incident Commander skills objectives in this chapter.

You Are the HazMat Responder

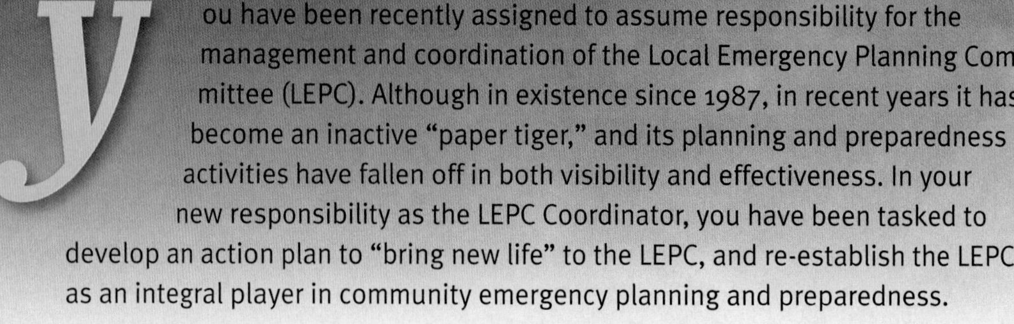

You have been recently assigned to assume responsibility for the management and coordination of the Local Emergency Planning Committee (LEPC). Although in existence since 1987, in recent years it has become an inactive "paper tiger," and its planning and preparedness activities have fallen off in both visibility and effectiveness. In your new responsibility as the LEPC Coordinator, you have been tasked to develop an action plan to "bring new life" to the LEPC, and re-establish the LEPC as an integral player in community emergency planning and preparedness.

1. What local agencies and organizations should serve as members of the LEPC?
2. Should the LEPC be limited to only hazardous materials-related activities, or should its scope and activities be expanded to have more of an "all hazards" focus?
3. What would be your top three priorities to reconstitute the LEPC as a viable organization that will positively affect community preparedness?

Introduction

This is a text about **hazardous materials** (hazmat) incident response, including weapons of mass destruction (WMD). It is designed to provide public and private sector emergency response personnel with a logical, building block system for responding to and managing hazmat/WMD emergencies. It is not a chemistry-oriented text. In fact, it assumes that most of the first-arriving emergency responders will have little or no formal chemistry training. It is designed to begin at the point where responders recognize that they are, in fact, dealing with a hazardous materials emergency, even when the specific hazardous materials have not yet been identified. Otherwise, normal fire, rescue, and emergency medical services (EMS) guidelines will be followed.

Our primary target audience includes Hazardous Materials Technicians, the on-scene **incident commander**, the Hazmat Officer (aka Group Supervisor or Branch Director), and members of organized hazmat response teams (HMRTs). Other special operations teams, such as bomb squads and technical rescue teams, will also find specific sections of interest (e.g., section on "hazard and risk assessment").

Role of the Hazardous Materials Technician

Hazardous Materials Technicians (HMTs) are individuals who respond to hazardous materials and WMD incidents using a risk-based response process by which they analyze a problem involving hazmat and WMD, select applicable decontamination procedures, and control a release using specialized protective clothing and control equipment (Source: NFPA 472).

The role of the HMT is significantly different from the roles of awareness-level personnel and operations-level responders. HMTs assume a more aggressive (i.e., proactive) role in controlling a hazardous materials incident, including the need to approach the point of a release to isolate or control the release. HMTs are often a member of a **Hazardous Materials Response Team (HMRT)**, which NFPA 472 defines as an organized group of trained response personnel operating under an emergency response plan and applicable standard operating procedures, who perform hazardous material technician-level skills at hazmat/WMD incidents. While not all HMRT members must be trained to the HMT-level, the HMRT must have the capability of delivering the HMT level of operational capabilities.

This text includes additional information to assist the reader in meeting the knowledge and skill requirements of the *Occupational Safety and Health Administration (OSHA) 1910.120(q) regulation* and the *National Fire Protection Association (NFPA) 472—Standard for Competence of Responders to Hazardous Materials/Weapons of Mass Destruction Incidents* competencies for the HMT and the On-Scene Incident Commander (2018 edition).

This text follows the NFPA 472 operational philosophy that emergency response operations to a terrorism or criminal scenario involving hazardous materials are based on the fundamental concepts of hazardous materials response. In other words, responders cannot safely and effectively respond to a terrorism or criminal scenario involving WMD and chemical, biological, radiological, nuclear and explosive (CBRNE) materials if they do not first understand hazardous materials response.

What Is A Hazardous Material?

You might assume everyone knows what a hazardous material is, or at least knows one when they see one. However, if we were to review the various state and federal regulations that govern the manufacture, transportation, storage, use, and clean-up of chemicals in the United States, the number of terms, definitions, and lists would be overwhelming. When one includes the terms associated with the criminal use of hazardous materials, WMD, and CBRNE materials, the challenges become even more pronounced. Table 1-1 lists key definitions from the various U.S. federal governmental agencies that regulate hazardous materials.

Each term in Figure 1-1 has its applications and limitations. In reality, we must recognize that hazmats can be found almost anywhere—in industry, in transportation, in the workplace, and even in the home. In addition, hazardous materials can also be used as a weapon for criminal or terrorist purposes—so a broad definition is necessary to cover all the variables.

Figure 1-1 Hazardous Materials—any substance that jumps out of its container when something goes wrong and hurts or harms the things it touches.

Table 1-1 Hazardous Materials Definitions

Hazardous materials—Any substance or material in any form or quantity that poses an unreasonable risk to safety, health, and property when transported in commerce (Source: U.S. Department of Transportation [DOT], 49 Code of Federal Regulations (CFR) 171).

Hazardous materials—A substance (either matter—solid, liquid, or gas—or energy) that when released is capable of creating harm to people, the environment, and property, including WMD as defined in 18 U.S. Code, Section 2332a, as well as any other criminal use of hazardous materials, such as illicit labs, environmental crimes, or industrial sabotage (Source: NFPA 472).

Hazardous substances—Any substance designated under the Clean Water Act and the Comprehensive Environmental Response, Compensation, and Liability Act (CERCLA) as posing a threat to waterways and the environment when released (Source: U.S. Environmental Protection Agency [EPA], 40 CFR 302). Note: Hazardous substances as used within OSHA 1910.120 refers to every chemical regulated by EPA as a hazardous substance and by DOT as a hazardous material.

Extremely Hazardous Substances (EHS)—Chemicals determined by the EPA to be extremely hazardous to a community during a spill or release as a result of their toxicities and physical/chemical properties (Source: EPA 40 CFR 355).

Hazardous chemicals—Any chemical that would be a risk to employees if exposed in the workplace (Source: OSHA, 29 CFR 1910).

Hazardous wastes—Discarded materials regulated by the EPA because of public health and safety concerns. Regulatory authority is granted under the Resource Conservation and Recovery Act (RCRA). (Source: EPA, 40 CFR 260–281).

Dangerous goods—In international transportation, hazardous materials are commonly referred to as "dangerous goods."

Weapons of Mass Destruction (WMD)—(1) Any destructive device, such as any explosive, incendiary, or poison gas bomb, grenade, rocket having a propellant charge of more than four ounces, missile having an explosive or incendiary charge of more than one quarter ounce (7 grams), mine, or device similar to the above; (2) any weapon involving toxic or poisonous chemicals; (3) any weapon involving a disease organism; or (4) any weapon that is designed to release radiation or radioactivity at a level dangerous to human life (Source: 18 U.S. Code, Section 2332a).

Regardless of the cause of the incident, risk-based hazmat emergency response operations primarily focus on the interaction of the hazmat and its container. Therefore, for the purposes of this text, we will use the definition of a hazardous material developed by Ludwig Benner, Jr., a former hazardous materials specialist with the National Transportation Safety Board (NTSB) in Washington, D.C.:

> Hazardous materials: Any substance that jumps out of its container when something goes wrong and hurts or harms the things it touches.

Benner's definition was developed in the 1970s, and it can still be applied today to all hazmats in all circumstances. The definition recognizes that emergency response is as much a container behavior problem as it is a chemical problem Figure 1-1. A hazardous materials incident can then be defined as the release, or potential release, of a hazardous material from its container into the environment.

Hazmat Laws, Regulations, and Standards

Operations involving the manufacture, transport, and use of hazardous materials, as well as the response to hazardous materials incidents, are affected by a large body of laws, regulations, and voluntary consensus standards. For most of us, discussing rules and regulations is like watching paint dry—it's boring! But as a professional, it's important to understand what the law requires and be able to back up what you think is the right thing to do on the street. These rules are important to responders because they influence virtually every facet of the hazardous materials business.

Because of their importance to emergency planning and response operations, hazmat program managers must have a working knowledge of how the regulatory system works. First, what is the difference between a law, regulation, and standard? These three terms are sometimes used interchangeably, but they do have distinctly different meanings.

Laws are primarily created through an act of Congress, by individual state legislatures, or by local government bodies. Laws typically provide broad goals and objectives, mandatory dates for compliance, and established penalties for noncompliance. Federal and state laws enacted by legislative bodies usually delegate the details for implementation to a specific federal or state agency. For example, the U.S. Occupational Safety and Health Act enacted by Congress delegates rule-making and enforcement authority on worker health and safety issues to OSHA.

Regulations, sometimes called rules, are created by federal or state agencies as a method of providing guidelines for complying with a law that was enacted through legislative action. A regulation permits individual governmental agencies to enforce the law through audits and inspections, which may be conducted by federal and/or state officials. Regulations directly affect the administrative and operational elements of a hazardous materials response program.

Voluntary consensus standards are normally developed through professional organizations or trade associations as a method of improving the individual quality of a product or system. Within the emergency response community, the National Fire Protection Association (NFPA) is recognized for its role in developing consensus standards and recommended practices that affect fire and electrical safety and hazmat operations. In the United States, standards are developed primarily through a democratic process whereby a committee of subject matter specialists representing varied interests writes the first draft of the standard. The document is then submitted to either a larger body of specialists or the general public, who then may amend, vote on, and approve the standard for publication. Collectively, this procedure is known as the Consensus Standards Process.

When a consensus standard is completed, it may be voluntarily adopted by government agencies, individual corporations, or organizations. Many hazmat consensus standards are also adopted by reference in a regulation. In effect, when a federal, state, or municipal government adopts a consensus standard by reference, the document becomes a regulation. An example of this process is the adoption of NFPA 30—*The Flammable and Combustible Liquids Code*, and NFPA 58—*The Liquefied Petroleum Gas Code*.

Responder Tip

One challenge for emergency responders is the extensive timeframe required for federal regulations to be revised and updated. For example, OSHA 1910.120 has not been revised since it was originally promulgated in 1989. Meanwhile, the risks encountered by hazardous materials emergency response community have changed significantly over the last 25-plus years.

Voluntary industry consensus standards can directly or indirectly affect the administrative and operational elements of a hazardous materials response program. This is especially true when they evolve and are eventually accepted as "Best Practices." Once a standard becomes widely adopted by its users, it establishes a new benchmark upon which the **Standard of Care** is established. Standard of Care is discussed in more detail later in this chapter.

Federal Hazmat Laws

Hazmat laws have been enacted by Congress to regulate everything from finished products to hazardous waste. Because of their lengthy official titles, many simply use abbreviations or acronyms when referring to these laws. The following summaries outline some of the more important laws affecting hazmat emergency planning and response.

- **RCRA—The Resource Conservation and Recovery Act (1976).** This law establishes a framework for the proper management and disposal of all waste materials (i.e., solid, medical, hazardous), including treatment, storage, and disposal facilities. It also establishes installation, leak prevention, and notification requirements for underground storage tanks.
- **CERCLA—The Comprehensive Environmental Response, Compensation, and Liability Act (1980).** Known as Superfund, this law addresses hazardous substance releases into the environment and clean-up of inactive hazardous waste disposal sites. It also requires those individuals responsible for the release of the hazardous materials (commonly referred to as the "responsible party") above a specified "reportable quantity (RQ)" to notify the National Response Center (NRC), which is the single point of contact for spill reporting to the federal government.
- **SARA—Superfund Amendments and Reauthorization Act of 1986.** SARA has had the greatest effect on hazmat emergency planning and response operations. As the name implies, SARA amended and reauthorized the Comprehensive Environmental Response, Compensation, and Liability Act of 1980 (CERCLA, or Superfund). While many of the amendments pertained to hazardous waste site clean-up, SARA's requirements also established a national baseline with regard to hazmat planning, community right-to-know, preparedness, training, and response.

  Title I of this law required OSHA to develop health and safety standards covering numerous worker groups who handle or respond to chemical emergencies, which led to the development of 29 CFR OSHA 1910.120, *Hazardous Waste Operations and Emergency Response (HAZWOPER)*.

  Most familiar to the emergency response community is SARA, Title III. Also known as the Emergency Planning and Community Right-to-Know Act (EPCRA), SARA, Title III led to the establishment of the State Emergency Response Commissions (SERCs) and the LEPCs.

Responder Tip

SARA, Title III, is the primary federal legislation that directly affects the local hazardous materials emergency preparedness program. Responders should identify the local agency responsible for the coordination of the LEPC, as well as those organizations that make up the local emergency response community.

- **CAA—The Clean Air Act.** This law establishes requirements for airborne emissions to help protect the environment. The Clean Air Act Amendments of 1990 addressed emergency response and planning issues at certain facilities with processes using highly hazardous chemicals. This included the establishment of a National Chemical Safety and Hazard Investigation Board, EPA's promulgation of 40 CFR Part 68—*Risk Management Programs for Chemical Accidental Release Prevention*, and OSHA's promulgation of 29 CFR 1910.119—*Process Safety Management of Highly Hazardous Chemicals, Explosives and Blasting Agents*. In addition, some facilities are required to make certain information available to the general public regarding the manner in which chemical risks are handled within their facility.
- **OPA—Oil Pollution Act of 1990.** Commonly referred to as OPA, this law amended the Federal Water Pollution Control Act. Its scope covers both facilities and carriers of oil and related liquid products, including deepwater marine terminals, marine vessels, pipelines, and railcars. Requirements include the development of emergency response plans, regular training and exercise sessions, and verification of spill resources and contractor capabilities. The law also requires the establishment of Area Committees and the development of Area Contingency Plans (ACPs) to address oil and hazardous substance spill response in coastal zone areas.

Hazmat Regulations

Laws delegate certain details of implementation and enforcement to federal, state, or local agencies, which are then responsible for writing the actual regulations that enforce the legislative intent of the law. Regulations will either (1) define the broad performance required to meet the letter of the law (i.e., performance-oriented standards); or (2) provide very specific and detailed guidance on satisfying the regulation (i.e., specification standards).

Federal Regulations

There are many important federal regulations that apply to the hazardous materials industry. Federal regulations are published in a series of manuals called The Code of Federal Regulations (CFR). The CFR is a massive publication containing all the rules and regulations enforced by the various federal departments and agencies.

The following summary includes several of the more significant federal regulations that affect hazmat emergency planning and response.

Hazardous Waste Operations and Emergency Response (29 CFR 1910.120). Also known as HAZWOPER, this federal regulation was issued under the authority of SARA, Title I. The regulation was written and is enforced by OSHA in those 23 states and 2 territories with their own OSHA-approved occupational safety and health plans. In the remaining 27 "non-OSHA" states, public sector personnel will be covered by a similar regulation enacted by the EPA (40 CFR Part 311).

The regulation establishes important requirements for both industry and public safety organizations that respond to hazmat or hazardous waste emergencies. This includes firefighters, law enforcement and EMS personnel, hazmat responders, and industrial Emergency Response Team (ERT) members. Requirements cover the following areas:

- Hazmat Emergency Response Plan
- Emergency Response Procedures, including the establishment of an Incident Management System (IMS), the use of a buddy system with back-up personnel, and the establishment of a Safety Officer
- Specific training requirements covering instructors and both initial and refresher training
- Medical Surveillance Programs
- Post-emergency termination procedures

Of particular interest to hazmat program managers and responders are the specific levels of competency and associated training requirements identified within OSHA 1910.120(q)(6). Table 1-2.

Individuals seeking further information on the application of OSHA standards to hazardous materials emergency response situations should consult the OSHA website at http://www.osha.gov. Specific attention should be paid to (1) the HAZWOPER Preamble, (2) OSHA interpretations of the HAZWOPER Standard, and (3) OSHA Directive Number CPL 02-02-073—*Inspection Procedures for the Hazardous Waste Operations and Emergency Response Standard, 29 CFR 1910.120 and 1926.65, Paragraph (q): Emergency Response to Hazardous Substance Releases* (August 27, 2007).

Community Emergency Planning Regulations (40 CFR parts 300 through 399). This requirement is the result of SARA, Title III and mandates the establishment of both state and local planning groups to review or develop hazardous materials response plans. The state planning groups are referred to as the State Emergency Response Commission (SERC). The SERC is responsible for developing and maintaining the state's emergency response plan. This includes ensuring that planning and training are taking place throughout the state, as well as providing assistance to local governments, as appropriate. States generally provide an important source of technical specialists, information, and coordination. However, they typically provide only limited operational support to local government in the form of equipment, materials, and personnel during an emergency.

Table 1-2 OSHA 1910.120 Levels of Emergency Responders

First Responder at the Awareness Level. These are individuals who are likely to witness or discover a hazardous substance release and who have been trained to initiate an emergency response notification process. The primary focus of their hazmat responsibilities is to secure the incident site, recognize and identify the materials involved, and make the appropriate notifications. These individuals would take no further action to control or mitigate the release.

First Responder–Awareness personnel shall have sufficient training or experience to demonstrate objectively the following competencies:

1. An understanding of what hazardous materials are and the risks associated with them in an incident.
2. An understanding of the potential outcomes associated with a hazardous materials emergency.
3. The ability to recognize the presence of hazardous materials in an emergency and, if possible, identify the materials involved.
4. An understanding of the role of the First Responder–Awareness individual within the local Emergency Operations Plan. This would include site safety, security and control, and the use of the Emergency Response Guidebook (ERG).
5. The ability to realize the need for additional resources and to make the appropriate notifications to the communication center.

The most common examples of First Responder–Awareness personnel include law enforcement and plant security personnel, as well as some public works employees. There is no minimum hourly training requirement for this level; the employee would have to have sufficient training to demonstrate objectively the required competencies.

First Responder at the Operations Level. Most fire department suppression personnel fall into this category. These are individuals who respond to releases or potential releases of hazardous substances as part of the initial response for the purpose of protecting nearby persons, property, or the environment from the effects of the release. They are trained to respond in a defensive fashion without actually trying to stop the release. Their primary function is to contain the release from a safe distance, keep it from spreading, and protect exposures.

First Responder–Operations personnel shall have sufficient training or experience to demonstrate objectively the following competencies:

1. Knowledge of basic hazard and risk assessment techniques.
2. Knowledge of how to select and use proper personal protective clothing and equipment available to the operations-level responder.
3. An understanding of basic hazardous materials terms.
4. Knowledge of how to perform basic control, containment, and/or confinement operations within the capabilities of the resources and PPE available.
5. Knowledge of how to implement basic decontamination measures.
6. An understanding of the relevant standard operating procedures and termination procedures.

First responders at the operations level shall have received at least 8 hours of training or have had sufficient experience to demonstrate objectively competency in the previously mentioned areas, as well as the established skill and knowledge levels for the First Responder–Awareness level.

Hazardous Materials Technician. These are individuals who respond to releases or potential releases for the purposes of stopping the release. Unlike the operations level, they generally assume a more aggressive role in that they are often able to approach the point of a release in order to plug, patch, or otherwise stop the release of a hazardous substance.

HMTs are required to have received at least 24 hours of training equal to the First Responder–Operations level and have competency in the following established skill and knowledge areas:

1. Capable of implementing the local Emergency Operations Plan
2. Able to classify, identify, and verify known and unknown materials by using field survey instruments and equipment (direct reading instruments)
3. Able to function within an assigned role in the Incident Management System
4. Able to select and use the proper specialized chemical personal protective clothing and equipment provided to the HMT
5. Able to understand hazard and risk assessment techniques
6. Able to perform advanced control, containment, and/or confinement operations within the capabilities of the resources and equipment available to the HMT
7. Able to understand and implement decontamination procedures
8. Able to understand basic chemical and toxicological terminology and behavior

Many communities and facilities have personnel trained as Emergency Medical Technicians (EMTs), yet do not have the primary responsibility for providing basic or ALS medical care. Similarly, HMTs may not necessarily be part of an HMRT. However, if they are part of a designated team as defined by OSHA, they must also meet the medical surveillance requirements within OSHA 1910.120.

Hazardous Materials Specialists. These are individuals who respond with and provide support to HMTs. While their duties parallel those of the Technician, Specialists require a more detailed or specific knowledge of the various substances they may be called upon to contain. This individual would also act as the site liaison with federal, state, local, and other governmental authorities in regard to site activities.

Similar to the technician level, Hazardous Materials Specialists shall have received at least 24 hours of training equal to the technician level and have competency in the following established skill and knowledge areas:

1. Capable of implementing the local Emergency Operations Plan
2. Able to classify, identify, and verify known and unknown materials by using advanced field survey instruments and equipment (direct reading instruments)
3. Knowledge of the state emergency response plan
4. Able to select and use the proper specialized chemical personal protective clothing and equipment provided to the Hazardous Materials Specialist
5. Able to understand in-depth hazard and risk assessment techniques

(continued)

Table 1-2 **OSHA 1910.120 Levels of Emergency Responders (*continued*)**

6. Able to perform advanced control, containment, and/or confinement operations within the capabilities of the resources and equipment available to the Hazardous Materials Specialist
7. Able to determine and implement decontamination procedures
8. Able to develop a site safety and control plan
9. Able to understand basic chemical, radiological, and toxicological terminology and behavior

Whereas the HMT possesses an intermediate level of expertise and is often viewed as a "utility person" within the hazmat response community, the Hazardous Materials Specialist possesses an advanced level of expertise. Within the fire service, the Specialist will often assume the role of the Hazmat Group Supervisor or the Hazmat Group Assistant Safety Officer, while an industrial Hazardous Materials Specialist may be "product specific." Finally, the Specialist must meet the medical surveillance requirements outlined within OSHA 1910.120.

On-Scene Incident Commander. Incident Commanders, who will assume control of the incident scene beyond the First Responder–Awareness level, shall receive at least 24 hours of training equal to the First Responder–Operations level. In addition, the employer must certify that the Incident Commander has competency in the following areas:

1. Know and be able to implement the local Incident Management System
2. Know how to implement the local Emergency Operations Plan (EOP)
3. Understand the hazards and risks associated with working in chemical protective clothing
4. Knowledge of the state emergency response plan and of the Federal Regional Response Team
5. Know and understand the importance of decontamination procedures

Skilled Support Personnel. These are personnel who are skilled in the operation of certain equipment, such as cranes and hoisting equipment, and who are needed temporarily to perform immediate emergency support work that cannot reasonably be performed in a timely fashion by emergency response personnel. It is assumed that these individuals will be exposed to the hazards of the emergency response scene.

Although these individuals are not subject to the HAZWOPER training requirements, they shall be given an initial briefing at the site prior to their participation in any emergency response effort. This briefing shall include elements such as instructions in using personal protective clothing and equipment, the chemical hazards involved, and the tasks to be performed. All other health and safety precautions provided to emergency responders and on-scene workers shall be used to ensure the health and safety of these support personnel.

Specialist Employees. These are employees who, in the course of their regular job duties, work with and are trained in the hazards of specific hazardous substances, and who will be called upon to provide technical advice or assistance to the Incident Commander at a hazmat incident. This would include industry responders, chemists, and related professional or operations employees. These individuals shall receive training or demonstrate competency in the area of their specialization annually.

The coordinating point for both planning and training activities at the local level is the LEPC. Ideally, among the LEPC membership are representatives from the following groups:

- Elected state and local officials
- Fire Department
- Law Enforcement
- Emergency Management
- Public health officials
- Hospital
- Industry personnel, including facilities and carriers
- Media
- Community organizations

The LEPC is specifically responsible for developing and/or coordinating the local emergency response system and capabilities. A primary concern is the identification, coordination, and effective management of local resources. Among the primary responsibilities of the LEPC are the following:

- Develop, regularly test, and exercise the Hazmat Annex of the Emergency Operations Plan.
- Conduct a hazards analysis of hazmat facilities and transportation corridors within the community.
- Receive and manage hazmat facility reporting information. This includes chemical inventories, Tier II reporting forms required under SARA, Title III, safety data sheets (SDSs) or chemical lists, and points of contact.
- Coordinate the Community Right-to-Know aspects of SARA, Title III.

In several communities, the LEPC has expanded its scope and responsibilities to adopt an all-hazards approach to emergency planning and management. Individuals desiring more information on both hazmat and all-hazards planning should consult the following websites:

- EPA Emergency Management—http://www.epa.gov/emergencies/index.htm
- Federal Emergency Management Agency (FEMA)—http://www.fema.gov
- International Association of Emergency Managers (IAEM)—http://www.iaem.com
- U.S. National Response Team—http://www.nrt.org
- U.S. Chemical Safety and Hazard Investigation Board—http://www.csb.gov. The Chemical Safety Board website also provides links to a number of CSB video clips developed as part of fixed facility incident investigations and which can be used as part of formal and refresher hazardous materials training programs.

Federal installations and military bases are also required to follow the EPA's right-to-know and pollution prevention regulations under Executive Order 12856 (August 3, 1993).

Risk Management Programs for Chemical Accidental Release Prevention (40 CFR Part 68). Promulgated under amendments to the Clean Air Act, this regulation requires that facilities that manufacture, process, use, store, or otherwise handle certain regulated substances above established threshold values develop and implement risk management programs (RMPs). The regulation is similar in scope to the OSHA Process Safety Management (OSHA 1910.119) standard, with the primary focus being community safety as compared to employee safety **Table 1-3**.

Risk management programs consist of three elements:

1. *Hazard assessment* of the facility, including the worst-case accidental release and an analysis of potential off-site consequences.
2. *Prevention program*, which addresses safety precautions, maintenance, monitoring, and employee training. The EPA believes that the prevention program should adopt and build upon the OSHA Process Safety Management standard.
3. *Emergency response considerations,* including facility emergency response plans, informing public and local agencies, emergency medical care, and employee training.

Hazard Communication (HAZCOM) Regulation (29 CFR 1910.1200). HAZCOM is a federal regulation that requires hazardous materials manufacturers and handlers to develop written Safety Data Sheets (SDS) on specific types of hazardous chemicals. These SDSs must be made available to employees who request information about a chemical in the workplace. Examples of information on SDSs include known health hazards, the physical and chemical properties of the material, first aid, firefighting and spill control recommendations, protective clothing and equipment requirements, and emergency telephone contact numbers. Under the HAZCOM requirements, hazmat health exposure information should be provided to emergency responders during the termination phase, and all exposures should be documented.

Under the Globally Harmonized System of Classification and Labeling of Chemicals (GHS), the MSDS term changed to the term **Safety Data Sheet (SDS)** as the impacts of global harmonization become more integrated into the regulatory and environmental health and safety communities. Additional information on global harmonization and its effects on chemical classification and labeling can be found on the OSHA website at http://www.osha.gov/dsg/hazcom/ghs.html.

Hazardous Materials Transportation Regulations (49 CFR 100–199). This series of regulations is issued and enforced by the U.S. DOT. The regulations govern container design, chemical compatibility, packaging and labeling requirements, shipping papers, transportation routes and restrictions, and so forth. The regulations are comprehensive and strictly govern how all hazardous materials are transported by highway, railroad, pipeline, aircraft, and by water.

Pipeline Regulations (49 CFR Part 190-199). The DOT Pipeline and Hazardous Materials Safety Administration (PHMSA) establishes rules and regulations governing the design, construction, operation, safety and maintenance of interstate pipelines. These include oil, gas, liquefied natural gas (LNG) and other hazardous liquid pipelines and facilities. Part 194, which outlines the oil spill emergency planning requirements for onshore oil pipelines, is of particular importance to local emergency response agencies.

National Contingency Plan or NCP (40 CFR 300, Subchapters A through J). This plan outlines the policies and procedures of the federal agency members of the National Oil and Hazardous Materials Response Team (also known as the National Response

Table 1-3 Process Safety Management Hazards Analysis Techniques

Hazards analysis is also an integral element of the Process Safety Management (PSM) process required by OSHA 1910.119—Process Safety Management of Highly Hazardous Chemicals, Explosives and Blasting Agents and EPA Part 68—*Risk Management Programs for Chemical Accidental Release Prevention.* Both regulations impact industries that manufacture, store, and use highly hazardous chemicals and explosives, including refineries and chemical and petrochemical manufacturing plants.

Hazards analysis methods commonly used by safety professionals within industry include the following:

- *What if Analysis.* This method asks a series of questions, such as, "What if Pump X stops running?" or "What if an operator opens the wrong valve?" to explore possible hazard scenarios and consequences. This method is often used to examine proposed changes to a facility.
- *HAZOP Study.* This is the most popular method of hazard analysis used within the petroleum and chemical industries. The hazard and operability (HAZOP) study brings together a multidisciplinary team, usually of five to seven people, to brainstorm and identify the consequences of deviations from design intent for various operations. Specific guide words ("No," "More," "Less," "Reverse," etc.) are applied to parameters such as product flows and pressures in a systematic manner. The study requires the involvement of a number of people working with an experienced team leader.
- *Failure Modes, Effects, and Criticality Analysis (FMECA).* This method tabulates each system or unit of equipment, along with its failure modes, the effect of each failure on the system or unit, and how critical each failure is to the integrity of the system. Then the failure modes can be ranked according to criticality to determine which are the most likely to cause a serious accident.
- *Fault Tree Analysis.* A formalized deductive technique that works backward from a defined accident to identify and graphically display the combination of equipment failures and operational errors that led up to the accident. It can be used to estimate the likelihood of events.
- *Event Tree Analysis.* A formalized deductive technique that works forward from specific events or sequences of events that could lead to an incident. It graphically displays the events that could result in hazards and can be used to calculate the likelihood of an accident sequence occurring. It is the reverse of fault tree analysis.

Under the original rule making, RMP results were to be made available to the public to ensure that the community was aware of the potential risks posed by specific facilities and chemicals. However, given the potential for this same information to be used as background and intelligence for a terrorist attack, its distribution and release to the general public is now being more closely controlled. Individuals desiring the latest information should consult the EPA's Emergency Management website at http://www.epa.gov/emergencies/index.com.

Team, or the NRT). The regulation provides guidance for emergency responses, remedial actions, enforcement, and funding mechanisms for federal government response to hazmat incidents. The NRT is chaired by the EPA, while the vice-chairperson represents the U.S. Coast Guard (USCG).

Each of the 10 federal regions also has a Regional Response Team (RRT) that mirrors the make-up of the NRT. RRTs may also include representatives from state and local government and Indian tribal governments.

When the NRT or RRT is activated for a federal response to an oil spill, hazmat, or terrorism event, a Federal On-Scene Coordinator (FOSC) will be designated to coordinate the overall response. For hazmat incidents, the FOSC will represent either the EPA or the USCG based upon the location of the incident. If the release or threatened release occurs in coastal areas or near major navigable waterways, the USCG will usually assume primary OSC responsibility. If the situation occurs inland and away from navigable or major waterways, the EPA will serve as the FOSC. Local emergency responders should contact the EPA or USCG personnel within their region to determine which agency has primary responsibility and will act as the federal OSC for their respective area. The lead state environmental agency will typically function as the State On-Scene Coordinator (SOSC).

If the incident is a terrorism-related event, the Federal Bureau of Investigation (FBI) will assume the role as federal OSC during the emergency response phase.

Facility and Modal Security Regulations. In the aftermath of the terrorist attacks of September 11, 2001, numerous regulations and programs pertaining to the security and protection of hazardous materials critical infrastructure and key resources (CI/KR) have been enacted. Examples include the DHS Chemical Facility Anti-Terrorism (CFAT) standards, as well as modal-specific security programs. Enforcement is spread among a number of federal agencies, based upon the mode of transportation and the nature of the facility operations. In addition to DHS, EPA, and DOT, other federal agencies involved in security planning, inspections, and enforcement include the U.S. Department of Energy, USCG, and the Transportation Security Administration (TSA).

State and Local Regulations

Each of the 50 states and the U.S. territories maintains an enforcement agency that has responsibility for hazardous materials. The three key players in each state usually consist of the State Fire Marshal, the State Occupational Safety and Health Administration, and the State Department of the Environment (sometimes known as Natural Resources or Environmental Quality). While there are many variations, the fire marshal is typically responsible for the regulation of flammable liquids and gases due to the close relationship between the flammability hazard and the fire prevention code, while the state environmental agency would be responsible for the development and enforcement of environmental safety regulations.

While known by various titles, most states have a government equivalent of the federal OSHA. Approximately 23 states have adopted the federal OSHA regulations as state law. This method of adoption has increased the level of enforcement of hazardous materials regulations, such as the Hazardous Waste and Emergency Response regulation described previously.

State governments also maintain an environmental enforcement agency and environmental crimes unit that usually enforces the federal RCRA, CERCLA, and CAA laws at the local level. Increased state involvement in hazardous waste regulatory enforcement has significantly increased the number of hazardous materials incidents reported. This increase is expected to continue in the future and will continue to generate more fire service activity at the local level.

Some state and local governments have a special set of codes and regulations that apply to pipeline operations. The statues under which the U.S. DOT's Federal Office of Pipeline Safety (OPS) operates provide for state assumption of part or all of the intrastate pipeline regulatory and enforcement responsibility through annual certifications and agreements. Under these agreements, OPS provides funding to defray the costs of their overall pipeline safety program and inspections, while the state must adopt the federal OPS regulations. A state or local government may adopt additional or more stringent regulations, provided they are consistent and compatible with the OPS regulations.

Voluntary Consensus Standards

Standards developed through the voluntary consensus process play an important role in increasing both workplace and public safety. Historically, a voluntary standard improves over time as each revision reflects recent field experience and adds more detailed requirements. As users of the standard adopt it as a way of doing business, the level of safety gradually improves over time.

Consensus standards are also updated more frequently than governmental regulations and can usually be developed more quickly to meet issues of the day. For example, in response to the need for emergency response personnel operating at terrorism incidents involving dual-use industrial chemicals or chemical or biological agents, NFPA approved *NFPA 1994—Protective Ensembles for Chemical/Biological Terrorism Incidents* to provide guidance in the selection and use of protective clothing and equipment.

In many respects, a voluntary consensus standard provides a way for individual organizations and corporations to self-regulate their business, discipline, or profession. Among the standards development organizations that have developed consensus standards for the hazmat and homeland security emergency response community are the NFPA, the American Society of Testing and Materials (ASTM), and the National Institute of Justice (NIJ).

Among the most important consensus standards used within the hazmat response community are the following:

NFPA Technical Committee on Hazardous Materials Response Personnel. This Technical Committee is responsible for documents on the requirements for professional qualifications, professional competence, training, procedures, and equipment for emergency responders to hazardous materials/WMD incidents. The Committee is responsible for the following documents:

- NFPA 472—*Standard for Competence of Responders to Hazardous Material / Weapons of Mass Destruction Incidents.* Additional information is provided below.
- NFPA 473—*Standard for Professional Competence of EMS Personnel Responding to Hazardous Material Incidents.* Additional information is provided below.

Responder Tip

Historically, the voluntary consensus standards community has the ability to react to changing events and situations much quicker than the regulatory process. For example, shortly after the 1995 Tokyo, Japan subway attack involving sarin, the NFPA's Hazardous Materials Technical Committee issued a Tentative Interim Amendment (TIA) that published recommended training requirements for chemical agents.

- NFPA 475—*Recommended Practice for the Organization and Management of a Hazardous Materials/Weapons of Mass Destruction Emergency Response Program.* This document is currently under development with an anticipated NFPA publication date of 2016. The scope of this recommended practice is to establish a common set of criteria for the organization, management, and deployment of personnel, resources, and programs for those public or private entities that are responsible for the hazardous materials/WMD emergency preparedness function.
- NFPA 1072—*Standard on Hazardous Materials Response Personnel Professional Qualifications.* This document has a publication date of 2017. This will be a hazardous materials ProQual standard that is Job Performance Requirements (JPR)-based for the following response levels: Awareness, Operations Core, Operations Mission Specific, HMT, and Incident Commander.

NFPA 472—Standard for Competence of Responders to Hazardous Material / Weapons of Mass Destruction Incidents. The purpose of NFPA 472 is to specify minimum competencies for those who will respond to hazardous materials/WMD incidents. The overall objective is to reduce the number of accidents, injuries, and illnesses during response to hazmat incidents and to prevent exposure to hazmats to reduce the possibility of fatalities, illnesses, and disabilities affecting emergency responders.

It is important to recognize that NFPA 472 applies to all emergency responders who respond to the emergency phase of a hazardous materials/WMD incident, regardless of the individual's response discipline. The standard is based upon the philosophy that emergency responders should be trained to perform their expected tasks, regardless of their discipline and organizational affiliation. Technical committee membership represents all elements of the hazardous materials response community, including the fire service, law enforcement, EMS, hazardous materials manufacturing and transportation sectors, military and federal government agencies.

Although there are commonalities between OSHA 1910.120 and NFPA 472, there are also distinct differences. This relationship can be summarized as follows:

- OSHA 1910.120 is the LAW of the land. NFPA 472 is a VOLUNTARY consensus standard. However, if you meet the requirements of NFPA 472 you will exceed the OSHA HAZWOPER emergency response training requirements (paragraph q.6).

Responder Tip

The NFPA hazardous materials standards and recommended practices are incorporated by reference into a number of other state and federal regulations and guidance documents, including OSHA 1910.120 and the FEMA Resource Typing Definitions.

- OSHA 1910.120 has not been revised since its initial final rule promulgation in 1989. In contrast, NFPA 472 (2013 edition) is in its fifth edition of the standard.

The NFPA 472 levels of training parallel those listed within OSHA 1910.120 for the primary response categories, with the exception that the Hazardous Materials Specialist has been deleted and the Private Sector Specialist Employee has been expanded upon. In addition, NFPA 472 includes a number of additional levels of training that are optional in scope and have been developed to meet specific hazardous materials training and response needs.

Table 1-4 summarizes the levels of training between OSHA 1910.120(q)(6) and NFPA 472.

NFPA 473—Standard for Professional Competence of EMS Personnel Responding to Hazardous Material Incidents. The purpose of NFPA 473 is to specify minimum requirements of competence and to enhance the safety and protection of response personnel and all components of the EMS system. The overall objective is to reduce the number of EMS personnel accidents, exposures, injuries, and illnesses resulting from hazmat incidents.

NFPA 473 is based on the following principles:

1. Emergency responders will be protected first, and then patients would be managed.
2. No advanced life support (ALS) measures would be performed in the hot zone.
3. EMS responders at hazardous materials incidents would essentially be the "first receivers of the first receivers," taking patient care from the scene to definitive hospital care.
4. No distinction is made in NFPA 473 between the agencies that "own" the EMS personnel, i.e., no distinction is made between a fire, rescue, law enforcement, or military EMS provider. The standard focuses on establishing the competency of the responder.
6. NFPA 473 provides mission-specific competencies for EMS personnel missions beyond the competencies of the ALS responder, including:
   - ALS Responder Assigned to an HMRT
   - ALS Responder Assigned to Provide Clinical Interventions at a Hazmat/WMD Incident

Responder Tip

NFPA standards are voluntary and consensus-based, which means that they can be adopted entirely or in part by the authority having jurisdiction (AHJ). In contrast, OSHA is the LAW of the land, and compliance is mandatory.

Table 1-4 OSHA 1910.120(q)(6)—NFPA 472 Levels of Training Comparison

OSHA 1910.120(q)(6) Responder Training Levels	NFPA 472 Responder Training Levels
First Responder at the Awareness Level	Awareness-Level Personnel
First Responder at the Operations Level	Core Competencies for Operations-Level Responders
No OSHA 1910.120(q) Equivalent	Operations-Level Responders Assigned Mission-Specific Competencies ■ Personal Protective Equipment ■ Mass Decontamination ■ Technical Decontamination ■ Evidence Preservation & Sampling ■ Product Control ■ Air Monitoring & Sampling ■ Victim Rescue & Recovery ■ Response to Illicit Laboratory Incidents ■ Improvised WMD Dispersal Device Disablement/Disruption & Operations at Improvised Explosive Incidents
Hazardous Materials Technician	Hazardous Materials Technician (HMT)
No OSHA 1910.120(q) Equivalent	■ HMT with a Tank Car Specialty ■ HMT with a Cargo Tank Specialty ■ HMT with an Intermodal Specialty ■ HMT with a Marine Tank and Non-Tank Vessel Specialty ■ HMT with a Flammable Liquids Bulk Storage Specialty ■ HMT with a Flammable Gas Bulk Storage Specialty ■ HMT with a Radioactive Material Specialty
Hazardous Materials Specialist	No NFPA 472 Equivalent
On-Scene Incident Commander	Incident Commander
Specialist Employee	Specialist Employee ■ Specialist Employee C ■ Specialist Employee B ■ Specialist Employee A
Skilled Support Personnel	No NFPA 472 Equivalent
No OSHA 1910.120(q) Equivalent	Hazardous Materials Officer (i.e., Hazmat Group Supervisor)
No OSHA 1910.120(q) Equivalent	Hazardous Materials Safety Officer (i.e., Assistant Safety Officer–Hazardous Materials)

- ALS Responders Assigned to Treatment of Smoke Inhalation Victims

NFPA Technical Committee on Hazardous Materials Protective Clothing and Equipment (NFPA 1991, 1992, 1994). This technical committee is responsible for the development of standards and documents pertaining to the use of personal protective clothing and equipment (excluding respiratory protection) by emergency responders at hazardous materials incidents. The committee scope includes personal protective equipment (PPE) selection, care, and maintenance.

Three hazmat protective clothing standards have been developed:

1. NFPA 1991—*Standard on Vapor-Protective Ensembles for Hazardous Materials Emergencies*
2. NFPA 1992—*Standard on Liquid Splash-Protective Ensembles and Clothing for Hazardous Materials Emergencies*
3. NFPA 1994—*Standard on Protective Ensembles for First Responders to CBRN Terrorism Incidents*

Other Standards Organizations

There are many other important standards-writing bodies that affect the hazardous materials emergency response community, including the American National Standards Institute (ANSI), the American Society for Testing and Materials (ASTM), the Compressed Gas Association (CGA), the Safety Equipment Institute (SEI), and the American Petroleum Institute (API). Each of these organizations approves or creates standards ranging from hazardous materials container design to personal protective clothing and equipment.

Individuals desiring more information on the standards-writing organizations discussed here should consult the following websites:

- American National Standards Institute (ANSI)—http://www.ansi.org/
- American Petroleum Institute (API)—http://api-ec.api.org/
- American Society for Testing and Materials (ASTM)—http://www.astm.org/
- The Chlorine Institute—http://www.chlorineinstitute.org/
- Compressed Gas Association (CGA)—http://www.cganet.com
- National Fire Protection Association (NFPA)—http://www.nfpa.org/
- Safety Equipment Institute (SEI)—http://www.seinet.org/

Standard of Care

"Standard of Care" is a widely accepted practice or standard that is followed by the majority of U.S. emergency response organizations. It represents the *minimum* accepted level of hazardous materials emergency service that should be provided regardless of location or situation.

Standard of care is established by existing laws and regulations, as well as voluntary consensus standards and recommended practices (e.g., NFPA 472 and 473). Many emergency responders are surprised to learn that the standard of care is also determined by local protocols and practices and what has been accepted in the past (i.e., precedent). So, when you consider the broad definition of *standard of care,* you can see how important it is for an Incident Commander or Hazmat Program Manager to be well versed in regulatory requirements, as well as what other response organizations are doing to manage hazardous materials responses.

Standard of care is also influenced by legal findings and case law precedents established through the judicial system. This, in turn, allows your actions to be judged based upon what is expected of someone with your level of training and experience acting in the same or similar situation.

Standard of care is a dynamic element and historically has improved over time. Looking back in history, consider the hazardous materials "washdowns" of the 1970s that were standard practices. We just flushed the spill into the storm sewer system and "made it go away." Today, that same procedure is viewed as a poor operating practice and a potential environmental crime.

Hazmat Program Managers should become familiar with the OSHA General Duty Clause (also known as Section 5(a)(1) of the OSH Act), which covers areas of occupational safety and health not covered by a specific OSHA regulation. Historically, OSHA inspectors have issued citations using industry voluntary consensus standards as the basis of the citation for a wide range of hazards. Given that HAZWOPER was initially promulgated in 1989 and has not been updated, Hazmat Program Managers should recognize the potential of voluntary consensus standards, such as NFPA 472 and 472 and other related standards, to potentially be used in the future as the basis for regulatory citations.

How do you know whether you're meeting the standard of care? Here are some questions that you should be asking:

1. *Our operations must be legal and within the requirements of the law.* Responders must be familiar with existing laws and regulations, including 29 CFR 1910.120 or HAZWOPER. Anyone who works in the hazmat response business should know this regulation well.
2. *Our actions and decisions must be consistent with voluntary consensus standards and recommended practices.* Most of what we do in hazmat planning and response is guided by national consensus standards developed by organizations such as NFPA, ANSI, and others. Command personnel working in the emergency response profession should be intimately familiar with not only what the standard requires, but the basis for that requirement. Don't just read the standard—you should also be familiar with the rationale and application of the requirement. Consulting publications such as the NFPA's *Hazardous Materials Handbook* can be helpful.
3. *Our actions and decisions to control a problem should have a technical foundation.* This is especially important when conducting hazard and risk evaluation. Responders must have a solid background in risk assessment and both basic chemistry and physical science.
4. *Our actions and decisions must be ethical.* Most of us learned what is right and wrong and fair or unfair from individuals we respect (e.g., parents, guardians, or mentors). If the decision you are making doesn't feel right and your internal "ethics meter" moves over into the red zone, you need to reevaluate your actions and decisions.

Responder Tip

Emergency responders must recognize that (1) a standard of care exists; and (2) that the "highbar" is constantly moving upward. Training and continuing education are among the best ways to ensure you will be able to provide the standard of care mandated by both society and the hazmat profession over time.

The Hazardous Materials Management System

The fire problem in the United States has traditionally been managed by fire suppression operations at the expense of prevention activities. In 1973, Congress issued *America Burning*, a historical report on the nation's fire problem. The report significantly influenced the way we manage the fire problem today. Community master planning, public education, residential sprinklers, improved fire code enforcement, and fire protection engineering are some examples of the changes influenced by this landmark report. What we accept as the standard approach to fire prevention and protection today primarily did not exist prior to 1970.

The lessons learned in analyzing the U.S. fire problem can also be applied to managing the hazardous materials problem. If we want to be effective in managing and controlling hazmat risks, we must approach the problem from a larger, integrated orientation that views the issue from a systems perspective.

There are four key elements in a hazardous materials management systems approach: (1) planning and preparedness, (2) prevention, (3) response, and (4) clean-up and recovery.

If the community or a facility is performing its responsibilities within the planning and prevention functions, you should see a reduction in the number and severity of response and clean-up activities.

Planning and Preparedness

Planning is the first and most critical element of the system. Planning is fundamentally a process to manage risk. Risk management is a process by which context is defined, risks are identified and assessed, and courses of action for managing those risks are analyzed, decided upon, implemented, monitored, and evaluated. In short, planning is a tool that allows for systematic risk management to reduce or eliminate future risks.

The ability to develop and implement an effective hazmat management plan depends upon two elements: hazards analysis and the development of a hazmat emergency operations plan.

- **Hazards analysis**—Analysis of the hazardous materials present in the community or facility, including their location, quantity, specific physical (i.e., how they behave) and chemical (i.e., how they harm) properties, previous incident history, surrounding exposures, and risk of release. Hazards analysis is the foundation of the planning process.
- **Contingency (emergency) planning**—A comprehensive and coordinated response to the hazmat problem. This response builds upon the hazards analysis process

Voices of Experience

As a technical specialist on a regional hazmat response team, I learned early on that providing timely, well-vetted information on dynamic variables and their effect on an active or potential hazardous materials release is essential for both the preplanning (Hazard Analysis) as well as the response phase (Contingency Planning) for hazmat incidents. Applying technician-level training—in conjunction with resources such as commodity flow studies, Tier II chemical inventory reports, and knowledge of unique hazards posed by terrain and weather phenomena within a given response area—should result in a safer and more efficient hazardous materials response based on sound risk management principles.

As an example, our team was dispatched late one afternoon to a rollover accident involving an MC 331 propane delivery truck that failed to navigate a corner on a steep and icy mountain road. Being January in the Midwest with a storm system settled in the area, the weather was cold and overcast with light snowfall. After mapping the location, the first piece of information passed to the team leader was—due to the combination of road grade, corner radiuses, and icy conditions—we would not be able to respond with our normal command and equipment vehicle. Having preplanned for such conditions, a reference kit containing a ruggedized laptop pre-loaded with current chemical response databases along with essential printed reference materials was loaded into a four-wheel drive pickup truck along with minimal personnel and response gear. The reference materials, staffing requirements, and needed equipment had all been predesignated in a product-specific, winter-condition force package developed through our hazard analysis and contingency planning processes. While it would be prohibitive on many levels to develop product-specific force packages for every chemical identified in a response area, we had used commodity flow studies and historical response data to determine that 90% of our calls were for hydrocarbon fuels and 17% of those calls were for propane and natural gas. Given the frequency of incidents involving these specific products, additional time and energy was invested in advanced planning specific to these products.

Upon arrival, our team was assigned as the Hazmat Branch under the ICS structure established by the first-due fire department. Our site characterization revealed that a chassis-mounted 3000 gallon MC 331 had gone over a guardrail and rolled one and a half times down a steep 40' rocky embankment, coming to rest upside down in a ravine. There was a steady stream of liquid propane leaking from the valves at the rear of the overturned truck, which, due in part to the 8°F air temperature, resulted in a visible pool of liquid propane under the valving cage. The damage assessment revealed several dents and gouges in the steel tank with up to 2" of deformation. The information passed to the Hazmat Branch Director and IC was that the atmospheric conditions were very stable, due to the combination of low ground temps, cold air, and very little wind. Because of these conditions we could expect poor dispersion of the propane vapor and that the vapor plume would most likely stay intact for a significant amount of time and distance—which might necessitate aggressive ignition control tactics and LEL measurements near some residential structures located at the downhill mouth of the ravine. In addition, a 10 degree increase in air temperature forecasted for early the following morning would result in a 1.5% expansion of the liquid propane still contained within the MC 331, creating an unacceptable pressure increase in an already significantly damaged tank.

After deploying LEL meters in areas of concern, a mobile weather station was used to develop a plume model using ALOHA software. Technicians dressed in bunker gear with SCBAs, made entry, placed a remote camera on the 331's pressure gauge, and eventually were able to stop the flow of liquid propane by shutting off a valve. Contingency planning, and the associated training and drills that are developed based on those contingency plans, allowed a small number of technicians to gather the pertinent data, provide atmospheric monitoring, and offer mitigation capabilities to Command in order to reach a successful resolution to the incident. In the final stages of this incident, ICS components were used to coordinate the efforts of multiple response agencies as well as private industry to offload and recover the damaged MC 331.

One last lesson from this response occurred when the private propane company attempted to offload the propane in an extremely unsafe manner simply because the operators did not want to be out in the cold. Command was informed of this immediately, which brought the dangerous practice to a quick halt.

In the end, contingency plans developed through a comprehensive hazard analysis and a solid ICS structure, along with timely and relevant information, led to a safe and successful outcome. With the air temp the following day still well below 20°F, the flaring operation to remove the last of the propane was a welcome bit of warmth.

Captain Dustin Willett
Rapid City Regional Hazardous Materials Emergency Response Team
Rapid City, South Dakota

and recognizes that no single public or private sector agency is capable of managing the hazmat problem by itself. For private sector organizations this plan may be a separate, standalone Hazmat Emergency Operations Plan or may be an element within an Integrated Contingency Plan (ICP). In the public sector, hazardous materials issues may be a functional or hazard-specific annex to the community Emergency Operations Plan.

The data and information generated by these activities will allow emergency managers to assess the potential risk to the community or facility and to develop and allocate resources as necessary. Many communities and facilities have adopted an "all-hazards" approach to emergency planning that develops a basic framework that is then applied to any emergency scenario, including natural hazards (e.g., floods, storms), technological hazards (e.g., utility outages), and terrorist events.

Hazards Analysis

Hazards analysis is the foundation of the planning process. It should be conducted for every location designated as having a moderate or high probability for a hazmat incident. In addition to risk evaluation, vulnerability—what is susceptible to damage should a release occur—must also be examined.

A hazards analysis provides the following benefits:

- It lets emergency responders know what to expect.
- It provides planning for less frequent incidents.
- It creates an awareness of new hazards.
- It may indicate a need for preventive actions, such as monitoring systems, remote isolation, and process modifications.
- It offers an opportunity to evaluate mitigation strategies using reduced chemical inventories or alternative chemicals to lower the consequences of an event.
- It increases the chance for successful emergency response operations.

An evaluation team familiar with the facility or response area can facilitate the hazards analysis process. For example, within a facility this team may include safety, environmental, and industrial/process engineering professionals. Similarly, fire officers and members from each battalion or district, as well as representatives from the prevention and hazmat section, would be appropriate for a fire department. The primary concern here is geographic, as most responders are very familiar with their "first due" area.

There are four components of a hazards analysis program:

1. *Hazards identification*—Provides specific information on situations that have the potential for causing injury to life or damage to property and the environment due to a hazardous materials spill or release. Hazards identification will initially be based upon a review of the history of incidents within the facility or industry, or evaluating those facilities that have submitted chemical lists and reporting forms under SARA, Title III and related state and local right-to-know legislation. Information should include the following:
   - Chemical identification
   - Location of facilities that manufacture, store, use, or move hazardous materials
   - The type(s) and design of chemical container or vessel
   - The quantity of material that could be involved in a release
   - The nature of the hazard associated with the hazmat release (e.g., fire, explosion, toxicity)
   - The presence of any fixed suppression or detection systems
   - The level of physical security to prevent theft by criminals or terrorists
2. *Vulnerability analysis*—Identifies areas that may be affected or exposed and what facilities, property, or environment may be susceptible to damage should a hazmat release occur. A comprehensive vulnerability analysis provides information on:
   - The size/extent of vulnerable zones—specifically, what size area may be significantly affected as a result of a spill or release of a known quantity of a specific hazmat under defined conditions
   - The population, in terms of numbers, density, and types—for example, facility employees, residents, special occupancies (e.g., schools, hospitals, nursing homes)
   - Private and public property that may be damaged, including essential support systems (e.g., water supply, power, communications) and transportation corridors and facilities
   - The environment that may be affected, and the impact of a release on sensitive environmental areas and wildlife
3. *Risk analysis*—Assesses (1) the probability or likelihood of an accidental release, and (2) the actual consequences that might occur. The risk analysis is a judgment of incident probability and severity based upon the previous incident history, local experience, and the best available hazard and technological information.

   In today's environment, we must also consider the threat of hazardous materials and WMD being used for criminal or terrorist activity. Emergency planners should coordinate and share risk analysis findings with law enforcement officials in their local jurisdiction. The U.S. Department of Homeland Security provides a number of resources and training courses specific to the threat and risk analysis process.
4. *Emergency response resources evaluation*—Based upon the potential risks, considers emergency response resource requirements. These would include personnel, equipment, and supplies necessary for hazmat control and mitigation, EMS, protective actions, traffic control, and inventories of available equipment and supplies and their functional status. For example, are Class B firefighting foam supplies adequate to control and suppress vapors from a gasoline tank truck rollover? What are our capabilities to initiate mass decon operations? What level of personal protective clothing is provided to law enforcement personnel?

Time and resources will dictate the depth and extent to which the hazards analysis can be completed. The focus is on the hazards created by the most common and most hazardous substances. Even the simplest plan will be better than no plan at all.

Responder Tip

A completed hazards analysis should allow emergency managers and planners to determine what level of response to emphasize, what resources will be required to achieve that response, and what type and quantity of mutual aid and other support services will be required.

Contingency and Emergency Operations Planning

Hazardous materials management is a multi-disciplined issue that goes beyond the resources and capabilities of any single agency or organization. There will be a variety of "players" responding to a major hazmat emergency, and the emergency operations plan and related procedures will establish the framework for how the emergency response effort will operate. To manage effectively the overall hazmat problem within the community, a comprehensive planning process must be initiated. This effort is usually referred to as contingency planning or emergency planning.

There are many federal, state, and local requirements that apply to emergency planning. In addition to FEMA's *Comprehensive Preparedness Guide (CPG) 101—Developing and Maintaining Emergency Operations Plans*, the one that most directly affects hazardous materials emergency planning is Title III of the Superfund Amendments and Reauthorization Act of 1986. SARA, Title III requires the establishment of SERCs and LEPCs. Title III also outlines specific requirements covering factors such as EHSs, threshold planning quantities, make-up of LEPCs, as well as the dissemination of the planning, chemical lists, and SDS information to the general public, facility inventories, and toxic chemical release reporting.

Figure 1-2 provides an overview of the emergency management planning process, including the following:

1. **Form a Collaborative Planning Team.** Planning requires community involvement throughout the process. Experience has shown that plans prepared by only one person or agency are doomed to failure. Emergency response requires trust, coordination, and cooperation. The diversity of the planning team leads to a better planning product. Remember, there is no single agency (public or private) that can effectively manage a major hazmat emergency alone.
2. **Understand the Situation.** Effective risk management is dependent upon a consistent evaluation of the hazards a facility or community may face. Components of this risk-based process should include (1) identifying the threats and hazards, and (2) assessing the level of risks. This should include an analysis of current prevention and response capabilities.
3. **Determine Goals and Objectives.** Based upon the output of the risk assessment process, determine operational priorities, goals, and objectives.
4. **Plan Development.** Elements of the plan development process would include the development and analysis of courses of action; the identification of resources and estimated current operational capabilities; and the identification of information and intelligence needs.
5. **Plan Preparation, Review, and Approval.** Depending upon the AHJ, there are two approaches to this step: (1) Develop or revise a hazmat appendix or a section of a multi-hazard emergency operations plan, or (2) develop or revise a single-hazard plan specifically for hazardous materials. Once the plan is written, the respective planning groups involved must review and approve the document.
6. **Plan Implementation and Maintenance.** Every emergency plan must be evaluated and kept up to date through the review and analysis of actual responses, simulation exercises, and the regular collection of new data and information.

Although emergency planning is essential, the completion of a plan does not guarantee that the facility or community is actually prepared for a hazmat incident. Planning is only one element of the total hazmat management system.

Other Planning Documents

There are other planning related documents that support the overall emergency preparedness process. These can include:

- *Pre-Incident Plans.* These are typically built around a specific hazardous materials/WMD target hazard or operational scenario. Preplanning allows planners and responders to bring a higher level of tactical, operational, and safety focus upon specific threats and hazards that may exist within the community. SARA, Title III reporting facilities and LEPC information are typically a good starting point for determining the priority of pre-incident plan development. Pre-incident plans can be found in a range of formats, including paper-based and electronic-based.
- *Standard Operating Procedures/Guidelines (SOP/SOG).* In the emergency response community, SOP/SOGs tend to be more tactical focused and provide the specific details of the preferred method(s) of performing a single operational function or a number of inter-related functions. In short, SOP/SOGs bring consistency to emergency response operations.
- *Field Operations Guides (FOG).* The FOG is a short-form version of the SOP/SOG and serves as a resource document. They are typically packaged as a durable pocket guide or as a Smart Phone application.
- *Job Aids.* These are checklists or other materials that help users perform a specific task. They may range from checklists and reporting templates to specific operating instructions and task lists.

CAER and TRANSCAER

Two highly successful programs that have been established to facilitate public and private sector planning efforts are the Community Awareness and Emergency Response (CAER) and Transportation Community Awareness and Emergency Response (TRANSCAER) programs. Both programs are voluntary national outreach efforts that focus on assisting communities to prepare for and respond to a possible hazardous materials incident. CAER programs are primarily focused upon local fixed facilities, while the TRANSCAER program focuses on the safe transportation and handling of hazardous materials. Additional information on TRANSCAER and its regional efforts can be found at http://www.transcaer.com.

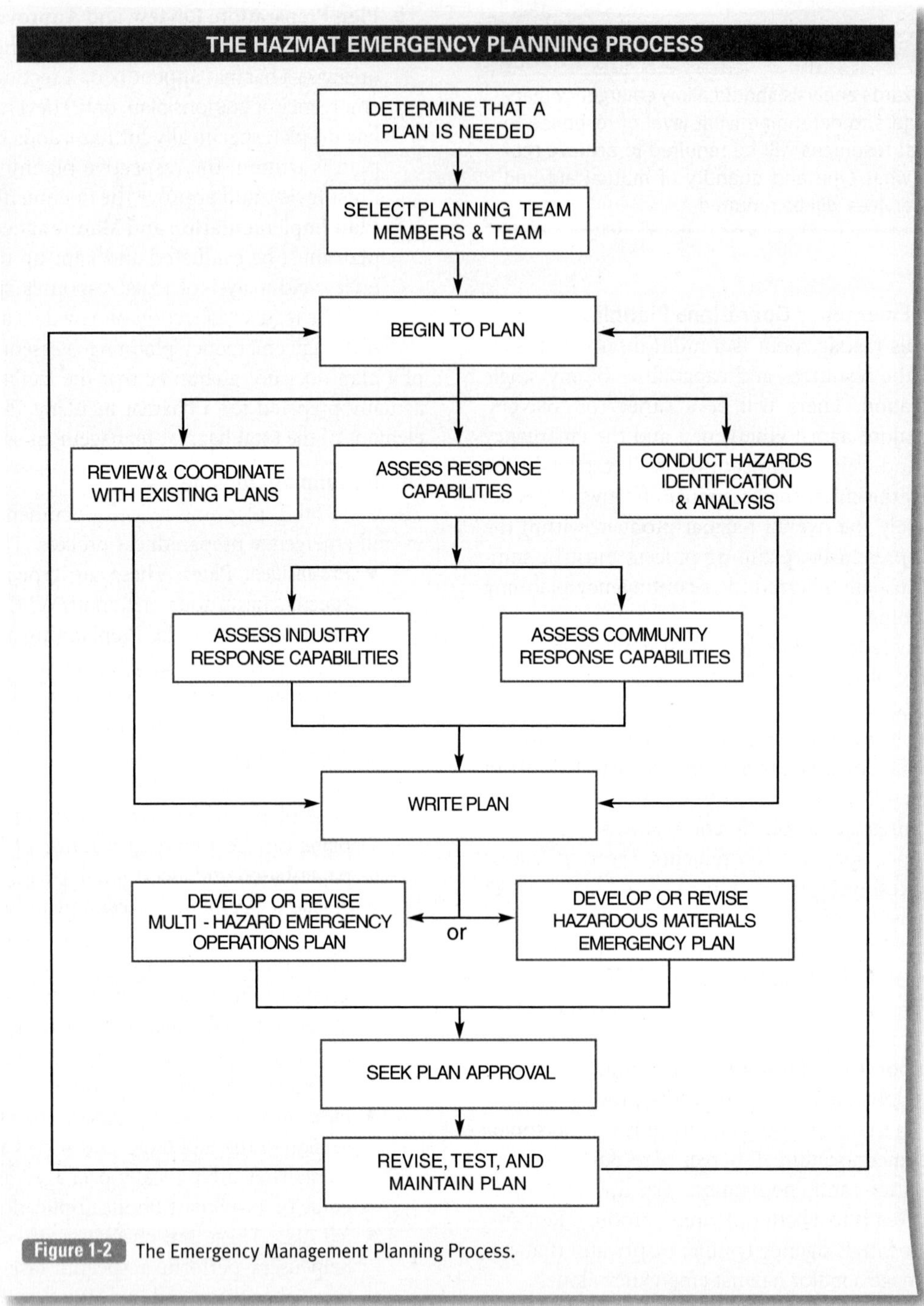

Figure 1-2 The Emergency Management Planning Process.

Prevention

The responsibility for the prevention of hazmat releases is shared between the public and private sectors. Because of their regulatory and enforcement capabilities, however, public sector agencies generally receive the greatest attention and often "carry the biggest stick."

Hazmat Process, Container Design, and Construction Standards

Almost all hazardous materials facilities, containers, and processes are designed and constructed to some standard. This standard may be based upon voluntary consensus standards, such as those developed by the NFPA and ASTM, or on government regulations. Many major petrochemical companies, hazardous materials companies, and industry trade associations (e.g., Chlorine Institute, Compressed Gas Association) have also developed their own respective engineering standards and guidelines.

All containers used for the transportation of hazardous materials are designed and constructed to both specification and performance regulations established by U.S. DOT. These regulations can be referenced in CFR Title 49. In certain situations, hazardous materials may be shipped in non-DOT specification containers that have received a DOT exemption.

A recent addition to the voluntary consensus standards community has been the development of NFPA 400—*The*

Hazardous Materials Code. This standard applies to the storage, use, and handling of hazardous materials in all occupancies and facilities, and brings a number of historically separate NFPA documents into a single standard. Hazardous materials classes falling within the scope of NFPA 400 include:

- Ammonium nitrate solids and liquids
- Compressed gases and cryogenic fluids
- Flammable solids
- Water-reactive solids and liquids
- Pyrophoric solids and liquids
- Unstable (reactive) solids and liquids
- Oxidizers—solids and liquids
- Organic peroxide formulations
- Toxic and highly toxic solids and liquids
- Corrosive solids and liquids

It should be noted that flammable liquids continue to be covered by NFPA 30—*The Flammable and Combustible Liquids Code*, while liquefied petroleum gases remain under NFPA 58—*The Liquefied Petroleum Gas Code*. In addition, Article 80 of the Uniform Fire Code is widely used throughout the western U.S. states.

Inspection and Enforcement

Fixed facilities, transportation vehicles, and transportation containers are normally subject to some form of inspection process. These inspections can range from comprehensive and detailed inspections at regular intervals to visual inspections each time a container is loaded. Fixed facilities will commonly be inspected by state and federal OSHA and EPA inspectors, in addition to state fire marshals and local fire departments. It should be recognized that many of these inspections will focus upon fire safety and life safety issues and may not adequately address either the environmental or process safety issues.

Transportation vehicle inspection is generally based upon criteria established within Title 49 CFR. The enforcing agency is often the state police, but this will vary according to the individual state, the hazardous materials being transported, and the mode of transportation. Some local fire departments routinely perform inspections of hazardous materials cargo tank trucks.

Among the U.S. DOT agencies with hazardous materials regulatory responsibilities are the following:

- *Office of Hazardous Materials Safety (OHMS)* of the *Pipeline and Hazardous Materials Safety Administration (PHMSA)*. Responsible for all hazardous materials transportation regulations except bulk shipment by ship or barge. Includes designating and classifying hazardous materials, container safety standards, label and placard requirements, and handling, stowing, and other in-transit requirements. Serves as the DOT representative to the National Response Team, supports the National Response Center operation in coordination with the USCG, and serves as the DOT liaison with FEMA on hazmat transportation issues. Also serves as a focal point for hazardous materials studies and reports by the National Transportations Safety Board (NTSB).
- *Office of Hazardous Materials Enforcement (OHME)* of the Office of Hazardous Materials Safety (OHMS). Inspection and enforcement staff determine compliance with safety standards by inspecting entities that offer hazardous materials for transportation; manufacture, rebuild, repair, recondition, or retest packaging used to transport hazardous materials; and handle intermodal transfers of hazardous materials. Responsible for the management and coordination of PHMSA's hazardous materials inspection and enforcement program.
- *Office of Pipeline Safety (OPS)* of the *Pipeline and Hazardous Materials Safety Administration (PHMSA)*. Administers DOT's national regulatory program to ensure the safe transportation of natural gas, petroleum, and other hazardous materials by pipeline. OPS develops regulations and other approaches to risk management to ensure safety in design, construction, testing, operation, maintenance, and emergency response of pipeline facilities. Provides technical and resource assistance for state pipeline safety programs to ensure oversight of intrastate pipeline systems and educational programs at the local level. Serves as a DOT liaison with the Department of Homeland Security and FEMA on matters involving pipeline safety.
- *Federal Railroad Administration (FRA)*. Responsible for enforcement of regulations relating to hazardous materials carried by rail or held in depots and freight yards.
- *Federal Aviation Administration (FAA)*. Responsible for the enforcement of regulations relating to hazardous materials shipments on domestic and foreign carriers operating at U.S. airports and in cargo handling areas.
- *U.S. Coast Guard (USCG)*. Responsible for the inspection and enforcement of regulations relating to hazardous materials in port areas and on domestic and foreign ships and barges operating in the navigable waters of the United States. The Coast Guard also plays a major role in providing Anti-Terrorism/Force Protection (AT/FP) to ensure the security of our major ports along the navigable waterways of the United States. The Coast Guard also supports the operation of the National Response Center in Washington, D.C.

Since 9/11 there has been increased networking between public safety HMRTs and USCG marine safety and port security personnel on a range of response scenarios. Some of the more obvious environmental scenarios include marine-based oil spills from vessels, pipelines, petroleum exploration, and production facilities (e.g., 2010 Deepwater Horizon incident). Other scenarios include maritime interdiction operations potentially involving hazmat/WMD and water rescue scenarios involving coordinated water and shored-based activities (e.g., the 2009 landing of US Airways Flight 1549 into the Hudson River, New York City).

Responder Tip

One of PHMSA's major programs that directly affects and benefits the emergency response community is the publication of the *Emergency Response Guidebook* (ERG). Historically, the ERG is revised and published on a three-year cycle.

Public Education

Hazmat safety is a concern not only for industry but also for the community. The average homeowner contributes to this problem by improperly disposing of substances such as used motor oil, paints, solvents, batteries, and other chemicals used in and around the home. As a result, many communities have initiated full-time household chemical waste awareness, education, and disposal programs. In other instances, communities have established used motor oil collection stations and chemical clean-up days in an effort to reclaim and recycle these materials. FEMA offers a web-based, free of charge independent study course entitled *IS-55: Household Hazardous Materials* that is available to the general public and can be accessed at http://www.training.fema.gov/is/.

An example of a highly successful public education program is the Wally Wise Guy™ program developed by the Deer Park, Texas LEPC. An example of industry and government partnership, Wally Wise Guy is a friendly turtle who teaches children and their parents how to shelter-in-place in case of a chemical emergency. The program has been produced and delivered in both English and Spanish and has been a highly successful educational tool. Further information on the Wally Wise Guy program can be obtained by consulting the Deer Park LEPC website at http://www.wally.org/.

Some hazardous materials industries are mandated by regulation to provide public awareness and education programs. For example, *American Petroleum Institute (API) Recommended Practice RP-1162—Public Awareness for Pipeline Operators*, was developed to help the pipeline industry meet the requirements of the Federal Pipeline Safety Improvement Act of 2002 (Section 5). Under the Act's requirements, pipeline operators must provide regular training to emergency responders located along the pipeline right-of-way. In addition, pipeline operators have extensive public education programs pertaining to the use of the One-Call System and the Dig Safely© program.

Handling, Notification, and Reporting Requirements

The guidelines for handling, notification, and reporting requirements actually act as a bridge between planning and prevention functions. There are many federal, state, and local regulations that require those who manufacture, store, or transport hazardous materials and hazardous wastes to comply with certain handling, notification, and reporting rules. Key federal regulations include CERCLA (Superfund), RCRA, and SARA, Title III. There are also many state regulations that are similar in scope and which often exceed the federal standard requirements.

In addition to the federal incident reporting requirements previously noted under CERCLA, state and local governments may also have equivalent spill reporting or emergency notification requirements. These requirements fall upon the facility owner or transportation carrier (i.e., the "Responsible Party") and are not the responsibility of public safety emergency responders. Also, reporting an emergency via 911 or the local equivalent may NOT satisfy the federal or state spill reporting requirements. Facility owners and transportation carriers should be familiar with the spill reporting requirements for all states where they may operate.

Response

When the prevention and enforcement functions fail, response activities begin. Since it is impossible to eliminate all risks associated with the manufacture, transportation, storage, and use of hazmats, the need for a well-trained, effective emergency response capability will always exist.

Response activities and operational capabilities should be based on the risk-based information and probabilities identified during the planning process. Response activities must be based upon an evaluation of the facility or local hazmat problem. While every community should have access to a technician-level hazmat response capability, that capability does not always have to be provided by either local government or the fire service. Numerous states and regions have established both statewide and regional HMRT systems that ensure the delivery of a competent and effective capability in a timely manner.

Response Groups

The emergency response community consists of various agencies and individuals who respond to hazmat incidents. They can be categorized based upon their knowledge, expertise, and resources. These responders can be compared to the levels of capability found within a typical EMS system. In that system, an injury such as a fractured arm can be effectively managed by a First Responder or Emergency Medical Technician, while a cardiac emergency will require the services of an EMT (Advanced Intermediate) or a Paramedic.

In the same way, a diesel fuel spill can usually be contained by First Responder–Operations-level personnel with the appropriate mission-specific competencies, such as a fire department engine company using absorbents or applying a Class B firefighting foam. An accident involving a Class 2.3 poison gas or a Class 5.1 oxidizer will, however, require the on-scene expertise of an HMT or HMRT. In simple terms, planners try to match the nature of the problem with the capabilities of the responders. Figure 1-3 illustrates this comparison.

Levels of Incident

Fortunately, not every incident is a major emergency. Response to a hazmat release may range from a single-engine company responding to a natural gas leak in the street to a railroad

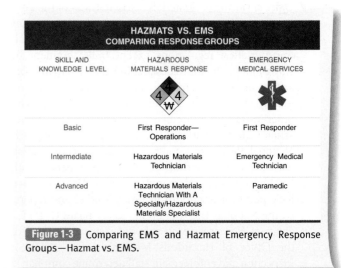

HAZMATS VS. EMS COMPARING RESPONSE GROUPS		
SKILL AND KNOWLEDGE LEVEL	HAZARDOUS MATERIALS RESPONSE	EMERGENCY MEDICAL SERVICES
Basic	First Responder—Operations	First Responder
Intermediate	Hazardous Materials Technician	Emergency Medical Technician
Advanced	Hazardous Materials Technician With A Specialty/Hazardous Materials Specialist	Paramedic

Figure 1-3 Comparing EMS and Hazmat Emergency Response Groups—Hazmat vs. EMS.

derailment involving dozens of government and private agencies and their associated personnel.

The National Incident Management System (NIMS) categorizes incidents into five types, based on the scope, impact, and resource requirements of the incident. A Type 1 incident is the most significant incident with national-level impact, while Type 5 incidents are the smallest, most common, and handled at the local jurisdiction level. Within the NIMS incident typing system, the overwhelming majority of hazardous materials incidents are Type 4 and Type 5 incidents, which are handled at the local or community level. If a hazardous materials incident exceeds local emergency response capabilities, is increasingly large or complex, and may extend into multiple operational periods, it may be typed as a Type 3 incident.

All emergency response operations start at the local level. As a result, communities and hazardous materials facilities may also develop levels of incidents based upon local planning, hazards and risks, and resource requirements. Under this process, levels of incidents can be categorized based on the severity of the incident and the level of local or company resources they require. **Table 1-5A and B** are examples of such response levels.

Hazmat Response Team (HMRT)

To respond effectively and efficiently to hazmat emergencies, many facilities and communities have established HMRTs. NFPA 472 defines an HMRT as an organized group of trained response personnel operating under an emergency response plan and applicable standard operating procedures, who perform hazardous material technician-level skills at hazardous materials/WMD incidents. The HMRT members respond to releases for the purpose of control or stabilization of the incident. Among the specialized equipment carried by an HMRT are technical reference libraries, computers and communications equipment, personal protective clothing and equipment, direct-reading detection and monitoring equipment, control and mitigation supplies and equipment, and decontamination supplies and equipment.

In evaluating the need for an HMRT, consider the following points:

- There is no single department or agency that can effectively manage the hazmat issue by itself. Regardless of what agency operates the HMRT, it must work closely with other local, state, and federal governmental agencies. A unified command organization is a necessity.
- Every community does not require an HMRT. However, every community should have access to an HMRT capability through local, regional, state, or contractor resources. In addition, the HMRT capability does not have to be provided through a single agency, but may be assembled by bringing together individuals and operational capabilities from multiple response agencies and organizations.
- An HMRT will not necessarily solve the hazmat problem. Remember the hazardous materials management system—planning, prevention, response and recovery.
- Like all special operations units, there are numerous constraints and requirements associated with developing an effective HMRT capability. These include legal, insurance, and political issues; both initial and continuing funding sources; resource determination and acquisition; personnel and staffing, and initial and continuing training requirements.
- Successful HMRT response programs are those that truly understand what services an HMRT can provide at all emergencies, not just those involving hazardous materials. For example, no organization within the emergency response community better understands and routinely practices (1) the process of risk evaluation; (2) the use of air monitoring and detection equipment and the interpretation of its results; and (3) the fundamentals of safe operating practices in a field environment better than HMRT personnel.

> **Responder Tip**
>
> The HMRT is *not* a chemical mitigation resource; it is a risk evaluation and health and safety resource with capabilities that can be used in a variety of response scenarios, including hazardous materials, confined space, structural collapse, aircraft accidents, acts of terrorism and criminal events involving the use of hazardous materials, and other significant fires and emergencies.

Regional and statewide hazardous materials response systems have been developed in many areas of the country, including North Carolina, Massachusetts, and California. A number of these response systems have different types or levels of HMRTs based upon their own specific staffing and equipment inventory. In states that have a mature intrastate mutual aid system, HMRTs are often reviewed and typed based upon the NIMS Resource Typing Definitions. A tiered response concept is also used in a number of metropolitan fire departments, where the HMRT is supported by a number of ladder or rescue companies designated as Hazardous Materials Support Companies. These Hazmat Support Companies are trained to handle the lower-risk incidents as well as provide support to the HMRT for a "working" hazardous materials response.

HMRTs typically function as a group or branch within the Incident Command System (ICS) under the direct control of a Hazardous Materials Officer typically functioning as a Hazmat Group Supervisor within the incident command organization. Based upon their assessment of the hazardous materials problem, the HMRT, through the Hazmat Group Supervisor, provides the Incident Commander or Operations Section Chief with a list of options and a recommendation for mitigation of the hazardous materials problem. However, the final decision always remains with the Incident Commander. This topic is discussed further in the section on "managing incidents."

HMRT members must be properly trained and participate in a medical surveillance program based upon the requirements of 29 CFR 1910.120. The two primary information sources for establishing an HMRT training program are OSHA 1910.120(q) and NFPA 472. While OSHA 1910.120 outlines the regulatory requirements (what you have to do), NFPA 472

Table 1-5A Levels of Hazardous Materials Incidents—Community

LEVELS OF HAZARDOUS MATERIALS INCIDENTS—COMMUNITY

RESPONSE LEVEL	DESCRIPTION	RESOURCES	EXAMPLES
 I Potential Emergency Conditions	An incident or threat of a release which can be controlled by the first responder. It does not require evacuation, beyond the involved structure or immediate outside area. The incident is confined to a small area and poses no immediate threat to life and property.	Essentially a local level response with notification of the appropriate local, state, and federal agencies. Required resources may include: • Fire Department • Emergency Medical Services (EMS) • Law Enforcement • Public Information Officer (PIO) • Chemtrec • National Response Center	• 500-gallon fuel oil spill • Inadvertent mixture of chemicals • Natural gas leak in a building
 II Limited Emergency Conditions	An incident involving a greater hazard or larger area than Level I which poses a potential threat to life and property. It may require a limited protective action of the surrounding area.	Requires resources beyond the capabilities of the initial local response personnel. May require a mutual aid response and resources from other local and state organizations. May include: • All Level I Agencies • Hazmat Response Teams • Public Works Department • Red Cross • Regional Emergency Management Staff • State Police • Public Utilities	• Minor chemical release in an industrial facility • A gasoline tank truck rollover • A chlorine leak at a water treatment facility
 III Full Emergency Conditions	An incident involving a severe hazard or a large area which poses an extreme threat to life and property and which may require a large-scale protective action.	Requires resources beyond those available in the community. May require the resources and expertise of regional, state, federal, and private organizations. May include: • All Level I and II Agencies • Mutual Aid Fire, Law Enforcement, and EMS • State Emergency Management Staff • State Department of Environmental Resources • State Department of Health • Environmental Protection Agency (EPA) • U.S. Coast Guard • Federal Emergency Management Agency (FEMA)	• Major train derailment with fire • Explosion or toxicity hazard • A migrating vapor cloud release from a petrochemical processing facility

Table 1-5B Levels of Hazardous Materials Incidents—Petrochemical Industry

LEVELS OF HAZARDOUS MATERIALS INCIDENTS—PETROCHEMICAL INDUSTRY			
RESPONSE LEVEL	DESCRIPTION	RESOURCES	EXAMPLES
 I Facility Incident	Minimal danger to life, property, and the environment. Problem is limited to immediate work area and public health, safety, and environment are not affected.	Handled by On-Shift Emergency Response Team (ERT) with no off-shift or mutual aid response.	• Minor spills and releases less than 55 gallons • Small pump seal fire • Minor vapor release during product transfer operations
 II Facility Serious Incident	Moderate danger to life, property, and the environment on the plant site. Problem is currently limited to plant property, but has the potential for involving additional exposures or migrating off-site and affecting public health, safety, and environment for a short period of time.	Requires On-Shift ERT response. Additional assistance required from off-duty ERT personnel and/or mutual aid units. Plant EOC may be activated. Corporate Crisis Emergency Plan may be activated.	• Large release of flammable, corrosive, or toxic vapors • Releases of over 100 gallons of hazardous material • Large spill fire or a seal fire on a floating roof tank
 III Facility Crisis Situation	Extreme danger to life, property, and the environment. Problem goes beyond the refinery property and can impact public health, safety and the environment or a large geographic area for an indefinite period of time.	Requires multi-organizational response from plant, local fire service, industrial mutual aid units, and public safety resources. Plant EOC is activated. Corporate Crisis Emergency Plan activated.	• Process unit fire or explosion • Major release of flammable, corrosive or toxic vapors • Shipboard fire, major oil spill, or HM release which can impact major waterways

spells out the specific training and educational competencies for the training program (how you can do it).

Both OSHA 1910.120 and NFPA 472 recommend that HMRT personnel be trained to the HMT level. According to OSHA 1910.120(q), the HMT requires a minimum of 24 hours of initial training at the First Responder Operations level. A survey of HMT training courses would find that they range from 24 to 200+ hours. Industry emergency response team members are regularly trained to the HMT-level in a 24- to 40-hour course because they respond to a limited number of chemicals and response scenarios within the confines of their facility. In contrast, given the wide range of potential scenarios that may occur in the community, public safety technician-level training requirements will most likely be in the 80 to 200+ hour area. This would be in addition to the basic-level fire, rescue, EMS, or law enforcement training a first responder may need to perform their day-to-day jobs.

Clean-Up and Recovery

Clean-up and recovery operations are designed to (1) clean up or remove the hazmat spill or release, and (2) restore the facility and/or community back to normal as soon as possible. In many instances, chemicals involved in a hazmat release will be eventually classified as hazardous wastes.

Clean-up operations fall under the guidelines of HAZWOPER, CERCLA (Superfund), and RCRA. Clean-up activities can be classified as follows:

- **Short term**—Those actions immediately following a hazmat release that are primarily directed toward the removal of any immediate hazards, reducing effects on critical infrastructure, and restoring vital support services (i.e., reopening transportation systems, drinking water systems, etc.) to minimum operating standards. Short-term activities may last up to several weeks.

  These activities are normally the responsibility of the "responsible party"—usually the facility owner/operator or transportation carrier (e.g., railroad or truck company). In situations where the responsible party has not been identified or does not have sufficient financial resources, clean-up operations may be assumed by state or federal environmental agencies.
- **Long term**—Those remedial actions that return vital support systems back to normal or improved operating levels. Examples would include groundwater treatment operations, the mitigation of both aboveground and underground spills, and the monitoring of flammable and toxic contaminants. These activities may not be directly related to a specific hazmat incident but are often the result of abandoned industrial or hazardous waste sites. These operations may extend over months or years.

One point of discussion during clean-up operations might be "how clean is clean." Depending on the nature of the incident and its physical location, this discussion can involve a number of organizations and entities, including both state and local environmental, health and safety (EH&S) representatives, EPA or USCG, and state and federal law enforcement agencies (e.g., illicit laboratory incidents). In addition, local government and political leaders may also have significant contributions to the discussion.

Recovery operations focus on restoring the facility, the community, and/or emergency response organization to normal operating conditions. Tasks include restocking all supplies and equipment, compilation and documentation of resources purchased and/or used during the emergency response, and financial restitution, where appropriate.

The economic cost of supporting and delivering a hazmat response capability can be extremely high, especially for "working" or long duration incidents. Post-incident issues will often include identifying a responsible party (RP) and recovering allowable costs associated with the incident response. To provide a legal foundation for pursuing cost recovery, many local and state jurisdictions have enacted legislation to allow response agencies the ability to seek reimbursement where a responsible party is identified. Hazmat incident cost recovery options may include:

- Establishment of a local cost recovery ordinance/legislation and supporting fee structures.
- Voluntary efforts through the RP's insurance carrier.
- EPA Local Government Reimbursement (LGR) Program. This was enacted under the original SARA, Title III legislation and can provide up to $25,000 per incident to local governments that do not have funds available to pay for response actions. However, the LGR program can only be used after other reimbursement strategies have been exhausted and a number of specific requirements have been satisfied. Additional information can be referenced at http://www.epa.gov/oem/content/lgr/ regarding civil actions through the judicial system.

Role of Emergency Responders During Clean-Up Operations

Many plant-level industrial responders are also responsible for the clean-up of minor spills and releases so that facility operations may continue. In contrast, public safety response personnel are usually not directly responsible for the final clean-up and recovery of a hazardous materials release. Depending upon the nature of the incident, however, they may continue to be responsible for site safety until all risks are stabilized and the emergency phase is terminated. Once the emergency phase is terminated, there must be a formal transfer of command from the lead response agency (e.g., fire department) to the lead agency responsible for post-emergency response operations (e.g., responsible party, state or federal EPA).

At short-term operations immediately following an incident, the Incident Commander should ensure that the work area is closely controlled, that the general public is denied entry,

Responder Tip

Although emergency responders are typically not responsible for clean-up and recovery operations, they are responsible for ensuring that there is a safe and effective transfer of command to the lead agency responsible for post-emergency response operations.

and that the safety of emergency responders and the public is maintained during clean-up and recovery operations. When interfacing with both industry responders and contractors, the Incident Commander should ensure that they are trained to meet the requirements of OSHA 1910.120 and/or NFPA 472.

Long-term clean-up and recovery operations do not normally require the continuous presence of the fire service. Depending upon the size and scope of the clean-up, a contractor or government official (Remedial Project Manager, or RPM) will be the central contact point. Emergency responders should be familiar with the clean-up operation, including its organizational structure, the OSC/RPM, work plan, time schedule, and site safety plan.

Clean-up operations should conform to the general health and safety requirements of state and federal EPA and OSHA standards, as well as local requirements. Although emergency responders generally do not have the authority to conduct inspections or issue citations at clean-up operations, they can bring specific concerns to the attention of the state or federal regulatory agency having jurisdiction.

Wrap-Up

Chief Concepts

- The primary target audience for this text includes HMTs, the On-scene Incident Commander, the Hazmat Officer (aka Group Supervisor or Branch Director) and members of organized HMRTs.
- Hazardous materials: Any substance that jumps out of its container when something goes wrong and hurts or harms the things it touches (Benner).
- A hazardous materials incident can then be defined as the release, or potential release, of a hazardous material from its container into the environment (Benner).
- SARA, Title III is the primary federal legislation that directly affects the local hazardous materials emergency preparedness program. Responders should identify the local agency responsible for the coordination of the LEPC, as well as who represents the emergency response community.
- OSHA Hazardous Waste Operations and Emergency Response (29 CFR 1910.120) is the primary federal law that directly affects hazmat emergency response and training activities in both the public and private sector.
- OSHA 1910.120 is the LAW of the land. NFPA 472 is a VOLUNTARY consensus standard. However, if you meet the requirements of NFPA 472 you will exceed the OSHA HAZWOPER emergency response training requirements (paragraph q.6).
- The NFPA publishes a number of voluntary consensus standards that directly influence hazmat/WMD emergency response, training, and certification throughout North America. These include NFPA 472 (training), 473 (EMS training), 1991 (chemical vapor protective clothing), 1992 (chemical splash protective clothing) and 1994 (chemical protective clothing—CBRNE terrorism scenarios).
- Standard of Care represents the *minimum* accepted level of hazardous materials emergency service that should be provided regardless of location or situation. It is established by existing laws and regulations, and is also influenced by legal findings and case law precedents.
- The two key elements of planning and preparedness activities are the hazards analysis and contingency planning processes.
- Hazmat prevention activities include (1) hazmat process, container design, and construction standards; (2) inspection and enforcement; (3) public education; and (4) handling, notification, and reporting requirements.
- Hazardous Materials Response Teams (HMRT) are defined as an organized group of trained response personnel, operating under an emergency response plan and applicable standard operating procedures, who perform hazardous material technician-level skills at hazardous materials/WMD incidents. The HMRT members respond to releases for the purpose of control or stabilization of the incident. Among the specialized equipment carried by an HMRT are reference libraries, computers and communications equipment, personal protective clothing and equipment, direct-reading detection and monitoring equipment, control and mitigation supplies and equipment, and decontamination supplies and equipment (NFPA 472).
- Although emergency responders are typically not responsible for clean-up and recovery operations, they are responsible to ensure there is a safe and effective transfer of command to the lead agency responsible for post-emergency response operations.

Hot Terms

Contingency (emergency) Planning A comprehensive and coordinated response to the hazmat problem, which builds upon the hazards analysis process and recognizes that no single public or private sector agency is capable of managing the hazmat problem by itself.

Hazardous Materials Any substance or material in any form or quantity that poses an unreasonable risk to safety and health and property when transported in commerce (Source: U.S. Department of Transportation, 49 Code of Federal Regulations 171).

Hazardous Materials Response Team (HMRT) An organized group of trained response personnel, operating under an emergency response plan and applicable standard operating procedures, who perform hazardous material technician-level skills at hazardous materials/WMD incidents.

Hazardous Materials Technician (HMT) An individual who responds to hazardous materials and WMD incidents using a risk-based response process

Hazards Analysis Analysis of the hazardous materials present in the community or facility, including their location, quantity, specific physical and chemical properties, previous incident history, surrounding exposures, and risk of release.

<u>Incident Commander (IC)</u> The individual responsible for establishing and managing the overall incident action plan (IAP). This process includes developing an effective organizational structure, developing an incident strategy and tactical action plan, allocating resources, making appropriate assignments, managing information, and continually attempting to achieve the basic command goals.

<u>Safety Data Sheet (SDS)</u> A document which contains information regarding the chemical composition, physical and chemical properties, health and safety hazards, emergency response, and waste disposal of the material as required by 29 CFR 1910.1200.

<u>Standard of Care</u> The minimum accepted level of hazmat service to be provided as may be set forth by law, current regulations, consensus standards, local protocols and practice, and what has been accepted in the past (precedent).

<u>Voluntary Consensus Standards</u> Standards and recommended practices developed through an open, democratic, consensus-based process. Examples of voluntary standards development organizations include NFPA, API, and ASTM.

HazMat Responder *in Action*

Your agency has decided to create a HazMat inspection group under the Fire Prevention division. This newly formed group will be responsible for enforcement of and compliance with federal rules and regulations within your agency's jurisdiction. This is the first day of the program and that you have been assigned a group of potential hazardous materials facilities within your response jurisdiction.

Questions

1. As a part of the rules and regulations, you are responsible for developing a hazard analysis of hazmat facilities along transportation corridors within the community. Which rule, law, regulation, and/or standard does this fall under?
   A. Risk Management Program for Chemical Accidental Release Prevention
   B. Hazardous Materials Transportation Regulations
   C. Hazardous Waste Operations and Emergency Response Rule
   D. Community Emergency Response Planning Regulations
2. Which phrase is defined in EPA 40 CFR 355?
   A. Dangerous goods
   B. Clean Air Act
   C. Extremely Hazardous Substances
   D. Local Emergency Planning Committee
3. Which of the following can assist with your documentation of operating policies and procedures?
   A. NFPA
   B. FEMA
   C. ASTM
   D. A and C
4. How can a hazmat public education program assist you with code enforcement in your newly formed unit?
   A. It will help increase the number of citations if the community volunteers information on code violations
   B. It will reduce the number of unsafe hazmat disposal or illegal dumping cases over time
   C. It will strengthen your legal case when prosecution of violators is required
   D. It will improve your department's pre-planning process

References and Suggested Readings

1. Callan, Michael, *Street Smart Hazmat Response,* Chester, MD: Red Hat Publishing (2002).
2. Federal Emergency Management Agency, *Comprehensive Preparedness Guide (CPG) 101—Developing and Maintaining Emergency Operations Plans* (Version 2.0), Washington, DC: FEMA (November 2010).
3. Fire, Frank L., Grant, Nancy K. and Hoover, David H., *SARA Title III—Intent and Implementation of Hazardous Materials Regulations*, Tulsa, OK: Fire Engineering Books and Videos (1990).
4. Lesak, David M., *Hazardous Materials Strategies and Tactics*, Upper Saddle River, NJ: Brady/Prentice Hall (1999).
5. National Fire Protection Association, *Hazardous Materials Response Handbook* (6th edition), Quincy, MA: National Fire Protection Association (2013).
6. National Fire Protection Association, *Fire Protection Handbook*, (20th Edition), Volume-II, *Public Fire Protection and Hazmat Management*, by Michael S. Hildebrand and Gregory G. Noll, pages 13-81–13-99, Quincy, MA: National Fire Protection Association (2008).
7. National Institute for Occupational Safety and Health (NIOSH), *Occupational Safety and Health Guidance Manual for Hazardous Waste Site Activities*, Washington, DC: NIOSH, OSHA, USCG, EPA (1985).
8. National Response Team, *Hazardous Materials Emergency Planning Guide* (NRT-1), Washington, DC: National Response Team (2001).
9. U.S. Environmental Protection Agency, *Hazmat Team Planning Guidance*, Washington, DC: EPA (1990).
10. U.S. Environmental Protection Agency, *Hazardous Materials Planning Guide*, Washington, DC: EPA (2001).
11. U.S. Environmental Protection Agency et al., *Technical Guidance for Hazards Analysis—Emergency Planning for Extremely Hazardous Substances*, Washington, DC: EPA, FEMA, DOT (1987).
12. U.S. Occupational Safety and Health Administration, *OSHA Directive Number CPL 2-2/59A—Inspection Procedures for the Hazardous Waste Operations and Emergency Response Standard, 29 CFR 1910.120 and 1926.65, Paragraph (Q): Emergency Response to Hazardous Substance Releases*, Washington, DC: OSHA (April 4, 1998).

Health and Safety

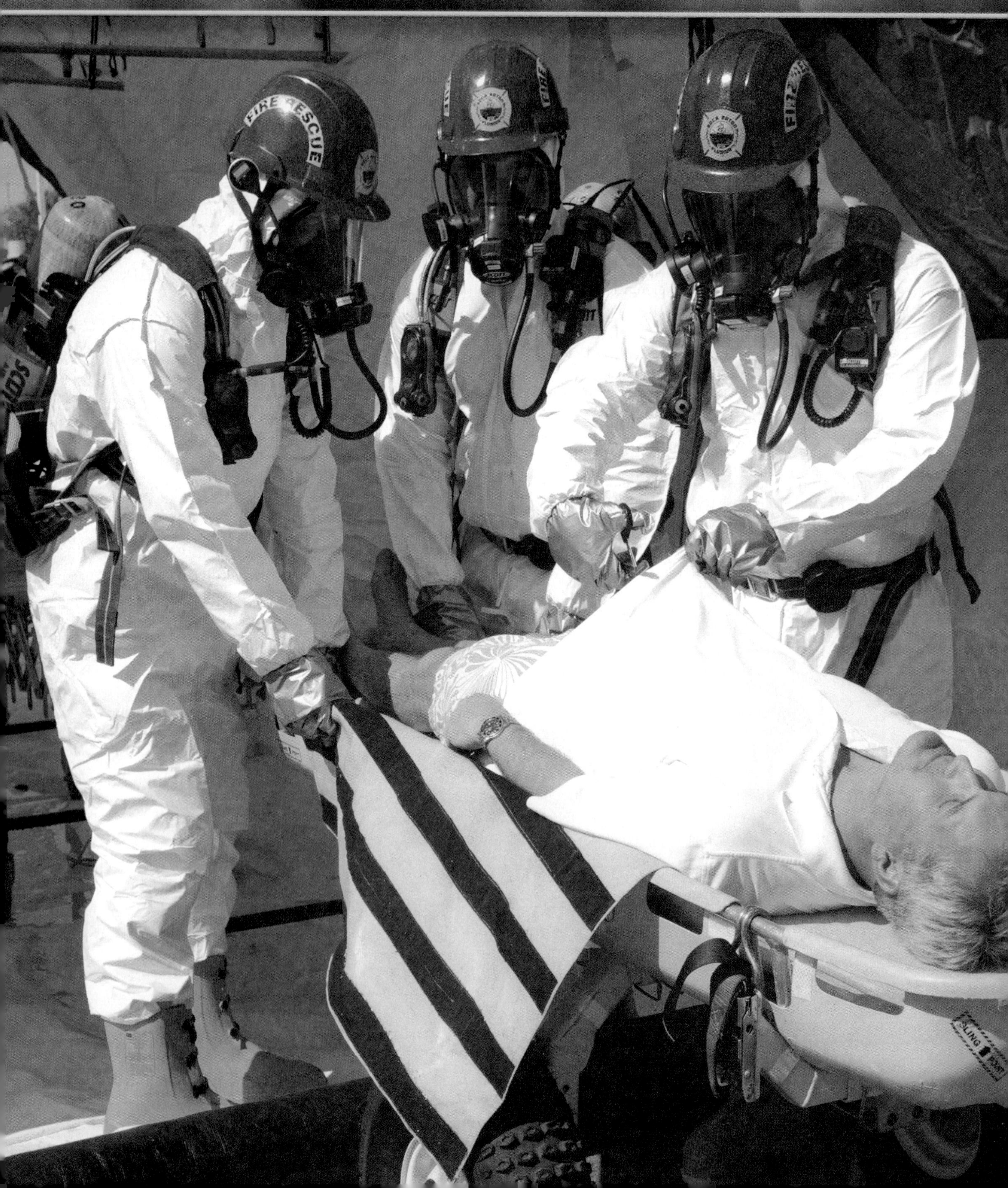

CHAPTER 2

Hazardous Materials Technician

Knowledge Objectives

After reading this chapter, you will be able to:

- Describe basic toxicological principles, including:
  - Exposure (p. 32)
  - Toxicity (p. 32)
  - Acute and chronic exposures (p. 33)
  - Acute and chronic effects (p. 33)
  - Routes of exposure to hazardous materials (p. 33, 34)
  - Dose/response relationship (p. 34)
  - Local and systemic effects (p. 34)
  - Target organs (p. 35)
- Identify the seven types of harm created by exposure to hazardous materials and their effects upon the human body. (p. 35)
- Identify the signs, symptoms, and emergency care procedures for handling heat stress emergencies. (p. 46, 47)
- Identify procedures for reducing the effects of heat stress on responders at a hazmat incident. (p. 47–49)
- Identify procedures for reducing the effects of cold on responders at a hazmat incident. (p. 49)
- Identify procedures for protecting responders against excessive noise levels at a hazmat incident. (p. 50)
- Identify the components of a medical surveillance program for hazmat responders as outlined in OSHA 1910.120(q). (p. 50–55)
- Identify the components of a personal protective equipment (PPE) program. (p. 55)
- Describe the components of a site safety plan for operations at a hazmat incident. (p. 56–60)
- Describe the procedures and components for conducting pre- and post-entry medical monitoring for response personnel operating at a hazmat incident. (p. 58)
- Describe the procedures for establishing and operating a Rehabilitation Area at a hazmat incident. (p. 60)

Skills Objectives

There are no Hazardous Materials Technician skills objectives in this chapter.

Hazardous Materials Incident Commander

Knowledge Objectives

After reading this chapter, you will be able to:

- Describe basic toxicological principles, including:
  - Exposure (p. 32)
  - Toxicity (p. 32)
  - Acute and chronic exposures (p. 33)
  - Acute and chronic effects (p. 33)
  - Routes of exposure to hazardous materials (p. 33, 34)
  - Dose/response relationship (p. 34)
  - Local and systemic effects (p. 34)
  - Target organs (p. 35)
- Identify procedures for reducing the effects of heat stress on responders at a hazmat incident. (p. 46–50)
- Describe the components of a site safety plan for operations at a hazmat incident. (p. 56–60)

Skills Objectives

There are no Incident Commander skills objectives in this chapter.

You Are the HazMat Responder

You are the Health and Safety Coordinator for a chemical manufacturing and distribution facility and have been assigned the task of developing a medical surveillance program for the members of the facility Hazardous Materials Response Team (HMRT). Facility hazards include flammable liquids and gases, corrosive liquids, and poisonous liquids.

The facility HMRT is trained to the "Hazardous Materials Specialist Employee A" level, as defined in NFPA 472, and can provide a Technician-level response capability for the products and risks found within the plant. HMRT members include personnel from the operations, maintenance, safety, and administration departments.

Your management team has asked for your recommendations on several specific issues pertaining to the health and safety of HMRT members.

1. What should be the components of the HMRT medical surveillance program?
2. What criteria should be used for the initial entry medical monitoring?
3. How could the facility coordinate its plant emergency response program with the local hospital?

Introduction

The health and safety of all emergency responders is a critical issue for both management and labor. Regardless of size or nature, every hazmat incident presents responders with a potentially hostile environment. Although preventing exposure to hazardous materials is always a primary concern, command personnel must also evaluate the physical working conditions, work intervals, and the stress of working in personal protective clothing and equipment. Our goal is simple—we want every responder to come home in the same physical, mental, and emotional condition as upon arrival at the incident.

Hazmat incidents are characterized by work environment hazards, which may pose an immediate danger to life and health but which may not be immediately obvious or identifiable. These hazards can vary according to the tasks being performed and a responder's location at the incident site and can change as response activities progress.

Protecting the health and safety of emergency response and support personnel, as well as the general public, must always be the Incident Commander's (IC's) primary concern. In this chapter, we will examine these health and safety concerns in detail.

Toxicology

Toxicology is the study of chemical or physical agents that produce adverse responses in the biological systems with which they interact. Chemical agents include gases, vapors, fumes, dusts, and so on, while physical agents include radiation, hot and cold environments, noise, and so forth.

Toxicity is defined as the ability of a substance to cause injury to a biological tissue. In humans this generally refers to unwanted effects produced when a chemical has reached a sufficient concentration at a particular location within the body.

> **Responder Safety Tip**
>
> To properly protect those personnel operating at a hazmat incident, both Command and Hazmat Group personnel must be able to understand the basic concepts of toxicology, as well as toxicological and exposure terms, and interpret toxicity and exposure data.

Several factors determine the toxicity of a chemical and the potential harm that may result. A simple way of understanding this concept and its potential for harm is the health hazard equation:

Exposure + Toxicity = Health Hazard

1. **Exposure** means you have had contact with the chemical. Common methods are inhalation, ingestion, skin absorption, or direct contact.

2. Dose is the concentration or amount of a material to which the body is exposed over a specific period of time. In simple terms, dose = concentration × time.
3. Toxicity refers to the ability of the chemical to harm your body once exposure has occurred. The level of toxicity will depend on the nature of the chemical, the route of entry, its ability to do harm to the body, and the dose. Different people may react differently to the same level of exposure. This principle is discussed further in the section on the "**dose/response relationship**."

Exposure Concerns

Exposures and Health Effects

Chemical exposures and their health effects are commonly described as acute or chronic. **Acute exposures** describe an immediate exposure, such as a single dose that might occur during an emergency response. **Chronic exposures** are low exposures repeated over time, such as responding to a number of hazmat emergencies while serving as a member of a public safety HMRT or conducting on-scene investigations of hazmat incidents over many years.

Acute effects result from a single dose or exposure to a material, such as a single exposure to a highly toxic material like hydrogen cyanide or arsine, or a large dose of a less toxic material, like nitrogen and some solvents. Signs and symptoms from the exposure may be immediate or may not be evident for 24 to 72 hours after the exposure. Some chemicals may produce both immediate and delayed effects. It is also possible that signs and symptoms of a chemical exposure may be confused with illnesses such as the common cold or a stomach virus. The potential for delayed or "masked" effects of exposure illustrate the importance of medical debriefings as part of the incident termination phase.

Chronic effects result from a single exposure or from repeated doses or exposures over a relatively long period of time. Although chronic exposures are usually correlated with long-term worker exposures in an industrial environment, they can also result from chemical exposures at long-term remedial clean-up operations or from breathing smoke or other elements resulting from firefighting operations.

Exposure to various types and doses of hazardous materials over a period of years is also associated with chronic health effects. To monitor such exposures, responders should be provided with baseline medical profiles, participate in a medical surveillance program, and document all exposures.

Routes of Exposure

Responders can be exposed to hazardous materials by the following means Figure 2-1:

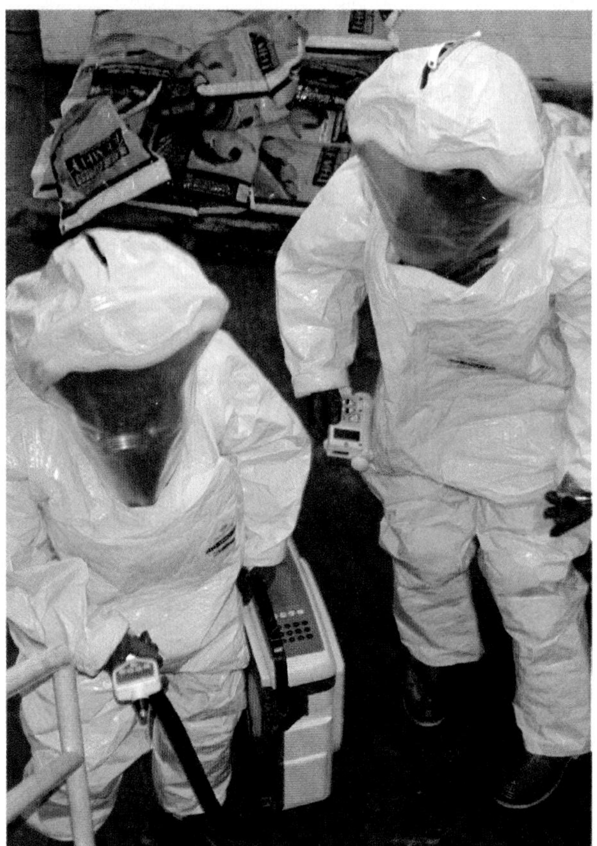

Figure 2-1 Responders can be exposed to hazmat by inhalation, skin absorption, ingestion, direct contact, and injection.

Inhalation

The introduction of a chemical or toxic product of combustion into the body by way of the respiratory system. Inhalation is the most common exposure route and often the most damaging. Remember that a material does not have to be a gas in order to be inhaled—solid materials may generate fumes or dusts in a dry, powdered form, while high vapor pressure liquid chemicals can generate vapors, mists, or aerosols that can be inhaled. Once within the body, toxins may be absorbed into the bloodstream and carried to other internal organs, may paralyze the respiratory system, reduce the ability of the blood to transport oxygen, or affect the upper and/or lower respiratory tract. Inhalation exposures are much more likely to involve a large number of civilian exposures.

Skin absorption

The introduction of a chemical or agent into the body through the skin. Skin absorption can occur with no sensation to the skin itself. Do not rely on pain or irritation as a warning sign of absorption (e.g., delayed effects of mustard agent). Some poisons are so concentrated that a few drops placed on the skin may result in death (e.g., parathion, VX nerve agent).

Skin absorption is enhanced by abrasions, cuts, heat, and moisture. This can create critical problems when working at incidents involving chemical or biological agents. Anyone with large, open cuts, rashes, or abrasions should be prohibited from working in areas where they may be exposed. Smaller cuts or abrasions should be covered with a nonporous dressing.

The rate of skin absorption can vary depending on the body part exposed and, to some degree, the way in which the exposure occurred and the chemical properties of the substance. For example, assuming the area of skin exposed and the duration of exposure are equal, the rate of skin absorption through the scalp or genitals area will be considerably faster than through the forearm. Absorption through the eyes is one of the fastest means of exposure since the eye has a high absorbency rate. This type of exposure may occur when a chemical is splashed directly into the eye, when a chemical is "carried" on toxic smoke particles into the eyes from a fire, or when gases or vapors are absorbed through the eyes. Likewise, this may be seen as an early warning signal for either PPE or respiratory protection failures.

Ingestion

The introduction of a chemical into the body through the mouth or inhaled chemicals trapped in saliva and swallowed. Exposed personnel should be prohibited from smoking, eating, or drinking except in designated rest and rehab areas after being decontaminated.

Direct contact

Direct skin contact with some chemicals, such as corrosives, will immediately damage skin or body tissue on contact. Acids have a strong affinity for moisture and can create significant skin and respiratory tract burns. However, the injury process also creates a clot-like barrier that blocks deep skin penetration. In contrast, caustic or alkaline materials dissolve the fats and lipids that make up skin tissue and change the solid tissue into a soapy liquid (think of how drain cleaners using caustic chemicals dissolve grease and other materials in sinks and drains!). As a result, caustic burns are often much deeper and more destructive than acid burns.

Injection

Other chemicals may be injected directly through the skin and into the bloodstream. Mechanisms of injury include needle stick punctures at medical emergencies and the injection of high pressure gases and liquids into the body similar to the manner in which flu shots are injected with pneumatic guns.

Dose/Response Relationship

A fundamental relationship exists between the chemical dose (i.e., dose = concentration × time) and the response produced by the human body. Given the broad range of toxicities any substance might cause, the wisdom of Paracelsus (1493–1541) becomes clear: "All substances are poisons; there is none which is not a poison. The right dose differentiates a poison and a remedy." Translation: Dose makes the poison.

The dose/response concept is based on the following assumptions:

- The magnitude of the response is dependent upon the concentration of the chemical at the biological site of action (i.e., target organ).
- The concentration of the chemical at the biological site of action is a function of the dose administered.
- Dose and response are essentially a cause/effect relationship.

Remember that dose is the concentration or amount of material to which the body is exposed over a specific time period. In simple terms, dose = concentration × time. The human body's response to a dose is either toxic or nontoxic. Human response may also be influenced by the age, gender, state of health, and nutritional state of the individual. Typically,

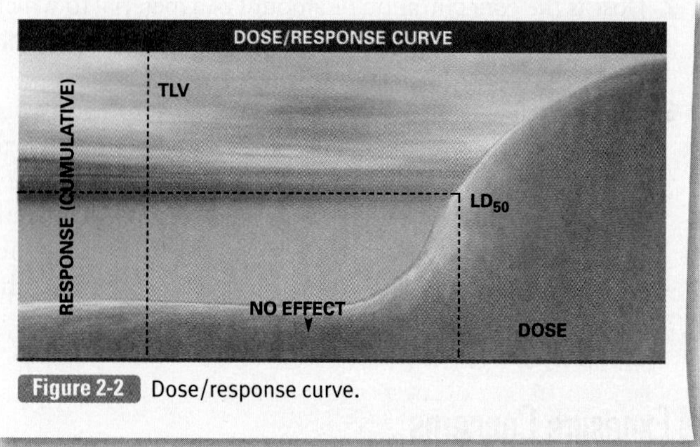

Figure 2-2 Dose/response curve.

as the size of the dose increases, the potential for a toxic response also increases. For example, one gram of aspirin is an accepted dose for medicinal purposes, yet quantities in excess of 10 grams (approximately 25 tablets) have caused death.

With many chemicals, increasing moderate doses will cause no apparent biological damage. However, increasing the dose rate will eventually reach a point defined as the threshold level. At this concentration and above, chemical exposure can result in biological harm. These effects may range from slight irritation to death, depending on the dose and properties of the chemical. Figure 2-2 illustrates a typical dose/response curve, and Table 2-1 illustrates the dose/response relationship using alcohol as an example.

Effects of Hazardous Materials Exposures

Health effects of a hazardous materials exposure can be described in terms of how a hazmat attacks the body. A local effect implies an effect at the point of contact—for example, a corrosive burn to the skin; eye irritation, and so on. A systemic effect occurs when a chemical enters the bloodstream and attacks target organs and internal areas of the human body. Multiple systemic effects are also a distinct possibility. Hydrofluoric acid is an example of a chemical that can be absorbed through the skin and ultimately cause systemic effects such as cardiac problems.

Table 2-1 Dose/Response Relationship

Dose	Acute Effect	Chronic Effect
1 oz. of Bourbon Consumed in 60 mins.	Minimal	None
1 qt. of Bourbon Consumed in 60 mins.	Illness or Death	Minimal
1 oz. of Bourbon Consumed every 60 mins. for 12 hrs. each day, 365 days a year	Minimal	Brain / Liver Damage
1 qt. of Bourbon Consumed over a year	None	None

Systemic effects often show up at target organs. Target organs are organs/tissues where a toxin exerts its effects; it is not necessarily the organ/tissue where the toxin is most highly concentrated. For example, over 90% of the lead in the adult human body is in the skeleton, but lead exerts its effects on the kidneys, the central nervous system, and the blood system.

Table 2-2 lists examples of target organs that may be affected by a systemic poison:

The human body can be subject to seven types of harm events:

- **Thermal**—Events related to temperature extremes. High temperatures are common at fire-related emergencies involving flammable liquids, gases, and solids, as well as explosions. Thermal harm resulting in frostbite can be found as a result of exposures to the extremely low temperatures associated with liquefied gases and cryogenic materials.
- **Mechanical**—Events resulting from direct contact with fragments and blast effects scattered because of a container failure, explosion, bombing, or shock wave. Remember that even small fragments and shrapnel can cause significant damage to the body.
- **Poisonous**—Events related to exposure to toxins. Some chemicals affect the body by causing damage to specific internal organs or body systems. Examples include hepatotoxins and nephrotoxins. Other chemicals, such as benzene and phenol, are considered blood toxins because of their effects on the circulatory system. Similarly, neurotoxins—such as organophosphate and carbamate pesticides—affect the central nervous system. A number of flammable liquids have some form of toxicity well before they become fire hazards (e.g., toluene, methyl alcohol).
- **Corrosive**—Events related to chemical burns and/or tissue damage from exposure to corrosive chemicals. Corrosives are divided into two chemical groups: acids and bases. Acids—such as strong inorganic acids like nitric, sulfuric, hydrochloric, and hydrofluoric—can cause severe tissue burns to the skin and permanent eye damage. Bases, or caustic or alkaline materials, break down fatty skin tissue and penetrate deeply. Sodium and potassium hydroxide are examples. Acids generally cause greater surface tissue damage while bases produce deeper, slower healing burns.

  Inhaled corrosive gases and vapors can also cause acute swelling of the upper respiratory tract and chemically induced pulmonary edema. High water soluble materials, such as anhydrous ammonia, will affect the upper respiratory tract, while low water soluble materials, such as nitrogen dioxide and phosgene, will affect the lower respiratory tract. Lower respiratory tract injuries can lead to chemically induced pulmonary edema and may be delayed for 24 to 72 hours after the exposure.
- **Asphyxiation**—Events related to oxygen deprivation within the body. Asphyxiants can be categorized as simple or chemical. Simple asphyxiants act on the body by displacing or reducing the oxygen in the air for normal breathing. Examples include carbon dioxide, nitrogen, and natural gas. Chemical asphyxiants disturb the normal body chemistry processes that control respiration. They can range from chemicals that inhibit oxygen transfer from the lungs to the cells (carbon monoxide) or prevent respiration at the cellular level (hydrogen cyanide) to those that incapacitate the respiratory system (hydrogen sulfide in large concentrations).
- **Radiation**—Events related to the emission of radiation energy. Radiation energy is defined as the waves or particles of energy emitted from radioactive sources. It may be in the form of alpha or beta particles or gamma waves, depending on the intensity of the source material. Examples include Cesium 137, tritium, plutonium and uranium hexafluoride. Nonionizing radiation, such as microwaves and lasers, may also create potential harm in certain emergency situations.
- **Etiological**—Events created by uncontrolled exposures to living microorganisms. Etiological/biological harm is normally associated with diseases such as typhoid fever and tuberculosis, with bloodborne pathogens such as the hepatitis (A, B, or C) virus, and with weapons-grade biological agents (e.g., anthrax). It is often difficult to detect when and where the physical exposure to the biological agent occurred and the route(s) of exposure.

Table 2-2 Target Organs Examples

Term	Target Organ	Examples
Hepatotoxins	Liver	Carbon tetrachloride, vinyl chloride monomer and nitroamines
Nephrotoxins	Kidneys	Halogenated hydrocarbons and mercury
Neurotoxins	Central Nervous System (CNS)	Lead, toluene, nerve agents and organophosphate pesticides
Respiratory Toxins	Lungs and Pulmonary System	Asbestos, chlorine and hydrogen sulfide
Hematotoxins	Blood System	Benzene, chlordane, and cyanides
Skeletal System	Bones	Hydrofluoric acid and selenium
Dermatotoxins	Skin (NOTE: Act as irritants, ulcers, chloracne, and/or cause skin pigmentation disturbances)	Halogenated hydrocarbons, coal tar compounds, and high levels of ultraviolet light
Teratogens	Fetus	Lead and ethylene oxide
Mutagens	Genetic damage to cells or organisms	Radiation, lead, and ethylene dibromide

Toxicity Concerns

All of the discussion up to this point has focused on the hazmat exposure side of the health hazard equation. Now let's turn our attention to the toxicity side of the equation. Specifically, toxicity depends on the chemical's ability to enter the cells of the body and the manner in which it reacts chemically with the cells' structure.

Toxicologists list four categories of factors that influence toxicity:

1. **Concentration or dose.** Usually, but not always, the speed and magnitude of the hazmat's action is in direct proportion to the concentration and the duration of exposure (dose = concentration × time). Remember—dose makes the poison!
2. **Rate of absorption.** The faster the chemical enters the body and into the bloodstream, the faster its effects will be manifested. There is often a direct relationship between the rate of absorption and the route of exposure. In order of decreasing rate of absorption, recognized routes of exposure are intravenous, inhalation, intraperitoneal (into the abdominal cavity), intramuscular, subcutaneous (beneath the skin), oral, and cutaneous (onto the skin). While this order applies to many chemicals, recognize there are also some exceptions.
3. **Rate of detoxification.** The human body has natural defenses that can break down the toxic components of a chemical. This process is possible only as long as the rate of absorption into the body is less than the rate at which the body cells can destroy or neutralize and eliminate the toxin.
4. **Rate of excretion.** Rate at which the chemical is removed or excreted from the cells and the body. Remember that there are some chemicals that cannot be excreted and will start to accumulate within the body. These include polychlorinated biphenyls or PCBs (in body fat) and hydrogen fluoride (in bones).

Other miscellaneous factors can influence the toxicity of a chemical, including age, weight, and sex, previous health problems, the individual's dietary and smoking habits, and prescription (or other) drugs within the bloodstream at the time of exposure.

Measuring Toxicity

Toxicity is measured in terms of the ability of a material to injure living tissue. Since only limited data is available from accidental human exposure, toxicologists rely on data from animal experiments to establish toxicological values for humans. Two units of measurement are commonly cited for determining the relative toxicity of a chemical substance or compound:

- **Lethal dose, 50% kill (LD_{50})**—The concentration of an ingested, absorbed, or injected substance which results in the death of 50% of the test population. LD_{50} is an oral or dermal exposure expressed in terms of weight—mg/kg. It is expressed as the milligrams of the substance per the kilograms of body weight or mg/kg. For example, a material with an LD_{50} of 1 mg/kg means that a toxic dose for a 150 pound (lb) (68 kg) person would be 68 mg. To convert pounds to kilograms, just remember that 1 kg equals 2.2 lb.

  Other lethal dose percentages may also be found in reference guidebooks, including LD_{Lo}, LD_1, LD_{10}, LD_{30}, and LD_{99}. LD_{Lo} or LDL is the lowest dose of a substance reported to have caused death in humans or animals. Always remember—the lower the dose, the more toxic the substance.
- **Lethal concentration, 50% kill (LC_{50})**—The concentration of an inhaled substance that results in the death of 50% of the test population in a specific time period (usually 1 hour). LC_{50} is an inhalation exposure expressed in terms of parts per million (ppm) for gases and vapors, and milligrams per cubic meter (mg/m^3) or micrograms per liter (μg/liter) for dusts and mists, as well as gases and vapors. (The lower the concentration, the more toxic the substance.)

  The lethal concentration low percentage (LC_{Lo}) may also be found in reference guidebooks. LC_{Lo} is the lowest concentration of a substance in air reported to have caused death in humans or animals. Always remember—the lower the concentration, the more toxic the substance.

U.S. Versus Metric Measurements

As noted, most toxicological measurements use the metric scale. The ability to move from U.S. to metric scales of measurement when discussing lethal dose and lethal concentration values may be confusing, so remember the basic rules of thumb listed in **Table 2-3**.

Table 2-3 US versus Metric Measurements

For Solids

1 kilogram (kg) = 1 million milligrams (mg)

Therefore, 1 mg/kg = 1 part per million (ppm)

1 kilogram (kg) = 1 billion micrograms (μg)

Therefore, 1 μg/kg = 1 part per billion (ppb)

For Liquids

1 liter (1) of water weighs exactly 1 kg

Therefore, 1 mg/l = 1 part per million and 1 μg/l = 1 part per billion

1 kg = about 2.2 pounds

1 liter = about 1 quart

For Gases and Vapors

mg/m³ can be converted to parts per million by volume by using the following formula:

$$\text{parts per million} = \frac{\text{mg/m}^3 \times 24.45}{\text{molecular weight}}$$

Parts per million can be converted to mg/m³ by the following formula:

$$\text{mg/m}^3 = \frac{\text{parts per million} \times \text{molecular weight}}{24.45}$$

Table 2-4 Parts per Million vs. Parts per Billion

Abbreviation	Numerical Equivalent	Common Application	Example
PPM (Parts per Million)	0.000,001 (10^{-6})	Harm to Humans	1 teaspoon per 10,000 gallons 1 ounce in 32 tons 1 car in bumper-to-bumper traffic between Washington, DC to Sacramento, CA 1% concentration in air = 10,000 ppm 10% concentration in air = 100,000 ppm
PPB (Parts per Billion)	0.000,000,001 (10^{-9})	Harm to Humans or the Environment	1 teaspoon per 1 million gallons 1 sheet of toilet paper in a roll stretching from Washington, DC to Frankfurt, Germany 1% concentration in air = 10,000,000 ppb

Table 2-4 shows a more detailed explanation of the relationship between parts per million and parts per billion.

Various terms are used to describe the toxicity of materials on safety data sheets (SDS) and other reference sources. These terms were originally published by H. C. Hodge and J. H. Sterner in the *American Industrial Hygiene Association Quarterly* in 1949 and have become commonly accepted for describing toxicity. The Hodge–Sterner table is shown in Table 2-5.

Exposure Values and Guidelines

The obvious question for emergency responders is, how do we limit our exposure to the "bad stuff" at the incident scene? In addition, what data are available to provide guidance in determining such factors as isolation distances, the size and location of hazard control zones, and protective action recommendations?

Numerous exposure values have been developed to provide guidance in protecting employees in the industrial workplace. Although not specifically developed for emergency response applications, exposure values are useful in understanding the degree of hazard. Remember—exposure values are only guidelines; they are NOT absolute boundaries between safe and dangerous conditions. Also, responders will not be able to measure concentrations of specific chemicals unless the appropriate monitoring instruments are available.

Table 2-5 Hodge Sterner Table

Experimental LD_{50} (mg/kg of Body Wt)	Degree of Toxicity	Probable LD_{50} for a 70 kg Man (150 lb)
<1.0 mg	Dangerously Toxic	A Taste
1–50 mg	Seriously Toxic	A Teaspoonful
50–500 mg	Highly Toxic	An Ounce
0.5–5 gm	Moderately Toxic	A Pint
5–15 gm	Slightly Toxic	A Quart
>15 gm	Extremely Low Toxicity	More Than a Quart

The Hodge Sterner table indicates the relative degree of toxicity of chemicals based on the lethal dose, 50% kill value.

Common exposure values and guidelines are listed below and are summarized in Table 2-6:

- Threshold Limit Value/Time Weighted Average (TLV/TWA)—The maximum airborne concentration of a material to which an average healthy person may be exposed repeatedly for 8 hours each day, 40 hours per week without suffering adverse effects. They are based upon current available data and are adjusted on an annual basis by the American Conference of Governmental Industrial Hygienists (ACGIH). Because TLVs involve an 8-hour exposure, they are difficult to adapt for ERP use. TLV/TWAs are expressed in ppm and mg/m^3. (The lower the TLV/TWA, the more toxic the substance.)
- Permissible Exposure Limit (PEL) and Recommended Exposure Levels (REL)—The maximum time-weighted concentration at which 95% of exposed, healthy adults suffer no adverse effects over a 40-hour work week; these levels are comparable to ACGIH's TLV/TWA. PELs are used by OSHA and are based on an 8-hour, time-weighted average concentration. RELs are used by the National Institute of Occupational Safety and Health (NIOSH) and are based on a 10-hour, time-weighted average concentration. Unless otherwise noted, both are expressed in either ppm or mg/m^3.

  PELs were originally based on ACGIH TLV/TWA values that were in place in 1971; however, many have not been revised since that time and do not reflect the revisions made to the ACGIH exposure guidelines as new toxicological and exposure data have become available during this time period. PELs are used by OSHA in evaluating workplace exposures. (The lower the PEL, the more toxic the substance.)

Table 2-6 Health Exposure Guidelines

Exposure Guideline	Target Group	Sponsoring Organization	Definition	Exposure Duration
Threshold Limit Value (TLV)	Workers	ACGIH	Occupational exposure for 8-hour time-weighted concentration	8 hours/day 20 to 30 years
Permissible Exposure Limit (PEL)	Workers	OSHA	Occupational exposure for 8-hour time-weighted concentration	8 hours/day 20 to 30 years
Recommended Exposure Limit (REL)	Workers	NIOSH	Occupational exposure for 10-hour time-weighted concentration	8 hours/day 20 to 30 years
Short-Term Exposure Limit	Workers	ACGIH	Occupational exposure for 15-minute time-weighted concentration	15 minutes
Immediately Dangerous to Life or Health (IDLH)	Workers	NIOSH	Concentration poses a dangerous to immediate threat to life or from which escape is possible without permanent damage	No exposure duration
10% (1/10th) of IDLH	General public	EPA / FEMA	Level of Concern (LOC)	30 minutes
Emergency Response Planning Guideline (ERPG)	General public	AIHA	Three-tiered emergency planning guideline for emergency response estimate based on 1/10th of the published IDLH	1 hour
Acute Emergency Exposure Guideline (AEGL)	General public	National Research Council—Committee on Toxicology	Three-tiered emergency guideline for emergency response for five different exposure durations	10 minutes 30 minutes 1 hour 4 hours 8 hours

- **Short-Term Exposure Limit (STEL)**—or threshold limit value/short-term exposure limit (TLV/STEL). The 15-minute, time-weighted average exposure that should not be exceeded at any time, nor repeated more than four times daily with a 60-minute rest period required between each STEL exposure. These short-term exposures can be tolerated without suffering from irritation, chronic or irreversible tissue damage, or narcosis of a sufficient degree to increase the likelihood of accidental injury, impairing self-rescue, or reducing efficiency. TLV/STELs are expressed in ppm and mg/m^3. (The lower the TLV/STEL, the more toxic the substance.)

  For some substances, ACGIH could not find sufficient toxicological data to develop the TLV/STEL for chemicals that had already been assigned a TLV/TWA. In these instances, ACGIH recommends that short-term exposures not exceed three times the TLV/TWA for more than a total of 30 minutes during the day. The STEL is found as ST in the *NIOSH Pocket Guide to Chemical Hazards*.
- **Threshold Limit Value/Ceiling (TLV/C)**—The maximum concentration that should not be exceeded, even instantaneously. (The lower the TLV/C, the more toxic the substance.) For some substances, ACGIH could not find sufficient toxicological data to develop the TLV/C for chemicals that had already been assigned a TLV/TWA. In these instances, ACGIH recommends that five times the TLV/TWA be used in place of the TLV/C.
- **Threshold Limit Value/Skin (TLV/Skin)**—Indicates possible and significant exposure to a material by way of absorption through the skin, mucous membranes, and eyes by direct or airborne contact. This attention-calling designation is intended to suggest appropriate measures to minimize skin absorption so that the TLV/TWA is not exceeded.
- **Immediately Dangerous to Life or Health (IDLH)**—An atmospheric concentration of any toxic, corrosive, or asphyxiant substance that poses an immediate threat to life, or would cause irreversible or delayed adverse health effects, or would interfere with an individual's ability to escape from a dangerous atmosphere. IDLH values are expressed in ppm and mg/m^3. (The lower the IDLH, the more toxic the substance.)

  There are three general IDLH atmospheres: toxic, flammable, and oxygen deficient. In the absence of an IDLH value for toxic atmospheres, emergency responders may consider using an estimated IDLH of 10 times the TLV/TWA. IDLH values for flammable atmospheres are typically 10% of the lower explosive limit, while an

IDLH oxygen-deficient atmosphere is 19.5% oxygen or lower. IDLH was not originally designed as an exposure level for evaluating protective actions. However, EPA has noted in the *Technical Guidance for Hazard Analysis* that using one-tenth (10%) of the IDLH value may be an acceptable level of concern for evaluating hazmat release concentrations and public protective options.

- Emergency Response Planning Guidelines (ERPG)—ERPGs are air concentration guidelines for single exposures to hazardous materials. Developed by the American Industrial Hygiene Association (AIHA) as an emergency planning tool for public protective action options, ERPGs have been developed for approximately 100+ chemicals, most of which are extremely hazardous substances (EHS) with airborne hazards (e.g., chlorine, anhydrous ammonia, hydrogen sulfide **Table 2-7**).

  Although there are three ERPG tiers, the ERPG-2 level is the most commonly cited guideline when using public protective action decision-making support tools. The ERPG-2 is defined as the maximum airborne concentration below which it is believed that nearly all individuals could be exposed for up to one hour without experiencing or developing irreversible or other serious health effects or symptoms that could impair an individual's ability to take protective action. Additional information can be obtained at the AIHA website at http://www.aiha.org.
- Acute Emergency Exposure Guidelines (AEGL)—Referred to as "eagles" and currently being developed through a U.S. EPA National Advisory Committee, these are intended to provide uniform exposure guidelines for the general public for a single short-term exposure. The Committee's objective is to define AEGLs for the 300+ EHS materials listed in SARA, Title III, as well as for chemical warfare agents.

  Three tiers of AEGLs have been developed covering five exposure periods: 10 minutes, 30 minutes, 1 hour, 4 hours, and 8 hours. You can find additional information on the EPA website at http://www.epa.gov.

Table 2-7 Exposure Values for Common Chemicals

Chemical Name	TLV/TWA (ppm)	IDLH (ppm)	ERPG-2 (ppm)
Sarin	0.000017	0.03	N/A
Chlorine	0.5	10	3
Nitric Acid	2	25	6
Anhydrous Ammonia	25	300	150
Carbon Monoxide	35	1200	350
Toluene	50	500	300
Acetone	250	2500	N/A

Responder Safety Tip

Exposure values should be regarded only as guidelines, not absolute boundaries between safe and dangerous conditions. In addition, they should not be used when combinations of materials are involved. To be most effective, exposure values must be combined with monitoring instrument readings and interpreted by response personnel who are familiar with their proper application and limitations.

When evaluating the establishment of hazard control zones at hazmat emergencies, the various TLV and the IDLH values are generally the most informative. When evaluating public protective action options, the ERPG, AEGL or one-tenth of the IDLH value are useful. Remember—the lower the reported concentration, the more toxic the material.

Controlling Personnel Exposures

The primary objective of using these various exposure guidelines is to minimize the potential for both public and responder exposures. One method for understanding these exposure guidelines and applying them at a hazmat emergency is the concept of safe, unsafe, and dangerous. Originally developed by Mike Callan and Frank Docimo, there are three basic atmospheres at an incident involving hazardous materials:

1. **Safe atmosphere**—No harmful hazmat effects exist, which allows personnel to handle routine emergencies without specialized PPE.
2. **Unsafe atmosphere**—Once a hazmat is released from its container, an unsafe condition or atmosphere exists. If one is exposed to the material long enough, some form of either acute or chronic injury will often occur.
3. **Dangerous atmosphere**—These are environments where serious irreversible injury or death may occur.

Understanding this system will assist responders in establishing hazard control zones, selecting skin and respiratory protection levels, evaluating public protective action options, and determining the end of the emergency phase of the incident. Of course, it is virtually impossible to quantify an atmosphere and determine the type of atmosphere present without the use of air monitoring instruments (see Scan Sheet 2-A).

What Is Safe?

Go back to our previous discussion on TLVs. All of these guidelines have one thing in common—remain below these values and the exposure is considered safe to the average healthy adult by all information known by today's health and safety professionals.

However, recognize that these are dynamic values and have a tendency to be lowered over time as additional toxicological research and studies are completed (e.g., benzene). In addition, exposure to multiple chemicals may have additive or synergistic effects. If the chemical concentration exceeds the TLV or PEL exposure values, you should assume you are now in an unsafe condition. Stay exposed long enough and some form of harmful effects may occur.

What Is Unsafe?

A general rule for responders should be—if the material has been released from its container, assume that an unsafe atmosphere may exist and some form of PPE is required. Prolonged exposures at high concentrations can lead to injury; however, acute injuries may not be lethal (e.g., headaches, nausea, irritation to the eyes, nose, or throat). Unsafe atmospheres do not become seriously dangerous unless the exposure continues or the concentration of contaminants increases.

The unsafe atmosphere is an area where some responders may ignore the signs and symptoms of overexposure. The fire service was often guilty of this during the days when "eating smoke" was viewed as an acceptable risk and part of the firefighting culture. In many respects, that culture continues today

Scan Sheet 2-A

Toxicity and Exposure Terminology

The terms used to describe chemical toxicity and exposures can seem complicated, and some have similar meanings, further complicating the issue. All relate to how long an individual can safely work in a chemical or hazardous atmosphere. There is a safety factor incorporated into each of these values, as each person can react differently based upon their age, sensitivity, and pre-existing medical conditions. All of these terms and values are based on a person wearing NO skin or respiratory protection.

Remember: HEALTH HAZARD = EXPOSURE + TOXICITY

"Poison lines" are an excellent tool for visually illustrating the significance and inter-relationship between health exposures and other hazards. Examine the following poison lines for chlorine (poison gas), methanol (flammable liquid with health hazards), and anhydrous ammonia (corrosive gas that is toxic by inhalation and flammable in certain atmospheres). Note that (1) as concentrations rise above the TLV, harm occurs; and (2) as the concentration increases, harm increases. At some point, the IDLH is reached.

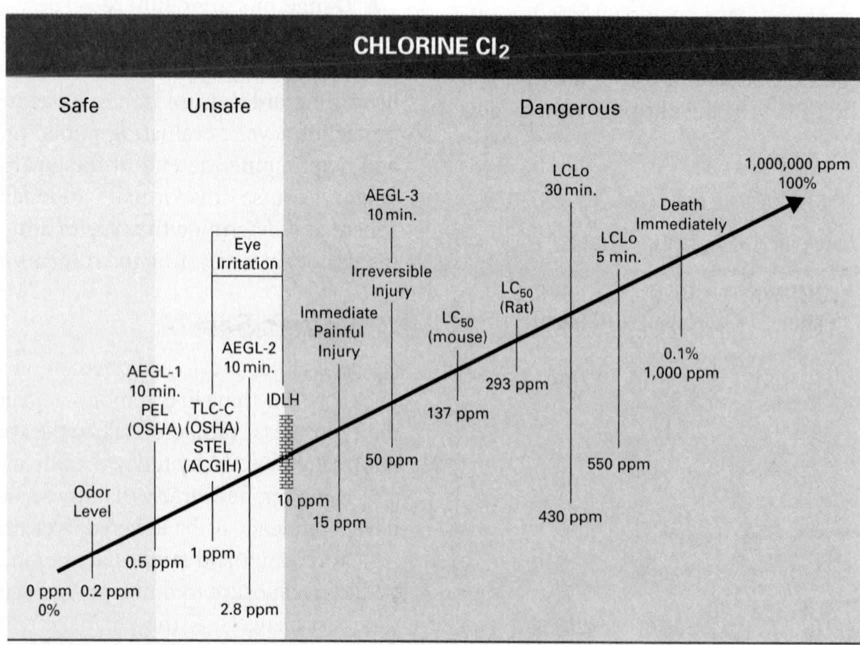

Adapted from original artwork by Michael Callan.

Scan Sheet 2-A (continued)

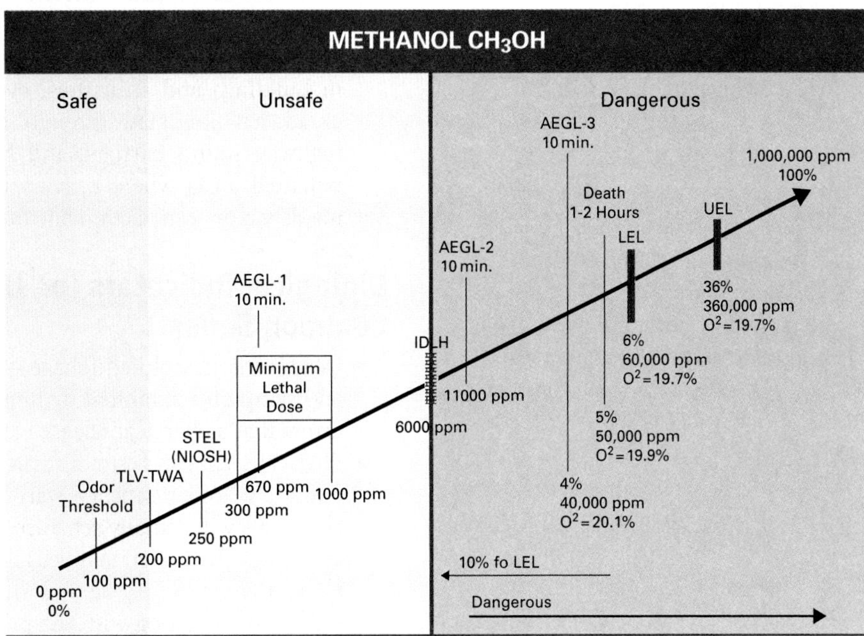

Adapted from original artwork by Michael Callan.

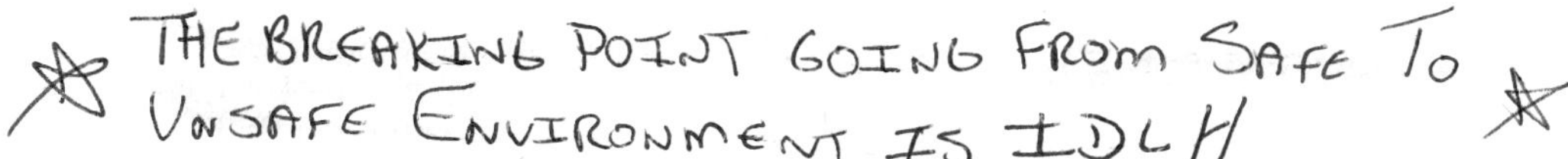

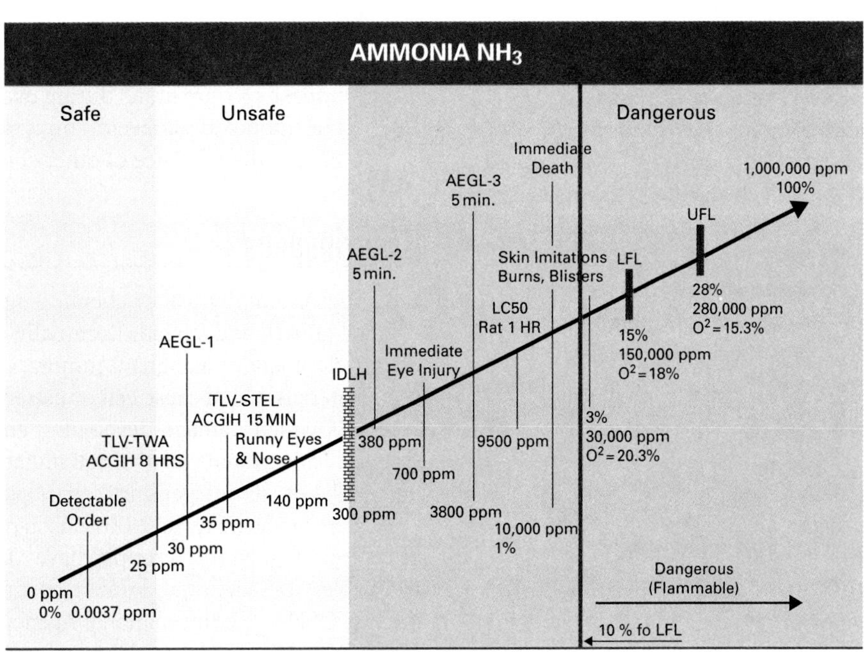

Adapted from original artwork by Michael Callan.

when fire fighters remove their SCBA masks once a fire is extinguished, even though many invisible gases and contaminants are still present. Unsafe conditions may also occur when emergency responders ignore unusual or unexplained air monitoring readings without confirming their cause. When dealing with a potential inhalation hazard, response personnel must use air-supplied respirators (e.g., SCBA) until the IC determines through the use of air monitoring that a decreased level of respiratory protection will not result in a hazardous exposure.

What Is Dangerous?

When concentrations continue to increase above unsafe levels, there is a high potential for life-threatening injuries or death to occur. This concentration level is the IDLH. There are three general IDLH atmospheres: toxic, flammable, and oxygen deficient. As we previously noted, in the absence of an IDLH value for toxic atmospheres, emergency responders may consider using an estimated IDLH of 10 times the TLV/TWA. IDLH values for flammable atmospheres are 10% of the lower explosive limit, while an IDLH oxygen-deficient atmosphere is 19.5% oxygen or lower. An oxygen-enriched atmosphere contains 23.5% oxygen or higher. While these atmospheres can cause harm in various ways, the degree of harm is similar.

Unfortunately, responders may not always have immediate access to monitoring instruments; therefore, what are some of the physical indicators of likely IDLH conditions?

Outside or Open Air Environment

- **Visible vapor cloud**—Large vapor clouds obviously indicate large concentrations of contaminants. Avoid entering a vapor cloud at all costs. Colored vapor clouds are not good! For example: greenish yellow = chlorine; orange brown = fuming nitric acid.
- **Release from a bulk container or pressure vessel**—Bulk containers can release large quantities of liquid and gas products. Bulk-pressurized containers, such as horizontal tanks and spheres containing liquefied gases and cryogenic liquids, will pose the greatest hazards because of their tremendous liquid to vapor expansion ratios.
- **Large liquid leaks**—All liquids give off vapors when released. High vapor pressure liquids, such as many solvents and fuming corrosives, and pooled liquefied gases, such as chlorine, sulfur dioxide, anhydrous ammonia, and anhydrous hydrogen fluoride, are particularly dangerous.

Inside or Limited Air Environment

- **Below grade rescues or releases**—Small amounts of heavier-than-air vapors can accumulate in low-lying areas. Remember—most gases and vapors are heavier than air. Such areas should be avoided unless personnel are properly protected.
- **Confined spaces**—Any enclosed area where there is poor ventilation can result in either an oxygen-deficient, toxic, or flammable atmosphere, depending on the hazmats involved. NIOSH historical statistics have shown that 60% of all fatalities at confined space emergencies are personnel acting in a rescuer capacity. Confined space hazards and risks are discussed in more detail in the section on "implementing response objectives."
- **Artificial or natural barriers**—Any time vapors can be trapped they will accumulate and potentially increase in concentration. Vapors can remain trapped in buildings and structures, even long after a vapor cloud has migrated through an area. Tank dikes, highway sound barriers, and high vertical walls are potential areas where heavier-than-air toxic or flammable vapors can accumulate outdoors.

Biological Indicators (or, Using Your Common Sense!)

- Dead birds, discolored foliage, sick animals, and signs and symptoms exhibited by humans are all good indicators that a chemical release may have occurred.
- Physical senses and "street smarts"—Be aware of strong odors and other sensory warnings; remember, some chemicals will actually sensitize or deaden the sense of smell; e.g., hydrogen sulfide. Likewise, you don't have to be a chemist to look at a situation and determine something is wrong with the picture. Don't underestimate your "sixth sense"—if the situation doesn't look or feel right, it probably isn't.
- Hazmats with a potential for quick and rapid harm, such as poison gases, explosives, some oxidizers, and materials with very low IDLH values, should dictate the use of extreme caution.
- Firefighting overhaul operations. Studies have clearly shown air contaminant concentrations exceed occupational exposure limits during overhaul operations. Carbon monoxide concentrations should not be used to predict the presence of other contaminants.

Carcinogens

Carcinogens are physical or chemical agents that cause abnormal cell growth and spread. Eventually, this action can lead to the development of malignant tumors. Carcinogens, mutagens (i.e., materials which cause cell mutation), and teratogens (i.e., materials which damage developing embryos) are similar in that each causes some form of cell mutation. Studies show that up to 90% of all mutagens are carcinogens.

There are as many as 2,000 substances that various scientific and regulatory groups have labeled as "suspect," "probable," or "definite" human carcinogens. These groups include OSHA, the International Agency for Research on Cancer (IARC), and the National Toxicology Program (NTP). In the 2011 ACGIH Threshold Limit Value (TLV) and Biological Exposure Indices (BEI) booklet, approximately 20+ substances, including asbestos, vinyl chloride, and benzene have been proven through human epidemiological studies to increase cancer rates. The remainder received their carcinogenic classification based on animal studies.

Scan Sheet 2-B

Fire Smoke: It's Bad—Don't Breathe It!

Smoke production is dependent on several factors, including the chemical make-up of the burning material, temperature of the combustion process, oxygen content supporting combustion, and presence or absence of ventilation. Simply put, fire is a complex process, and the smoke produced is an intricate collection of particulates, superheated air, and toxic chemical compounds (especially in closed compartment residential structure fires). The most important thing for you to remember is smoke is bad and you shouldn't breathe it!

The extensive commercial and residential use of synthetic materials (plastics, nylons and polymers such as Styrofoam® and polyurethane foam) has a significant impact on combustion and fire behavior, as well as the smoke produced during a structure fire. Most of these materials are carbon-based and bonded with various atoms like hydrogen, nitrogen, chlorine, and sulfur. Synthetic substances ignite and burn fast, causing rapidly developing fires and toxic smoke and making structural firefighting more dangerous than ever before.

Some of the compounds liberated during the combustion process are irritants such as hydrogen chloride and ammonia, which can cause eye irritation or airway problems during smoke exposure. Other common chemicals liberated in the combustion process include several oxides of nitrogen such as nitric oxide and nitrogen dioxide, along with ozone, acrolein, and many different types of aldehydes. These compounds and many others are present during the typical combustion process, creating a toxic "soup" that assaults the respiratory system, central nervous system, and cardiovascular system. In short, smoke attacks the entire human body and could leave behind such legacy illness as Parkinson's disease, various kinds of cancer, and long term respiratory compromise. When these irritants and toxins are looked at from a hazmat standpoint, it would be pretty hard to convince an entry team they wouldn't need a high level of respiratory protection.

From a health and safety perspective, structural firefighting should be thought of with more of a "hazmat eye," meaning that while you approach fires and hazmat incidents from a different tactical perspective, wearing your PPE and reducing the exposure to toxins should be approached as if you were contemplating a Level A entry! Think about this—carbon monoxide levels can reach 10,000 ppm in a working structure fire. Is doing an entry without an SCBA, or leaving someone short on air supply, a good plan for a long and happy life?

Other compounds, like carbon monoxide and cyanide, are toxic when inhaled. Carbon monoxide is created when carbon and hydrogen bond and is partly responsible for incapacitating a smoke inhalation victim. Cyanide, formed by carbon-hydrogen-nitrogen bonding during the combustion process, disrupts the body's ability to use oxygen and causes asphyxia at the cellular level. Recent studies conclude that cyanide, along with concurrent carbon monoxide poisoning, is responsible for many smoke related deaths and injuries.

Sofas, cabinets, drapes, blankets, and carpeting all produce various levels of cyanide and other common toxins as by-products of combustion. Vehicle fires are also capable of generating cyanide, along with almost everything found in garage or dumpster fires.

To that end, it is incumbent on fire fighters, EMS providers, hazmat responders, and members of the law enforcement community to have at least a basic understanding of the hazards of fire smoke and how to recognize the signs and symptoms of a smoke exposure.

It is becoming more common to see fire fighters performing gas detection at fire scenes in order to try to identify some of these toxins and determine when it is safe to remove SCBA. This may seem like a straightforward task, but trying to identify what's in the smoke in a hot, humid particulate-laden environment is challenging at best. Should you use standard gas detection devices with electrochemical sensors, or maybe colorimetric tubes? At what level should you be concerned about the irritants and toxins? The answers to some of these questions are beyond the scope of this text, and in some cases, industry standards are still being developed. You can do something today, however, to make your job much safer—something that doesn't cost the agency anything, doesn't take any more time or effort on your part, and is something you already know how to do: Wear your SCBA! Check it out before you wear it, manage the air you have, and wear the mask and breathe the air from the tank.

(Continued)

Scan Sheet 2-B (continued)

Responder Safety Tip

Here are several exposure values for two of the most common "bad actors" found in fire smoke—carbon monoxide and hydrogen cyanide.

- Carbon Monoxide (CO)
  - IDLH = 1,200 ppm
  - Recommended Exposure Limit (REL) = 25 ppm
- Hydrogen Cyanide (HCN)
  - IDLH = 50 ppm
  - REL = 5 ppm
  - HCN at concentrations above 270 ppm can be fatal in 6–8 minutes

Wear and use your SCBA!

Manifestations of Cyanide Poisoning

Early Indications of Exposure to Low Inhaled Concentrations:

- Anxiety · Headache · Tachypnea
- Drowsiness · Impaired judgment · Vertigo
- Dyspnea · Tachycardia

Inhalation of Moderate to High Concentrations:

- Cardiovascular collapse · Hypotension · Respiratory depression or arrest · Smell of almonds on the breath (sometimes undetectable)
- Cardiac dysrhythmia · Markedly altered level of consciousness · Seizure

Manifestations of Carbon Monoxide Poisoning

Early Indications of Exposure to Low Inhaled Concentrations:

- Difficulty with balance · Fatigue · Headache · Palpitations

Inhalation of Moderate to High Concentrations:

- Altered level of consciousness · Nausea and vomiting · Seizure · Shock and death
- Cardiac dysrhythmia · Respiratory arrest · Severe headache · Syncope

Responder Safety Tip

Although some carcinogens may have a TLV value (e.g., benzene), many do not. It is assumed that there is no threshold value below which these materials are "safe." In addition, carcinogens will not have an IDLH value.

Radioactive Materials

Radiation is the emitting of energy from an atom in the form of either particles or electromagnetic waves. Radiation can be classified into two types:

- **Nonionizing radiation—**Characterized by its lack of energy to remove electrons from atoms. Infrared waves, radio waves, and visible light are all nonionizing radiation. The amount of energy in these waves is small compared to ionizing radiation. Microwaves and lasers are common examples of nonionizing radiation.
- **Ionizing radiation—**Characterized by its ability to create charged particles, or ions, in anything it strikes. Exposure to low levels of ionizing radiation can produce short- or long-term cellular changes with potentially harmful }effects, such as cancer and leukemia. X-rays are a familiar form of ionizing radiation.

There are four types of ionizing radiation:

- **Alpha particles—**Largest of the common radioactive particles, alpha particles have extremely limited penetrating power. They travel only 3 to 4 inches in air and can be stopped by a sheet of paper or a layer of human skin. Alpha radiation is primarily an internal hazard and the greatest health hazard exists when alpha particles enter the body, such as through inhalation or ingestion.
- **Beta particles—**A particle that is the same size as an electron and can penetrate materials much farther than large alpha particles. Depending on the source, beta particles can travel several yards in air and penetrate paper and human skin but cannot penetrate internal organs. Depending on their energy, beta particles represent both an internal and external radiation hazard. They can be shielded by plastic, glass or metal foil.

- **Gamma rays**—Most dangerous form of common radiation because of the speed at which it moves, its ability to pass through human tissue, and the great distances it can cover. The range of gamma waves depends on the energy of the source material. Gamma radiation penetrates most materials very well, and it is considered a whole body hazard as internal organs can be penetrated and damaged.
- **Neutron particles**—Another form of high-speed particle radiation, neutron radiation consists of a "neutron" emitted at a high speed from the nucleus of a radioactive atom. There are few natural emitters of neutron radiation; the natural background of neutron radiation comes from cosmic rays from outer space interacting with gas molecules in the atmosphere. Neutrons are considered a whole body hazard.

Half-life is the time it takes for the activity of a radioactive material to decrease to one-half of its initial value through radioactive decay. The half-life of known materials can range from a fraction of a second to millions of years. Although not normally a critical factor in incident mitigation, responders should still note the half-life. Some extremely active radioactives used for radionucleocides and other medicinal purposes have relatively short half-lives (e.g., thallium).

In dealing with a radioactive material emergency, responders must understand the difference between exposure and contamination. Exposure means the human body has been subjected to radiation emitted from a radioactive source. Contamination means the actual radioactive material has come in direct contact with one's body or clothing (i.e., the person is "dirty"). As long as the material remains in contact with one's body or clothing, they are contaminated until decontamination measures are performed. Contamination and cross-contamination are discussed in detail in the section on "decontamination."

Exposure guidelines for radioactive materials are related to the old adage—"time, distance, and shielding." These basic site safety concepts can also apply to any hazmat release:

- **Time**—The shorter the exposure time, the less the exposure. Radiation exposures are additive in their effects upon the body or any other subject—site safety and control procedures to monitor all entry operations are critical.
- **Distance**—The closer you are, the greater the exposure. The energy emitted from a radioactive source declines as one moves further away from the source. The inverse square law is a simple tool for applying this safety concept. Simply stated, if you double the distance from a point radiation source, the radiation intensity is lowered by one-fourth. If you increase the distance 10 times, the radiation intensity is one-hundredth of the original value.
- **Shielding**—Although personal protective clothing can offer protection against alpha particles, it will provide limited protection against beta particles and no protection against gamma or neutron radiation. Therefore, dense materials must be kept between emergency responders and higher radiation sources. Common shielding materials include lead, cement, and even water. Any shielding material is better than none!

Responder Safety Tip

Radiation that doesn't hit anybody doesn't hurt anybody! Always keep the acronym "ALARA" in mind when considering radiation exposures—keep the exposure "As Low as Reasonably Achievable."

Units of Measurement for Radiation

Radiation source activity, exposure, and dose are measured in two systems of units. Within the United States, the English system is most commonly used, while the international system (SI) is commonly used in other parts of the world. The SI units of measurement are also used by the U.S. Department of Defense (DOD). **Table 2-8** summarizes the units of measurement and their respective conversion factors.

Radioactive Dose Limits for Emergency Response Operations

Table 2-9 summarizes the U.S. and voluntary consensus standard guidance on radioactive dose limits for personnel performing emergency duties from the following sources:

- *EPA Manual of Protective Action Guides and Protective Actions for Nuclear Incidents* (EPA 400-R-92-001)

Table 2-8 Radiation Dose and Exposure

	Source Activity (disintegrations per unit time)	Exposure	Dose
U.S. System (old system)	Curie (Ci) = 37 × 10^9 dps (dps—disintegrations per second)	Roentgen (R)	rem—Roentgen equivalent man RAD or rad—Radiation absorbed dose
SI Units (new system)	Becquerel (Bq) 1 Bq = 1 dps 1 Ci = 37 GBg (or 10^9 GBq)	Gray (Gy) 1 Gy = 100 rad 1 rad = 1 cGy (centi Gray)	Sievert (Sv) 1 Sv = 100 rem 1 rem = 10 mSv (milli Sievert)

Table 2-9 Radioactive Materials Dose Limits for Workers Performing Emergency Services (Offensive Operations)

Total Effective Dose Equivalent Guideline	Activity	Condition	Source
≤ 5 rem (0.05 Sv)	All Occupational Exposures	All reasonably achievable actions have been implemented to minimize dose	EPA 400 FEMA 2008 ASTM E2601-08
10 rem (0.1 Sv)	Protecting Valuable Property Necessary for Public Welfare (e.g., power plant)	All reasonably achievable actions have been implemented; exceeding 5 rem (0.05 Sv) is unavoidable and on a voluntary basis. Responders fully informed of risks	EPA 400 FEMA 2008 ASTM E2601-08
25 rem (0.25 Sv)	Lifesaving or Protection of Large Populations	All reasonably achievable actions have been implemented; exceeding 5 rem (0.05 Sv) is unavoidable and on a voluntary basis. Responders fully informed of risks	EPA 400 FEMA 2008 ASTM E2601-08
50 rem (0.5 Sv)	Lifesaving activities in catastrophic incidents (e.g., Improvised Nuclear Device – IND)	All reasonably achievable actions have been implemented; exceeding 5 rem (0.05 Sv) is unavoidable and on a voluntary basis. Responders fully informed of risks	ASTM E2601-08

- *FEMA Application of Protective Action Guides for Radiological Dispersal Device (RDD) and Improvised Nuclear Device (IND) Incidents* (2008)
- *ASTM E2601-08—Standard Practice for Radiological Emergency Response* (2008)

In applying these exposure values at a radioactive material incident, responders should always keep these basic points in mind:

- Emergency duties should be limited to nonpregnant adults.
- These limits apply to doses incurred over the duration of an emergency and should be treated as a once-in-a-lifetime exposure.
- The dose to the individual responder should always be managed to the lowest possible level. ALARA principles should always prevail; if only defensive operations are required, responder dose should not require management beyond the 5 rem (0.05 Sv) guideline for any emergency activity.
- Emergency responders exceeding the 5 rem (0.05 Sv) guideline should understand both the acute and long term (i.e., cancer) effects of radiation.

Exposure to Environmental Conditions

During an incident, the physical working environment must constantly be evaluated. The human body tolerates a limited range of thermal environments. Exposure to either hot or cold weather conditions over a sustained period of time can adversely affect both the physiological and psychological conditions of response personnel. Continued exposures may result in physical discomfort, loss of efficiency, and a higher susceptibility to accidents and injuries.

Factors that influence an individual's susceptibility to environmental conditions include:

- Lack of physical fitness
- Lack of acclimatization to the elements
- Age
- Level of hydration
- Obesity
- Alcohol and drug use (including prescription drugs)
- Infection
- Allergies
- Chronic disease

Heat Stress

In many respects, the human body is similar to a machine. Like a vehicle, the body must take fuel and mix it with oxygen to produce work. A natural by-product of this work is heat given off into the environment. Also like a vehicle, the body has an excess heat transfer system. Where a vehicle uses a radiator with water, air, and coolant lines, the human body uses skin with sweat, air, and veins to dissipate heat buildup. However, wearing impermeable, "nonbreathing" clothing will significantly reduce the ability of the body to transfer this additional heat. The net result is increased heat levels within the deep inner body core.

Experience over the last 30 years has shown that responders wearing chemical protective clothing (CPC) are more likely to be injured as a result of heat stress than a chemical exposure. Heat stress is a concern when wearing any type of impermeable protective clothing. When wearing normal clothing, excess body heat can normally escape to the atmosphere. However, while CPC is designed to protect the user from a hostile environment and prevent the passage of harmful substances into the "protective envelope," CPC also reduces the body's ability to discard excess heat and perform natural body ventilation.

A key indicator of body heat levels is the body's core temperature or the body's internal temperature. If the body heat cannot be eliminated, it will accumulate and elevate the core temperature. In addition to CPC, conditions that promote heat stress include high temperatures, high humidity, high radiant heat, and strenuous physical activities. Even after cooling is initiated, the body's core temperature can take several hours to return to normal resting levels.

Heat stress can also be enhanced by certain medically prescribed drugs and over-the-counter medicines. Responders should receive guidance from their personal physician or medical director.

Physical reactions to heat include the following:

- **Heat rash**—An inflammation of the skin resulting from prolonged exposure to heat and humid air and often aggravated by chafing clothing. Heat rash is uncomfortable and decreases the ability of the body to tolerate heat.
- **Heat cramps**—A cramp in the extremities or abdomen caused by the depletion of water and salt in the body. However, there is no observed increase in the body's core temperature. Usually occurs after physical exertion in an extremely hot environment or under conditions that cause profuse sweating and depletion of body fluids and electrolytes. The onset of heat cramps may be delayed several hours after the heat exposure. When the cramps occur in the abdomen, they may be confused with a surgical abdominal emergency, such as appendicitis.
- **Heat exhaustion**—A mild form of shock caused when the circulatory system begins to fail as a result of the body's inadequate effort to give off excessive heat. Although not an immediate life-threatening condition, the individual should be immediately removed from the source of heat, rehydrated with electrolyte solutions, and the body kept cool. If not properly treated, heat exhaustion may evolve into heat stroke.
- **Heat stroke**—A severe and sometimes fatal condition resulting from the failure of the temperature regulating capacity of the body. It is caused by exposure to the sun and/or high temperatures. Reduction or cessation of sweating may be an early symptom. Body temperature of 105°F (40.5°C) or higher; rapid pulse; hot, dry, or moist skin; headache; confusion; unconsciousness; and convulsions may occur.

Heat stroke is a true medical emergency requiring immediate transport to a medical facility. Any means available should be used to cool the victim; the body temperature should be taken at five-minute intervals and should not go below 101°F (38.3°C). Normal intravenous saline solution may be required to rehydrate and balance electrolyte levels. However, the patient is susceptible to pulmonary edema, and careful monitoring is essential.

The signs and symptoms of heat stress and the related emergency care procedures are outlined in **Table 2-10** *and* **Table 2-11**.

Table 2-10 Heat Stress Emergencies Signs And Symptoms

Signs and Symptoms	Condition		
	Heat Cramps	**Heat Exhaustion**	**Heat Stroke**
Muscle Cramps	Yes	No	No
Breathing	Varies	Rapid Shallow	Deep Then Shallow
Pulse	Varies	Weak	Full Rapid
Weakness	Yes	Yes	Yes
Skin	Moist-Warm No Change	Cold Clammy	Dry-Hot
Perspiration	Heavy	Heavy	Little or None
Loss of Consciousness	Seldom	Sometimes	Often

Table 2-11 Emergency Care of Heat Stress Emergencies

Heat Cramps

- Move patient to a nearby cool place, such as in shade or an air-conditioned ambulance, and loosen clothing
- Drink water or commercial fluid replacements (i.e., low sugar drinks, sports drinks)
- If cramps are severe or don't go away, seek medical attention.

Heat Exhaustion

- Move the patient to a nearby cool place, such as an air-conditioned ambulance.
- Remove enough clothing to cool the patient without chilling him or her (watch for shivering).
- Use active cooling (e.g., forearm immersion, misting fans or cold towels) to lower the body core temperature.
- If responsive and not nauseated, have the individual drink water or other fluids as specified by local medical protocols. If nauseous or vomiting, give nothing by mouth.
- If patient is unconscious, fails to recover rapidly, has other injuries, or has a history of medical problems, transport as soon as possible.
- Responders exhibiting the signs and symptoms of heat exhaustion should not return to service at the incident.

Heat Stroke

- Immediately transport to a medical facility. This is a life threatening emergency.
- Cool the patient—using active cooling techniques, move the patient out of the sun or away from the heat source. Use the maximum setting on the air conditioner in the patient compartment of the ambulance.
- Remove the patient's clothing, as appropriate. If cold packs or ice bags are available, apply them to the groin.

Table 2-11 Emergency Care of Heat Stress Emergencies (continued)

Heat Stroke

- Keep the skin wet by applying water with sponges or wet towels, or by wrapping the person in sheets soaked in cool water.
- Give the individual nothing by mouth.
- Monitor vital signs. Should transport be delayed, find a tub or container and immerse the responder up to the neck in tepid or cool water.
- If not managed appropriately, heat stroke will likely result in the responder's death. Early recognition of the condition and rapid cooling are essential to survival.

Source: Edward T. Dickinson and Michael A. Weidner, Emergency Incident Rehabilitation, Upper Saddle River, NJ: Brady/Prentice Hall (2000)

Figure 2-3 Heat stress should be managed through a series of administrative controls and through the use of PPE cooling devices.

To minimize the effects of heat stress, EMS personnel should be on-site to monitor and screen response personnel. Heat stress should be managed through a series of both administrative controls and through the use of PPE cooling devices Figure 2-3. Administrative controls, including the need for acclimatization or conditioning the body to working in hot environments, work/rest scheduling, rehab, and fluid replacement are outlined below. PPE cooling options are outlined here:

- **Air-cooled jackets and suits**—Consists of small airlines attached to either vests, jackets, or CPC to provide convective cooling of the user by blowing cool air over the body inside the suit. Cooling may be enhanced by the use of a vortex cooler or by refrigeration coils and a heat exchanger. Although sometimes found at remediation operations, they are typically not used for emergency response applications.

  These units require an airline and large quantities of breathing air (10 to 25 cubic feet per minute) and are not as effective as the active and passive cooling units in controlling the body's core temperature.
- **Ice-cooled vests**—These vests consist of frozen ice or synthetic coolant packs that are part of a vest or jacket. This passive cooling system operates on the principle of conductive heat cooling. Although not as effective as the full-body cooling suit in controlling the body's core temperature, studies have shown ice vests are better than both the air-cooled units and water-cooled jackets. In addition to the physiological advantages, there may also be psychological benefits; however, people may say they "feel better" even though their body's core temperature is actually increasing.

  Ice cooled units are relatively inexpensive and lightweight, improve worker comfort, decrease lens fogging, and are "user friendly" (that means responder-proof!). On the negative side, some vests require frozen coolant packs or an ice source at the scene of the emergency (unfortunately, the frozen coolant packs often leave the hazmat unit for someone's lunch box or cooler) and may add bulk underneath the CPC.

  These vests can also be used with heat packs for operations in extremely cold working environments.
- **Liquid-cooled vests and suits**—These units consist of a heat transfer garment (vest or full-body suit) and a cooling unit. The cooling unit normally consists of a battery or power source, a pump, and an ice/water or cooling agent (e.g., liquid nitrogen) container. Essentially, the cooling agent is circulated throughout the garment and operates on the principle of conductive heat transfer.

  These units often add both weight and bulk to the CPC. As a result, they tend to be more prevalent in the hazmat clean-up and remediation industry, where longer work times are required. Studies have shown the full-body suits are substantially more effective than vests alone and, overall, are the most effective method for controlling the body's core temperature. This same technology can also be used as part of a full-body warming system for use under cold work conditions.
- **Phase Change Cooling Technology**—These units consist of vest packs and neck/head wraps that contain a unique phase-liquid change material (PCM) that functions as a heat sink. PCM produces a cooling effect as it is transformed from a solid state to a liquid state when the temperature rises above a certain activation point (e.g., 82°F [27.7°C] to 90°F [32°C]). Worn under PPE, PCM operates under the principle of conductive heat cooling in which the PCM provides long-lasting, temperature-specific cooling relief at approximately 58°F (3°C).

  Originally developed for military applications, phase change cooling technology is widely used in both emergency response that involves the use of specialized protective clothing (e.g., hazmat, bomb suits) and industrial applications with long-duration cooling requirements. The PCM material is nontoxic, nonflammable, reusable, and never fully freezes.

Advantages include constant temperature (approximately 58°F [14°C]), long duration of cooling capability (up to 3 hours), usage options (e.g., vest, neck wrap, head wrap), and ease of reuse. Both packs and wraps can be reused by simply inserting them into ice water for up to 45 minutes or placing them in a freezer until solid.

Cold Temperature Exposures

Hazardous materials such as liquefied gases and cryogenic liquids expose personnel to the same hazards as those created by cold weather environments. In addition, conducting decon operations in cold weather environments will provide numerous challenges, especially if technical or mass decon operations are required. Exposure to severe cold even for a short period of time may cause severe injury to body surfaces, especially the ears, nose, hands, and feet.

Two factors significantly influence the development of cold injuries—the ambient temperature and wind velocity. Since still air is a poor heat conductor, emergency responders working in areas of low temperature and little wind can endure these conditions for long periods, assuming their clothing is dry. However, when a low ambient temperature is combined with an active airflow, a dangerous condition described as "wind chill" is created. For example, an ambient temperature of 20°F (–6.6°C) with a wind of 15 miles per hour (mph) is equivalent in its chilling effect to still air at –5°F (–20.5°C). Generally, the greatest increase in wind chill effect occurs when a wind of only 5 mph increases to 10 mph.

Response personnel should also understand the term *water chill,* which is body heat lost through conduction. Wet clothing extracts heat from the body up to 240 times faster than dry clothing. In a worst-case scenario, it may also lead to hypothermia, in which the body temperature falls below 95°F (35°C). Hypothermia is a true medical emergency.

Regardless of the ambient temperature, personnel will perspire heavily in impermeable chemical protective clothing. When this clothing is removed in the decontamination process, particularly in cold environments, the body can cool rapidly.

Usually, injuries from cold exposures will be local or confined to a small area of the body. Response operations in low-temperature environments, when the wind chill factor is low or when working in wet clothing, can impair your ability to safely work and possibly lead to frostbite or hypothermia. Ensure all personnel are wearing appropriate clothing (preferably layered) and have warm shelters or vehicles such as buses available.

It is essential that the layer next to the skin, especially socks, be dry. Trenchfoot can result when wet socks are worn at long-term emergencies in cool (not cold) environments.

Reducing Heat Stress

Educate your responders on heat stress, including its prevention. Physical conditioning, prehydration, and acclimatization are critical. The degree to which the body physiologically adjusts to work in hot environments affects its ability to actually do the work.

- Provide plenty of liquids, including prehydration with 8 to 16 ounces of fluids. Don't rely on thirst as an indicator of the need to hydrate. If you are thirsty you waited too long to begin prehydration! Replace body fluids (water, electrolytes, and simple carbohydrates); use water or commercial electrolyte mixes per local protocols.
- Pre- and post-entry medical monitoring procedures are critical to ensure the physical status of entry personnel is properly monitored and evaluated.
- Body cooling devices (e.g., cooling vests) can be used to aid natural body ventilation. These devices may add weight or bulk, may have to be worn a considerable time to impact the body's core temperature prior to the entry, and must be carefully balanced against responder inefficiency.
- During exceptionally hot and/or humid weather conditions, consider the installation of mobile showers, portable hose-down facilities, and misting devices to reduce internal CPC temperatures and to cool protective clothing.
- Rotate personnel on a shift basis. Many response organizations recommend CPC be worn for a maximum work time of 20 to 30 minutes, based on the ambient temperature.
- Establish a rehab area, and provide shelter or shaded areas to protect personnel during rest periods. Entry personnel should be placed in cool areas (e.g., air-conditioned ambulances) before and after entry operations.
- Active cooling devices, such as forearm immersion, misting fans and towels soaked in water, provide for greater cooling and a more rapid decline in body temperature after entry operations.
- When dealing with heat stress emergencies, apply sponges, wet towels, or ice packs around the neck and under the armpits for maximum body cooling effectiveness.
- At remediation operations, conduct nonemergency response activities in the early morning or evening when ambient temperatures are cooler.

Figure 2-4 Fire fighters conduct firefighting and search operations during a snowstorm as a result of a commuter train collision.

Carefully schedule work and rest periods and monitor physical working conditions. It is especially important to have warm shelters available for protective clothing donning and doffing activities Figure 2-4.

Noise

Hazardous materials incidents can often involve excessive noise levels. Examples include actuation of pressure relief devices, generators, pumps, and the use of heavy equipment on the scene. Excessive noise can create safety hazards at an incident scene through distraction and interruption of responders performing critical tasks in hazardous environments. The effects of excessive noise levels can include the following:

- Personnel being annoyed, startled, or distracted
- Physical damage to ears, pain, and temporary and/or permanent hearing loss
- Interference with communications, which may limit the ability of emergency responders to warn of danger or enforce proper safety precautions (verbal and radio)

Standardized hand signals should be developed for situations where excessive noise levels make verbal or radio communications impossible. Effective noise and ear protection should always be provided. This usually consists of ear plugs or ear muffs. OSHA regulations require that hearing protection be provided whenever noise levels exceed 85 dBA (decibels on the A-weighted scale).

Up to this point in section of the text, we have discussed the health and safety problems created by both the physical and chemical environment at a hazmat incident. The following section will review the components of a health and safety management program.

Health and Safety Management Program

Personnel involved in hazmat emergency response operations can be exposed to high levels of both physiological and psychological stress. Routine activities may expose them to both chemical and physical hazards. They may develop heat stress while wearing protective clothing or while working under temperature extremes, not to mention the possibility of facing life-threatening emergencies such as fires and explosions.

A health and safety management program should be an integral element of any emergency response organization. The components of a health and safety management system for hazmat responders are outlined in OSHA 1910.120, *Hazardous Waste Site Operations and Emergency Response* (HAZWOPER). Key areas within the regulation include medical surveillance, personal protective equipment, and site safety practices and procedures.

Medical Surveillance

A **medical surveillance program** is the cornerstone of an effective employee health and safety management system. The primary objectives of a medical surveillance program are (1) to determine that an individual can perform his or her assigned duties (i.e., "fit for duty"), including the use of personal protective clothing and equipment; and (2) to detect any changes in body system functions caused by physical and/or chemical exposures.

A medical surveillance program should provide surveillance (pre-employment screening, periodic and follow-up medical examinations where appropriate, and a termination examination), on-scene evaluation, treatment (emergency and nonemergency), recordkeeping, and program review. A sample medical surveillance program is summarized in Table 2-12.

Medical Surveillance Program Success Factors

The success of any medical program depends on management support and employee involvement. Occupational health physicians and specialists, emergency medicine physicians, safety professionals, local or regional poison control center specialists, as well as advanced EMS personnel should be consulted for their expertise. Many hazmat units have a physician with a background or interest in hazardous materials who serves as the medical director for their unit.

Confidentiality of all medical information is paramount. Prospective employees and new hazmat response team members must provide a complete, detailed occupational and medical history so a baseline profile can be established. Responders should be encouraged to document any suspected exposures, regardless of the degree, along with any unusual physical or physiological conditions. Training programs must emphasize that even minor complaints (e.g., headaches, skin irritations) may be important.

Additional information on medical surveillance programs can be found at the OSHA website at www.osha.gov, and NFPA 1582, *Standard on Comprehensive Occupational Medical Program for Fire Departments*.

Pre-Employment Screening

The objectives of pre-employment screening are to determine an individual's fitness for duty, including respirator and protective

Scan Sheet 2-C

Cold Weather Decontamination Operations

As part of the DOD Improved Response Program to terrorism incidents involving chemical weapons and agents, the U.S. Army's Edgewood (MD) Chemical Biological Center provided guidance on cold weather mass decontamination procedures. The report, *Guidelines for Cold Weather Mass Decontamination During a Terrorist Chemical Agent Incident*, outlines information that can be useful to responders operating at any hazmat emergency in a cold weather environment.

Important points of the report pertaining to cold weather operations include the following:

- Cold weather experts involved in the report recommended an ambient temperature of 65°F (18.3°C) as a "breakpoint" for outdoor decontamination. Comfort, rather than physiological effects, was the overriding consideration for this recommendation.
- At ambient temperatures below 65°F (18.3°C), individuals are likely to be much less willing to participate in decon showering because of the significant perception of discomfort.
- As ambient air temperatures decrease, the risk of serious health complications from exposure to cold water increases for some people. Encouraging people to enter the shower stream gradually and allowing the body to adjust to the cold water can minimize this risk.
- When the ambient air temperature approaches freezing, the risk of accidents from frozen shower water and equipment must be considered.
- Whenever possible, some form of shelter or dry clothing should be provided.
- Victims decontaminated with water should be observed for signs of shivering. Shivering generates body heat and is an indication of normal bodily response to the cold environment. Should shivering stop in such situations, medical attention should be sought immediately since cold weather injury could be imminent.
- Cold shock is the sudden manifestation of physiological responses triggered by a cold water exposure, and can result in sudden death in susceptible individuals. Cold shock occurs almost immediately and must be considered by responders. The risk is greatest for those with preexisting conditions, such as heart disease and the aged. Cold shock is more likely to cause serious medical problems than hypothermia during mass decon operations. Cold shock can be minimized by inquiring about preexisting medical conditions before decon (if the situation permits), and by encouraging people to gradually get wet, rather than being instantaneously deluged in cold water.
- Hypothermia is not a significant risk for most people undergoing mass decon in cold weather. Most individuals can tolerate 55°F (12.8°C) and although they would shiver severely and experience great discomfort, they would not be in an immediate life-threatening situation due to hypothermia. If an individual is cold, wet, and not shivering, prompt medical attention should be sought since some people are not able to shiver and are at greater risk of developing hypothermia.
- Regardless of the ambient temperature, people who have been exposed to a known life-threatening level of chemical contamination should disrobe; undergo decon with copious amounts of low-pressure, high-volume water or alternative decon methods; and be sheltered as soon as possible.

Additional information on the SBCCOM report can be referenced from the Edgewood Chemical Biological Center (ECBC) website at http://www.ecbc.army.mil.

Table 2-12 Medical Surveillance Program
Medical surveillance is an essential element of a comprehensive employee health and safety program.

Component	Recommendation
Pre-employment screening	▪ Medical History ▪ Occupational History ▪ Physical Examination ▪ Determination of fitness to wear protective equipment (e.g., respirator fit testing per OSHA 1910.134) ▪ Baseline monitoring for specific exposures
Periodic medical examination	▪ Annual update of medical and occupational history; annual physical examination; testing based on (1) examination results, (2) exposures, and (3) job class and task ▪ More frequent testing may be required based upon specific exposures ▪ Exams may be biannual based on a physician's recommendation.
Emergency treatment	▪ Provide medical care on site. ▪ Develop liaison with local hospital and medical specialists. ▪ Arrange for decontamination of victims. ▪ Arrange in advance for the transport of victims. ▪ Provide for the transfer of medical records; provide details of the incident and medical history to the next-care provider. ▪ Provide for postincident surveillance of potentially exposed emergency responders and civilians.
Non-emergency treatment	▪ Develop a mechanism for nonemergency health care.
Recordkeeping and review	▪ Maintain and provide access to medical records in accordance with OSHA, state, and provincial records. ▪ Report and record occupational injuries and illnesses. ▪ Review medical surveillance program periodically. ▪ Focus on current site hazards, exposures, and industrial hygiene standards.

clothing use, and to provide baseline data for future medical comparisons. The screening should focus on the following areas:

- **Occupational and medical history**—This questionnaire should be completed with attention toward prior exposures to chemical and physical hazards. Also note previous illnesses, chronic diseases, hypersensitivity to specific substances, ability to use PPE, family history, and general lifestyle habits such as smoking and drug use. The OSHA respirator medical evaluation questionnaire (OSHA 1910.134—Appendix C) should also be included as part of this package.
- **Physical examination**—Complete a comprehensive physical examination focusing on the pulmonary, cardiovascular, and musculoskeletal systems. Additional tests that can help gauge the capacity to perform emergency response duties while wearing protective clothing include pulmonary function, electrocardiograms (EKG), hearing, and cardiac "stress tests."
- **Baseline laboratory profile**—This verifies the effectiveness of protective measures and determines whether the responder is adversely affected by previous exposures. The profile may include medical screening and biological monitoring tests based on potential exposures, such as liver, renal, and blood forming functions. Pre-employment blood and serum specimens may be frozen for later testing and comparison.

Periodic Medical Examinations

Periodic "fit for duty" exams must be used in conjunction with pre-employment screening. Their comparison with the pre-employment baseline physical may detect trends and early warning signs of adverse health effects. Under the OSHA 1910.120 requirements, such exams shall be administered annually, and no longer than every 2 years if the attending physician believes a longer interval is appropriate. In addition, more frequent intervals may be required depending on the nature of potential or actual exposures, type of chemicals involved, and the individual's medical and physical profile.

If an individual develops signs or symptoms indicating possible overexposure to hazardous substances or health hazards or has been injured or exposed to substances above accepted exposure values in an emergency, medical examinations and consultations shall be provided as necessary. Periodic screening exams can include medical history reviews that focus on health changes, illness and exposure-related symptoms, physical examinations, and specific tests such as pulmonary function, audiometric, blood, and urine.

To ensure the completion of a comprehensive medical profile, a medical exam is required to be given to all personnel when they are removed from active duty as a hazmat responder and at the termination of their employment or membership.

Emergency Treatment

EMS personnel and units must be available at each hazmat incident. OSHA 1910.120 (q)(3)(vi) requires that "advanced first-aid support personnel, as a minimum, shall stand-by with medical equipment and a transportation capability at hazmat emergencies." The level of emergency medical support may be influenced by the nature of the incident, risks involved, tasks to be performed, and the intensity and/or duration of the tasks. Informal interpretations by OSHA have indicated that the following factors will be considered in determining whether responders or a facility are in compliance with this requirement:

- Advanced first-aid personnel are considered as individuals who have been trained to the 2010 International

Voices of Experience

As an employee of the Indiana Department of Homeland Security, in the hazmat section, I have responded to many large-scale incidents. One that has particular bearing on the importance of monitoring devices was a large plastics fire in an old heavy-timber building, which produced huge volumes of black, acrid smoke.

Building and Contents Description

The building was of heavy timber construction with a limited number of windows and doors. In the past it had been used by a distillery. As is quite common, the building contents did not conform to occupancy/suppression standards. The building was approximately 100' x 100' and stood six stories tall. Every floor in this building was full of plastic materials and eventually became involved with fire.

Suppression Efforts

With site access severely limited and a lack of suitable windows, there was very little on-scene fire departments could do as far as fire suppression. With the limited access it was quite apparent that this fire was going to burn for a long time. And, with the huge volume of plastics in the building. great amounts of black, acrid smoke was being produced.

Monitoring Requirements

With the wind being very light the smoke carried downwind for about 2 blocks and then started settling, impacting residential, commercial, and high hazard occupancies, including a hospital and a nursing home. Hazard Control Zones had to be determined on a priority basis. The only accurate and defendable way to establish these hazard zones was by using the correct, properly calibrated monitoring device. This monitoring was performed using PIDs, 4-gas, and the appropriate colorimetric tube (formaldehyde, ammonia, and benzene). By using exposure limits referenced in the NIOSH we were able to determine when the hospitals and nursing homes HVAC air intakes had to be shut down. Monitoring was conducted continuously for 2 days. Because the air quality monitoring was performed correctly not one person reported any ill effects from this incident.

Tim Thomas
Indiana Department of Homeland Security
Indianapolis, Indiana

Consensus Guidelines for Cardiopulmonary Resuscitation (CPR) and Emergency Cardiac Care (ECC) for Advanced First-Aid level or higher (e.g., Emergency Medical Responder, Emergency Medical Technician, etc.) and are capable of providing basic medical care.

- Medical equipment is not required to be on scene but must be available for immediate response. As a general rule, medical treatment should be provided within 3 to 4 minutes of the incident, while a transportation capability should be on-site in approximately 15 to 20 minutes.

Although most public safety response agencies will have an EMS unit on scene as part of the response, that is not always the case with facility Emergency Response Teams (ERTs) who typically deal with small spills of hazardous materials commonly found or used within the facility Figure 2-5.

When possible, an EMS responder with a background in hazmat operations should be in charge of the EMS operations and coordinate closely with the Hazmat Group Supervisor. At "working" incidents the IC may establish a Medical Group to coordinate all EMS activities. Specific responsibilities of EMS or Medical Group personnel include the following:

- Provide technical assistance to responders in the development and analysis of EMS-related data and information. This shall include signs and symptoms of exposure, medical treatment procedures, antidote information, patient handling guidelines, transportation recommendations, and medical resource requirements Figure 2-6.
- Designate a treatment and triage area in proximity to the decontamination area.
- Perform pre-entry and post-entry medical monitoring of all entry and backup team personnel, as appropriate.
- Coordinate and supervise all patient handling activities, including decontamination, treatment, handling and transportation of contaminated victims. This should include recommendations for the protection of all EMS personnel.
- Communicate and coordinate with local hospitals and specialized treatment facilities, including the poison control center, as necessary.

Figure 2-5 EMS personnel must be on scene at hazmat emergencies.

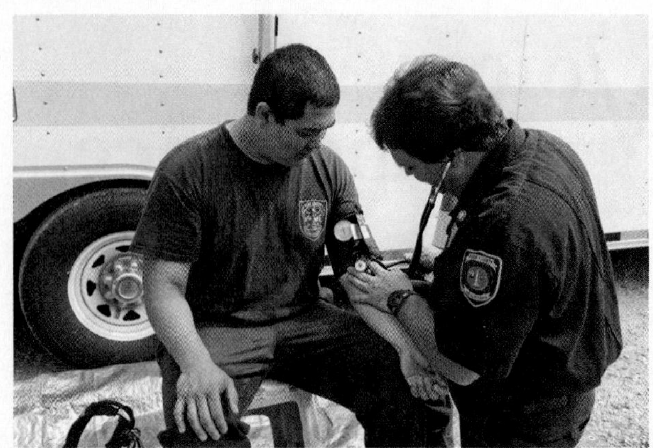
Figure 2-6 Medical monitoring is the systematic, ongoing evaluation of individuals at risk of suffering adverse effects to heat, stress or hazardous materials.

Some regions and industrial facilities provide either advanced life support (ALS) units or individuals specially trained and equipped for hazardous materials emergencies. They often carry drugs and antidotes for chemicals commonly found within the plant or community, along with medical information and baseline profiles of all hazmat response team personnel. Hazmat training competencies for EMS personnel can be found in NFPA 473, *Competencies for EMS Personnel Responding to Hazardous Materials Incidents.* Other good sources of training and information are the *Advanced Life Support Response to Hazardous Materials Incidents* (R247) offered by the National Fire Academy (see http://www.usfa.dhs.gov/index.shtm) and the Advanced Hazmat Life Support course offered by the University of Arizona Health Services Center (see http://www.ahls.org).

Standard Operating Procedures (SOPs) for the clinical management and transportation of chemically contaminated patients must be developed as part of the planning process. In addition, the specific roles, responsibilities, and capabilities of hazmat personnel, EMS personnel, and local medical facilities need to be determined.

The handling of chemically contaminated patients will be addressed in the section on "decontamination" elsewhere in this text; however, keep these basic principles in mind:

- Always ensure EMS personnel are properly protected—both skin and respiratory.
- When dealing with victims in a contaminated environment, determine whether it is a rescue operation or a body recovery operation.
- Although certain situations may exist where decontamination may aggravate victim care or further delay priority treatment, as a rule of thumb all victims should receive gross decontamination.
- The ABCs can be administered to a contaminated victim if rescuers and EMS personnel are protected. It's a much better option than having a fully decontaminated but dead patient.
- Always coordinate with your local medical facilities.

Nonemergency Treatment

The signs and symptoms of certain exposures may not be present for 24 to 72 hours after exposure. This is particularly true when dealing with some biological agents and chemicals that have delayed effects. In many instances, those exposed may already be off duty and out of contact. Personnel operating at an incident should be medically evaluated before being released. In addition, the termination procedure should provide for a briefing for all emergency responders on the signs and symptoms of exposure, documentation, and completion of health exposure logs or forms, post-incident points of contact, and how to get immediate treatment, if necessary.

Recordkeeping and Program Review

Recordkeeping is an important element of the medical surveillance program. Time intervals between initial exposures and the appearance of possible chronic effects may take years to develop. Individual records should be kept for all personnel. OSHA requires that exposure and medical records be maintained for at least 30 years after the employee retires. These records must be made available to all affected employees and their representatives upon request. Procedures should also be developed for emergency access to individual records if personnel are hospitalized and their medical surveillance records are requested.

Individual medical records should include all medical exams completed, their purpose (e.g., baseline, periodic), the examining physician's observations and recommendations, and if they were a result of a specific exposure. A copy of the incident report should also be maintained in the file. Any injuries sustained during line-of-duty operations should be noted, and follow-up treatment and personal exposure logs should be maintained.

Regular evaluation of the overall medical surveillance program is important to ensure its continued effectiveness. Review the following elements on an annual basis:

- Ensure each accident/illness is promptly investigated to determine its root cause and update health and safety procedures, as necessary.
- Evaluate the effectiveness of medical testing in light of potential and confirmed exposures.
- Add or delete specific medical tests as recommended by the medical director and by current industrial hygiene and environmental health data.
- Review all emergency care protocols.

Critical Incident Stress

Although not an element of the medical surveillance program, critical incident stress should be recognized as an issue that can potentially impact the health and welfare of responders. Hazmat emergencies and the potential health risks from exposures can create high levels of psychological stress for both responders and their families. Unknown or perceived exposures can generate as much emotional stress as an actual documented high-level exposure. Both the IC and the Hazmat Program Manager must recognize that these personal stressors exist and must be managed.

Medical debriefings as part of the incident termination phase are essential elements in reducing the level of stress. The debriefing should review the hazmats involved in the incident, the signs and symptoms of exposure, exposure documentation procedures, and the procedures to follow in the event an individual starts to show these signs and symptoms over the next several days. Employee Assistance Programs (EAP) and Critical Incident Stress Management (CISM) teams can be an effective post-incident resource and should be used as necessary. Post-incident debriefings are discussed elsewhere in this text.

Personal Protective Equipment Program

The objectives of a PPE program are to protect personnel from both chemical and physical safety and health hazards, and to prevent injury to the user from the incorrect use and/or malfunction of protective clothing or equipment.

A comprehensive PPE program should include the following:

- **Hazard assessment**—The selection and purchase of PPE should be based on a hazard assessment of those chemicals stored, transported, or used within the facility or the community.
- **Medical monitoring of personnel**—Medical surveillance, respirator fit testing, and medical monitoring results are among the criteria for the initial selection and continuing certification of "fitness for duty" of response personnel. Other issues include the effects of heat stress and environmental temperature extremes.
- **Equipment selection and use**—Recognize the relationship between the environment being encountered (e.g., flammable versus toxic), the response objectives (defensive versus offensive), the PPE user, and the PPE ensemble used.
- **Training program**—An effective PPE program cannot exist without a comprehensive training program.
- **Inspection, maintenance, and storage program**—These are key elements of a PPE program, and the absence of any one of these elements may lead to injury. This would include inspection procedures prior to, during, and after use, methods of storage, maintenance guidelines, and capabilities. Tool inspection and maintenance procedures should follow the recommendations of the product's manufacturer and the guidelines of the AHJ.

A written PPE program outlining these elements is required under OSHA 1910.120(q)(5). It should include a policy statement with the guidelines and procedures from the preceding list. Copies should be made available to all employees. Technical data concerning equipment, maintenance manuals, relevant regulations, and other essential information should also be made available as necessary. PPE training is also required under OSHA 1910.120. It should allow the user to become familiar with the equipment in a nonhazardous environment, which builds user confidence. Understanding the capabilities and limitations of this equipment also improves the safety and efficiency of emergency operations. Finally, a well-rounded PPE training program often reduces associated maintenance expenses.

A written respiratory protection program (RPP) is also required when necessary to protect the health of employees from workplace contaminants or when the employer requires the use of respirators, per OSHA 1910.134(c)(1). Additional information on RPP requirements can be found on the OSHA website at www.osha.gov.

Certain personal features and lifestyle choices of your responder personnel may jeopardize their safety while using PPE during emergency response operations, including the following:

- **Facial hair (beards) and long hair**—Interferes with the use of respiratory equipment. Per the requirements of OSHA 1910.134(g)(1)(i), no facial hair should pass between the face and the facepiece surface. Testing has documented that even a few days' growth of facial hair may allow contaminants to penetrate the facepiece.
- **Eyeglasses with conventional temple pieces**—Interferes with facepiece seal. A spectacle kit should be installed into the facepiece of those who require corrected vision.
- **Gum and tobacco chewing**—May cause ingestion of contaminants and compromise facepiece fit.
- **Prescription drug use**—Certain prescription drugs may place responders at greater risk for environmental illness or chemical toxicity. Individual prescription medications should be evaluated as part of medical surveillance examinations. Changes in medications or medications that are used on an "as needed" basis should be evaluated by a physician who is knowledgeable in hazmat response operations.

Site Safety Practices and Procedures

Safety Issues

Safety is not simply an organizational rule or a government regulation. Safety is an attitude, a behavior, and a culture. Safety MUST be an inherent part of all operations from the development of SOPs to the selection and purchase of PPE. The operating philosophy of every emergency response organization should be, "If we cannot do this safely, then we will not do it at all."

There are two phases of an incident where the potential for responder injury and harm is greatest—first, during initial response operations where the "fog of war" can complicate the sizing-up process, and second, when the incident shifts gears from the emergency phase to the clean-up and recovery phase. You cannot manage a hazmat incident if you do not have control of the scene—in fact, you will quickly find out that the incident is actually managing you!

Gaining and maintaining control of the incident scene is one of the most difficult tasks faced by the IC. A continuous problem will be everyone—including facility managers, plant personnel, the media, the general public, and even responders—wanting to get as close as possible to the action. There will also be situations where site safety becomes lax or even nonexistent. This is especially true at "campaign" incidents extending over hours or days. As time passes it is quite easy for responders to become bored and careless in their attitudes and actions.

Considering the validity of these observations, recognize the importance of the following safety truths:

"What occurs during the initial 10 minutes will dictate what will occur for the next hour, and what occurs during the first hour will dictate what will occur for the initial eight hours of the incident."

Translation: If first-on-the-scene responders become part of the problem, it will shift initial strategic goals and probably take hours to undo their mistakes and get the system on track. If you don't know what to do, isolate the area, deny entry, and call for help. It may take a while, but eventually someone will show up who knows what to do!

"There is nothing wrong with taking a risk. However, always remember that there are good risks and bad risks—if there is much to be gained, then perhaps much can be risked. Of course, if there is little to be gained, then little should be risked."

Translation: Life safety is always our number one priority—including the life safety of all emergency responders. There is a significant difference between a rescue and body recovery operation when hazmats are involved. Considering the average response and operational set-up time for most HMRTs, rescue is seldom a strategic priority except in criminal or terrorism scenarios involving the use of hazardous materials or weapons of mass destruction (WMD) where large numbers of casualties may be occur without rapid intervention.

"Safety must be more than a policy or procedure . . . it is both an attitude and a responsibility that must be shared by all responders."

Translation: Safety must be an integral element of all hazmat operations. Yes, the Incident Safety Officer does play a critical role in ensuring that all operations are conducted both safely and effectively. Ultimately, however, safety must be the responsibility of every responder.

"Protective clothing is not your first line of defense but is your last line of defense."

Translation: The selection of strategies and tactics to minimize any direct exposure to the materials involved should always be your first line of defense. Remember—evaluate defensive strategies first, then offensive. If you minimize the potential for exposure, you reduce the potential of relying solely upon your protective clothing. Of course, that assumes you ARE wearing your protective clothing!

"Final accountability always rests with the Incident Commander."

Translation: Although emergency responders are normally not responsible for product transfer and removal (except in some industrial facilities), site safety is still the IC's responsibility. Do not become lax during this phase of the emergency—site safety procedures must be continuously enforced throughout the clean-up and recovery phase.

Site Safety Plan

A site safety plan is required under OSHA 1910.120, paragraph b, as a mechanism for ensuring the health and safety of personnel operating at clean-up and hazardous waste site operations. Although a site safety plan is not required under OSHA 1910.120, paragraph q, site safety must be an integral element of on-scene response operations. SOPs and checklists should be used both to verify and document that safety elements are addressed during the course of the emergency. These checklists are usually divided by positions within the Incident Command System and the Hazmat Group (e.g., Assistant Safety Officer–Hazardous Materials, Decon Unit Leader).

Some of the advantages of using operational checklists to meet the site safety requirements are the ability to ensure that specific organizational guidelines and SOPs are followed, the ability to track activities and performance, and the ability to document the plan of action and decision-making process.

Under OSHA 1910.120, paragraph b, components of a site safety plan should include site map or sketch, hazard and risk analysis of the identified hazardous materials, site monitoring, establishment of hazard control zones, site safety practices and procedures, communications, implementation of an incident management organization, and the location of the incident command post, decontamination practices, EMS support, and other relevant topics. Standard site safety practices and procedures are outlined in Table 2-13.

Table 2-13 Standard site safety practices

- Minimize the number of personnel operating in the contaminated area.
- Suspected contaminated surfaces. Avoid walking through any suspected releases or placing equipment on contaminated surfaces.
- Advise all entry personnel of all site control policies including entry and egress points, decon layout and procedures, and working times.
- Always have an escape route. Ensure that everyone knows the emergency evacuation signals.
- Ensure that all tasks and responsibilities are identified before attempting entry. If necessary, practice unfamiliar operations prior to entry.
- Use the buddy system for all entry operations. Always ensure that properly staffed and equipped back-up crews are in place.
- Maintain radio communications between entry, backup crews, and the Safety Officer (whenever possible).
- Prohibit drinking, smoking, and any other practices that increase the possibility of hand-to-mouth transfer in all contaminated areas.
- Follow decontamination and personal cleanliness practices before eating, drinking, or smoking after leaving the contaminated area.

Safety Officer and Safety Responsibilities

Under the OSHA HAZWOPER regulation, the safety function must be addressed at every incident in which hazardous materials are involved. At small, NIMS Level 5 incident scenarios the safety function can be easily managed by the IC. However, as the scope and complexity of the hazmat problem increase, it will be necessary to designate an Incident Safety Officer.

At incidents where an HMRT is operating, safety responsibilities will often be divided into two areas—first, the safety of all units operating within the incident scene and under the control of the Incident Safety Officer; and second, the safety of those operating within the ICS Hazmat Group and under the control of the Assistant Safety Officer–Hazardous Materials (aka the Hazmat Group Safety Officer).

Although the Assistant Safety Officer–Hazardous Materials is subordinate to the Incident Safety Officer, he or she has certain responsibilities within the Hazmat Group that may circumvent the normal chain of command. In either case, both the Incident Safety Officer and all Assistant Safety Officers must have the authority to stop any operations they deem unsafe.

Among the primary responsibilities of the Safety Officer are the following:

Overall Site Safety

- Ensure the Safety Officer is identified to all personnel. The IC should advise all operating personnel, as appropriate. The use of command vests for identification is recommended.
- Ensure all personnel and equipment are positioned in a safe location. Remember the basics—upwind, uphill—and always have an escape route and predesignated withdrawal signal. Consider having vehicles and apparatus back into the incident scene for a quick exit.
- Ensure hazard control zones are identified, established, constantly monitored, and their locations communicated to all personnel. Consider the location of the Incident Command Post, Hazmat Group, and Staging Area in relation to the hazard control zones and the potential worsening of the emergency.
- When necessary, designate a security officer to maintain overall site security. Delegate to plant security or law enforcement whenever possible.
- Ensure all personnel in controlled areas are in the proper level of personal protective clothing.
- Use a Personal Accountability System to track all responders throughout the incident. Conduct regular personal accountability reviews (PAR).

Entry Operations

- Coordinate with the Medical Unit Leader to ensure pre-entry medical monitoring has been conducted.
- Hold a pre-entry safety briefing prior to recon or entry operations. This may be provided by either the Assistant Safety Officer–Hazardous Materials or other Hazmat Group personnel (e.g., Entry Unit Leader). All entry and backup personnel must be familiar with the objectives, tasks, and procedures to be followed. Topics should

include objectives of the entry operation, potential safety issues with mitigation actions, a review of all assignments, verification of radio procedures (designated channels) and emergency signals (both hand signals and audible), emergency escape plans and procedures, protective clothing requirements, immediate signs and symptoms of exposure, and the location and layout of the decon area.

- Coordinate entry operations with backup crews and the Decon Unit. The Entry Team should be permitted to enter the hot zone only when backup crews are in place and the decon area is prepared.
- Monitor entry operations and advise entry personnel and the IC of any unsafe practices or conditions.
- During the termination phase, advise all personnel of the possible signs and symptoms of exposure and ensure that health exposure forms are documented.

Hazmat Medical Monitoring and Support

Medical monitoring is defined as an ongoing, systematic evaluation of individuals at risk of suffering adverse effects of stress or exposure to heat, cold, or hazardous environments. The objectives of medical monitoring are (1) to obtain baseline vital signs, and (2) to identify and preclude from participation individuals who are at increased risk for sustaining injury or illness.

It should be emphasized that medical monitoring is one element of a comprehensive medical surveillance program that starts with the baseline and periodic "fit for duty" medical examination. Consider the following points:

- Firefighting is equally as strenuous as wearing chemical protective clothing; however, the fire service does not conduct pre-entry medical monitoring at firefighting operations due to the obvious time constraints.
- Hazmat emergency responders should not view pre-entry medical monitoring as an absolute, especially in those scenarios where immediate life safety and rescue requirements may exist.
- To complement the medical surveillance program, some career fire department HMRTs conduct pre-entry medical monitoring at the start of each shift to facilitate the rapid response concept.
- Not providing a pre-entry evaluation should not preclude the need for post-entry medical monitoring and support.

Pre- and post-entry medical monitoring is normally the responsibility of the Medical Group supporting hazmat operations. Assuming that the response organization has a medical surveillance program in-place, pre-entry exams for hazmat emergency responders normally focuses upon vital signs (i.e., blood pressure, pulse, respiratory rate, body temperature and weight) and level of hydration.

If the response organization is using external specialist employees and other outside personnel, additional elements may be incorporated into the pre-entry exam, including:

- Skin evaluation, with an emphasis on rashes, lesions, and open sores or wounds.
- Lung sounds, including wheezing, unequal breath sounds, and so on.
- Mental status (alert and oriented to time, location, and person).
- Recent medical history, including medications, alcohol consumption, any new medical treatment or diagnosis within the last 2 weeks, and symptoms of fever, nausea, diarrhea, vomiting, or coughing within the past 72 hours.
- A 10-second EKG rhythm strip or a 12-lead EKG may also be taken. Preferred by many physicians, the 12-lead EKG can be interpreted by a paramedic and compared to baseline traces to indicate possible cardiac abnormalities.

Criteria should be established for evaluating responders prior to entry operations. These criteria should be reviewed by the HMRT Medical Director or an occupational health physician/specialist who is familiar with the duties and tasks of hazmat responders.

The following exclusion criteria are used by a number of Hazmat Medical Groups and are provided as an example; however, they should not supersede any existing criteria established by the local medical control. Entry shall be denied if the following criteria are not satisfied:

- Blood pressure—BP exceeds 100 mm Hg diastolic.
- Pulse—Greater than 70% maximum heart rate (>115) or irregular rhythm not previously known.
- Respirations—Respiratory rate is greater than 24 per minute.
- Temperature—Oral temperature less than 97°F or exceeds 99.5°F. Core temperature less than 98°F or greater than 100.5°F.
- Body weight—No pre-entry exclusion.
- EKG—Dysrhythmias not previously detected must be cleared by medical control.
- Mental status—Altered mental status (e.g., slurred speech, clumsiness, weakness).
- Other criteria, including:
  Skin—Open sores, large skin rashes, or significant sunburn.
  Lungs—Wheezing or congested lung sounds.
- Medical history—Recent onset of heart or lung problems, hypertension, diabetes, etc. Experienced nausea and vomiting, diarrhea, fever, or heat exhaustion within the last 72 hours. Use of prescription medication and over-the-counter medicines (e.g., decongestants, antihistamines, etc.) must be cleared through local medical control. Heavy alcohol consumption within the previous 24 hours or any alcohol within the past 2 hours.

Post-entry medical monitoring is performed following decontamination to determine if the responder has suffered any immediate effects from heat stress or a chemical exposure, and to determine the individual's health status for future assignment during or after the incident. Components of the post-entry exam should include the following:

- Any signs or symptoms of chemical exposures, heat stress, or cardiovascular collapse. EMS personnel should be aware of the symptomology for specific toxidromes for the hazards involved.
- Vital signs, including blood pressure, pulse, respiratory rate, temperature, and body weight. If available, a 10-second EKG rhythm strip may also be taken.

Scan Sheet 2-D

Emergency Rescue Capabilities at Interior Firefighting Operations "Two-In / Two-Out"

An emergency rescue capability must be provided when operations are conducted in an IDLH atmosphere. In permit-required, confined space entry operations where the IDLH atmosphere has been characterized and controlled, this requirement is usually satisfied by having one or more stand-by persons who are trained and equipped to provide an effective emergency rescue (see OSHA 1910.146—*Permit-Required Confined Space Entry Operations*). Similarly, emergency rescue requirements for hazmat incidents are outlined in OSHA 1910.120(q).

The release of the OSHA Respiratory Protection Standard (1910.134) brought the question of emergency rescue capabilities to interior firefighting operations. Commonly referred to as the "two-in/two-out rule," it requires a backup and rescue capability be provided for interior structural fires. OSHA Compliance Directive 2-0.120—*Inspection Procedures for the Respiratory Protection Standard*, makes the following notes:

- It is the IC's responsibility, based on training and experience, to judge whether a fire is an interior structural fire and how it will be attacked.
- There must always be at least two fire fighters stationed outside during interior structural firefighting, and they must be trained, equipped, and prepared to enter, if necessary, to rescue the fire fighters inside. However, the IC has the responsibility and flexibility to determine when more than two outside fire fighters are necessary given the circumstances of the fire. The two-in/two-out rule does not require an arithmetic progression for every fire fighter inside (i.e., the Standard should not be interpreted as 4-in/4-out, 8-in/8-out, etc.).
- Life-saving activities in interior structural firefighting are not precluded by the OSHA standard. There is an explicit exemption in the standard that if life is in jeopardy, fire fighters have the discretion to perform the rescue, and the "two-in/two-out" requirement is waived. There is no violation of the standard under such life-saving rescue circumstances.
- The two-in/two-out provision is not intended as a staffing requirement. The two-in/two-out rule is a worker safety practice requirement, not a staffing requirement.
- The standard allows one of the standby fire fighters to have other duties such as serving as the IC, Safety Officer, or fire apparatus operator. However, one of the outside fire fighters must actively monitor the status of the inside fire fighters and may not be assigned additional duties. The second outside fire fighter may be involved in a wide variety of activities. Both of the outside personnel must be able to provide support and assistance to the two interior fire fighters; any assignment of additional duties for one of the outside fire fighters must be weighed against the potential for interference with this requirement.
- The two fire fighters (buddies) entering an IDLH atmosphere to perform interior structural firefighting must maintain visual or voice communication at all times. Electronic methods of communication such as the use of radios shall not be substituted for direct visual contact between the team members in the danger area. However, reliable electronic communication devices are not prohibited and certainly have value in augmenting communication and may be used to communicate between inside team members and outside standby personnel.

Since its promulgation, the OSHA two in/two out requirement has generated much discussion on rapid intervention and rescue of emergency responders operating in IDLH environments. An extensive research project was conducted by the Phoenix, Arizona Fire Department (PFD) after a 2001supermarket fire, which resulted in a fire fighter fatality. PFD conducted 200 rapid intervention drills that evaluated the department's ability to remove two fire fighters in trouble in a similar occupancy. The results showed that rapid intervention may not be rapid and revealed three consistent ratios: (1) It takes 12 fire fighters to rescue one; (2) 1 in 5 rescuers will get into trouble themselves; and (3) a 3,000-psi air cylinder has approximately 18.7 minutes of air (+ or – 30%).

How do these experiences relate to a hazmat emergency response? As a result of the rapid intervention drills, the Phoenix Fire Department's HMRT conducted informal drills on removing a downed hazmat responder from a contaminated environment. The results showed the following:

- For hazmat incidents outdoors in an open-air environment, the two-in/two-out back-up procedure is still effective.
- For hazmat incidents inside a structure, a two-in/four-out back-up team procedure is required.
- Incidents that require the backup team to remove downed entry personnel via a stairwell will significantly complicate the timing and effectiveness of any rescue operation. Operations may be further complicated if entry personnel are wearing chemical vapor (EPA Level A) protective clothing.

In summary, the nature of the hazmat entry operation should dictate what the back-up capability looks like. While two in/two out is a regulatory minimum, the exact make-up and composition of the back-up team should be based upon a number of factors, including the number of entry personnel, indoor vs. outdoor entry operations, visibility of

(Continued)

entry operations, above-grade vs. below-grade entry access, and the presence of any obstructions. It is entirely possible that the hazmat back-up team may need to be a crew, comparable to the Rapid Intervention Team (RIT) concept, which is regularly employed within the fire service.

For further explanation, refer to the preamble of the OSHA *Respiratory Protection Standard* and the *Respirator Question and Answer* document (August 3, 1998). Both documents can be found at the OSHA Web site at http://www.osha.gov.

- Skin evaluation, with an emphasis on rashes, lesions, and open sores or wounds.
- Lung sounds, including wheezing, unequal breath sounds, and so on.
- Mental status (alert and oriented to time, location, and person). One cognitive test often used in the field is to have the individual spell any five-letter word backward.
- Hydration—Provide plenty of liquids. Replace body fluids (water and electrolytes); use a 0.1% salt solution or commercial electrolyte mixes.

Vital signs should be monitored every 5 to 10 minutes, with the victim resting, until they return to approximately 10% of the baseline. If vital signs do not return to normal, it may be necessary to transport the individual to a medical facility. Medical control should be consulted for direction and recommendations, as necessary.

Emergency Incident Rehabilitation

Incident scene rehabilitation (or rehab) is an excellent risk management tool. The IC should consider the circumstances of the incident and make adequate provisions early in the incident for the rest and rehabilitation of all personnel operating at the scene. This is particularly critical for "campaign" emergencies that extend over a period of hours. The IC may establish a Rehabilitation Group to coordinate rest and rehab activities, or the rehab function may be coordinated through the Medical Group.

In addition to coordinating for EMS support, treatment, and monitoring, rehab is responsible for providing food and fluid replenishment, mental rest, and relief from the extreme environmental conditions associated with the incident. The Rehabilitation Area should meet the following parameters:

- Be in a location that provides physical rest by allowing the body to recuperate from the hazards and demands of the emergency. It should also be located to allow for prompt reentry back into the emergency operation upon complete rehabilitation.
- Be located in a safe location within the cold zone (see text section on "site management") so personnel can remove their protective clothing and be afforded physical and mental rest from the stress of the emergency operation. It should also be easily accessible by EMS units.
- Provide suitable protection from the prevailing environmental conditions. During hot weather, it should be located in a cool, shaded area. In cold weather, it should be in a warm, dry area. In addition, it should be free of vehicle exhaust fumes.
- Be large enough to accommodate multiple crews at the same time, based on the size of the incident.

An obvious safety question is how long should emergency responders remain in the Rehab Area before being released? Unfortunately, there is a lack of standards regarding heat stress monitoring and rehab periods while wearing CPC at a hazmat emergency. Most guidelines, such as those from the EPA's *Standard Operating Safety Guides,* are directed primarily towards hazardous waste remediation operations.

Additional information on rehabilitation can be culled from NFPA 1584—*Standard for the Rehabilitation Process for Members During Emergency Operations and Training Exercises.*

Wrap-Up

Chief Concepts

- Personnel protection is the number one priority at any hazmat incident. Toxicology is the health and safety concern of every emergency responder, including study of chemical and physical agents that produce adverse responses in the biological systems with which they interact. Chemical agents include gases, vapors, fumes, and dusts, while physical agents include radiation, hot and cold environments, noise, and so forth.
- Toxicity is defined as the ability of a substance to cause injury to a biological tissue. In humans this generally refers to unwanted effects produced when a chemical has reached a sufficient concentration at a particular location within the body.
- Exposure + Toxicity = Health Hazard
- Chemical exposures and their health effects are commonly described as acute or chronic. Acute exposures describe an immediate exposure, while chronic exposures are low exposures repeated over time.
- Common methods of exposure are inhalation, ingestion, skin absorption, direct contact, or injection.
- Health effects of a hazardous material can be described in terms of how a hazmat attacks the body. A local effect implies an effect at the point of contact—for example, a corrosive burn to the skin, eye irritation, etc. A systemic effect occurs when a chemical enters the bloodstream and attacks target organs and other internal areas of the human body.
- The human body can be subject to seven types of harm events—thermal, mechanical, poisonous, corrosive, asphyxiation, radiation, and etiological.
- Units of measurement for determining the relative toxicity and health exposure of a chemical substance or compound are lethal dose and lethal concentration.
- Exposure values are only guidelines and interpretations—NOT absolute boundaries between safe and dangerous conditions. Examples of common exposure values are threshold limit value (TLV), permissible exposure limit (PEL), immediately dangerous to life and health (IDLH), and emergency response planning guidelines (ERPG).
- Radiation that doesn't hit anybody doesn't hurt anybody! Always keep the acronym "ALARA" in mind when considering radiation exposures—keep the exposure As Low As Reasonably Achievable.
- Heat stress is a significant concern when wearing any type of impermeable protective clothing. Physical reactions to heat include heat rash, heat cramps, heat exhaustion, and heat stroke. Heat stroke is a true medical emergency.
- Heat stress can be managed through both administrative controls (e.g., acclimatizing or conditioning the body to working in hot environments, work/rest scheduling, rehab, and fluid replacement) and the use of PPE cooling options.
- Medical surveillance is the cornerstone of an effective employee health and safety management system and site safety practices and procedures. Objectives include (1) to determine whether an individual can perform his or her assigned duties (i.e., "fit for duty"), including the successful use of personal protective clothing and equipment; and (2) to detect any changes in body system functions caused by physical and/or chemical exposures.
- A written PPE program is required under OSHA 1910.120(q)(5) and should include hazard assessment; medical monitoring; equipment selection and use; training; and inspection, maintenance, and storage. A written respiratory protection program is also required under OSHA 1910.134(c)(1).
- Safety is an attitude, a behavior, and a culture, and MUST be an inherent part of all operations from the development of SOPs to the selection of PPE. The fundamental operating philosophy of every emergency response organization should be, "If we cannot do this safely, then we will not do it at all."
- Establishing an Incident Safety Officer and developing a written Site Safety Plan are key elements in ensuring the health and safety of personnel operating at a hazmat incident.
- Medical monitoring is defined as an ongoing, systematic evaluation of individuals at risk of suffering adverse effects from stress and/or exposure to heat, cold, or hazardous environments. The objectives of medical monitoring are (1) to obtain baseline vital signs, and (2) to identify and preclude from participation all individuals at increased risk for sustaining injury or illness. It is one element of a comprehensive medical surveillance program that starts with a baseline and periodic "fit for duty" medical examination.

Hot Terms

Acute Effects Results from a single dose or exposure to a material. Signs and symptoms may be immediate or may not be evident for 24 to 72 hours after the exposure.

Acute Exposure An immediate exposure, such as a single dose that might occur during an emergency response.

Acute Emergency Exposure Guidelines (AEGL) Developed by The National Research Council's Committee on Toxicology to provide uniform exposure guidelines for the general public.

Chronic Effects Results from a single exposure or from repeated doses or exposures over a relatively long period of time.

Chronic Exposure Low exposure repeated over time, such as responding to a number of hazmat emergencies while serving as a member of a public safety HMRT or conducting on-scene investigations of hazmat incidents over many years.

Dose/Response Relationship The cause/effect relationship relating to a substance absorbed by the body and the body's response to that substance.

Emergency Response Planning Guidelines (ERPG) Air concentration guidelines for single exposures to hazardous materials.

Exposure Contact with a chemical. Common methods are inhalation, ingestion, skin absorption, or direct contact.

Immediately Dangerous to Life and Health (IDLH) An atmospheric concentration of any toxic, corrosive, or asphyxiant substance that poses an immediate threat to life, or would cause irreversible or delayed adverse health effects, or would interfere with an individual's ability to escape from a dangerous atmosphere.

Medical Surveillance Program The cornerstone of an effective employee health and safety management system. The primary objectives are (1) to determine that an individual can perform his or her assigned duties and (2) to detect any changes in body system functions caused by physical and/or chemical exposures.

Permissible Exposure Limit (PEL) The maximum time-weighted concentration at which 95% of exposed, healthy adults suffer no adverse effects over a 40-hour work week, based on an 8-hour, time-weighted average concentration; used by OSHA.

Recommended Exposure Levels (REL) The maximum time-weighted concentration at which 95% of exposed, healthy adults suffer no adverse effects over a 40-hour work week, based on a 10-hour, time-weighted average concentration; used by NIOSH.

Short-Term Exposure Limit (STEL) The 15-minute, time-weighted average exposure that should not be exceeded at any time, nor repeated more than four times daily with a 60-minute rest period required between each STEL exposure.

Threshold Limit Value/Ceiling (TLV/C) The maximum concentration that should not be exceeded, even instantaneously. The lower the value, the more toxic the substance.

Threshold Limit Value/Skin (TLV/Skin) Indicates a possible and significant contribution to overall exposure to a material by absorption through the skin, mucous membranes, and eyes by direct or airborne contact.

Threshold Limit Value / Time Weighted Average (TLV/TWA) The maximum airborne concentration of a material to which an average healthy person may be exposed repeatedly for 8 hours each day, 40 hours per week without suffering adverse effects.

Toxicity The ability of a substance to cause injury to a biological tissue.

Toxicology The study of chemical or physical agents that produce adverse responses in the biological systems with which they interact.

HazMat Responder *in Action*

It is summertime and you have been called by Law Enforcement to a house that has a fatality and an individual who is described as "acting strange." When you arrive you learn that the incident is a suspected illicit laboratory (i.e., drug lab), and drug enforcement agents are on scene processing evidence. The agents tell you that the deceased person is in the back bedroom and the other perpetrator is now in custody, physically in the back of the patrol car behind which your team is parked.

Questions

1. Assuming that both perpetrators were in the same location over the same time frame, why has one died and the other not?
   - **A.** Routes of exposure are different
   - **B.** It is a function of the dose/response relationship
   - **C.** Toxicity of one chemical was greater than another
   - **D.** All the above
2. The Law Enforcement Commander approaches you and states that one of the SWAT entry team members is starting to act in a fashion similar to the second perpetrator who is in custody. What is your primary line of questions going to be?
   - **A.** What personal protective equipment was the officer wearing at the time of entry?
   - **B.** Is the officer taking any medications?
   - **C.** Is this normal for the officer after a drug raid?
   - **D.** Does the officer have a history of doing this?
3. The Law Enforcement Commander states that they have found a substance that is labeled Paraoxone. You look it up in your reference library and find that it is toxic at 14mg/kg. What does this information tell you about potential hazards present?
   - **A.** This chemical is moderately hazardous
   - **B.** It has an extremely low toxicity and is not to be concerned with
   - **C.** It is a seriously toxic substance
   - **D.** It is slightly toxic
4. Which exposure guideline should you use to achieve a safe environment for responders to work in?
   - **A.** LD50
   - **B.** IDLH
   - **C.** PEL
   - **D.** STEL

References and Suggested Readings

1. American Society of Testing and Materials, ASTM E2601-08—*Standard Practice for Radiological Emergency Response,* West Conshohocken, PA (2008).
2. Berger, M., W. Byrd, C.M. West, and R.C. Ricks, *Transport of Radioactive Materials: Q & A About Incident Response.* Oak Ridge, TN: Oak Ridge Associated Universities (1992).
3. Bolstad-Johnson, Dawn M., et. al., "Characterization of Firefighter Exposures During Fire Overhaul." *American Industrial Hygiene Association Journal* (September/October, 2000), pages 636–641.
4. Bowen, John E., "Understanding Chemical Toxicity." *Fire Engineering* (August, 1987), pages 19–28.
5. Brunacini, Alan V., and Nick Brunacini, *Command Safety.* Peoria, AZ: Across the Street Productions (2004).
6. Callan, Michael, *Street Smart Hazmat Response.* Chester, MD: Red Hat Publishing, Inc. (2001).
7. Cameron, Mark, "Health and Safety Concerns for Law Enforcement Personnel Investigating Clandestine Drug Labs." *Chemical Health & Safety* (January/ February, 2002), pages 6–9.
8. Dickinson, Dr. Edward T., and Michael A. Wieder, *Emergency Incident Rehabilitation.* Upper Saddle River, NJ: Brady/Prentice Hall (2000).
9. U.S. Department of Homeland Security, Federal Emergency Management Agency—U.S. Fire Administration, *Emergency Incident Rehabilitation.* Washington, DC: FEMA/USFA (2008).
10. U.S. Department of Homeland Security, Federal Emergency Management Agency, *Special Operations Program Management Course.* Emmitsburg, MD: FEMA/USFA/NFA (2010).
11. Jones and Bartlett Learning, *Live Fire Training—Principles and Practices.* Burlington, MA: Jones and Bartlett Learning (2012).
12. National Fire Protection Association, *Hazardous Materials Response Handbook* (6th edition). Quincy, MA: National Fire Protection Association (2013).
13. National Fire Protection Association, *NFPA 473—Standard for Competencies for Ems Personnel Responding to Hazardous Materials/Weapons of Mass Destruction Incidents.* Quincy, MA: National Fire Protection Association (2013).
14. National Fire Protection Association, *NFPA 1500—Fire Department Occupational Safety and Health Program Handbook.* Quincy, MA: National Fire Protection Association (2002).
15. National Fire Protection Association, *NFPA 1582—Standard on Comprehensive Occupational Medical Program for Fire Departments.* Quincy, MA: National Fire Protection Association (2013).
16. National Fire Protection Association, *NFPA 1584—Standard on the Rehabilitation Process for Members During Emergency Operations and Training Exercises.* Quincy, MA: National Fire Protection Association (2008).
17. National Institute for Occupational Safety and Health (NIOSH), *Occupational Safety and Health Guidance Manual for Hazardous Waste Site Activities.* Washington, DC: NIOSH, OSHA, USCG, EPA (1985).
18. Occupational Safety and Health Administration, *OSHA Technical Manual.* Washington, DC: OSHA. Information can be accessed through the OSHA website at http://www.osha.gov.
19. Pollak, Andrew N., MD, *FAAOS, Emergency Medical Responder* (5th Edition). Sudbury, MA: Jones and Bartlett Learning (2011).
20. Skinner, Lars, "Responding to Radiation Incidents: Concepts and Tactics for Hazmat Teams, Part I." *Firefighting.Com* (February 2001).
21. U.S. Department of Energy, Office of Transportation and Emergency Management, *Module Emergency Response Radiological Transportation Training (MERRTT).* Washington, DC: DOE (2000).
22. U.S. Department of Homeland Security, Federal Emergency Management Agency. *Application Of Protective Action Guides for Radiological Dispersal Device (Rdd) and Improvised Nuclear Device (Ind) Incidents.* Washington, DC: DHS/FEMA (2008).
23. U.S. Department of Justice, Office of Justice Programs/Office for Domestic Preparedness (OJP/ODP), *Weapons of Mass Destruction (Wmd) Radiation/Nuclear Course for Hazardous Materials Technicians.* Mercury, NV: Department of Energy (DOE) and Bechtel Nevada (2002).
24. U.S. Environmental Protection Agency, *Computer Assisted Management of Emergency Operations (CAMEO) Emergency Software Program.* Washington, DC: EPA (2010).

25. U.S. Environmental Protection Agency, *Manual of Protective Action Guides and Protective Actions for Nuclear Incidents* (EPA 400-R-92-001). Washington, DC: EPA (1992).
26. U.S. Environmental Protection Agency, et al., *Technical Guidance for Hazards Analysis—Emergency Planning for Extremely Hazardous Substances.* Washington, DC: EPA, FEMA, DOT (1987).
27. Veghte, James H., "Physiologic Field Evaluation of Hazardous Materials Protective Ensembles." *U.S. Fire Administration Report FA-109* (September 1991).
28. Williams, Philip L., and James L. Burson, *Industrial Toxicology.* New York: Van Nostrand Reinhold (1985).
29. Wray, Thomas K., "Mutagens, Teratogens and Carcinogens." *Hazmat World* (March, 1991), pages 88–89.
30. Wray, Thomas K., "Risk Assessment." *Hazmat World* (January, 1993), pages 45–47.

Managing the Incident: Problems, Pitfalls, and Solutions

CHAPTER 3

Hazardous Materials Technician

Knowledge Objectives

After reading this chapter, you will be able to:

- List the categories of players and participants at a hazardous materials incident. (p. 68–72, 80–84)

Skills Objectives

There are no Hazardous Materials Technician skills objectives in this chapter.

Hazardous Materials Incident Commander

Knowledge Objectives

After reading this chapter, you will be able to:

- List the categories of players and participants at a hazardous materials incident. (p. 80–84, 87)
- Identify the key organizational elements of the Incident Command System. (p. 72–80)

Skills Objectives

There are no Incident Commander skills objectives in this chapter.

You Are the HazMat Responder

A new chemical manufacturing facility is being constructed in your community and the county Hazardous Materials Response Team (HMRT) has been invited to an initial meeting to discuss the operational capabilities and proposed organizational relationships between the facility and emergency responders in the event of an emergency. The facility uses several chemical-based processes to develop high-pressure petrochemical products. One of the agenda items will be the application and use of an incident command organization that will allow the various response organizations to work within a single, integrated response organization.

1. How could unified command be applied in a manner that recognizes the jurisdictional responsibilities of public safety agencies with the process/product knowledge and expertise of the facility owner and operator?
2. How could facility Emergency Response Team (ERT) members, who are trained to the Hazmat Technician level for the products within the facility, be integrated with county HMRT personnel to provide a single, integrated Hazmat Group operation?

Introduction

Direct and effective command and control operations are essential at every type of incident. However, hazardous materials incidents place a special burden on the command system because they often involve communications among separate agencies and the coordination of many different functions and personnel assignments—from public protective actions and the use of specialized personal protective equipment (PPE) to investigative efforts involving the potential criminal use of hazardous materials.

This chapter reviews the fundamental concepts of incident management and its application at a hazmat incident. Primary topics include the various players who characteristically appear at an incident, the elements of the Incident Command System (ICS), the functions and responsibilities of the Hazmat Group, and "street smart" tips. This chapter is intended to *complement* the information the reader should have already gained by completing the ICS-100—*Introduction to the Incident Command System* and ICS-200—*ICS for Single Resources and Initial Action Incidents* level training courses. Where the ICS-100 and 200-level courses provide the ICS "science" associated with managing hazardous materials incidents, this chapter reinforces that information by looking at the "art" of managing hazmat incidents.

Managing the Incident: The Players

A hazmat incident requires different skill sets to safely stabilize the emergency and bring it to closure. Each response discipline brings its own agendas, organizational structures, and priorities to the incident scene. The seeds of a successful response start by having a pre-existing relationship with your peers and other players well before you meet at the incident scene. The 0200-hours meeting on Interstate 66 should not be the first time you meet your public safety response partners.

> **Responder Tip**
>
> The formula for a safe and effective response is having a coordinated incident command structure in which all of the players integrate their resources to make the problem go away in a safe and effective manner. In simple terms, you need a system, and the ICS provides the structure you need.

The basic ICS organization that must be created to bridge these potential gaps and problems includes the following:

- **The <u>Incident Commander (Command or IC)</u>**—The individual responsible for establishing and managing the overall <u>incident action plan (IAP)</u>. This process includes developing an effective organizational structure, developing an incident strategy and tactical action plan, allocating resources, making appropriate assignments, managing information, and continually attempting to achieve the basic command goals in a safe and competent manner. Everyone working at the event reports through the chain of command to this individual.
- <u>Unified Commanders (UCs)</u>—Command-level representatives from each of the primary responding agencies who present their agency's interests as a member of a

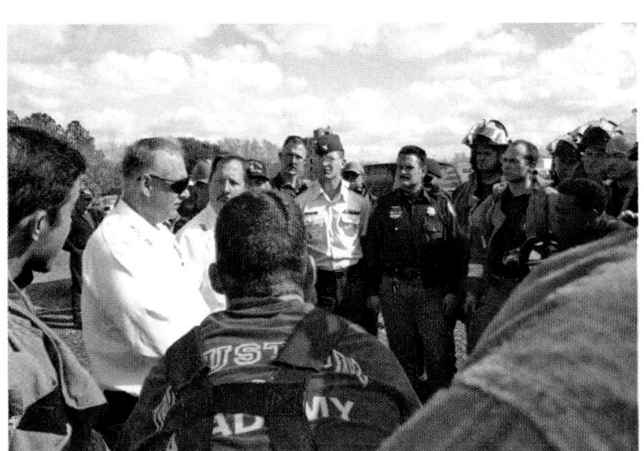

Figure 3-1 Hazardous materials incidents can bring many different players and agencies to the scene.

unified command organization. Depending on the scenario and incident timeline, they may be the lead IC or play a supporting role within the command function. The UCs manage their own agency's actions and make sure all efforts are coordinated through the unified command process Figure 3-1.

- **ICS General Staff**—ICS provides a mechanism to divide and delegate tasks and develop a management structure to handle the overall control of the incident. Section Chiefs are members of the IC's general staff and are responsible for the broad response functions of operations, planning, logistics, and finance/administration. Individuals below the section level are the front-line supervisors who implement tactical objectives to meet the strategies established by the IC within a branch, group, or division (e.g., Hazmat Group Supervisor).
- **ICS Command Staff**—Those individuals appointed by and directly reporting to the IC. These include the Safety Officer, the Liaison Officer (LNO), and the Public Information Officer (PIO).

The Players

Regardless of who they are and how they materialize on the scene, the IC must also be able to quickly identify and categorize the various players and participants who will interact within the ICS organization. These include the following:

- **Fire/Rescue/EMS Companies**—Provide resources for fire suppression, rescue, medical triage, treatment, and transport. They implement assigned tasks, provide support to specialized assets, and help to coordinate overall response efforts. Examples include fire fighters and fire officers, EMS personnel, and other knowledgeable responders on scene.
- **Law Enforcement Officers**—Resources for ensuring scene safety (i.e., scene and traffic control), implementing public protective actions, perpetrator arrest or control, evidence preservation and collection, incident investigation, etc. They implement assigned tasks, provide support to specialized assets, and help coordinate overall response efforts. Examples include police and security personnel in both government and industry who provide fundamental law enforcement services.
- **Emergency Response Team (ERT)**—Crews of specially trained personnel used within business and industrial facilities for the control and mitigation of emergency situations. They may consist of full-time personnel, shift personnel with ERT responsibilities as part of their job assignment (e.g., plant operators), or volunteer members. ERTs may be responsible for any combination of fire, hazmat, medical, and technical rescue emergencies, depending on the nature, size, and operation of their facility.
- **Hazardous Materials Response Teams (HMRTs)**—An organized group of trained response personnel operating under an emergency response plan and applicable standard operating procedures who perform hazardous materials technician-level skills at hazardous materials/WMD incidents (NFPA 472 definition). They may be staffed with members from emergency services, private industry, governmental agencies, environmental contractors, or any combination. They generally perform more complex and technical hazmat response functions than fire, rescue, and EMS companies.
- **Special Operations Teams**—Highly trained and equipped response teams who deliver a highly specialized response service and capability. Examples include urban search and rescue (US&R) teams, technical rescue, bomb squads, explosive ordinance disposal (EOD) units, water rescue, and law enforcement tactical units (e.g., SWAT—Special Weapons and Tactics). In recent years with the increasing threat of terrorism, there has been a trend towards an integrated Special Operations Team among the response disciplines.
- **Communications Personnel**—The central communications function for emergency services. In the public sector, is commonly referred to as the Public Safety Answering Point (PSAP). They receive "911" calls for assistance and dispatch appropriate units and resources to the incident locations. Communications personnel (sometimes known as "comms") provide a crucial link for those working on scene by dispatching additional needed resources, including technical specialists, and providing response data and information to field units. Some large industrial complexes may have a central communications center that monitors plant operations and alarms and would forward any facility emergency alarms to the local public safety communications center (or PSAP).
- **Responsible Party (RP)**—Organization legally responsible under government environmental laws for the clean-up of a hazmat release. Depending on the scenario, this may be a transportation carrier or a facility owner.
- **Facility Managers**—Individuals who normally do not have an on-scene emergency response function, but who are key players within the plant environment. In the

event of a plant emergency, they report to the plant Emergency Operations Center (EOC) and are responsible for providing overall plant command and logistical support to field emergency response units and for coordinating external issues, including community liaison, media relations, and agency notifications.

- **Support Personnel**—Individuals who provide important support services at the incident. Water and utility company employees, heavy equipment operators, and food service/rehab personnel are some examples. Within OSHA 1910.120 (q)(4), these individuals are referred to as Skilled Support Personnel and provide immediate emergency support work that cannot be reasonably performed by emergency responders and who may be exposed to the hazards at the incident scene.
- **Technical Specialists**—Individuals who provide specific expertise to the IC and emergency responders either in person, by telephone, or through other means. They usually are product and/or container specialists representing the manufacturer or shipper or are familiar with the chemicals, containers, or problems involved. CHEMTREC™ and the Local Emergency Planning Committee (LEPC) often provide an excellent way to reach information resources in any given area of expertise. May also be referred to as Allied Professionals in NFPA 472 **Figure 3-2**.
- **Environmental Clean-up Contractors**—Individuals who provide mitigation and support services at the incident. Capabilities may include spill control, product transfer operations, site clean-up and recovery, and remediation operations. They are usually retained by the responsible party (RP), the IC, or government environmental agencies (e.g., EPA, State Dept. of Environmental Quality, etc.). Personnel should be trained to meet the training requirements of OSHA 1910.120, paragraphs (b) through (o).

Figure 3-2 Technical specialists and environmental clean-up contractors can be key players in the safe and successful management of a hazmat incident.

- **Government Officials**—Individuals who normally do not have an emergency response function but bring a lot of political clout to the incident. Examples include mayors, city/county managers, or other elected officials who may be involved. For large-scale events they may play a policy and coordination role, or they may delegate this responsibility to an emergency manager. Failure to professionally address their questions and concerns within the ICS organization can have significant political and other impacts both during and after the response.
- **News Media**—Individuals representing various elements of the media who work to inform the public of major happenings within their community or region. Because of the unusual and often frightening nature of hazardous materials incidents, it is very important that the public be accurately informed quickly and regularly of the incident. Television and radio, as well as social media and other online options, are excellent methods with which to coordinate and manage large-scale public protective actions activities. On-scene media may also be willing to provide quality video equipment and aerial shots for emergency service use in exchange for coverage.
- **Investigators**—Individuals who are responsible for determining the origin and cause of the hazmat release, including any related evidence collection and preservation. A hazmat incident is not concluded until the investigation is complete. Future legal proceedings, possible regulatory citations or criminal charges, and financial reimbursement for the time, equipment, and supplies of emergency services may well depend on investigators' efforts. Certain types of incidents require interaction between investigators on the federal, state, and local levels, as well as in the private sector.
- **The Customer**—The injured, affected, or victims. Individuals who may be exposed, contaminated, displaced, injured, or killed as a result of the hazardous materials/WMD incident. Once emergency responders begin to provide treatment and care, they themselves may become patients. Special care should be given to their welfare and to informing them of potential short- and long-term signs and symptoms of exposure.
- **Spectators**—Curious, usually well-meaning members of the facility and/or general public who arrive at the scene to assist or watch the unfolding of events. Since they are often difficult to control—especially during campaign incidents—spectators need to be monitored and managed constantly to ensure their safety.
- **The Bad Guy**—Also referred to as suspect, perpetrator, criminal, terrorist, or whatever else the current politically correct term is. A potentially harmful person (or group of persons) who is intent on bringing harm, destruction, injury, or death to both the general public and/or public safety professionals. At those incidents where hazardous materials are being used for criminal purposes or WMD are being used, there may be one or

Scan Sheet 3-A

Military Assets in the Hazmat Response Community

In the post 9/11 environment, a number of resources and assets have been developed by the U.S. military to deal with domestic homeland security scenarios. These include the following:

- *Weapons of Mass Destruction Civil Support Teams (WMD CST)*. The CST mission is to support civil authorities at a domestic CBRNE incident site with identification and assessment of hazards, advice to civil authorities, and facilitating the arrival of follow-on military forces during emergencies and incidents of WMD terrorism, the release of CBRN materials, and natural or man-made disasters in the United States. The CSTs are designed to complement and enhance, but not duplicate, state CBRNE response capabilities. There are currently approximately 55 CSTs located in each state and U.S. territory.

  Each CST is comprised of 22 full-time Army and Air National Guard personnel. Each unit is divided into six sections, including command, operations, communications, administration/logistics, medical/analytical, and survey. CSTs can be activated through the governor or the adjutant general and can deploy within 3 hours of notification to an incident site. Operational vehicles and resources include a command vehicle, operations trailer, a communications platform known as the Unified Command Suite (provides secure communications capabilities), an Analytical Laboratory System (contains chemical, biological, and radiological analysis equipment), and several general-purpose vehicles.

- *Chemical, Biological, Radiological, Nuclear and High-Yield Explosive (CBRNE) Enhanced Response Force Package (CERFP)*. The CERFP mission is to provide an immediate response capability to a state governor, including search capability of damaged buildings, rescuing trapped casualties, providing decontamination, and performing medical triage and initial treatment to stabilize patients for transport to medical facilities. There are currently 17 states with a CERFP Team, with a 6–12 hour response posture.

  A CERFP is staffed with approximately 170 personnel from National Guard units and comprises four elements:

  1. Search and extraction, which is assigned to an Army National Guard Engineering Company
  2. Decontamination, which is assigned to an Army National Guard Chemical Company
  3. Medical, which is assigned to an Air National Guard Medical Group, and
  4. Command and control, which directs the overall activities and coordinates with the Joint Task Force—state and the Incident Commander.

- *Homeland Response Force (HRF)*. In 2010 the military response concept to a domestic CBRN incident was revised to provide a bridge between the initial civilian community first response and the follow-on federal military response to a CBRN event. There is one HRF assigned to each of the 10 federal regions, with an operational time of 18 hours after notification within its respective fezderal region.

  The HRF is a 570+ member unit composed of the following elements:

  - Brigade Command and Control element—consists of approximately 200 personnel. The element is capable of providing C2 for up to 5 CERFPs and 9 CSTs. It brings its own logistics section and is capable of self-sustaining the HRF for 72 hours.
  - Security elements—consist of approximately 200 members, who provide security to the HRF and its subordinate elements. All security element personnel are trained to the Hazmat Operations level and are equipped with personnel protective clothing and equipment (EPA Level C).
  - CERFP elements—consist of approximately 170 members, as outlined above. In addition, this element can be augmented by a Fatality Search and Recovery Team (FSRT) that can coordinate with other related assets, such as the local medical examiner or a federal Disaster Mortuary Operational Response Team (DMORT).

many Bad Guys at the scene, and they may conceal their identity and blend in with victims and spectators after an attack.

- **The Hazardous Material**—A potentially harmful substance or material that has escaped or threatens to escape from its container. It should be considered an active, mobile opponent that must be carefully monitored at all times. Whenever its container is stressed or it has already escaped, the hazardous material should be considered a threat to the other players. From an emergency response perspective, it really doesn't matter whether the material is classified as a hazardous material, hazardous substance, or weapon of mass destruction.

Managing the Incident: The Incident Command System

Why so much interest in incident command? One primary reason is the OSHA 1910.120(q) requirement that both public safety and industrial emergency response organizations use a "nationally recognized Incident Command System for emergencies involving hazardous materials." Another driver is the Homeland Security Presidential Directive (HSPD) 5—*Management of Domestic Incidents*, which was released in February 2003 and establishes a single, comprehensive national incident management system. But beyond regulatory and governmental requirements, experience has shown that the normal, day-to-day business organization is not well-suited to meeting the broad demands created by "working" hazmat incidents.

The information presented in this chapter is based on the **National Incident Management System (NIMS)**. NIMS is the baseline incident management system established under HSPD-5, which is used by federal, state, and local governments, as well as private sector organizations throughout North America. There are five major components within NIMS:

- Preparedness
- Communications and Information Management
- Resource Management
- Command and Management
- Ongoing Management and Maintenance

This section will be limited to the ICS component of the Command and Management section of NIMS. ICS is an organized system of roles, responsibilities, and procedures for the command and control of emergency operations. It is a procedure-driven, all-risk system based upon the same business and organizational management principles that govern organizations on a daily basis. As is the case with the day-to-day management of any organization, ICS has both technical and political aspects that must be understood by the key players.

In the past, these players were concerned primarily with the technical or operational aspects of an emergency. Today, however, the playing field has changed significantly. "Working" hazmat incidents have political, legal, and financial effects on how both the public and the corporate shareholders view the performance of emergency response professionals. Recent incidents have taught us that an emergency can have a favorable technical or operational outcome and still be a "political" disaster. In some instances, solving the operational problems may be the easiest piece of the command equation. These political issues will be reviewed later in this chapter.

Incident Management vs. Crisis Management

Crisis management is an integral element of corporate and industrial organizations. Experience has shown there is a direct relationship between incident management and crisis management concerns as well as the organizational structure for managing each event.

What is the difference between an incident and a crisis? In the broadest sense, *an incident can be defined as an occurrence or event, either natural or human made, which requires action by emergency response personnel to prevent or minimize loss of life or damage to property and/or natural resources.* Essentially, an incident interrupts normal procedures, has limited and definable characteristics, and has the potential to precipitate a crisis. Examples of incidents may include fires, hazmat, medical, and rescue emergencies. In short, an incident does not necessarily mean that an organization or a community has a crisis.

> **Responder Tip**
>
> The transition from incident management to crisis management may not be easy. The scope or severity of an incident is not the sole factor that determines its potential to develop into a crisis. Other factors may come into play, including the occurrence of recent incidents in the area, the occurrence of similar incidents nationally or within the region, and the political environment in which the incident occurs.

The definition of a crisis will vary significantly depending on the type of organization (e.g., public vs. private sector) and situations anticipated. A good starting point is the following definition: *A crisis is an unplanned event that can exceed the level of available resources and has the potential to impact significantly an organization's operability, credibility, and reputation, or pose a significant environmental, economic, or legal liability.*

Crisis management builds on the philosophy of incident management. If ICS is not used on a regular basis for the "more routine" incidents, it will be very difficult to implement it successfully during a major incident or crisis situation.

ICS Lessons Learned

ICS must be a foundational element of any successful hazmat response program. Response experience has provided us with the following lessons learned:

1. **A variety of different players will respond to a working hazmat incident. Therefore, what occurs during the planning and preparedness phase will establish the framework for how the emergency response effort will operate.**

   Title III of the Superfund Amendments and Reauthorization Act of 1986 (SARA, Title III) and OSHA

Figure 3-3 Local Emergency Planning Committees (LEPCs) can play an important role in the hazmat planning process at the community level.

1910.120 focused much attention on the need for hazmat planning and the implementation of an ICS organization **Figure 3-3**. Communities and facilities are now required to develop planning documents to meet pre-established regulatory criteria. Unfortunately, the result has sometimes been emergency response plans (ERP) that look good on paper but don't really work on the street. In the rush to develop ERPs that meet the letter of the law, some have lost sight of the importance of the ERP's operational utility.

Too much emphasis has been placed on "Do we meet the requirements of the law?" as compared to "Can our personnel perform the mission requirements?" In fact, a subjective assessment of emergency planning programs could be summarized as follows: The majority of industrial emergency response programs are compliance oriented, where the letter of the law is satisfied, but the performance of facility emergency responders may be untested and not verified through either exercises or actual experience. In contrast, some public safety response programs are "operationally oriented," where personnel are able to perform the expected emergency response tasks, but sometimes lack the required regulatory documentation.

Planning and preparedness establish the framework for how the emergency response effort will function. While ERPs must satisfy minimum regulatory requirements, the response plan must be both operationally oriented and representative of actual personnel and resource capabilities. Response plans that are not regularly updated and user friendly do not get used. In short, response plans that are poorly understood by the players are poorly executed in the field.

2. **There is no single agency that can effectively manage a major emergency alone.**

A major incident will require the resources and expertise of various organizations and agencies within the region. All emergency response organizations bring their own agendas to the emergency scene. Each of these agendas represent real, valid, and significant concerns. Problems are often created, however, when there is no communication prior to the emergency (i.e., you don't know the key players and their emergency response mission) and everyone feels that his or her specific agenda or interest is the most important. Remember, the ability to mount a safe and effective response builds on what is accomplished during planning and preparedness activities. The real issue is not only command, it is also coordination.

In the absence of planning, organizational relationships are often based on perceptions, and perceptions are often based on our experiences with one individual or one incident. If that experience was positive, we tend to view the respective organization in a positive light until proven otherwise. Similarly, if that experience was negative . . . well, you know the rest!

There is no single organization that can effectively manage a major hazmat incident. Organizations that attempt to maintain the normal organization or bureaucracy in managing a major event will have inherent problems in implementing a timely and effective emergency response. There is no excuse for not knowing who the key players are within your area and with whom you are going to interact with on scene.

3. **Many special operations teams, including HMRTs, tend to be people-dependent programs.**

Emergency response programs can be categorized as being either "people-dependent" or "system-dependent." Special operations teams are often very people-dependent when they are initially formed. These organizations rely on the experience of a few key individuals and can result in failed emergency response efforts if these key individuals are not present at an incident **Figure 3-4**.

In contrast, a system-dependent organization has clearly defined objectives, specific duties, and responsibilities that are spelled out in standard operating procedures (SOPs) and checklists, and resources based upon probable response scenarios. A system-dependent response allows individuals to assume different roles in an emergency regardless of their daily activities. Written procedures, operational checklists, and an effective training and critique program ensure that less experienced personnel can get the job done with an acceptable level of safety and efficiency.

In essence, a system-dependent organization delivers a consistent level of quality and service, regardless of personnel or location. For example, when you order a Big Mac from any McDonalds throughout the world, there will

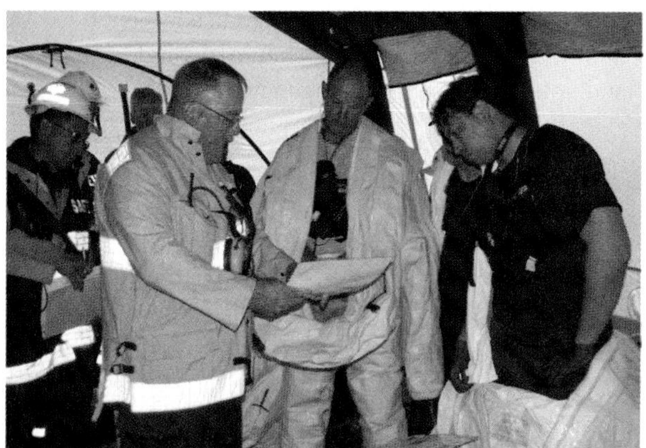

Figure 3-4 Special operations units that rely on the experience of a few key individuals can result in failed emergency response efforts if these key individuals are not present at an incident.

be little difference in either quality or taste. How do they do it several billion times? By emphasizing common and consistent procedures and personnel training. In short, McDonalds is the epitome of a system-dependent organization.

An organizational philosophy and management goal of emergency response programs should be the development of operational procedures that will bring consistency to emergency operations. The components of this system include:

- A hazard and risk analysis that provides the foundation for assessing the likely threats and hazards and determining the desired operational capability
- The development of SOPs that outline key tactical-level functions and tasks
- Training all personnel in the scope, application, and implementation of the SOPs
- The execution of the SOPs on the emergency scene
- Post-incident review and critique of their operational effectiveness
- Revision and updating of SOPs on a regular basis

This standard management cycle helps build an organization with the ability to self-improve over time. This is critical, as the accepted standard of care keeps rising over time. What was considered an adequate emergency response program 5 years ago may be viewed as inadequate by today's standards.

4. **In those cases where ICS has not resulted in the operational improvements expected, the problems are typically associated with planning, training, and the organization buying into the ICS program, as compared to the ICS system itself.**

   ICS is not a panacea but is an organizational process and a resource management tool. As with any new effort, it will be necessary to establish and communicate a policy regarding the application and use of ICS for incident management and crisis management purposes. This policy must be established and supported by the highest levels of management and communicated throughout the organization. If the boss doesn't buy in to the program, neither will the troops!

5. **The management and control of routine, day-to-day incidents establishes the framework for how the larger, more significant events will be managed.**

   ICS should be the basic operating system for the management of all emergencies and significant events. The incident management structure should expand as the nature and complexity of the emergency expands, resulting in a smooth transition with minimum organizational changes or adjustments.

   If ICS is not used for all routine emergencies, don't expect the organizational structure to function and adapt efficiently and safely when a major emergency occurs. The routine establishes the foundation on which the nonroutine must build. The more routine decisions made prior to the emergency, the more time the IC and subordinates will have to make critical decisions during the emergency.

ICS Elements

ICS has common characteristics that permit different organizations to work together safely and effectively in order to bring about a favorable outcome to the emergency. ICS is predicated on basic management concepts, including the following:

- Division of labor—Work is assigned based on the functions to be performed, the equipment available, and the training and capabilities of those performing the tasks.
- Lines of authority—are clearly defined, including the delegation of authority and responsibility, as appropriate. However, ultimate responsibility always rests with the Incident Commander.
- Unity of command—Every person reports to only one supervisor (and only one IC). This provides for a proper chain of command, helps to eliminate confusion and freelancing, and facilitates personnel and organizational accountability.
- Span of control—For most emergency service applications, a span of control of five individuals (range of three to seven) is recommended. The actual span-of-control should be based upon the tasks to be performed, the associated danger or difficulty, and the level of delegated authority. For example, in the special operations world, high-risk tasks will typically have a lower span-of control than basic-level tasks performed in non-IDLH environments. When span-of-control problems arise, they can be addressed by expanding the organization in a modular fashion.
- Establishment of both line and staff functions within the organization. Line functions are directly associated with the implementation of incident operations to "make the problem go away," while staff functions support incident operations.

Common Terminology

A hazmat response program must be built around an ICS organization that uses standardized terminology for organizational functions, resource elements, and incident facilities. This is particularly critical at multiagency and multijurisdictional incidents. Basic ICS organizational terms include the following:

1. *Incident Commander (IC)*: The individual responsible for the management of on-scene emergency response operations. The IC must be thoroughly trained to assume these responsibilities and is not automatically authorized to perform these activities by virtue of his/her position within the organization.

   The IC will be located at the emergency scene and will operate from a designated incident command post (ICP) location. As the scope of the incident escalates and senior officers and managers are activated, command and/or coordination may be transferred from the emergency scene to a Multi-Agency Coordination Center, such as a facility or community EOC **Figure 3-5**. In this instance, on-scene command will be under the direction of the Operations Section Chief or the On-Scene Commander, depending on local terminology. In simple terms, life in the field doesn't change much; what does change is the establishment of a broader response management organization that can (1) rapidly support the field response and (2) effectively deal with the external world effects and support issues caused by the incident.

   A single command structure is used when one response agency has total responsibility for the overall incident. Some hazmat emergencies will require that command be unified

Figure 3-5 Facility and community Emergency Operations Centers (EOC) can play a significant role in coordinating the efforts of multiple organizations during a working or significant hazmat incident.

or shared between several organizations. Unified command is described in greater detail later in this chapter.

2. *Sections*—The organizational level with functional responsibility for primary functions of emergency incident operations. Sections and their respective unit-level positions are only activated when their respective functions are required by the incident.

   Sections are part of the IC's general staff and represent the broad functional areas of operations, planning, logistics, and administration/finance. Section Chiefs report directly to Command.

   Primary responsibilities of each respective section include the following:

   - Operations Section—Delivers the required tactical-level "services" in the field to make the problem go away, including fire, hazmat, oil spill, technical rescue, and emergency medical operations. Until the Operations Section is established, the IC has direct control of all tactical resources. (*Note:* This ICS terminology should not be confused with an industrial facility's Operations Department, and does not specifically refer to process/operations personnel or activities.)
   - Planning Section—Typically regarded as the information management and research and development (R&D) arm of the ICS process. The Planning Section conducts assessments and identifies the future needs and then develops the plans required to support the response. In the early phase of an incident, planning will focus on what the requirements are for 1 or 2 hours into the future. Once the incident stabilizes, planning begins to develop plans for the next operational period (NOP) in the future. At major incidents, this may be 12+ hours or multiple days into the future.

     Another important role of the Planning Section is to develop plans that maximize the resources available. For example, an event involving a large geographic area or multiple locations must have someone thinking about how scarce resources will be allocated and operations coordinated. Failure to establish a Planning Section for a moderate to large incident is similar to a military general officer attempting to manage the battlefield without any reconnaissance or military intelligence support. Planning units include situation status, resource status, documentation and demobilization, and technical specialists. Planning also plays a critical role in the scheduling and facilitating of incident meetings and briefings.

     The Planning Section plays a critical role in providing specialized expertise based on the needs of the incident, such as health and safety, industrial hygiene, environmental, WMD, and process engineering.
   - Logistics Section—Once staged resources are exhausted, the response effort becomes extremely dependent on the availability of the resupply and logistics effort. Logistics is responsible for providing all incident support needs, including facilities, services, and materials. This would encompass communications, rehabilitation, and food unit concerns within the Service Branch, and supply, facilities, and ground support unit concerns within the Support Branch.

     The Logistics Section should not be viewed as an "order taker" to support the response effort. To be effective, the Logistics Section must be proactive and anticipate future needs, be knowledgeable of both the emergency and project procurement process, understand communications, and be able to facilitate the transition from the emergency response to the project phase of a long-term incident.
   - Administration/Finance Section—Responsible for all costs and financial actions of the incident. The primary responsibility is to get funds where they are needed; ensure that adequate, yet simple, financial controls are in place; and keep track of all funds. Primary financial units would include time, procurement, cost, and compensation/claims. In some corporate organizations, the legal unit may also be assigned to the Finance Section.

     Some incidents may be large enough that a federal disaster is declared, thereby allowing the local jurisdiction to be reimbursed for some of its incident response costs. To receive reimbursement, expenses must be tracked and accounted for. Unfortunately, experience has shown that many costs are not tracked or documented during the initial stages of a campaign event.

3. *Branch:* The organizational level having functional or geographic responsibility for major segments of incident operations. The branch level is organizationally between the section and division/group levels and is supervised by a Branch Director. Branches are often established when the number of divisions/groups or sectors exceeds the recommended span of control. Hazmat operations may be combined into a Hazmat Branch at a large Type 3 or 4 incident.

4. *Division/Groups:* Divisions are the organizational level having responsibility for operations within a defined geographic area. A building floor, plant location, or process area may be designated as a division, such as Division 4 (e.g., 4th floor area) or the Alky Division (e.g., Alkylation Process Unit). Divisions are under the direction of a Supervisor.

   Groups are the organizational level responsible for a specified functional assignment at an incident. Hazmat units may operate as a Hazmat Group at smaller Type 4 or 5 hazmat incidents. Groups are under the direction of a Supervisor and may move between divisions at an incident.
5. *Command Staff Officers:* Based on response requirements, the IC may assign personnel to serve on the Command Staff to provide Information, Safety, and Liaison services for the entire organization. Although there is only one Command Staff position for each of these functions, the Command Staff may have one or more assistants, as necessary. For example, hazmat incidents will often have an Assistant Safety Officer (ASO)–Hazmat, who functions as a member of the tactical unit (e.g., Hazmat Group Safety Officer).
   - Safety Officer—Responsible for the safety of all personnel, including monitoring and assessing safety hazards, unsafe situations, and developing measures for ensuring personnel safety. The Incident Safety Officer (ISO) has the authority to terminate any unsafe actions or operations and is a required function based upon the requirements of OSHA 1910.120 (q). Depending on the geographic layout of the incident and the tactics being employed, there may be multiple ASO/Hazmat personnel. Oil spills with both on-shore and off-shore impacts are an example of an incident where multiple ASOs may be found **Figure 3-6**.
   - Liaison Officer (LNO)—On large incidents or events, representatives from other agencies (usually called Agency Representatives) may be assigned to the incident to coordinate their agency's involvement. The LNO serves as a coordination point between the IC and any assisting or coordinating agencies who have responded to the incident, but who are not part of unified command or represented at the ICP. In many respects, Liaison can often become the IC's "political officer," who may need to run interference for the IC.
   - Public Information Officer (PIO)—The PIO will be the point of contact for the media and other organizations seeking information directly from the incident or event. In today's world of online communications, they may also have expanded responsibilities in monitoring communications and interfacing with the general public through the myriad of social media networks.

Figure 3-6 The Incident Safety Officer and Assistant Safety Officers are responsible for developing measures to ensure personnel safety.

In those incidents where multiple Incident Officers from different agencies are present, a Joint Information System and a Joint Information Center (JIC) should be established and all of the individual agency Information Officers work jointly and cooperatively from one location. JICs may be established at various levels of government or incident sites or can be components of Multi-Agency Coordination Systems. A single JIC location is preferable, but the system is flexible and adaptable enough to accommodate virtual or multiple JIC locations, as required. Incidents that may cover a large geographic area, such as the 2010 Deepwater Horizon oil spill which affected a number of states along the Gulf of Mexico, are examples where multiple JICs may be established.

Modular Organization

The ICS organizational structure develops in a modular fashion based on the size and nature of the incident. The system builds from the top down, with initial responsibility and performance placed upon the IC. At the very least, an IC must be identified on all incidents, regardless of their size. As the need exists, separate sections can be developed, each with several divisions/groups that may be established **Figure 3-7**.

The specific ICS organizational structure will be based on the management needs of the incident. For example, at a small incident, such as a 30-gallon carboy spill of sulfuric acid on a loading dock, personnel are not required to staff each major functional ICS area. The operational demands and the number of resources do not require delegation of management functions. However a working incident, such as a major train derailment or a large toxic vapor cloud release, may require staffing all sections to manage each major functional area and delegate management functions.

Predesignated Incident Facilities

Emergencies require a central point for communications and coordination. Depending on the incident size, several types of predesignated facilities may be established to meet ICS requirements. These are as follows:

1. *Incident Command Post (ICP):* The on-scene location where the IC develops goals and objectives, communicates with subordinates, and coordinates activities between various agencies and organizations. The ICP is the field office for on-scene response operations and requires access to communications, information, and both technical and administrative support **Figure 3-8**. It may range from the front seat of a Suburban to a mobile command unit that could pass as a bus for a rock star. It should be identified so that arriving command and general staff personnel can easily find its location (e.g., green flag or flashing light).

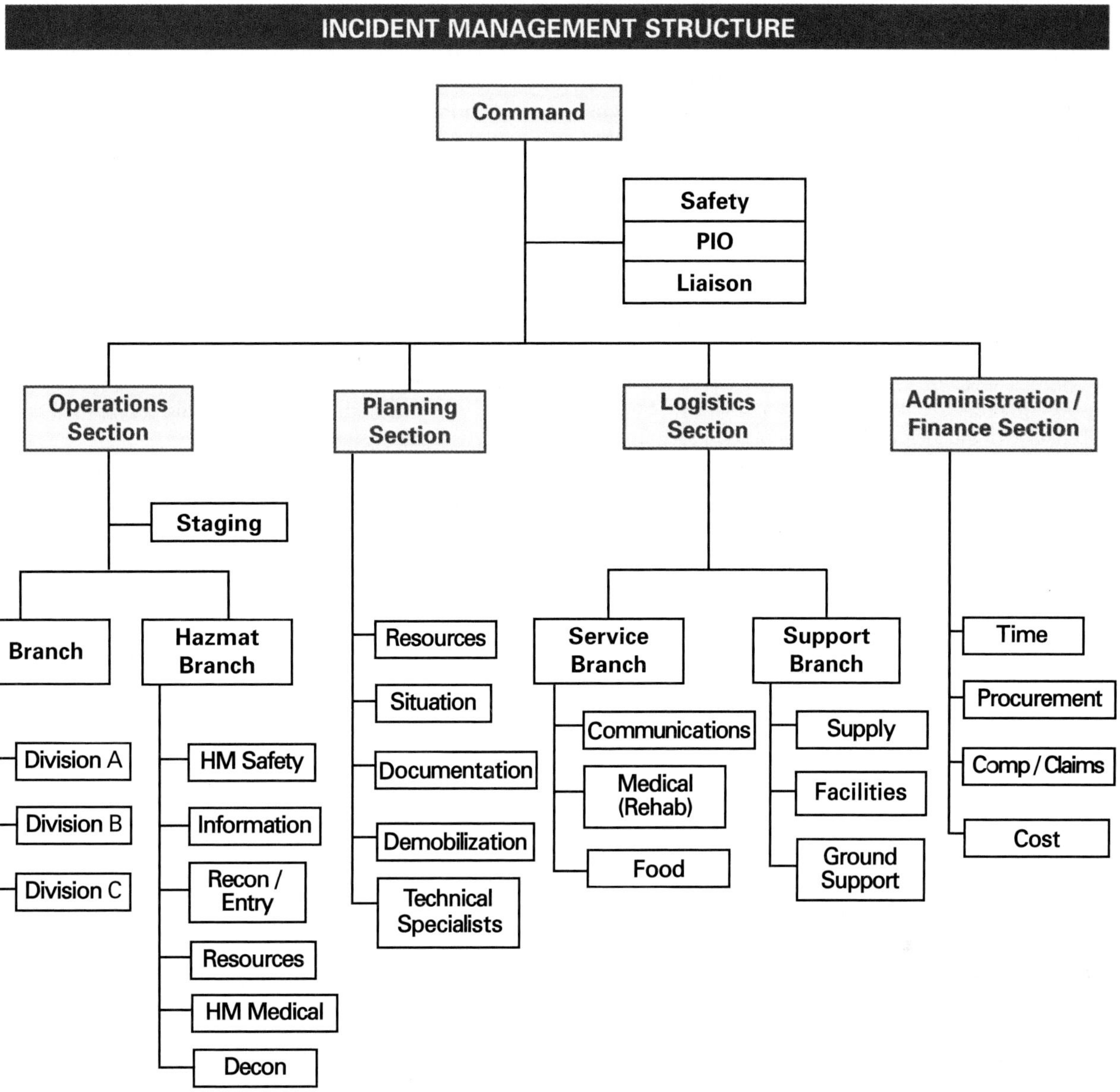

INCIDENT MANAGEMENT STRUCTURE

- Command
  - Safety
  - PIO
  - Liaison
  - Operations Section
    - Staging
    - Branch
      - Division A
      - Division B
      - Division C
    - Hazmat Branch
      - HM Safety
      - Information
      - Recon / Entry
      - Resources
      - HM Medical
      - Decon
  - Planning Section
    - Resources
    - Situation
    - Documentation
    - Demobilization
    - Technical Specialists
  - Logistics Section
    - Service Branch
      - Communications
      - Medical (Rehab)
      - Food
    - Support Branch
      - Supply
      - Facilities
      - Ground Support
  - Administration / Finance Section
    - Time
    - Procurement
    - Comp / Claims
    - Cost

Figure 3-7 The Incident Command System provides a modular-based organization that can be expanded or contracted based on the size and nature of the incident.

Figure 3-8 The Incident Command Post (ICP) is the field office for on-scene response operations and requires access to communications, information, and both technical and administrative support.

An ICP should provide command with the following:

- A place safe from the hazardous material(s) or problem and which is easily identifiable to the players
- A (relatively) quiet place in which to think, discuss, and decide
- A vantage point from which to see (when possible)
- A primary and secondary source of power (e.g., generator)
- Inside lighting
- A place to write and record
- Protection from the weather
- Protection from the media
- Staff space

The IC should remain at the ICP so that he or she is readily accessible to all personnel. As noted, a number of emergency response organizations have specialized command vehicles that function as a mobile ICP. However, the necessary equipment for an ICP can be prepackaged into an ICP kit. Minimum equipment should include the following:

- Radio capability to communicate with responders, mutual aid units, and facility maintenance/operations personnel. This should include mutual aid radios, programmable scanner radios for monitoring emergency radio frequencies, and access to the NOAA weather frequencies, where appropriate. Agency interoperability is critical!
- Cellular telephone capability. Remember, unless you have an encryption capability, all cell phone communications can be monitored. Also, responder cell phone usage can be adversely affected by increased cell phone usage by the general public based on the nature of the incident and its impact on the public.
- Copies of appropriate emergency response guidebooks and other reference sources. These should include the ERP and a response folder/booklet containing copies of all ICS checklists and worksheets.
- Technical and administrative support, including internet access, the use of laptop computers, and related electronic equipment.
- ICS command vests
- Tactical command chart
- Pair of binoculars

2. *Emergency Operations Center (EOC):* The ICP is the nerve center of on-scene operations and is usually located near the scene of the emergency. However, if the scope of the incident increases, the plant or community EOC could then be activated. In this situation, overall command and/or coordination would be transferred to the EOC, while on-scene response operations would continue to be managed from the ICP. All communications with the media and outside agencies would now be coordinated through the EOC.

Based on physical needs and safety requirements, the EOC is normally remote from the emergency scene. The EOC should provide phone, radio, and fax communications, internet access, information technology (IT) resources, and the ability for a large number of personnel to work in a comfortable and secure area. These elements become essential as the number of players increases and the incident stretches into days as opposed to hours.

It is important to understand the differences and relationship between the ICP and the EOC when both are operating simultaneously at a major emergency. The ICP is primarily oriented towards tactical control issues pertaining to the on-scene response, while the EOC deals with both strategic and external world issues and *coordinates* all logistical and resource support for on-scene operations.

Emergencies have occurred where the EOC was impacted by the incident and could not be used. Perhaps the most vivid example was the loss of the New York City EOC in Building 7 of the World Trade Center complex on September 11, 2001. As a result, many facilities and some communities have identified both a primary and alternate EOC location. Both EOCs should be similarly equipped and provided with comparable information and resource capabilities. In evaluating potential EOC locations, consideration should be given to the impact of potential fire, hazardous materials spills, vapor releases, or terrorist attacks on the site. It should be noted that control rooms are typically poor options for an EOC within high-risk process control industries.

An EOC should be equipped with the following:

- Radio, phone, and fax communications. This should include mutual aid radios, programmable scanner radios for monitoring emergency radio frequencies, and access to the NOAA weather frequencies, where appropriate. Sufficient telephone lines should be available for both incoming and outgoing calls.
- Detailed copies of area and facility maps, site plot plans, emergency preplans, hazard analysis documentation, and other related information.
- Copies of appropriate emergency response guidebooks and other reference sources. These should include the Emergency Response Plan, the LEPC Plan, safety data sheets (SDS), and other pertinent plans and procedures.
- General administrative support, including writing boards, incident status and documentation boards, telefax, and copying machines.
- Electronic communication capabilities, including the use of computers and e-mail, the Internet and intranet, and the development of incident or agency-specific Web sites. Always consider the security of your electronic communication system.
- Television sets and AM/FM radios to monitor local and national news coverage.
- Backup emergency power capability to support EOC lighting, telephone, and radio base stations.

3. *Staging Area*—The designated location where emergency response equipment and personnel are assigned on an immediately available basis until they are needed. Staging is effective when the IC anticipates that additional resources may be required and orders them to respond to a predesignated area approximately three minutes from the scene. In simple terms, staging is the IC's tool box.

The Staging Area should be clearly identified through the use of signs, color-coded flags or lights, or other suitable means. The exact location of the Staging Area will be based on prevailing wind conditions and the nature of the emergency. Emergency response personnel are then assigned to the emergency scene from the Staging Area as needed. Staging ensures that resources are close by but not in the way.

Staging becomes an element within the Operations Section. The Staging Area Manager reports to the Operations Section Chief and is responsible to account for all incoming emergency response units, to dispatch resources to the emergency scene at the request of the IC, and to request additional emergency resources, as necessary.

Integrated Communications

Communications are critical to safe and efficient incident management. Ideally, the IC should be able to communicate directly with all on-scene units and support personnel. However, the more players at the incident, the less likely that they will share the same radio frequency or have the ability to directly communicate with each other. The more people using the same frequency, the more crowded it becomes.

Communications are managed most effectively through the use of a common communications center and network. Communications interoperability is the key. Where common or mutual aid radio frequencies are unavailable, the IC should request that a designated individual report to the ICP with a radio from his or her organization. Using these individual radios, the IC can ensure that communications flow horizontally and vertically within the command structure. When common radio channels do not exist between all on-scene units, it may be more effective to have all companies operating on a common radio frequency to work together as a group or division, rather than dividing their resources between functions or areas.

Whenever a situation is encountered that could immediately cause or has caused injuries to emergency response personnel, the term *emergency traffic* should precede the radio transmission. This will be given priority over all other radio traffic. If an "emergency traffic" message is issued by the Communications Center, it may be preceded by a radio alert tone.

Although difficult, it is critical that the IC be provided with regular and timely progress reports throughout the course of the emergency. Information is power; when the Operations Section becomes a "black hole" for all incident information, it takes away the ability of the ICS structure to support the operational response effort and address the myriad of external issues in a timely and effective manner.

Communications of a sensitive nature should not be given over nonsecure cellular telephones or radios, which can be monitored.

Unified Command

Hazmat incidents often involve situations where more than one organization shares management responsibility or where the incident is multijurisdictional in nature. A unified command structure simply means that the key agencies that have jurisdictional authority or functional responsibility for an incident jointly contribute to the process of:

- Determining common set of incident objectives and strategies
- Developing a single IAP
- Maximizing use of all assigned resources
- Resolving conflicts between the players

The sooner a unified command structure is established, the better **Figure 3-9**. Unified command is *not* management by committee; there will always be a lead agency or one agency that has 51% of the vote as compared to the other players. A representative of the lead agency should serve as the IC and should be supported by the senior officers from the other agencies involved. For example, during a chemical agent terrorism scenario where rescue and medical operations are underway, the lead IC will most likely be a fire department officer. However, once these life safety issues are completed and the focus switches to crime scene management and incident investigation, the lead IC should transfer to a law enforcement representative.

Figure 3-9 Unified command is NOT management by committee!

As in a single-agency command structure, the Operations Section Chief will have responsibility for implementation of the IAP. When multiple agencies are involved in the response, the selection of the Operations Section Chief must be made by the mutual agreement of the unified command team. This may be done on the basis of greatest jurisdictional involvement, number of resources involved, existing statutory authority, or by mutual acknowledgement of the individual's qualifications.

If the players aren't coordinated and don't play as a team, the result is often ugly. While the members of unified command should only perform command-level tasks and avoid getting involved in direct activities at the incident, they can and often are supported by aides in the field (i.e., Field Observers) who are observing on-scene operations and serve as their "eyes and ears."

Consolidated Plan of Action

Every incident needs some form of an IAP. The IAP consists of incident priorities, strategic goals, tactical objectives, and resource requirements. For a small hazmat incident, the plan is usually simple and can be verbally communicated directly to the individuals carrying out the IC's orders. However, on large-scale incidents or campaign events this may not be practical, and a formal, written IAP may be required.

Emergencies involving multiple organizations or jurisdictions working within a unified command structure require consolidated action planning. As more organizations arrive at the emergency scene, they bring with them individual agendas and objectives. These may be driven by:

- Facility responsibilities
- Legally mandated requirements
- Financial interests

- Contractual responsibilities
- Specific mission goals and charters

For example, local law enforcement agencies arriving at an incident may be primarily interested in traffic management and safety, while the EPA and the State Department of Environmental Quality (DEQ) are both interested in the environmental impact of the release. All three agencies have a legal right to be involved in the emergency, but none of the groups has exactly the same objective.

A consolidated action plan is used to ensure that:

- Everyone works together toward a common emergency response goal; that is, protecting life safety, the environment, and property.
- Individual response agendas are coordinated so that personnel and equipment are used effectively and in a spirit of cooperation and mutual respect.
- Everyone works safely at the scene of the emergency.

The most effective way to ensure a consolidated plan of action is implemented is to have the senior representative of each major player at the incident present at the ICP and/or EOC at all times. Command runs the incident like a business meeting, making sure that every organization has its say and that the entire group works toward a resolution of the emergency as quickly and as safely as possible. The IC should remain focused on realistic objectives and ensure that each entity or special interest has input into the plan.

Comprehensive Resource Management

The Incident Commander must analyze overall incident resource requirements and deploy available resources in a well-coordinated manner. In the case of a major hazmat incident or WMD event, resources will include personnel, equipment, and supplies. The bottom line is simple—the proper type and level of resources must be available to support emergency operations in a timely manner **Figure 3-10**.

Logistics and resource management have been the Achilles' heel of many responses. Once the initial supply of resources is exhausted, responders are dependent on the ability of the ICS organization, specifically the Logistics Section, to provide a steady and reliable flow of resources. Purchasing procedures that work well for day-to-day operations may be totally inadequate in an emergency or crisis environment.

Figure 3-10 Implementing an Incident Action Plan without sufficient resources is like writing a check with insufficient funds in the bank.

Managing the Incident: Hazmat Group Operations

Depending on the scope and complexity of an incident, special operations (e.g., hazmat, bomb squad, technical rescue) may be managed as either a Branch or Group within the ICS organization. This section provides a brief overview on the application and use of a **Hazardous Materials Group** at a hazmat incident.

The Hazardous Materials Group is normally under the command of a senior Hazmat Officer (known as the Hazardous Materials Group Supervisor), who, in turn, reports to the Operations Section Chief or the IC. The Hazardous Materials Group is directly responsible for all tactical hazmat operations that occur in the hot and warm zones of an incident. Tactical operations outside of these hazard control zones (e.g., public protective actions) are not the responsibility of the Hazardous Materials Group.

Although a number of resources may be assigned to the Hazardous Materials Group, they are typically HMRT personnel and resources. The scope and nature of the problem will determine which roles are staffed.

Primary functions and tasks assigned to the Hazardous Materials Group include the following:

- **Safety function**—Primarily the responsibility of the Hazardous Materials Group's Safety Officer (i.e., ASO–Hazmat). Responsible for ensuring that safe and accepted practices and procedures are followed throughout the course of the incident. Possesses the authority and responsibility to stop any unsafe actions and correct unsafe practices.

Responder Tip

Among the resource management lessons learned as a result of previous incidents and exercises are the following:

- Get it done rather than argue about whose problem it is.
- It is easier to "gear down" operations than it is to play "catch-up." If you think you will need it, call for it!
- Overreact until the emergency situation is fully assessed and completely understood. React to the incident potential, not the existing situation. If you have to play "catch-up," you'll usually come in second place.
- Accept help from others. Mrs. Smith doesn't really care from where an asset or resource is obtained ... neither should you. Remember, our customer just wants the problem to go away!
- Treat outside personnel with the same respect and care you provide your own personnel.
- Don't pay for stuff you don't need—downsize once the emergency has been stabilized and it is safe to do so. Demobilization can be as important as the initial ramp-up and mobilization, and should start as early as possible.

- **Entry/back-up function**—Responsible for all entry and backup operations within the hot zone, including reconnaissance, monitoring, sampling, and mitigation.
- **Decontamination function**—Responsible for the research and development of the decon plan, set-up, and operation of an effective decontamination area capable of handling all potential exposures, including entry personnel, contaminated victims, and equipment. If necessary, will include the coordination of a Safe Refuge Area.
- **Site access control function**—Establish hazard control zones, establish and monitor egress routes at the incident site, and ensure that contaminants are not being spread. Monitor the movement of all personnel and equipment between the hazard control zones. Manage the Safe Refuge Area, if established.
- **Information/research function**—responsible for gathering, compiling, coordinating, and disseminating all data and information relative to the incident. This data and information will be used within the Hazardous Materials Group for assessing hazards and evaluating risks, evaluating public protective options, selecting PPE, and developing the IAP.

Secondary support functions and tasks that may be assigned to the Hazardous Materials Group will include the following:

- **Medical function**—Responsible for pre- and post-entry medical monitoring and evaluation of all entry personnel, and provides technical medical guidance to the Hazardous Materials Group, as requested.
- **Resource function**—Responsible for control and tracking of all supplies and equipment used by the Hazardous Materials Group during the course of an emergency, including documenting the use of expendable supplies and materials. Coordinates, as necessary, with the Logistics Section Chief (if activated).

Figure 3-11 illustrates the standard ICS positions found at a typical hazardous materials incident. Remember, ICS is a modular organization. Only those functions necessary for the control and mitigation of the incident would be activated, as appropriate.

Hazardous Materials Group Staffing

Hazardous Materials Group Supervisor. Responsible for the management and coordination of all functional responsibilities assigned to the Hazardous Materials Group, including safety, site control, research, entry, and decontamination. The Hazardous Materials Group Supervisor must have a high-level of technical knowledge and be knowledgeable of both the strategic and tactical aspects of hazardous materials response.

The Hazardous Materials Group Supervisor will be trained to the Hazardous Materials Technician level and will normally be filled by either the HMRT Team Leader or Hazmat Officer. Depending on the scope and nature of the incident, the Hazardous Materials Group Supervisor will usually report to either the Operations Section Chief or the IC.

Based on the IC's strategic goals, the Hazardous Materials Group Supervisor develops the tactical options to fulfill the hazmat portion of the IAP and is responsible for ensuring the following tasks are completed:

- Establish and monitor hazard control zones.
- Conduct site monitoring to determine the presence and concentration of contaminants.
- Develop and implement a Site Safety Plan for Hazmat Group operations.
- Establish tactical objectives for the Hazardous Materials Entry Team within the limits of the team's training and equipment limitations.
- Coordinate all hot zone operations with the Operations Section Chief or IC to ensure tactical goals are being met.

Hazardous Materials Group Safety Officer (i.e., Assistant Safety Officer—Hazmat). The Hazardous Materials Group Safety Officer reports to the Hazardous Materials Group Supervisor and is subordinate to the ISO. This individual is responsible for coordinating safety activities within the Hazardous Materials Group but also has certain responsibilities that may require action without initially contacting the normal chain of command. The ISO is responsible for the safety of all personnel operating at the incident, while the ASO–Hazardous Materials is responsible for all operations within the Hazardous Materials Group and within the hot and warm zones. This includes having the authority to stop or prevent unsafe actions and procedures during the course of the incident.

The Assistant Safety Officer–Hazardous Materials must have a high level of technical knowledge to anticipate a wide range of safety hazards. This should include being hazmat trained, preferably to the Hazardous Materials Technician level, and being knowledgeable of both the strategic and tactical aspects of hazmat response. This position is typically filled by a senior HMRT officer or member. While it is not the ASO–Hazardous Materials job to make tactical decisions or to set goals and objectives, it is his or her responsibility to ensure that operations are implemented in a safe manner.

Specific functions and responsibilities of the ASO include the following:

- Advise the Hazardous Materials Group Supervisor of all aspects of health and safety, including work/rest cycles for the Entry Team.
- Coordinate site safety activities with the ISO, as appropriate.
- Possess the authority to alter, suspend, or terminate any activity that may be judged to be unsafe.
- Participate in the development and implementation of the Site Safety Plan.
- Ensure the protection of all Hazardous Materials Group personnel from physical, chemical, and/or environmental hazards and exposures.
- Identify and monitor personnel operating within the hot zone, including documenting and confirming both "stay times" (i.e., time using air supply) and "work times" (i.e., time within the hot or warm zone performing work) for all entry and decon personnel.
- Ensure EMS personnel and/or units are provided with necessary resources, and coordinate with the Hazardous Materials Medical Leader.

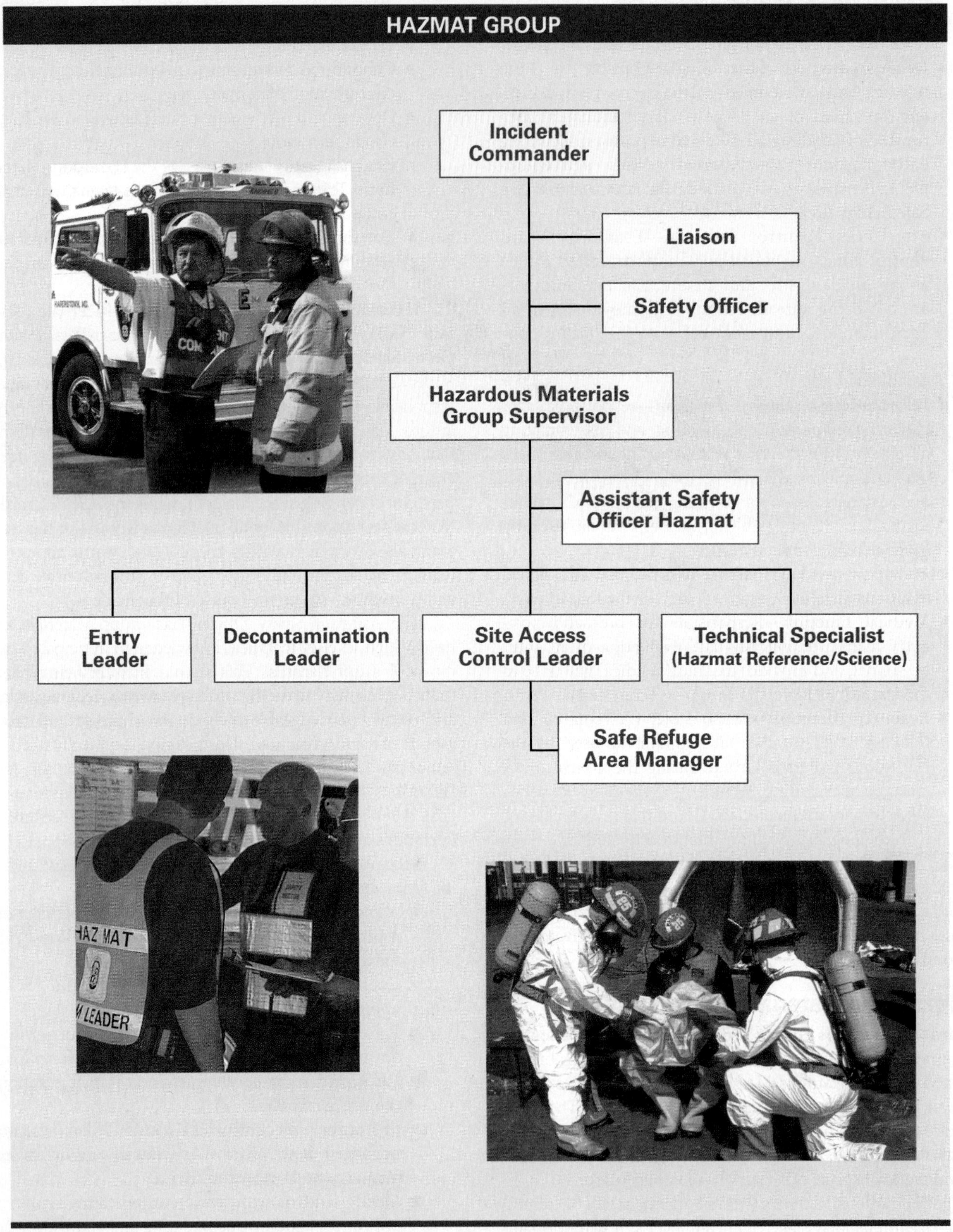

HAZMAT GROUP

- Incident Commander
  - Liaison
  - Safety Officer
  - Hazardous Materials Group Supervisor
    - Assistant Safety Officer Hazmat
    - Entry Leader
    - Decontamination Leader
    - Site Access Control Leader
      - Safe Refuge Area Manager
    - Technical Specialist (Hazmat Reference/Science)

Figure 3-11 The Hazmat Group is responsible for all tactical-level hazmat response operations.

- Ensure health exposure logs and records are maintained for all Hazardous Materials Group personnel, as necessary.

Entry Team. The Entry Team is managed by the Entry Unit Leader (aka Entry Officer). This individual is responsible for all entry operations within the hot zone and should be in constant communication with the Entry Team. The Entry Team and the Entry Leader are responsible for the following:

- Recommend actions to the Hazardous Materials Group Supervisor to control the emergency situation within the Hot Zone.
- Implement all offensive and defensive actions, as directed by the Hazardous Materials Group Supervisor, to control and mitigate the actual or potential hazmat release.
- Direct rescue operations within the hot zone, as necessary.
- Coordinate all entry operations with the Decon, Hazmat Information, Site Access, and Hazardous Materials Medical Units.

Personnel assigned to the Entry Team will include the entry and back-up teams, and personnel assigned for entry support. The Entry Team consists of all personnel who will enter and operate within the hot zone to accomplish the tactical objectives specified within the IAP. Entry Teams will always operate using a buddy system.

The Back-Up Team is the safety team that will extract the entry team in the event of an emergency. Depending upon the scenario, the back-up capability may be provided by a Rapid Intervention Team or "RIT," as commonly used within the fire service. The Back-Up Team must be in-place and ready whenever entry personnel are operating within the hot zone. Entry support personnel (may also be known as the Dressing Team) are responsible for the proper donning and outfitting of both the Entry and Back-Up Teams.

Decontamination Team. The Decon Team is managed by the Decon Leader (aka Decon Officer) and are responsible for the following:

- Determine the appropriate level of decontamination to be provided.
- Ensure proper decon procedures are used by the Decon Team, including decon area set-up, decon methods and procedures, staffing, and protective clothing requirements.
- Coordinate decon operations with the Entry Leader, Site Access Control, and other personnel within the Hazardous Materials Group.
- Coordinate the transfer of decontaminated victims requiring medical treatment and transportation with the Hazardous Materials Medical Group.
- Ensure that the decon area is established before any entry personnel are allowed to enter the hot zone. If rapid rescue operations are required, establish an emergency decon capability until a formal decon area can be set up.
- Monitor the effectiveness of decon operations.
- Control all personnel entering and operating within the decon area.

Site Access Control. The Site Access Control Unit is managed by the Site Access Control Leader and is responsible for the following:

- Monitor the control and movement of all people and equipment through appropriate access routes at the incident scene to ensure the spread of contaminants is controlled.
- Based upon recommendations from the Entry, Decon, and Info/Research Units, oversee the placement of the hazard control zone lines.
- As necessary, establish a safe refuge area and appoint a Safe Refuge Area Manager. This would include coordinating with Decon on decon and medical priorities for contaminated victims.
- Ensure injured or exposed individuals are decontaminated prior to departure from the incident scene.

Hazardous Materials Information/Research Team. The Hazardous Materials Information/Research Team is managed by the Information Leader (may also be known as Research or Science). Depending upon the level of the incident and the number of hazardous materials involved, the Information Team may consist of several persons or teams. The Hazardous Materials Information Team and the Information Leader are responsible for the following:

- Provide technical support to the Hazardous Materials Group.
- Research, gather, and compile technical information and assistance from both public and private agencies.
- Provide and interpret environmental monitoring information, including the analysis of hazardous materials samples and the classification and/or identification of unknown substances.
- Provide recommendations for the selection and use of protective clothing and equipment.
- Project the potential environmental impacts of the hazardous materials release.

Hazardous Materials Medical Unit. Medical support services may be provided by either a Hazardous Materials Medical Unit or Medical Group within the ICS organization. Hazardous Materials Medical Unit personnel will be located in the Entry Team dressing area and in the Rehabilitation Area. The Hazardous Materials Medical Unit and Hazardous Materials Medical Leader are responsible for the following:

- Provide pre-entry and post-entry medical monitoring of all entry and backup personnel.
- Provide technical assistance for all EMS-related activities during the course of the incident.
- Provide emergency medical treatment and recommendations for ill, injured, or chemically contaminated civilians or emergency response personnel.
- Provide EMS support for the Rehab Area.

The Hazardous Materials Medical Unit will conduct post-entry medical monitoring, cooling and rehydration of entry and backup personnel in the Rehab Area. All operating personnel should not be given anything to eat or drink unless approved by Hazardous Materials Medical personnel. Medical findings and personal exposure forms should be forwarded to the ASO/Hazardous Materials and/or the Entry Team Leader.

Hazardous Materials Resource Unit. At some "working" incidents, a hazardous materials resource function may be established to support Hazardous Materials Group activities and is directed by the Hazardous Materials Resource Leader. This unit will be located in the cold zone and will be responsible for acquiring all supplies and equipment required for Hazardous Materials Group operations, including protective clothing, monitoring instruments, leak control kits, and so on. In addition, the Hazardous Materials Resources Leader will also be responsible for documenting all supplies and equipment expended as part of the emergency response effort. The Hazardous Materials Resources Unit and Leader must work closely and coordinate with the Logistics Section Chief.

Managing the Incident: Street Smarts

Have you ever watched Olympic figure ice skating? You would find that the contestants are judged in two broad areas: (1) technical merit, which accounts for the technical precision and quality of their program; and (2) artistic impression, which accounts for how the program is choreographed and the relationship between the skater's performance and music. In simple terms, a poor score in one area can offset a great performance in the other.

Does hazmat response resemble figure ice skating? It shouldn't! However, emergency response operations are increasingly being judged in how responders perform in two similar areas. Consider the following comparison:

- *Technical merit*—In simple terms did responders make the "problem" go away. While we can often argue about safety and response efficiencies, most would agree that the emergency response community is very effective! At the end of the day, the problem usually goes away . . . sometimes because of a lack of fuel, but the problem still goes away!
- *Artistic impression*—Our version of artistic impression is how well we manage the "external world" effects of the problem outside of the response community. This could cover everything from media communications to community relations and intergovernmental relationships.

To restate this concept in another manner, the overall performance of a hazmat response program will be based on two interrelated factors: (1) the implementation of a timely, well-trained, and equipped emergency response effort; and (2) the effective management of the interpersonal and organizational dynamics created by the event, particularly those dealing with external groups and audiences (e.g., the media, government agencies, and the public at large). An effective field response effort can be compromised or completely negated by poor management of the political and external issues. Or to put it another way, an incident can become a crisis when the political and external issues are not effectively addressed.

Remember, perceptions are reality. Emergency response efforts that do not address external and political concerns will often be perceived by the public, elected officials, and management as a failed response, regardless of how effective the on-scene emergency response effort actually is. This section will examine some of the factors that may influence the political and external impact of an emergency response effort.

Command and Control

Experienced officers often regard the hazmat problem in enemy-oriented, pessimistic terms. In brief, the hazardous materials involved are your "enemy" and will behave in both known and unknown ways. Command and control efforts must be applied toward achieving results—that is, defeating the "enemy." Confident ICs and hazmat officers refuse to be overwhelmed as they assume command. As soon as possible, they delegate certain responsibilities and empower their subordinates to make decisions and do the job they have been assigned. Effective ICs recognize that a few minutes spent establishing effective command and control at the beginning may save hours in the course of a long-term incident.

A fire service colleague once mentioned that the best officers he had worked with during his career had one thing in common—their presence was so strong that it inspired people to perform well at the emergency scene. If you look like you know what you are doing, it sets a tone for the management of the emergency. Many refer to this as command presence. If command presence is not strong, both individual and organizational "freelancing" and on-scene "rubber-necking" can result.

Constant reassessment and possible revision of tactical operations are needed to maximize response effectiveness. Both Command and the Hazmat Group Supervisor must be able to integrate evaluation and revision into the overall management approach. However, review and evaluation are useful only when the tactical plan is kept open-ended. Both Command and the Hazmat Group Supervisor need to plan ahead and operate with a backup plan, constantly asking questions such as these Figure 3-12:

- Where will the incident be in 30 minutes? 60 minutes?
- Given the current problem, what is our worst-case scenario?
- What is Plan B and Plan C if Plan A doesn't work?

Responder Safety Tip

Don't accept a bad situation; on the emergency scene, things go right and things go wrong. As an officer, you must be willing to admit when a mistake has been made or that conditions have changed and then modify the placement and deployment of available personnel and resources accordingly. Failure to constantly evaluate and adjust your plan, when necessary, can eventually place the IAP and the safety of responders at a great disadvantage.

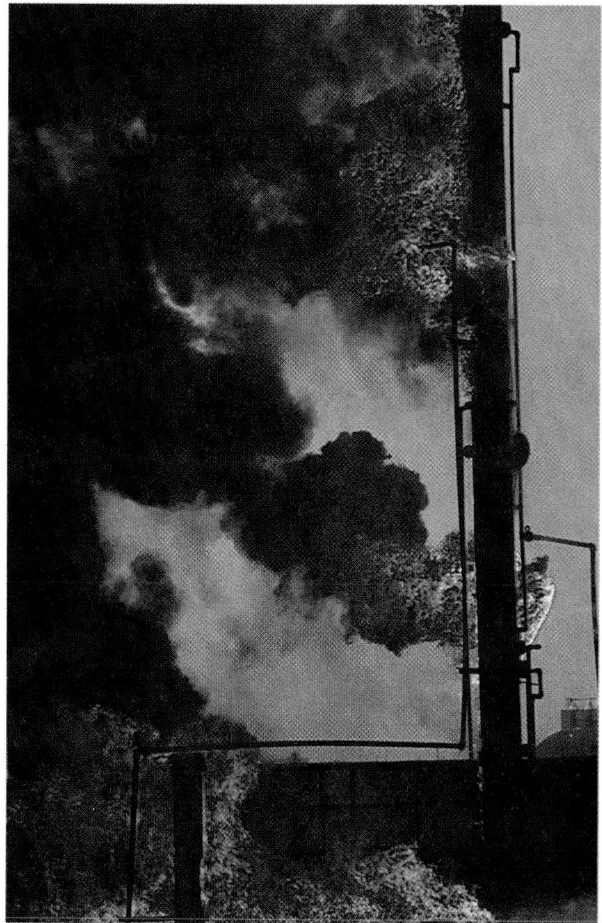

Figure 3-12 What will the incident be in 30 minutes? What is Plan B? Plan C?

Incident Potential

The development of incident objectives and the related strategies and tactics to defeat the enemy must be based on the assessment of incident potential. Elements of incident potential can include incident severity, magnitude and duration of the event, as well as the nature and degree of incident impacts. In addition, don't forget about the "artistic" side of the equation, including community impacts, external world and media affairs impacts, and legal concerns.

Situational awareness and assessment of incident potential go hand-in-hand. What are some of the causes of a delayed assessment of incident potential? They include the following:

- Responders get so focused on the tactical problem that they fail to consider the big picture or strategic aspects of the incident.
- Response agencies that act like they can do it all.
- Response agencies that believe a request for mutual aid or assistance might involve acknowledging a mistake.
- Responders define a problem down to a manageable level.
- Responders are inexperienced.
- There is a lack of information.

Responder Tip

You cannot defeat the enemy if you only focus on where the problem is now—always consider where the enemy is going and how it may change (i.e., incident potential).

Decision Making

The decision-making process begins with both Command and the Hazmat Group Supervisor recognizing the need to avoid dead-end decisions. Whenever possible, decisions must be open-ended, allowing for expansion or reversal. Having to make quick decisions may worry the inexperienced officer, but they become easier to make once you recognize the following:

- *Distinguish between assumptions and facts.* Response operations must sometimes be based upon incomplete or assumed information. Factual information is often not available or incomplete, particularly in the initial stages of an emergency. Realize that both information and decision-making will have the opportunity to improve as the incident grows older.
- *Maintain a flexible approach to decision making.* The overall IAP must be constantly updated as more and better information is received from operations units and other outside sources. Regular feedback allows for revisions to the general strategy, specific tactics, and all major decisions.
- *Shift to a management role after initiating action.* An IC cannot make all ongoing response decisions. The efficiency of command decisions will improve once the IC delegates tactical responsibility; otherwise Command will quickly become overwhelmed with both people and information. The same concept holds true within the Hazmat Group.

The IC must quickly prioritize problems and develop solutions. This requires the ability to effectively gather, record, and organize information.

Information Is Power

Emergency scene information and intelligence can rapidly supply Command and the Hazmat Group with random data and information. Both functions must develop and use an information-gathering and processing routine that functions within their limitations. Tactical worksheets can help facilitate this process. However, without effective information management, mental overload can quickly occur and decision making will suffer.

Unfortunately, the quality and quantity of information during the initial stages of an incident is often poor and incomplete. In other instances, time is required to gather the information required to develop an initial IAP. The term *fog of war* is not limited to the military battlefield but is present at most working incidents, especially during the early stages.

Equally critical is the ability to manage and disseminate information in a timely manner. Most information has a "half-life," in that it is valuable for a limited period of time. If critical

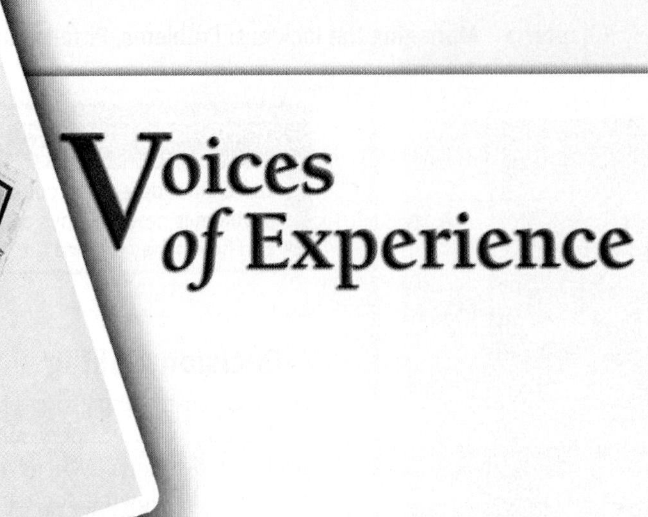

Voices of Experience

It was a beautiful Saturday morning just 4 days before Christmas, and I was looking forward to a weekend packed with family activities. Having been recently promoted to battalion chief/fire marshal, everything seemed to be coming up roses. Little did I suspect Murphy's Law was about to kick me in the head.

Around 8:47 am, dispatch called to request my presence at a commercial structure fire. The fire was consuming a 150,000 square foot warehouse near the interstate highway and would prove to be the largest fire in my city's history. Additionally, as Murphy would have it, most of the administrative staff was enjoying overdue vacation time, and I was covering several positions during this (normally) slow time of year. The Operations Division initially assumed command, and I acted as Fire Marshal/PIO. After the first operational period, I became the Incident Commander.

The fire took several days to extinguish, required extensive mutual aid, and was suspected to be an act of arson. Sections of the roof collapsed, creating a river of toxic sludge (a combination of massive water streams and damaged product from the warehouse), which flowed into the streets. In addition, a criminal investigation was underway, and the media clamored for attention . . . Thank you Murphy!

Fortunately for me, our fire department consistently trained in the principles of emergency management and practiced those disciplines during routine incidents. So, when this "Big One" occurred, we had the safety net of previous training and experience to rely on. A command structure, established by the first arriving units, evolved as the incident shifted from one phase into another. Using internal and external resources, we addressed site safety, fire suppression, hazardous materials, incident stabilization, fire investigation, and media relations. The totality of the circumstances required me to continue as site manager for several weeks after my first arrival.

Eventually, paperwork became my best friend. I relied upon a series of documents (i.e., incident action plans, site safety plans, and personnel logs) to record events, regulate activities, and serve as action reminders. The disciplined and systematic organization of this complex incident proved invaluable throughout the event, the subsequent criminal prosecution, and the eight years of civil actions that followed.

Consequently, my best advice is to learn constantly, train continually, and stay organized. Make it a habit to practice proper Site Management (at least in your head) during any event to which you respond. You may never know when Murphy will pay you a visit, but you can be prepared to deal with him if he does decide to drop by.

Kevin Hammons
Revelations, Inc.
Aurora, Colorado

information is received after a tactical window of opportunity has closed, it will not help in defeating the enemy. How many times have you heard the statement, "It would have been nice to have this information 15 minutes ago?"

Occasionally there is reluctance for on-scene personnel to provide regular and timely updates to the ICP or EOC. This is a critical problem, as both Command and the EOC cannot support tactical response operations and address the external issues (e.g., media relations, community notifications, etc.) if they do not have a regular and reliable flow of information. If they don't know the "true" picture, they may begin to "make it up" based on what they think is happening.

In the heat of battle, the Operations Section may start to reduce the flow of information to the remainder of the ICS organization. In essence, it becomes an information "black hole"—questions flow to the Operations Section but nothing comes out. Ways to avoid this problem include assigning a communications aide to the Operations Section Officer and establishing procedures that require that updates and status reports be provided at timely intervals (e.g., every 10 minutes).

Information is power. Whoever controls the flow of information controls the incident **Figure 3-13**.

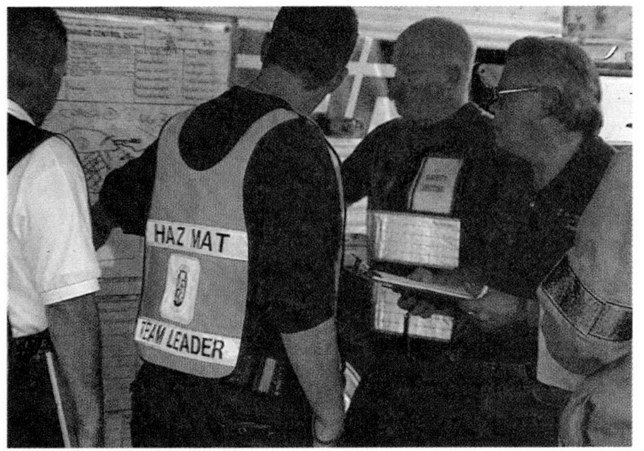

Figure 3-13 Regular situation status reports must flow from the Operations Section to the Incident Commander.

The Rules of Engagement

HMRTs are often requested to respond on a mutual aid basis into other areas or jurisdictions to provide technician-level response services. At one time in our response history, these requests were "far and few"—local ICs often stated that "... there is no way any hazmat team is going to come into my town/plant and run my incident." Today, some ICs are willing to give majority command and control of the incident to an HMRT, often making comments like "Please make my problem go away" or "If you need me I'll be over there."

Law enforcement, special ops, and military operations typically lay out rules of engagement (ROE) for all of the players prior to initiating actions. The ROE provides the structure for engaging the enemy, including the chain of command, decision-making authority, accountability, and responsibility. In a similar manner, the ROE may need to be established for both hazmat and special operations events, particularly those where the special ops unit is operating in someone else's jurisdiction.

The best scenario is one where the ROE are outlined and agreed upon as part of the planning process. If the incident is a "home game," departmental policies and procedures will usually address this issue. However, if the incident is an "away game," the ROE should be reviewed and established between Command and the HMRT before hazmat response operations are initiated.

What should the ROE look like? At a minimum, they should include the following:

- Clarify the relationship between the IC and the HMRT. In simple terms, hazmat should always work for the IC. Don't allow a well-meaning and well-intentioned IC to saddle you with command and control of the entire operation.
- Clearly identify what objectives, tasks, or areas are the responsibility of the Hazmat Group or HMRT.
- HMRTs are often reliant on other response units for various support personnel and resources, such as people, water, and supplies. In other cases, unusual or specialized resources may be required, including heavy equipment. Command must understand if you (i.e., the HMRT) ask for it, you need it.
- Finally, always have a "get out of jail" card when playing at an away game. Be prepared for those occasions where the local IC may reject the recommendations of the HMRT and pursue tactical options with unacceptable risks to the hazmat responders. Although not common, this scenario has tremendous potential career, political, and legal liabilities.

Liaison Officer

Hazmat incidents pose a multitude of technical, managerial, and political issues. As previously noted, emergencies that have been effectively managed from a technical perspective have been perceived to be poorly managed by the public and governmental agencies because the political issues and external impacts of the incident were not adequately addressed.

ICS is an effective and necessary tool for ensuring that the internal, technical, and external/political aspects of an emergency are all addressed. A key member of the ICS command staff is the Liaison Officer (LNO). The LNO is sometimes viewed as the "Political Officer" who serves as the initial point of contact for all assisting and cooperating agencies not represented within the unified command structure. Assisting agencies are those organizations which provide personnel, services, and other resources to the agency with direct responsibility for incident management. An assisting agency provides tactical resources. In contrast, a cooperating agency provides assistance other than direct operational or support functions or resources to the incident management effort.

This ICS position allows the IC and general staff to focus on problem resolution while ensuring that political sensitivities are still addressed. The Liaison Officer's ability to effectively coordinate, handle, and, if necessary, "stroke" these individual agencies and representatives will have a significant impact on how the incident will be perceived from both a political and external perspective.

What You See Is Not Necessarily What You Get

Underestimating the significance of a hazmat emergency can increase the level of risk to both responders and the public. Human tendency is to downplay the potential of "minor" emergencies; however, there have been instances where some individuals have not acted on early warnings for fear of "screwing up" or because they were not sure if they had the authority to take action. Of course, consistently overreacting is also a problem and can damage the credibility of emergency responders.

The absence of physical indicators of a hazard (i.e., smoke, vapors, odors, etc.) can influence both public and political perception of an emergency. If visible physical indicators are present, fewer individuals will initially question responder decisions. However, in the absence of such indicators, all bets are off.

Of course, the opposite can be true when there is not an emergency but physical indicators are present that may be perceived as a problem. For example, an unusually large or smoky flare stack at a refinery may be perceived by the public as an on-site problem. If the facility does not notify local public safety agencies, the public perception may be that "they're hiding something." Facilities with good community relationships have procedures for notifying the community when there are both actual emergencies and nonevents that may be perceived as an emergency by the community.

The duration of the incident can also be an influencing factor, even in the presence of a physical indicator. Hazmat releases often create significant operational and community disruptions until the incident is controlled and terminated. Public intolerance is directly proportional to the length of time citizens are inconvenienced Figure 3-14.

Working with Technical Specialists

When gathering information, public safety responders are often the true nonbelievers. They have been lied to so often they automatically question most information when it is initially presented to them. This is not necessarily a bad trait, as long you use a structured procedure to guide you through the information-gathering process. In many respects, the role of responders during the information-gathering process is similar to that of a detective.

A likely source of hazard information will be product or container specialists, or other individuals who have specific knowledge needed by responders. When evaluating these sources and the information they provide, consider the following observations:

- Individuals who are specialists in a narrow, specific technical area may not have an understanding of the broad, multi-disciplined nature of emergency response. For example, a technical specialist may provide extensive data on process engineering and design, yet may be unfamiliar with basic site safety and personal protective clothing practices.
- Some technical specialists have knowledge totally based upon dealing with a chemical or process in a structured and controlled environment. When faced with the same chemical or process in an uncontrolled

Figure 3-14 Public intolerance of response operations is directly proportional to the amount of time people are inconvenienced.

or emergency response situation, they may have a more limited framework for analyzing the problem.

- There are no experts, but only information sources! Each individual source will have its own advantages and limitations. A colleague once provided this response after being referred to an expert: "We aren't experts, but we do have good judgment. Recognize, however, that good judgment is based upon experience, and experience is often the product of bad judgment." Translation—we screwed up enough to know what will probably work and what probably won't.
- Sometimes responders interact with individuals with whom they have had no previous contact. Before relying upon their recommendations, ascertain their level of expertise and job classification by asking specific questions. Remember these two points: (1) Technical smarts is not equivalent to street smarts! Having an alphabet behind one's name (e.g., PhD., M.S., CSP, etc.) does not automatically mean an individual necessarily understands the world of emergency response and operations in a field setting; and (2) Twenty years of experience may actually be one year of experience repeated twenty times.
- Questioning information sources is an art and requires the skills of both a detective and a diplomat. While this is certainly not an interrogation process, you must be confident of the Technical Specialist's authority and expertise. Always conduct the interview with respect to the person's rank or position within the field or organization. One method is to ask questions for which you already know the answer in order to evaluate that person's competency and knowledge level. Remember, final accountability always rests with Command.
- A time-tested rule for minimizing political vulnerability is the "rule of threes." Simply, when faced with significant or politically sensitive decisions, consult at least three independent reference sources. The more politically sensitive the incident, the greater the need for the reference sources to be respected and reputable individuals. For example, using a single emergency response guidebook as the sole technical justification for evacuating 5,000 people is only asking for well-justified technical and political criticism of the emergency response effort, even if it was the correct decision. To minimize your political vulnerability, also seek input from CHEMTREC™, the shipper, or local technical information specialists.

Finally, remember that everybody brings their own agenda and scorecard to a hazmat incident. Don't assume that your concerns (1) are the most important, and (2) are always going to be the same as "their" concerns.

Everybody Has the Answer to Your Problem

Major incidents attract a great deal of attention. They also bring a number of entrepreneurs, salespeople, managers, and "do gooders" to the attention of the IC, many of them professing to have the answer to your problem. The reality of incident command is that while these people can be helpful, they are usually a major distraction, especially at long-term emergencies or if they have the ear of a local political or governmental official.

The IC may have to designate the LNO to initially address these external contacts and serve as the "gate keeper" into the ICS organization, as they occasionally do provide useful information or resources and can influence both public and political opinion. At the least, failure to seriously address these people may generate bad publicity and political backlash.

Everyone Has a Boss

Regardless of one's position within an organization, everyone has a boss. If the bosses don't have a clearly defined role within the organization's incident management/crisis management program, you may have problems as challenging as the incident itself. Don't wait until the incident to address this issue.

The Eternal Optimists

Be aware of the eternal optimists within your own ranks. Don't allow your own people to "walk you down the yellow brick road" without exploring all viable options and alternatives. Remember, initial observations often underestimate the significance of a problem. Command should always be prepared to implement an alternative action plan if the current plan fails. Remember the PACE Model for planning: have a **P**rimary Plan, an **A**lternate Plan, a **C**ontingency Plan, and an **E**mergency Plan.

Experience has shown there can sometimes be an inadequate flow of timely and accurate information from the hot zone or the incident scene. When asking subordinates for progress reports, don't ask questions that can be answered with only a simple "yes" or "no." If you only ask if everything is okay, don't be surprised when your people consistently say it is.

Responder Tip

Both the IC and the ASO/Hazardous Materials have a view of the incident that many of the other players do not. Remember the difference between "street smarts" and "technical smarts"—you don't need a degree in chemistry to know when something just doesn't look right.

Long-Term Incidents and Planning

The majority of hazmat incidents are "high intensity—short duration" events that are terminated in 8 hours or less. Activities are typically oriented towards planning ahead of current events and ensuring that the IAP is adequately staffed and resourced. Future events and needs are often assessed using a "what if" approach.

Campaign incidents extending over a period of days or weeks create different challenges for emergency responders. Issues such as developing a shift schedule, determining short-term and long-term logistical requirements, and establishing a formal IAP development flow and process are foreign to many responders.

Scan Sheet 3-B

Crew Resource Management: An ICS Tool

The term "Crew Resource Management" (CRM) was originally defined in 1977 by aviation psychologist Dr. John Lauber as "using all available resources—information, equipment and people—to achieve safe and efficient flight operations." Dr. Lauber noted that "CRM is the stuff that takes two competent, proficient pilots and turns them into one competent, proficient crew."

CRM was an outgrowth of aviation accident investigations that showed a significant percentage of losses (70%) were due to flight crew failures rather than technical problems. These failures included a lack of situational assessment and awareness, failure to follow or complete SOPs, failure to adapt to unusual or emergency situations, and interpersonal communications and coordination problems. CRM focuses upon the cognitive and interpersonal skills needed to manage the flight within an organized aviation system. These skills include the mental processes for gaining and maintaining situational awareness, for solving problems, and for making decisions.

Experience shows there are many parallels between in-flight decision making and emergency services decision making. A number of emergency response professionals are now looking at opportunities for integrating the concepts of CRM into the emergency response community. For example, a CRM training approach has been used since the early 1990s for training Offshore Installation Managers (OIMs) and their respective teams who operate on offshore oil production platforms in the North Sea. The U.S. Department of the Interior has incorporated the basic concepts of CRM into its ICS and leadership training curricula as a result of wildland fire incidents that resulted in fire fighter injuries and fatalities.

The underlying tenets of CRM are leadership and resource management.

Leadership

Leadership can directly influence the quality of the ICS organization. While the exercise of leadership is typically associated with the IC, every ICS supervisor has a certain amount of authority, responsibility, and opportunity to exercise leadership. Leadership styles will vary depending upon the situation.

CRM elements that can be used to improve leadership effectiveness include problem definition, inquiry, and advocacy.

- **Problem Definition.** The first and most critical step in problem solving is defining the problem ("A problem well-defined is half-solved."). Once the problem is understood, alternative solutions can be evaluated. Involving the ICS organization in the "diagnosis" phase increases both decision-making effectiveness and organizational "buy in."
- **Inquiry.** Human error is inevitable. Unfortunately, the cost of human error at a hazmat incident is often measured in lives. Inquiry is a mental process that involves constantly evaluating and re-evaluating everything that can be anticipated during an incident. Commonly known as "playing the devil's advocate" or "what ifs," this process can help the IC differentiate between what actually is happening or about to happen and what should be occurring. Inquiry is the responsibility of every individual within the ICS organization.
  - **Advocacy.** Advocacy is the responsibility of personnel to speak out in support of a course of action different from the one currently being planned or implemented (e.g., "I'm not comfortable with this" or "Let me push back ..."). Reasons may include technical concerns, safety issues, risk versus gain, etc. Likewise, advocacy also includes listening carefully to viewpoints that might be contradictory to one's position.

Inquiry and advocacy are essential to each other. Inquiry, which results in the detection of potential safety problems, is of little use unless the information is advocated so others can react. Advocacy is constructive questioning and should not be seen as undermining authority.

Effective ICs and ICS supervisors should work to create an environment where subordinates are encouraged to raise concerns or ask questions about a particular course of action. The input received may be used to enhance the problem definition and decision-making process. Subordinates must remember, however, that while advocacy is important, the final decision still remains with the IC or the ICS supervisor.

Resource Management

Resource management includes the effective use of emergency responder skills, knowledge, and expertise. Communications, coordination, conflict resolution, and critique are critical CRM elements that can enhance the effectiveness of both individuals and the overall ICS organization.

- *Communications* within the ICS organization is an essential prerequisite for good CRM. In addition to transferring information, it is the vehicle for both situational awareness and problem solving. Communication involves both talking and listening. To talk when no one is listening is worthless; to have deaf ears to what is being said is equally ineffective.

(continued)

Scan Sheet 3-B (continued)

One must always be sensitive to the "atmosphere" of the discussion. When the atmosphere is open, it allows for a free flow of communication and encourages input.

- *Coordination* is the process by which information, plans, and operational activities are considered and shared throughout the ICS organization. Coordination minimizes the likelihood of confusion because the players understand the IAP. It also reduces the potential for error because of overlooked or disregarded information.
- *Conflict Resolution.* Conflict is inevitable. Differences in thoughts, opinions, values, or actions—whether actual or perceived—can lead to disagreements and disputes. Differences in personality alone may create conflict. Conflict is not necessarily negative or destructive. What makes conflict "bad" is the inability to constructively cope with it, such as when the conflicting positions are passively given up or aggressively suppressed. The concept of "legitimate avenue of dissent" is critical for clearing the air and maintaining lines of communication. When effectively channeled by the IC, conflict can be transformed into an effective comparison of viewpoints, problem definitions, options, and sound solutions.
- *Critique.* Many have said, "experience is the best teacher." This may be true, but only if one takes full advantage of the lessons learned. Critique involves studying a plan of action before, during, and after its implementation. Feedback is at the heart of a critique, but don't confuse critique with criticism.

In summary, CRM is a training tool that can improve teamwork and performance when operating at an emergency. In 1989, United Airlines Captain Al Haynes, in coordination with his crew and an air traffic controller, was able to land a DC-10 aircraft damaged by an in-flight explosion in Sioux City, Iowa despite having limited control of the aircraft. He later acknowledged the role of CRM in this accomplishment:

> *"I am firmly convinced that CRM played a very important part in our being able to land at Sioux City with any chance of survival. I also believe that its principles apply no matter how many crew members are in the cockpit."*

Where can you go for help? This is the focus of the ICS 300—*Intermediate ICS for Expanding Incidents for Operational First Responders Course.* In addition, a number of regions and states have also established concepts such as Incident Support Teams (ISTs) and regional or state-based IMTs (Type 3 and 4) to assist local ICs with the management of complex or long-term emergencies.

Some Final Thoughts

The following are some random thoughts and comments pertaining to the management of a hazmat incident:

- Command must be aware of all major decisions and operations made under his/her jurisdiction. If the IC is unsure or uncomfortable with any part of the IAP or any of the information received, the strategy and tactics should be put on hold until the IC is satisfied. The IC is a risk evaluator and a resource manager.
- From the time of arrival on scene, the IC must prioritize problems and develop solutions by collecting information. The effective IC will:
  1. Seek out data that is current, accurate, and specific.
  2. Delegate information retrieval.
  3. Know where to find reference data and how to use it.
  4. Collect the right information in the right order.
  5. Use a wide variety of sources of information.
- Managing a major hazmat incident is no different than taking an army to war—the emergency response effort (Operations Section) will be no better than the information, forecasting, and technical expertise available (Planning Section), the physical and personnel resources available for timely response and support (Logistics Section), and the financial and administrative support provided (Administrative/Finance Section).
- Solicit opinions and ideas—they foster both individual and organizational "buy-in" into the decision-making process. Allow everyone (through the ICS organization) to voice opinions, particularly when dealing with situations where the hazards are exceptionally high. Remember, the people being asked to take the risks should have a voice in the decision-making process. This collective input will strengthen the decision-making

process, as well as present the IC with more options. Be careful to avoid "group-think," as voiced by those "yes-men" who surround every IC.

- Emergency responders often seek direction on how to handle a decision-making situation where the IC or a senior officer does not agree with a subordinate's recommendation, or where a senior plant manager has no on-scene responsibility and does not belong on the emergency scene. There are no easy answers here other than these two fundamental points: (1) always be professional in this situation (i.e., don't make your boss look bad!); and (2) always try to have one-to-one conversations away from the rest of the troops.
- Never say never, particularly when dealing with a long-term, campaign operation. History is full of incidents where tactical options that appeared totally unrealistic during the first hour eventually looked good and were actually implemented during the twentieth hour.
- Consider the art of communications. Effective communications is one part talking and ten parts listening. Beware of individuals whose hearing is affected by management position or promotion, as well as the "yes men" who show up at major emergencies and often flock around the IC.
- When an incident goes bad or is particularly politically sensitive, anticipate being the scapegoat. In order to minimize political vulnerability, the IC must continuously (1) consult and build a consensus within the IAP; (2) document; and (3) assume nothing.

Wrap-Up

Chief Concepts

- The successful management of a hazmat incident is directly linked to the rapid development of an effective incident management process and organization. From the arrival of the first emergency responder, the IC must match and balance the size and structure of the ICS organization with the number of units and organizations present on the incident scene. A functional hazmat command system must allow the IC to use the standard ICS elements to establish and maintain control, to be responsive to the unique hazards and risks of each incident, and, finally, to apply the same ICS system and process to every incident, regardless of its nature and size.
- A variety of different players will respond to a working hazmat incident. Therefore, what occurs during the planning and preparedness phase will establish the framework for how the emergency response effort will operate.
- There is no single organization that can effectively manage a major hazmat incident. Organizations that attempt to maintain their usual structure or bureaucracy while managing a major event will have inherent problems in implementing a timely and effective emergency response.
- Emergency response programs can be categorized as being either "people-dependent" or "system-dependent." Special operations teams are often very people-dependent when they are initially formed. These organizations rely on the experience of a few key individuals and can result in failed emergency response efforts if these key individuals are not present at an incident.
- If ICS is not used for all routine emergencies, don't expect the ICS structure to function and adapt effectively and safely when a major emergency occurs. The routine establishes the foundation on which the nonroutine must build. The more routine decisions made prior to a major emergency, the more time the IC will have to make the critical decisions during the emergency.
- The Hazardous Materials Group is normally under the command of a senior Hazmat Officer (often known as the Hazardous Materials Group Supervisor), who, in turn, reports to the Operations Section Chief or the IC. The Hazardous Materials Group is directly responsible for all tactical hazmat operations that occur in the hot and warm zones of an incident.
- The Hazardous Materials Group Supervisor is responsible for the management and coordination of all functional responsibilities assigned to the Hazardous Materials Group, including safety, site control, research, entry, and decontamination.
- The ASO/Hazardous Materials is responsible for coordinating safety activities within the Hazardous Materials Group but also has certain responsibilities that may require action without initially contacting the normal chain of command. The ISO is responsible for the safety of all personnel operating at the incident, while the ASO/Hazardous Materials is responsible for all operations within the Hazardous Materials Group and within the hot and warm zones. This includes having the authority to stop or prevent unsafe actions and procedures during the course of the incident.
- Your emergency response performance will be evaluated on two interrelated factors: (1) the implementation of a timely, well-trained and equipped emergency response effort in the field, and (2) the effective management of the interpersonal, organizational, and external impacts created by the incident. An effective response effort can be compromised or completely negated by poor management of the political and external issues.

Hot Terms

Hazardous Materials Group Normally under the command of a senior Hazmat Officer and the Operations Section Chief, the Hazardous Materials Group is directly responsible for all tactical hazmat operations that occur in the hot and warm zones of an incident.

Incident Action Plan (IAP) A plan that includes developing an effective organizational structure, developing an incident strategy and tactical action plan, allocating resources, making appropriate assignments.

Incident Commander (Command or IC) The individual responsible for establishing and managing the overall incident action plan (IAP).

National Incident Management System (NIMS) The baseline incident management system established under HSPD-5, which is used by federal, state, and local governments, as well as private sector organizations throughout North America.

Unified Command (UC) Command-level representatives from each of the primary responding agencies who present their agency's interests as a member of a unified command organization.

HazMat Responder *in Action*

You are rewriting the policy and procedure manual for your agencies HazMat Response team. You are looking for compliance with national standards and national documentation. Your primary task is to review and revise the section on the Incident Command System. Your findings raise some questions:

Questions

1. In years past, your department has used an assortment of terminology to describe Incident Management structure. What are the four main sections within the ICS plan?
   A. Division/Group/Sectors/Elements
   B. Operations/Planning/Logistics/Administration
   C. Command/HazMat Branch/Sector/Operations
   D. Decontamination/Medical/Entry/Safety
2. What are the main positions under the HazMat group?
   A. Site safety/Logistics/planning/Sector control/Incident Action planning
   B. Assistant Safety Officer/Hazmat/Entry/Decontamination/Site Access/Hazmat Reference/Science
   C. HazMat Group/Technical Safety/Entry Sector/Planning Group
   D. Safety Officer/Entry/Decontamination/Liaison/Technical Specialist
3. What position is the point of contact for assisting and cooperating with external representatives?
   A. Safety Officer
   B. Incident Commander
   C. Assistant Incident Commander
   D. Liaison Officer
4. Which command style is defined by a lead agency or is an agency that has 51% of the vote as compared to other players at a HazMat incident?
   A. Consolidated Incident Command
   B. Incident Command System
   C. Unified Command Structure
   D. Lead Agency Command

References and Suggested Readings

1. Air Force Civil Engineer Support Agency (AFCESA) and PowerTrain, Inc., *Hazardous Materials Incident Commander Emergency Response Training CD*. Tyndall Air Force Base, FL: Headquarters AFCESA (2010).
2. BP Exploration (Alaska) Inc., *BP Seals*—Skills Enhancement and Leadership Seminar Materials. Anchorage, AK: BP Exploration (Alaska)—Crisis Management Center (2009).
3. Brunacini, Alan V., and Nick Brunacini, *Command Safety*. Phoenix, AZ: Across the Street Productions (2004).
4. Brunacini, Alan V., *Fire Command* (2nd edition), Quincy, MA: National Fire Protection Association (2002).
5. ConocoPhillips Inc., *TIGER—Training for Integrated Group Emergency Response Materials*. Anchorage, AK: ConocoPhillips (Alaska)—Crisis Management Center (2009).
6. Daimler, Gottlieb, and Karl Benz Foundation et al. *Workshop On Group Interaction In High Risk Environments*. Zurich, Switzerland: Gottleib Daimler and Karl Benz Foundation (July 5–6, 2001).
7. Department of Homeland Security (DHS), *The National Response Framework*. Washington, DC: Department of Homeland Security (2008).
8. Emergency Film Group, *Industrial Incident Management* (videotape). Plymouth, MA: Emergency Film Group (2009).
9. Flin, Rhona, *Sitting In the Hot Seat—Leaders and Teams for Critical Incident Management*. New York: John Wiley & Sons (1996).
10. Flin, Rhona, et al, *Decision Making Under Stress—Emerging Themes and Applications*. Aldershot, England: Ashgate Publishing Ltd (1997).
11. LeSage, Paul, Jeff T. Dyar, and Bruce Evans, *Crew Resource Management: Principles and Practices*. Sudbury, MA: Jones and Bartlett Learning (2011).
12. National Fire Protection Association, *Hazardous Materials Response Handbook* (6th edition). Quincy, MA: National Fire Protection Association (2013).
13. National Fire Protection Association, *NFPA 1500—Fire Department Occupational Safety and Health Program Handbook*. Quincy, MA: National Fire Protection Association (2007).
14. National Fire Protection Association, *NFPA 1561—Technical Standard On Emergency Services Incident Management Systems*. Quincy, MA: National Fire Protection Association (2008).
15. National Fire Service Incident Management System Consortium Model Procedures Committee, *IMS Model Procedures Guide For Hazardous Materials Incidents*. Stillwater, OK: Fire Protection Publications, Oklahoma State University (2000).
16. National Institute for Occupational Safety and Health (NIOSH), *Occupational Safety and Health Guidance Manual for Hazardous Waste Site Activities*. Washington, DC: NIOSH, OSHA, USCG, EPA (1985).
17. Royal Aeronautical Society—Crew Resource Management Standing Group, "A Paper on Crew Resource Management." London, England: Royal Aeronautical Society (October 1999).
18. Tippett, John, *Crew Resource Management—A Positive Change for the Fire Service*. Fairfax, VA: International Association of Fire Chiefs (2002).

The Eight Step Process©: An Overview

CHAPTER 4

Knowledge Objectives

After reading this chapter, you will be able to:

- Describe the Eight Step Process© and its application as a tactical incident management tool for managing on-scene operations at a hazardous materials incident. (p. 98–107)
- Describe the critical success factors in managing the first hour of a hazardous materials incident. (p. 98–99)

Skills Objectives

There are no skills objectives in this chapter.

You Are the HazMat Responder

You are a regional Hazmat Officer who has been requested to respond to an incident involving a suspected illicit drug laboratory. Both fire service and law enforcement personnel are already on-scene and a suspected lab operator is now under arrest. The suspected laboratory is located on the 3rd floor of a 4-story multi-family (i.e., garden apartment) building. At this time, no entry operations have taken place and emergency responders are in the process of developing their Incident Action Plan (IAP). As the Hazmat Officer, both fire and law enforcement leadership are looking for recommendations on how to proceed.

1. What are your incident priorities?
2. What are your tactical priorities?
3. Identify any additional resources or specialized units that might be required as part of entry and control operations.

Introduction

On-scene response operations must always be based on a structured and standardized system of protocols and procedures. Regardless of the nature of the incident and response, a reliance on standardized procedures will bring consistency to the tactical operation. If the situation potentially involves hazardous materials or weapons of mass destruction (WMD), this reliance on standardized tactical response procedures will help to minimize the risk of exposure to all responders.

This chapter will introduce the reader to a system for the tactical management of incidents or emergencies that may involve hazardous materials or related special operations scenarios. This includes hazardous materials, hazardous substances, the criminal use of hazardous materials, and materials capable of being used as WMD agents.

The **Eight Step Process©** is widely used throughout the country by governmental and private sector hazardous materials response teams for the tactical management of hazardous materials emergencies. It also serves as an example of a structured system that can be used by law enforcement and special operations personnel at incidents involving hazardous substances and materials.

In eight chapters of this text, we have dedicated one chapter to each step in the Eight Step Process©. This material is further expanded upon with structured checklists and procedures included in the companion *Field Operations Guide* (*FOG*). The FOG also includes an incident commander's tactical worksheet to guide the user through the incident management process.

Making the Transition

Experience has shown that the critical success factors in hour 1 of a hazmat response will typically be (1) the ability to establish command and control in a timely manner; (2) the ability of responders to recognize "clues" that indicate the incident may involve hazardous materials; and (3) the ability of responders to quickly gain control of the incident scene and separate bystanders from the problem.

The ability of emergency responders initially to size-up and assess the clues that hazardous materials may be involved starts with the quantity and quality of information provided by Communications (or Dispatch). What do responders need to know while en route? Basic considerations during the alerting, notification, and initial response phase include the following:

- Where is the location of the incident? Is the incident at a target occupancy or target hazard event? Is there a pre-incident plan for the location?
- Based on available clues (e.g., container shapes, markings, placards), are hazmats involved? If unsure, are hazmats manufactured, stored, or used at the location?
- Are there reports or physical clues of any unusual odors? Explosions? Smoke or vapor clouds? Hazardous materials?
- Are any injuries or casualties involved? Are the reasons known or unknown?
- Are any initial responders down?
- Is there any suspicious activity in the area? Are Law Enforcement units also on scene?

These factors will assist responders and the Incident Commander (IC) in determining whether to follow "normal" or hazmat response protocols.

The Eight Step Process©

The Eight Step Process© outlines the basic tactical functions to be evaluated and implemented at incidents involving hazardous materials. Like all SOPs, the Eight Step Process© should be viewed as a flexible guideline and not as a rigid rule. Individual departments and agencies should decide what works best for them.

The Eight Step Process© offers several benefits. First, it recognizes that the majority of incidents involving hazardous materials are minor in nature and generally involve limited quantities. It also builds on the action of first responding units and facilitates identifying the roles and responsibilities of each level of response. The Eight Step Process© provides a flexible management system that expands as the scope and magnitude of the incident grows and, finally, it provides a consistent management structure, regardless of the classes of hazardous materials involved.

Essentially, there are eight basic tactical-level functions that must be evaluated at emergencies involving, or suspected of involving, hazmats or WMD agents. These eight functions typically follow an implementation timeline at the incident:

1. Site Management and Control
2. Identify the Problem
3. Hazard Assessment and Risk Evaluation
4. Select Personal Protective Clothing and Equipment
5. Information Management and Resource Coordination
6. Implement Response Objectives
7. Decontamination (Decon) and Clean-Up Operations
8. Terminate the Incident

Step 1: Site Management and Control

FUNCTION: Site management and control involves managing and securing the physical layout of the incident Figure 4-1. The operational reality is you cannot safely and effectively manage the incident if you do not have control of the scene. As a result, site management and control is a critical benchmark in the overall success of the response and is the foundation on which all subsequent response functions, strategies, and tactics are built.

Figure 4-1 Step 1 involves managing and securing the physical layout of the incident.

GOAL: Establish the playing field so that all subsequent response operations can be implemented safety and effectively.

CHECKLIST:

- ❑ During the initial approach to the incident scene, avoid committing or positioning personnel in a hazardous position. Assess the situation first and strive to have an escape route out of the area if conditions should deteriorate suddenly.

> CAUTION: Emergency responders must be aware that certain chemical releases may travel throughout the scene and impact response routes. In addition, some chemicals may produce vapor clouds that may be mistaken for fog or other normal weather and environmental conditions (e.g., anhydrous ammonia, liquefied petroleum gas or LPG). The danger area may extend beyond the visible vapor cloud.

- ❑ Establish command of the incident and establish an Incident Command Post (ICP). If other public safety units are already on scene, ensure that operations are coordinated and unified.
- ❑ Establish a Staging Area (Level I, II) for additional responding equipment and personnel.
- ❑ Establish an isolation perimeter (i.e., outer perimeter) to isolate the area and deny entry. Establish access control and traffic control points—restrict emergency site access to authorized essential personnel only; all nonessential personnel should be isolated from the problem. Isolation perimeters should include land, water, and air areas.
- ❑ Establish a hot zone or inner perimeter as the "playing field." The location of the inner hot zone should be identified and communicated to all personnel operating at the site. Methods of identifying the restricted area include visible landmarks, barricade tape, traffic cones, and so on.
- ❑ Do not attempt to enter the area without the appropriate level of respiratory and skin protection, based on the hazards present. If victims are down, SCBA will be considered the minimum level of respiratory protection for initial emergency response operations. Depending on the scenario and materials involved, structural firefighting clothing may not provide sufficient personal protection.

Responder Tip

- Site Management establishes the playing field for the players (responders) and the spectators (everyone else).
- The initial 10 minutes of the incident will determine operations for the next 60 minutes, and the first 60 minutes will determine operations for the next 8 hours.
- Don't try to control more real estate than you can effectively isolate and control. Smaller and tighter may be better than bigger and looser.
- Remember the basics. The more time, distance, and shielding between you and the material, the lower the risk will be. Don't get the stuff on you!
- Designate an emergency evacuation signal and identify rally points if emergency evacuation is necessary.
- Remember the first law of hot zone operations when dealing with hazardous materials: To play in the game you must:
  - Be trained to play
  - Be dressed to play
  - Have a buddy system with backup personnel (minimum of 2 in/2 out)
  - Have decon established
  - Coordinate with Command and Safety

- ❑ If civilians are injured and personal contamination is suspected, isolate all personnel until emergency decon can be established.
- ❑ Initiate public protective actions (PPA).

Step 2: Identify the Problem

FUNCTION: Identify the scope and nature of the problem. This includes:

- Recognition, identification, and verification of the hazardous materials/WMD involved in the incident
- Type of container, as applicable
- Exposures

Methods of identification include analyzing container shapes, markings, labels and placards, and facility documents (e.g., pre-incident plans, Tier 2 reporting forms, Safety Data Sheets or SDS); using monitoring and detection equipment; and identifying by the senses (i.e., physical observations, smell, reports from victims, etc.). Responders should remember that even when the hazardous substances have been initially identified, the information should always be verified **Figure 4-2**.

Figure 4-2 The goal of Step 2 is to identify the scope and nature of the problem, including the type and nature of hazardous materials involved.

GOAL: Identify the scope and nature of the problem, including the type and nature of hazardous materials involved.

CHECKLIST:

- ❑ Survey the incident—identify the nature and severity of the immediate problem, including the recognition, identification, and verification of the material(s) involved, type of container involved, and any potential or existing life hazards. If multiple problems exist, prioritize them and make independent assignments.
- ❑ Clues for determining the identity of the materials involved include:
  - Occupancy and location
  - Container shapes
  - Markings and colors
  - Placards and labels
  - Shipping papers or facility documents
  - Monitoring and detection equipment
  - Senses, including physical observations, smell, and so on.
- ❑ Factors to consider in assessing the type of container involved include:
  - Bulk versus nonbulk
  - Pressurized versus nonpressurized
  - Number of compartments
  - Material(s) of construction
  - Container valves, piping, and pressure relief devices
- ❑ Conduct offensive or defensive reconnaissance, as necessary, to gather intelligence (or intel) on the situation.

Responder Tip

- A problem well defined is half solved.
- Assume initial information is not correct. Always verify your initial information.
- Conduct recon operations, as necessary.

Defensive Recon. Objective is to obtain information on site layout, containers, physical hazards, access, and other related conditions from beyond the inner perimeter. This is normally obtained through threat assessments, interviews, physical observations, and so on.

Offensive Recon. Objective is to obtain intel and incident information by physically entering the inner perimeter. At incidents that involve both hazmat and explosive risks, this may be a joint-entry operation with both bomb squad and hazmat personnel to conduct monitoring, sampling, and video or photo documentation for analysis or evidence.

- Be alert for the presence of secondary events and improvised explosive devices (IEDs). IED clues can include:
  - Abandoned container out of place for the surroundings
  - Obvious explosive device components, such as batteries, timers, blasting caps, charges, and wires taped to containers.
  - Trip wires, motion sensors, or partially exploded devices
  - Unusual or foreign devices attached to hazmat containers, especially liquefied and compressed gas cylinders, flammable liquid containers, and bulk storage tanks and vessels
  - Unattended vehicles not appropriate to the immediate environment

Step 3: Hazard Assessment and Risk Evaluation

FUNCTION: This is the most critical function that public safety personnel perform. The primary objective of the risk evaluation process is to determine whether or not responders should intervene, and what strategic objectives and tactical options should be pursued to control the problem **Figure 4-3**. You can't get this wrong. If you lack the expertise to do this function adequately, get help from someone who can provide that assistance, such as local HMRTs and product/container technical specialists.

Figure 4-3 Step 3, assessing hazard and risk, is the most critical function that public safety personnel perform.

GOAL: Assess the hazards present, evaluate the level of risk, and establish an Incident Action Plan (IAP) to make the problem go away.

CHECKLIST:

- ❑ Assess the hazards posed by the problem (health, physical, chemical, weapons, other).
- ❑ Collect, prioritize, and manage hazard data and information from all sources, as appropriate, including:
  - Technical reference manuals
  - Technical information sources
  - Hazmat databases
  - Technical information specialists
  - Safety Data Sheets (SDS)
  - Monitoring and detection equipment
- ❑ Primary technical information centers available to public safety personnel include:
  - CHEMTREC—(800) 424-9300
  - National Response Center (NRC), which serves as the federal single point of contact for accessing federal assistance
  - State single point of contact
  - Product or container specialists
  - State and regional poison control centers
- ❑ Air monitoring and the *General Hazmat Behavior Model* are critical in implementing a "risk-based response." Understand the relationship between the container and the hazmat(s) involved:
  - Stress
  - Breach
  - Release
  - Engulf
  - Impingement/Contact
  - Harm
- ❑ Based on the risk evaluation process, develop your IAP. Determine whether the incident should be handled offensively, defensively, or by nonintervention.

Strategy	Offensive	Defensive	Nonintervention
Rescue	X		
Public protective actions	X	X	X
Spill control	X	X	
Leak control	X		
Fire control	X	X	
Clean-up and recovery	X	X	

Responder Tip

- Focus on those things you can change and that will have a positive impact on the outcome.
- Every incident will have an outcome, with or without your help. If you can't change the outcome in a positive manner, why get involved? Doing nothing and letting the incident take its course without intervention may be a viable option.
- There's nothing wrong with taking a risk. If there is much to be gained, there is much to be risked. If there is little to be gained, then little should be risked.
- Public safety personnel should view their roles as risk evaluators, rather than risk takers, where hazardous materials are involved. Bad risk takers get buried; good risk evaluators come home.
- Hour 1 priorities within the IAP are as follows:
  - Establish Site Management and Control.
  - Determine the materials/agents involved.
  - Ensure the safety of all personnel from all hazards.
  - Ensure PPE is appropriate for the hazards.
  - Initiate tactical objectives to accomplish initial rescue, decon, medical, and public protective action needs.
  - If criminal activities are involved (e.g., terrorism incidents), maintain the integrity of potential evidence.
- Remember the PACE model of planning—Primary plan, Alternate plan, Contingency plan(s), and Emergency plan.

Remember that offensive tactics increase the risks to emergency responders.

- **Offensive tactics**—Require responders to control/mitigate the emergency from within the area of high risk.
- **Defensive tactics**—Permit responders to control/mitigate the emergency remote from the area of highest risk.
- **Nonintervention tactics**—Pursue a passive attack posture until the arrival of additional personnel or equipment, or allowing the fire to burn itself out.

Step 4: Select Personal Protective Clothing and Equipment

FUNCTION: Based on the results of the hazard and risk assessment process, emergency response personnel will select the proper level of personal protective clothing and equipment Figure 4-4. Two primary types of personal protective clothing are commonly used at hazmat incidents: (1) structural firefighting protective clothing, and (2) chemical protective clothing.

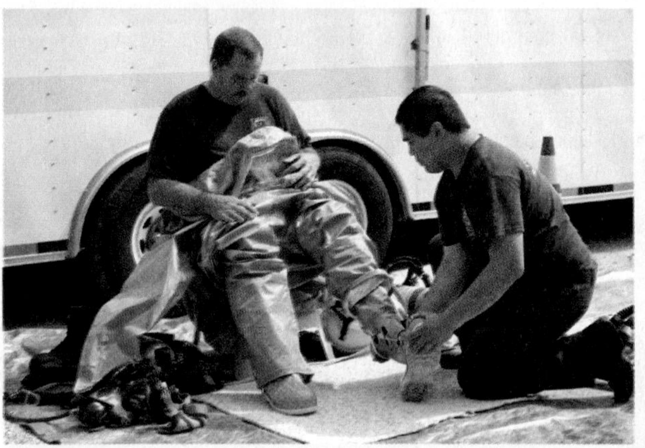

Figure 4-4 The goal of Step 4 is to ensure all emergency response personnel have the appropriate level of PPE for the expected tasks.

GOAL: Ensure all emergency response personnel have the appropriate level of personal protective clothing and equipment (skin and respiratory protection) for the expected tasks.

CHECKLIST:

- ❑ The selection of personal protective clothing will depend on the hazards and properties of the materials involved and the response objectives to be implemented (i.e., offensive, defensive, or nonintervention). In evaluating the use of specialized protective clothing, the following factors must be considered:
  - The hazard to be encountered, including the specific tasks to be performed. If victims are present, their signs and symptoms can provide clues that will influence the selection of PPE.
  - The tasks to be performed (e.g., entry, decon, support)
  - The level and type of specialized protective clothing to be used
  - The capabilities of the individual(s) who will use the PPE in a hostile environment. Remember—specialized protective clothing places a great deal of both physiological and psychological stress on an individual.
- ❑ The following levels of personal protective clothing are typically utilized by emergency responders at hazmat incidents, as appropriate:
  - **Structural firefighting clothing**—Includes helmet, fire retardant hood, turnout coat and pants, personal alert safety system (PASS device), and gloves. Positive-pressure SCBA should be considered the minimum level of respiratory protective clothing.
  - **Chemical vapor protective clothing**—This is specialized chemical protective clothing, which when used in conjunction with air supplied respiratory protection devices offers a sealed, integral level of full-body protection from a hostile environment. It is primarily designed to offer protection from both chemical gases and vapors, as well as total body splash protection. It may also be referred to as EPA Level A chemical protective clothing or NFPA 1991 compliant clothing.
  - **Chemical splash protective clothing**—This is specialized protective clothing that protects the wearer against chemical liquid splashes but not against chemical vapors or gases. It is primarily designed to provide personal protection against liquid splashes, solids, dusts, and particles. It can be found in both single- and multipiece garment ensembles and may be referred to as EPA Level B chemical protective clothing when air supplied respiratory protection is provided or EPA Level C chemical protective clothing when air-purifying respirators (APR) are provided. Based on the garment, its design and chemical resistance may also be referred to as NFPA 1992 or NFPA 1994 compliant clothing.
- ❑ Ensure all emergency response personnel are using the proper protective clothing and equipment equal to the hazards present. If the incident involves CBRN agents, ensure both skin and respiratory protection can provide the required level of personal protection.
- ❑ Do not place personnel in an unsafe emergency situation or location.

Responder Tip

- Remember that structural firefighting protective clothing is not designed to provide protection against chemical hazards.
- With some notable exceptions, there is no one single barrier that will effectively combine both chemical and thermal protection.
- Wearing any type and level of impermeable protective clothing creates the potential for heat stress injuries.
- Personal protective clothing is NOT your first line of defense . . . it is your last line of defense!

- ❑ Order additional personnel and other specialized equipment and technical expertise early in the incident. If you are unsure of your requirements, always call for the highest level of assistance available. Do not wait to call for emergency assistance—it's always better to have too much assistance than not enough.

Step 5: Information Management and Resource Coordination

FUNCTION: Refers to proper management, coordination, and dissemination of all pertinent data and information within the ICS in effect at the scene. In simple terms, this function cannot be effectively accomplished unless a unified ICS organization is in place. Of particular importance is the ability to determine the incident factors involved, which functions of the Eight Step Process© have been completed, what additional information must be obtained, and what incident factors remain unknown.

GOAL: Provide for the timely and effective management, coordination, and dissemination of all pertinent data, information, and resources among all of the players.

CHECKLIST:

- ❑ Confirm the ICP is in a safe area.

> *NOTE:* Personnel not directly involved in the overall command and control of the incident should be removed from the ICP area.

- ❑ Confirm a unified command organization is in place and all key response and support agencies are represented directly or through the Liaison Officer.
- ❑ Expand the ICS and create additional branches, divisions, or groups, as necessary.
- ❑ Ensure all appropriate internal and external notifications have been made. Coordinate information and provide briefings to other agencies, as appropriate.
- ❑ Ensure safety procedures are part of all tactical briefings. Confirm emergency orders and follow through to ensure they are fully understood and correctly implemented. Maintain strict control of the situation.
- ❑ Make sure there is continuing progress toward solving the emergency in a timely manner. Do not delay in calling for additional assistance if conditions appear to be deteriorating.
- ❑ If activated, provide regular updates to the local/facility Emergency Operations Center (EOC).

Responder Tip

- Consider the security of the ICP and all other incident response areas (e.g., Staging, Rehab) of the incident.
- Don't look stupid because you didn't have a plan.
- Bad news doesn't get better with time. If there's a problem, the earlier you know about it the sooner you can start to fix it.
- Don't allow external resources to "freelance" or do an "end run" and develop their own plan.
- Don't let your lack of a Planning Section become the Achilles' heel of your response. Establish it early, particularly if the incident has the potential to become a "campaign event."
- Operations may win battles, but logistics wins wars. Incidents, which have specialized or substantial resource needs, require the Logistics Section to be activated as early as possible.

Step 6: Implement Response Objectives

FUNCTION: The phase where emergency responders implement the best available strategic goals and tactical objectives, which will produce the most favorable outcome. If the incident is in the emergency phase, this is where we "make the problem go away." Common strategies to protect people and stabilize the problem include rescue, public protective actions, spill control, leak control, fire control, and recovery operations. In simple terms, these strategies are typically implemented by fire, rescue, and hazardous materials units with law enforcement responsible for all security and criminal-related issues.

If the incident is in the post-emergency response phase, the focus of response personnel will likely become scene safety, clean-up, evidence preservation (as appropriate), and incident investigation. Specific tasks will include (1) initial site entry and monitoring to determine the extent of the hazards present; (2) an evaluation of the scene to locate evidence that can be used to reconstruct the events leading up to the incident; (3) identification of the contributing factors that caused the incident; (4) interviewing on-scene personnel and witnesses to corroborate the information obtained and opinions formed based on the available data; and (5) documentation of preliminary results Figure 4-5.

Figure 4-5 The goal of Step 6 is to ensure the incident priorities are accomplished in a safe, timely, and effective manner.

GOAL: Ensure the incident priorities (i.e., rescue, incident stabilization, environmental and property protection) are accomplished in a safe, timely, and effective manner.

CHECKLIST:

- ❑ Implement response objectives. Remember that offensive tactics increase the risks to emergency responders; evaluate the risks of offensive control tactics before sending emergency response crews into the hazard area.
  - **Offensive tactics**—Require responders to control/mitigate the emergency from within the area of high risk.
  - **Defensive tactics**—Permit responders to control/mitigate the emergency remote from the area of highest risk.
  - **Nonintervention tactics**—Pursue a passive attack posture until the arrival of additional personnel or equipment, or allowing the fire to burn itself out.

Strategy	Offensive	Defensive	Nonintervention
Rescue	X		
Public protective actions	X	X	X
Spill control	X	X	
Leak control	X		
Fire control	X	X	
Clean-up and recovery	X	X	

NOTE: Rapidly changing incident conditions may require using multiple tactics simultaneously or switching from one tactic to another. Defensive tactics are always desirable over offensive tactics if they can accomplish the same objectives.

- ❑ Ensure properly equipped backup personnel wearing the appropriate level of personal protective clothing are in-place before initiating operations.
- ❑ Ensure any detection or monitoring equipment is operational and properly calibrated.
- ❑ Ensure entry teams have been briefed prior to being allowed to enter the hot zone. For hazardous materials emergencies, this should include the following:
  - All watches, jewelry, and personal valuables (e.g., wallets, cell phones) have been removed
  - Objectives of the entry operation
  - Known hazards within the area, including physical hazards (e.g., downed powerlines, stairwells, open pits)
  - Safety procedures, including medical, radio communications, SCBA, and PPE checks
  - Emergency procedures, including placement of backup teams, escape signals, and escape corridors
  - Decon area location, set-up, and procedures
- ❑ Conduct regular monitoring of the hazard area to determine if conditions are changing.

CAUTION: Decon should be set up prior to initiating entry operations. For situations where civilians are down and chemical exposures are suspected, emergency decon must be established as soon as possible.

Responder Tip

- Positive pressure ventilation (PPV) can be used to significantly reduce chemical vapor levels within a building and increase the safety of emergency responders who must enter a structure to perform rescue operations. Remember to keep exposure hazards downwind, if applicable.
- Never touch or handle anything in a clandestine lab operation until the area has been evaluated and cleared by bomb squad personnel who have proper training in identifying booby traps.
- What will happen if I do nothing? Remember—this is the baseline for hazmat decision making and should be the element against which all strategies and tactics are compared.
- There is a very fine line between explosives, oxidizers, and organic peroxides. All are capable of releasing tremendous amounts of energy.
- Surprises are nice on your birthday but not on the emergency scene. Always have a Plan B in case Plan A doesn't work!

Step 7: Decon and Clean-Up Operations

FUNCTION: **Decontamination** (decon) is the process of making personnel, equipment, and supplies "safe" by reducing or eliminating harmful substances (i.e., contaminants) that are present when entering and working in contaminated areas (i.e., hot zone or inner perimeter). Although decon is commonly addressed in terms of "cleaning" personnel and equipment after entry operations, in some instances, due to the nature of the materials involved, decon of clothing and equipment may not be possible and these items may require disposal.

All personnel trained to the First Responder Operations level should be capable of performing or delivering an emergency decon capability. At most "working" hazmat inci-

Voices of Experience

On a warm Saturday summer afternoon, my department was dispatched to a truck fire on a very busy state highway. According to our Standard Operating Procedures, this type of assignment requires a response of two engine companies. As the first due engine (E1101) approached the scene, the Captain stated in his "First In Report" that there was a single axle MC-331 propane cargo truck off the side of the road with the engine hood open and no smoke or fire showing. The engineer positioned E1101 to protect the scene and awaited further orders should handlines need to be deployed.

The Captain maintained a mobile command and ordered the second due engine (E1110) to position behind E1101 should a resupply be necessary. All firefighters were in their full personal protective equipment. As the Captain approached the propane truck he detected a slight smell off propane gas and the driver standing next to the truck. The closer he got to the tank truck, the stronger the smell of propane. The Captain then cleared the area around the truck, determined there were no ignition sources, and ordered a 1¾-inch handline be advanced upwind from the incident as a safety precaution. He then requested additional resources and that the highway be shut down to all traffic.

As the Assistant Chief and Department Training Officer, I responded to the scene based on the request for additional resources. Once I arrived on-scene the Captain gave me a full briefing of the incident and what had been done. I assumed Command and placed the Captain in charge of Operations. We then met with the company owner of the propane truck to try to determine the source of the leak. He informed us that a valve was replaced a few days earlier and this could be the source of the leak.

A state Department of Public Safety Hazardous Materials Technician arrived on-scene and was briefed as well. We developed an Incident Action Plan for isolating the leaking valve, incorporating all resources we had on-scene. After successfully closing the main internal valve on the truck, along with constant air monitoring, it was determined our plan was effective. Our next task was to conduct air monitoring in the low areas around the truck as propane is heavier than air. Our readings turned up normal and we determined that we successfully mitigated the hazard.

By using the Eight Step Process© we were able to mitigate the incident quickly and ensure life safety throughout. It is important to use all your resources when dealing with a potential hazardous material incident. Both engine crews had expected this incident to be nothing more than a truck fire, and yet it turned out to be a Hazardous Materials incident.

Jeffry J. Harran
Buckskin Fire Department
Parker, Arizona

dents, technical decon services will be provided by HMRTs or fire and rescue units working under the direction of a Hazmat Technician. Questions pertaining to disposal methods and procedures should be directed to environmental officials and technical specialists, based on applicable federal, state, and local regulations.

GOAL: Ensure the safety of both emergency responders and the public by reducing the level of contamination on scene and minimizing the potential for secondary contamination beyond the incident scene.

CHECKLIST:

- ❑ Ensure the technical decon operations are coordinated with tactical operations. This should include the following tasks:
  - The decon area is properly located within the warm zone, preferably upslope and upwind of the incident location.
  - The decon area is well marked and identified.
  - The proper decon method and type of personal protective clothing to be used by the Decon Team have been determined and communicated, as appropriate.
  - All decon operations are integrated within the ICS organization.
- ❑ Incidents involving large numbers of contaminated or potentially contaminated individuals will require the establishment of a mass decon operation. Basic principles of mass casualty decon include the following:
  - Anticipate a five to one ratio of unaffected to affected casualties (in simple terms, for every victim who is physically contaminated there will be five others who may be exposed or who think they are contaminated). Bottom line—lots of folks may be seeking attention.
  - Decontaminate ASAP. It may be necessary to "corral" those contaminated until a mass decon corridor is established.
  - Disrobing is decon. Top to bottom, more is better.
  - Water flushing for a period of 3 minutes is generally the best mass decon method (Note: SBCCOM findings for chemical agents).
  - After a known exposure to a hazardous material, responders must decon ASAP to avoid serious effects.
- ❑ Ensure proper decon of all personnel before they leave the scene. For example, flammable gases and some toxic and corrosive gases can saturate protective clothing and be carried into "safe" areas.
- ❑ Establish a plan to clean up or dispose of contaminated supplies and equipment before cleaning up the site of a release. Federal and state laws require proper disposal of hazardous waste.

Responder Tip

- Establishing an emergency decon capability should be part of the Incident Action Plan for any incident where hazardous materials are involved (e.g., clandestine drug labs, SWAT tactical operations).
- Decon involving large numbers of people will be a challenge and will make for a long day. Remember the basics—separate people from the problem and keep them corralled until mass decon is established. It won't be easy and it might be ugly!
- Permeation can occur with any porous material, not just PPE. That includes shoes, belts, fire hose, holsters, gun stocks, and so on.
- Sodium hypochlorite (i.e., bleach) may be used as a decon agent for equipment when dealing with chemical and biological materials. Be aware that bleach will degrade some materials (e.g., Kevlar™) and will shorten the life of the material.
- Degradation chemicals, such as bleach, should never be applied directly to the skin.
- Never transport contaminated victims from the scene to any medical facility without conducting field decon. At best, healthcare workers will be mad at you; at worst, you will endanger other people and completely shut down the emergency room of the hospital.
- Law enforcement operations at incidents where hazmats are involved create some unique challenges:
  - Be aware of "Bad Guys" possibly being mixed in with civilians when conducting mass decon operations.
  - Establish a weapons safety officer as part of decon to ensure all weapons are properly handled and rendered safe.
  - Ensure security procedures are in place when conducting decon of contaminated prisoners.

Step 8: Terminate the Incident

FUNCTION: This is the termination of emergency response activities and the initiation of **post-emergency response operations (PERO)**, including investigation, restoration, and recovery activities. This would include the transfer of command to the agency that will be responsible for coordinating all post-emergency activities.

GOAL: Ensure overall command is transferred to the proper agency when the emergency is terminated and all post-incident administrative activities are completed per local policies and procedures.

CHECKLIST:

- ❑ Account for all personnel before securing emergency operations.
- ❑ Conduct incident debriefing session for on-scene response personnel. Provide background information necessary to ensure the documentation of any exposures.
- ❑ Ensure command is formally transferred from the lead response agency to the lead agency for all post-emergency response operations.

- ❑ Ensure public safety units are aware of any hazards remaining at the incident scene in the event emergency responders must return to the scene.
- ❑ Ensure the following elements are documented:
  - All operational, regulatory, and medical phases of the emergency, as appropriate.
  - All equipment and supplies used during the incident.
  - The names and telephone numbers of all key individuals. This should include contractors, public officials, and members of the media.
- ❑ Ensure all emergency equipment is decontaminated, reserviced, inspected, and placed back in service. Provide a point of contact for all post-incident questions and issues.
- ❑ Conduct a critique of all major and significant incidents based on local protocols.

Responder Tip

- Although every organization has a tendency to develop its own critique style, never use a critique to assign blame; public meetings are the worst time to discipline personnel.
- Organizations must balance the potential negatives against the benefits gained through the critique process. Remember—the reason for doing the critique in the first place is to improve your operations.
- Most critiques fall into one of three categories: (1) We lie to each other about what a great job we just did; (2) we beat up each other for screwing up; or (3) we focus on the lessons learned and the changes/improvements that must be made to our response system.

Wrap-Up

Chief Concepts

- Emergency response operations at incidents involving hazardous materials must always be based on a structured and standardized system of response protocols and procedures. Regardless of the nature of the incident, the nature of the response, or the personnel involved, a reliance on standardized procedures will bring consistency to the tactical operation and facilitate the delivery of a risk-based operational response capability. If the situation potentially involves hazardous materials or WMD agents, this reliance on standardized tactical response procedures will help minimize the risk of exposure to all responders.
- The Eight Step Process© is a tool used for the tactical management of hazardous materials emergencies. It serves as an example of a structured system that can be used by response personnel at incidents involving hazardous substances and materials.
- Although the level of equipment, training, and personnel may vary among organizations, there are fundamental functions and tasks that must be evaluated and implemented on a consistent basis. The Eight Step Process© provides a necessary framework to translate planning and preparedness into the delivery of an effective system for responding to and investigating incidents where hazardous materials and WMDs may be involved.
- The eight functions within the Eight Step Process© are:
  1. Site Management and Control
  2. Identify the Problem
  3. Hazard Assessment and Risk Evaluation
  4. Select Personal Protective Clothing and Equipment
  5. Information Management and Resource Coordination
  6. Implement Response Objectives
  7. Decon and Clean-Up Operations
  8. Terminate the Incident

Hot Terms

Decontamination The process of making personnel, equipment, and supplies "safe" by reducing or eliminating harmful substances (i.e., contaminants) that are present when entering and working in contaminated areas (i.e., hot zone or inner perimeter).

Defensive Recon Obtaining information on site layout, containers, physical hazards, access, and other related conditions from beyond the inner perimeter of the incident.

Eight Step Process© A tool for the tactical management of hazardous materials emergencies; it can be used by response personnel at incidents involving hazardous substances and materials.

Offensive Recon Obtaining intelligence and incident information by physically entering the inner perimeter of an incident.

Post-Emergency Response Operations (PERO) All post-emergency activities, including investigation, restoration, and recovery activities.

Site Management and Control Managing and securing the physical layout of the incident; the first step in the Eight Step Process©.

HazMat Responder *in Action*

The Hazardous Materials response team is en route to an accident involving a car and a tractor trailer. The car has crashed under the belly of the trailer and there is product leaking from the trailer. You are setting up for containment and control of the product, and you are using the Eight Step process.

Questions

1. In initiating offensive tactics, which step will reduce or stop the flow of product?
   A. Step 4
   B. Step 2
   C. Step 6
   D. Step 8
2. Which step is a critical benchmark in the overall success of a response and is the foundation on which all subsequent response functions and tactics are built?
   A. Step 2
   B. Step 4
   C. Step 3
   D. Step 1
3. You need to ensure that site workers are using the proper protective equipment and clothing appropriate for the hazards present. Which step is this?
   A. Step 1
   B. Step 2
   C. Step 3
   D. Step 4
4. Which steps refers to the proper management, coordination and dissemination of all pertinent data and information?
   A. Step 3
   B. Step 5
   C. Step 1
   D. Step 8

Site Management

CHAPTER 5

Hazardous Materials Technician

Knowledge Objectives

After reading this chapter, you will be able to:

- Define site management and control. (p. 114)
- List and describe the six major tasks that must be implemented as part of the site management and control process. (p. 114)
- Define the following terms and describe their significance in controlling emergency response resources at a hazmat incident:
  - Staging (p. 113–114)
- Describe the procedures for establishing scene control through the use of an isolation perimeter at a hazmat incident. (p. 115, 116)
- Describe the procedures for establishing scene control through the use of Hazard Control Zones at a hazmat incident. (p. 117–119)
- Describe three criteria for evaluating Protection-in-Place as a public protective action option. (p. 123)
- Describe three criteria for evaluating evacuation as a Public Protective Action option. (p. 123)

Skills Objectives

There are no Hazardous Materials Technician skills objectives in this chapter.

Hazardous Materials Incident Commander

Knowledge Objectives

After reading this chapter, you will be able to:

- Define site management and control. (p. 114)
- List and describe the six major tasks that must be implemented as part of the site management and control process. (p. 114)
- Describe the procedures for initially establishing command at a hazmat incident. (p. 114–119)
- Describe the guidelines for the safe approach and positioning of emergency response personnel at a hazmat incident. (p. 115)
- Define the following terms and describe their significance in controlling emergency response resources at a hazmat incident:
  - Staging (p. 113–114)
- Describe the procedures for establishing scene control through the use of an isolation perimeter at a hazmat incident. (p. 115, 116)
- Describe the procedures for establishing scene control through the use of Hazard Control Zones at a hazmat incident. (p. 117–119)
- Describe the role of on-scene security and law enforcement personnel in establishing perimeters at a hazmat emergency. (p. 119)
- Define the following terms and describe their significance in protecting the public at a hazmat incident: (p. 119–127)
  - Public protective actions (p. 120)
  - Protection-in-Place (p. 121)
  - Evacuation (p. 123)
- Describe three criteria for evaluating Protection-in-Place as a public protective action option. (p. 123)
- Describe the guidelines and procedures for implementing Protection-in-Place at a hazmat incident. (p. 123, 129)
- Describe three criteria for evaluating evacuation as a Public Protective Action option. (p. 123)

- List the four criteria for implementing a limited scale evacuation. (p. 123–125)
- Define "sick building syndrome" and list three indicators of a sick building. (p. 125)
- List four situations that may justify a full-scale evacuation. (p. 126, 127)
- List four critical issues that must be addressed by the IC to effectively manage a full-scale evacuation. (p. 127)

Skills Objectives

There are no Incident Commander skills objectives in this chapter.

You Are the HazMat Responder

You are a member of the Local Emergency Planning Committee (LEPC) and have been appointed as the chairperson of a subcommittee charged with developing recommendations for improving evacuation procedures within your county.

Currently, the fire department is responsible for issuing an evacuation order, the police department is responsible for alerting the public, the school district is responsible for providing buses used for evacuating people with critical transportation needs, and local schools and churches are used as evacuation shelters. The Emergency Management Agency is responsible for evacuation planning and shelter management. The agreements to provide these services have been long-standing but there are no formal and legally binding documents in place.

The high hazard areas the LEPC has identified within the community include an anhydrous ammonia storage facility, a water treatment plant which uses chlorine, several farm chemical facilities, an LPG bulk plant, a four-lane state highway, and a major east/west railway corridor.

1. Based on the hazards identified in the community, which ones will place the community at greatest risk in terms of the need for public protective actions?
2. Which hazards identified in the community will most likely require local evacuation plans?
3. What are the major elements that will need to be addressed in your community's evacuation plan?

Introduction

Site management is the first step in the Eight Step Process©. The major emphasis of site management is establishing control of the incident scene by assuming command of the incident and isolating people from the problem by establishing an isolation perimeter and Hazard Control Zones. Site management and control provide the foundation for the response. Responders cannot safely and effectively implement an incident action plan (IAP) unless the playing field is clearly established and identified for both emergency responders and the public.

The actions taken by the first arriving units in the first 10 minutes of the incident usually dictate how well the next hour will go. Addressing site management tasks early in the incident ultimately helps the IC manage the biggest problem first—people. People too close to the hazmat scene can become potential rescues. The IC should clear the most hazardous areas first and use an isolation perimeter and Hazard Control Zones to protect responders and spectators; however, keep in mind this action can be trickier than you think—isolate what you need but don't take real estate than you can not control. It is also easier to make a Hazard Control Zone smaller once you have control as opposed to trying to make a smaller zone larger after the fact.

When the IC launches into tactical operations without first addressing basic site management tasks, safety issues will continue to arise throughout the course of the incident and become a problem. The IC should not begin extended operations until the Hazard Control Zones have been identified and the isolation perimeter is secured.

Terminology and Definitions

To understand the material in this chapter better, let's quickly review some terminology. The following terms are listed in the order they will be discussed in this chapter:

Staging—The National Incident Management System (NIMS), defines staging as the location where resources can be placed while awaiting a tactical assignment. The Operations Section is responsible for the management of the Staging Area, and depending upon the incident geography and resource requirements, there may be multiple staging areas.

Staging areas are typically established at hazmat incidents under the following conditions:

- Initial response operations—The first-arriving unit responds to the incident scene, while all other units are ordered to stage at a safe location close to, but away from the scene, from where they can deploy in a safe, timely, and effective manner. Historically, this was known as Level I staging.
- Sustained response operations—As an incident grows or escalates, the IC designates a fixed location where resources responding beyond the initial response can be placed until given a tactical assignment. This tool is usually employed for large, complex, or lengthy hazmat operations. Historically, this was often known as Level II staging.

Isolation Perimeter—The designated crowd control line surrounding the Hazard Control Zones. The isolation perimeter is always the line between the general public and the Cold Zone (definition below). Law enforcement personnel may also refer to this as the "outer perimeter."

Hazard Control Zones—Designation of areas at a hazardous materials incident based on safety and the degree of hazard. These zones are defined as the hot, warm, and cold zones and are located inside the isolation perimeter.

Hot Zone—The control zone immediately surrounding a hazardous materials incident that extends far enough to prevent adverse effects from hazardous materials releases to personnel outside the zone.

Warm Zone—The control zone at a hazardous materials incident site where personnel and equipment decontamination (decon) and Hot Zone support takes place.

Cold Zone—The areas at a hazardous materials incident that contains the incident command post and other support functions necessary to control the incident.

Area of Refuge—A holding area within the Hot Zone where personnel are controlled until they can be safely decontaminated, treated, or removed.

Public Protective Actions—The strategy used by the IC to protect the general population from the hazardous material by implementing a strategy of either (1) Protection-in-Place, (2) Evacuation, or (3) a combination of Protection-In-Place and Evacuation.

Protection-In-Place (shelter-in-place)—Directing people to go inside a building, seal it up as effectively as possible, and remain inside the building until the danger from a hazardous materials release has passed.

Evacuation—The controlled relocation of people from an area of known danger or unacceptable risk to a safer area or one in which the risk is considered to be acceptable.

Site Management Tasks

Site management can be divided into six major tasks:

1. Assuming command and establishing control of the incident scene.
2. Ensuring safe approach and positioning of emergency response resources at the incident scene.
3. Establishing staging as a method of controlling arriving resources.
4. Establishing an isolation perimeter around the incident scene.
5. Establishing Hazard Control Zones to ensure a safe work area for emergency responders and supporting resources.
6. Sizing up the need for immediate rescue and implementing initial public protective actions (PPA).

Life safety is the highest strategic priority for any IC. There will always be situations where initial size-up warrants that responders move directly into rescue operations (e.g., a driver who is alive and trapped in the cab of an overturned tank truck leaking a life-threatening product). However, even under the most extreme situations, implementing initial Site Management tasks will save lives. Don't let a bad situation become worse by allowing emergency responders to charge into rescue situations without following safe operating procedures. Rescue is discussed in more detail in the Implementing Response Objectives chapter.

Command and Control

The chapter that covers "managing the incident" discusses incident command in detail. This section provides a brief review of the Incident Command process as it relates to Site Management.

Hazmat incidents require strong, centralized command. Without it, the scene will usually degenerate into an unsafe, disorganized group of freelancers (people running around the incident scene doing their own thing with no clue what the response objectives are).

The success or failure of emergency operations will depend on the manner in which the first-arriving officer or responder establishes command. Regardless of job title or rank, this individual should initiate the following command functions:

- Correctly assume command. The person assuming command should be the highest ranking or most experienced person on-scene and who also meets the requirements of OSHA 1910.120q.
- Confirm command. Confirm that all personnel on the scene and en route have been notified of the command structure. Responders need to know (1) who is in command, (2) how to contact command, and (3) where to find the location of the Incident Command Post (ICP).
- Select a stationary location for the ICP. An experienced commander only gives up the advantage of a stationary command post when it is absolutely necessary for the IC to personally provide one-on-one direction to responders operating in forward positions. In either case, the IC must maintain a command presence on the radio. The IC should also be readily identifiable using a command vest or by other means.
- Establish a staging area. Ensure staging is in an easily accessible location and has been announced over the radio for incoming personnel and apparatus. Staging is discussed in more detail later in this chapter.
- Request necessary assistance. Do the levels of resources required to mitigate the incident match the scope and

nature of the hazmat problem? The IC should be familiar with the capabilities of mutual aid resources before they are needed. Resource Typing of Hazardous Materials Response Teams is an effective way to ensure the mutual aid hazmat capability you need is what shows up at the scene.

Approach and Positioning

Safe approach and positioning by the initial emergency responders is critical to how the incident will be managed. Emergencies that start bad because of poor positioning sometimes stay bad. If initial emergency responders become "part of the problem," the IC has to change the IAP to deal with the new circumstances. For example, if emergency responders become contaminated, the IAP shifts from protecting the public to rescuing and decontaminating the responders.

Good approach and positioning of personnel and apparatus should follow basic safe operating principles. Some general guidelines include the following:

- **Approach from uphill and upwind.** We recognize that highway engineers don't always build roads upwind from hazmat incidents, but if the weather and topography are on your side, take advantage of it to make the incident scene safer. If you do find yourself approaching from the downwind side, then use distance from the incident to your maximum advantage or switch to self-contained breathing apparatus (SCBA). Keep in mind your local weather patterns. An initial command post location may not remain viable if you have unpredictable wind shifts or other unexpected changes in the environment.
- **Look for physical hazmat clues.** Avoid wet areas, vapor clouds, spilled material, and so on. Use some common sense; if birds are flying in one side of the vapor cloud and not coming out the other side, you probably have a problem.

Conditions can change quickly at hazmat incidents. The wind can shift and the vapor cloud can overrun your position. Don't position too close until a proper size-up has been completed. The DOT *Emergency Response Guide* can serve as a good reference to establish initial evacuation zones.

Ask yourself: Where is the hazmat being released from now? Where is it going? Where am I in relation to where it is going? How fast is it moving? What will the hazmat do to me when it gets to where I am? If the answer is you don't know or will be hurt, you are in the wrong position and need to move! This thought process should be ongoing throughout the incident.

Staging Areas

Staging procedures facilitate safety and accountability by allowing for the orderly, systematic, and deliberate deployment of responders. The staging area is the designated location where emergency response resources (people, equipment, and supplies) are assigned until they are needed.

Staging becomes an element within the operations section command structure (see the chapter on "managing the incident"). The staging area manager (STAM) accounts for all incoming emergency response units, assigns resources from the staging area to their tactical assignment at the request of the IC, and requests additional emergency resources as necessary.

The ideal staging area is close enough to the isolation perimeter to reduce response time significantly, yet far enough away to allow units to remain highly mobile for tactical assignments. The staging area should be easily accessible to responding apparatus and sufficiently large (expandable) for the anticipated requirements. Staging is effective when the IC anticipates that additional resources may be required and orders them to respond to a predesignated area several minutes from the scene.

Complex incidents may require that extensive resources be brought to the scene at different times throughout the emergency. If resources will not be required for some time, the IC should consider establishing primary and secondary staging areas.

Staging should be used as part of initial response operations for any multiple unit response to a hazmat incident. The first unit arriving at the scene of the emergency assumes command and begins site management operations. All other responding units stage at a safe distance away from the scene until ordered into action by the IC. Using staging as part of the initial response places a response unit on scene ASAP and allows the first responding officer to establish command, size up the problem, and begin to formulate an IAP. It also provides the IC with the most options in assigning the remainder of the response.

Normally, staging as part of initial response operations takes place when units stop short of the scene (approximately one typical city block in their direction of travel) or remain outside of a facility's main gate. Staging should be in a safe, upwind location. Obviously, you should not drive through an unsafe location just to take up a position in a safer location.

Staging is employed as a resource management tool when an incident escalates past the capability of the initial response and for sustained response operations, especially for large, complex, or lengthy hazmat operations. As additional resources arrive near the emergency, they are staged together in a designated location under the command of a Staging Area Manager until given a tactical assignment. The location of the Staging Area will be based on the nature of the emergency, available geographic locations, and prevailing wind conditions.

Isolation Perimeter

The isolation perimeter is the designated crowd control line surrounding the incident scene to maintain the safety and security of the spectators and the responders. Designating and establishing the isolation perimeter is an Incident Command responsibility. Maintaining the perimeter throughout the incident is usually the responsibility of law enforcement and security professionals. In simple terms, the isolation perimeter separates the responders from the spectators. Who makes up each group?

Spectators—The general public. This includes the media and anyone else who is not a Player. Spectators should always be kept outside the isolation perimeter.

Responders—Emergency responders and other support personnel who are part of the team brought to the incident scene to safely make the problem go away. The IC controls the Responders at the incident scene by establishing Hazard Control Zones.

Responder Safety Tip

The isolation perimeter should be flexible and it may become larger or smaller depending on the hazards and risks. Big hazards and risks = Big isolation perimeter.

The objective in setting up the isolation perimeter is to immediately limit the number of civilian and public safety personnel exposed to the problem. This process begins by using existing geographic or building features whenever practical. For example, when confronted with an incident inside a structure, the best place to begin controlling people is at the points of entry, such as the main entrance doors. Once the doorways are secured and the entry of unauthorized personnel is denied, crews can begin to isolate above and below the hazard. Proper protective clothing and equipment must be worn **Figure 5-1**.

The same concept applies for outdoor scenarios. First, use the existing terrain and roadway features to secure the main entry points into the area and then establish an isolation perimeter around the hazard. Begin by controlling intersections, on/off ramps, service roads, or any other access routes to the scene. Controlling traffic flow early in the incident helps to reduce gridlock. When this task has been completed, a reconnaissance team can begin a size-up. Injured civilians are still a top priority, but highways and access points can quickly choke up and totally restrict any type of access to the scene. Surrounding the problem with gridlocked, occupied vehicles will quickly compound rescue operations and restrict access to the incident **Figure 5-2**.

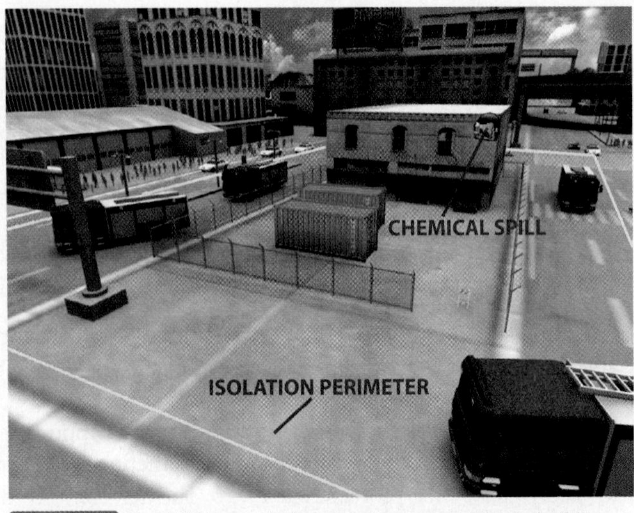

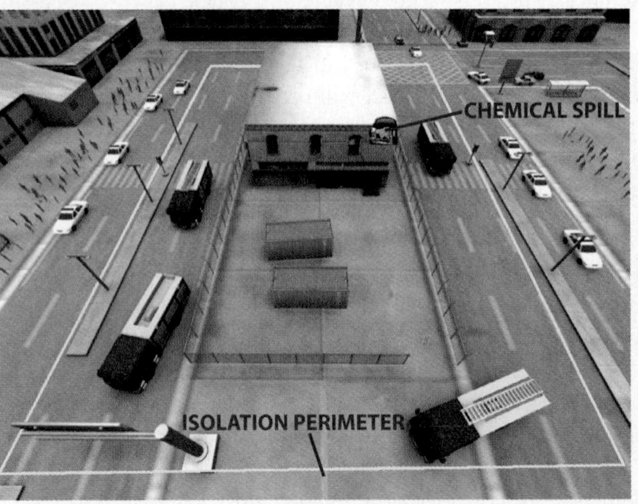

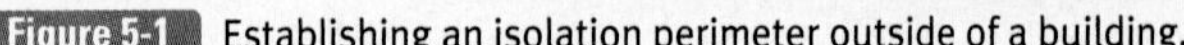

Figure 5-1 Establishing an isolation perimeter outside of a building.

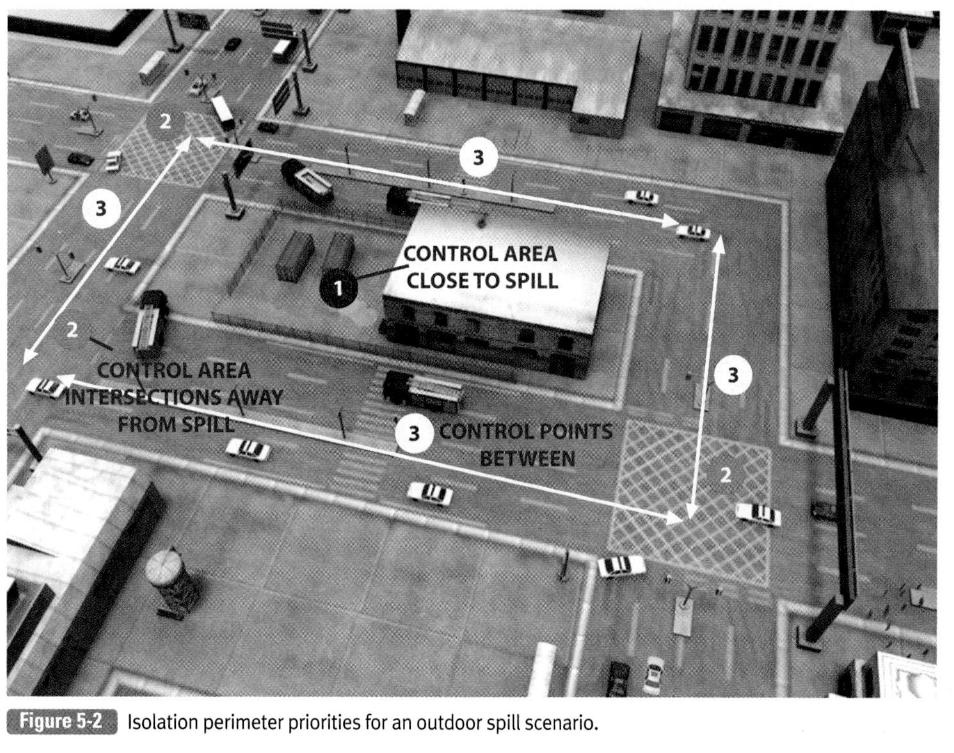

Figure 5-2 Isolation perimeter priorities for an outdoor spill scenario.

Hazard Control Zones

Hazard control zones are designated areas at a hazardous materials incident based on safety and the degree of hazard.

Defining Hazard Control Zones

Hazard control zones are designated restricted areas within the isolation perimeter based upon their degree of hazard. Hazard control zones are designated from the most hazardous to least hazardous as hot, warm, and cold (hot = greatest risk and cold = least risk).

The primary purpose of establishing three different Hazard Control Zones within the isolation perimeter is to provide the highest possible level of control and personnel accountability for all responders working at the emergency scene. Defined control zones help ensure that responders do not inadvertently cross into a contaminated area or place themselves in locations that could be quickly endangered by explosions, migrating vapor clouds, or structural instability.

As a general rule, the public and the news media should be located outside of the isolation perimeter, the ICP and support personnel should be located in the Cold Zone, emergency responders supporting the tactical hazmat response operation should be positioned in the Cold and Warm Zones, and the entry team(s) should be located in the Hot Zone. To work in the hot zone or inner perimeter, remember the First Law of Hot Zone Operations: To play in the game safely, you should:

- Be trained to play
- Be dressed to play (i.e., proper level of PPE based on the hazards present)
- Be medically fit to participate
- Have a backup capability (i.e., OSHA Two-In/Two-Out rule.)
- Have decon established
- Have an entry plan coordinated with the IC and safety officer

Emergency responders can be challenged by scenarios where civilians and the public are exposed to a hazmat prior to their arrival. In this scenario, an Area of Refuge (i.e., Safe Refuge Area) should be established within the Hot Zone to control any exposed personnel until they can be safely managed (e.g., a decon area has been established). The Hot Zone should be large enough to provide one or more areas of refuge as necessary Figure 5-3. The reality is that sometimes people with exposures do not wait around for emergency responders to arrive and go directly to the hospital, which creates additional challenges.

Identifying Hazard Control Zones

Hazard Control Zones should be established through air monitoring, and clearly marked and posted on the IC's tactical worksheet. A Tactical Worksheet for Site Management functions is included in the *Hazardous Materials: Managing the Incident Field Operations Guide.*

In outdoor situations, Hazard Control Zones can be designated by using key geographical reference points, such as a storage tank dike wall, fence line, or street name. Geographic areas should be communicated verbally by radio or in a face-to-face briefing between the IC and his/her subordinates. The Hot Zone can be indicated with colored banner tape, color-coded traffic cones, color-coded light sticks, or temporary fencing for long-term operations.

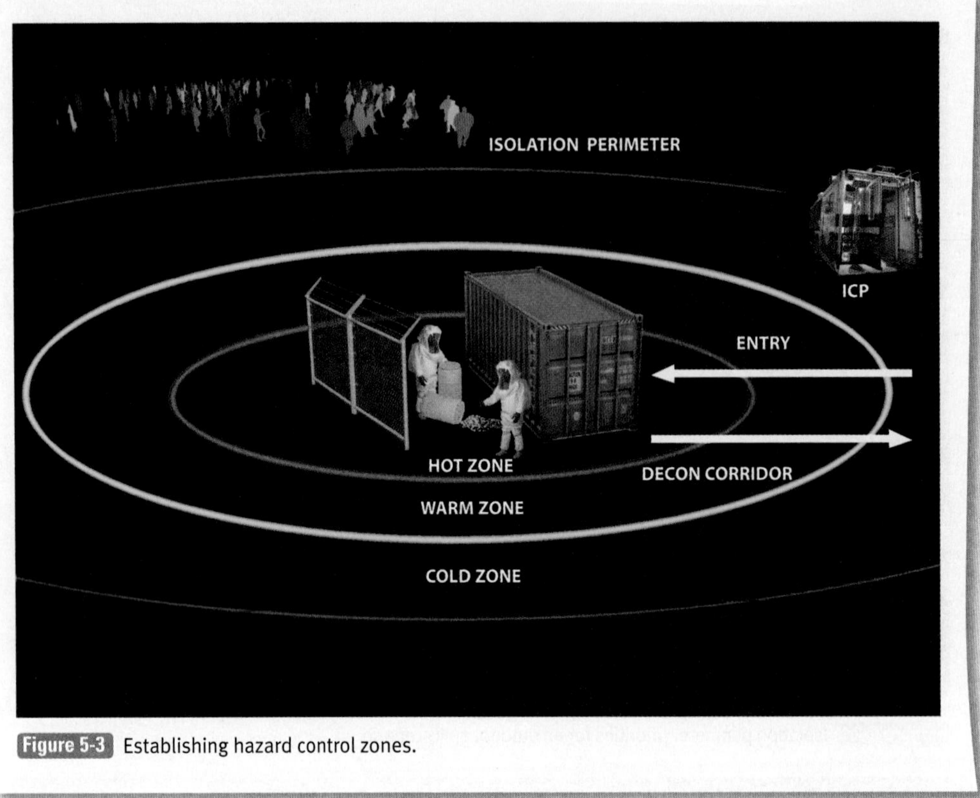

Figure 5-3 Establishing hazard control zones.

When the hazard is inside of a building, these zones can be denoted by their location within the structure. For example, a spill in room 321 may dictate that rooms 320 to 322 would be a Hot Zone, the rest of the building would be the Warm Zone, and the area outside of the building itself would be designated the Cold Zone (Figure 5-2).

Although it is acceptable to estimate the size of the Hazard Control Zones early in the incident based on visible clues, the IC should move toward a more definitive assessment using monitoring instruments.

As a guideline, any area with a measurable concentration of a contaminant using available monitoring should initially be considered within the hot zone until additional data are obtained and evaluated. Initial monitoring efforts should concentrate on determining if Immediately Dangerous to Life and Health (IDLH) concentrations are present. Decisions regarding the size of Hazard Control Zones should be based on the following:

1. **Flammability**—Flammability is usually measured in percent of the atmosphere using a combustible gas indicator that displays its results in percentage. The IDLH action level is 10% of the lower explosive limit (LEL). Any areas above this concentration are clearly inside the Hot Zone.
2. **Oxygen**—Oxygen is measured in percent of the atmosphere using an oxygen meter and displays in %. An IDLH oxygen-deficient atmosphere is 19.5% oxygen or lower, while an oxygen-enriched atmosphere contains 23.5% oxygen or higher. When evaluating an oxygen-deficient atmosphere, consider that the level of available oxygen may be influenced by contaminants that are present. Areas containing atmospheres that are either oxygen deficient or enriched should be designated as the Hot Zone.
3. **Toxicity**—Toxicity is measured in parts per million and is normally measured by a Photo Ionization Device (PID) with substances that are organic. Unless a published action level or similar guideline is available such as Emergency Planning Guidelines (ERPG), the STEL or IDLH values should initially be used. If there is no published IDLH value, consider using an estimated IDLH of 10 times the TLV/TWA. Initial Hazard Control Zones can be established for toxic materials using the following guidelines:
   - **Hot Zone**—Monitoring readings above STEL or IDLH exposure values.
   - **Warm Zone**—Monitoring readings equal to or greater than TLV / TWA or PEL exposure values.
   - **Cold Zone**—Monitoring readings less than TLV / TWA or PEL exposure values.
4. **Radioactivity**—Any positive reading two times above background levels or alpha and/or beta particles that are 200 to 300 counts per minute (cpm) above background would confirm the existence of a radiation hazard and should be used as the basis for establishing a Hot Zone. The chapter on "hazard assessment and risk evaluation" includes a more detailed discussion concerning the application and interpretation of monitoring instruments as it relates to hazard and risk assessment.

Safe operating procedures should strictly control and limit the number of personnel working in the Hot Zone. Most Hot Zone operations can be accomplished with a minimum of four personnel working for specified time periods using the Buddy

System following the OSHA Two-In/Two-Out Rule. The purpose of the Buddy System is to provide rapid assistance in the event of an emergency. This system organizes emergency response personnel into work groups so that each person in the work group is designated to be observed by at least one other entry person.

The size of Hazard Control Zones should change over time as the risks increase or decrease. The zones might expand or contract depending on the size of the incident and the nature of the hazards and risks. Retaining large Hazard Control Zones during prolonged operations without good safety and technical reasons will eventually create problems with property owners and the affected community. This is especially true at incidents involving critical infrastructure or industrial facilities. On longer-duration incidents, holding onto large pieces of real estate may generate political problems that can erode the IC's credibility. Property owners and law enforcement personnel should be briefed on how and why you have established the Hazard Control Zones as you have.

Although a large perimeter surrounding the incident is desirable, a common mistake is to seal off more area than can be effectively controlled. Given limited personnel, it is better to secure a smaller area completely and expand the perimeter outward as additional resources become available. On the other hand, it is much easier to contract a zone as the hazards decrease than to try to expand one when things go bad.

Roles of Security and Law Enforcement Personnel

The IC should make isolation perimeter assignments as soon as possible. This requires close coordination with security or law enforcement supervisors at the ICP. These individuals will become key players in establishing inter-agency communications and determining areas that will be controlled first, and how they will be managed throughout the incident.

Security and law enforcement officers involved in establishing a perimeter or in securing buildings need to know exactly what the potential hazards and risks appear to be. If there is a risk these officers may be exposed to the hazard as the isolation area expands, they must be provided with proper PPE.

Law enforcement personnel are best used where traffic and crowd control will involve large groups of people on public property. Key law enforcement functions include staffing entry control points and patrolling the perimeter for unauthorized entry into the control zones.

In today's threat environment, law enforcement also plays a very important role in providing security for the ICP and emergency responders within the Isolation Perimeter where they may become the target of other security-related threats; e.g., active shooters, snipers, suicide bombers. Police are better trained for perimeter security than firefighters **Figure 5-4**.

When operating on a private facility such as an industrial plant, the on-site security force assumes similar duties as law enforcement. Industrial security officers are very familiar with the site, its hazards, and the available resources. They often know the employees inside the plant by sight and can provide specifics on evacuation plans, emergency procedures, and availability of special tools. The IC can usually rely on security to handle perimeter control issues within the plant while local law enforcement officers control the areas outside of the plant. Working together under a unified command system, law enforcement and the security team can be a valuable asset to the IC.

Figure 5-4 Law enforcement officers are essential for controlling the isolation perimeter.

Rescue and Initial Public Protective Actions

Scenarios may occur where immediate response actions are required in order to rescue or remove victims from a hazardous environment. First responders will be the most likely personnel to be faced with the decision to "go or not go" in order to attempt a rescue and favorably change the outcome. If the hazards are known and the risks can be readily evaluated, initial rescue operations may be implemented at an acceptable level of risk to responders. In contrast, rescue scenarios that may be encountered at a terrorism incident involving chemical agents will pose many unknowns and present first responders with significantly higher risk levels.

The reality of most tactical hazmat response operations is that they are not well suited for rapid entry and extrication operations. For rescue operations in a hazmat environment to be effective, first responders must have: (1) the proper training, (2) pre-established standard operating procedures, and (3) appropriate personal protective clothing and equipment, which must be immediately available on-scene and rapidly donned.

Rescue as a strategy will be discussed further in the chapter on "implementing response objectives."

Responder Safety Tip

The reality of most tactical hazmat response operations is that they are not well suited for rapid entry and extrication operations. Make sure you are performing a rescue to save a life before you take extreme risks. All decisions made by the incident commander should be based on a calculated risk using a risk vs. benefit thought process.

Initiating Public Protective Actions

Public Protective Actions (PPAs) are the strategy used by the IC to protect the general population from the hazardous material by implementing (1) Protection-In-Place, (2) evacuation, or (3) a combination of Protection-In-Place and evacuation.

PPA strategies are implemented after the IC has established an isolation perimeter and defined the Hazard Control Zones for emergency responders.

There are no clear benchmarks for the PPA decision-making process. A combination of factors must be evaluated to select the best strategy based on incident conditions. Some of the more important factors include what has been released, how much, the hazards of the material(s) involved, the population density, time of day, weather conditions, type of facility, and the availability of air-tight structures. As you can see, there is a great deal to consider both before and during an incident.

One common misunderstanding about PPA strategies is that the IC must use one strategy versus the other (e.g., either Protection-in-Place or evacuation). This is not an either/or choice. Sometimes evacuation is the best strategy, sometimes Protection-In-Place is the best choice, and sometimes the best option is to use both strategies at the same time (e.g., evacuate the area close to the hazards while directing others further away to Protect-In-Place).

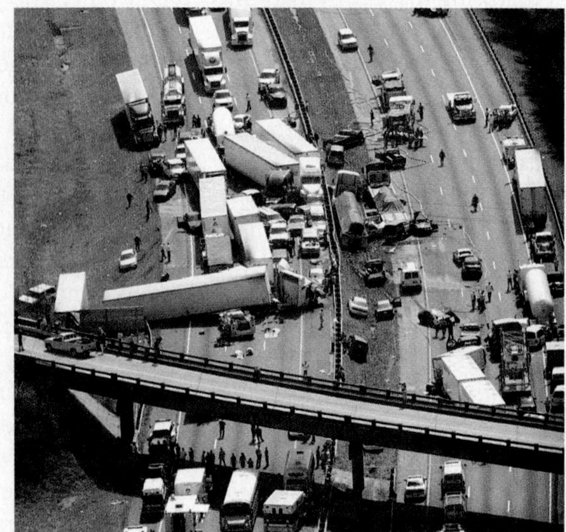

Figure 5-5 For short-term releases, remaining inside a structure may be the best decision.

Initial PPA Decision Making

PPA decisions are incident specific and require the use of the IC's judgment and experience. If a release occurs over an extended period of time, or if there is a fire that cannot be controlled within a short time, evacuation is typically the preferred option for nonessential personnel. Evacuation may not always be necessary during incidents involving the airborne release of extremely hazardous substances, such as anhydrous ammonia, chlorine, and hydrofluoric acid. Airborne materials can move downwind so rapidly that there may be no time to evacuate a large number of plant employees or the surrounding community. In other situations, evacuating people may actually expose them to greater potential risk. For short-term releases, the most prudent course of action may be to remain inside of a structure **Figure 5-5**.

The IC's decision to either evacuate or protect-in-place should be based on the following factors:

- Hazardous material(s) involved, including their characteristics and properties, quantity, concentrations, physical state, and location of release.
- The population at risk, including facility personnel and the general public. In addition, the IC must consider the resources required to implement the recommended protective action, including notification, movement/transportation, and possible relocation shelters and the time required to open and staff the shelters.
- The time factors involved in the release. Consideration must be given to the rate of escalation of the incident, the size and observed or projected duration of the release, the rate of movement of the hazardous material, and the estimated time required to implement protective actions.
- The effects of the present and projected meteorological conditions on the control and movement of the hazardous materials release. These would include atmospheric stability, temperature, precipitation, and wind conditions.
- The capability to communicate with the population at risk and emergency response personnel prior to, during, and after the emergency.
- The availability and capabilities of hazmat responders and other personnel to implement, control, monitor, and terminate the protective action. This should include a size-up of the structural integrity and infiltration rates of structures potentially available for Protection-in-Place throughout the area.

Prior knowledge of the hazmat or the facility through planning information or computer dispersion models acquired through the hazards analysis process can also assist the IC in this evaluation.

Regardless of the tactic used, achieving PPA objectives translates into gaining control of a specified area beyond the isolation perimeter, securing and clearing that area, and then controlling a second downwind or adjacent area. In this manner, more and more threatened areas can be secured as more resources become available to the IC.

It is imperative the IC use a systematic and structured approach to clearing the public away from the hazard area. Establishing priorities and communicating the plan for PPA tactics are important from the beginning and should be updated on a map at the ICP as new PPA zones are identified and controlled.

In the early stages of an incident, the IC is often preoccupied with size-up and rescue activities and can easily overlook people problems in the immediate area. Be aware that if the situation deteriorates rapidly, exposures within 1,000 feet (304 meters) may be contaminated. In other words, everyone inside the isolation perimeter is a potential rescue.

University of Alberta Research

In the late 1980s the Department of Mechanical Engineering at the University of Alberta, Edmonton, Canada, conducted extensive tests on Canadian and American homes to determine whether protection-in-place tactics would actually work. Test findings revealed that for an accidental toxic gas release that occurs over several minutes to half an hour, even a very leaky building contains a sufficient reservoir of fresh air to provide effective sheltering-in-place. However, for longer-duration releases of 1 to 3 hours, the average indoor concentration may reach 80% or more of the outdoor average during a steady and continuous release.

For releases that have a long duration of an hour or more, the choice between shelter and evacuation is difficult to make. Typical air exchange rates in a house are about 0.5 air changes per hour (ACH). For a 3-hour release, this exchange rate causes the air in the house to be replaced 1.5 times during the event. See Figure 5-6. After this air exchange, the indoor concentration is about 80% of the average outdoor value. Obviously, the more energy efficient or tight the building is, the slower the air exchange rate will be.

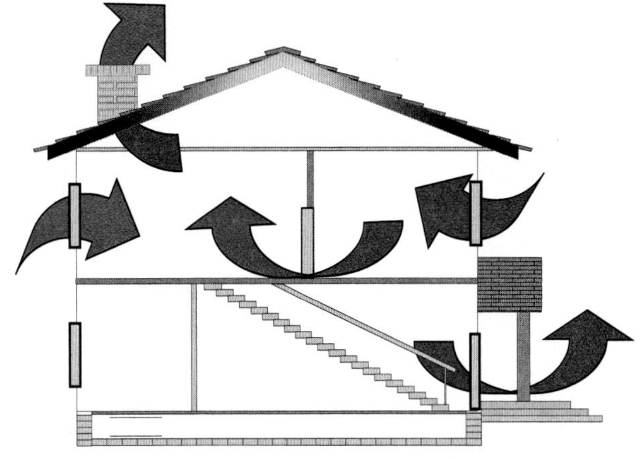

Figure 5-6 Typical air exchange rates in a house are about 0.5 air changes per hour (ACH).

[*Source:* "Effectiveness of Indoor Sheltering During Long Duration Toxic Gas Releases," by D. J. Wilson, Department of Mechanical Engineering, University of Alberta, Edmonton, Alberta, T6G 2G8.]

Areas that should receive immediate attention by the IC include the following:

- Locations within 1,000 feet (304 meters) of the incident that will be rapidly overtaken by the hazmat. This is especially a concern when a flammable or toxic gas or vapors are drifting downwind.
- Locations near the incident where people are reasonably safe from the hazmat. People near the hazmat should be alerted to keep clear of the hazard and remain indoors until given other instructions.
- Key locations that control the flow of traffic and pedestrians into the hazard area (for example, doorways, on-ramps, and grade crossings).
- Special high-occupancy structures such as schools.
- Structures containing sick, disabled, or incarcerated persons.

The *Emergency Response Guidebook* (ERG) is a good resource document to guide the IC in making quick initial judgment calls on which PPA option to implement. The *Guidebook* also provides some basic guidelines concerning the size of the initial isolation zone based on the type of hazardous material and size of container. The IC should be thoroughly familiar with how to use the *Emergency Response Guidebook*. The instructions to the ERG provide some useful background information on PPA decision making.

There is a fine line between isolation objectives and evacuation. For our purposes, isolation requires quick action to protect the public and first responders from an immediate, life-threatening situation. Isolation is a necessity; failure to act when people are outside, in exposed locations, will result in injuries. In contrast, evacuation implies a prolonged, precautionary removal from the affected location.

Protection-In-Place

The Protection-in-Place/Shelter-In-Place strategy involves directing people to go inside of a building, seal it up as effectively as possible, and remain there until the danger from a hazardous materials release has passed. Protection-in-place (sheltering in place) is a concept that is a familiar part of people's daily activities. We routinely shelter ourselves indoors when we close windows to keep out noise or dust or to keep the house cool on a hot summer day when the air conditioning is running. The concept of protection-in-place as applied to a hazmat release is identical except the objective is to prevent toxic vapors from entering the structure.

Research and accident investigations clearly indicate that staying indoors may provide a safe haven during toxic vapor releases. The following summarizes some of the key research that has been conducted over the last 30 years.

Los Alamos National Laboratory Research

Studies conducted at the Los Alamos National Laboratory in the late 1990s by Michael Brown and Gerald Streit indicate that "for an outdoor plume release, closing the windows and doors and staying inside may be initially safer than remaining outdoors. Due to infiltration (e.g., leaks, ventilation) and the eventual passage of the bulk of the plume, there may come a time when it is actually safer to be outdoors rather than indoors." Figure 5-7

Emergency response decision makers responsible for public announcements should be aware that safe zones may change from inside to outside buildings during the course of the event, so that strict guidelines for when to go outside may be difficult to derive.

Figure 5-7 Staying inside may be initially safer than going outdoors.

[*Source:* Brown, Michael, J., and Gerald E. Streit, "Emergency Responders' Rules-of-Thumb for Air Toxics Releases in Urban Environments." U.S. Department of Energy, Publication LA-UR-98-4539, Los Alamos National Laboratory, Los Alamos, NM (1998), page 21.]

National Institute of Chemical Studies

In June 2001 the National Institute of Chemical Studies (NICS) in Charleston, West Virginia released a research report entitled "Sheltering in Place as a Public Protective Action." NICS reviewed a number of case studies to determine whether it is better to evacuate or protect-in-place during a chemical release. The report concluded as follows:

- Sheltering-in-place is an appropriate public protection tool in the right circumstances. For chemical releases of limited duration, it is faster and usually safer to shelter in place than to evacuate.
- For all case studies examined during this study, there were no fatalities associated with sheltering in place.

The body of evidence from the research referenced in this chapter suggests that if there is insufficient time to complete an evacuation, or the chemical leak will be of limited duration, or conditions would make an evacuation more risky than staying in place, then sheltering in place is a good way to protect the public during chemical emergencies.

Even though sheltering is effective in most situations, it must also be recognized that sustained continuous releases may eventually filter into a structure and endanger the occupants. Protection-in-place may not be the best option if the nature of the incident involves a prolonged release of a toxic material. It is also not the best option when the release involves a flammable gas when the gas can permeate the structure and create an internal explosion hazard.

The IC may have to make critical decisions based on weather conditions and forecasts. High humidity and warm air can force vapors toward the ground. Air ventilation and air conditioning ducts may force toxic vapors into the building before the public is warned and the order to take shelter is issued by the IC.

Protection-in-place is usually the best option for the following types of situations:

- The hazardous material has been totally released from its container and is dissipating. As the clock ticks, the situation is getting better and the air is getting cleaner.
- The released material forms a "puff" or migrating plume pattern (e.g., vapor clouds that will quickly disperse and is not originating from a fixed, continuous source from the leak).
- A fast-moving toxic vapor cloud will quickly overrun exposed people. While evacuation might be the best option, time may be working against you and the only option may be to move indoors where there is a better chance for survival.
- Short-duration solid or liquid leaks are present. These types of incidents usually threaten a very limited geographic area.
- Leaks that can be rapidly controlled at their source by engineered suppression or mitigation systems or through responder containment and confinement tactics.

When protection-in-place is the best course of action, the IC must make sure the public is provided with clear instructions on what they should do. The following are examples:

- Close all doors to the outside and close and lock all windows (windows seal better when locked). Seal any obvious gaps around windows, doors, vents, and so on with tape, plastic wrap, wet towels, or other materials.
- Turn off all heating, ventilating, and air conditioning systems (HVAC) systems and window unit air conditioners. If applicable, place inlet vents in the closed position.
- Close fireplace dampers. If there is a fire in the fireplace, let it burn down without closing the dampers.
- Turn off and cover all exhaust fans.
- Close as many internal doors as possible.
- Pick one room in the house or structure to use as a shelter room. A bathroom can be a good choice because water and a toilet are available, if needed. A master bedroom can also be a good choice if it has a bathroom and a phone. An upstairs room may be a better choice than a basement room, as many chemicals are heavier than air and tend to sink near the ground.

- Monitor the local radio or television stations for further information.

The protect-in-place option only works if the public is compliant with the IC's recommendations. Compliance with an order to shelter-in-place will be dependent on the following factors:

- Receipt of a timely and an effective warning message. The public needs to get the word rapidly and the instructions must be clear. For example, a warning message broadcast in English in a Spanish-speaking neighborhood may be misunderstood. A warning to simply "take shelter" may be misunderstood and interpreted to mean evacuate.
- Clear rationale for the decision to protect-in-place as compared to an evacuation. If the instructions are not consistent with the public's perception of the danger, they probably won't be implemented. If the public feels threatened (they can see smoke, fire, or are watching the situation unfold on live television), they may want to evacuate even though the safest course of action would be to stay where they are.
- Credibility of emergency response personnel with the general public. If the public's perception of your public safety agency is high, you will have more credibility and trust when you ask them to do something in an emergency.
- Previous training and education by fixed facility personnel and the public on the application and use of protection-in-place. If the public has never even heard the words "protect-in-place," don't be surprised when they ignore your order. Advance training of the public in what to do when a chemical emergency occurs makes a big difference in establishing your credibility. Citizens are more comfortable following protect-in-place orders if they understand that the community has a well thought-out plan and they are familiar with the protection-in-place procedure.

Additional shelter-in-place information can be obtained at the NICS website at http://www.nicsinfo.org.

Wally Wise Guy™

An excellent example of an effective public education effort is the "Wally Wise Guy™" program. Originally developed by the Deer Park, Texas, LEPC, Wally is a turtle that "Knows it's wise to go inside his shell whenever there's danger." This costumed character is a mascot designed to teach children and their parents how to shelter-in-place in case of a chemical emergency.

Wally is designed to reach children from kindergarten to fourth grade, so Wally makes personal appearances at elementary schools, day care centers, community parades, and other civic events.

The Wally Wise Guy™ program has been a very effective tool to market the protection-in-place concept. The program has spread across the United States through LEPCs. Wally's message has been projected on everything from coloring books (in Spanish and English) to refrigerator magnets. Wally's image has even been used on fire apparatus compartment doors in industrial areas as a marketing tool. For more information, go to http://www.wally.org.

Evaluating Structures for Protection-In-Place

The age and construction of a building greatly influence how successful protection-in-place will be. Pre-incident plans and hazards analysis surveys should incorporate this type of information. As a general rule, the older a building, the less likely it will provide a safe refuge for periods longer than one hour.

Hazmat responders should be familiar with the types of structures within the community and consider the following variables:

- Age of the building—Older buildings do not seal well. Structures with leaky windows, doors, and poke-throughs may not be good places to seek refuge during a toxic gas release.
- Prevailing wind direction—Most communities have a predictable prevailing wind direction that is seasonal. You should be familiar with the local "wind rose," which indicates the direction the wind blows the majority of the time for the time of year in the area you live in. This information is available from the National Oceanic and Atmospheric Administration, National Weather Service.
- Building height—The geometry of high-rise buildings in urban environments, especially in high-rise building areas, can significantly affect the movement of airborne toxic gases. Canyon and tunnel effects in high-rise areas can make toxic gas plumes go in directions against the prevailing wind that are unpredictable.

Evacuation

Evacuation is the controlled relocation of people from an area of known danger or unacceptable risk to a safer area or one in which the risk is considered to be acceptable. Evacuation of both industrial fixed facility personnel and the general public is an attempt to avoid their exposure to any quantity of the released hazardous material. Under ideal conditions, evacuation will remove these individuals from any exposure to the released hazmat for a given length of time. For our purposes, we will describe evacuations in terms of limited-scale and full-scale evacuations.

Limited-Scale Evacuations

Limited-scale evacuations are implemented by the IC when the incident affects one or two buildings in the vicinity of the incident. The majority of the evacuations required at hazmat incidents affect a small number of people.

Case Study 1

I-610 at Southwest Freeway, Houston, Texas—May 11, 1976.

On a bright sunny day at 11:08 am on May 11, 1976, a Transport Company of Texas tractor–semitrailer tank truck transporting 7,509 gallons of anhydrous ammonia struck and broke through a bridge rail on a ramp connecting I-610 with the Southwest Freeway (U.S. 59) in Houston, Texas. The truck left the ramp, struck a bridge support column of an adjacent overpass, and fell onto the Southwest Freeway, approximately 15 feet below.

The tank truck breached immediately on impact, releasing most of its contents into the atmosphere. At the time of the accident there were about 500 people within 1/4 mile of the release. The released ammonia fumes rapidly penetrated automobiles and buildings. When their occupants left to escape the fumes during the early minutes of the release, many were exposed to fatal doses of ammonia Figure 5-8.

The temperature at the time of release was in the low 80s (27° C). The released ammonia immediately vaporized and the 7-mph wind gradually decreased the vapor concentration at ground level. Witnesses reported the white ammonia vapor cloud initially reached a height of 100 feet before being carried by the 7-mph wind for approximately 1/2 mile. After 5 minutes, most of the liquefied ammonia had boiled off and the vapor cloud was completely dispersed.

Seventy-eight of the 178 victims who were within 1,000 feet of the release point were hospitalized and treated for symptoms of ammonia inhalation. More than 100 persons were treated for less severe injuries. Five of the six fatalities were due to ammonia exposure. Because all fatalities were within 200 feet of the estimated release, it is estimated that the ammonia concentration within this distance was greater than 6,500 ppm for at least two minutes. The IDLH for ammonia is 500 ppm.

Figure 5-8 The tank truck breached immediately on impact, releasing most of its contents into the atmosphere.

A detailed investigation of this incident conducted by the U.S. National Transportation Safety Board (NTSB) in 1979 revealed there were significant differences in the degree of injury among the exposed victims who evacuated buildings and those who protected-in-place.

The Board's conclusion was that the protection offered survivors by the vehicles and buildings demonstrated there were alternatives to simply running away from the released hazmat. A detailed investigation conducted by the Board showed that people who sheltered and remained inside buildings received no harm from the ammonia. Also, people who remained inside of their automobiles generally received less severe injuries than those who left their cars and tried to escape the ammonia.

Although there have been other investigations conducted with similar conclusions, the Houston case was the first the authors are aware of concerning protection-in-place issues documented by forensic science. As a result of NTSB's report, many emergency response and safety professionals began to rethink whether evacuation was always the best tactical option.

The investigation conducted by the Board also documented that the actions taken by the Houston Fire Department (HFD) saved lives. Within 10 minutes the HFD dispatched 14 emergency rescue units and 4 pieces of fire apparatus. The Incident Command System was used to coordinate EMS, fire, and police agencies. As a result, many of the contaminated and injured victims were quickly located by search and rescue teams and escorted to safety, where they received medical treatment.

A copy of the final report can be obtained at http://www.ntsb.gov. NTSB Number: HAR-77/01, NTIS Number: PB-268251.

The U.S. Department of Health and Human Services, Agency for Toxic Substances Disease Registry (ATSDR), conducts a biennial study involving participation from state health departments. The most recent study tracked hazmat incidents reported by 14 state health departments from 2007 to 2008. The study found that evacuations were ordered in 1,155 of 15,462 events for which evacuation status was reported. Of this sample, the majority (75.9%) was from the affected building or affected areas of one or more buildings.

The number of persons evacuated was recorded for 802 (69.4%) of the 1,155 events and ranged from 0 to 3,800 persons, with a median of 20. The median duration of evacuation was 2 hours (range: 0 hours to 18 days).

[*Source:* U.S. Department of Health and Human Services, Agency for Toxic Substances Disease Registry, "Hazardous Substances Emergency Events Surveillance 2007–2008, Biennial Report," Atlanta, Ga. (2010).]

A limited-scale evacuation may be the best option for the IC under the following conditions:

- Whenever the building is on fire
- Whenever the hazmat is leaking inside the building and the material is flammable or toxic or the state of the hazmat is such that it will be hard to control
- Whenever explosives or reactive materials are involved and can detonate or explode, producing flying glass or structural collapse
- Whenever the building occupants show signs or symptoms of acute illness and there is a known hazmat spill inside the structure

Sick Building Syndrome

Emergency responders are sometimes called to incidents where several people are exhibiting signs and symptoms of illness but there is no apparent source (e.g., there is no reported hazmat leak or spill inside or outside the building). This may be indicative of sick building syndrome (SBS).

SBS is a situation in which occupants of a building experience acute health effects that seem to be linked to time spent in a building, but no specific illness or cause can be identified. The complaints may be localized in a particular room or zone or may be widespread throughout the building.

Persons with SBS may exhibit the following symptoms:

- Anxiety
- Hyperventilation
- Headache
- Itchy skin
- Dizziness and nausea
- Nose or throat irritation
- Difficulty in concentrating
- Intermittent muscle spasms or twitching (tetany)
- Severe shortness of breath

In cases involving a single victim, there may be complaints about sensitivity to odor. For example, the odor emitted from fresh paint, new carpet, or other furnishings. Atmospheric testing may reveal no harmful levels in the ambient air.

Indicators of a sick building may include the following:

- There are none of the usual indicators that there is a hazardous materials release inside the building (e.g., no smoke, strange odors, or unusual meter readings). Other than the symptoms being displayed by people in the building everything else appears normal.
- The structure is a "tight building." Sick buildings are usually office buildings that house many people working in close proximity. The building is energy efficient and designed to reduce heat loss from windows that do not open. Heating and air conditioning ducts originate from a single source.
- Most of the complainants report relief of their symptoms soon after leaving the building.

Sick building problems may also result from the building being operated or maintained in a manner that is inconsistent with its original purpose of design or prescribed operating procedures. For example, improper modifications to ventilation systems can result in a build up of carbon monoxide. If you suspect you have a sick building, always conduct a test for carbon monoxide.

Sometimes indoor air problems are a result of remodeling activities like painting, using glues and maskings for flooring and wall coverings, or the recent installation of new synthetic carpet or plastic materials. Other sources of SBS problems include copy machines, cleaning agents, recently applied pesticides (roach control), and chemicals that may release volatile organic compounds (VOCs), including formaldehyde.

SBS problems may also be caused by biological contaminants inside the building. Biological contaminants include pollen, bacteria, viruses, and molds. These contaminants can breed in stagnant water accumulated in humidifiers, drain pans, and ducts, or where water has collected on ceiling tiles, insulation, carpet, or outside the building near air intakes. Biological contaminants can cause fever, chills, cough, chest tightness, muscle aches, and allergic reactions. One indoor air bacterium, Legionella pneumophila, has caused both Pontiac fever and Legionnaire's disease.

[*Source:* Environmental Health Center, National Safety Council, Washington, D.C. (http://www.nsc.org). Also see U.S. Environmental Protection Agency, Washington, D.C., *Indoor Air Facts No. 4* (revised): *Sick Building Syndrome* (http://www.epa.gov)].

The IC should consider initiating a limited-scale evacuation of a suspected SBS incident whenever the signs and symptoms of the affected people become acute. Be aware that many SBS complaints also have a psychological component that can result in a "mass hysteria" reaction with psychogenic symptoms.

Evacuation of Fixed Industrial Facilities

Unlike their public safety counterparts, industrial emergency response teams have the advantage of knowing exactly what types of hazardous materials are in their facilities. This prior knowledge allows evacuation decisions to be more specific.

OSHA requires fixed facilities to have written evacuation procedures (OSHA 1938[a]). Well-written PPA procedures can provide useful guidelines to supervisors and the facility IC concerning whether a limited- or full-scale evacuation is necessary. Many facilities have developed a tiered approach to implementing PPAs. One method used is to define three levels of PPA that are associated with the Levels of Incidents described in the chapter covering "hazardous materials management systems."

Levels of Protective Actions

Fixed industrial facilities should provide written guidelines to employees and contractors concerning when it is appropriate to evacuate or protect-in-place. Industrial facilities are familiar with the hazards and risks of the products they manufacture and can develop specific guidelines. The Protective Action example in **Table 5-1** was prepared by a gas plant that processes natural gas with high hydrogen sulfide levels.

Full-Scale Evacuations

Full-scale evacuations involve the relocation of large populations from a hazardous area to a safe area. Full-scale evacuations present two major problems for the IC:

1. **Life Safety**—In some cases, evacuations may endanger the lives of the people being evacuated. Traffic accidents, stress-induced heart attack, and accidental exposure to the hazmat being released are real-world examples. You evacuate them, you own the decision you made!
2. **Expense**—Full-scale evacuations are expensive. One study conducted by the Battelle Human Affairs Research Center for the Atomic Industrial Forum indicated that the cost to individual households for evacuation would be expected to be almost seven times the cost of protection-in-place. Costs to the public sector are approximately three times as high and fifteen times as high for the manufacturing sector. An expensive evacuation operation might be very hard to justify after the fact. There will be no shortage of critics the day after the evacuation (even if you made the right decision).

Fortunately, most ICs will never be faced with the decision to evacuate thousands of people, but that should not be an excuse for being unprepared. When emergency responders are having their worst day, they usually don't "rise to the occasion," but instead default to their level of planning and training.

Some of the worst days in modern history that required large, full-scale evacuations include:

- **Japanese Earthquake and Tsunami**—On March 11, 2011 Japan was hit with a 9.0 magnitude earthquake. The northeastern shore of Japan was struck by a devastating tsunami which reached 80 feet high in some places. The giant wave severely damaged several nuclear power plants located in Fukushima. The subsequent nuclear crisis at the Fukushima Daiichi plant resulted in the evacuation of 80,000 residents who lived within a 12-mile evacuation zone around the damaged nuclear plants **Figure 5-9**.
- **The World Trade Center**—The terrorist attacks of September 11, 2001 and the subsequent collapse of the New York City's World Trade Center towers was the worst building disaster in modern history, killing some 2,900 people. More than 350 emergency response personnel died in the line of duty and are credited with saving approximately 25,000 lives through their extreme heroism when they directly supervised the evacuation of the buildings' occupants. It is estimated that between 300,000 and 1 million people evacuated Manhattan Island by powerboat, ferry, barge, or tugboat. One vessel, the Staten Island Ferry's Samuel I. Newhouse, evacuated 6,000 people **Figure 5-10**.

Table 5-1 Levels of Protective Action

Level I Incident Protective Action—A Level 1 Protective Action requires all employees and contractors to evacuate the work site in an upwind direction and report to their pre-designated briefing area as defined in the site-specific emergency plan. Personnel will be accounted for and emergency work assignments will be issued by the on-site supervisor. Individual employees who are working alone at a remote location and must evacuate a work site should contact the Control Room Operator and report that they are leaving the area. Examples of incidents requiring a Level 1 protective action would include a small flammable or toxic leak from a valve or flange that produces concentrations in the 10 ppm range on company property. The leaking gas is not likely to go beyond company property. The leak is easily repairable using safety equipment on-site. There is no risk to the public, but the fire department is notified immediately that there is a Level I Incident in progress.

Level II Incident Protective Action—A Level II incident will require personnel to evacuate the immediate work site and meet at the pre-designated assembly area for accountability and emergency work assignment as specified in the site-specific emergency plan. The public surrounding the work site should be protected-in-place. Examples of incidents requiring a Level II Incident Protective Action would include a moderate release of hydrogen sulfide that is producing atmospheres of up to 300 ppm on company property with atmospheres of 10 ppm immediately adjacent and downwind of company property. The leak can be rapidly repaired, and the hydrogen sulfide gas will rapidly disperse. The fire department is alerted via 911, and a chief officer assumes command of the incident.

Level III Incident Protective Action—A Level III incident is a major emergency which requires the total evacuation of all company personnel and the surrounding public to a pre-designated location(s) outside the immediate area. Depending on the nature of the incident and the potential for the problem to migrate off-site, there will also be PPA instructions given to the surrounding community. Concentrations of hydrogen sulfide are rapidly exceeding 300 ppm immediately adjacent to company property, and the off-site concentrations will reach 50 to 100 ppm. The leak cannot be immediately repaired, atmospheric concentrations are unfavorable for rapid dispersing of hydrogen sulfide gas, and homes within the H_2S release area are not of energy-efficient construction.

Figure 5-9 The March 11, 2011 9.0 earthquake and tsunami severely damaged nuclear power plants located in Fukushima, Japan.

Figure 5-10 It is estimated that between 300,000 and 1 million people evacuated Manhattan Island on 9/11.

- **Mississauga, Ontario**—On November 10, 1979 at 11:53 p.m., a Toronto-bound Canadian Pacific train (No. 54) derailed at Mississauga, Ontario (a suburb of Toronto) spilling hazardous materials. Police evacuated 218,384 residents from the area due to the threat presented by chlorine vapors.
- **Three Mile Island**—On March 28, 1979 at 4:00 a.m., a malfunction occurred in the secondary cooling circuit at the Three Mile Island Nuclear Plant near Harrisburg, Pennsylvania, which initiated the largest nuclear power facility accident in U.S. history. The accident led to 145,000 people being evacuated from an area that extended about 15 miles from the plant.

The decision to commit most emergency response resources to a full-scale evacuation should be initially determined by the IC based on the specific conditions of the emergency. Some situations that may justify a full-scale evacuation include the following:

- Large leaks involving flammable and/or toxic gases from large-capacity storage containers and process units.
- Large quantities of materials that could detonate or explode, damaging additional process units, structures, and storage containers in the immediate area.
- Leaks and releases that are difficult to control and could increase in size or duration.
- Whenever the IC determines the release cannot be controlled and facility personnel and/or the general public are at risk.

When the decision is made to commit to a full-scale public evacuation, four critical issues must be addressed and managed effectively in order for the operation to succeed.

1. **Alerting and Notification**—People must be alerted that an evacuation order has been issued and they must be told where to go and what to do when they get there. People do not normally panic; they usually make rational decisions based on the risks they perceive. That said, rational decisions are not necessarily good ones if they are based on bad information. Getting clear instructions to people that need to be evacuated is the key to getting buy-in.

   Using only one method of alerting and notification is a losing strategy. You must use a variety of tools to be successful (e.g., radio and television, sirens, telephone and email alerts, use of social media, door-to-door visits by public safety officers). Alerting methods are discussed later in this chapter.
2. **Transportation**—If you live in a metro area it may not seem obvious when you are driving to work in bumper-to-bumper traffic, but not everyone owns an automobile and not everyone has more than one vehicle. Transportation must be provided to move lots of people from where they are now to where you want them to go. Sound simple? It's not. Think about the Fukushima, Mississauga, and Three-Mile Island scenarios.
3. **Relocation Facility**—Once people have been removed from harm's way, they need shelter and to be taken care of. People need to be safe, comfortable, and provided with the basic things everyone needs (e.g., food, water, ice, and a safe place to sleep, prescription medication, medical interventions). Did we mention bathrooms and showers, beds, pet kennels, and something to keep the kids out of trouble? Keep in mind that all of the problems (i.e., medical, physical, criminal) a community experiences on a "normal" day will likely be replicated in a shelter environment. The longer people must remain in a temporary facility, the greater their needs will be. See Figure 5-12. Consideration for pets should also be factored into the evacuation process and relocation facility.
4. **Information**—You must keep displaced people informed about your progress. People need to know what is going on and when they might be able to go home. They also want to know if their family and friends are safe. Nobody likes not knowing what is going on. Lack of information creates concern, and concern breeds anger, which leads to outrage, which gets the attention of elected officials, who will eventually get your attention. Keep evacuees in the information loop.

Case Study 2

Train Derailment and Full-Scale Evacuation, Graniteville, South Carolina—January 6, 2005

On January 6, 2005, at 2:39 a.m., a northbound Norfolk Southern Railway Company freight train (NS train 192), while traveling about 47 mph through Graniteville, South Carolina, encountered an improperly lined switch that diverted the train from the main line onto an industry track, where it struck an unoccupied, parked train (NS train P22). The collision derailed both of the locomotives and 16 of the 42 freight cars of train 192, as well as the locomotive and 1 of the 2 cars of train P22. Among the derailed cars from train 192 were three tank cars containing chlorine, one of which was breached, releasing 60 tons of poisonous liquefied chlorine gas. The train engineer and eight other people died as a result of chlorine gas inhalation **Figure 5-11**.

About 554 people complaining of respiratory difficulties were taken to local hospitals. Of these, 75 were admitted for treatment. Six firefighters were treated and released; one firefighter was admitted to the hospital and remained there for several days. Two sheriff's department officers were also treated and released.

Because of the chlorine release, approximately 5,400 people within a 1-mile radius of the derailment site were evacuated for several days. Total damages exceeded $6.9 million.

When train 192 struck train P22, both locomotives and the first 16 cars of train 192 derailed. The ninth car from 192, containing 90 tons of chlorine, was punctured during the derailment and released chlorine gas. Winds were light at the time of the accident, and the chlorine vapor cloud settled in the low-lying valley along the tracks.

Based on emergency responder observations and the locations of those receiving fatal injuries, the cloud extended at least 2,500 feet to the north, 1,000 feet to the east, 900 feet to the south, and 1,000 feet to the west. The sudden release and expansion of the escaping gas caused the product remaining in the tank to auto-refrigerate and remain in the liquid state, slowing the release of additional gas.

[*Source:* Railroad Accident Report, Collision of Norfolk Southern Freight Train 192 With Standing Norfolk Southern Local Train P22 With Subsequent Hazardous Materials Release at Graniteville, South Carolina, January 6, 2005, NTSB/RAR-05/04, National Transportation Safety Board, Washington, D.C.]

Figure 5-11 60 tons of chlorine were released from this rail car, requiring an evacuation of 5,400 people within a 1-mile radius.

Figure 5-12 Relocation facilities need to address the safety, health, and welfare of people. Food, refrigeration, sanitation are just some of the services that need to be available.

Alerting and Notification

For an evacuation to be successful, the IC must ensure that people are quickly alerted of an emergency in progress. The methods of notification will vary depending on the location of the emergency, the type of plan and hardware in place, and the time of day.

The location of the general population and the time of day should always be a factor in the IC's decision-making process. Rush-hour traffic and the time of day are both significant factors in deciding whether or not to evacuate. Some studies conducted by the nuclear power industry show that with good planning and traffic control assistance from police agencies, many urban highways are capable of handling large traffic flows created by a full-scale evacuation. These same studies, however, also point out that high-density traffic jams can be created at critical traffic arteries when large crowds attempt to evacuate locations like athletic stadiums and concert halls. Table 5-2 provides a more detailed picture of principal locations of the population.

Table 5-2 **Average 24 Hour Population Location**

Population Locations		
Location/ Activity	**Hours Per Day**	**Percent of Time**
Home	16.6	69.2
School or work	4.7	19.6
Commuting	1.2	5.0
Outdoors	1.5	6.2

A variety of communication technologies exist to assist the IC with the warning process. There are advantages and disadvantages of each of these systems, and it is important to recognize one is not necessarily better than another. Each one is a different type of tool and must be selected based on local conditions.

A brief summary of the different alerting methods is as follows:

- **Door-to-door**—Responders are required to go door-to-door to advise residents in the affected area. This is usually completed by law enforcement agencies. This initial contact has a multiplier effect as residents contact neighbors and relatives, alerting them that there is a problem.
- **Loudspeakers/public address systems**—Loudspeakers on emergency vehicle siren systems are an effective way of alerting people outside in public areas such as parks. Sound from public address systems on emergency vehicles travels 500 feet, which means the vehicle has to stop every 1,000 feet to broadcast the message. This is a slow process. Public address systems inside shopping malls and public assembly buildings may also be used.
- **Tone-alert radios**—Some fixed facilities, such as chemical plants and oil and gas plants, have special tone-alerts for the radios of residents living near their facility. These operate on the same principle as a volunteer firefighter's pager or radio. A radio signal is sent from the control room at the plant that sets off an alerting tone inside each home's radio. A live, real-time message can then be broadcast. Special weather radios are also available that are activated by the National Weather Service (NWS) for severe storm warnings. This system can be used to issue special warnings if prior arrangements have been made with the NWS.
- **Emergency Alerting System**—The Emergency Alerting System, or EAS, is an effective method of alerting people in buildings and automobiles. The Federal Emergency Management Agency and the NWS operate the EAS. The system allows emergency messages to be broadcast by cable, satellite, and other services through

Voices of Experience

As a Hazmat responder for 18 years I have been on some well-controlled incidents and others that have significantly lacked proper Site Management. A few memorable incidents have taught me that no matter how obvious a hazard is, how secure a facility is, or how small an area you are trying to control, Site Management is a key element of the Eight Step Process©. At the most basic level, Site Management protects responders and the public from hazards, and a lack of Site Management can lead to freelancing, injuries, contamination spread, and litigation.

Early in my career I responded to a vapor release at a local oil refinery. The plant personnel activated the local refinery siren system and had their security force secure the facility gates. They believed that by activating the siren and locking the facility perimeter gates that Site Management was accomplished. Approximately 10 minutes into the incident, a plant employee drove through the incident scene where the vapors were being released. He never heard the siren because he was driving in a vehicle, and the lack of on-scene control (no Hazard Control Zones established) allowed him to pass through the scene before anyone could stop him. Fortunately, the vapors never ignited; however, a follow-up vapor cloud release study revealed that, had the vapors ignited, the overpressure (explosion) would have injured or killed eight personnel (including me).

Basic scene control is second nature to emergency responders. We do a great job of identifying Isolation Perimeters (i.e., traffic control zones) to keep the public away from the response. However, setting the Hazard Control Zones is sometimes overlooked. The Hazard Control Zones need to be established early on to protect the responders from the hazards. Like most emergency responders, I have been on a number of incidents that had poor Hazard Control Zones. Three incidents stick in my mind:

- While on a fuel tanker rollover we watched in horror as a local ambulance squad drove through the fuel spill not just once, but twice, and a number of their crew walked through the fuel oil. The Incident Commander had established control of the traffic on the highway (Isolation Perimeter), but did not establish the Hazard Control Zones. This lack of control lead to a huge risk to the responders and helped spread the fuel spill outside of the original contaminated area.
- Our statewide HMRT responded to a local fire department's "white powder" call at a nearby post office. Our entry team entered the post office to retrieve the envelope and decontaminate the mail slot. Local law enforcement was controlling the Isolation Perimeter and the Incident Commander was controlling the Cold and Warm zones, but somehow a local resident strolled past law officers and walked right into the post office. The resident literally came face to face with one of our entry team members, which as you can imagine, startled both of them.
- Our statewide HMRT was dispatched to a fire at a local auto parts/paint store. Seven area fire departments had been battling the fire for over 4 hours, trying to extinguish the fire and protect an adjacent exposure. The HMRT was requested because fire fighters were beginning to complain of skin and respiratory irritation. Command tasked the HMRT with identifying the materials involved and the chemicals the fire fighters were exposed to. When arriving on-scene, I immediately observed a complete lack of scene control. Fire fighters were walking through significant contamination from motor oils, paints, battery acids, firefighting runoff water, etc. The contamination was being spread to surrounding ditches, roadways, and adjacent properties. Fire hoses, tools, and SCBA were being staged on contaminated asphalt, most of which were completely oil soaked. Without any Hazard Control Zones established, no one was controlling entry into and out of the Hot Zone, and there was no decontamination area established, which lead to contamination being carried elsewhere, including back to the fire stations.

These examples show that a failure to establish the First Step in the Eight Step Process© can easily negate all of the following Steps. If you are not controlling the incident scene, no amount of PPE, problem identification, and hazard risk assessment will protect the public or your response personnel.

Tyler Bones
Fairbanks North Star Borough Hazmat Response Team
Fairbanks, Alaska

radio and television stations in the United States. By law, the EAS must be available to the President of the United States within 10 minutes using normal activation procedures from any area in the United States for national emergencies. State Emergency Management agencies can provide local emergency notification using the EAS communications backbone.

- **Personalized Localized Alerting Network**—The Personalized Localized Alerting Network or PLAN is an expansion of the Federal Communications Commission's EAS. PLAN is managed by the Federal Emergency Management Agency. The system has the ability to send emergency text messages over participating commercial wireless networks through cell towers to PLAN-enabled mobile devices. Three types of alerts can be sent out over PLAN: messages issued by the President of the United States, alerts involving imminent threats to life safety, and Amber Alerts.
- **Weather Radios**—The National Oceanographic and Atmospheric Administration (NOAA) weather radio system covers a major portion of the population within the country. The station broadcasts continuously and can be used to warn people of special atmospheric emergencies such as migrating toxic vapor clouds. This is especially useful for boaters who may be downwind on lakes or rivers. The U.S. Coast Guard may also contact boaters directly by issuing special broadcasts on marine channel 16.
- **Commercial Television and Radio**—Television Capture Systems can be used as alerting tools in communities where cable television systems are used. Emergency services can "capture" the cable station and transmit a message called a "crawler" across the bottom of the viewer's screen. The media may also break into normal radio and television programming with a special broadcast. Some emergency management agencies have special agreements with local radio stations that allow them to break into programming and broadcast an emergency message live from the Emergency Operations Center.
- **Smart Phone Applications**—Hand-held data and voice communications devices, such as "Smart Phones," are widely used and are now almost household items. There are many phone applications available that provide news feeds and flash alerts by text message, e-mail, or auto-dialing a recorded voice message.
- **Social Networking**—Emergency managers are employing social networking systems to stay connected with the public. This method of alerting has a multiplier effect as various networks of individuals share information extremely quickly.
- **Sirens and Alarms**—These may include the community Emergency Management Agency sirens or special sirens installed in areas around fixed facilities like chemical plants, refineries, and weapons depots. Some of these devices can also function as a public address system.
- **Aircraft**—Helicopter loudspeakers can be an effective method of alerting people in outdoor and remote areas (e.g., flying over parks, campgrounds, and hunting areas). Most law enforcement helicopters have voice air-to-ground capability.
- **Electronic Billboards**—Many interstate highways and commuter routes have electronic message signs for alerting drivers of local traffic conditions, kidnappings, terrorist threats, and so on. Some bus and train stations have similar electronic bulletin boards.
- **Computerized Telephone Notification Systems**—CT/NS can reach a potentially large number of people by telephoning hundreds of phone numbers simultaneously by computer. Prerecorded emergency messages provide instructions to residents.
- **Low-power AM Radio Systems**—These systems use a low-power radio transmitter in the AM band to broadcast traffic and weather information 24 hours a day. Some systems also broadcast continuous public safety messages. Telephone books, flashing roadside signs, and other public education literature direct the citizen to switch to the designated frequency whenever a public emergency occurs (e.g., "Switch your radio to AM 1600"). These systems are very effective when they regularly broadcast useful information on a daily basis like the current weather. People become used to going to the public safety radio program to get reliable information about what is going on in their community.

A good community alerting and notification system is based on a variety of systems that are described in the Community Emergency Response Plan. The plan should spell out who has the authority and responsibility to activate each system. Each system component should be tested on a regular basis. A layered approach to alerting and notification is required to reach the majority of the targeted population. A good alerting strategy attempts to notify the populations at risk who are closest to the danger area first and then expand outward Table 5-3.

Many locations have improved their alerting systems by providing detailed PPA instructions in the local telephone directory. Alerting messages received by the EAS direct residents to a specific page number in their telephone book where detailed instructions are provided.

Alerting Systems in Fixed Facilities

At a fixed facility such as a refinery or chemical plant, the notification process normally occurs by activation of sirens or by use of an on-site public address system. One of the most frequent problems encountered in fixed facilities is the confusion created by a single warning tone that may also be used to indicate the beginning or ending of a work shift, a fire, toxic gas release, and so on. The same tone is used with one, two, or three blasts on the horn, which have different meanings. As a general rule, evacuation alarms should be unique and distinctly different from any other type of alarm in the facility.

Table 5-3 Estimates of Alerting Times in Minutes

Estimates of Warning Times for Alternative Systems

Warning System	Percent of Population Warned 25	Percent of Population Warned 50	Percent of Population Warned 75	Percent of Population Warned 90
Media	20–30	45–60	80–120	180–240
Door to door	40–45	60–80	100–120	150–180
Route alert	25–35	40–50	60–70	90–150
Tone alert radio or auto telephone	2–3	4–5	7–10	10–15
Siren/media	5–10	12–15	15–20	20–35
Siren/fixed response	1–2	2–3	14–15	10–15

As part of the pre-incident planning process, fixed facility personnel should have special knowledge of the following:

- Methods by which all personnel are notified of an emergency evacuation, including the specific sound of the alarm system.
- Instructions on where personnel should report to and assemble when the evacuation alarm is sounded.
- Facility evacuation routes and corridors.
- Safe havens where personnel can seek refuge if evacuation is not possible or advisable.
- Ability to communicate with facility personnel at evacuation assembly areas. This is critical in accounting for evacuees and initiating search and rescue operations.
- Location of both primary and alternate assembly locations and a method of employee and visitor accountability.

From a tactical perspective, the IC should be aware that an activated evacuation alarm at a fixed facility does not necessarily mean that all occupants have either been protected-in-place or evacuated. Many manufacturing plants with high noise levels have "dead spots" where alarms cannot always be heard. Additional direction and assistance from emergency response personnel may be required to complete the evacuation Figure 5-13.

Figure 5-13 An activated evacuation alarm at an industrial facility does not mean that all occupants have either been protected in place or evacuated.

Once notified and evacuated, a head count should be taken of facility personnel, contractors, and visitors to ensure all personnel are accounted for. Supervisors are usually responsible for coordinating all personnel accountability activities. Information regarding any missing personnel and their previously known location within the facility should be relayed to the IC, so search and rescue operations can begin. However, the IC should be aware that initial reports of people missing based on head counts are usually not correct.

Transportation

Experience in large-scale evacuations indicates that the majority of the affected population will leave the hazardous area by use of their own automobile or by catching a ride with a friend or neighbor. However, a significant portion of the population cannot drive or do not own an automobile. This is especially the case in urban areas, where many people use public transportation as their only means of getting around.

In any major city there will be a large number of people with **Critical Transportation Needs (CTN)**. CTNs are defined as people who cannot self-evacuate by any means of transportation. This group includes people too young or old to drive, people with medical issues or disabilities that prohibit them from driving, or people who simply do not own an automobile. Some of these people have special needs that make mobility and passenger loading difficult Figure 5-14.

Commercial motor coaches are the only practical and cost-effective way of moving large numbers of people with CTN. Initially this may involve using local transit buses or school buses, but these vehicles are often already committed to other responsibilities in a crisis (e.g., moving children from affected schools to a safe area). Emergency management

Figure 5-14 People with critical transportation needs sometimes require assistance in loading and unloading a motor coach for large-scale evacuations.

agencies in areas of high population density should have an Evacuation Plan. When large numbers of CTNs are known to exist the Evacuation Plan should be supported by a Motor Coach Evacuation Plan Annex. For example, if the CTN population in your town is estimated to be 10,000 people, it would take about 222 motor coaches to move them. Bus contracting, passenger loading, diesel fuel supply, and a policy on allowing pets on the buses are just a few of the many details that would need to be outlined in your Motor Coach Evacuation Plan. The evacuation of CTNs from southeast Louisiana after Hurricane Katrina required nearly 600 buses.

Special Emergency Extraction

Special emergency extraction may be required to transport people from areas close to the hazardous materials release. Some organizations have addressed this problem by purchasing Emergency Escape Packs, also known as Emergency Breathing Apparatus (EBA). These devices typically have 5 to 10 minutes of breathing air in their cylinders. The facepiece is a simple plastic hood placed over the head to provide a fresh-air breathing supply for limited duration. This is usually adequate to move someone from inside their home and into an awaiting vehicle, where they will be transported to a safe area.

EBAs may be carried on Special Air Units operated by the fire department or be staged in strategic locations for use by Emergency Response Personnel. They may be found at or near special project sites, such as oil and gas wells, where hydrogen sulfide is present, or toxic waste dump remediation projects.

Relocation Facilities

Relocation facilities, also known as shelters, are used to temporarily house people displaced during an evacuation. One study suggests approximately 65% of evacuees do not stay at public relocation shelters. They may check into a hotel or stay with friends and relatives outside of the evacuated area. Using these estimates, this means the remaining 35% of the population will require some form of public shelter. In a town of 25,000 that would be about 8,700 people—still a lot of people that must be cared for!

Relocation facilities are typically located at schools, sports complexes, and other similar structures with large, open floor areas. In order for relocation shelters to be effective, they need the following elements in place:

- **Safe Building**—Relocation facilities may be in service for several hours or for several days. The building must be safe, have a food service area, adequate restrooms and bathing facilities, and an area large enough for temporary sleeping furniture such as cots. The building must be air conditioned or heated and have adequate security to protect residents. Relocation shelters should always be energy efficient in case the occupants need to protected-in-place.
- **Shelter Manager**—An individual trained in shelter management techniques should be assigned to each Relocation Facility. The shelter manager organizes and supervises shelter activities and is the single point of contact between the shelter and the field command post. Many jurisdictions have a cooperative arrangement with the local Red Cross chapter or religious and civic organizations like the Salvation Army. The local Emergency Preparedness Agency usually coordinates the shelter program, provides the funding, and trains the shelter managers, while the cooperating organizations provide the personnel to staff the shelter.
- **Shelter Support Staff**—If the relocation shelter will be operating for an extended period, a shelter staff should be provided to assist with its operation. Examples include receptionists to document who arrives and departs the shelter, EMTs or nurses to attend to the sick and disabled, food service personnel to prepare meals, building engineers or maintenance personnel, and counselors. It is also advisable to assign a police officer at each shelter for security. Direct radio contact with the Emergency Operations Center is recommended.

If shelters will remain in operation for an extended period, arrangements must be made for around-the-clock staffing and provisions.

Information

The IC should ensure that displaced people are kept informed of the actions being taken by emergency responders to mitigate the problem. Failure to keep people informed often creates political and public affairs issues.

Relocation facility shelter managers should receive regular situation briefings from the ICP or the Emergency Operations Center. This can be accomplished by direct radio or telephone contact on an hourly basis or by wireless networks set up at each shelter location.

When the incident covers a period of days, the IC should consider issuing a written progress report, which can be posted twice daily in the shelter. If the displaced population is not kept informed, they will often quickly form a negative opinion of your operation and make their feelings known to the rest of the world through the media.

The news media can be a powerful tool in confirming the initial evacuation was well handled and is still necessary for public safety. It is important the IC project to the media an image of professionalism and control during the evacuation. The IC and PIO should hold regularly scheduled joint press briefings with senior representatives of the media present. For example, the Emergency Management, Police, and Fire agencies should conduct their briefings as a team and project unity in their decision-making.

Wrap-Up

Chief Concepts

- Site Management is the first step in the Eight Step Incident Management Process©. Its major focus is on establishing control of the incident scene by assuming command of the incident and isolating people from the problem by establishing an Isolation Perimeter and Hazard Control Zones.
- Site management and control provide the foundation for the response. Responders cannot safely and effectively implement an IAP unless the playing field is clearly established and identified for both emergency responders and the public.
- Safe approach and positioning by the initial emergency responders is critical to how the incident will be managed. Emergencies that start bad because of poor positioning sometimes stay bad.
- Staging procedures facilitate safety and accountability by allowing for the orderly, systematic, and deliberate deployment of responders. The Staging Area is the designated location where emergency response resources (people, equipment, and supplies) are assigned until they are needed.
- The isolation perimeter is the designated crowd control line surrounding the incident scene to maintain the safety and security of the spectators and the responders. Designating and establishing the isolation perimeter is an ICs responsibility.
- Hazard Control Zones are designated restricted areas within the isolation perimeter based upon their degree of hazard. Hazard Control Zones are designated from the most hazardous to least hazardous as Hot, Warm, and Cold. (Hot = Greatest Risk and Cold = Least Risk.)
- Safe operating procedures should strictly control and limit the number of personnel working in the Hot Zone. Most Hot Zone operations can be accomplished with a minimum of four personnel working for specified time periods using the Buddy System following the OSHA Two-In/Two-Out Rule.
- Public Protective Actions (PPAs) are the strategy used by the IC to protect the general population from the hazardous material by implementing either (1) Protection-in-Place, (2) Evacuation, or (3) Combination of Protection-in-Place and Evacuation.

Hot Terms

Area of Refuge A holding area within the Hot Zone where personnel are controlled until they can be safely decontaminated, treated, or removed.

Cold Zone The areas at a hazardous materials incident that contains the incident command post and other support functions necessary to control the incident.

Critical Transportation Needs (CTN) People who have no means of transportation.

Evacuation The controlled relocation of people from an area of known danger or unacceptable risk to a safer area or one in which the risk is considered to be acceptable.

Hazard Control Zones Designation of areas at a hazardous materials incident based on safety and the degree of hazard. These zones are defined as the hot, warm, and cold zones and are located inside the isolation perimeter.

Hot Zone The control zone immediately surrounding a hazardous materials incident that extends far enough to prevent adverse effects from hazardous materials releases to personnel outside the zone.

Isolation Perimeter The designated crowd control line surrounding the Hazard Control Zones. The isolation perimeter is always the line between the general public and the Cold Zone (definition below). Law enforcement personnel may also refer to this as the "outer perimeter."

Protection-In-Place (shelter-in-place) Directing people to go inside a building, seal it up as effectively as possible, and remain inside the building until the danger from a hazardous materials release has passed.

Public Protective Actions The strategy used by the IC to protect the general population from the hazardous material by implementing a strategy of either (1) Protection-in-Place, (2) Evacuation, or (3) a combination of Protection-In-Place and Evacuation.

Staging The National Incident Management System (NIMS), defines staging as the location where resources can be placed while awaiting a tactical assignment. The Operations Section is responsible for the management of the Staging Area, and depending upon the incident geography and resource requirements, there may be multiple staging areas.

Warm Zone The control zone at a hazardous materials incident site where personnel and equipment decontamination (decon) and Hot Zone support takes place.

HazMat Responder *in Action*

Your HazMat Captain has just finished with a review and has rewritten the department's Emergency Response Plan. As the Chief of Operations, you would like to test this plan and analyze the initial response. You have set up an exercise to conduct this evaluation. You are specifically concerned with the initial public protective actions.

Questions

1. Your plan states that a crowd control line surrounding the hazard control zones is critical in protecting bystanders. Which term best describes this requirement?
   A. Hazard Control zone
   B. Protection in place
   C. Isolation perimeter
   D. Area of Refuge
2. Based on incident safety, the first arriving units are considered to have arrived at:
   A. Level I staging
   B. Level II staging
   C. Level III staging
   D. Level IV staging
3. The overall concept and strategy to protect the general public is located in what section of the Emergency Response Plan?
   A. Area of Refuge
   B. Public Protective Actions
   C. Evacuation Protocols
   D. Staging procedures
4. The primary purpose of establishing hazard control zones within the isolation perimeter is to provide:
   A. Evacuation boundaries
   B. Control and personnel accountability
   C. Effective allocation of resources
   D. Effective communications

References and Suggested Readings

1. Agency for Toxic Substances and Disease Registry, *Hazardous Substances Emergency Events Surveillance,* 2007–2008 Biennial Report. Atlanta, GA: ATSDR (2010).
2. Brown, Michael, J., and Gerald E. Streit, "Emergency Responders' Rules-of-Thumb for Air Toxics Releases in Urban Environments." U.S. Department of Energy Publication LA-UR-98-4539. Los Alamos, NM: Los Alamos National Laboratory (1998).
3. Federal Emergency Management Agency—U.S. Fire Administration, *CSX Tunnel Fire,* Baltimore, MD (USFA-TR-140). Washington, DC: FEMA (July 2001).
4. Emergency Film Group, *Site Management and Control.* Plymouth, MA: Emergency Film Group (2004).
5. National Fire Service Incident Management System Consortium Model Procedures Committee, *IMS Model Procedures Guide for Highway Incidents.* Stillwater, OK: Fire Protection Publications, Oklahoma State University (2003).
6. National Institute for Chemical Studies, *Sheltering in Place as a Public Protective Action.* Charleston, WV: National Institute for Chemical Studies (June 2001).
7. National Institute for Occupational Safety and Health, *Guidance for Protecting Buildings from Airborne Chemical, Biological or Radiological Attacks.* Cincinnati, OH: NIOSH (May 2002).
8. National Transportation Safety Board, "Special Investigation Report: Survival in Hazardous Materials Transportation Accidents" (Report NTSB/HZM-79-4). Washington, DC: National Transportation Safety Board (December 1979).
9. Nordin, John, "Technical Dialogue—Evacuate or Shelter in Place," *The First Responder Newsletter* (May 2003).
10. Sherman, M. H., D. J. Wilson, and D. E. Kiel, "Variability in Residential Air Leakage." A technical paper presented at the ASTM Symposium on Measured Air Leakage Performance in Buildings, Philadelphia, Pennsylvania (April 1984).
11. Sorenson, John, H., "Evaluation of Warning and Protective Action Implementation Times for Chemical Weapons Accidents." A report prepared for the Aberdeen Proving Ground, MD by Oak Ridge National Laboratory (Report No. ORNL/TM-10437). Oak Ridge TN: Oak Ridge National Laboratory (April 1988).
12. Wilson, D. J., "Accounting for Peak Concentrations in Atmospheric Dispersion for Worst Case Hazard Assessments," Department of Mechanical Engineering, University of Alberta, Edmonton, Alberta (May 1991).
13. Wilson, D. J., "Effectiveness of Indoor Sheltering during Long Duration Toxic Gas Releases," Department of Mechanical Engineering, University of Alberta, Edmonton, Alberta (May 1991).
14. Wilson, D. J., "Wind Shelter Effects on Air Infiltration for a Row of Houses," Department of Mechanical Engineering, University of Alberta, Edmonton, Alberta (September 1991).
15. Wilson, D. J., "Variation of Indoor Shelter Effectiveness Caused by Air Leakage Variability of Houses in Canada and the USA." Prepared by the Department of Mechanical Engineering, University of Alberta, Edmonton, Alberta. Presented at the U.S. EPA/FEMA Conference on the Effective Use of Sheltering-in-Place as a Potential Option to Evacuation During Chemical Release Emergencies, Emmitsburg, MD (December 1988).

Identifying the Problem

CHAPTER 6

Hazardous Materials Technician

Knowledge Objectives

After reading this chapter, you will be able to:

- Describe the principles of recognition, identification, classification, and verification as they apply to hazardous materials emergencies. (p. 143)
- List and describe the seven basic methods of identifying hazardous materials. (p. 144)
- Identify the following railroad cars by name and specification, identify the typical contents by name and hazard class, and identify the basic design and construction features, including closures:
  - Cryogenic liquid tank cars (p. 165)
  - Nonpressure tank cars (p. 165)
  - Pneumatically unloaded hopper cars (p. 166)
  - Pressure tank cars (p. 165)
- Identify the following intermodal tanks by name and specification, identify the typical contents by name and hazard class, and identify the basic design and construction features, including closures:
  - Nonpressure intermodal tanks:
    - IM-101 portable tanks (IMO Type 1 internationally) (p. 156)
    - IM-102 portable tank (IMO Type 2 internationally) (p. 156)
- Pressure intermodal tank (DOT Spec. 51; IMO Type 5 internationally, p 156)
- Specialized intermodal tanks (p. 157)
  - Cryogenic intermodal tanks (DOT Spec. 51; IMO Type 7 internationally) (p. 157)
  - Tube modules (p. 157)
- Identify the following cargo tanks by name and specification, identify the typical contents by name and hazard class, and identify the basic design and construction features, including closures:
  - Compressed gas tube trailers (p. 161)
  - Corrosive liquid tanks (p. 160)
  - Cryogenic liquid tanks (p. 161)
  - Dry bulk cargo tanks (p. 161)
  - High-pressure tanks (p. 160)
  - Low-pressure chemical tanks (p. 160)
  - Nonpressure liquid tanks (p. 161)
- Identify the following facility storage tanks by name, identify the typical contents by name and hazard class, and identify the basic design and construction features, including closures:
  - Cryogenic liquid tank (p. 171)
  - Nonpressure tank (p. 168)
  - Pressure tank (p. 169–170)
- Identify the following nonbulk packaging by name, identify the typical contents by name and hazard class, and identify the basic design and construction features, including closures:
  - Bags (p. 148)
  - Carboys (p. 150)
  - Cylinders (p. 151)
  - Drums (p. 149)
- Identify the following radioactive material container/package by name, identify the typical contents by name, and identify the basic design and construction features, including closures:
  - Excepted (p. 166)
  - Industrial (p. 166)
  - Type A (p. 166, 167)
  - Type B (p. 166, 167)
  - Type C (p. 166, 167)

- Identify the following packaging by name, identify the typical contents by name and hazard class, and identify the basic design and construction features, including closures:
  - Intermediate Bulk Container (IBC) (p. 153)
  - Ton container (pressure drum) (p. 152)
- Identify the approximate capacity facility and transportation containers, and using the markings on the container and other available resources, identify the capacity (by weight or volume) of the following transportation vehicles and facility containers:
  - Cargo tanks (p. 158, 159)
  - Tank cars (p. 163–166)
  - Tank containers (p. 155–157)
  - Cryogenic liquid tank (p. 171)
  - Nonpressure tank (p. 168)
  - Pressure tank (p. 169–170)
- Identify the type or category of a radioactive material label, contents, activity, transport index, and criticality safety index as applicable for the label, and describe the radiation dose rates associated with label (p. 178–179).
- Describe the basic design and construction features of a pipeline and how a liquid petroleum liquid pipeline can carry different products, and identify the following:
  - Ownership of the line (p. 173, 174)
  - Procedures for checking for gas migration (p. 173)
  - Procedures for shutting down the line or controlling the leak (p. 174)
  - Type of product in the line (p. 174)

Skills Objectives

There are no Hazardous Materials Technician skills objectives in this chapter.

Hazardous Materials Incident Commander

Knowledge Objectives

There are no Incident Commander knowledge objectives in this chapter.

Skills Objectives

There are no Incident Commander skills objectives in this chapter.

You Are the HazMat Responder

At approximately 1600 hours on a cold February weekday, you are the senior officer at the scene of a box tractor-trailer incident involving a leaking unknown liquid substance. The incident was reported by the truck driver who noticed liquid leaking from his trailer while driving down an adjacent interstate highway. The driver pulled off the interstate and then parked his vehicle on a side street in a heavily congested area and dialed 911 for assistance.

Initial responders have isolated the area and have interviewed the driver who is very cooperative but has limited knowledge of his cargo. Shipping papers show the trailer contains approximately 60 55-gallon drums of mixed hazardous waste; most of the drums contain different pesticide solutions. Based on the physical location of the vehicle, significant traffic impact, and the amount of surrounding commercial exposures, this is starting to look like a very ugly rush hour!

1. What methods of identification could hazmat personnel use to confirm that the cargo is hazardous waste? How could hazmat personnel identify the unknown liquid substance coming from the trailer?
2. How could hazmat personnel identify the constituents of the individual hazardous waste drums?

At 1730 hours, local TV media crews are going to provide "live" reports from the scene. Your department Public Information Officer (PIO) is going to conduct the initial press briefing, however, you inform the PIO that recent hazmat incidents along the interstate highway corridor have generated many citizen complaints concerning the need to close down the highway. In short, there is a lot of sensitivity to this issue and the PIO should anticipate questions that may be raised.

1. What are the three key points you want to convey in the 1730 initial briefing? At approximately 1735 hours while the press briefing is taking place, hazmat personnel confirm through sampling that the unknown liquid is actually water. Further discussions with the shipper verify the drums were stored outdoors and were covered with snow and ice.
2. How would you brief the media after you discovered the leaking material was water? Would you specifically tell the media that the unknown material was water and not hazardous waste?

Introduction

In 1971, a railroad derailment in the Houston, Texas metropolitan area caused a breach in a pressure tank car transporting propane, which subsequently ignited. The propane fire then impinged on an adjoining tank car of vinyl chloride. After approximately 45 minutes of exposure to fire, the vinyl chloride tank car violently ruptured. As a result, a fire department training officer photographing the incident was killed and a number of emergency response personnel and civilians were injured. Because of the nature of the incident, the National Transportation Safety Board (NTSB) conducted an investigation and concluded that the following factors contributed to the severity of the accident:

- The lack of adequate training, information, and documented procedures for on-scene identification
- The lack of adequate assessment of threats to safety
- Reliance on firefighting recommendations that did not take into consideration the full range of hazards

The American philosopher George Santayana (1863–1952) stated, "Those who do not remember the past are condemned to relive it." Although the Houston incident occurred over 40 years ago, the lessons learned are timeless. Timely identification and verification of the hazmats involved are critical to the safe and effective management of a hazmat incident.

This chapter discusses the second step in the Eight Step Process©—Identify the Problem. Although this text is directed toward the Hazardous Materials Technician and the On-Scene Incident Commander, the material in this chapter has broad application to all first responders.

The hazardous materials recognition and identification process actually starts as soon as emergency responders are notified and dispatched to the incident. Remember the Eight Step Process©—Problem Identification cannot be safely and effectively accomplished if responders have not first controlled the incident scene. Likewise, strategic goals and tactical response objectives cannot be formulated if the nature of the problem is not defined. Remember—a problem well-defined is half-solved!

We will also review the basic principles of problem identification and methods of hazmat recognition, identification, and classification. The authors would especially like to recognize Charles J. Wright, Manager of Hazardous Materials Training for the Union Pacific Railroad (retired) and original member of the NFPA 472 Technical Committee, for his life-long efforts in developing many of the recognition and identification tools and training materials referenced in this chapter.

Basic Principles

Knowing the Enemy

Managing a hazmat incident is much like trying to manage a war. Neither effort will be very successful unless you have good intelligence and learn as much as possible about the "enemy"—where it can be found and its general tendencies and behaviors. Among the most critical tasks in managing a hazmat incident are surveying the incident scene to detect the presence of hazmats, identifying the nature of the problem and the materials involved, and identifying the type of hazmat container and the nature of its release. This effort is made more difficult by the number and variety of hazardous materials found in society and the increasing likelihood of hazmats being used for both criminal and terrorism events.

Historically, numerous public and private sector studies have evaluated hazmat transportation, hazmat incidents, the materials involved, and the source of their release Table 6-1.

Table 6-1 Top 10 Commodities Ranked by High Impact Casualties for 2000–2009.

Commodity Name	High Impact Casualties	Fatalities	Major Injuries	Incidents
1. Chlorine	92	9	83	48
2. Gasoline	49	30	19	1,306
3. Propylene	22	1	21	15
4. Diesel Fuel	19	12	7	573
5. Liquefied Petroleum Gas (LPG)	17	1	16	473
6. Sodium Hydroxide Solution	10	0	10	2,239
7. Sulfuric Acid	9	2	7	1,269
8. Ammonia, Anhydrous	8	1	7	317
9. Corrosive Liquids, Toxic, n.o.s.	8	0	8	511
10. Carbon Dioxide, refrigerated liquid	6	3	3	51

Source: U.S. Department of Transportation—Pipeline and Hazardous Materials Safety Administration, "Top Consequence Hazardous Materials by Commodities & Failure Modes—2005–2009," Washington, DC: U.S. DOT-PHMSA (September 2011)

Although the specific numbers may vary, these studies clearly show some trends and patterns, including the following:

- Approximately 75% of releases occur in facilities that produce, store, manufacture, or use chemicals; the remaining 25% occur during transportation.
- The majority of hazmat emergencies involve flammable and combustible liquids (e.g., gasoline-blended fuels, diesel fuel, and fuel oils) and compressed gases (e.g., chlorine, anhydrous ammonia, natural gas, propane, butane). The next most common hazard class is corrosive materials, specifically materials such as sulfuric acid and sodium hydroxide.
- Trailers on flat cars (TOFC) and containers on flat cars (COFC) account for the largest number of rail car movements.
- The top hazardous materials transported by rail include liquid petroleum gas (LPG), sodium hydroxide, sulfuric acid, anhydrous ammonia, chlorine, gasoline, and blended motor fuels.

Responder Safety Tip

Although responders may deal with thousands of chemicals, a rather small list of hazmats accounts for the majority of our problems. Be practical and realistic—the routine establishes the foundation on which the nonroutine must build. If you don't "have your act together" managing high-probability flammable liquid and gas emergencies, it's unlikely you'll perform well trying to manage an incident involving a more exotic hazardous material. These classes of materials have a long history of causing emergency responder injuries and fatalities. See the "flammable liquids and gases" section elsewhere in this text for additional information.

This background information is no substitute for conducting a hazard analysis and developing an Emergency Response Plan for your plant or community; however, it does support the point that, with proper analysis and planning, your hazmat emergency response program can become more effective and efficient.

Surveying the Incident

The identification process starts with a survey of the incident site and surrounding conditions. Responders should complete the following tasks: (1) Identify the hazmats involved; (2) identify the presence and condition of the containers involved; and (3) assess the conditions at the incident site, e.g., injuries, exposures, topography, and environment.

The identification process is built on the following basic elements:

1. *Recognition*—Recognize the presence of hazardous materials. Positive recognition automatically shifts the response into the hazmat mode. Basic recognition clues include occupancy and location, container shapes, markings and colors, placards and labels, shipping and facility documents, monitoring and detection instruments, and senses.
2. *Identification*—Identify the hazardous materials involved and the nature of the problem. Primary hazmat identification clues include container markings, shipping and facility documents, and monitoring and detection instruments. Regardless of the method of recognition or identification, always verify the information obtained. Don't take anybody's word at face value, including other responders. Always verify, verify, verify!
3. *Classification*—Determine the general hazard class or chemical family of the hazardous materials involved. A rule of thumb is, "If you can't identify, then try to classify." Basic classification clues include occupancy and location, container shapes, placards and labels, monitoring and detection instruments, and senses.

When dealing with unknown substances, responders should rely on monitoring instruments and chemical analytical kits, which use a systematic process to determine the identity and hazards. Although responders may not always be able to identify the hazmat(s) involved, they will usually be able to determine the hazard class or chemical family. **Table 6-2** shows the relationship between the clues and the hazmat/weapons of mass destruction (WMD) identification process.

In recent years there has been increasing emphasis on the potential use of hazardous materials and WMDs in terrorism or criminal events (e.g., illicit labs). These scenarios bring a

Table 6-2 Clues and the Identification Process

Clue	Hm Recognize	Hm Identify	Hm Classify	Weapons of Mass Destruction
Occupancy & Location	X		X	X
Container Shapes	X		X	X
Markings & Colors	X	X		
Labels & Placards	X		X	
Shipping Papers & Facility Documents	X	X		
Monitoring & Detection Equipment	X	X	X	X
Senses	X		X	X

Responder Safety Tip

Being able to determine what an unknown material *isn't* can be nearly as important as identifying what a material *is*.

number of new issues, including the use of weapons and armed assaults, secondary devices, and booby traps. However, from a health and safety perspective, the "bad stuff" is still a hazardous material and the basic concepts of recognition, identification, and verification still apply. You can't play in the WMD world if you don't first understand the hazmat response world!

Identification Methods and Procedures

In some situations, responders will initiate offensive response operations without immediately realizing hazardous materials are involved. In other cases, they may initiate aggressive, offensive operations before the material(s) are positively identified, verified, and evaluated. In either case, responders have placed themselves at an unacceptable risk.

Emergency responders rely on seven basic clues as part of their identification process. Look for hazmats in every incident; then identify or at least classify the material. The seven clues are:

1. Occupancy and location
2. Container shapes
3. Markings and colors
4. Placards and labels
5. Shipping papers and facility documents
6. Monitoring and detection equipment
7. Senses

These clues are also listed in relation to their distance from the hazmat **Figure 6-1**. The closer you are to the release site, the greater the likelihood you can accurately identify the material(s). The downside is that you also face a substantially higher risk of exposure and/or contamination. Your initial objective should be to learn as much as possible about the problem from as far away as possible.

To ensure personnel safety during size-up, many emergency responders rely on binoculars or spotting telescopes for identification. These tools have many applications, including surveying outdoor and indoor incidents to verify information such as container labels, markings, and other clues from a distance. Although they provide a narrow field of vision, telescopes can also be a useful tool for the identification process. In many facilities and some communities, closed circuit cameras and webcameras can also be used to facilitate the identification process.

Occupancy and Location

Hazardous materials surround us every day—not only in transportation and industrial facilities but in stores, hospitals, supermarkets, warehouses, garages, and even in our homes. These potential locations can be categorized into four basic areas—production, transportation, storage, and use **Figure 6-2**. Occupancy and location is the first clue for the recognition and classification of the hazmat/s involved.

The key for determining these potential sites is through the hazard analysis process. SARA, Title III and state/local "Right-to-Know" legislation also requires facilities to notify the fire department, the Local Emergency Planning Committee (LEPC), and other government agencies when on-site quantities exceed established threshold values (e.g., 500 pounds, 10,000 pounds, etc.). Hazard analysis information should include a list of the hazmats on site, their quantity and location, and hazards. In addition, safety data sheets (SDSs) or comparable information should be available. Many local jurisdictions require fixed facilities to use Knox Boxes™ or similar devices at the front gate or main entrance for the storage of all pertinent facility information. In other instances, response information is provided electronically through the facility intranet or web-based systems for rapid access.

Experienced responders are able to associate certain hazmats with different types of occupancies. Do you know what's in your

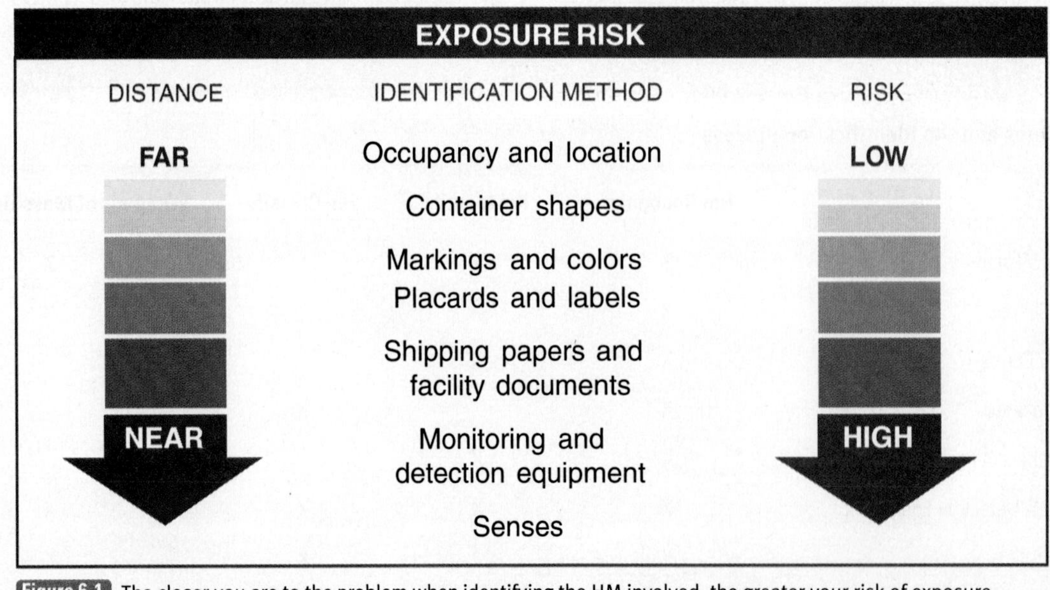

Figure 6-1 The closer you are to the problem when identifying the HM involved, the greater your risk of exposure.

Figure 6-2 Hazardous materials locations: the local emergency planning committee (LEPC) and the hazard analysis process are good tools for identifying hazardous materials locations within the plant and community.

community? Check it out—you might be surprised! Consider the following examples:

- You respond to a person down in a metal plating firm. What types of hazmats could you encounter? If you said sodium cyanide, potassium cyanide, and strong inorganic acids (e.g., nitric acid, sulfuric acid, chromic acid, etc.), you've got the right idea.
- Water treatment plants are found in almost every community, and the vast majority use chlorine in the treatment/disinfection process. Chlorine is typically transported and stored in these facilities in either 1-ton containers or 90-ton railroad tank cars, and represent both a hazmat and terrorist target hazard. In a worst-case scenario, a major release from a 90-ton chlorine tank car could generate a plume that could impact downwind areas over 10 miles away. Some water treatment plants are also using chlorine alternatives, such as sodium hypochlorite solutions.
- You have a facility that produces the metal bases for light bulbs, a nonhazardous product. No hazmat problems, right? A facility walk-through would show that anhydrous ammonia is used as a fuel source—"you mean that same anhydrous ammonia, which is nonflammable?" The anhydrous ammonia is piped to a dissociator, where it is split into hydrogen and nitrogen; the hydrogen is then used as a fuel source for burners and the nitrogen is released. Other hazmats found in this facility include sulfuric and hydrochloric acid, caustic cleaners, and oils, solvents and lubricants.
- You are involved in the planning process for a raid on a suspected illicit methamphetamine lab. What types of hazards may be present? First, there are the obvious concerns with the perpetrator's potential use of weapons and booby traps. From a hazmat perspective, there are flammable liquids and solvents, corrosives, reactives, and possibly anhydrous ammonia depending on the cooking process. Would you want to take a lab down while the bad guys are cooking?

Responders must recognize that hazmat exposures are no longer limited to industrial or transportation emergencies. The use of hazardous materials for criminal or terrorist purposes is a real threat. Not every emergency is a hazmat emergency; however, responders have the potential of being exposed to hazmats at ANY emergency.

Container Shapes

All hazardous materials are controlled as long as they remain within their container. The size, shape, and construction features of a container/packaging are the second clue to the standard hazmat identification process and can be used as a clue for both hazmat recognition and classification. The U.S. Department of Transportation (DOT) defines **packaging** as a receptacle, which may require an outer packaging and any other components or materials necessary for the receptacle to perform its containment function and to ensure compliance with minimum packaging requirements.

Packaging used for transporting hazardous materials is regulated by the DOT. Other types of containers are used only

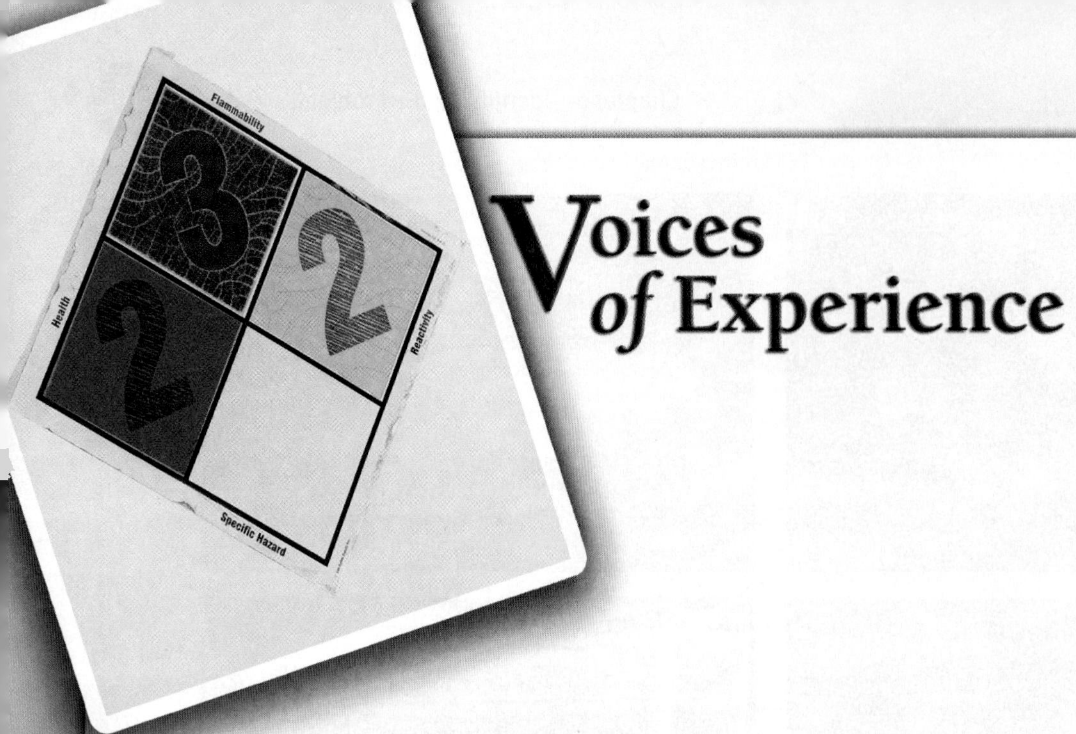

Voices of Experience

I would like to describe briefly a hazardous materials incident I encountered early in my fire service career, when I responded to a call for a propane gas grill on fire in a residential backyard. To most responders, including myself, this seemed rather ordinary. Most gas grill fires either have a broken hose to the burners or a defective valve. In this particular case, the hose was defective.

The grill itself was portable and on wheels and had a storage cabinet below the burner assembly where two twenty pound cylinders could be stored. Only one cylinder could be hooked-up at a time and it was the one that was hooked-up that had the leaking hose on fire. Noticing that the fire was coming from the defective hose and was starting to impinge on the spare cylinder, my first thought was to remove the spare cylinder and eliminate it as part of the problem.

As I approached the fire, I kept low to the ground by crawling. The fire was not large and, in my mind, it was still a routine incident and I didn't expect any problems. My plan was to remove the spare cylinder, not knowing whether it was empty or full. As I grabbed the spare cylinder and pulled it from the cabinet, to my surprise the attached cylinder fell over and shot a fire ball just above me. Fellow fire fighters thought I was engulfed in the fire ball, but, fortunately, the fire ball went right over me as I lay on the ground. I had not realized nor saw that the spare tank was actually holding up the burning cylinder.

At this incident, a hazard assessment and risk evaluation should have been given more consideration. Things are not always what they appear to be. As stated earlier, this was considered a routine call. We should all realize there is nothing routine about the way we conduct business or fight fires. Every situation is unique, and if we don't play our "A" game, we lose.

Chief Richard Kosmoski
President, New Jersey Volunteer Fire Chiefs Association

at fixed facilities, such as process towers, piping systems, and reactors. Packaging used for production, storage, and use is usually built according to nationally recognized consensus standards, such as those provided by the American Society of Mechanical Engineers (ASME), the American Society of Testing and Materials (ASTM), and the American National Standards Institute (ANSI). For example, welded steel atmospheric pressure petroleum storage tanks are constructed according to specifications of the American Petroleum Institute (API), Standard 650. This standard is, in turn, referenced by the NFPA *Technical Standard on Flammable and Combustible Liquids (NFPA 30)*.

Packaging is divided into three general groups: nonbulk packaging, bulk packaging, and facility containment systems **Table 6-3**.

Nonbulk Packaging

Nonbulk packaging will hold solid, liquid, or gaseous materials. DOT provides the following definitions:

- Liquid—capacity of 119 gallons (450 liters) or less
- Solid—net mass of 882 pounds (400 kg) or less for solids, or capacity of 119 gallons (450 liters) or less
- Compressed Gas—water capacity of 1,001 pounds (454 kg) or less

Table 6-3 Examples of Nonbulk, Bulk and Facility Containment Systems

Transportation Containment Systems		Facility Containment Systems
Nonbulk	**Bulk**	
Bags	Bulk Bags	Buildings
Bottles	Bulk Boxes	Machinery
Boxes	Cargo Tanks	Open Piles (outdoors & indoors)
Carboys	Covered Hopper Cars	Piping
Cylinders	Freight Containers	Reactors (chemical & nuclear)
Drums	Gondolas	Storage bins, cabinets or shelves
Jerricans	Pneumatic Hopper Trailers	
Multicell	Portable Tanks and Bins	
Wooden Barrels	Protective Overpacks for Radioactive Materials	
	Tank Cars	
	Ton Containers	
	Van Trailers	

Source: Chemical Manufacturers Association and Association of American Railroads, *Technical Bulletin on Packaging for Hazardous and Non-Hazardous Materials.* Washington, DC: Chemical Manufacturers Association (1989).

Nonbulk packaging may consist of single packaging (e.g., drum, carboy, cylinder) or combination packaging—one or more inner packages inside of an outer packaging (e.g., glass bottles inside a fiberboard box, infectious disease sample containers). Nonbulk packaging may be palletized or placed in overpacks for transport in vehicles, vessels, and freight containers. Examples include bags, boxes, carboys, cylinders, and drums.

Bulk Packaging

Bulk packaging is any packaging, including transport vehicles, that meets the following DOT definition:

- Liquid—capacity greater than 119 gallons (450 liters).
- Solid—net mass greater than 882 pounds (400 kg) for solids, or capacity greater than 119 gallons (450 liters).
- Compressed Gas—water capacity greater than 1,001 pounds (454 kg).

Bulk packages can be an integral part of a transport vehicle (e.g., cargo tank trucks, tank cars, and barges) or packaging placed on or in a transport vehicle (e.g., intermodal portable tanks, ton containers, intermediate bulk containers).

Facility Containment Systems

Facility containment systems refer to packaging, containers, and containment systems that are part of a fixed facility's operations. Depending on the nature of the facility and its operations, the types of containment systems can vary greatly. Examples include pressurized and nonpressurized storage tanks, process towers, chemical and nuclear reactors, piping systems, pumps, storage bins and cabinets, dryers and degreasers, machinery, and so forth.

Nonbulk Packaging

Nonbulk packaging is constructed to performance or specification standards mandated by DOT. Performance tests for nonbulk packaging include a drop test, leak-proof test, hydrostatic test, stacking test, and vibration standard test.

The type of material and the end use of the product will determine the type of packaging. For example, packaging for household or consumer commodities is usually different from industrial applications, even for the same material. Industrial grade materials are usually more concentrated, while the household version is often diluted with other materials, such as water or a solvent.

The type of container is typically a good clue to its hazards and contents. In general, the more substantial or durable the container, the greater the hazards of the material, especially when dealing with drums. **Figures 6-3 through 6-10** depict the design, construction features, and contents found in the common types of nonbulk packaging.

Bulk Packaging

Bulk packaging may be placed on or in a transport vehicle or may be an integral part of a transport vehicle (e.g., cargo tank truck, tank car). **Figures 6-11 through 6-24** provide additional information on the design and construction features, as well as the contents found in both bulk transportation and storage containers:

- Bulk Packaging Placed on or in a Transport Vehicle **Figures 6-11 through 6-21**
- Cargo Tank Trucks **Figure 6-22**
- Railroad Tank Cars **Figure 6-24**

NONBULK PACKAGING

BAGS

Stiched bag

Folded and glued bag

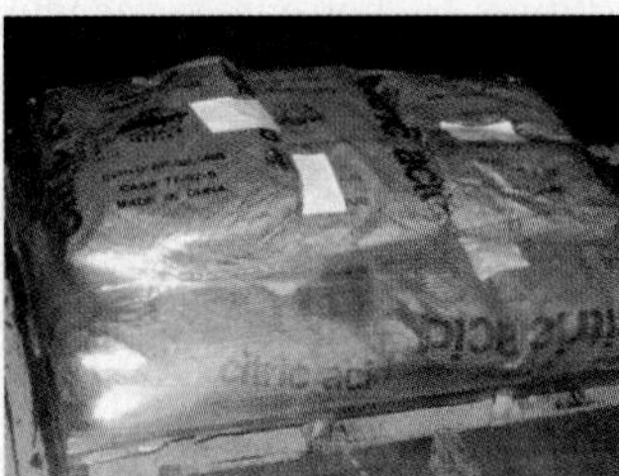

Shrink-wrapped bags

DESIGN AND CONSTRUCTION

- Flexible packaging constructed of cloth, burlap, kraft paper, plastic, or a combination of these materials.
- Closed by folding and gluing, heat sealing, stitching, crimping with metal, or twisting and tying.
- Typically contain up to 100 lbs. of material, although large tote bags weighing up to 500 lbs. may be found.
- Tote bags may be palletized or hung inside of boxed vehicles.

CONTENTS/HAZARD CLASS

- Solid materials, including explosives, flammable solids, oxidizers and organic peroxides, poisons and corrosives.
- Common examples include fertilizers, pesticides, caustic powders, and many non-hazardous materials (e.g., food products).
- Large reinforced bladders may be used for the transport of non-hazardous materials.

Figure 6-3 Nonbulk Packaging: Bags.

BOTTLES

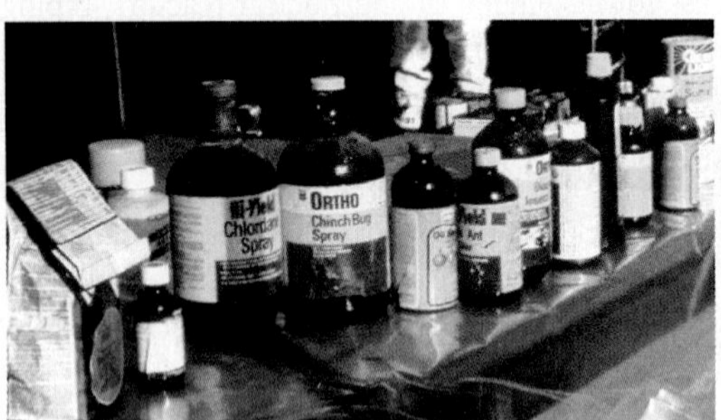

Protected bottles

Plastic bottles

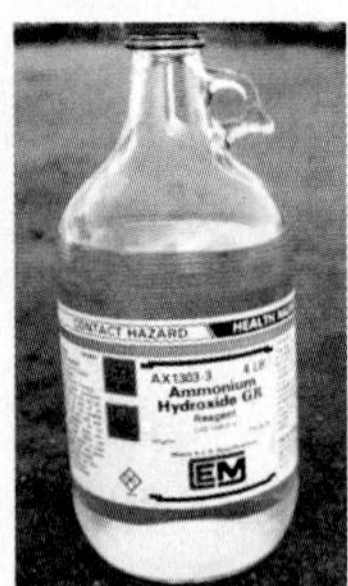

Glass bottle

DESIGN AND CONSTRUCTION

- Constructed of glass, plastic and metal, although ceramic may be found. Some newer generation of glass bottles may be encased in an outer plastic coating.
- Closed head with threaded caps or stoppers.
  Range from several ounces to >20 gal.
- Usually placed within an outer package for transport, such as a box.
- May be referred to as jugs or jars.

CONTENTS/HAZARD CLASS

- Liquid and solid materials, including flammable and combustible liquids, poison liquids, and corrosives.
- Common examples include laboratory reagents and corrosive liquids.
- Generally, brown glass bottles are used used for light sensitive and reactive materials, such as ethers and organic peroxides.

Figure 6-4 Nonbulk Packaging: Bottles.

DRUMS

5 gallon drum

Closed head stainless steel drums

Overpack drum

Closed head plastic drums

Open head plastic drum

Open head fiber drum

DESIGN AND CONSTRUCTION

- A cylindrical package constructed of metal, plastic, fiberboard, or other suitable materials.
- Typical capacity is 55 gallons, although smaller (e.g., pails) and larger drums can be found.
- May be open-head (i.e., drum lid comes off) or closed-head (i.e., drum lid is fixed and cannot be removed). Closed head drums tend to be used for the more hazardous materials, and usually contain 2 openings—2-inch and 3/4-inch diameter plugs called "bungs." Bungs can be a source of leaks.
- Chimes and seams can be a source of leaks on rusty steel drums.
  - Overpack container construction.
  - Open-head drum constructed of steel with an inner coating or polyethylene. Lid may be attached by either a locking ring or screw-type design.
  - Sizes range from lab packs to 110-gallon overpacks.
  - Should be certified for use as a DOT salvage drum. Overpacks used by emergency response units should comply with UN Packing Group 1 (X-rating) requirements.
  - Overpack must be compatible with the contents of the leaking nonbulk container.

CONTENTS/HAZARD CLASS

- Liquids, solids and mixtures, including flammable and combustible liquids, flammable solids, oxidizers, organic peroxides, poisons, corrosives, and radioactive materials.
- As a general rule (with a lot of exceptions!), liquids are found in closed head drums, while solids are typically found in open head drums.
- Steel Drums—commonly used for flammable and combustible liquids, poisons, mild corrosives, and liquids used in food production. Stainless Steel Drums—commonly used for more hazardous and reactive liquids, such as nitric acid or oleum (super concentrated sulfuric acid @ 120–160% concentrations).
- Aluminum Drums—hold materials that would react with rust or with steel, and cannot be shipped in a poly drum. Contents are often combustible or toxic, such as some pesticides. Caustic corrosives would NOT be shipped in an aluminum drum.
- Plastic (Poly) Drums—commonly used for corrosive liquids, some flammable or combustible liquids, and food production liquids.
- Fiberboard Drums—commonly hold solid materials, such as powders, granules or pellets. May be toxic or corrosive, or may present little or no risk. Are usually lined with a poly liner; when lined, may also hold gels and some low hazard liquids.
- Overpack Drums—hold liquids, solids, and mixtures. Used for transporting damaged or leaking nonbulk containers of all types and sizes, as well as hazardous waste materials (e.g., contaminated PPE, spill control materials, etc.).

Figure 6-5 Nonbulk Packaging: Drums.

BOXES

Fiberboard box

Wooden box

Divided fiber board box

DESIGN AND CONSTRUCTION

- Rigid packaging that completely encloses the contents.
- Commonly used as the outside packaging for other nonbulk packages. Inner packaging may be surrounded with absorbent materials.
- Constructed of fiberboard, wood, metal, plywood, plastic, or other suitable materials.
- Fiberboard boxes may contain up to 65 lbs. of material; wooden boxes up to 550 lbs.

CONTENTS/HAZARD CLASS

- Liquid and solid materials. Almost any hazmat can be found inside a box.
- Combination packaging using an outer box with inner packaging commonly used for infectious disease samples and radioactive materials.

Figure 6-6 Nonbulk Packaging: Boxes.

MULTI-CELL PACKAGING

Base element

DESIGN AND CONSTRUCTION

- Consist of form-fitting, expanded polystyrene box encasing one or more bottles.
- When transporting certain hazmats, DOT limits the capacity to no more than 6 bottles and a maximum bottle capacity of 4 liters.

CONTENTS/HAZARD CLASS

- Liquids, including flammable and combustible liquids, poison liquids, and corrosives.
- Common examples include laboratory reagents, specialty chemicals, solvents, and corrosive liquids.

Figure 6-7 Nonbulk Packaging: Multi-cell Packaging.

CARBOYS

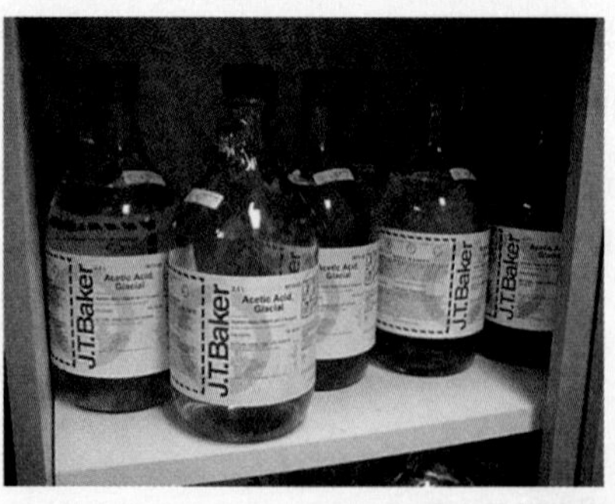

DESIGN AND CONSTRUCTION

- Glass or plastic bottle-like containers encased in an outer packaging, such as a polystyrene box, wooden crate, etc.
- Range in capacity up to 20 gallons.
- Typically found in laboratories and chemical manufacturing facilities.

CONTENTS/HAZARD CLASS

- Liquids, including flammable and combustible liquids, and corrosives.
- Common examples include acids and caustics.

Figure 6-8 Nonbulk Packaging: Carboys.

CYLINDERS

DESIGN AND CONSTRUCTION

- Three basic types of cylinders—aerosol containers, uninsulated cylinders, and cryogenic (insulated) cylinders.
- May have an outer packaging.
- Common service pressures can range from <100 psi to approximately 5,000 psi, although pressures as high as 10,000 psi can be found.
- Majority are equipped with one or more pressure relief devices, such as a relief valve (pressure actuated), rupture disc (pressure actuated) or fusible plug (temperature actuated).
- Protective bonnet often covers the valve(s) during transportation.
- If in-service, will have a regulator attached to control the rate of discharged gas.

CONTENTS/HAZARD CLASS

- Pressurized, liquefied, and dissolved gases, including flammable and combustible liquids and gases, poisons, and reactive liquids and gases.
- Although there are several voluntary color-code systems, none are mandatory. The only reliable method to identify cylinder contents is to check the attached DOT label.

Aerosol Containers

DESIGN AND CONSTRUCTION

- Small cylinders constructed of metal, glass or plastic.
- Transported in boxes.
- Propellant may be a flammable (e.g., propane) or inert gases.

CONTENTS/HAZARD CLASS

- Commonly used for consumer commodities, such as lubricants, paints, toiletries, and pesticides.

Uninsulated Containers

DESIGN AND CONSTRUCTION

- Constructed of steel, although aluminum or fiberglass aluminum combinations may be found.
- Range of sizes, from small lecture cylinders containing electronic gases (e.g., phosphine, silane) to 420 lb. propane cylinders.
- Do not have uniform taper on the cylinder head, and cylinder valve inlet and outlet threads will vary depending upon contents.
- Valve may be either right-handed or left-handed thread to close.
- Protective bonnet often covers the valve(s) during transportation.

CONTENTS/HAZARD CLASS

- Flammable and non-flammable pressurized gases, including nitrogen, hydrogen, and oxygen.
- Flammable and toxic liquefied gases (e.g., butane, chlorine, and anhydrous ammonia).
- Dissolved gases (e.g., acetylene in acetone or dimethylformamide).
- Pyrophoric or reactive liquids and gases (e.g., aluminum alkyls, diborane, phosphine).

Cryogenic (Insulated) Containers

DESIGN AND CONSTRUCTION

- Constructed of an insulated metal cylinder within an outer protective metal jacket.
- Space between the inner container and outer jacket may be under a vacuum.
- Designed for a specific range of service pressures and temperatures, depending upon the cryogenic liquids being stored.
- Venting of vapors is a normal occurrence for liquid nitrogen cylinders.
- Commonly found in laboratories and many industrial facilities.

CONTENTS/HAZARD CLASS

- Flammable and non-flammable cryogenic liquids, including liquid nitrogen, liquid hydrogen, liquid oxygen (LOX), and liquefied natural gas (LNG).

Figure 6-9 Nonbulk Packaging: Cylinders.

CYLINDERS (*continued*)

Dewars Containers

DESIGN AND CONSTRUCTION

- Open head, non-pressurized, vacuum jacketed vessels used for handling cryogenic liquids. Valving may be installed on the container head.
- Sizes range from 1 liter to >1,000 liters depending upon working requirements.
- Commonly found in laboratories. May be fixed or wheeled units, and will often have dispensing tubes attached.

CONTENTS/HAZARD CLASS

- Flammable and non-flammable cryogenic liquids, such as liquid helium and liquid hydrogen.

Figure 6-10 Nonbulk Packaging: Cylinders.

BULK PACKAGING PLACED ON/IN TRANSPORT VEHICLES

TON CONTAINERS

DESIGN AND CONSTRUCTION

- Cylindrical pressure tanks approximately 3 ft. diameter and 8 ft. long with concave heads. Designed to lay horizontally on its side.
- Transport 1 ton (2,000 lbs) of product; actual cylinder weighs approximately 800 to 1,300 lbs. empty.
- Two identical container valves are found on one end under a detachable protective hood. Valves should be in the 12 o'clock (vapor) and 6 o'clock (liquid) position for normal operations.
- Chlorine and sulfur dioxide containers have pressure relief devices (fusible plugs); phosgene containers have no pressure relief devices.
- Commonly transported on flat-bed vehicles. Multi-unit tank cars (i.e., railroad flatcar transporting 15 one ton containers) may also be found, although use is limited.

CONTENTS/HAZARD CLASS

- Liquefied gases, including chlorine, sulfur dioxide, phosgene, and some nonflammable fluorocarbon refrigerant gases.
- Containers with convex design usually contain refrigerant gases.
- Containers with concave design usually contain chlorine, phosgene, etc.

Figure 6-11 Bulk Packaging Placed On/In Transport Vehicles/Ton Containers

INTERMEDIATE BULK CONTAINERS (IBCs)

Flexible Containers or "Super Sacks"

DESIGN AND CONSTRUCTION

- Pre-formed packaging constructed of flexible materials (e.g., polypropylene, cloth or woven fabric). Available with coatings or liners dependent upon materials being handled. Often tied at the top with a bottom discharge.
- Flexible bag may also be found inside a heavy cardboard box, which may then be shrink-wrapped.
- Standard sizes range from 15 to 85 ft.3, with capacities of 500 to 5,000 lbs.
- Transported in both open and closed transport vehicles. Bulk bags may be palletized or hung inside of boxed vehicles.

CONTENTS/HAZARD CLASS

- Solid materials, including explosives, flammable solids, oxidizers and organic peroxides, poisons and corrosives.
- Common examples include fertilizers, pesticides, caustic powders, water treatment chemicals, and Class 9 materials.
- Are not authorized for liquid hazardous materials except under a DOT exemption. Large reinforced rubber bladders may be used for the transport of non-hazardous materials.

Figure 6-12 Bulk Packaging Placed On/In Transport Vehicles: Intermediate Bulk Containers (IBCs)/Flexible Containers or "Super Sacks".

RIGID CONTAINERS—POLYETHYLENE AND STEEL TANKS OR "COMPOSITE IBCs"

DESIGN AND CONSTRUCTION

- Consist of high-density polyethylene (HDPE) tank inside of a rigid steel frame, with a top tank fill opening and top or bottom discharge piping. Some may have a rigid polyethylene frame.
- Approximately 4 ft. square and 6 ft. high, with a capacity of 300 to 500 gallons.
- May have a "volume marking" allowing responders to see how much product is in the container.
- Polyethylene tank may separate from the steel framing, creating mechanical stress and container failure.
- While many IBCs are approved for the transportation of Class IB and IC flammable liquids, most fire codes prohibit the storage of any flammable liquid in these containers.

CONTENTS/HAZARD CLASS

- Liquid materials, including flammable and combustible liquids, oxidizers, corrosives and Class 9 materials.
- Used for Packing Group II and III products.

Figure 6-13 Bulk Packaging Placed On/In Transport Vehicles: Intermediate Bulk Containers (IBCs): Rigid Containers—Polyethlene and Steel Tanks or "Totes"

RIGID CONTAINERS—PORTABLE TANKS OR "TOTES"

DESIGN AND CONSTRUCTION

- Circular or rectangular metal tank constructed with a top fill opening, and often with a bottom discharge piping or opening. Materials include steel, stainless steel, aluminum, and may be lined.
- Approximately 4 ft. square and 6 ft. high, with a capacity of 300 to 500 gallons.
- May have low pressures, depending upon materials being transported. Any pressure relief devices are typically incorporated into the container fill lid.
- **WARNING:** The bottom discharge valve is often a quarter-turn ball valve. Unlike typical valves, when the valve handle is in the "in-line" or parallel position, it is CLOSED; when in the cross or perpendicular position, it is OPEN. Look for labels around the valve that should clearly indicate valve position guidelines.
- Designed to be moved with forklifts and cranes in the marine industry.

CONTENTS/HAZARD CLASS

- Liquid materials, including flammable and combustible liquids, oxidizers, and corrosives.

Figure 6-14 Bulk Packaging Placed On/In Transport Vehicles: Intermediate Bulk Containers (IBCs).

PORTABLE BINS

DESIGN AND CONSTRUCTION

- Constructed of steel or stainless steel, and designed to be moved by forklift vehicles or overhead cranes.
- Approximately 4 ft. square and 6 ft. high. Appear similar to IBC portable tanks, except are used for the transport of solids.
- Depending upon contents, may weigh 3+ tons.
- Usually transported on flat bed vehicles or rail cars.

CONTENTS/HAZARD CLASS

- Solid materials, including flammable solids, oxidizers, and corrosives.
- Common examples include ammonium nitrate fertilizers, and calcium carbide.

Figure 6-15 Bulk Packaging Placed On/In Transport Vehicles: Portable Bins.

INTERMODAL PORTABLE TANK CONTAINERS (i.e., Tank Containers or "Iso-Tanks")

Intermodal tank containers are built to both U.S. and international standards. Tank containers are designed for multi-modal use, including marine, rail and highway. In addition, they may be integrated into fixed facility process operations. Although the tank itself may vary, there are common design and construction features that can be found on all tank containers. Key construction features include:

- Usually consists of a single, non-compartmented vessel within a sturdy metal supporting frame that allows the unit to be lifted by specialized handling cranes and vehicles. Most tanks are built to ASME pressure vessel standards.
- Majority of containers are 20 ft. long, 8 ft. wide, and 8 to 9-1/2 ft. high.
- Two basic types of supporting frames: (1) box type that encloses the container in a cage-like framework with continuous side rails; and (2) beam-type that relies upon the inherent strength of the tank as a beam.
- Corner castings used to secure the tank and lift it with standard container handling equipment.
- Emergency remote shut-off device for containers with bottom outlet "foot" valves.
- May be lined, insulated and/or jacketed, have refrigeration unit, or be heated.

Figure 6-16 Bulk Packaging Placed On/In Transport Vehicles: Intermodal Portable Tank Containers/Tank Containers or ISO-Tanks.

Tank Container Markings can be used to gain knowledge about the tank design and construction features. Additional technical, approval, and operational data can also be found on the dataplate that is permanently attached to the tank or frame. Tank container markings include:

- *Reporting Marks and Number.* Intermodal portable tank containers are registered with the International Container Bureau in France, and must be marked with reporting marks and a tank number. The initials indicate ownership of the tank and the tank number identifies the specific tank. These markings are generally found on the right-hand side of the tank (as you face it from either side), and on both ends.
- *Specification Marking.* The specification marking indicates the standards to which a portable tank was built. These markings will be on both sides of the tank, generally near the tank's reporting marks and number. Examples of specification markings are IM-101, IM-102, and Spec. 51.
- *DOT Exemption Marking.* Exemptions are sometimes authorized from DOT regulations. In these cases, the outside of each package/container must be plainly and durably marked "DOT-E" followed by the exemption number assigned (e.g., DOT-E8623). On intermodal tanks, these markings must be in 2-inch letters.
- *AAR-600 Marking.* For interchange purposes in rail transportation, intermodal containers should conform to the requirements of Section 600—"Specification for Acceptability of Tank Containers" of the Association of American Railroads (AAR) Specifications for Tank Cars. Tanks meeting these requirements will display the "AAR 600" marking in 2-inch letters on both side's near the tank's reporting marks and number. The "AAR 600" marking indicates tanks that can be used for regulated materials. An "AAR-600NR" marking indicates the container can only be used for non-regulated materials.
- *Country, Size, and Type Markings.* The tank will display a size/type code (see Figure 6-21—BER-2275). The country code (two or three letters) indicates the tank's country of registry.

The four digit size/type code follows the country code. The first two numbers jointly indicate the container's length and height. The second pair of numbers is the type code that indicates the pressure range of the tank.

Common Size Codes

20 = 20 ft. (8 ft. high)
22 = 20 ft. (8 ft. 6 inches high)
24 = 20 ft. (>8 ft. 6 inches high)

Common Type Codes—Maximum Allowable Working Pressure

Nonhazardous Commodities

T0 (70) = <0.44 (6.4 psig) Bar test pressure
T1 (71) = 0.44 (6.4 psig) to 1.47 (21.3 psig) Bar test pressure
T2 (72) = 1.47 (21.3 psig) to 2.94 (42.6 psig) Bar test pressure
T3 (73) = spare

Hazardous Commodities (NOTE: 1 Bar = 14.5 psi)

T4 (74) = <1.47 (21.3 psig) Bar test pressure
T5 (75) = 1.47 (21.3 psig) to 2.58 (37.4 psig) Bar test pressure
T6 (76) = 2.58 (37.4 psig) to 2.94 (42.6 psig) Bar test pressure
T7 (77) = 2.94 (42.6 psig) to 3.93 (57.0 psig) Bar test pressure
T8 (78) = >3.93 (57.0 psig) Bar test pressure
T9 (79) = spare

NOTE: The "T" designation is the new type code, but many of the older "7" designations still exist in the international tank container fleet.

Figure 6-17 Bulk Packaging Placed On/In Transport Vehicles. Intermodal Portable Tank Markings. *(Continued)*

Figure 6-17

NON-PRESSURE TANK CONTAINERS (IM 101 And IM 102—Also Known As IMO Type 1 And IMO Type 2 Internationally) (UN Portable Tank T Codes T1 – T-22)

DESIGN AND CONSTRUCTION

- Although classified as non-pressure, can have maximum allowable working pressures (MAWP) up to 100 psig.
- Normal capacities range from 5,000 to 6,340 gal. Capacity can be determined from tank dataplate or tank markings.
- Two basic types of non-pressure containers: IM-101 (also known as IMO Type 1—International Maritime Organization) and IM-102 (aka IMO Type 2).
- IM 101—MAWP range of 25.4 (1.75 bar) to 100 psig (6.8 bar).
- IM 102—MAWP range of 14.5 (1.0 bar) to 25.4 psig (1.75 bar).
- Valves and fittings are found in the top spillbox, and may include manway, eduction pipe or "dip tube," dipstick, airline connections, and pressure/vacuum relief valves.
- May be top or bottom loaded/unloaded, depending upon product and facilities.

CONTENTS/HAZARD CLASS

- IM 101 tanks transport hazardous and non-hazardous liquid materials, including flammable liquids with flash points <32° F (0° C), poisons and environmentally harmful liquids, and corrosives.
- IM 102 tanks transport hazardous and non-hazardous liquid materials, including flammable liquids with a flash point ranging from 32° F (0° C) to 140° F (60° C), whiskey, alcohols, poison liquids (e.g., pesticides), foodgrade commodities, and some corrosives.
- IM-101 and IM-102 account for over 90% of the total number of tank containers worldwide.

Figure 6-18 Bulk Packaging Placed On/In Transport Vehicles. Intermodal Portable Tank Containers.

PRESSURE TANK CONTAINERS (DOT Spec 51 Or IMO Type 5) (UN Portable Tank T Code 50)

DESIGN AND CONSTRUCTION

- Pressurized containers within a box or beam supporting frame.
  Normal capacities range from 4,500 to 5,000 gallons, although Spec. 51 containers as small as 100 gallons may be found for specialized materials (e.g., pyrophoric liquids). Capacity can be determined from tank dataplate or tank markings.
- Service pressures range from 100 (6.8 bar) to 500 psi (34 bar).
- Valves and fittings are usually enclosed within one end under a cover and can include liquid and vapor valves, gauging devices, and thermometer wells.
- Pressure relief devices mounted on top of the container; usually found in pairs. A sun shade may also cover the vapor space area on top of the tank container.

CONTENTS/HAZARD CLASS

- Liquefied gases (e.g., anhydrous ammonia, LPG, refrigerant gases).
- Other highly regulated materials, such as UN Packaging Group 1 materials, high vapor pressure flammable liquids, highly toxic poison liquids or gases, and pyrophoric liquids (e.g., aluminum alkyls).

Figure 6-19 Bulk Packaging Placed On/In Transport Vehicles. Intermodal Portable Tank Containers.

Cryogenic Liquid for Refrigerated Gases (UN Portable Tank IMO Type 7 or UN Portable Tank T Code 75)

DESIGN AND CONSTRUCTION

- Pressurized containers within a box or beam supporting frame. Must meet DOT Spec 51 standards.
- Normal capacities range from 4,500 to 5,000 gallons. Capacity can be determined from tank dataplate or tank markings.
- Consist of a tank within a tank design, with an insulating space between the inner tank and outer shell.
- Insulating space is normally maintained under vacuum.

CONTENTS/HAZARD CLASS

- Cryogenic liquids and refrigerant gases, including liquid nitrogen, liquid argon, liquid oxygen, and liquid helium.

Figure 6-20 Specialized Tank Containers–Cryogenics.

Multiple Element Gas Containers for Non-Refrigerated Compressed Gases at High Pressures (Based on types of cylinders used in tube modules 3AX, 3AAAX, or 3T)

DESIGN AND CONSTRUCTION

- Consist of seamless steel cylinders from 9 to 48 inches in diameter, permanently mounted inside an open frame.
- Service pressures up to 2,400 psi and higher.
- Box-like compartment at one end enclose all valves.
- May also find larger 35 ft. modules.

CONTENTS/HAZARD CLASS

- Pressurized gases, such as oxygen, hydrogen, nitrogen, helium, ethane and silane.

Figure 6-21 Specialized Tank Containers–Tube Modules.

CARGO TANK TRUCKS

Cargo tank trucks also have various markings that, when combined with container shapes, can be important clues in the overall identification process. These include company names and logos, vehicle identification numbers, and the manufacturers specification plate.

Manufacturers Specification Plate. Mounted on the front third of the tank on the driver's side near the landing gear. For tanks constructed prior to July 1, 1985, the plate may be found on the right side near the landing gear or on the nose of the trailer. Key information on the plate includes:

- Name of the manufacturer, manufacturer's serial number, and date of manufacture.
- DOT container specification number (e.g., MC-306/DOT-406)
- Materials of construction, including head, shell, weld and lining material (if applicable).
- Maximum allowable working pressure - psig (MAWP)
- Nominal compartments capacity (in gallons) of each compartment, front to rear.
- Water capacity of the tank in pounds. This will be found on pressure vessels, such as MC-331 and MC-338 cargo tanks. Remember – 1 gallon of water weighs 8.35 lbs.
- Maximum product load (in pounds).
- Maximum temperature is the maximum temperature at which the cargo tank will safely carry material without failure.
- Date of manufacture.

Materials of construction and lining materials are critical information if the product must be transferred to another tank or container. Tank shell material codes include the following:

AL	Aluminum
CS	Carbon Steel
HSLA	High Strength Low Alloy Steel
HSLA-QT	High Strength Low Alloy, Quenched and Tempered Steel
MS	Mild Steel
SS	Stainless Steel

For example, an DOT-406 AL specification would indicate that the cargo tank is an DOT-406 (low pressure cargo tank) constructed of aluminum.

Some cargo tanks are designed to multiple container specifications that allow the unit to transport more than one type of commodity. The most common multi-purpose configurations are the combination DOT-407 and DOT-412 units. In addition to the manufacturer's specification plate, these tanks have a second "multi-purpose" plate that identifies the specification under which the cargo tank is being operated. These plates are either color-coded or stamped as such, as are the fittings that are added to make the cargo tank meet the respective specifications. A sliding shield exposes the specification plate currently in use. However, these plates may move as a result of an accident or may not be properly positioned by the vehicle driver. The respective color codes are:

MC-306 or DOT-406	Red Plate and Fittings
MC-307 or DOT-407	Green Plate and Fittings
MC-312 or DOT-412	Yellow Plate and Fittings
Non-Specification Tank	Blue Plate and Fittings

ASME Containers. Cargo tanks that are pressurized and that are built to the requirements of the American Society of Mechanical Engineers (ASME) Pressure Vessel Code will have this information stamped on the specification plate or will carry a separate plate outlining the ASME requirements. The ASME plate is easily identified by an embossed "U" in the upper left corner of the plate. Examples of tanks subject to the ASME requirements include MC-331 high pressure and MC-338 cryogenic liquid cargo tanks.

Figure 6-22 Cargo Tank Trucks: Specification Plate and Inspection Markings. *(Continued)*

Inspection Markings. Cargo tanks are subject to various inspections, and inspection markings can be found on the front head or tank shell near the specification plate. These markings will indicate the date and type of inspection conducted. For example, MC-331 cargo tanks are required to be inspected annually and be tested every 5 years; an example of a marking and abbreviations are listed below.

NQT
TESTED 7 -02 P I
INSPECTED 9 -02 V K

ABBREVIATION	TEST INSPECTION TYPE
V	External Visual Inspection
I	Internal Visual Inspection
K	Leakage Test
L	Lining Inspection
P	Pressure Test
T	Thickness Test (tanks transporting corrosives)

MC-331 Tank Shell Markings. MC-331 cargo tanks will also be stenciled with the letters "QT" or "NQT" near the specification plate. During manufacture, the tank shell steel is hardened through different heat treatment or quenching processes—quench tempering (QT) or non-quench tempering (NQT). Understanding these points can become a critical consideration if an MC-331 cargo tank is involved in a severe rollover incident.

Responders should understand the following basic principles regarding QT and NQT tanks:

- QT tanks are heated to a high temperature, then rapidly cooled. This produces a metal that is quite hard and strong, but can also be somewhat brittle. The brittleness of a metal can become an issue if the cargo tank is subjected to extensive mechanical stress, such as when the tank is involved in a serious rollover or strikes a concrete abutment.
- NQT tank metals are heated to specified temperature for a specified period of time, then slowly cooled through the use of some type of heated media (e.g., melted salt). This produces a metal that is hardened, but not as hard as QT metal. In addition, NQT metals are not as brittle as QT metals.

Tank Color. MC-331 specification tanks are required to have the upper two-thirds of the tank painted a light reflective color. Corrosive tank trucks (MC-312/DOT-412) will often have a contrasting color band on the tank in line with the dome covers and overturn protection. This band is usually a corrosive-resistant paint or rubber material.

Figure 6-22

MC-306 / DOT-406

Atmospheric Pressure Cargo Tank Truck

DESIGN AND CONSTRUCTION

- Oval cross-section on most indicates an atmospheric pressure tank (less than 4 psi).
- Normally constructed of aluminum. May also find round cross-section tanks constructed of stainless steel and aluminum.
- Maximum of 6 compartments per trailer.
- Maximum capacity of approximately 9,000 gallons in U.S. transportation system; 13,800 gallons in Canada and border states. Capacity can be determined from manufacturers specification plate, or by compartment capacity markings along the overturn rail.
- Valves can be air or cable operated,
- Safety features include internal safety valves with shear protection; fusible links, nuts and plugs; emergency remote shutoff; pressure / vacuum relief protection; vapor recovery systems; and overturn protection.

CONTENTS/HAZARD CLASS

- Flammable and combustible liquids, including gasoline, diesel fuel, and fuel oils.
- Round cross-section, stainless steel units often transport flammable solvents and chemicals.

Figure 6-23 Cargo Tank Trucks *(Continued)*

	DESIGN AND CONSTRUCTION	CONTENTS / HAZARD CLASS
MC-307 / DOT-407 **Low Pressure Chemical Cargo Tank Truck** 	• Circular cross-section or noninsulated version with visible strengthening rings. Stainless steel shell is most common. • Horseshoe shape or insulated versions with an outer jacket (usually light gauge sheet metal) that conceals rings and insulation. • Up to 5 compartments with overturn protection and crash boxes. • Maximum allowable working pressure (MAWP) up to 40 psi. • Total capacity of 5,000 to 8,000 gallons. Capacity can be determined from manufacturers specification plate. • Valves can be air, hydraulic or cable operated. • Safety features include internal safety valves; fusible links, nuts and plugs; emergency remote shutoff (air, cable or hydraulic); pressure/vacuum relief protection; and overturn protection.	• Flammable and combustible liquids, poisons, mild corrosives, and chemicals with a vapor pressure of 18 psi at 100° F (37.8° C) or greater, but not more than 40 psi at 170° F (4.4° C). • Workhorse of the chemical industry.
MC-312 / DOT-412 **Corrosive Cargo Tank Truck** 	• Narrow circular cross-section on noninsulated version with visible strengthening rings. • Horseshoe shape on insulated versions with an outer jacket (usually light gauge sheet metal) that conceals rings and insulation. • Normally a single compartment cargo tank with spill box in the rear. Capacity of 3,300 to 6,300 gallons, depending upon the product. Capacity can be determined from manufacturers specification plate. • Multiple spill boxes may indicate that the tank is a top-unloading version; not necessarily an indicator of the total number of compartments. • Normal maximum allowable working pressure (MAWP) between 35 and 50 psi; rare versions may operate at pressures up to 100 psi. • Valves can be air, hydraulic or manually operated. • Safety features include internal safety valves, pressure and vacuum relief protection, and overturn protection. • Some versions of top unloading MC-312 tanks may not have an emergency shut-off device. However, cutting off the air supply during a transfer can reduce the flow of product.	• Corrosives and high density liquids.
MC-330 / MC-331 **High Pressure Cargo Tank Truck** 	• Usually a single compartment unit constructed of top quality seamless or welded steel; circular cross-section with rounded ends or heads. Steel may be quench tempered (QT) or non-quenched tempered (NQT). • Minimum design pressure of 100 psi and maximum of 500 psi. • Single shell, normally noninsulated tank. • May be found as a bobtail or a transport. Bobtail capacities range from 2,300 to 3,000 gallons; transports range from 9,000 to 14,500 gallons. Capacity can be determined from manufacturers specification plate. • Transports as large as 17,500 gallons are allowed in some states. • Valves can be air, cable or hydraulic operated. • Safety features include pressure relief devices, internal safety valves with sheer protection; excess flow valves; fusible links and nuts; emergency remote shutoff; and off-truck emergency remote shutdown devices.	• Liquefied gases, primarily propane, butane and anhydrous ammonia. Also transport high hazard or high vapor pressure Class 3, 4, 6 and 8 materials.

Figure 6-23 *(Continued)*

DESIGN AND CONSTRUCTION | **CONTENTS / HAZARD CLASS**

MC-338
Cryogenic Liquid Cargo Tank Truck

Design and Construction

- Large, well insulated "thermos bottle" design. Inner container holds the product, and the vacuum-sealed outer shell is filled with insulation.
- Many versions include a work box "dog house" at the rear of the unit and evaporator coils underneath the belly of the tank. Work box includes discharge and fill piping, valves and pump.
- Single compartment unit with working pressures ranging from 25 to 500 psi. Capacity of 8,000 to 10,000 gallons.
- Safety features include pressure relief devices on both the inner tank and the outer tank, excess flow valves, fusible links and nuts, and emergency remote shutoff.
- MC-338 units transporting non-flammable gases that occur naturally in the atmosphere (e.g., nitrogen, oxygen) are designed to regularly vent off vapors during transport.

Contents / Hazard Class

- Cryogenic liquids, including liquid oxygen, liquid nitrogen, liquid carbon dioxide, liquefied natural gas and ethylene.
- Atmospheric gases, such as nitrogen and oxygen, are usually transported in Compressed Gas Association (CGA) 341 cargo tanks, which are similar in design to the MC-331 cargo tank. Flammable and toxic cryogenic liquids are transported in the MC-338 cargo tank.

Compressed Gas Trailer

Design and Construction

- Consist of 2 to approximately 20 seamless steel cylinders permanently mounted together. Often referred to as a "tube trailer."
- Service pressures range up from 2,000 to 5,000 psi.
- Cylinder valves, piping and controls are typically at the back of the unit. These units operate similar to a cascade system used to fill SCBA cylinders.
- Safety features include pressure relief devices.

Contents / Hazard Class

- Pressurized gases, such as oxygen, hydrogen, nitrogen, and helium.
- Units are commonly found at construction and industrial sites. When empty, the entire unit is removed and replaced with a similar full unit.

Heated Material Cargo Tank Trucks

Design and Construction

- Constructed of mild steel or aluminum.
- Tank is covered with insulation and an outer jacket (usually aluminum) that conceals the insulation.
- Typically does not have pressure relief protection, but may have an open vent into the vapor space to prevent pressure build-up.
- May be provided with steam coils to keep the product heated and fluid (e.g., molten sulfur).
- Some asphalt tanks have burner tubes and may carry propane cylinders to fuel the burners.
- Required to be placarded as a "HOT" material. Molten sulfur cargo tanks are required to have the contents stenciled on both sides and ends of the tank.
- Molten sulfur tank trucks frequently have a large breathing air cylinder attached near the trailer's ladder at or near the rear of the unit; provides supplied air for the vehicle operator while transferring molten sulfur to/from the unit.

Contents / Hazard Class

- Heated materials, such as tar, asphalt, molten sulfur, and #7 and #8 fuel oils.

Dry Bulk Tanktruck

Design and Construction

- Bottom hopper design for off-loading the product. Air pressure is generally used for product transfer; is not pressurized during transport. May have multi-hoppers.
- Carry very heavy loads; centrifugal force the cause of many rollovers.
- Static charge build-up is a common hazard when transferring product.

Contents / Hazard Class

- Solid products, such as fertilizers (e.g., ammonium nitrate), oxidizers, cement, dry caustic soda, plastic pellets, flour, etc.

Figure 6-23

RAILROAD TANK CARS

Railroad Tank Car Nomenclature. The railroad industry uses tank car test pressure as the criterion for differentiating between pressurized and on-pressurized tank cars. Non-pressure tank cars have a test pressure of 100 psig or less, while pressure tank cars have a test pressure greater than 100 psig.

When describing a tank car, the "B-end" is used as the initial reference point. The B-end is where the hand brake wheel is located; numbers 1 through 4 indicate the wheels.

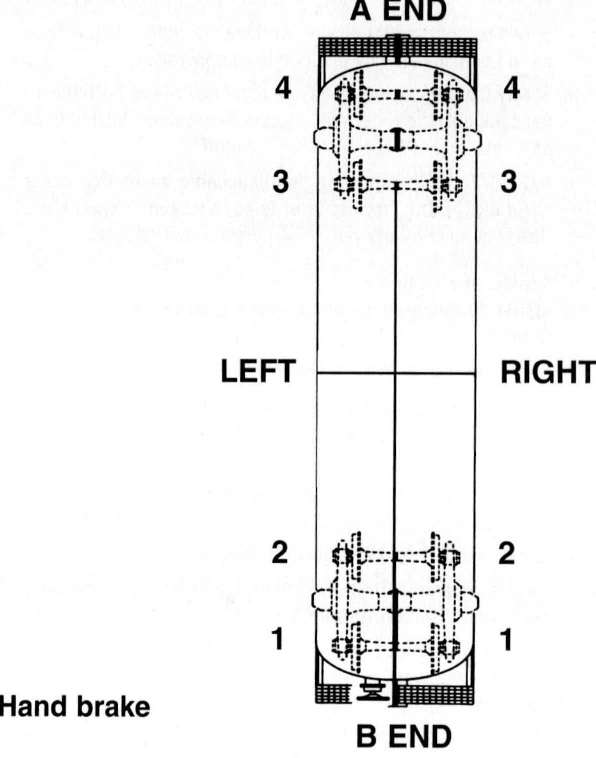

Railroad Tank Car Markings can be used to gain knowledge about the tank itself and its contents. This information would be useful in evaluating the condition of the container. These markings include the following:

- *Commodity Stencil.* Tank cars transporting anhydrous ammonia, ammonia solutions with more than 50% ammonia, Division 2.1 material (flammable gas), or a Division 2.3 material (poison gas) must have the name of the commodity marked on both sides of the tank in 4-inch (102 mm) minimum letters.
- *Reporting Marks and Number.* Railroad cars are marked with a set of initials and a number (e.g., GATX 12345) stenciled on both sides (left end as one faces the tank) and both ends of the car. These markings can be used to obtain information about the contents of the car from the railroad, the shipper or CHEMTREC. The last letter in the reporting marks has special meaning:
  - "X" indicates a rail car is not owned by a railroad (for a rail car, the lack of an "X" indicates railroad ownership).
  - "Z" indicates a trailer.
  - "U" indicates a container.

  Some shippers and car owners also stencil these markings on top of tank cars to assist in identification in an accident or derailment scenario. New tank cars are also stenciled on top for tank car verification during loading operations, as plant personnel can easily verify the car initial and number without climbing down from the rack.

Figure 6-24 Railroad Tank Cars: Markings (*Continued*)

- *Capacity Stencil.* Shows the volume of a tank car in gallons (and sometimes liters), as well as in pounds (and sometimes kilograms). These markings are found on the ends of the car under the reporting marks. For certain tank cars (e.g., DOT-105, DOT-109, DOT-112, DOT-114 and DOT-111A100W4), the water capacity / water weight of the tank car is stenciled near the center of the car.

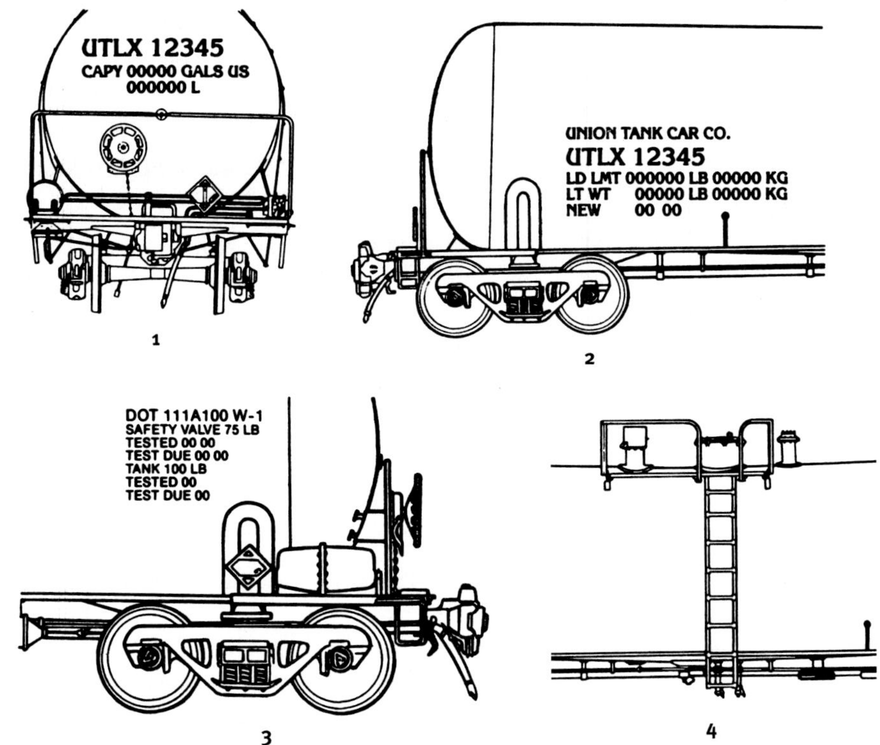

- *Specification Marking.* The specification marking indicates the standards to which a tank car was built. These markings will be on both sides of the tank car (right end as one faces the tank). The specification marking is also stamped into the heads of the tank, where it is not readily visible. The markings provide the following information:

  Approving Authority (e.g., DOT – Department of Transportation, AAR – Association of American Railroads, ICC – Interstate Commerce Commission (authority to DOT in 1966), CTC – Canadian Transport Commission, TC – Transport Canada)

  Class Number – three numbers which follow the approving authority designation

  Separator/Delimiter Character (significant in certain tank cars)

  Tank test pressure

  Type of material used in construction—most tank cars are carbon steel and no designation appears. When other construction materials are used (e.g., aluminum, nickel, alloy steel), designations are used.

  Type of weld used

  Fittings/material/lining

- *Specification Plate.* Tank cars ordered after 2003 will have a plate on the A-end right bolster and the B-end left bolster (i.e., structural cross-member which cradles the tank) that provides information about the tank car's characteristics. Although not designed as an emergency response tool, the plate will provide the following information:

  Car Builder's Name

  Builder's Serial Number

  Certificate of Construction/Exemption

  Tank Specification

  Tank Shell Material/Head Material

  Insulation Materials

  Insulation Thickness

  Underframe/Stub Sill Type

  Date Built

Figure 6-24 (*Continued*)

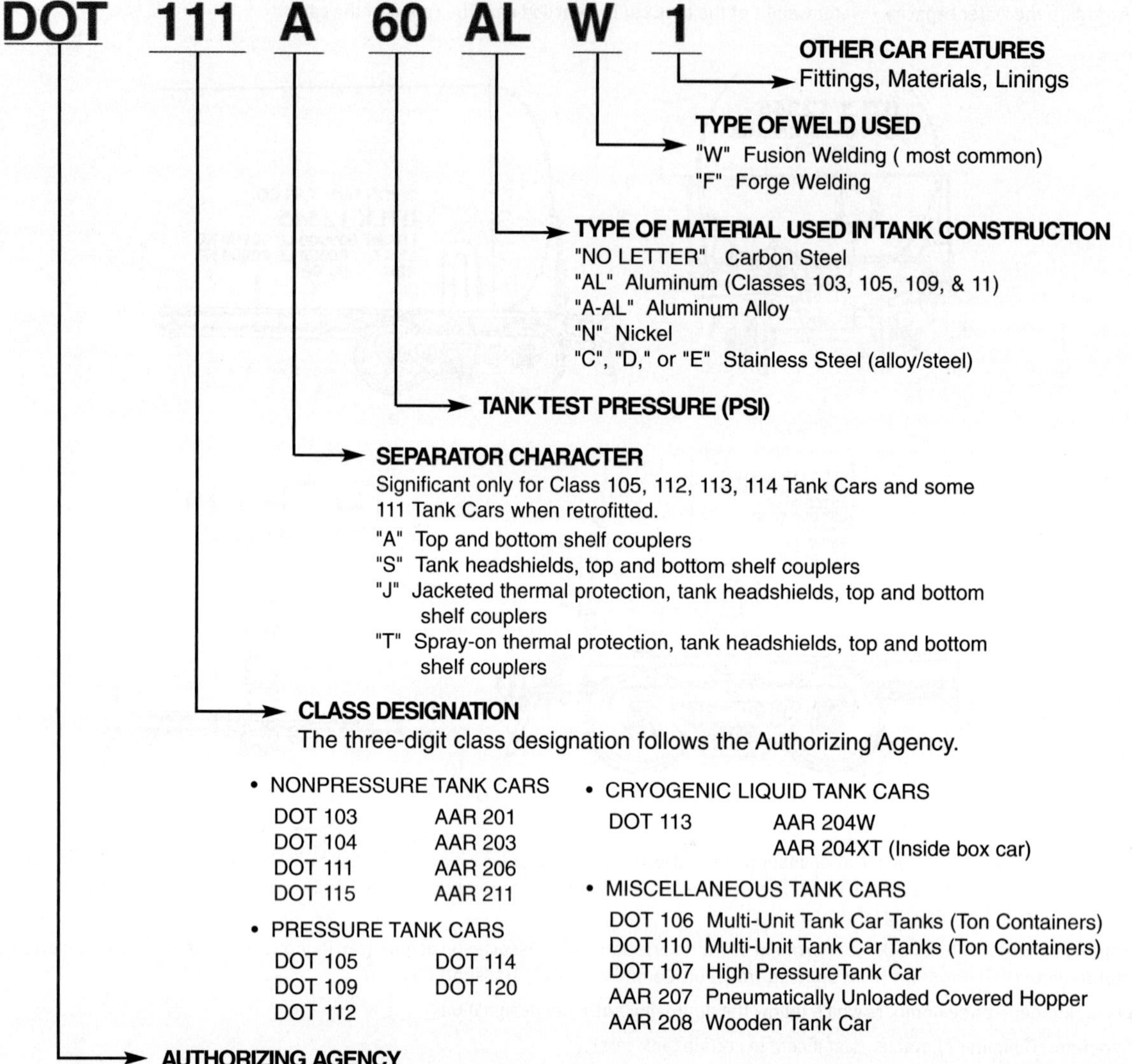
DOT SPECIFICATION MARKINGS FOR RAILROAD TANK CARS
DOT 111 A 60 AL W 1
OTHER CAR FEATURES
Fittings, Materials, Linings
TYPE OF WELD USED
"W" Fusion Welding (most common)
"F" Forge Welding
TYPE OF MATERIAL USED IN TANK CONSTRUCTION
"NO LETTER" Carbon Steel
"AL" Aluminum (Classes 103, 105, 109, & 11)
"A-AL" Aluminum Alloy
"N" Nickel
"C", "D," or "E" Stainless Steel (alloy/steel)
TANK TEST PRESSURE (PSI)
SEPARATOR CHARACTER
Significant only for Class 105, 112, 113, 114 Tank Cars and some 111 Tank Cars when retrofitted.
"A" Top and bottom shelf couplers
"S" Tank headshields, top and bottom shelf couplers
"J" Jacketed thermal protection, tank headshields, top and bottom shelf couplers
"T" Spray-on thermal protection, tank headshields, top and bottom shelf couplers
CLASS DESIGNATION
The three-digit class designation follows the Authorizing Agency.
• NONPRESSURE TANK CARS
DOT 103 AAR 201
DOT 104 AAR 203
DOT 111 AAR 206
DOT 115 AAR 211
• PRESSURE TANK CARS
DOT 105 DOT 114
DOT 109 DOT 120
DOT 112
• CRYOGENIC LIQUID TANK CARS
DOT 113 AAR 204W
AAR 204XT (Inside box car)
• MISCELLANEOUS TANK CARS
DOT 106 Multi-Unit Tank Car Tanks (Ton Containers)
DOT 110 Multi-Unit Tank Car Tanks (Ton Containers)
DOT 107 High PressureTank Car
AAR 207 Pneumatically Unloaded Covered Hopper
AAR 208 Wooden Tank Car
AUTHORIZING AGENCY
Tank Car specifications start with three letters designating the agency under whose authority the specification was issued
• DOT DEPARTMENT OF TRANSPORTATION
• AAR ASSOCIATION OF AMERICAN RAILROADS
• ICC INTERSTATE COMMERCE COMMISSION (Regulatory authority assumed by DOT in 1966)
• CTC CANADIAN TRANSPORT COMMISSION
• TC TRANSPORT CANADA (replacing CTC)

Figure 6-24

Non-Pressure Tank Car

DESIGN AND CONSTRUCTION

- Horizontal tank with flat or nearly flat ends (NOTE: Only jacketed tank cars will appear to have flat ends). May also be referred to as general service or low pressure tank cars.
- Fittings and valves visible on top of tank car. However, some non-pressure cars may be found with a protective housing around all fittings.
- Older cars will have an expansion dome with visible fittings. These were last built in 1973 and are seldom encountered.
- Although classified as non-pressure, actually have tank test pressures at 60 and 100 psi.
- Capacities of 4,000 to 45,000 gallons. Capacity can be determined from tank car markings.
- May have up to 6 compartments, each having its own set of fittings.
- Often has bottom unloading valve.
- Non-pressure tank cars transporting strong corrosives may have a wide band of corrosion-resistant paint running vertically at the manway, or may be identified by staining or corrosion around the manway area.
- Safety features can include pressure and vacuum relief valves, shelf couplers, head shields, and thermal protection.
- Non-pressure car classes

DOT 103	AAR 201
DOT 104	AAR 203
DOT 111	AAR 206
DOT 115	AAR 211

CONTENTS/HAZARD CLASS

- Transports liquids and solids with vapor pressures below 25 psig at 105 – 115° F.
- Examples include flammable and combustible liquids, flammable solids, reactive liquids and solids, oxidizers, organic peroxides, liquid poisons, and corrosives.
- Non-hazardous liquids include food grade liquids, such as corn syrup and vegetable oils.

Pressure Tank Car

DESIGN AND CONSTRUCTION

- Cylindrical steel or aluminum tank with rounded ends. May be insulated or thermally protected (either jacketed or sprayed on). Those without insulation or jacketed thermal protection must have the top two-thirds painted white.
- Fittings and valves enclosed in a protective housing with a cover.
- Tank test pressures range from 100 to 600 psi.
- Capacities range from 4,000 to 34,500 gallons. Capacity can be determined from tank car markings.
- Safety features can include thermal protection, pressure relief protection, excess flow valves, shelf couplers, and head shields.
- Pressure car classes

DOT 105	DOT 114
DOT 109	DOT 120
DOT 112	

CONTENTS/HAZARD CLASS

- Transports flammable, non-flammable and poisonous compressed gases. May also transport flammable liquids.
- Examples include LPG, chlorine, anhydrous ammonia, and anhydrous hydrogen fluoride.
- Anhydrous hydrogen fluoride may be found in a tank car with an orange/red stripe running the middle of the car. These color and design schemes are an industry standard and are not DOT requirements.

Cryogenic Liquid Tank Car

DESIGN AND CONSTRUCTION

- Well-insulated tank-within-a-tank or "Thermos Bottle" design. Inner stainless steel tank holds the product, and the space between the inner and outer tanks is filled with insulation and under a vacuum.
- Transports low pressure refrigerated liquids (pressures 25 psig or lower). Capacity can be determined from tank car markings.
- Tank test pressure can range up to175 psi.
- Absence of any top fittings or protective housing.
- Loading / unloading fittings and pressure relief device often found in cabinets on both sides or at one end of the tank car.
- Cryogenic tank car classes

DOT 113	AAR 204W

CONTENTS/HAZARD CLASS

- Gases that are liquefied through refrigeration rather than pressurization. Pressures usually below 25 psig at product temperatures <−130° F.
- Examples include liquid oxygen, liquid hydrogen, and liquid argon. LNG and ethylene may be found at slightly higher pressures.
- Combination of insulation and vacuum protects the contents from ambient temperatures for only 30 days. Shippers and railroads closely track these "time sensitive" shipments.

Figure 6-25 Railroad Tank Cars: DOT Specification Markings (*Continued*)

	DESIGN AND CONSTRUCTION	CONTENTS/HAZARD CLASS
Pneumatically Unloaded Covered Hopper Car	• Covered hopper car that is loaded through a series of hatches on top of the car and unloaded pneumatically. • Although non-pressure cars, tank test pressures range from 20 to 80 psi. Pressure is only on the car during unloading operations. • The movement of finely divided solids and air during transfer operations can generate static charges. Dust explosions may be a possibility if the solid is combustible. • Tank car class AAR 207.	• Transports dry, solid materials. • Examples include pellets, caustic flake, flour, resin powder, and corn starch. • Phosphorous pentasulfide and calcium carbide shipped with an inert gas blanket; will be a separate container for inert gas.

Figure 6-25

Radioactive Material Packaging

Federal regulations place strict controls on the transportation of radioactive materials. The transport of radioactive materials is based on the philosophy that (1) safety should be primarily focused on the package (i.e., packaging is the first line of defense), and (2) package integrity should be directly related to the degree of hazard of the material it contains. Radiological packaging may be made of any number of materials, depending on the level of radioactivity involved.

More than two-thirds of radioactive material shipments are of human-made radio-isotopes used in medicine, industry, agriculture, and scientific research. These are typically shipped in their most stable forms, which is usually as a solid material. The majority of radioactive material shipments are made by truck, while rail is often used to carry large volumes of low-level radioactive waste or to transport packages too heavy for over-the-road transport.

You should be familiar with five basic types of radioactive material packaging:

1. Excepted Packaging—Used to transport material with low levels of radioactivity. Excepted Packaging does not have to pass any performance tests but must meet the general requirements for any hazardous materials package as spelled out in DOT regulations. Key design and construction features include the following:
   - Authorized for limited quantities of radioactive material that would pose a very low hazard if released in an accident. Examples include consumer goods, such as smoke detectors and lantern mantles.
   - Are exempted from specific radioactive material packaging, marking, labeling, and shipping paper requirements.

   Included under Excepted Packaging is Strong Tight Packaging. Strong Tight Packaging is allowed for shipping Low Specific Activity (LSA) and Surface Contaminated Objects (SCOs) transported domestically in "exclusive use" shipments. These must meet only the general requirements for any hazardous materials package.
2. Industrial Packaging—Used in certain shipments of LSA material and SCOs, which are typically categorized as radioactive waste. Most low-level radioactive waste, such as contaminated protective clothing and handling materials, is shipped in secured packaging of this type. DOT regulations require that these packagings allow no identifiable release of the material into the environment during normal transport and handling.
3. <u>Type A Packaging</u>—Used to transport small quantities of radioactive material with higher concentrations of radioactivity than those shipped in Industrial Packaging. Key design and construction features include the following:
   - Typically constructed of steel, wood, or fiberboard and have an inner containment vessel made of glass, plastic, or metal surrounded with packing material made of polyethylene, rubber, or vermiculite.
   - Designed to ensure the package retains its containment integrity and shielding under normal transport conditions. However, they are not designed to withstand the forces of an accident.

Type A

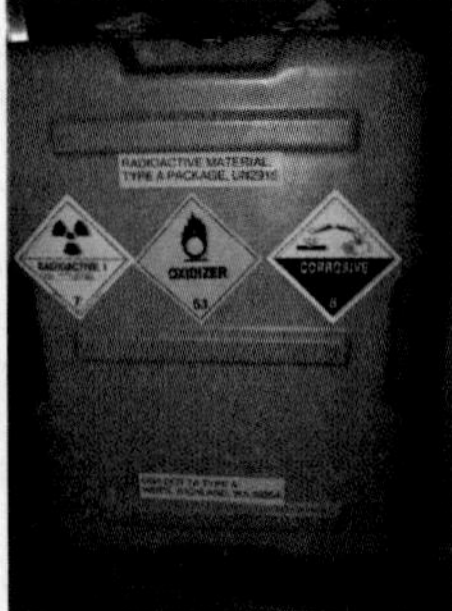

Type B

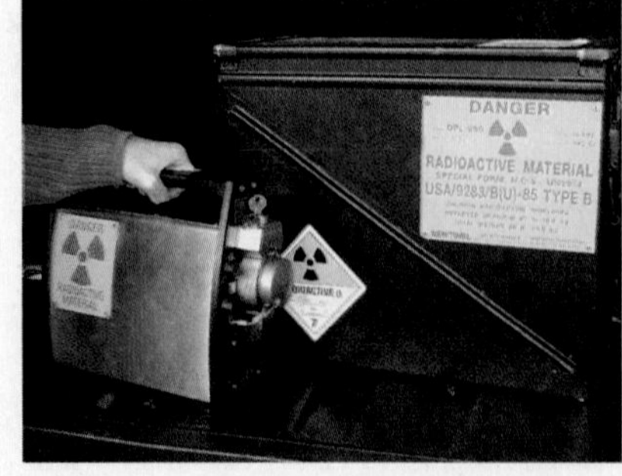

Type C

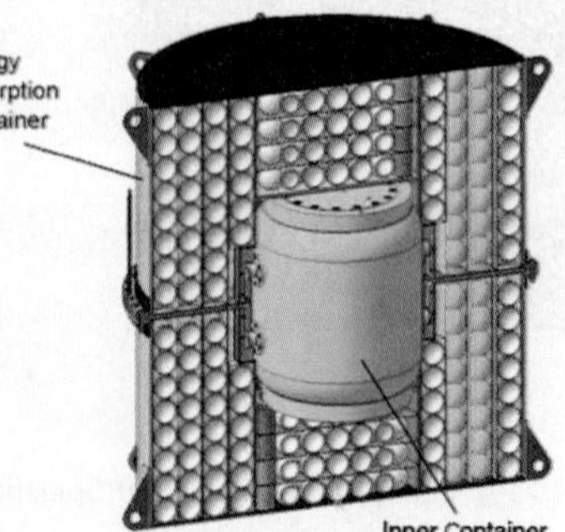

Figure 6-26 Responders must be familiar with the basic types of radioactive material packaging.

- The consequences of a release of material from a Type A package would not be major since the quantity of radioactive material transported in this package is limited.
- Are only used to transport non-life-endangering amounts of radioactive material, such as radiopharmaceuticals, radioactive waste, and radioactive sources used in industrial applications.

4. Type B Packaging—Used to transport radioactive material with the highest levels of radioactivity, including potentially life-endangering amounts that could pose a significant risk if released during an accident. Key design and construction features include the following:
   - Must meet all of the Type A requirements, as well as a series of tests that simulate severe or worst-case accident conditions. Accident conditions are simulated by performance testing and engineering analysis.
   - Range from small, hand-held radiography cameras to heavily shielded steel casks that weigh up to 125 tons. Examples of materials transported in Type B packaging include spent nuclear fuel, high-level radioactive waste, and high concentrations of other radioactive materials such as cesium and cobalt.
   - In the 50+ year history of transporting radioactive material, there has never been a release from a certified Type B package or an injury or death resulting from the release of a radioactive material in a transportation incident.
5. Type C Packaging—Used to transport by aircraft high-activity radioactive materials that have not been certified as "low dispersible radioactive material" (including plutonium). They are designed to withstand severe accident conditions associated with air transport without loss of containment or significant increase in external radiation levels.

Fundamental design and construction features include the following:

- The Type C packaging performance requirements are significantly more stringent than those for Type B packaging.
- Type C packaging is not authorized for domestic use but can be authorized for international shipments of high-activity radioactive material consignments.
- Type C packaging performance requirements include those applicable to Type B packaging, but with enhancements on some tests significantly more stringent than those for Type B packaging. For example, a 200 mph (321.8 km/hr) impact onto an unyielding target is required instead of the 30 ft. (9.1 m) drop test required of a Type B packaging; a 60-minute fire test is required instead of the 30-minute test for Type B packaging; and a puncture/tearing test is required. These stringent tests are expected to result in packaging designs that will survive more severe aircraft accidents than Type B packaging designs.

Facility Containment Systems

Facility containment systems are packaging, containers, and/or associated systems that are part of a fixed facility's operations. Examples may include storage tanks, process towers, chemical and nuclear reactors, piping systems, pumps, storage bins and cabinets, dryers and degreasers, and machinery.

Figure 6-27 through 6-31 provide further information on shapes, descriptions, and contents of common facility storage tanks and vessels:

- Atmospheric and Low-Pressure Liquid Storage Tanks
- Pressurized Storage Vessels

ATMOSPHERIC PRESSURE LIQUID STORAGE TANKS

	DESIGN AND CONSTRUCTION	CONTENTS/HAZARD CLASS
Cone Roof Tank 	• Welded steel tank with vertical cylindrical walls supporting a fixed bottom and a flat or conical roof. However, older riveted tanks still exist. • Typically 20 to 100 ft. diameter, but can be as large as 200+ ft. Depending upon tank diameter and height, capacities can range from 1,000 to 100,000+ barrels (NOTE: 1 barrel = 42 gallons). • Capacity can be determined from tank spec plate or pre-incident planning (if not marked). • Have a vapor space between the liquid level and the roof. Operates at atmospheric pressures, and uses a pressure-vacuum valve for normal tank "breathing" during transfer operations. • Emergency venting provided by a weak roof-to-shell seam designed to fail in case of fire or explosion (if tank is constructed to API 650 specifications). • May be insulated, particularly for fuel oil or asphalt service. • Tanks in chemical service may be lined; high vapor pressure solvents may have an inert blanket (e.g., nitrogen) over the product.	• Flammable and combustible liquids, solvents, oxidizers, and corrosives. • At petroleum storage facilities, will primarily store combustible liquids due to environmental emissions and economic considerations. • At chemical facilities, may store virtually any hazardous or non-hazardous liquid.

Figure 6-27 Atmospheric Pressure Liquid Storage Tanks (*Continued*)

ATMOSPHERIC PRESSURE LIQUID STORAGE TANKS

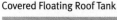

	DESIGN AND CONSTRUCTION	CONTENTS/HAZARD CLASS
Open Floating Roof Tank 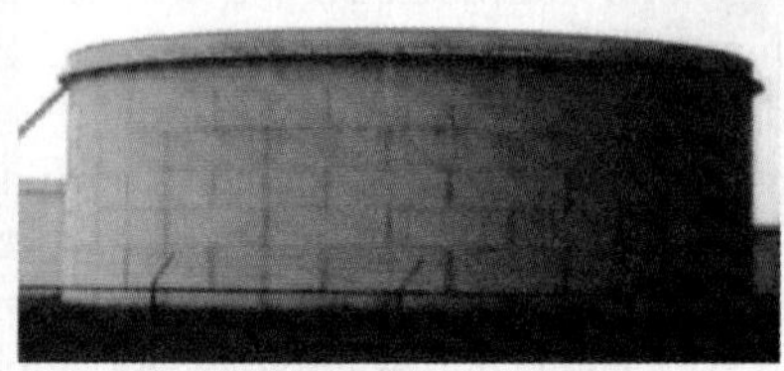	• Steel tank with vertical cylindrical walls and a roof that floats on the surface of the product. There is a seal area between the tank shell and floating roof. • Have no vapor space, except when the product is at lowest levels and the floating roof is sitting on its roof supports. • Identified by the wind girder around the top of the tank shell and the roof ladder. • Floating roof may be a pontoon or honeycomb design, and have a drainage system to carry normal rainwater off the roof and to the ground. • Typically 40 to 120 ft. diameter, but can be as large as 300+ ft. at refineries and crude oil terminals. Depending upon tank diameter and height, capacities can range from 1,000 to 500,000+ barrels (NOTE: 1 barrel = 42 gallons). • Capacity can be determined from tank spec plate or pre-incident planning (if not marked). • Most common storage tank found in the petroleum industry; most common scenario is a seal fire. • When installed, semi-fixed suppression systems are generally designed to protect only the seal area.	• Flammable and combustible liquids. • At petroleum storage facilities, will primarily store flammable liquids and higher vapor pressure products due to environmental emissions and economic considerations.
Covered Floating Roof Tank   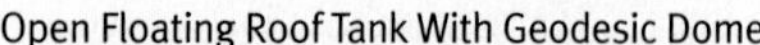	• Welded steel tank with vertical cylindrical walls supporting a fixed flat or conical roof, with an internal floating roof or pan that floats on the surface of the product. • Is essentially a cone roof tank with an internal floating roof. • Have no vapor space, except when the product is at lowest levels and the floating roof/pan is sitting on its roof supports. • Identified by the large "eyebrow" vents at the top of the tank shell. • Typically 40 to 120 ft. diameter, but can be as large as 300+ ft. at refineries and crude oil terminals. Depending upon tank diameter and height, capacities can range from 1,000 to 500,000+ barrels (NOTE: 1 barrel = 42 gallons). • Capacity can be determined from tank spec plate or pre-incident planning (if not marked). • Fires can be difficult to extinguish through portable application devices.	• Flammable and combustible liquids. • At petroleum storage facilities, will primarily store flammable liquids and higher vapor pressure products due to environmental emissions and economic considerations.
Open Floating Roof Tank With Geodesic Dome 	• Is essentially an open floating roof tank with a lightweight aluminum geodesic dome installed over the top. • Geodesic roof structure is installed for environmental emissions and economic considerations.	• Flammable and combustible liquids. • At petroleum storage facilities, will primarily store flammable liquids and higher vapor pressure products due to environmental emissions and economic considerations.

Figure 6-27

LOW PRESSURE LIQUID STORAGE TANKS

Vertical Storage Tank

- Steel tank with vertical cylindrical walls supporting a fixed bottom and a flat or conical roof. Typically welded construction, although older riveted tanks are still in-service. Also referred to as dome roof tanks.
- Smaller diameter (approximately 25 ft. and less). Capacities range from 100 to 10,000 barrels (NOTE: 1 barrel = 42 gallons).
- Capacity can be determined from tank spec plate or pre-incident planning (if not marked).
- Working pressures range from 2.5 to 15 psi.
- Uses a pressure-vacuum valve for normal tank operations. If designed, emergency relief protection provided by a weak roof-to-shell seam. However, the roof may not always fail as designed!
- Tanks in chemical service may be lined; high vapor pressure solvents may have an inert blanket (e.g., nitrogen) over the product.
- Smaller fiberglass tanks with capacities up to 25,000 gallons may also be found in chemical service.

- Flammable and combustible liquids, solvents, oxidizers, poison liquids, and corrosives.
- At petroleum storage facilities, will primarily store combustible liquids due to environmental emissions and economic considerations.
- At chemical facilities, may store virtually any hazardous or non-hazardous liquid.

Horizontal Storage Tank

- Wide range of horizontal tanks in service. Range from welded or riveted steel plate construction to double-walled steel tanks encased in fire retardant materials with its own containment box (i.e., tank within a tank design).
- Capacities typically range from 300 gallons to 10,000 gallons, although larger tanks may be found. Working pressures are at atmospheric pressure.
- Capacity can be determined from tank spec plate or pre-incident planning (if not marked).
- Structural integrity of tank supports is critical in a fire scenario. Under NFPA standards, horizontal tanks storing flammable liquids must be protected by materials having a fire resistance of not less than 2 hours.
- Uses a pressure-vacuum valve for normal tank operations.

- Flammable and combustible liquids, solvents, oxidizers, poison liquids, and corrosives.
- Newer generation of tank within a tank design is widely used for motor fuels applications (e.g., gasoline, diesel fuel).

Figure 6-28 Low Pressure Liquid Storage Tanks

Underground Storage Tank

- EPA defines an Underground Storage Tank System as a tank and any underground piping connected to the tank that has at least 10% of its combined volume underground.
- Horizontal tank constructed of steel, fiberglass, or steel with a fiberglass coating. Tank capacities can range up to 25,000+ gallons.
- Leak detection, overfill protection, and cathodic protection are required under EPA regulations. Tanks may be single or double-walled to meet requirements.
- Visible clues are vents, fill points, and potential occupancy and locations (e.g., service station, fleet maintenance).

- Flammable and combustible liquids.
- At chemical facilities, may store virtually any hazardous or non-hazardous liquid.

Figure 6-29 Atmospheric And Low Pressure Liquid Storage Tank: Underground Storage Tank

HIGH PRESSURE STORAGE TANKS

DESCRIPTION	DESIGN AND CONSTRUCTION	CONTENTS/HAZARD CLASS
High Pressure Horizontal Tank 	• Single shell, non-insulated welded storage tanks constructed to ASME standards. • Pressures range from 100 to 500 psi, and are protected with pressure relief valves. • Capacities can vary – residential tanks range from 100 to 1,000 gallons; industrial and distribution facilities range from 1,000 to 120,000 gallons. • Capacity can be determined from tank spec plate or pre-incident planning (if not marked). • Majority are horizontal, although some smaller vertical tanks may be found. Some tanks may be mounded, where most of the tank surface is covered by a earthen surface. • Piping systems will include both a liquid line and a vapor line. Liquid lines are generally larger than vapor lines, and may be painted different colors. • Some storage tanks may be protected with water spray systems	• Flammable and non-flammable liquefied gases, including LPG, anhydrous ammonia, chlorine, and vinyl chloride. • Bulk LPG storage facilities can also include cylinder and bobtail filling stations, and truck and rail car unloading racks.
High Pressure Spherical Tank 	• Single shell, non-insulated welded storage tanks constructed to ASME standards. Can be up to 90+ ft. diameter. • Pressures range from 100 to 500 psi, and are protected with pressure relief valves. • Capacities can be as large as 600,000 gallons. Capacity can be determined from tank spec plate or pre-incident planning (if not marked). • Piping systems will include both a liquid line and a vapor line. Liquid lines are generally larger than vapor lines, and may be painted different colors. • Some storage tanks may be protected with water spray and deluge systems	• Flammable and non-flammable liquefied gases, including LPG, anhydrous ammonia, and vinyl chloride. • Found at large gas distribution facilities and petrochemical manufacturing facilities.
High Pressure Underground Tank 	• Single shell, non-insulated welded storage tanks constructed to ASME standards. Outer shell coated with a mastic material and provided with cathoidc protection. • Capacities can range from 500 to 10,000+ gallons, with pressures ranging from 100 to 500 psi, • Most underground propane tanks under 2,000 gallons (water capacity) have a riser with a combination valve containing the filler valve, service valve, pressure relief valve and other appurtenances. The protective shroud covering the riser may be the only identification clue! • Some areas store LPG gases in earth covered domes, underground caverns and salt mines	• Liquefied petroleum gases (propane, butane and mixtures)

Figure 6-30 High Pressure Storage Tanks

CRYOGENIC LIQUID AND REFRIGERATED STORAGE TANKS

DESCRIPTION	DESIGN AND CONSTRUCTION	CONTENTS / HAZARD CLASS
Cryogenic Liquid Storage Tank 	▪ Large, well insulated "thermos bottle" design. Inner container constructed of stainless steel or metal suitable for low temperature service, while the outer shell is typically a gastight carbon steel jacket, with the space filled with insulation and under vacuum. Constructed to ASME standards. ▪ Normally have vaporizing unit and pumps to warm the product and convert it to a gas, as well as pressurize the container and "move" the gas. ▪ Liquid oxygen tanks will always be over a concrete pad; never over asphalt or grass. ▪ Capacities range from 500 to 20,000 gallons (most are in the range of 1,500 to 11,000 gallons), with design working pressures up to 250 psi. ▪ Capacity can be determined from tank spec plate or pre-incident planning (if not marked). ▪ Safety features include pressure relief devices on both the inner tank and the outer tank, excess flow valves, fusible links and nuts, and emergency remote shutoff.	▪ Cryogenic liquids, including liquid oxygen, liquid nitrogen, liquid carbon dioxide, liquid argon, and liquid helium. ▪ Often find liquid oxygen (LOX) at hospitals and some industries, and liquid nitrogen at hospitals, universities, and industry.
Refrigerated Storage Tank 	▪ Insulated and refrigerated cylindrical bulk storage tank. Depending upon product and service, may be a combination of single steel wall, double steel wall, or a concrete exterior and double steel interior wall combined with insulation. ▪ Products are stored at temperatures near their boiling point. ▪ Design working pressures <15 psi. ▪ Typically 50 to 150 ft. diameter, but can be as large as 200+ ft. Depending upon tank diameter and height, capacities can range from 20,000 to 200,000+ barrels (NOTE: 1 barrel = 42 gallons). ▪ Capacity can be determined from tank spec plate or pre-incident planning (if not marked).	▪ Liquefied gases, including LPG, LNG, ethylene and anhydrous ammonia. ▪ Found at pipeline terminals, LNG peak shaving plants, and utility plants.

Figure 6-31 Cryogenic liquid and refrigerated storage tanks

Markings and Colors

Markings and colors on hazmat packaging or containment systems are the third clue in the standard identification process. These clues may include color codes, container specification numbers, signal words, or even the content's name and associated hazards. At facilities, clues may include Hazard Communication markings, piping color code systems, and specific signs and/or signal words (e.g., "Hydrofluoric Acid Area").

Markings and colors can be used as a clue for hazmat recognition, identification, and classification. We will evaluate these systems based on the nature and use of the package.

Nonbulk Package Markings

Agricultural Chemicals and Pesticide Labels. Individual nonbulk packages, particularly those storing pesticides and agricultural chemicals (ag chems), will display useful information for identification and hazard assessment. These container markings include the following:

- *Toxicity Signal Word*—The signal word indicates the relative degree of acute toxicity. Located in the center of the front label panel, it is one of the most important label markings. The three toxic categories are high toxicity, moderate toxicity, and low toxicity **Figure 6-32**.
- *Statement of Practical Treatment*—Located near the signal word on the front panel, it is also referred to as the "first-aid statement" or "note to physician." It may have precautionary information as well as emergency procedures for exposures. Antidote and treatment information may also be added.
- *Physical or Chemical Hazard Statement*—A statement displayed on a side panel, as necessary. It will list special flammability, explosion, or chemical hazards posed by the product.
- *Product Name*—The brand or trade name is printed on the front panel. If the product name includes the term "Technical," as in Parathion-Technical, it generally indicates

TOXICITY SIGNAL WORDS	
LEVEL OF TOXICITY	SIGNAL WORD
HIGH	DANGER, POISON Skull and Crossbones Symbol
MODERATE	WARNING
LOW	CAUTION

Figure 6-32 Toxicity Signal Words Found on AG Chem and Pesticide Containers Indicate the Relative Degree of Acute Toxicity of the Contents

a highly concentrated pesticide with 70% to 99% active ingredients.

- *Ingredient Statement*—All pesticide labels must have statements that break down the chemical ingredients by their relative percentages or as pounds per gallon of concentrate. "Active ingredients" are the active chemicals within the mixture. They must be listed by chemical name, and their common name may also be shown. Inert ingredients have no pesticide activity and are usually not broken into specific components, only total percentage. A number of agricultural chemical products, particularly those used in the home, use flammable products (e.g., propane used for aerosols and diesel fuel or solvents used with liquids) only as the inert ingredients and may not be easy to identify.
- *Environmental Information*—The label may provide information on both the storage and disposal of the product, as well as environmental or wildlife hazards that could occur. This information can be most useful when planning clean-up and disposal after a fire or spill.
- *EPA Registration Number*—This number is required for all ag chems and pesticide products marketed in the United States. It is one of three ways to positively identify a pesticide. The other ways are by the product name or by the chemical ingredient statement. The registration number will appear as a two- or three-section number (e.g., EPA Reg. No. 239-2491-AA) and indicates the manufacturer, specific product, and locations where the product may be used. When relaying this number to CHEMTREC® or other sources, make sure you include each dash. A U.S. Department of Agriculture number may appear on products registered before 1970.
- *EPA Establishment Number*—The location where the product was manufactured. This number has little significance to responders.

Responder Tip

When faced with an incident involving poisonous or toxic products where there are injuries or exposures, obtain an uncontaminated container or label for reference whenever possible. When victims are transported to a medical facility as a result of a suspected exposure, ensure proper identification and medical care by also forwarding an uncontaminated label or container.

Chemical Abstract Service (CAS) Number. Often found on hazardous materials containers, as well as SDSs, the CAS number is often used by state and local right-to-know regulations for tracking chemicals in the community and workplace. Sometimes referred to as a chemical's "social security number," sequentially assigned CAS numbers identify specific chemicals and have no chemical significance.

Cylinder Color Codes. A number of gases are stored and transported in compressed gas cylinders. Although there are several voluntary color schemes, none are mandatory. The only reliable method to identify cylinder contents is to check the DOT label attached to the cylinder head. Unfortunately, if cylinders are exposed to fire, the labels may burn off and further hamper identification.

Bulk Packaging and Transportation Markings

Identification Number. Four-digit identification numbers are assigned to a hazardous material or group of hazardous materials. They are used to determine the name of the material and to obtain hazard and response information from emergency response guidebooks. They are required on shipping papers and on the following:

- Nonbulk packages of hazardous materials (except limited quantities) printed adjacent to the required labels on the package
- Shipments of nonbulk packages when loaded at one location of more than 8,820 lbs. of a single material
- Bulk packages of hazardous materials, including cargo tanks, rail cars, portable tanks, and railroad hopper cars

Acceptable methods of displaying this marking are shown in **Figure 6-33**. Identification numbers are prohibited on DOT explosives, dangerous, radioactive, and subsidiary hazard placards. The identification number must be displayed on the supplemental orange panel for these materials.

When viewing shipping papers or SDSs, the identification number may be found with the prefix "UN" or "NA." Those

Figure 6-33 The four-digit identification number may be displayed on a bulk container in one of three ways.

identification numbers preceded by the letters "UN" (i.e., United Nations) are appropriate for domestic and international transportation. Those preceded by the letters "NA" (i.e., North America) are not recognized for international transportation, except to and from Canada.

Inhalation Hazard Markings. The words *Poison—Inhalation Hazard* or *Inhalation Hazard* indicate that the material is considered toxic by inhalation (e.g., chlorine).

Marine Pollutant Markings. These markings are displayed on both sides and both ends of bulk packages of materials designated on the shipping papers as a marine pollutant, except when the container is properly placarded (e.g., Poison, Flammable Liquid, etc.). The mark must appear when the package moves by water.

Elevated Temperature Materials. Elevated temperature materials are materials that, when transported in a bulk container, are:

- Liquids at or above 212°F (100°C).
- Liquids with a flash point at or above 100°F (37.8°C) that are intentionally heated and are transported at or above their flash point.
- Solids at a temperature at or above 464°F (240°C).

Except for a bulk container transporting molten aluminum or molten sulfur (which must be marked "MOLTEN ALUMINUM" or "MOLTEN SULFUR" respectively), the container must be marked on each side and each end with the word "HOT" in black or white lettering on a contrasting background. The word "HOT" may be displayed on the bulk packaging itself or in black lettering on a white square-on-point configuration similar in size to the placard **Figure 6-34**.

Pipelines. Pipelines and piping systems are the safest and second largest hazmat transportation mode within the United States. While emergency responders are familiar with pipeline systems used for both intrastate and interstate transportation, pipelines are also used for transporting products between industrial facilities, transferring raw materials and finished products within oil, chemical and petrochemical facilities, and delivering liquid and gas fuels directly to the consumer.

From a design and construction viewpoint, all piping systems are based upon the following principles:

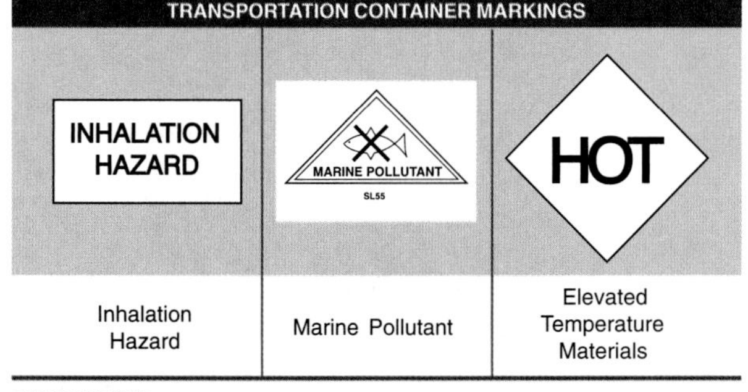

Figure 6-34 Transportation container markings for special situations

1. *A material is inserted or injected into a pipe.* Although slurries may be transported in a pipeline, liquid and gas products are most common.
2. *The product is moved from this origination point to a prespecified destination.* The product is physically moved as a result of gravity, the pressure of the product, energy created through the use of pumps or compressors, or a combination of all of the above. In addition, various valves and manifolds may be used to control and direct the flow of the product.
3. *The product is ultimately removed from the pipeline at its destination point.* Depending upon the type of pipeline and the location, the product may be transferred to another mode of transportation (e.g., marine, rail, highway), placed into a container for storage (e.g., tank, underground cavern, etc.), or used.

Pipelines often cross over or under roads, waterways, and railroads. At each of these crossover locations, a marker should identify the pipeline right-of-way. Although its format and design may vary, all markers are required to provide the pipeline contents (e.g., natural gas, propane, liquid petroleum products, etc.), the pipeline operator, and an emergency telephone number **Figure 6-35**.

The pipeline emergency telephone number goes to a control room, where an operator monitors pipeline operations and can start emergency shutdown procedures. It should be stressed that even when a ruptured pipeline is immediately shut down, product backflow may continue for several hours until the product drains to the point of release.

Most gas pipelines are dedicated to one product (e.g., natural gas, butadiene, anhydrous ammonia). However, liquid petroleum transmission pipelines may carry several different petroleum products simultaneously. **Figure 6-36** shows a typical liquid products shipping cycle. Liquid pipeline personnel normally refer to product flows in terms of "barrels" rather than gallons (*Note:* 1 barrel equals 42 gallons).

For liquid petroleum pipelines, there is usually no physical separator (e.g., sphere or pig) between different products. Rather, the products are allowed to "co-mingle." This interface can range from a few barrels to several hundred, depending on the pipeline size and products involved. Verification of the shipment arrival is made by examining a sample of the incoming batch for color, appearance, and/or chemical characteristics.

Product flows through many transmission pipeline systems are monitored through a computerized pipeline SCADA

Responder Tip

Emergency response personnel should NEVER attempt to isolate any pipeline valves on transmission or distribution pipelines unless under the direction of pipeline operations personnel. Operation of pipeline valves and systems must be under the full command and direction of the pipeline operator. Failure to do so may create additional problems far worse than the original event.

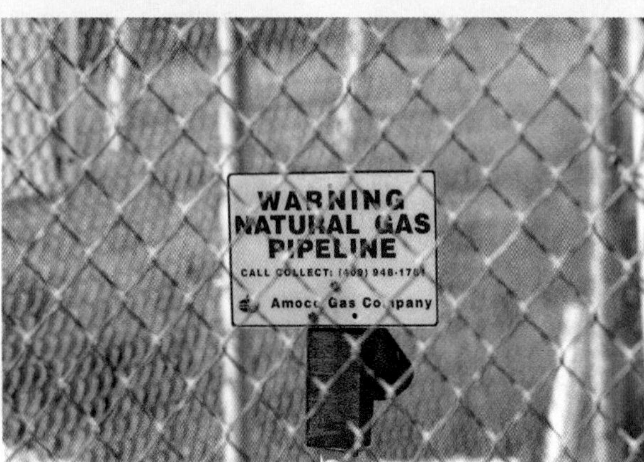

Figure 6-35 Pipeline markers must provide the pipeline contents, the pipeline operator, and an emergency telephone number.

System (Supervisory Control and Data Acquisition System). The exact injection date and time of the particular product into the pipeline is noted and its delivery date/time is projected. As the product gets close to its destination, a sensor in the line signals the arrival of the shipment. The SCADA System provides pipeline personnel with the ability to monitor pipeline flows and pressures and initiate emergency shutdown procedures in the event of a release.

Readers may reference the *Pipeline Emergencies* (2nd edition) textbook and related Internet-based training curriculum at *http://www.pipelineemergencies.com* for additional information. The curriculum is a result of a cooperative agreement between the U.S. Department of Transportation's Pipeline and Hazardous Materials Safety Administration (PHMSA) and the National Association of State Fire Marshals (NASFM).

Figure 6-36 Typical product cycle for a liquid petroleum pipeline

Figure 6-37 Emergency responders should NEVER attempt to isolate any pipeline valves on transmission or distribution pipelines unless under the direction of pipeline operations personnel.

Responder Tip

Emergency responders should know the following about pipelines:

1. The location of pipelines in your area. The National Pipeline Mapping System (https://www.npms.phmsa.dot.gov) is an excellent tool for learning about certain types of pipelines in a given area.
2. The operator of the pipeline.
3. What is being transported by the pipeline.
4. The shut-off valve locations. This can help emergency responders to determine how quickly the release can be stopped. Remember—emergency responders should NEVER attempt to isolate any pipeline valves on transmission or distribution pipelines unless under the direction of pipeline operations personnel.
5. The worst-case scenario.

Facility Markings

NFPA 704 System. Many state and local fire codes mandate the use of the NFPA 704 marking system at all fixed facilities in their jurisdiction, including tanks and storage areas. The NFPA 704 system, which is shown in Figure 6-38, is not used on transport vehicles.

Hazard Communication Marking Systems. There are several hazard communication systems used within industry to comply with the OSHA Hazard Communication System (29 CFR 1910.1200). They may be known by several titles, including the Hazardous Materials Information System (HMIS) and the Hazardous Materials Identification Guide (HMIG). Similar to NFPA 704, these systems provide a standardized hazard rating scale from 0 (minimal hazard) to 4 (extreme hazard) for health, flammability, and reactivity. In addition, there are alphabetical designations for the required level of personal protective clothing. These markings may be found at facility entrances, room entrances, and directly on containers.

In 2013 OSHA updated the Hazard Communication Standard and adopted the Global Harmonization System (GHS) to harmonize U.S. hazard communication (hazcom) regulations with those of the international community in order to promote and facilitate international trade Figure 6-39. There are three main elements of the GHS:

- Classification System—GHS identifies physical, chemical, and environmental hazards, and provides objective criteria for their classification within each hazard class. The GHS is generally regarded as a more accurate system than the older OSHA system.
- Labeling—GHS requires that the following information appear on a label: the product name; company's name, address and phone number, standardized hazard statement, a pictogram, and a precautionary statement. A prescribed hazard statement is assigned to each hazard category along with a signal word requirement ("Danger" or "Caution").
- Safety Data Sheets (SDSs)—previously known as a material safety data sheet (MSDS), there is now a standardized 16-section format for all SDSs.

Under the GHS process, the same chemical from different manufacturers should now carry the same hazard information. In addition, hazard classes and definitions will now be aligned between U.S. federal agencies (e.g., U.S. DOT and U.S. EPA). The classification, labeling, and SDS requirements go into full effect on June 1, 2015, although distributors may continue to ship products using older hazcom labels until December 1, 2015.

It should be emphasized that NFPA 704 is intended for emergency response, while hazard communication marking systems are intended for facility employees and personnel. Although not a response information tool, both can give clues to the identity or hazard class of a hazardous material.

Polychlorinated Biphenyls (PCBs). In the 1980s a great deal of attention was focused on the hazards of polychlorinated biphenyls (PCBs). Although severely restricted and eliminated from most industries, PCB-laden equipment may still be found in use. PCB transformers and equipment are required to be marked with black-on-yellow warning labels. Unfortunately, these labels are small (6 inches by 6 inches or less) and often cannot be seen from a safe distance Figure 6-40.

Only transformers and equipment containing 50 ppm or greater concentration must be marked. As a general rule, regard all transformer incidents as PCB-related until tests prove otherwise.

There are situations in which high concentrations of PCBs can be encountered with no warning labels. Examples include fluorescent light ballasts manufactured prior to 1977, hydraulic and vacuum pump systems, some adhesives, microscope immersion oils, and the manufacture of "carbonless" carbon paper. Although the hazards of PCB fluids must be recognized, greater risks are created if the PCB fluids burn and break down into more toxic dioxin and furan compounds.

U.S. Military Marking System. This marking system will be found primarily on both structures and containers at U.S. military facilities. The system consists of both fire and chemical hazard symbols Figure 6-41. The four fire division symbols parallel Division 1.1 through 1.4 explosive materials.

Other Marking Systems. Bulk liquid petroleum marketing and storage facilities that move more than one product typically use a color code and symbol system that denotes the different grades of gasoline and fuel oils. These markings are typically found at the loading rack and at transfer valves and connections.

The American Petroleum Institute (API) has developed a recommended uniform marking system published in API 1637—The API Color Symbol System. This system classifies the many hydrocarbon fuels and blends into leaded and unleaded gasoline (regular, premium, super), gasoline additives, and distillates and fuel oils. This marking system has also been adopted by several states for use at liquid petroleum facilities and service stations for identifying piping and connections at loading racks, as well as the fill point connections for service station tanks.

Color codes are also used at bulk LPG facilities. Liquid product lines will often be color-coded either dark blue or orange while vapor lines are light blue or yellow. Liquid lines will always be larger in diameter than vapor lines. Fire protection piping will normally be painted red.

THE NFPA 704 MARKING SYSTEM

The NFPA 704 Marking System distinctively indicates the properties and potential dangers of hazardous materials. The following is an explanation of the meanings of the Quadrant Numerical Codes:

HEALTH (Blue)

IN GENERAL, HEALTH HAZARD IN FIREFIGHTING IS THAT OF A SINGLE EXPOSURE WHICH MAY VARY FROM A FEW SECONDS UP TO AN HOUR. THE PHYSICAL EXERTION DEMANDED IN FIREFIGHTING OR OTHER EMERGENCY CONDITIONS MAY BE EXPECTED TO INTENSIFY THE EFFECTS OF ANY EXPOSURE. ONLY HAZARDS ARISING OUT OF AN INHERENT PROPERTY OF THE MATERIAL ARE CONSIDERED. THE FOLLOWING EXPLANATION IS BASED UPON PROTECTIVE EQUIPMENT NORMALLY USED BY FIREFIGHTERS:

4 MATERIALS TOO DANGEROUS TO HEALTH TO EXPOSE FIREFIGHTERS. A FEW WHIFFS OF THE VAPOR COULD CAUSE DEATH OR THE VAPOR OF LIQUID COULD BE FATAL ON PENETRATING THE FIREFIGHTER'S NORMAL FULL PROTECTIVE CLOTHING. THE NORMAL, FULL-PROTECTIVE CLOTHING AND BREATHING APPARATUS AVAILABLE TO THE AVERAGE FIRE DEPARTMENT WILL NOT PROVIDE ADEQUATE PROTECTION AGAINST INHALATION OR SKIN CONTACT WITH THESE MATERIALS.

3 MATERIALS EXTREMELY HAZARDOUS TO HEALTH, BUT AREAS MAY BE ENTERED WITH EXTREME CARE. FULL-PROTECTIVE CLOTHING INCLUDING SELF-CONTAINED BREATHING APPARATUS, COAT, PANTS, GLOVES, BOOTS AND BANDS AROUND LEGS, ARMS, AND WAIST SHOULD BE PROVIDED. NO SKIN SURFACE SHOULD BE EXPOSED.

2 MATERIALS HAZARDOUS TO HEALTH, BUT AREAS MAY BE ENTERED FREELY WITH FULL-FACE MASK AND SELF-CONTAINED BREATHING APPARATUS WHICH PROVIDES EYE PROTECTION.

1 MATERIALS ONLY SLIGHTLY HAZARDOUS TO HEALTH. IT MAY BE DESIRABLE TO WEAR SELF-CONTAINED BREATHING APPARATUS.

0 MATERIALS WHICH WOULD OFFER NO HAZARD BEYOND THAT OF ORDINARY COMBUSTIBLE MATERIAL UPON EXPOSURE UNDER FIRE CONDITIONS.

FLAMMABILITY (Red)

SUSCEPTIBILITY TO BURNING IS THE BASIS FOR ASSIGNING DEGREES WITHIN THIS CATEGORY. THE METHOD OF ATTACKING THE FIRE IS INFLUENCED BY THIS SUSCEPTIBILITY FACTOR.

4 VERY FLAMMABLE GASES OR VERY VOLATILE FLAMMABLE LIQUIDS. SHUT OFF FLOW AND KEEP COOLING WATER STREAMS ON EXPOSED TANKS OR CONTAINERS.

3 MATERIALS WHICH CAN BE IGNITED UNDER ALMOST ALL NORMALTEMPERATURE CONDITIONS. WATER MAY BE INEFFECTIVE BECAUSE OF THE LOW FLASH POINT.

2 MATERIALS WHICH MUST BE MODERATELY HEATED BEFORE IGNITION WILL OCCUR. WATER SPRAY MUST BE USED TO EXTINGUISH THE FIRE BECAUSE THE MATERIAL CAN BE COOLED BELOW ITS FLASH POINT.

1 MATERIALS THAT MUST BE PREHEATED BEFORE IGNITION CAN OCCUR. WATER MAY CAUSE FROTHING IF IT GETS BELOW THE SURFACE OF THE LIQUID AND TURNS TO STEAM. HOWEVER, WATER FOG GENTLY APPLIED TO THE SURFACE WILL CAUSE A FROTHING WHICH WILL EXTINGUISH THE FIRE.

0 MATERIALS THAT WILL NOT BURN.

REACTIVITY (STABILITY) (Yellow)

THE ASSIGNMENT OF DEGREES IN THE REACTIVITY CATEGORY IS BASED UPON THE SUSCEPTIBILITY OF MATERIALS TO RELEASE ENERGY EITHER BY THEMSELVES OR IN COMBINATION WITH WATER. FIRE EXPOSURE WAS ONE OF THE FACTORS CONSIDERED ALONG WITH CONDITIONS OF SHOCK AND PRESSURE.

4 MATERIALS WHICH (IN THEMSELVES) ARE READILY CAPABLE OF DETONATION OR OF EXPLOSIVE DECOMPOSITION OR EXPLOSIVE REACTION AT NORMAL TEMPERATURES AND PRESSURES. INCLUDES MATERIALS WHICH ARE SENSITIVE TO MECHANICAL OR LOCALIZED THERMAL SHOCK. IF A CHEMICAL WITH THIS HAZARD RATING IS IN AN ADVANCED OR MASSIVE FIRE, THE AREA SHOULD BE EVACUATED.

3 MATERIALS WHICH (IN THEMSELVES) ARE CAPABLE OF DETONATION OR OF EXPLOSIVE DECOMPOSITION OR EXPLOSIVE REACTION BUT WHICH REQUIRE A STRONG INITIATING SOURCE OR WHICH MUST BE HEATED UNDER CONFINEMENT BEFORE INITIATION. INCLUDES MATERIALS WHICH ARE SENSITIVE TO THERMAL OR MECHANICAL SHOCK AT ELEVATED TEMPERATURES AND PRESSURES OR WHICH REACT EXPLOSIVELY WITH WATER WITHOUT REQUIRING HEAT OR CONFINEMENT. FIREFIGHTING SHOULD BE DONE FROM AN EXPLOSIVE-RESISTANT LOCATION.

2 MATERIALS WHICH (IN THEMSELVES) ARE NORMALLY UNSTABLE AND RAPIDLY UNDERGO VIOLENT CHEMICAL CHANGE BUT DO NOT DETONATE. INCLUDES MATERIALS WHICH CAN UNDERGO CHEMICAL CHANGE WITH RAPID RELEASE OF ENERGY AT NORMAL TEMPERATURES AND PRESSURES OR WHICH CAN UNDERGO VIOLENT CHEMICAL CHANGE AT ELEVATED TEMPERATURES AND PRESSURES. ALSO INCLUDES THOSE MATERIALS WHICH MAY REACT VIOLENTLY WITH WATER OR WHICH MAY FORM POTENTIALLY EXPLOSIVE MIXTURES WITH WATER. IN ADVANCE OR MASSIVE FIRES, FIREFIGHTING SHOULD BE DONE FROM A SAFE DISTANCE OR FROM A PROTECTED LOCATION.

1 MATERIALS WHICH (IN THEMSELVES) ARE NORMALLY STABLE BUT WHICH MAY BECOME UNSTABLE AT ELEVATED TEMPERATURES AND PRESSURES OR WHICH MAY REACT WITH WATER WITH SOME RELEASE OF ENERGY BUT NOT VIOLENTLY. CAUTION MUST BE USED IN APPROACHING THE FIRE AND APPLYING WATER.

0 MATERIALS WHICH (IN THEMSELVES) ARE NORMALLY STABLE EVEN UNDER FIRE EXPOSURE CONDITIONS AND WHICH ARE NOT REACTIVE WITH WATER. NORMAL FIREFIGHTING PROCEDURES MAY BE USED.

SPECIAL INFORMATION (White)

MATERIALS WHICH DEMONSTRATE UNUSUAL REACTIVITY WITH WATER SHALL BE IDENTIFIED WITH THE LETTER W WITH A HORIZONTAL LINE THROUGH THE CENTER (W).

MATERIALS WHICH POSSESS OXIDIZING PROPERTIES SHALL BE IDENTIFIED BY THE LETTERS OX.

MATERIALS POSSESSING RADIOACTIVITY HAZARDS SHALL BE IDENTIFIED BY THE STANDARD RADIOACTIVITY SYMBOL

Figure 6-38 NFPA 704 Marking System

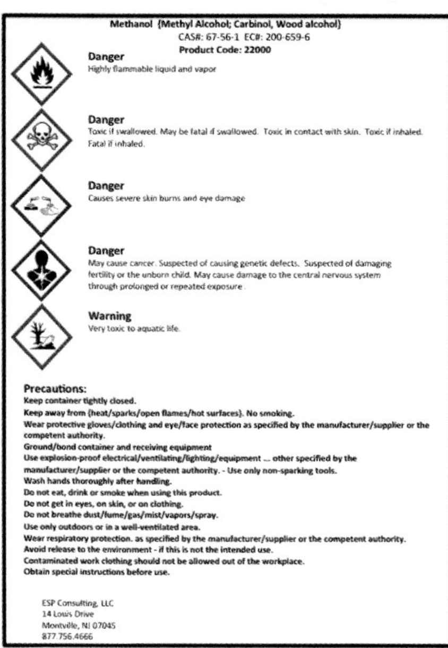

Figure 6-39 Example of a label in compliance with the OSHA's Global Harmonization Standard and Hazard Communication standard requirements

The OSHA Hazard Communication Standard and many state and local right-to-know regulations require the marking of hazardous materials storage vessels, piping, and process units within industry. Although the format may vary, the marking should clearly spell out the hazard class or specific contents of the container. ANSI A13.1—*Scheme for Identification of Piping Systems* outlines a common system of identification often used in business and industry.

Placards and Labels

Placards and labels are the fourth clue in the standard identification process and can be used as a clue for hazmat recognition and classification. This is a basic First Responder Awareness and Operations-level competency, and should be well-understood by this point in your hazmat response career. Although designed for transportation purposes, placards and labels may also be found on both bulk and nonbulk packaging at fixed locations.

Figure 6-40 PCB-contaminated transformers and equipment containing more than a 50 ppm concentration must be marked with this label.

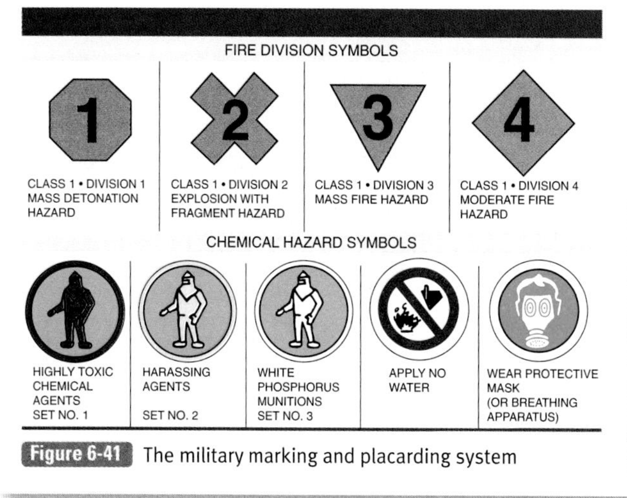

Figure 6-41 The military marking and placarding system

DOT Hazardous Materials Regulations outline the hazmat placarding and labeling requirements within the United States. The backbone of DOT's hazmat regulations is the Hazardous Materials Table or HMT (49 CFR 172.101). The HMT governs the transportation of hazmats by all modes of transportation. For each material listed, the HMT identifies its hazard class or specifies that the material is forbidden in transportation. It provides the proper shipping name of the material or directs the user to the preferred proper shipping name. In addition, it specifies or references requirements pertaining to labeling and placarding, packaging, quantity limits aboard aircraft, and stowage of hazmats aboard vessels.

Placards and labels provide recognition and general hazard classification by way of:

- Colored background
- Respective hazard class symbol
- Hazard class/division number (found at the bottom of the placard or label)
- Hazard class description wording or the four-digit identification number (found in the center of the placard)

Labels are approximately 4-inch (100 mm) square markings applied to individual hazardous materials packages. They are generally placed or printed near the contents names or are printed on the manufacturing label. When labels cannot be applied directly to the container because of its nonadhesive surface, they are placed on tags or cards attached to the package.

The proper placard and label(s) are determined by the product's hazard class **Figure 6-42**. Some hazmats have more than one hazard and multiple placards or labels may be required. For example, red fuming nitric acid nonbulk packages are labeled "CORROSIVE," "OXIDIZER," and "POISON" to indicate that it is a poison inhalation hazard (PIH) material.

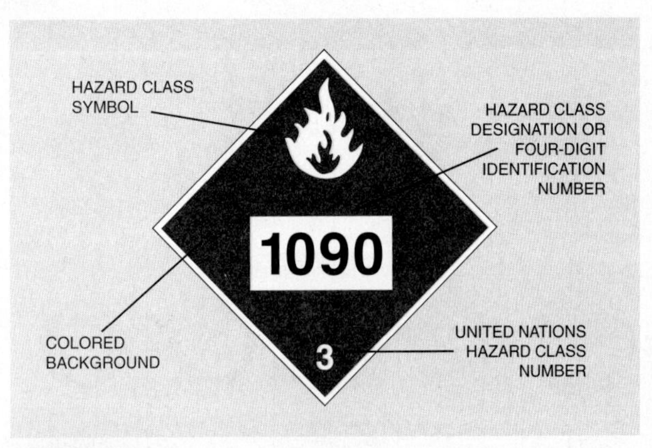

Figure 6-42 Placards and labels are the fourth clue in the standard identification process and provide hazard communication information in several manners.

Placards are approximately 10.75-inch (273 mm) square markings applied to both ends and each side of freight containers, cargo tanks, and portable tank containers. Factors such as the individual package labels, the size of individual packages, and the total quantity of the product will determine the correct placard to be used.

Radioactive Material Labels and Placards. Radioactive packages will be labeled based on the type and quantity of material being shipped and associated levels of radiation. Three different labels are used on radioactive material packaging—Radioactive White–I, Radioactive Yellow–II, and Radioactive Yellow–III **Table 6-4**.

Radioactive label markings are essential in assessing the integrity of the packaging during an incident and will include:

- *Contents*—The radioactive content of the packaging (e.g., Am-241 Americium, Cs-137 Cesium).
- *Activity*—The rate of disintegration or decay of a radioactive material. Listed in the appropriate SI units (e.g., Becquerels [Bq], Terabecquerels [TBq], etc.) or in both SI units and the appropriate customary units (Curies [Ci], millicuries [mCi] microcuries [uCi], etc.). Activity indicates how much radioactivity is present and not how much material is present.
- *Transport Index (TI)*—Is used by the shipper to control the number of packages in a shipment. It is obtained by taking the maximum radiation level in millirems per hour (mrem/hour) at 1 meter (3 ft) from an undamaged package. The TI can be an indicator for determining the external radiation hazard of an undamaged package and can be a starting point for determining whether or not damage has occurred.
- *Criticality Safety Index (CSI)*—assigned to a package, overpack, or freight container containing fissile material, for the transport of Class 7 material. This label will appear with one of the three Class 7 labels (Rad White I, Rad Yellow II, Rad Yellow III). The CSI number is used to provide control over the accumulation of packages, overpacks, or freight containers containing fissile material.

Responder Safety Tip

Placards and labels are simply another clue and should not be considered as a definitive source of hazmat information. Although enforcement measures are constantly improving, experience shows that a number of vehicles and containers are still improperly labeled or placarded.

Shipping Papers and Facility Documents

Shipping Paper Requirements

Hazmat shipping papers and related documents are the fifth clue in the standard identification process. Shipping papers are required to accompany each transport vehicle. Responders must be familiar with the information noted on shipping papers, their location on each transport vehicle, and the individual responsible for them. **Table 6-5** summarizes this information for each transportation mode.

Basic Description. Each transport mode has its own terms for shipping papers. All shipping papers are required to contain the following entries, known as the hazardous material's basic description. All information shall be printed in English:

- *Proper Shipping Name.* Identifies the name of the hazmat as found in the HMT. The word "WASTE" will precede the proper shipping name for those shipments classified as hazardous wastes.
- *DOT Hazard Class/Division Number.* Indicates the category of hazard assigned to a hazardous material in the DOT Hazardous Materials Regulations. A division is a subset of a hazard class. Note: A hazmat may meet the criteria for more than one hazard class, but is assigned to only one hazard class.
- *Subsidiary Hazard Class.* Indicates a hazard of a material other than the primary hazard assigned.
- *Identification Number(s).* The four-digit identification number assigned to each hazardous material. The identification number may be found with the prefix "UN" (i.e., United Nations) or "NA" (i.e., North America). "UN" identification numbers can be used for both domestic and international shipment, while "NA" are not recognized for international transportation, except between the U.S. and Canada.
- *Packing Group.* DOT regulations require that all shipments meet basic DOT requirements. Packing group further classifies hazardous materials based on the degree of danger represented by the material. There are three groups:
  1. Packing Group I indicates great danger (PG I or I).
  2. Packing Group II indicates medium danger (PG II or II).
  3. Packing Group III indicates minor danger (PG III or III).

  Packing Groups may be shown as "PG I", etc. Packing Groups are not assigned to Class 1 materials (explosives), Class 2 materials (compressed gases), Division 5.2 materials (organic peroxides), Class 7 materials (radioactive materials), some Division 6.2 materials (infectious substances), and ORM-D materials.

Table 6-4 Radioactive Material Label Summary

LABEL	EXTERNAL RADIATION LEVELS	TRANSPORT INDEX (Dose at 1 Meter)
RADIOACTIVE WHITE—I LABEL 	■ Packages with extremely low or almost no levels of radiation. ■ Maximum contact radiation level of 0.5 mrem/hour.	No TI on this label.
RADIOACTIVE YELLOW—II LABEL 	■ Packages with low radiation levels. ■ Maximum contact radiation level ranging from 0.5 mrem/hour to 50 mrem/hour.	Maximum Allowable TI = 1
RADIOACTIVE YELLOW—III LABEL 	■ Packages with higher radiation levels. ■ Maximum contact radiation level ranging from 50 mrem/hour to 200 mrem/hour. ■ Also required for Fissile Class III or large quantity shipments, regardless of radiation level.	Maximum Allowable TI = 10
CRITICALITY SAFETY INDEX LABEL WITH YELLOW III LABEL 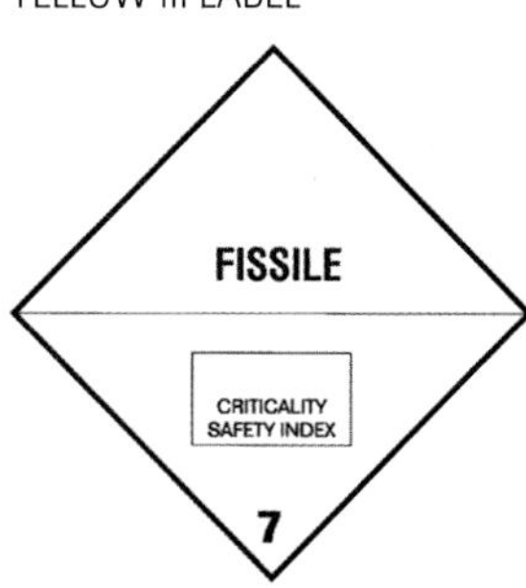	■ Packages containing fissile materials ■ Will appear with one of the 3 Class 7 labels	

Table 6-5 Shipping Paper Information

Mode of Transportation	Title of Shipping Paper	Location of Shipping Papers	Responsible Person(s)
Highway	Bill of Lading or Freight Bill	Cab of Vehicle	Driver
Railroad	Waybill and/or Consist, Switch List, Training List, Track List	With Train Crew (e.g., Conductor or Engineer)	Conductor
Water	Dangerous Cargo Manifest	Wheelhouse or Pipe-like Container on Barge	Captain or Master
Air	Airbill with Shipper's Declaration of Dangerous Goods	Cockpit (may also be found attached to the outside of packages)	Pilot

- *Total Quantity.* Indicates the quantity by net or gross mass, capacity, and so on. May also indicate the type of packaging. The number and type of packaging (e.g., 1 TC, 7 DRM) may be entered on the beginning line of the shipping description. Carriers often use abbreviations to indicate the type of packaging. The following are examples of packaging abbreviations that may be found:

BA = Bale	CH = Covered Hopper	KIT = Kit
BG = Bag	CL = Carload	KL = Container Load
BOX = Box	CY = Cubic Yard	PA = Pail
BC = Bucket	CYL = Cylinder	PKG = Package
CA = Case	DRM = Drum	SAK = Sack
CAN = Can	JAR = Jar	TB = Tube
CR = Crate	JUG = Jug	TC = Tank Car
CTN = Carton	KEG = Keg	TL = Trailer Load

- *Emergency Contact.* All shipping papers must also contain an emergency response telephone number. This is a telephone number for the shipper or shipper's representative that may be accessed 24 hours a day, 7 days a week in the event of an accident. The contact person must be either knowledgeable of the hazardous characteristics and emergency response information for the hazmat(s) listed on the shipping paper or have immediate access to someone who has that knowledge. If the shipper is registered with CHEMTREC® or comparable service, that phone number may be displayed as the emergency contact.

Shipping Papers—Additional Entries

Additional shipping paper entries may be required for some hazardous materials. They include the following:

- *Compartment Notation.* Identifies the specific compartment of a multi-compartmented rail car or cargo tank truck in which the material is located. On rail cars, compartments are numbered sequentially from the "B" end (the end where the hand brake wheel is located), while cargo tank trucks are numbered sequentially from the front.
- *Empty Packaging (Residue).* Identifies packaging that contains a hazmat residue and has not been cleaned and purged or reloaded with a material not subject to DOT Hazardous Materials Regulations. Residue is indicated by the words "Residue: Last Contained" before the proper shipping name. It is only used in rail transportation.
- *HOT.* Identifies elevated temperature materials other than molten sulfur and molten aluminum.
- *Technical Name.* Identifies the recognized chemical name currently used in scientific and technical handbooks, journals, and texts. Generic descriptions may be found provided they identify the general chemical group. With some exceptions, trade names may not be used as technical names. Examples of acceptable generic descriptions are organic phosphate compounds, tertiary amines, and petroleum aliphatic compounds.
- *Not Otherwise Specified (NOS) Notations.* If the proper shipping name of a material is an "NOS" notation, the technical name of the hazardous material must be entered in parentheses with the basic description. If the material is a mixture or solution of two or more hazardous materials, the technical names of at least two components that most predominantly contribute to the hazards of the mixture/solution must be entered on the shipping paper. For example, Flammable Liquid, Corrosive Liquid, NOS, (contains methanol, potassium hydroxide), 3, UN 2924, PGII.
- *Subsidiary Hazard Class.* Indicates a hazard of a material other than the primary hazard assigned.
- *Reportable Quantity (RQ) Notation.* Indicates the material is a hazardous substance by the EPA. The letters "RQ" (reportable quantity) must be shown either before or after the basic shipping description entries. This designation indicates that any leakage of the substance above its RQ value must be reported to the proper agencies (e.g., National Response Center). Regardless of which agencies are involved, the legal responsibility for notification still remains with the spiller.
- *Marine Pollutant.* Indicates the material meets the definition of a marine pollutant. If the basic description does not identify the component that makes the material a marine pollutant, the name of the component(s) must appear in parentheses.
- *EPA Waste Stream Number.* Indicates the number assigned to a hazardous waste stream by the U.S. EPA to identify that waste stream. *Note:* For all hazardous

waste shipments, a Uniform Hazardous Waste Manifest must be prepared in accordance with both DOT and EPA regulations.

- *EPA Waste Characteristic Number.* Indicates the general hazard characteristics assigned to a hazardous waste by the U.S. EPA. Waste characteristics include EPA corrosivity, EPA toxicity, EPA ignitability, and EPA reactivity.
- *Radioactive Material Information.* Should provide the following:
  - "Radioactive Material"—if not part of the proper shipping name
  - Proper shipping name and UN ID
  - Name of each radionuclicide (e.g., Cs-137)
  - Radioactivity level per package (will be listed as activity)
  - Category of label applied (e.g., White I, Yellow II, etc.)
  - Transport Index (for Yellow II and III labels)
  - Package type (e.g., type A, type B, etc.)
  - Physical/chemical form (if not special form)
  - "Fissile Excepted" or "Criticality Safety Index" (for fissile materials only)
  - "Exclusive Use" if shipment is being made under exclusive use provisions
  - Highway Route Controlled Quantity or "HRCQ" (if shipment is HRCQ)
- *Poison Notation.* Indicates a liquid or solid material is poisonous when the fact is not disclosed in the shipping name.
- *Poison-Inhalation Hazard (PIH)* or *Toxic-Inhalation Hazard (TIH) Notation.* Indicates gases and liquids that are poisonous by inhalation.
- *Hazard Zone.* Indicates relative degree of hazard in terms of toxicity (only appears for gases and liquids that are poisonous by inhalation):
  - Zone A—LC_{50} less than or equal to 200 ppm (most toxic)
  - Zone B—LC_{50} greater than 200 ppm and less than or equal to 1000 ppm
  - Zone C—LC_{50} greater than 1000 ppm and less than or equal to 3000 ppm
  - Zone D—LC_{50} greater than 3000 ppm and less than or equal to 5000 ppm (least toxic)
- *Dangerous When Wet Notation.* Indicates a material that, by contact with water, is liable to become spontaneously flammable or give off flammable or toxic gas at a rate greater than 1 liter per kilogram of the material, per hour.
- *Limited Quantity (LTD QTY).* Indicates a material being transported in a quantity for which there is a specific labeling and packaging exception.
- *Canadian Information.* Indicates information required for hazardous materials entering or exiting Canada in addition to that required in the United States (e.g., ERG reference number, the 24-hour Canadian emergency telephone number, Canadian class).
- *Placard Notation.* Indicates the placard applied to the container. Where placards are not required (e.g., ORM commodities or nonflammable cryogenic gases), the notation "MARKED" is followed by the four-digit identification number.
- *Trade Name.* A name that enables organizations, such as emergency responders and CHEMTREC®, to access the SDS for additional information.
- *DOT Exemption Notation.* If a hazmat shipment is made under a DOT exemption for specific packaging or shipping procedures, the shipping papers must include the letters "DOT—E" followed by the assigned exemption number. The exemption number must be placed so that it is clearly associated with the description to which the exemption applies.
- *Hazardous Materials STCC Number.* A seven-digit Standard Transportation Commodity Code (STCC) number will be found on all shipping papers accompanying rail shipments of hazmats. It will also be found when intermodal containers are changed from rail to highway movement. Look for the first two digits—"49"—as the key identifier for a hazmat. The STCC number will follow the notation "HAZMAT STCC."
- *Shipper Contact.* Indicates the identity of the producer or consolidator of the materials described.

Shipping Papers—Emergency Response Information

Emergency response information must also be included with shipping papers. An emergency telephone number for the shipper or shipper's representative is required on the shipping paper. Emergency response information must provide the following:

- Brief product description
- Emergency actions involving fire
- Emergency actions involving release only
- Personnel protective measures
- Environmental considerations, as appropriate
- First-aid measures

Several common sources of emergency response information requirements are an SDS, a copy of the *Emergency Response Guidebook* (ERG), a copy of the specific page from the ERG for the hazmat being transported, or railroad emergency response information sheets cross-referenced with the train consist. With most major railroads, the railroad emergency response information sheets will be part of the train consist.

Facility Documents

Various types of facility documents are available to assist in the information process. They can be a source for hazmat recognition, identification, and classification at an emergency.

The specific type and nature of information provided will vary based on pertinent federal, state, and/or local reporting requirements. Examples include hazmat inventory forms, shipping and receiving forms, Risk Management Plans and supporting documentation, SDSs, and Tier II reporting forms required to be submitted to the LEPC and the fire department under SARA, Title III.

Both the Risk Management Plans and the Tier II reporting forms can be used as part of the hazard analysis process. For example, the Tier II reporting forms provide information such as chemicals on-site that exceed the reporting thresholds, physical and chemical hazards, average and maximum amounts on site, and types of storage containers and location.

Monitoring and Detection Equipment

If the hazardous material cannot be identified from the previously discussed methods, monitoring and detection equipment can often provide data concerning the overall nature of the problem you face as well as the specific materials involved. Monitoring and detection equipment is essential for identifying, verifying, or classifying the hazmat(s) involved and is the sixth clue in the standard identification process.

Although considered here as an identification tool, monitoring and detection equipment are also critical tools for evaluating real-time data and developing a risk-based response. Monitoring helps responders to:

- Determine the appropriate levels of personal protective clothing and equipment.
- Determine the size and location of hazard control zones.
- Develop protective action recommendations and corridors.
- Assess the potential health effects of exposure.

The selection, application, and use of monitoring instruments are addressed in the chapter of the text that covers "hazard and risk evaluation."

Senses

Sometimes clues are not as obvious as a building occupancy, container shapes, or markings, and your senses must come into play. Senses are not a primary identification tool and are the final clue in the standard identification process. The only senses that offer some protection are visual and hearing; the use of other senses means you are at potential risk from the hazmat. In most cases, if you are close enough to smell, feel, or hear the problem, you are probably too close to operate safely.

Nonetheless, senses can be valuable assets and can offer immediate clues to the presence of hazardous materials. For our purposes, "senses" refers to any personal physiological reaction to or visual observation of a hazmat release. Smells, dizziness, unusual noises (i.e., relief valve actuations), and destroyed vegetation are some examples.

The inhalation (or smelling) of chemicals should always be avoided, but there are times when you may enter a situation and an odor will be present. Being able to characterize those odors is important to your safety so that you can quickly move to a safe area. Unfortunately, being able to detect the odor effectively also means you have most likely been exposed to potentially dangerous concentrations of the gas or vapors at some point.

The concept of *Immediately Dangerous to Life or Health (IDLH)* atmospheres was discussed in the "health and safety" section of this text. As a review, there are some basic street smart clues of IDLH atmospheres you should always remember:

- *Visible vapor clouds*—Avoid entering any vapor clouds, smoke, or mists. Fires are considered IDLH conditions. Vapor clouds with colors (e.g., green, orange, brown) are not good! Remember that the lack of a visible vapor cloud does not mean there is no hazard.
- *Releases*—Releases from bulk containers or a pressurized container are extremely hazardous because the more product available, the greater the risk. Most pressurized containers hold liquefied gases with large liquid to vapor expansion ratios, thereby presenting greater risk.
- *Large liquid leaks*—Avoid contact with any amount of released materials. The larger the leak, the more vapors that may be produced.
- *Below grade or confined spaces*—These can include artificial barriers, all of which can be oxygen deficient or accumulate toxic gases.
- *Dead birds, brown foliage, sick animals, and sick humans*—These are all biological indicators of a possible chemical release.
- *Physical senses and "street smarts"*—If a situation doesn't seem right, it probably isn't. Trust your instincts and back away from the problem until you can sort it out. Be aware of odors and sources of energy that can be released and harm you—mechanical, electrical, and kinetic. Remember—dealing with unknowns is always dangerous!

Note: An excellent textbook on basic street smart hazmat response issues is *Street Smart HazMat Response* by Michael Callan.

Scan Sheet 6-A

Hazardous Materials Incidents and Aircraft

When compared to other modes of hazardous materials transportation, the limited number of air incidents seems small and relatively insignificant. However, numerous case studies have clearly illustrated the challenges that such incidents can create for responders, flight crews and, when commercial passenger aircraft are involved, the passengers. Among the most notable is the 1996 Value Jet incident in the Everglades, Florida when a DC-9 with 110 passengers and crew crashed after an in-flight fire erupted in a forward compartment containing chemical oxygen generators, a Class 5.1 material.

Nearly every aircraft incident has the potential to be a hazmat incident from a health and safety perspective. Aircraft transport large amounts of fuel, use flammable metals in their construction, and plastics and composite materials that can emit toxic gases and carbon fibers when involved in fire. Aircraft systems include hydraulic fluids, oxygen, fuel, compressed air cylinders, and high-pressure wheel and tire assemblies filled with nitrogen. Military aircraft add another dimension to the problem, ranging from munitions, ejection systems, and special hazards such as liquid oxygen and hydrazine.

Aircraft incidents involving hazardous materials fall into three broad scenarios:

1. *Ground-based incidents* that occur with the aircraft on the ground, either parked or taxiing. Scenarios range from spills during fueling or defueling operations, to "unknown odors" in the cargo hold while an aircraft is being loaded. Tactically, this scenario is managed along the same lines as any other hazmat incident once the aircraft is in a safe location and ground travel is cleared through the airport control tower.
2. *In-flight emergencies* where the flight crew must take rapid actions to get the aircraft on the ground in a minimum amount of time. Scenarios can range from smoke or chemical odors within the aircraft fuselage to a loss of aircraft control due to the effects of a hazmat release upon the aircraft controls or systems. Tactical response considerations for in-flight emergencies include:
   - The aircraft landing may be a hard landing, with a maximum effort stop. Ensure you are positioned well clear of the runway prior to the landing.
   - The aircraft may have blown tires or hot brakes as a result of the landing and maximum effort stop. *Do not approach the landing gear from the sides due to the explosion hazard!*
   - Aircraft evacuation may be initiated as soon as the aircraft comes to a stop.
   - Be alert for numerous safety hazards, including engine intakes and exhaust, moving flight control surfaces, fuel spills, etc.
3. *Aircraft crash*, such as an off-airport crash or emergency landing scenario that results in the break-up of the aircraft.

Types of Hazardous Cargo. The shipment of hazardous materials or dangerous goods by air is governed by DOT and International Civil Aviation Organization (ICAO) regulations. ICAO is a United Nations governing body that exercises control over international air transportation. Most air cargo carriers also use the *Dangerous Goods Regulations* published annually by the International Air Transport Association (IATA), which provides additional information on packaging, labeling, loading, and maximum allowable quantities that can be transported on aircraft for airfreight carriers.

Categories of dangerous goods within the aviation industry include:

1. *Material Forbidden on Any Aircraft.* These materials are forbidden on any aircraft, in any quantity, and under any circumstances. Lists of these materials can be referenced in both DOT (i.e., the Hazardous Materials Table) and ICAO regulations.
2. *Materials That Can Only Be Transported on Cargo Aircraft.* These materials are forbidden on passenger aircraft and must have an orange and black "Cargo Aircraft Only" label attached to the container (see photo). Except for "Cargo Aircraft Only" packages specifically marked as radioactive materials, the flight crew is mandated to have direct access to these containers during the flight. The quantity of dangerous goods transported on cargo aircraft are generally limited by the maximum quantity per individual package, as referenced from DOT and ICAO regulations.

3. *Materials Allowed on Passenger or Cargo Aircraft in Inaccessible Cargo Holds.* These materials are generally carried in the lower cargo holds that are inaccessible during flight and can be loaded as follows:
   - Class 1, 3, 4, 5, 6, and 8 materials are limited to 55 lbs. per compartment.
   - In addition, 165 lbs. of a nonflammable compressed gas (Division 2.2) may be loaded in the same compartment.
   - Class 7 materials are limited by their combined Transport Index (TI) number.
   - There are no limits for Class 9 materials (except dry ice) or ORM-D materials aboard an aircraft.

Recognize, however, that exceptions to these regulations may be encountered, such as aircraft carrying both passengers and cargo in accessible cargo areas on the main deck.

4. *Air Carrier Company Materials (COMAT).* These are replacement items for installed aircraft equipment, serviceable items, or items removed for servicing and repair that are being shipped on an aircraft. Many of these materials are regulated as both hazardous materials under U.S. DOT regulations and dangerous goods under international regulations, and may include oxygen cylinders, batteries, life rafts, paint, and chemical oxygen generators. Only an airline with a U.S. Federal Aviation Administration's (FAA) approved Hazardous Materials/Dangerous Goods program may transport its own materials as COMAT.
5. *Undeclared and Unknown Dangerous Goods.* These are shipments about which the airline has no knowledge, such as individuals/organizations unaware of the regulations or who deliberately attempt to conceal information because of the additional costs or burdens of shipping hazardous materials.

Packaging. Hazardous materials may be shipped in a wide range of nonbulk or bulk packaging, providing they are permissible under DOT and ICAO regulations. Nonbulk packages are often placed in freight containers or aircraft unit load devices (ULDs). ULDs may be found in either a container or pallet configuration; pallets are secured by a net attached to the rim of the pallet. Containers are constructed of aluminum with an aluminum or fiberglass shell, with a volume ranging from 64 to 640 cubic ft. (1.8 to 18 cubic meters). ULDs should have the appropriate DOT labels, and at least one of the required labels must be displayed on or near the closure for the container.

A ULD tag, which indicates the type of dangerous goods loaded within the container or pallet, is often found. In general, all airlines have their own ULD tags with their own logo and requirements stated. The tag may have a red border and will be wired to or placed in a clear plastic holder on the exterior of the ULD. The ULD tag will list the hazard class and division of any dangerous goods being transported within the ULD. Some cargo airlines also use color codes or markings to specifically identify the ULD hazmat containers.

Shipping Papers. The shipper is responsible for properly packaging, labeling, and marking the individual package, as well as completing the "Shipper's Declaration of Dangerous Goods." This is similar to an individual waybill. Copies of this declaration should be affixed directly to the package (or one of the packages in a multi-package shipment from one shipper) in a clear plastic envelope, as well

as in or close to the flight deck. A red candy-striped border on any cargo paperwork will indicate the presence of hazardous materials; however, some computer-generated forms without the red border may also be found. Information found on the Shipper's Declaration includes:

- Basic description of the hazardous materials, per DOT regulations.
- Shipper and consignee.
- Departure airport and destination.
- Information pertaining to whether the material is radioactive, Cargo Aircraft Material Only, or allowed aboard a passenger aircraft.

An airbill should be in the possession of the flight crew or near the flight deck, and should indicate whether a material is a dangerous good. The crew will also have a "Pilot's Notification for Loading Restricted Articles" or "Dangerous Goods Load Notification to Captain" form. Information that must be provided to the flight crew by the aircraft operator or loading facility includes:

- Proper shipping name, hazard class, and four-digit identification number.
- Total number of packages.
- Location of dangerous goods aboard the aircraft.
- Telephone number for a point-of-contact not aboard the aircraft from whom information contained in the Pilot's Notification can be obtained.
- Confirmation that the package must be carried on "Cargo Aircraft Only" if its transportation aboard passenger-carrying aircraft is forbidden.

On cargo aircraft, a "Load Planning Sheet" indicating the weights and locations of the cargo at each position on the aircraft may also be found. The initials "DG" may be written in the position square on the form, indicating the location of the dangerous goods.

Wrap-Up

Chief Concepts

- The evaluation of hazards and the assessment of the risks build on the timely identification and verification of the hazardous materials involved. A problem well defined is half-solved.
- Among the most critical tasks in managing a hazmat incident are surveying the incident scene to detect the presence of hazmats, identifying the nature of the problem and the materials involved, and identifying the type of hazmat container and the nature of its release.
- The identification process is built on the basic elements of (1) recognition, (2) identification, and (3) classification of the materials involved.
- Identification and verification of the hazmats involved are critical to the safe and effective management of a hazmat incident. The seven basic clues for recognition, identification, and classification are:
  - Occupancy and location
  - Container shapes
  - Markings and colors
  - Placards and labels
  - Shipping papers and facility documents
  - Monitoring and detection equipment
  - Senses
- All hazardous materials are controlled as long as they remain within their container. Hazmat responders should be able to recognize the various container profiles and know the general hazmat class/division of materials that may be found within each type of container.
- Markings and colors on hazmat packaging or containment systems may include color codes, container specification numbers, signal words, or the content's name and associated hazards. At facilities, clues may include Hazard Communication markings, piping color code systems, and specific signs and/or signal words.
- Shipping papers are required to accompany each transport vehicle. Responders must be familiar with the information noted on shipping papers, their location on each transport vehicle, and the individual responsible for them.
- Various types of facility documents are available to assist in the information process and can be a source for hazmat recognition, identification, and classification at an emergency. Examples include hazmat inventory forms, shipping and receiving forms, Risk Management Plans and supporting documentation, SDSs, and SARA, Title III-Tier II reporting forms.
- Monitoring and detection equipment can provide data concerning the overall nature of the problem responders face as well as the specific materials involved. They are also critical tools for evaluating real-time data and developing a risk-based response.

Hot Terms

Bulk Packaging—Any packaging having a capacity meeting one of the following criteria:

- Liquid—capacity greater than 119 gallons (450 liters).
- Solid—net mass greater than 882 pounds (400 kg) for solids, or capacity greater than 119 gallons (450 liters).
- Compressed Gas—water capacity greater than 1,001 pounds (454 kg).

Bulk packages can be an integral part of a transport vehicle or packaging placed on or in a transport vehicle.

Container—Any vessel or receptacle that holds a material, including storage vessels, pipelines and packaging. Includes both bulk and non-bulk packaging, and fixed containers.

Excepted Packaging—Used to transport material with low levels of radioactivity. Excepted Packaging does not have to pass any performance tests, but must meet specific design requirements spelled out in DOT regulations.

Facility Containment Systems—Packaging, containers, and/or associated systems that are part of a fixed facility's operations.

Industrial Packaging—Used in certain shipments of Low Specific Activity (LSA) radioactive material and Surface Contaminated Objects, which are typically categorized as radioactive waste. Most low-level radioactive waste, such as contaminated protective clothing and handling materials, is shipped in secured packaging of this type. DOT regulations require that these packagings allow no identifiable release of the material into the environment during normal transportation and handling.

Nonbulk Packaging—Any packaging having a capacity meeting one of the following criteria:

- Liquid—internal volume of 119 gallons (450 L.) or less;
- Solid—capacity of 882 lb. (400 kg) or less; and Compressed Gas—water capacity of 1,001 lb. (454 kg) or less.
- Nonbulk packaging may consist of single packaging or combination packaging—one or more inner packages inside of an outer packaging.

Packaging—A receptacle, which may require an outer packaging and any other components or materials necessary for the receptacle to perform its containment function and to ensure compliance with minimum packaging requirements (U.S. DOT). Packaging is divided into three general groups: nonbulk packaging, bulk packaging, and facility containment systems.

Type-A Packaging—Used to transport small quantities of radioactive material with higher concentrations of radioactivity than those shipped in Industrial Packaging. Designed to ensure that the package retains its containment integrity and shielding under normal transport conditions. However, they are NOT designed to withstand the forces of an accident.

Type-B Packaging—Used to transport radioactive material with the highest levels of radioactivity, including potentially life-endangering amounts that could pose a significant risk if released during an accident. Must meet all of the Type A requirements, as well as a series of tests which simulate severe or "worst case" accident conditions. Accident conditions are simulated by performance testing and engineering analysis.

Type C Packaging—Used to transport by aircraft high-activity radioactive materials that have not been certified as "low dispersible radioactive material" (including plutonium). They are designed to withstand severe accident conditions associated with air transport without loss of containment or significant increase in external radiation levels.

HazMat Responder *in Action*

A community that is thirty miles from yours has just reported a train derailment and has asked for mutual aid response. While en route you have been receiving limited information over you Mobile Data Terminal. The information advises you of the stenciling found on several of the railroad cars—AAR 103 J 60 AL F 1, DOT 111 A 60 AL W 1—and a flat trailer with several medium-sized boxes that are have been identified as Type A packaging.

Questions

1. What do the first three letters on the two rail cars suggest?
   - **A.** These are the specification of the rail car
   - **B.** They identify the type of materials used in tank construction
   - **C.** They identify the Authorizing Agency under whose authority the specification was issued
   - **D.** They are a class designation of the construction and materials type of the rail car
2. The communication of Type A packaging identifies which hazard?
   - **A.** High-Grade poison
   - **B.** Highly corrosive materials
   - **C.** Flammable materials
   - **D.** Radioactive materials
3. On the second rail car that has been identified, what is the type of material used for container construction and what is its test pressure?
   - **A.** Nickel plate/111 psi
   - **B.** Aluminum/60 psi
   - **C.** Stainless steel/1 psi
   - **D.** Aluminum/1 psi
4. What is the difference between AAR 103 and DOT 111?
   - **A.** Both are non-pressure rail cars with the same authorizing agency
   - **B.** AAR 103 is pressurized while the DOT is not
   - **C.** Both are non-pressure tank cars with different authorizing agencies
   - **D.** Both are pressure tank cars with different authorizing agencies

References and Suggested Readings

1. Air Force Civil Engineer Support Agency (AFCESA) and PowerTrain, Inc., *Hazardous Materials Incident Commander Emergency Response Training CD.* Tyndall Air Force Base, FL: Headquarters AFCESA (2010).
2. Air Force Civil Engineer Support Agency (AFCESA) and PowerTrain, Inc., *Hazardous Materials Technician Emergency Response Training CD.* Tyndall Air Force Base, FL: AFCESA (2010).
3. Butters, Tim. "5 Things Every Fire Chief Should Know About Pipeline Incidents." FIRE CHIEF Magazine (July, 2012).
4. Callan, Michael. *Street Smart HazMat Response.* Chester, MD: Red Hat Publishing, Inc. (2002).
5. Carr, John and Les Omans, "Responding to Commercial Aircraft HazMat Incidents." *Fire Engineering* (November, 2001), pages 47–52.
6. Code of Federal Regulations, *Title 49 CFR Parts 100–199 (Transportation),* Washington, DC: U.S. Government Printing Office.
7. Compressed Gas Association, *Handbook of Compressed Gases* (4th Edition). Norwell, MA: Kluwer Academic Publishers (2003).
8. Emergency Film Group, *The Eight Step Process: Step 2—Identifying The Problem* (videotape). Edgartown, MA: Emergency Film Group (2004).
9. General American Transportation Corporation, *GATX Tank Car Manual* (6th edition). Chicago, IL: General American Transportation Corporation (1994).
10. Hawley, Chris, *Hazardous Materials Air Monitoring and Detection Instruments.* Albany, NY: Delmar—Thomson Learning (2002).
11. Hawley, Chris, *Hazardous Materials Incidents.* Albany, NY: Delmar—Thomson Learning (2002).
12. Hawley, Chris, Gregory G. Noll and Michael S. Hildebrand, *Special Operations for Terrorism and Hazmat Crimes.* Chester, MD: Red Hat Publishing, Inc. (2002).
13. Hildebrand, Michael S. and Gregory G. Noll, *Gasoline Tank Truck Emergencies: Guidelines and Procedures* (2nd Edition). Stillwater, OK: Fire Protection Publications—Oklahoma State University (1996).
14. Hildebrand, Michael S., Gregory G. Noll and Michael Donahue, *Hazardous Materials Emergencies Involving Intermodal Containers.* Stillwater, OK: Fire Protection Publications—Oklahoma State University (1995).
15. Hildebrand, Michael S. and Gregory G. Noll, *Propane Emergencies* (3rd Edition). Washington, Dc: National Propane Gas Association (2007).
16. Hildebrand, Michael S. and Gregory G. Noll, *Storage Tank Emergencies.* Chester, MD: Red Hat Publishing, Inc. (1997).
17. Maslansky, Carol J. and Steven P. Maslansky, *Air Monitoring Instrumentation.* New York, NY: Van Nostrand Reinhold (1993).
18. National Fire Protection Association, *Fire Protection Handbook* (20th edition). Quincy, MA: National Fire Protection Association (2008).
19. National Fire Protection Association, *Hazardous Materials Response Handbook* (6th Edition). Quincy, MA: National Fire Protection Association (2013).
20. National Fire Protection Association, *Liquefied Petroleum Gases Handbook.* Quincy, MA: National Fire Protection Association (2011).
21. National Fire Protection Association, *NFPA 30—National Flammable and Combustible Liquids Code.* Quincy, MA: National Fire Protection Association (2012).
22. NIOSH/OSHA/USCG/EPA, *Occupational Safety and Health Guidance Manual for Hazardous Waste Site Activities.* Washington, DC: U.S. Government Printing Office (1985).
23. Safe Transportation Training Specialists, "Cargo Tank Truck Construction And Emergency Response." Student handout presented at International Association of Fire Chiefs (IAFC) Hazardous Materials Conference, Hunt Valley, MD (June, 2011).
24. Scheffy, Joseph, P.E., "Testing the Plastic (IBC's)." *NFPA Fire Journal* (January/February 2009), pages 54–58.
25. Sea Containers Limited, *Inspection, Repair and Test Requirements for Tank Containers,* London, England: Sea Containers Group (undated).
26. Union Pacific Railroad Company, *A General Guide to Tank Cars,* Omaha, NE:
27. Union Pacific Railroad Company, Technical Training (February, 2009).
28. Union Pacific Railroad Company, *A General Guide to Tank Containers,* Omaha, NE: Union Pacific Railroad Company, Technical Training (September, 2006).
29. U.S. Department of Energy—Office of Transportation and Emergency Management, *Module Emergency Response Radiological Transportation Training (MERRTT),* Washington, DC: DOE.
30. *Radiation/Nuclear Course For Hazardous Materials Technicians.* Mercury, NV: Department of Energy (DOE) and Bechtel Nevada.

Hazard Assessment and Risk Evaluation

CHAPTER 7

Hazardous Materials Technician

Knowledge Objectives

After reading this chapter, you will be able to:

- Describe the concept of hazard assessment and risk evaluation. (p. 218–222)
- Identify and describe the components of the General Hazardous Materials Behavior Model (GHBMO). (p. 236, 237)
- Describe the factors that influence the underground movement of hazardous materials in soil and through groundwater. (p. 249)
- Identify the hazards associated with the movement of hazardous materials in the following types of sewer collection systems: (p. 250–253)
  - Storm sewers
  - Sanitary sewers
  - Combination sewers
- List five site safety procedures for handling an emergency involving a hydrocarbon spill into a sewer collection system. (p. 252)

Skills Objectives

After reading this chapter, you will be able to:

- Demonstrate the steps necessary to identify unknown substances. (p. 220)
- Demonstrate techniques to identify chemical hazards (corrosivity, flammability, oxidation potential, oxygen deficiency, radioactivity, toxicity, and pathogenicity) for a select group of instruments. (p. 221, 222)
- Demonstrate field maintenance and testing procedures for certain types of instrumentation. (p. 224)
- Demonstrate the ability to collect samples of solids, liquids and gases. (p. 232–234)
- Describe the use of radiation survey and monitoring equipment, to determine whether the integrity of any container has been breached. (p. 217–220)
- Identify the signs and symptoms of exposure to a variety of chemicals and the target organ effects of exposure to those materials. (p. 199–203)
- Identify resources and methods for dispersion pattern prediction and modeling. (p. 204–205)
- Determine applicable public protective response options and the geographic areas to be protected. (p. 218)
- Describe response objectives for a variety of simulated hazardous materials release scenarios. (p. 243–244)

Hazardous Materials Incident Commander

Knowledge Objectives

After reading this chapter, you will be able to:

- Describe the concept of hazard assessment and risk evaluation. (p. 199–208, 235)
- Identify and describe the components of the General Hazardous Materials Behavior Model (GHBMO). (p. 236, 237)
- Describe the factors that influence the underground movement of hazardous materials in soil and through groundwater. (p. 249)
- Identify the hazards associated with the movement of hazardous materials in the following types of sewer collection systems: (p. 250–253)
  - Storm sewers
  - Sanitary sewers
  - Combination sewers
- List five site safety procedures for handling an emergency involving a hydrocarbon spill into a sewer collection system. (p. 252)

Skills Objectives

After reading this chapter, you will be able to:

- Collect and interpret hazard and response information not available from the current edition of the DOT *Emergency Response Guidebook* or an SDS. (p. 199–206)
- Identify the methods available to the organization for obtaining local weather conditions and predictions for short-term future weather changes. (p. 236)

You Are the HazMat Responder

Your emergency response organization has been dispatched to a freight train derailment within your community. Twenty-one freight cars, including nine butane tank cars, have derailed. The nine butane tank cars came to rest adjacent to a chemical manufacturing plant, and butane released from two breached tank cars immediately caught fire with some extension into the plant. The fire ultimately required the evacuation of approximately 1,750 residents.

After public safety responders received and verified the consist information from the train crew, on-scene response operations were implemented in a timely and professional manner, resulting in the control and extinguishment of fires involving the butane tank car and chemical manufacturing plant.

With fire control operations addressed, incident action plan strategies now focus on the off-loading, uprighting, and removal of the derailed railroad tank cars and wreckage.

1. What should the incident command structure look like during this phase of emergency response operations? Specifically, who is in charge? What is the role of the railroad? What is the role of the Hazardous Materials Response Team (HMRT) to the Incident Commander (IC)?
2. Assess the hazards and evaluate the risks associated with the butane tank car offloading, uprighting, and removal operations.

This incident actually occurred in Akron, Ohio on February 26, 1989. Unfortunately, 2 days later, during derailment clean-up operations, a butane tank car rolled off its tracks and forced a second evacuation from the area. Ultimately, the incident was safely terminated with no additional community impact.

The subsequent National Transportation Safety Board (NTSB) report of this accident was one of the first to address the issue of hazard and risk evaluation from the perspective of emergency responders. Key findings and issues that were identified in the final NTSB accident investigation report included:

- Both the fire department and the city government depended on the expertise of the railroad (CSX) for the removal of wreckage from the derailment site. The Operations Section Chief considered it unsafe to offload the butane tank cars because of a continuing fire from one of the derailed tank cars. After discussions with CSX, he agreed with an action plan proposed by CSX to re-rail the tank cars, move them to the Akron Junction Yard where the tank cars would be more permanently secured, and finally transport them to Canton, Ohio, where the product would be offloaded. During the course of these conversations, however, the railroad did not discuss other alternatives with the fire department; nor did the railroad advise the fire department of the possible risks associated with re-railing the tank cars.
- On February 28, while the re-railed tank cars were being moved from the accident site, a butane tank car rolled off its tracks and forced the evacuation of approximately 25 families from the area. Only after this second derailment were alternative plans and the risks associated with each potential course of action thoroughly discussed with the fire department and city of Akron officials.
- The NTSB stated that it recognized the limited technical resources that may be available to local communities regarding train wreckage clearing operations and the communities' reliance on the railroad to take the appropriate course of action. Thus, NTSB believed that it is necessary for the railroad to discuss with local emergency response personnel (1) the severity of known damage to hazmat tank cars; (2) the relative dangers posed to public safety; (3) all possible courses of action; and (4) any associated risks.

Both the IC and the Hazmat Group Supervisor must play an active role during the hazard and risk assessment process.

Tasks should include seeking available information on the severity of known tank car damage and the dangers posed, potential alternatives and solutions, and the risks involved. Likewise, the IC will rely upon the guidance and recommendations of the Hazmat Group Supervisor and Hazmat Technicians in evaluating these options and coming to a decision.

Introduction

The evaluation of hazard information and the assessment of risks is the most critical decision-making point in the successful management of a hazardous materials incident. The decision to intervene, or more often to not intervene, is not easy. While most responders recognize the initial need for isolating the area, denying entry, and identifying the hazardous materials involved, failure to develop effective analytical and problem-solving skills can lead to injury, and in a worst-case scenario, death.

This chapter discusses the third step in the Eight Step Process©—Hazard and Risk Evaluation. This chapter is based on the premise that responders have (1) successfully implemented site management procedures, and (2) identified the nature of the problem and the materials potentially involved. It builds upon the basic tenets of a risk-based response process, outlining a systematic process by which emergency responders can assess the hazards, evaluate the potential consequences or risks, and determine appropriate response actions based upon facts, science, and the circumstances of the incident. Chapter topics will include understanding hazardous materials behavior, outlining the common sources of hazard information, evaluating risks, and determining response objectives.

Basic Principles

The concept of hazard assessment and risk evaluation is recognized as a critical benchmark in safe and successful emergency response operations Figure 7-1. Although our focus is directed toward hazmat response, the reality is that the risk management and evaluation process influences all disciplines within the emergency response community.

If we review incidents and case studies where emergency responders have been injured or killed, in most instances it is *not* due to their failure to assess and understand the hazard. In contrast, one of the most common root causes is our failure to adequately evaluate and understand the level of risk involved.

What are Hazards and Risks?

Hazards refer to a danger or peril. In hazardous materials response operations, hazards generally refer to the physical and chemical properties of a material. You can obtain hard data on hazards through emergency response guidebooks, computer and digital-based databases, and the like. Examples include flash point, toxicity levels (LC_{50}, LD_{50}), exposure values (TLVs), protective clothing requirements, and compatibility.

Risks refer to the probability of suffering harm or loss. Risks can't be determined from books or pulled from computerized models and databases—they are those intangibles that are different at every hazmat incident and must be evaluated by a knowledgeable IC and Hazmat Technicians. Although the risks associated with hazmat response will never be completely eliminated, they can be successfully managed. The objective of response operations is to minimize the level of risk to responders, the community, and the environment. Hazmat responders must see their role as risk evaluators, not risk takers.

Risk levels are variable and change from incident to incident. Factors that influence the level of risk include the following:

- **Hazardous nature of the material(s) involved.** This is normally based upon an assessment of the material's physical and chemical properties, such as flammability, reactivity, and toxicity.
- **Quantity of the material involved.** Risks will often be greater when dealing with bulk quantities of hazardous materials as compared to limited-quantity, individual containers. However, quantity must also be balanced against the hazardous nature of the material(s) involved—small quantities of highly toxic, reactive, and energetic materials can create significant risks.
- **Containment system and type of stress applied to the container.** Containers may be either pressurized or nonpressurized. Risks are inherently higher for pressurized containers as compared to low-pressure and atmospheric-pressure containers. In addition, the type of stressors involved (thermal, chemical, mechanical, or combination) and the ability of the container to tolerate that stress, will influence the level of risk.
- **Proximity of exposures.** This would include both distance and the rate of dispersion of any chemical release. The faster the hazmat escapes from its container, the greater the level of risk to responders in protecting exposures. Exposures include emergency response personnel, the community, property, and the environment.
- **Level of available resources.** The availability of resources and their response time will influence the level of risk. This includes both the training and knowledge level of responders for the scenario encountered, where risk may be inversely proportional to the level of training and experience for the hazmat/weapons of mass destruction (WMD) scenario being encountered.

This hazard and risk evaluation process will ultimately lead to the development of an Incident Action Plan (IAP) that should be designed to favorably change or influence the outcome.

In this chapter, the hazard and risk evaluation process will be viewed as three distinct yet inter-related tasks:

1. *Hazard Assessment*—Assessment of the hazards that may be involved in the incident, including the collection and interpretation of hazard and response information. Information pertaining to this task will be covered in the "Understanding the Enemy" and "Sources of Hazard Information" sections that follow.
2. *Risk Evaluation*—This is the process where responders analyze the problem and assess potential outcomes. Factors that will influence the level of risk include the hazardous materials involved, type of container and its integrity, and the environment where the incident has occurred. This task will be covered in the "Evaluating Risks" section.

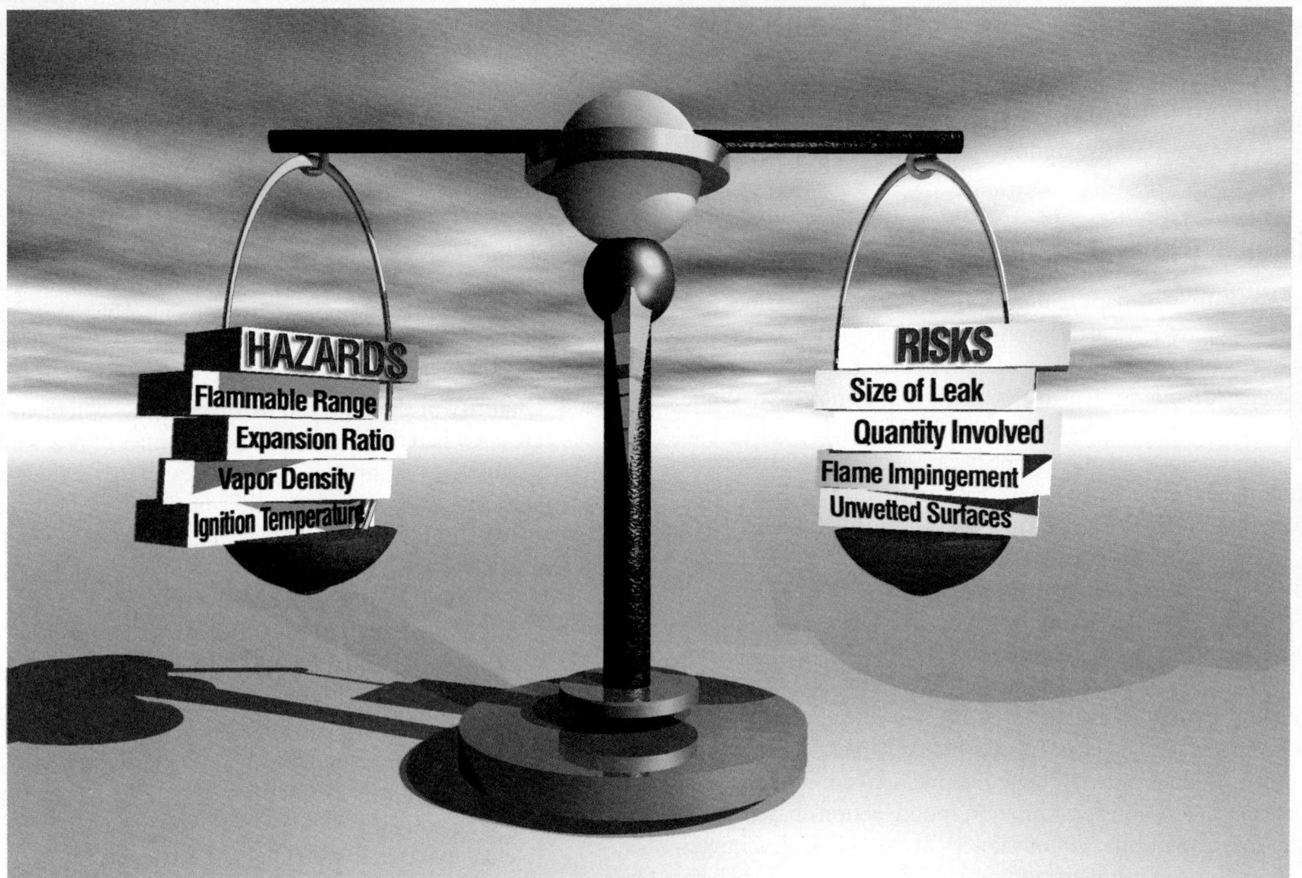

Figure 7-1 Responders must understand the differences between hazards and risks. Failure to understand these differences can result in responder injuries and death.

3. *Development of the IAP*—The output of the risk evaluation process is the implementation of strategies and tactics that will produce a favorable outcome. This will be covered in the section titled "Developing an Incident Action Plan." Detailed information on the implementation of response objectives is covered in the section on "implementing response objectives found elsewhere in this text.

Understanding the Enemy: Physical and Chemical Properties

To evaluate risks effectively, responders must be able to identify and verify the materials involved and determine their hazards and behavior characteristics. This includes collecting and interpreting available hazard data.

To be effective in combat, one must understand the tendencies and behavior of your enemy. In hazmat response, our enemy is the hazardous material. To mount a safe and effective hazmat response, responders must understand (1) how the enemy will behave (i.e., its physical properties), (2) how it can harm (i.e., its chemical properties), and (3) how these properties are influenced by the environment where the incident is occurring. In this section we review the key physical and chemical properties of hazardous materials and their role in the risk assessment process.

General Chemical Terms and Definitions

The following terms are commonly found on safety data sheets and in various emergency response references as part of a material's description or basic chemical make-up. Understanding these terms and definitions is critical in evaluating and predicting the behavior of hazardous materials and their containers during an incident.

- **Element**—Pure substance that cannot be broken down into simpler substances by chemical means. The Periodic Table breaks elements into groups and periods based upon their atomic number and chemical structure.
- **Compound**—Chemical combination of two or more elements, either the same elements or different ones, that is electrically neutral (e.g., table salt or NaCl). Compounds have a tendency to break down into their component parts, sometimes explosively (e.g., organic peroxides).
- **Mixture**—Substance made up of two or more elements or compounds, physically mixed together. Each element or compound in the mixture keeps its own properties.
- **Solution**—Mixture in which all of the ingredients are completely dissolved. Solutions are composed of a solvent (water or another liquid) and a dissolved substance (known as the solute).
- **Slurry**—Pourable mixture of a solid and a liquid.

- **Cryogenic liquid**—Gases that have been transformed into extremely cold liquids stored at temperatures below –130°F (–90°C) (U.S. DOT definition). Cryogenic liquids will have the following hazards: (1) extremely cold temperature, (2) tremendous liquid-to-vapor expansion ratio, and (3) hazards of the respective material (e.g., liquid oxygen is an oxidizer, while liquid hydrogen is flammable). When released from their container, cryogenic liquids will start to boil off rapidly due to their extremely low boiling point; for example, the expansion ratio for liquefied natural gas (LNG) is 600:1 and liquid oxygen is 875:1. Based upon the container and type of release scenario, auto-refrigeration can also occur during the release timeline.
- **Chemical change**—Substances are chemically altered and are changed into different substances with different physical and chemical properties after the change. A chemical change is irreversible. Examples of chemical change include neutralization, decomposition, and oxidation.
- **Chemical interactions**—Process that leads to the transformation of one set of chemical substances to another. Chemical reactions can be either spontaneous (i.e., require no input of energy) or non-spontaneous (i.e., occur following the input of some type of energy). Chemical interactions within a closed container can result in a build-up of heat that, in turn, causes an increase in pressure. Likewise, chemical interactions may alter the corrosivity of a product, resulting in a material more corrosive than the container is designed to handle.
- **Ionic bonding**—The electrostatic attraction of oppositely charged particles. Atoms or groups of atoms can form ions or complex ions.
- **Covalent bonding**—The force holding together atoms that share electrons.
- **Organic materials**—Materials that contain carbon atoms. Organic materials are derived from materials living or were once living, such as plants or decayed products. Most organic materials are flammable. Examples include methane (CH_4) and propane (C_3H_8).
- **Inorganic materials**—Compounds derived from other than vegetable or animal sources, which lack carbon chains but may contain a carbon atom (e.g., sulfur dioxide or SO_2).
- **Hydrocarbons**—Compounds primarily made up of hydrogen and carbon. Examples include LPG, gasoline, and fuel oils.
- **Saturated hydrocarbons**—A hydrocarbon possessing only single covalent bonds, and all of the carbon atoms are saturated with hydrogen. May also be referred to as alkanes. Examples include methane (CH_4), propane (C_3H_8), and butane (C_4H_{10}).
- **Unsaturated hydrocarbons**—A hydrocarbon with at least one multiple bond between two carbon atoms somewhere in the molecule. Generally, unsaturated hydrocarbons are more active chemically than saturated hydrocarbons and are considered more hazardous. May also be referred to as the alkenes and alkynes. Examples include ethylene (C_2H_4), butadiene (C_4H_6), and acetylene (C_2H_8).
- **Aromatic hydrocarbons**—A hydrocarbon containing the benzene "ring," which is formed by six carbon atoms and contains resonant bonds. Examples include benzene (C_6H_6) and toluene (C_7H_8).
- **Halogenated hydrocarbons**—A hydrocarbon with a halogen atom (e.g., chlorine, fluorine, bromine, etc.) substituted for a hydrogen atom. They are often more toxic than naturally occurring organic chemicals, and they decompose into smaller, more harmful elements when exposed to high temperatures for a sustained period of time.

Physical Properties

Physical properties provide information on the behavior of a material. These properties or characteristics of a material can be observed and measured and will provide responders with an understanding of how a material will behave both inside and after being released from its container.

- **Normal physical state**—The physical state or form (solid, liquid, gas) of a material at normal temperatures (68°F [20°C] to 77°F [25°C]). Determining the physical state of a material can allow responders to assess potential harm. Consider the following points:
  - Solids will generally cause limited harm, as the typical route of exposure is physical contact or ingestion. Exceptions to this are dusts (depending on the size of the particles) and radioactive substances.
  - Liquids present additional risks, as they are not as easily controlled, and may have the ability to evaporate, thereby creating an inhalation hazard. In addition, they have the ability to damage skin, and some are toxic through skin absorption. Of the 700 chemicals listed in the *NIOSH Pocket Guide to Chemical Hazards,* approximately 85 are listed as toxic through skin absorption.
  - Gases present the greatest risk, as they may be odorless, colorless, toxic, corrosive, and/or flammable. Gases and liquids with high vapor pressures pose among the greatest risks to emergency responders.
- **Temperature of product**—The temperature of the material within its container. A material's temperature will influence both the range of hazards and potential counter-measures. Hazards may also vary dependent upon the ambient temperature. Temperatures are usually measured in Fahrenheit (°F) or Celsius (°C).
- **Vapor pressure**—The pressure exerted by the vapor within the container against the sides of a container. This pressure is temperature dependent; as the temperature increases, so does the vapor pressure. Consider the following three points:
  1. The vapor pressure of a substance at 100°F (37.7°C) is always higher than the vapor pressure at 68°F (20°C).
  2. Vapor pressures reported in millimeters of mercury (mm Hg) are usually very low pressures. 760 mm Hg

is equivalent to 14.7 psi or 1 atmosphere. Materials with vapor pressures greater than 760 mm Hg are usually found as gases.

3. The lower the boiling point of a liquid, the greater vapor pressure at a given temperature.

Water has a vapor pressure of 25 mm Hg; materials with a vapor pressure above 25 mm Hg are producing vapors and can present a significant inhalation risk. Materials with a vapor pressure over 760 mm Hg will be gases under normal conditions.

- **Specific gravity**—The weight of a solid or liquid material as compared with the weight of an equal volume of water. If the specific gravity is less than one, the material is lighter than water and will float. If the specific gravity is greater than one, the material is heavier than water and will sink. Most insoluble hydrocarbons are lighter than water and will float on the surface. This is a significant property for evaluating spill control options and clean-up procedures for waterborne releases.
- **Vapor density**—The weight of a pure vapor or gas compared with the weight of an equal volume of dry air at the same temperature and pressure. If the vapor density of a gas is less than 1.0 the material is lighter than air and may rise. If the vapor density is greater than 1.0, the material is heavier than air and will collect in low or enclosed areas. Materials with a vapor density close to 1.0 (e.g., 0.8 to 1.2) will likely hang at mid-level and will not travel unless moved by wind or ventilation drafts. In the *NIOSH Packet Guide to Chemical Hazards,* vapor density is identified as RgasD or relative gas density. Vapor density is a significant property for evaluating exposures and determining where gases and vapors will travel.

  If a reference source does not provide a vapor density, you can calculate it by using the molecular weight of the material. The molecular weight of air is 29; materials with a molecular weight of < 29 will rise, and those with a molecular weight > 29 will sink. For example, anhydrous ammonia has a molecular weight of 17 and a vapor density of 0.59.

An easy way to remember common hazardous gases and simple asphyxiants lighter than air is the acronym 4H MEDIC ANNA:

H = Hydrogen (VD = 0.069)
H = Helium (VD = 0.14)
H = Hydrogen Cyanide (VD = 1.0)
H = Hydrogen Fluoride (VD = 0.70)
M = Methane (VD = 0.554)
E = Ethylene (VD = 0.97)
D = Diborane (VD = 0.96)
I = Illuminating Gas (10% ethane and 90% methane mixture – VD = 0.6)
C = Carbon Monoxide (VD = 0.97)
A = Anhydrous Ammonia (VD = 0.588)
N = Neon (VD = 0.7)
N = Nitrogen (VD = 0.96)
A = Acetylene (VD = 0.90)

Note that there are situations where some of these gases, if sufficiently chilled, could initially sink and be considered heavier than air. Examples include liquefied gases such as anhydrous ammonia, anhydrous hydrogen fluoride, and ethylene.

- **Boiling point**—The temperature at which a liquid changes its phase to a vapor or gas. The temperature at which the vapor pressure of the liquid equals atmospheric pressure. The lower the boiling point, the more vapors that are produced at a given temperature. The closer a material is to its boiling point, the more vapors produced.

  When evaluating flammable liquids, remember that flash point and boiling point are directly related. A low flash point flammable liquid will also have a low boiling point, which translates into greater amounts of vapors being given off.
- **Melting point**—The temperature at which a solid changes its phase to a liquid. This temperature is also the freezing point, depending on the direction of the change. For mixtures, a melting point range may be given. This is a significant property in evaluating the hazards of a material, as well as the integrity of a container (e.g., frozen material may cause its container to fail).
- **Sublimation**—The ability of a substance to change from the solid to the vapor phase without passing through the liquid phase (e.g., dry ice, naphthalene or moth balls). An increase in temperature can increase the rate of sublimation. Significant in evaluating the flammability or toxicity of any released materials that sublime. The opposite of sublimation is deposition (changes from vapor to solid).
- **Critical temperature and pressure**—Critical temperature is the temperature above which a gas cannot be liquefied no matter how much pressure is applied. Critical pressure is the pressure that must be applied to liquefy a gas at its critical temperature. Both terms relate to the process of liquefying gases. A gas cannot be liquefied above its critical temperature. The lower the critical temperature, the less pressure required to bring a gas to its liquid state.
- **Auto-Refrigeration**—Normally associated with liquefied gases and cryogenic liquids. Occurs during the rapid release (i.e., boiling) of a liquefied gas that causes it to rapidly cool, slow down the boiling process, and remain in a liquid state. Emergency responders may falsely assume that a container has emptied when auto-refrigeration occurs, until the liquefied gas resumes boiling and there is a subsequent release. This can be a common scenario when dealing with breaches of LPG cylinders.
- **Volatility**—The ease with which a liquid or solid can pass into the vapor state. The higher a material's volatility, the greater its rate of evaporation. Vapor pressure is a measure of a liquid's propensity to evaporate; the higher a liquid's vapor pressure, the more volatile the material. This is a significant property in that volatile materials will readily disperse and increase the hazard area.
- **Evaporation rate**—The rate at which a material will vaporize or change from liquid to vapor, as compared

to the rate of vaporization of a specific known material—*n*-butyl acetate. This is useful in evaluating the health and flammability hazards of a material. The relative evaporation rate of butyl acetate is 1.0, and other materials are then classified as shown in **Table 7-1**.

- **Expansion ratio**—The amount of gas produced by the evaporation of one volume of liquid at a given temperature. This is a significant property when evaluating liquid and vapor releases of liquefied gases and cryogenic liquids. The greater the expansion ratio, the more gas produced and the larger the hazard area.
- **Solubility**—The ability of a solid, liquid, gas, or vapor to dissolve in water or other specified medium. The ability of one material to blend uniformly with another, as in a solid in liquid, liquid in liquid, gas in liquid, or gas in gas. This is a significant property in evaluating spill control tactical options and for the selection of control and extinguishing agents, including the use of water and Class B firefighting foams.
- **Miscibility**—The ability of materials to dissolve into a uniform mixture. If a material is miscible in water, we mean it is infinitely dissolvable in water.
- **Degree of solubility**—An indication of the solubility and/or miscibility of the material.
  Negligible—less than 0.1%
  Slight—0.1 to 1.0%
  Moderate—1 to 10%
  Appreciable—greater than 10%
  Complete—soluble at all proportions
- **Viscosity**—Measurement of the thickness of a liquid and its ability to flow. High-viscosity liquids, such as heavy oils, must first be heated to increase their fluidity. A low-viscosity liquid will spread like water and increase the size of the hazard area.

Table 7-1 Evaporation Rates

Speed	Evaporation Rate	Examples
Fast	> 3.0	Methy ethyl ketone = 3.8 Acetone = 5.6 Hexane = 1.4
Medium	0.8 to 3.0	Naptha = 1.4 Xylene = 0.6 Isobutyl alcohol = 0.6
Slow	< 0.8	Water = 0.3 Mineral spirits = 0.1

Chemical Properties

Chemical properties are the intrinsic characteristics or properties of a substance described by its tendency to undergo chemical change. In simple terms, the true identity of the material is changed as a result of a chemical reaction, such as reactivity and the heat of combustion. Chemical properties typically provide responders with an understanding of how a material may harm.

Toxicity Hazards

Toxicity terms and definitions are covered in the chapter on "health and safety" found elsewhere in this text.

- **Dose**—Concentration of material to which the body is exposed over a specific time period. Dose = concentration × time.
- **Dose-Response**—Biological reaction caused by the dose in the body. The degree of harm is directly related to the dose (concentration and time) and its impact at the biological site of action (i.e., target organ). This can relate to chemical, biological, and radiological doses.

Flammability Hazards

- **Flash point**—Minimum temperature at which a liquid gives off sufficient vapors that will ignite and flash over, but will not continue to burn without the addition of more heat. Significant in determining the temperature at which the vapors from a flammable liquid are readily available and may ignite. Flash point is also linked to boiling point and vapor pressure; low flash point materials will typically have low boiling points and increasing vapor pressures.
- **Fire point**—Minimum temperature at which a liquid gives off sufficient vapors that will ignite and sustain combustion. It is typically several degrees higher than the flash point. In assessing the risk posed by a flammable liquids release, greater emphasis is placed on the flash point, since it is a lower temperature and sustained combustion is not necessary for significant injuries or damage to occur.
- **Ignition (autoignition) temperature**—The minimum temperature required to ignite gas or vapor without a spark or flame being present. Significant in evaluating the ease at which a flammable material may ignite. Materials with lower ignition temperatures generally have a greater risk of ignition.
- **Flammable (explosive) range**—The range of gas or vapor concentration (percentage by volume in air) that will burn or explode if an ignition source is present. Limiting concentrations are commonly called the lower flammable (explosive) limit and the upper flammable (explosive) limit. Below the lower flammable limit, the mixture is too lean to burn; above the upper flammable limit, the mixture is too rich to burn. If the gas or vapor is released into an oxygen-enriched atmosphere, the flammable range will expand. Likewise, if the gas or vapor is released into an oxygen-deficient atmosphere, the flammable range will contract. Chemical families with wide flammable ranges include alcohols, aldehydes, and ethers.
- **Toxic products of combustion**—The byproducts of the combustion process that are harmful to humans. Based on the burning material(s), the byproducts of combustion will vary as follows:
  - If there is incomplete combustion = carbon monoxide

- If nitrogen is present = nitrogen oxide gases. Incomplete combustion can produce hydrogen cyanide and ammonia.
- Organic materials containing a halogenated material (e.g., chlorine, bromine, fluorine) = hydrogen chloride, hydrogen bromide, or hydrogen fluoride in a fire.
- If sulfur is present = sulfur dioxide gas. Incomplete combustion = hydrogen sulfide and carbonyl sulfide.

Reactivity Hazards

- **Reactivity/Instability**—The ability of a material to undergo a chemical reaction with the release of energy. Reactivity can be initiated by mixing or reacting with other materials, the application of heat, physical shock, and so on. Reactive materials include materials that decompose spontaneously, polymerize, or otherwise self-react. Examples include oxidizers and organic peroxides, corrosives, pyrophoric (i.e., air reactive) materials, and water reactive materials. The term instability is often used inter-changeably with the term reactivity.
- **Oxidation ability**—The ability of a material to (1) either give up its oxygen molecule to stimulate the oxidation of organic materials (e.g., chlorate, permanganate, and nitrate compounds), or (2) receive electrons being transferred from the substance undergoing oxidation (e.g., chlorine and fluorine).
- **Water reactivity**—Materials that react with water and release a flammable gas (e.g., acetylene) or present a health hazard. Some materials may also react explosively with water (e.g., sodium).
- **Air reactivity (pyrophoric materials)**—Materials that ignite spontaneously in air without an ignition source (e.g., aluminum alkyls, white phosphorous).
- **Chemical reactivity**—A process involving the bonding, unbonding, and rebonding of atoms, that can chemically change substances into other substances. The interaction of materials in a container may result in a build-up of heat and pressure, and may cause container failure. Similarly, the combined materials may be more corrosive than the container was originally designed to withstand and lead to container failure.
- **Polymerization**—A reaction during which a monomer is induced to polymerize by the addition of a catalyst or other unintentional influences, such as excessive heat, friction, contamination. If the reaction is not controlled, it is possible to have an excessive amount of energy released.
- **Catalyst**—Used to control the rate of a chemical reaction by either speeding it up or slowing it down. If used improperly, catalysts can speed up a reaction and cause a container failure due to pressure or heat build-up.
- **Inhibitor**—Added to products to control their chemical reaction with other products. If the inhibitor is not added or escapes during an incident, the material will begin to polymerize, possibly resulting in container failure.
- **Maximum Safe Storage Temperature (MSST)**—The maximum storage temperature that an organic peroxide may be maintained, above which a reaction and explosion may occur.
- **Self-Accelerating Decomposition Temperature (SADT)**—The temperature at which an organic peroxide or synthetic compound will react to heat, light, or other chemicals and release oxygen, energy, and fuel in the form of an explosion or rapid oxidation. When this temperature is reached by some portion of the mass of an organic peroxide, irreversible decomposition will begin.

Corrosivity Hazards

- **Corrosivity**—A material that causes visible destruction of, or irreversible alterations to, living tissue by chemical action at the point of contact. Corrosive materials include acids and caustics or bases.
- **Dissociation**—the process by which acids and bases break down when dissolved in water to produce hydrogen (H+) ions or hydroxide (OH–) ions.
- **Acids**—Compound that forms hydrogen (H+) ions in water. These compounds have a pH < 7, and acidic aqueous solutions will turn litmus paper red. Materials with a pH < 2.0 are considered a strong acid.
- **Caustics (base, alkaline)**—Compound that forms hydroxide (OH–) ions in water. These compounds have a pH > 7, and caustic solutions will turn litmus paper blue. Materials with a pH > 12 are considered a strong base. Also known as alkali, alkaline, or base.
- **pH**—Acidic or basic corrosives are measured to one another by their ability to dissociate in solution. Those that form the greatest number of hydrogen ions are the strongest acids, while those that form the hydroxide ion are the strongest bases. The measurement of the hydrogen ion concentration in solution is called the pH (power of hydrogen) of the compound in solution. The pH scale ranges from 0 to 14, with strong acids having low pH values and strong bases or alkaline materials having high pH values. A neutral substance would have a value of 7.
- **Strength**—The degree to which a corrosive ionizes in water. Those that form the greatest number of hydrogen ions are the strongest acids (e.g., pH < 2), while those that form the greatest number of hydroxide ions are the strongest bases (pH > 12).
- **Concentration**—The percentage of an acid or base dissolved in water. Concentration is not the same as strength.

Radioactive Materials

- **Radioactivity**—The ability of a material to emit any form of radioactive energy. Radiation is the movement of energy through space or matter in the form of waves or particles.
- **Activity**—The rate of disintegration or decay of a radioactive material. Measured in curies (1 curie = 37 billion disintegrations per second), although it is usually expressed in either millicuries or microcuries. Activity indicates how much radioactivity is present and not how much material is present.
- **Counts per minute (CPM)**—Standard unit of measurement for alpha and beta radiation, and is also

commonly used to express background radiation in numerical terms. Contamination survey instruments typically measure in CPM or kCPM.

- **Dose**—A quantity of radiation or energy absorbed by the body, usually measured in millirems (mrem).
- **Dose rate**—The radiation dose delivered per unit of time (e.g., mrem/hour).
- **Half-life**—The time it takes for the activity of a radioactive material to decrease to one half of its initial value through radioactive decay. The half-life of known materials can range from a fraction of a second to millions of years.

Chemical and Biological Agents/Weapons

- **Biological agents and toxins**—Biological threat agents consist of pathogens and toxins. Pathogens are disease-producing organisms and include bacteria (e.g., anthrax, cholera, plague, e coli), and viruses (e.g., small pox, viral hemorrhagic fever). Toxins are produced by a biological source and include ricin, botulinum toxins, and T2 mycotoxins.
- **Incubation period**—The time from exposure to a biological agent to the appearance of symptoms in infected persons.
- **Infectious dose**—The amount of pathogen (measured in the number of organisms) required to cause infection in the host.
- **Chemical agents**—Chemical agents are classified in military terms based upon their effects on the enemy. The intent of using chemical weapons is to incapacitate and to kill. Categories of chemical agents are:
  1. Nerve agents (neurotoxins)
  2. Choking agents (respiratory irritants)
  3. Blood agents (chemical asphyxiants)
  4. Vesicants or blister agents (skin irritants)
  5. Antipersonnel agents (riot control agents)
     - **Nerve agents**—Chemical warfare agents that are the most toxic of the known chemical agents. Primarily consist of organophosphate agents that attack the central nervous system. Nerve agents are hazardous in both their liquid and vapor state, and can cause death within minutes after exposure. Examples of chemical warfare agents include tabun (GA), sarin (GB), soman (GD), and VX.
     - **Choking agents**—Chemical agents that can damage the membranes of the lung. Examples include phosgene and chlorine.
     - **Blood agents**—Chemical agents that consist of a cyanide compound, such as hydrogen cyanide (hydrocyanic acid or AC) and cyanogens chloride (CK). These agents are identical to their civilian counterpart used in industry.
     - **Vesicants (blister agents)**—Chemical agents that pose both a liquid and vapor threat to all exposed skin and mucous membranes. These are exceptionally strong irritants capable of causing extreme pain and large blisters upon contact. Examples include mustard (H), lewisite (L), and phosgene oxime (CX).
     - **Riot control agents**—Usually solid materials dispersed in a liquid spray and cause pain or burning on exposed mucous membranes and skin. Common examples include tear gas or "Mace" (CN) and pepper spray (i.e., capsaicin or OC).
- **Persistence**—Refers to the length of time a chemical agent remains as a liquid. A chemical agent is said to be "persistent" if it remains as a liquid for longer than 24 hours and nonpersistent if it evaporates within that time. Among the most persistent chemical agents are VX, tabun (GA), mustard (H), and lewisite (L).

Sources of Hazard Data and Information

Two primary tasks within the hazard and risk evaluation process are to (1) gather hazard data and information on the materials involved, and (2) to compile that data in a useful manner so that the risk evaluation process can be accomplished in a timely and efficient manner.

In this section, we discuss the various sources of hazard data and information and some methods for compiling that data into a logical and useful format.

For ease of discussion, hazard data and information sources can be broken into the following categories:

- Reference manuals and guidebooks
- Technical information centers
- Hazardous materials databases
- Technical information specialists
- Hazard communication and right-to-know regulations
- Monitoring instruments

Reference Manuals and Guidebooks

A wide variety of emergency response guidebooks and reference manuals exist. These range from small Field Operations Guides (or FOGs) that can fit into your pocket and multi-volume reference manuals that require considerable storage space, to digital-based reference sources that can be accessed on a smart phone. In the early days of hazmat response, it was not uncommon for responders to be hindered by the lack of adequate information during an incident. Today, one can find situations where responders are either (1) overwhelmed by the amount of data and information available from a multitude of reference sources, or (2) simply rely upon one digital-based reference source and do not verify the accuracy of the information. Compounding these problems can be situations where different sources (i.e., printed or digital versus verbal) provide conflicting information.

Despite the large number of resources available, most responders initially rely on three to five primary response guidebooks for most of their data and information. These sources are generally selected based on personal preferences and experience, ease of use, a large chemical listing, and a large, effective data bank. **Table 7-2** lists some of the most common reference guidebooks, including their target audience and type of information provided.

Table 7-2 Common Emergency Response Guidebooks and Manuals

	User / Customer					Hazard Information*					
Guidebook / Manual	**FRA / FRO**	**HMT / HMS**	**IC**	**Product Specialists**	**Other**	**General Response Info.**	**Health & Safety**	**Physical / Chemical Properties**	**Reactivity**	**Specific Product / Containers**	**Comments**
ACGIH Threshold Limit Values *American Council of Governmental Industrial Hygienists*		X		X	X	X	X				■ Provides exposure values (TLVs) for the chemicals studied by ACGIH. ■ Published annually; TLV values are reviewed more frequently than government published exposure values.
General Handling of Hazardous Materials in Surface Transportation *Association of American Railroads*	X	X	X			X					■ Provides initial emergency response information for broad range of HM. ■ Includes cross-reference list with UN/DOT identification numbers and Standard Transportation Commodity Codes (STCC)
Hazardous Materials Emergency Action Guidesheets *Association of American Railroads*	X	X	X	X	X	X	X	X		X	■ Provides list of common materials transported by rail. Excellent resource for HMRT. ■ Provides basic to advanced emergency response information, including guidance on clean-up and mitigation strategies.
Field Guide to Tank Cars *Association of American Railroads*	X	X	X	X	X					X	■ Divided into non-pressure, pressure and cryogenic tank car sections. ■ Provides information on tank car stenciling, as well as cutaways and schematics of typical valve arrangements. ■ Includes information on initial emergency response and tank car damage assessment.
Bretherick's Handbook of Reactive Substances *Leslie Bretherick and Peter Urben*		X		X	X				X	X	■ Widely used for its information on chemical interactions and reactions. ■ Provides cross-referencing of similar compounds not obviously related.
Emergency Care for Hazardous Materials Exposure *Alvin Bronstein and Phillip Currance*	X	X	X		X		X				■ Field reference for the health effects of hazardous materials exposures. ■ Format is similar to the NA ERG, with the addition of treatment and drug protocols.
Handbook of Compressed Gases *Compressed Gas Association*		X		X	X					X	■ Best source of information on compressed gases, including general properties, design and construction features, and information on specific compressed gases and mixtures.
Gardener's Chemical Synonym and Trade Names *Michael and Irene Ash, Editors*		X		X	X			X		X	■ Provides common synonyms for household products and industrial chemicals, and trade names, chemical compositions, applications and manufacturers. ■ Not designed for emergency response, but a good reference.
GATX Tank Car Manual *General American Transportation Corp.*		X		X	X					X	■ Provides information on tank car stenciling, as well as cutaways and schematics of railroad tank cars and valving.

Table 7-2 Common Emergency Response Guidebooks and Manuals (continued)

	User / Customer					Hazard Information*					
Guidebook / Manual	**FRA / FRO**	**HMT / HMS**	**IC**	**Product Specialists**	**Other**	**General Response Info.**	**Health & Safety**	**Physical / Chemical Properties**	**Reactivity**	**Specific Product / Containers**	**Comments**
Hawley's Condensed Chemical Dictionary *Richard L. Lewis Sr.*		X		X	X			X		X	■ Compiles technical data and descriptive information on thousands of chemicals. ■ Good initial reference for public inquiries on specific hazardous materials, and their application and use.
Jane's Chem-Bio Handbook *Frederick Sidell, et. al.*	X	X	X	X	X					X	■ Provides extensive information on chemical and biological weapons and agents. ■ Four primary sections: On-Scene Procedures, Chemical Agents, Biological Agents, and Appendix materials.
Medical Management of Biological Casualties Handbook *U.S. Army Medical Research Institute of Infectious Diseases*		X	X	X	X					X	■ Provides concise information on biological casualty management, with an emphasis upon military operations.
Medical Management of Chemical Casualties Handbook *U.S. Army Medical Research Institute of Chemical Defense*		X	X	X	X					X	■ Provides concise information on biological casualty management, with an emphasis upon military operations.
Medical Management of Radiological Casualties Handbook *Armed Forces Radiobiology Research Institute*		X	X	X	X					X	■ Provides concise information on radiological casualty management, with an emphasis upon military operations. ■ Topics inlcude acute radiation syndrime, biodosimetry, skin injury, internal contamination, psychological support, delayed effects and decontamination.
Crop Chemicals Handbook (formerly known as Farm Chemicals Handbook) *Meister Publishing Company*		X		X	X					X	■ Published for the agricultural industry; not designed for emergency response. ■ Combines Electronic Pesitice Dictionary into the CD version. ■ Best single source of information on pesticides, herbicides and insecticide background information.
Quick Selection Guide to Chemical Protective Clothing *Krister Forsberg and S. Z. Mansdorf*		X		X	X		X				■ Provides chemical compatibility information against 500+ chemicals for the most common CPC materials.
The Merck Index *Merck Publishing Group*		X		X	X		X	X		X	■ Encyclopedia of chemicals and pharmaceutical materials listed in alphabetical order. Each entry cites a monograph that can be referenced. ■ Provides basic chemical information but limited response information.

Table 7-2 Common Emergency Response Guidebooks and Manuals (continued)

Guidebook / Manual	User / Customer					Hazard Information*					Comments
	FRA / FRO	HMT / HMS	IC	Product Specialists	Other	General Response Info.	Health & Safety	Physical / Chemical Properties	Reactivity	Specific Product / Containers	
NFPA Fire Protection Guide on Hazardous Materials *National Fire Protection Association*	X	X	X	X		X	X	X	X		■ Combination of four NFPA standards—49, 325, 491 & 704. NFPA 491 is helpful when assessing chemical reactivity.
NFPA Hazmat Quick Guide *National Fire Protection Association*	X	X	X			X		X			■ Provides information on approximately 2,000 chemicals referenced from NFPA 49, 325, 491 and the NAERG. ■ Cross references provided by ID/UN number, CAS Number, or synonym
Patty's Industrial Hygiene Handbook *Robert L. Harris, Editor*		X		X	X		X			X	■ Four volume set primarily designed as a desk-top reference; not specifically used for emergency response operations.
CHRIS – Chemical Hazards Response Information System *U.S. Coast Guard*	X	X	X	X	X	X		X		X	■ Focuses on chemicals found in water transportation, with a focus upon water-based emergency response issues. ■ Good reference for physical and chemical properties on land and sea.
NIOSH Pocket Guide to Chemical Hazards *U.S. Department of Health & Human Services*		X	X	X			X	X			■ Good initial reference manual; provides quick overview of physical and chemical properties, and PPE. Must understand abbreviations. ■ Provides Ionization Potentials (IP) which are required when using a photionization detector.
Emergency Response Guidebook (ERG) *U.S. Department of Transportation*	X	X	X			X					■ The most common and widely used ERG. ■ Responders should be inherently familiar with this document, including the Table on Isolation and Protective Action Distances.
Dangerous Properties of Industrial Chemicals *Irving Sax*		X		X	X		X	X	X		■ Useful resource for chemicals used in industrial applications.
Firefighter's Handbook of Hazardous Materials *Charles Baker*	X	X	X			X					■ Provides data on flammability, thermal stability, extinguishing agents, and related data.

Hazard Information Notes

1. *General Response Information*—provides tactical level guidance, such as spill, leak and fire control recommendations
2. *Health & Safety*—provides information such as exposure values, PPE or CPC recommendations, decon guidance, medical information, etc.
3. *Physical / Chemical Properties*—provides information on the physical (i.e., how the material will behave) and chemical (i.e, how the material will harm) properties of hazardous materials.
4. *Reactivity*—provides information on the reactivity of a chemical, including mixtures.
5. *Specific Product or Container Information*—provides response information pertaining to a specific hazardous material or type of hazmat container.

Responder Safety Tip

Most responders will evaluate a minimum of two or three independent information sources and reference guidebooks (aka "Rule of Threes") before permitting personnel to operate within the Hot Zone. With experience, responders develop a preference for those manuals and information sources they feel most comfortable with.

As with all resources, guidebooks are an information tool with both advantages and limitations. Several operational considerations should be kept in mind when using them:

1. You must know how to use the reference material before the incident in order to use them effectively. Evaluate reference materials before use and make sure that all references are using the same definitions for hazard terms. A good guidebook should have a well-written "How to Use" section.
2. In some instances, there may be conflicting information between guidebooks. This is often due to the testing methods used (e.g., closed cup versus open cup test for flash point) or differences in terminology. Always select the most conservative values or recommendations to ensure the greatest margin of safety.
3. Be realistic in your evaluation of the data contained in the guidebooks. For example, the ambient temperature of a liquid and whether it is flammable or combustible, is much more critical than a slight discrepancy in flash points between references.
4. Always rely on the protective clothing compatibility charts provided by the clothing manufacturer. Likewise, conflicts may exist when chemical compatibility charts from several different sources are referenced. If in doubt, always choose the most conservative recommendation.
5. Although reference guidebooks contain data on those chemicals most commonly encountered during hazmat incidents, they are not a complete listing of all the chemicals found in your community. There is no replacement for hazard analysis and contingency planning at both the facility and community levels.
6. Electronic versions of most of the major emergency response guidebooks are also available in a wide range of formats and applications. However, don't bet the farm on going "totally electronic"; paper-based guidebooks work "first time, every time" if you know how to use the reference.

Technical Information Centers

A number of private and public sector hazardous materials emergency telephone "hotlines" exist. Their functions include (1) providing immediate chemical hazard information; (2) accessing secondary forms of expertise for additional action and information; and (3) acting as a clearinghouse for spill notifications. They include both public and subscription-based systems.

In the United States, the most recognized emergency information center is CHEMTREC®. Operated by the American Chemistry Council (ACC) in the National Capital Region, CHEMTREC® is a free public service that can be contacted 24 hours daily at (800) 424-9300 from anywhere within the United States, as well as Puerto Rico, the Virgin Islands, and Canada.

Callers outside of the United States and ships at sea can contact the Center using CHEMTREC's® international and maritime number at (703) 527-3887. (*Note:* Collect calls are accepted.) Non-English-speaking callers to CHEMTREC® are handled through the use of an interpreter service.

CHEMTREC® provides a number of emergency and non-emergency services, including the following:

1. *Emergency response information.* CHEMTREC® provides immediate information to callers anywhere on how to cope with hazardous materials involved in a transportation or fixed facility emergency. In addition to its extensive database, it can also access shippers, manufacturers, or other forms of expertise for additional and appropriate follow-up action and information. CHEMTREC® also has immediate access to medical professionals and toxicologists, who can provide medical information for incidents involving chemical exposures.

   CHEMTREC® Emergency Service Specialists have the capability to initiate conference phone calls between on-scene responders and company representatives as well as fax Safety Data Sheets (SDSs) from their database directly to on-scene personnel.
2. *Emergency communications.* CHEMTREC® helps hazmat shippers comply with DOT regulations (49 CFR 172.604) that require shippers to provide a 24-hour emergency point-of-contact. Manufacturers and shippers that register with CHEMTREC® must provide SDSs for their products, as well as emergency and administrative contacts in the event additional information or assistance is required.

   CHEMTREC® also serves as the point of contact for several product-specific industry mutual aid programs. These include products such as chlorine, hydrogen chloride, hydrogen peroxide, phosphorous, and compressed gases.
3. *Participation in drills and exercises.* CHEMTREC® is available to participate in local exercises and provide assistance as in an actual emergency. Requests for this service should be coordinated with CHEMTREC® at least 48 hours before the exercise. Further information on this program and CHEMTREC® operations can be found at http://www.chemtrec.com. *Note: CHEMTREC is a registered service mark of the American Chemistry Council (ACC) and is used with the permission of the ACC.*

Other emergency response and general information numbers that may be useful for both emergency planning and response purposes include the following:

- **CANUTEC (Canadian Transport Emergency Centre)** is operated by Transport Canada and can be contacted at (613) 996-6666 or *666 on a cellular phone. The Canadian counterpart of CHEMTREC®, it provides assistance in hazmat identification and establishing contact with shippers and manufacturers of hazardous materials that originate in Canada. CANUTEC is also the primary contact point for the Transport of Dangerous Goods Directorate on questions regarding dangerous goods transport regulations and chemical products. The general information number is (613)

992-4624. Additional information on CANUTEC can be found at http://www.tc.gc.ca/canutec.

- **SETIQ (Emergency Transportation System for the Chemical Industry)** is a service of the Mexico National Association of Chemical Industries and can be contacted at 01-800-00-214-00 in the Mexican Republic. For calls originating in Mexico City and its metro area, SETIQ can be contacted at 5559-1588. For calls originating elsewhere, call 0-11-52-5-559-1588. Additional information on SETIQ can be found at http://www.aniq.org.mx/setiq/setiq.htm.
- **U.S. Coast Guard and the Department of Transportation National Response Center** (NRC) at (800) 424-8802 or (202) 267-2675. The NRC is the federal government's central reporting point for all oil, chemical, radiological, biological, and etiological releases into the environment within the United States and its territories. In addition, the NRC receives reports via the toll-free number on potential or actual domestic terrorism and coordinates notifications and response with the FBI and the appropriate military command (e.g., NORTHCOM).

  The NRC must be notified by the responsible party (i.e., the spiller) if a hazardous materials release exceeds the reportable quantity (RQ) provisions of CERCLA. Additional information on the NRC can be found at http://www.nrc.uscg.mil.
- **The Agency for Toxic Substances and Disease Registry (ATSDR) at (800) 232-4636.** ATSDR is the leading federal public health agency for hazmat incidents and operates a 24-hour emergency number (770) 488-7100 for providing advice on health issues involving hazmat releases. If necessary, ATSDR can deploy an Emergency Response Team to provide on-scene assistance.

  ATSDR also has developed Medical Management Guidelines (MMGs) for Acute Chemical Exposures to aid emergency department physicians and other emergency healthcare professionals who manage acute exposures resulting from chemical incidents, as well as other emergency response resources for healthcare and public health professionals. Additional information on ATSDR can be found at http://www.atsdr.cdc.gov.
- **ASPCA Animal Poison Control Center at (888) 426-4435.** Operated by the American Society for the Prevention of Cruelty to Animals (ASPCA) as an allied agency of the University of Illinois College of Veterinary Medicine, this number provides 24-hour consultation in the diagnosis and treatment of suspected or actual animal poisonings or chemical contamination. Additional information on this center can be found at http://www.aspca.org.
- **National Pesticide Information Center (NPIC)** at (800) 858-7378. Operated by Oregon State University in cooperation with EPA, NPIC provides information on pesticide-related health/toxicity questions, properties, and minor clean-up to physicians, veterinarians, responders, and the general public. Hours of operation are 7:30 am to 3:30 pm (Pacific Time), Monday through Friday. Additional information on NPIC can be found at http://www.npic.orst.edu.

Emergency responders should develop a telephone roster of those individuals and agencies at the state and local level that can offer technical assistance. Examples include environmental and health departments, local chemical industry personnel, local hazmat spill cooperatives and clean-up contractors, and regional poison control centers. Your Local Emergency Planning Committee (LEPC) can be a good resource in this process.

Hazardous Materials Websites and Computer Databases

Portable computers, personal data assistants (PDAs), smart phones, CDs, and Internet access have revolutionized the ability of emergency responders to search and access hazard information from the field. Virtually all of the major emergency response guidebooks previously cited have an electronic equivalent that can be referenced through third-party software products, CDs, or websites. In addition, there are numerous public and commercial web-based tools that can facilitate the risk evaluation process.

Some public-based electronic resources include:

- **CAMEO® (Computer Assisted Management of Emergency Operations)** is the most widely used computer-based software tool used by hazmat responders. Developed by the EPA Chemical Emergency Preparedness and Prevention Office (CEPPO) and the National Oceanic and Atmospheric Administration (NOAA) Office of Response and Restoration, CAMEO has application as both a planning and response tool. It consists of four major elements that interact and communicate with each other either alone or as a "suite" of programs:

  The *CAMEO Chemicals database* contains over 6,000 hazardous chemicals, 80,000 synonyms, and product trade names. CAMEO Chemicals provides a powerful search engine linked to chemical-specific information on physical and chemical hazards, fire and explosive hazards, health hazards, firefighting techniques, clean-up procedures, and protective clothing. CAMEO Chemicals can be downloaded to a computer or used via a web connection.

  The *CAMEO FM* database is used to store information regarding facilities, their chemical inventories, contact information, site plans, etc. (i.e., SARA, Tier II information). It can also store information on emergency resources, transportation routes, and special locations, as well as link to outside documents. Information can be accessed through the LEPC or the State Emergency Response Commission (SERC).
- ***MARPLOT (Mapping Applications for Response, Planning, and Local Operational Tasks)*** is the mapping application that allows users to "see" their data (e.g., roads, facilities, schools, response assets), display this information on computer maps, and print the information on area maps. Areas contaminated by potential or actual chemical release scenarios also can be overlaid on the maps to determine potential impact.

MARPLOT can use maps available from the U.S. Census Bureau, ESRI "shape" files (available from most GIS sources), and aerial image files (available from U.S. Department of Agriculture or local GIS bureaus). It also "links" to CAMEO FM to display most information available in the CAMEO FM database, and to ALOHA to display "threat" zones of actual or potential releases.

- ***ALOHA (Area Locations of Hazardous Atmospheres)*** is an atmospheric plume dispersion model used for evaluating releases of hazardous materials vapors. ALOHA allows the user to estimate the downwind dispersion of a chemical cloud based on its physical and toxicological properties, atmospheric conditions, and specific circumstances of the release. Graphical outputs include a "threat zone" that can be plotted on maps with MARPLOT to display the location of other hazmat facilities and vulnerable locations (e.g., hospitals, schools). Specific information on these locations can be extracted from CAMEO information modules to help make decisions about the degree of risk posed by a release scenario. It can also provide information on release times, ambient concentrations, etc.

  Additional information can be referenced at http://www.epa.gov/emergencies and following the CAMEO link.
- **The CHEMTREC® and EPA Chemical Emergency Preparedness and Prevention Office (CEPPO)** websites provide good starting points for gathering hazard information. With their links, responders have a free, regularly updated starting point for accessing a number of both public and private websites that can provide both chemical or container information. The CHEMTREC® website can be found at http://www.chemtrec.org, while the EPA website can be found at http://www.epa.gov/swercepp.
- **The Wireless Information System for Emergency Responders (WISER)** is an electronic tool developed by the U.S. National Library of Medicine. WISER was specifically designed for emergency responders to provide a wide range of information on hazardous substances, including substance identification support, physical characteristics, health information, and mitigation advice. WISER is available for download as a stand-alone application for PCs, smart phones, and other electronic devices, and can be accessed online through "WebWiser." Additional information can be referenced at http://wiser.nim.nih.gov.
- **The NOAA Chemical Reactivity Worksheet** is an excellent tool that can be downloaded and used for determining the effects of various chemical mixtures. It includes reactivity information for more than 6,000 chemicals and can be referenced at http://response.restoration.noaa.gov/chemaids/react.html

  Nearly all of the major federal agencies with an emergency planning or response mission (e.g., DOT, OSHA, EPA, NRC) have websites with a range and variety of reference sources and links.

Web-based tools provide a wide range of options for accessing technical response information. In addition, the Internet provides response organizations with the capability to create an incident-specific website where response partners and parties not on-scene can have access to incident-specific information, including status reports, digital photos, video clips, and so forth.

When evaluating electronic-based information sources, consider the following criteria:

- How will the tool complement or improve your response operations and decision making?
- Costs, including any initial subscription and user fees.
- Hardware and software requirements, including communications technology, software upgrades, and technology half-life and upgrades as the "current" technology changes and is upgraded (e.g., transition from personal computers to PDAs to smart phones to computer tablets, etc.).
- Communications security (COMSEC) vulnerabilities, as appropriate.
- Ease of use and user friendliness.
- Technical support.

Technical Information Specialists

Personnel (aka Technical Specialists) who either work with the hazardous material(s) or their processing—or who have some specialized knowledge, such as container design, toxicology, or chemistry—are a common source of hazard information. When evaluating these product and container specialists and the information they provide, consider these observations and lessons learned:

- Many individuals who are specialists in a narrow, specific technical area may not have an understanding of the broad, multi-disciplinary nature of hazmat emergency response. For example, some information sources will provide extensive data on container design yet may be unfamiliar with basic emergency response principles.
- Each information specialist has their own strengths and limitations. It's a good idea to remove the term *expert* from your vocabulary; be wary of self-proclaimed experts without first verifying their background and knowledge.
- You will often interact with individuals with whom you have had no previous contact. Before relying on their recommendations, verify their level of expertise and job classification by asking specific questions. More than one responder has been disappointed or embarrassed to find that the "expert" they have been waiting for is actually a truck driver or a product marketing representative.

Responder Safety Tip

When questioning outside information sources, consider yourself as playing the role of a detective. Remember, final accountability always rests with the IC. Although this is certainly not an interrogation process, you must be confident of the specialist's expertise and authority. In some cases, responders will ask questions for which they already know the answer in order to evaluate that person's competency and knowledge level.

- Networking and relationships are everything! Local responders and facility personnel must get out into their communities and establish personal contacts and relationships with your response partners. These include state, regional, and federal environmental response personnel, law enforcement, clean-up contractors, industry representatives, wrecking and rigging companies, and so on.
- Investigate the existence of local and state "Good Samaritan" legislation that may provide legal liability protection for outside representatives as they assist you on the scene.

Hazard Communication and Right-To-Know Regulations

Numerous state and local worker and community right-to-know laws exist across the country. In addition, the OSHA Hazard Communication Standard (OSHA 29 CFR 1910.1200) has specific requirements pertaining to hazard markings, worker access to hazmat information, and worker exposures to chemicals in the workplace. In 2012, OSHA announced that the Hazard Communication Standard will be revised over a 3-year period to bring the U.S. regulation in line with the Globally Harmonized System of Classification and Labeling of Chemicals (GHS). Additional information on global harmonization and its impacts upon hazard communication chemical classification and labeling can be found on the OSHA website at http://www.osha.gov/dsg/hazcom/ghs.html.

Although the scope of these regulations may vary, most right-to-know laws provide emergency responders with access to SDSs and have specific requirements mandating the development of facility pre-incident plans and community hazardous materials plans.

Responders must be able to both read and interpret safety data sheets. Under the revised GHS requirements, OSHA will require that all SDS eventually be modified and formatted into 16 sections, including the following:

- *Section 1—Product and Company Identification.* Includes manufacturer's name, address, and emergency phone number.
- *Section 2—Hazards Identification.* GHS classification of the substance/mixture and any additional regional information.
- *Section 3—Composition/Information on Ingredients.* Chemical identity, including common name, synonyms, etc. Breaks out the active ingredients by percentage.
- *Section 4—First-Aid Measures.* Includes routes of exposure, signs and symptoms of exposure, emergency care, and special treatments.
- *Section 5—Firefighting Measures.* Includes extinguishing agents, specific hazards from the chemical, and personal protective equipment (PPE) and firefighting protection requirements.
- *Section 6—Accidental Release Measures.* Personal precautions, protective equipment and emergency procedures, including methods and materials for containment and clean-up.
- *Section 7—Handling and Storage.* Precautions for safe handling, and conditions for safe storage, including any incompatibilities.
- *Section 8—Exposure Controls/Personal Protection.* Occupational exposure values, required engineering controls, use of PPE, etc.
- *Section 9—Physical and Chemical Properties.* Listing of pertinent physical and chemical property values.
- *Section 10—Stability and Reactivity.* Chemical stability, including incompatible materials, possible hazardous reactions, decomposition products, and conditions to avoid.
- *Section 11—Toxicological Information.* Concise but complete and comprehensible description of the various health effects, available data, likely routes of exposure, symptoms, acute and chronic effects of exposure, and numerical exposure and toxicity.
- *Section 12—Ecological Information.* Ecotoxicity, persistence, and degradability, etc.
- *Section 13—Disposal Considerations.* Description of waste residues and information on safe handling and methods of disposal, including contaminated packaging.
- *Section 14—Transport Information.* Includes UN number, UN proper shipping name, transport hazard classes. Packing group, and any special precautions needed to ensure compliance with transportation.
- *Section 15—Regulatory Information.* Safety, health, and environmental regulations for the product in question.
- *Section 16*—Other Information. Other information including information on the preparation and revision of the SDS itself.

SDSs developed in accordance with these newly enacted OSHA and GHS regulations will provide many benefits to emergency responders. However, until the system is fully implemented in 2015 responders should still recognize that:

- Current SDSs have no uniform or consistent format or layout. The only regulatory requirements are that the specified data be provided.
- Computer-generated SDSs may be difficult to initially use and interpret because of their layout.
- There are no regulatory requirements concerning the language and terminology used. SDSs for the same chemical that are produced by different manufacturers may appear different and, in some instances, may use different terminology.

Monitoring Instruments

Monitoring and detection equipment are critical tools for evaluating real-time incident data to:

- Determine whether anything (i.e., a hazmat) is present.
- Classify or identify unknown hazards, including possible quantification of the hazard.
- Determine the appropriate levels of personal protective clothing and equipment.
- Determine the size and location of hazard control zones.
- Develop protective action recommendations and corridors.
- Assess the potential health effects of exposure.

- Determine when the incident scene is safe so that the public and/or facility personnel may be allowed to return.

Monitoring is an integral part of site safety operations and a cornerstone of a risk-based emergency response philosophy. Numerous response organizations have been issued regulatory citations for their failure to identify hazardous and IDLH conditions, to evaluate constantly the incident site for changes, and to verify the accuracy of hazard control zone locations.

Hazardous materials concentrations can be identified, quantified, and/or verified in two ways: (1) on-site use of direct-reading instruments, which provide readings at the same time that monitoring is being performed, and (2) laboratory analysis of samples obtained through several collection methods. Both tools are discussed in this section.

Responder Safety Tip

There is no single detection/monitoring device on the market that can do everything. Make sure you understand how an instrument will fit into your standard operating procedures and emergency response strategies. Anyone can use an instrument; the challenge is interpreting what the instrument is (and isn't) telling you and then making risk-based decisions to make the problem go away!

Selecting Direct-Reading Instruments

Direct-reading instruments provide information at the time of sampling, thereby allowing for rapid, on-scene risk evaluation and decision making. They are used to detect and monitor flammable or explosive atmospheres, oxygen-enriched and oxygen-deficient atmospheres, certain toxic and hazardous gases and vapors, chemical agents, certain biological agents, and ionizing radiation. The selection of types of monitoring instruments should be based on local/facility hazards and anticipated response scenarios **Figure 7-2**.

When evaluating survey instruments for emergency response use in the field, consider the following criteria:

- *Portability and user friendliness*—Easy to carry, easy to operate, weight, etc. Consider who will be using the instrument and their ability to consistently use it safely and correctly.
- Instrument Response Time—Also known as lag time, this is the period of time between when the instrument senses a product and when a monitor reading is produced. Depending on the instrument, lag times can range from several seconds to minutes. Variables will include the following:
  - Does the instrument have a pump? Monitors with a pump typically have a response time of 3–5 seconds; monitors without a pump and operating in a diffusion mode have response times of 30–60 seconds. Consult your manufacturer's operation manual for specific information.
  - Use of sampling tubing—add 1 to 2 seconds of lag time for each foot of hose.

  Monitors also have a recovery time, which is the amount of time it takes the monitor to clear itself of the sample. Recovery time is influenced by the properties of the sample, the amount of sampling hose, and the amount absorbed by the monitor.
- *Sensitivity and selectivity*—The ability of the instrument to select slight changes in product concentrations and select a specific chemical or group of chemicals that react similarly. Monitoring instruments are calibrated on specific materials (e.g., methane, pentane, isobutylene). Increased selectivity widens the relative response of an instrument and can increase its accuracy; however, it may not be possible to determine the exact contaminant present.

  Note: Amplifiers are used in some monitoring instruments to widen their response to more hazmats and increase accuracy. However, other electrical equipment in the area, including radios, other types of monitoring instruments, power lines, and transformers, may interfere with the amplifier in the instrument being used.
- Lower detection limit (LDL)—The lowest concentration to which a monitoring instrument will respond. The lower the LDL, the quicker contaminant concentrations can be evaluated. Many instruments have several scales of operation for monitoring both very low and very high concentrations.
- Calibration—The process of adjusting a monitoring instrument so that its readings correspond to actual, known concentrations of a given material. If the readings differ, the monitoring instrument can then be adjusted so that readings are the same as the calibrant gas. Conditions that will affect calibration include atmosphere, humidity, temperature, and atmospheric pressure.

  There are four types of calibration:
  1. *Factory calibration*—Instrument is returned to a certified factory/facility for testing and adjustment by certified instrument technicians.
  2. *Full calibration*—Instrument is shown a calibration gas and the readings are adjusted (automatically or manually) to the certified calibration gas values.
  3. *Field calibration*—Instrument is exposed to a known calibration gas and the user verifies that the readings correspond to +10% of the calibration gas. Field adjustments to the monitor may then be made, as appropriate.

Figure 7-2 Air monitoring and detection is an essential element of risk-based response process. No single instrument does it all!

4. *Bump test*—A field test in which an instrument is exposed to a known calibration gas and the sensors show a response or alarm. In simple terms, you are making sure the instrument will "see" the contaminant before being used in a hazardous area. If the instrument does not respond appropriately, then it should undergo a field or full calibration.

- *Correction factors (i.e., relative response curves)*—Relative response is the difference between a calibrated response and a response from a product for which the meter is not calibrated. This is a common issue with combustible gas indicators (CGIs) and photoionization detectors when dealing with known materials. The manufacturer should provide a comparison or response curve table to adjust the readings for the product being evaluated.
- *Inherent safety*—The inherent safety and the ability of the device to operate in hazardous atmospheres must be evaluated (see Scan Sheet 7-A). In addition, instruments should be certified by an approved testing laboratory for expected operating conditions.

In addition to the previous criteria, the following operational, storage, and use considerations should be evaluated:

- Where and in what type of storage container will the instruments be stored? This is especially critical when placing monitoring instruments inside or outside of vehicles. In addition, some storage containers may off-gas and contaminate sensors.
- Can field maintenance be easily performed? For example, are field calibration kits available and can sensors be easily changed in the field?
- Can buttons, switches, and so on be easily manipulated while wearing chemical gloves?
- How long does it take for the monitoring instruments to "warm up" before they can be used in the field?
- What types of alarms does the instrument have? Can meters be read easily while wearing protective clothing and respiratory protection? Is there a glare problem during daytime operations and a lighting problem for operations at night?
- What types of batteries are required for the instrument—off-the-shelf batteries or rechargeable batteries? How long will the unit operate with a full charge?

In recent years, there has been an explosive growth in the detection and monitoring industry, especially as it relates to monitoring technologies. The technologies presented in this text are based on those listed in NFPA 472 and represent the most current available instruments at the time of publication. While Hazmat Technicians must be proficient in the use of that equipment supplied by the Authority Having Jurisdiction (AHJ), they should be knowledgeable about all technologies in order to know what may be available for specific and unique emergency response purposes. Scan Sheet *7-B* lists the NFPA 472 list of instrument technologies and the types of hazards against which they may be employed.

Types of Direct-Reading Instruments

All direct-reading instruments have inherent limitations. Many detect and/or measure only specific classes of chemicals. As a general rule, they are not designed to measure and/or detect airborne concentrations below 1 ppm. Also, many direct-reading instruments designed to detect one particular substance may detect other substances (interference) and give false readings.

Figure 7-3 outlines the common types of direct-reading instruments used in the emergency response field.

When using direct-reading instruments, interpret instrument readings conservatively and consider the following guidelines:

- Conduct a daily check of your instruments, as well as before use.
- Use chemical correction factors when dealing with known materials, as appropriate.
- Remember that instrument readings have some limitations when dealing with unknown substances. Report readings of unknown contaminants as positive instrument response rather than specific concentrations (i.e., ppm). Conduct additional monitoring at any location where a positive response occurs.
- A reading of zero should be reported as no instrument response rather than clean, since quantities of chemicals may be present that cannot be detected by that particular instrument.
- The Rule of Threes that was originally developed for using emergency response guidebooks and manuals can also be applied to monitoring instrument technologies. When dealing with unknowns and suspected criminal scenarios involving hazardous materials, use several different types of detection technologies to classify or identify the hazard. Do not rely on one type of instrument technology, especially in those scenarios where your decisions have significant potential risks.
- After the initial survey, continue frequent monitoring throughout the incident.

Initial air monitoring and reconnaissance operations pose the greatest threat to emergency responders. Always remember these basic safety considerations:

- Air monitoring personnel have the greatest risk of exposure—protective clothing must be sufficient for the expected hazards. The situation should determine the level of protective clothing used.
- The air monitoring team should consist of at least two personnel, with a backup team wearing an equal level of protection.
- Protect the instruments, as appropriate. If they have the potential to become contaminated, the instrument may be wrapped in clear plastic to protect against contamination. However, make sure that the intake and exhaust ports are open.
- Approach the hazard area from upwind whenever possible. The initial site survey should begin upwind and then move to the flanks of the release. As soon as any positive indication is received, proceed with caution.
- Priority areas should include confined spaces, low-lying areas, and behind natural or artificial barriers (e.g., hills, structures, etc.) where heavier-than-air vapors can accumulate.

Scan Sheet 7-A

Hazard/Monitoring Technology Matrix

NFPA 472 TECHNOLOGY LIST	Corrosivity	Flammability	Oxidation Potential	Oxygen Deficiency	Pathogenicity	Radioactivity	Toxicity/ Chemical Agents
Biological Immunoassays					X		
Chemical Agent Monitors							X
Colorimetric Indicators	X	X	X				X
Combustible Gas Indicator (CGI)		X					
DNA Fluoroscopy					X		
Electro-chemical Cells				X			X
Flame Ionization Detector (FID)		X					X
Gas Chromatograph Mass Spectrometer (GC/MS)							X
Infrared Spectroscopy (FT-IR)		X			X		X
Ion Mobility Spectroscopy (IMS)							X
Gamma Spectrometer (RIID)						X	
Metal Oxide Sensor (MOS)							X
Photoionization Detector (PID)		X					X
Polymerase Chain Reaction (PCR)					X		
Radiation Detection & Measurement						X	
Raman Spectroscopy							X
Surface Acoustical Wave (SAW)							X
Wet Chemistry	X	X	X				X

TYPES OF DIRECT-READING MONITORING INSTRUMENTS

CORROSIVITY

HAZARD MONITORED	APPLICATION	INSTRUMENT TECHNOLOGY / METHOD OF OPERATION	GENERAL COMMENTS
Corrosivity—Acidity or Alkalinity	Measures the acidity or alkalinity of a corrosive material	■ pH paper—chemical reaction changes the color of the detection paper. Can detect corrosive liquids and some corrosive vapors in air, with a range of 0 to 14. Acids are normally shades of red and purple; bases/caustics are shades of blue. ■ pH meters use a probe that is inserted into the liquid after being buffered to neutral.	■ Readings <2 or >12 present a significant risk for injury. ■ Neither pH paper or pH meters will provide the specific concentration of the corrosive. ■ Using neutral water to wet pH paper will make it easier to detect corrosive vapors in air. ■ pH meters must be calibrated before each use. Probes must be thoroughly rinsed with distilled water before and after each calibration and use. They are typically not used extensively for emergency response purposes.

Figure 7-3 Types of direct-reading monitoring instruments. *(Continued)*

TYPES OF DIRECT-READING MONITORING INSTRUMENTS

FLAMMABILITY - COMBUSTIBLE GAS INDICATORS (LEL Meters)

HAZARD MONITORED	APPLICATION	INSTRUMENT TECHNOLOGY / METHOD OF OPERATION	GENERAL COMMENTS
Flammable Gases and Vapors 	Measures the concentration of a flammable gas or vapor in air	■ Basic principle is that a stream of sampled air passes through the senor housing, causing a heat increase thereby increasing the resistance in the electrical circuit, and subsequently showing an instrument reading (except for infrared sensor). ■ Three types of LEL sensors: catalytic bead (most common) metal oxide semi-conductors, and infrared. ■ Meter may read in either percent of the LEL, ppm, or percent of gas by volume.	■ Although commonly known as CGIs, they are used for detecting flammable gases and vapors. ■ Readings are relative to a calibrant gas (e.g., methane, pentane, hexane). Response curves required. ■ Catalytic bead sensors may have problems including: – Intended for use in normal atmospheres (i.e., not oxygen deficient or enriched). Will affect LEL readings if O_2 levels are deficient (will lower readings). – Sensor can be damaged by certain materials, including silicone, tetraethyl lead, and acid gases. Some problems can be reduced through use of filters and water traps between instrument and sampling tube. – Chronic exposure through high levels may saturate the sensor and cause it to be useless for a long period of time until purged and recalibrated. ■ 1% of the LEL=10,000 ppm. Just because you don't have a fire problem does not mean you don't have a health problem.

OXIDATION POTENTIAL & OXYGEN DEFICIENCY / ENRICHMENT

HAZARD MONITORED	APPLICATION	INSTRUMENT TECHNOLOGY / METHOD OF OPERATION	GENERAL COMMENTS
Oxidation Potential 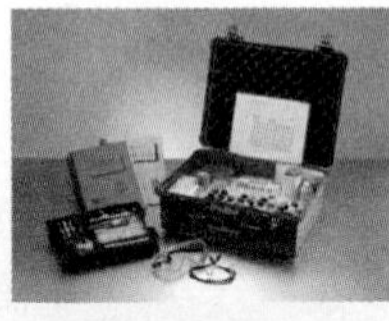	Test for the presence of oxidizers.	■ Wet chemistry: consists of applying a sample of the sample material to oxidizer test paper (e.g., KI paper)	■ Typically part of the HazCat® Chemical Identification System or a commercial test strip.
Oxygen Deficient and Enriched Atmospheres 	Measures the percentage of oxygen in air. Should measure both oxygen deficient (<19.5%) and enriched (>23.5%) atmospheres	■ Electrochemical cell: operates by a diffusion process, in which air diffuses into the sensor. Oxygen reacts with electrolytes in a cell, thereby generating a current flow in the meter. ■ May be either a passive sensor or have an internal pump that draws in the air sample ■ Meter will normally read percent of oxygen in the sample (e.g., 21% O_2)	■ Most O_2 sensors are combined with a multi-gas meter (e.g., LEL, × O_2 and 1 or 2 toxicity sensors). ■ Some materials (e.g., chlorine, fluorine) will indicate a high or normal level of O_2, when the actual atmosphere may be oxygen deficient. ■ Extreme cold temperatures often result in sluggish, delayed movement of the meter. ■ O_2 sensors are adversely affected by materials with a lot of oxygen in their chemical structure (e.g., chlorine, ozone, carbon dioxide). ■ A drop of 0.1% of O_2 means that 5,000 ppm of "something else" is in the atmosphere. ■ Must be calibrated prior to use to compensate for altitude and barometric pressure

Figure 7-3 *(Continued)*

TYPES OF DIRECT-READING MONITORING INSTRUMENTS

PATHOGENICITY

HAZARD MONITORED	APPLICATION	INSTRUMENT TECHNOLOGY / METHOD OF OPERATION	GENERAL COMMENTS
Hand-Held Immunoassays (HHA) 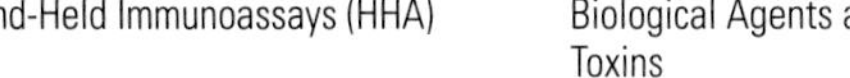	Biological Agents and Toxins	■ Similar to pregnancy tests, which tag/identify the antibodies present and provide a color change. Color change is read visually or through use of an instrument. ■ If threat is credible, sample should be sent to an accredited lab for analysis.	■ Some units use fluorescence to mark the antigen. ■ May be subject to false-positives and false negatives ■ Not a definitive tool to determine the presence of any bio pathogen or toxin. Should not be used to make decisions on patient management or prophylaxis. Can be useful in assessing credibility of the threat. ■ Examples: RAMP (Rapid Analyte Measurement Platform); Bio Threat Alert™ Test Strips and Guardian Reader System™; Smart Tickets
Polymerase Chain Reaction (PCR) Technology 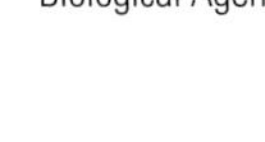	Biological Agents	■ DNA sample of a suspected bio-agent is put into a reaction and will go through a chain reaction process to amplify a specific DNA sequence ■ Cannot differentiate between live and dead organisms; for definitive results, test must be confirmed through lab analysis. ■ DNA fluoroscopy may be added as an enhancing technology to better verify the product being tested. Although this technology is currently used for bio-aerosol detectors for surveillance, it is primarily limited to laboratory applications.	■ Examples: RAZOR™ EX BioDetection System; R.A.P.I.D. System™ (Ruggedized Advance Pathogen Identification Device); Bio Seeq™

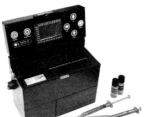

RADIATION DETECTION

HAZARD MONITORED	APPLICATION	INSTRUMENT TECHNOLOGY / METHOD OF OPERATION	GENERAL COMMENTS
Monitors—Ionizing Radiation (alpha and beta particles, gamma rays) 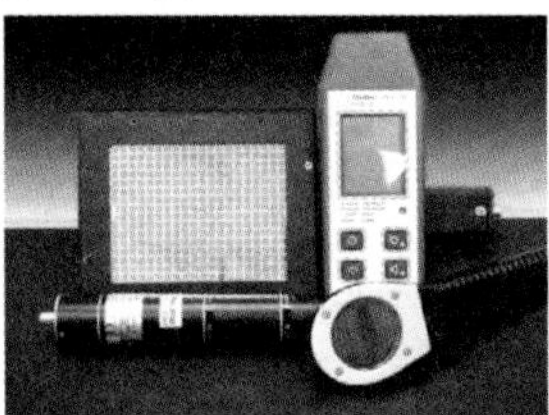	Detect and measure radiation levels	■ Ionization detectors that collect and count ions electronically ■ Three types of radiation detectors: scintillation, Geiger Mueller tubes, and ion chambers. Different probes can be used to measure for alpha, beta or gamma sources: – Scintillation: range of 0.02 to 20 mR/hr. – GM tubes: range of 0.2 to 20 mR/hr or 800 to 80,000 cpm (most common detection technology found in emergency response). – Ion chamber: range of 1 mR/hr to 500 R/hr. ■ Meter may read as counts per minute (CPM), and μR/Hr to R/hr. ■ Radiation pagers are used for detecting gamma and x-ray radiation, but may also detect high energy beta particles. Excellent tool when dealing with terrorism or criminal events when radioactive materials are suspected.	■ Radiation instruments should be selected based on: – Type of radiation to be detected – Energy of the radiation to be detected – Range high enough to measure the intensity of the source. ■ An instrument reliability check should be performed before each use. If the reading is not with ± 20% of the initial reading listed on the calibration sticker, the instrument should be recalibrated. ■ Meters using the International System (SI) for measuring radiation (e.g., Gray, Sievart) may be found at fixed facilities. ■ Electromagnetic fields can give "false positive" readings.

Figure 7-3 *(Continued)*

TYPES OF DIRECT-READING MONITORING INSTRUMENTS

HAZARD MONITORED	APPLICATION	INSTRUMENT TECHNOLOGY / METHOD OF OPERATION	GENERAL COMMENTS
		– Units are activated when radiation levels exceed background levels. – Can provide an audible, visual (LED light), or vibra alarm ■ Personal dosimeters monitor the accumulated dose received by an individual. Types include older style personal dosimeters that were visually read in the field, to sensors that must be read in a laboratory, to newer generation electronic dosimeters integrated into radiation pagers.	
Radiation Pager—Ionizing Radiation (gamma rays, neutron radiation)	Detects and measures radiation levels and measure radiation exposure	■ A small, personal radiation detector worn on a person or placed in a vehicle that provides visual, audio or vibration alarm on the presence of elevated radiation. Designed for early detection. ■ Typically relies on Geiger Mueller tube for radiation detection and responds only to highly penetrating ionizing radiation (i.e., gamma or neutron). Cannot detect alpha or beta radiation. ■ Can provide real-time exposure rates in physical units (μR/h or mR/h), numerical values (e.g., 1–9), or color warning lights	■ Widely distributed to emergency responders to alert them to the presence of possible radiological exposure ■ Useful for radiological terrorist scenarios, including Radiological Dissemination Device (RDD); Radiological Exposure Device (RED); and Improvided Nuclear Device (IND)
Radiation Isotope Identification Device (RIID) (beta particles, gamma rays and neutron radiation)	Identify man-made and natural radionuclides	■ Uses gamma spectroscopy to identify the radioactive source material. Each type of radioactive material produces different ratios of gamma rays to beta particles, which allows for the identification of the radionuclide. ■ Detector contains a crystalline material which, when hit by beta, gamma or neutron radiation, vibrates and emits a tiny flash of light. By analyzing the signal the radioactive material can be identified.	■ Useful for radiological terrorist scenarios, including Radiological Dissemination Device (RDD), Radiological Exposure Device (RED) and Improvised Nuclear Device (IND). ■ Useful for pre-event clearing/sweep of large areas
TOXICITY - COLORIMETRIC INDICATOR TUBES (Detector Tubes)			
Specific Gases and Vapors	Measures the concentration of specific gases and vapors in air. Used in hazard categorization systems for testing for unknowns; enable the user to classify the hazard class or chemical family of the unknown.	■ Glass tubes are filled with different reagents that react with the chemical being tested. When that chemical is present, the reagent may change color, or produce a colored stain that must be evaluated or measured using the reference points on the tube (ppm or % of material). ■ Ends of glass tube are first broken off; arrow on the tube should face *towards* the pump. Sample is then drawn through the tube using a bellows, piston, or thumb pump. ■ Readings can be directly read, or be interpreted based on the number of strokes or pumps. Wrong number of strokes/pumps can lead to incorrect readings.	■ Primarily used to determine if a specific chemical is present (or not present), as compared to specific quantitative results. ■ May be found in pre-packaged hazmat kits with a sampling matrix for the identification of unknowns or chemical warfare agents. ■ Read the tube instructions. ■ Operational issues include: – Greatest sources of error are (1) how the user judges the color stain or stain's endpoint, and (2) the tubes accuracy. Tubes may have error margin up to 35%. – Tubes have a limited shelf life (typically 2-3 years); can be affected by high humidity (tubes are calibrated at 50% humidity) and temperature extremes (most operate from 32° F [0° C] to 122° F [50° C]).

Figure 7-3 *(Continued)*

TYPES OF DIRECT-READING MONITORING INSTRUMENTS

HAZARD MONITORED	APPLICATION	INSTRUMENT TECHNOLOGY / METHOD OF OPERATION	GENERAL COMMENTS
 		■ Read instructions for each tube prior to use! ■ Draeger Chip Measurement System (CMS) uses a bar-coded chip that is inserted into a pump. Sampling is performed automatically and readings are provided on an LCD screen in ppm.	– Response times may vary greatly from chemical to chemical. – Tubes can have cross-sensitivities- many similar chemicals can interfere with the sampling and give "false positive" readings. However, this can also be useful in identifying an unknown. – Measured concentration of the same compound may vary between different manufacturers tubes. In addition, different manufacturer's tubes should not be used with another manufacturer's pump. WMD Agents ■ WMD colorimetric tubes may be used differently from some "regular" detector tubes. Read the manufacturer's instructions. ■ Examples: M8/M9 chemical agent detector papers. (*note*: can get false positives from pesticides and some petroleum products); M18A2 chemical agent detector kit; M256A1 chemical agent detector kit
TOXICITY - TOXIC GAS SENSORS			
Specific Gases and Vapors 	Designed to detect a specific chemical Common toxic gas sensors include carbon monoxide, hydrogen sulfide, chlorine, ammonia, sulfur dioxide	■ Sensor technology includes electrochemical sensors with two or more electrodes and a chemical mixture sealed in a sensor housing, and metal oxide semi-conductors (MOS). ■ Meter will normally read ppm of specific gas or vapor.	■ More accurate than detector tubes, but are limited to significantly fewer chemicals. ■ Hydrogen sulfide and carbon monoxide sensors are often combined into a multi-gas meter (e.g., LEL, × O_2 and toxicity sensors). ■ Toxic sensors can have a range of shelf-lives and a maximum exposure limit. High concentration exposures will shorten the life of the sensor. ■ Chemicals in the same chemical family as the sensor may give a reading that may be misinterpreted by the user.
Passive Dosimeter 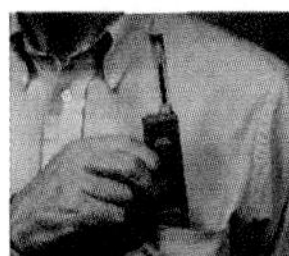	Designed to detect and measure concentrations of a specific chemical, typically in the workplace	■ Passive monitor that measures an individual's exposure to a specific chemical. ■ Provides either immediate results or is required to be sent to a laboratory for analysis.	■ Must acquire the specific dosimeter for the materials in question (e.g., organic vapors, mercury, ethylene oxide). ■ Commonly used to monitor for TLV/TWA and TLV/STEL exposures in non-emergency scenarios.
TOXICITY - IONIZATION DETECTORS			
***NOTE*:** There are two primary ionizing detectors used in the field: the photo-ionization detector (PID) and the flame ionization detector (FID). These are discussed below. Other types of ionizing detectors used in the weapons of mass destruction area will be covered later in this section under terrorism agent detection.			
A. PHOTO-IONIZATION DETECTORS (PID)			
Organic and Some Inorganic Gases and Vapors 	Detects total concentration of many organic and some inorganic gases and vapors (e.g., ammonia, arsine, phosphine, hydrogen sulfide). Can also detect some chemical agents (tabun, sarin, soman)	■ Gas/vapor sample is exposed to an ultraviolet (UV) lamp, which ionizes the sample. Ions are collected, amplified and produce a current which is read as total meter units.	■ General survey instrument that does *not* tell the user what is there; only that something is there. ■ IPs for gases and vapors are provided by the PID manufacturer; can also be referenced from the *NIOSH Pocket Guide to Chemical Hazards*.

Figure 7-3 (*Continued*)

TYPES OF DIRECT-READING MONITORING INSTRUMENTS

HAZARD MONITORED	APPLICATION	INSTRUMENT TECHNOLOGY / METHOD OF OPERATION	GENERAL COMMENTS
		■ PID reads from 0.1 to 2,000 ppm, although some units may read up to 10,000 ppm. There are also PIDs available that will read in the ppb range. All readings are expressed in terms of meter units for the calibration gas: iso-butylene. ■ The vapor or gas must be able to be ionized, known as its Ionization Potential (IP). IP is measured in electron volts (eV). ■ There are various UV lamps, with their intensity measured also measured in eV. Most common bulbs are 10.2, 10.6 and 11.7 eV. The 10.2 or 10.6 bulbs are typically used for emergency response purposes. ■ For the PID to read a gas/vapor, the IP of the sample must be less than the eV rating of the bulb. Response times are usually <5 seconds.	■ Operational issues include: – Lamps are affected dirt and dust, as well as high humidity and fog. – Higher levels of methane may suppress some of the ionizing potential of the lamp. Methane has an IP of approximately 13 eV. – PIDs cannot separate out mixed gases. WMD Agents ■ The 10.6 eV lamp will see most chemical agents, but the 11.7 eV lamp is the better to see the nerve agents. Given their low vapor pressures, none of the chemical agents will likely read very high on the PID. ■ Always use more than one detection technology when using a PID. ■ Available for monitoring point sources or for ambient air monitoring of a given area connected to a host controller unit ■ Difficult to calibrate after agent exposure

B. FLAME IONIZATION DETECTORS (FID)

HAZARD MONITORED	APPLICATION	INSTRUMENT TECHNOLOGY / METHOD OF OPERATION	GENERAL COMMENTS
Organic Gases and Vapors  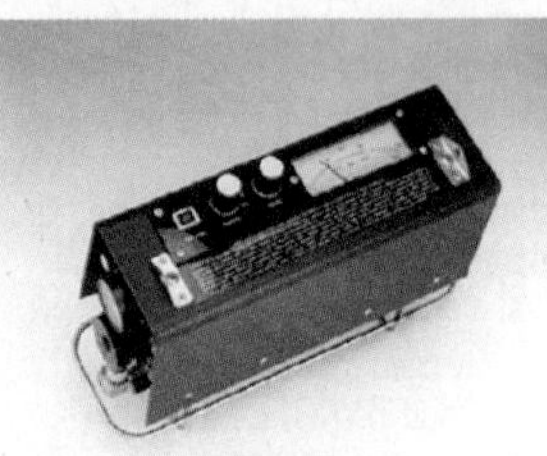 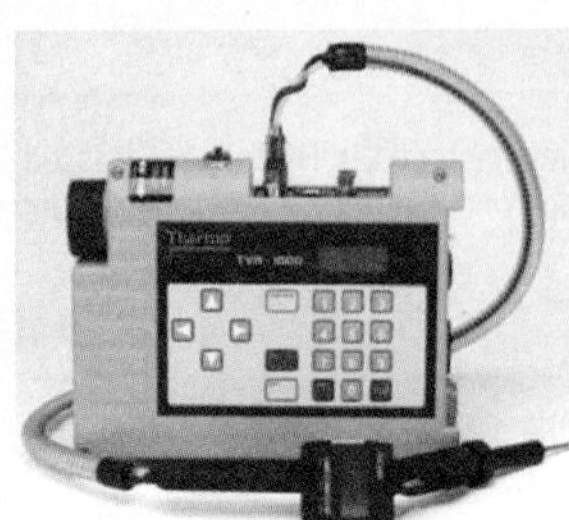	Detect total concentration of many organic gases and vapors Operate in either the survey mode or gas chromatograph mode	■ Gases/vapor sample is exposed to a hydrogen flame to ionize the sample. ■ Operates in two modes: – Survey mode: detects total concentration of many organic gases and vapors. All organic compounds are ionized and detected at the same time. – Gas chromatograph (GC) mode: identifies and measures specific organic compounds; volatile compounds are separated. ■ *Survey mode method of operation*: ionizes any chemical with an IP <15.4 eV. Sample is exposed to a hydrogen flame and is ionized; ions are then collected, amplified, and produce a current that is read on the display as total organic vapors present (in ppm). ■ *GC method of operation*: – Sample is drawn into a column with an inert material. Hydrogen gas is then passed through the tube and picks up various components of the sample. Each chemical takes a period of time to exit the tube. – The mass spectrometer (MS) is usually coupled with a GC and is the identifying portion of the device. As the sample passes into the detector, the energy is measured on a strip recorder as a peak. – The MS measures the relative mass of the molecular fragments of the sample and compares it to materials within the MS library, as each molecular fragment has a different weight.	■ Operational issues include: – Not all FID's are intrinsically safe. Must first establish that the operating area is safe (i.e., not in the explosive range). – FID is more accurate than the PID. – FID has the ability to read methane. – Response affected by temperatures below 40°F (4.4°C), as gases may condense in the pump and column. – High learning curve for using GCs and mass spectrometers in field settings for emergency response purposes. ■ The more extensive the MS library, the more likely one will be able to identify an unknown substance. ■ Many police department crime labs will have GC/MS capabilities.

Figure 7-3 *(Continued)*

TYPES OF DIRECT-READING MONITORING INSTRUMENTS

TOXICITY – CHEMICAL WARFARE AGENTS

HAZARD MONITORED	APPLICATION	INSTRUMENT TECHNOLOGY / METHOD OF OPERATION	GENERAL COMMENTS
WMD Chemical Agents	Designed to detect nerve and blister agents	■ Ion-Mobiity Spectrometry (IMS) is one of the most common technologies for chemical agents. Sample is ionized, usually through a radioactive source, "fingerprinted" based on its chemical make-up, and then compared to a library. ■ Can detect low levels, down to parts-per-billion levels. However, is slow to respond to low levels. ■ IMS is extremely sensitive to false positives (e.g., cleaning agents, floor wax).	■ Examples: APD 2000 (closed loop system), Chem Pro 100 (open loop system), Improved Chemical Agent Monitor (I-CAM). ■ APD-2000 can also detect irritants and riot control agents. ■ Can be used for both point or area monitoring ■ Technology can also be used to detect explosives, toxic industrial materials and street drugs.
WMD Chemical Agents	Designed to detect nerve agents and hydrogen mustard	■ Flame spectroscopy burns agent with hydrogen gas, which decomposes any chemical agents or TIMs in the sample. Decomposed agent turns into hydrogen phosphorous or sulfur which emit light and then read by optical fibers. ■ Technology is similar to the flame ionization detector (FID). ■ Generally used in conjunction with gas chromatograph	■ Provides a very quick response time, but cannot be used in an explosive environment ■ Primarily designed for military battlefield applications ■ Examples: Miniature CAM Monitor, AP2C/AP4C Detector
WMD Chemical Agents	Desinged to detect nerve, blister, blood and choking agents	■ Surface Acoustical Wave (SAW) technology relies on chemically selective coated piezoelectric crystals that absorb target gases. The absorption causes a change in the resonant frequency of the crystal that is measured by a microcomputer.	■ Can identify more than one agent at a time. Technology also detects toxic industrial materials. ■ Provides a low number of false positives versus IMS and flame spectroscopy. However, has a high detection threshold level. ■ Can be used for point or area monitoring ■ Subject to electro-magnetic radiation interference and false positives from solvents, household chemicals and aftershave ■ Examples: Hazmat CAD Plus, JCAD Chem Sentry

UNKNOWNS - FOURIER-TRANSFORM INFRARED SPECTROMETRY (FT-IR)

HAZARD MONITORED	APPLICATION	INSTRUMENT TECHNOLOGY / METHOD OF OPERATION	GENERAL COMMENTS
Organic and Some Inorganic Gases and Vapors  	Identification of (1) unknown solid, slurry and liquid substances, or (2) unknown gases and vapors. *NOTE*: These are two separate applications of the FT-IR technology	■ Sample is obtained and taken to the unit. FT-IR records the interaction of infrared radiation with the chemical sample, measuring the frequencies at which a sample absorbs radiation and the intensities of the absorptions. ■ As chemical functional groups are known to absorb light at specific frequencies, the FT-IR technology can allow for the specific identification of liquid and solid samples. ■ Reading is compared against a library of various hazardous and non-hazardous substances. If the material is in the library, identification is confirmed in <1 minute.	■ Libraries include hazardous materials, chemical agents, white powders, explosives, clan lab precursors, drugs, and other common liquid and solid materials. ■ Although they cannot identify biological materials, they can determine if a compound is of biological or non-biological origin. ■ FT-IR strengths complements Raman weaknesses, and vice versa. ■ Examples include: TravelIR HCI™; HazMat™ID; Tru Defender FT and FTi; Avatar™ ESP

Figure 7-3 *(Continued)*

TYPES OF DIRECT-READING MONITORING INSTRUMENTS

UNKNOWNS – RAMAN SPECTROSCOPY

HAZARD MONITORED	APPLICATION	INSTRUMENT TECHNOLOGY / METHOD OF OPERATION	GENERAL COMMENTS
Organic and Some Inorganic Gases and Vapors   	Identification of unknown solid and liquid substances, including explosives	■ Optical analysis technique that uses a laser as its light source. When placed against a transparent container, the Raman instrument emits light at a single focused wavelegth, then collects the light that is scattered by the sample, thereby creating a unique molecular fingerprint ■ Fingerprint is compared against a library of various hazardous and non-hazardous substances	■ Provides for non-destructive analysis of the sample material. Sample can remain inside of a glass container or plastic bag while being analyzed ■ Laser can cause injuries if used improperly. ■ Raman strengths complements FT-IR weaknesses, and vice-versa. New units which combine both the FT-IR and Raman technology are now available on the market (e.g., Thermo Gemini).

OTHER MISCELLANEOUS DETECTION DEVICES

HAZARD MONITORED	APPLICATION	INSTRUMENT TECHNOLOGY / METHOD OF OPERATION	GENERAL COMMENTS
HazCat® Chemical Identification System 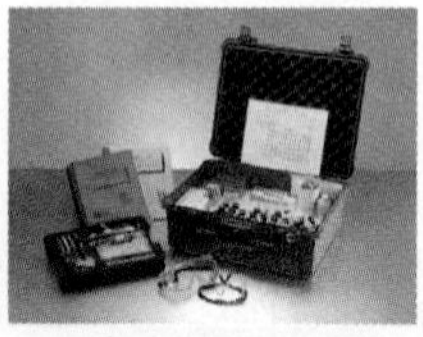	On-site identification or characterization of hazardous and non-hazardous materials	■ Field chemistry testing kit that provides a logic system for conducting specific chemical tests that will result in the identification or characterization of solids, liquids, and gases ■ System is based on the concept that chemicals within the same family react in the same manner	■ Requires specific training; can typically provide characterization of unknowns within 10–30 minutes, depending on the substance ■ Kit options are offered for methamphetamine labs and chemical/biological/radiological applications.
Chemical Test Strips 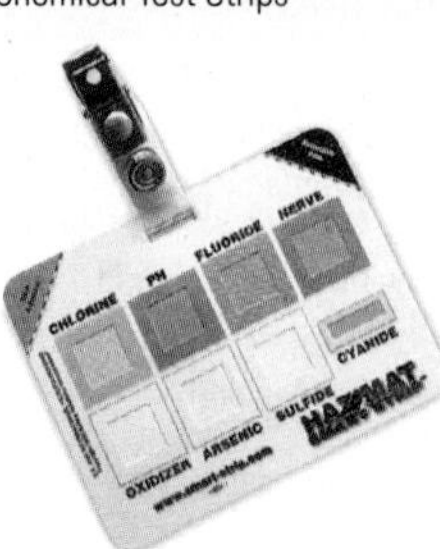	Detects the presence of specific chemicals or hazards	■ Requires a sample of the contaminant to come in contact with the test strip or chemical indicator paper. If the contaminant is present, it will trigger a change in color on the test strip or indicator paper.	■ May be part of a hazmat identification system; e.g., HazCat® Chemical Identification System. ■ Indicator papers include pH paper, and M-8/M-9 tape for chemical agents. ■ Chemical Classifier® Kits can test liquid spills for corrosivity, oxidizers, fluoride, organic solvents/petroleum distillates, and iodine/bromine/chlorine. ■ HazMat Smart-Strip® can test liquids and aerosols for chlorine, pH, sulfides, oxidizer, fluorides, nerve agents, cyanide, and arsenic. ■ Readings should be confirmed by other instruments, as appropriate (e.g., colorimetric detector tubes, chemical agent monitors).

Figure 7-3 *(Continued)*

TYPES OF DIRECT-READING MONITORING INSTRUMENTS

HAZARD MONITORED	APPLICATION	INSTRUMENT TECHNOLOGY / METHOD OF OPERATION	GENERAL COMMENTS
Mercury Detector	Detects mercury vapors in air	■ Mercury vapors collect on a gold foil, changing the electrical resistance and causing a corresponding display reading.	■ Can be susceptible to false readings if hand carried or moved around; place on a stable surface and bring the contaminant to the unit, if possible ■ If the unit must be moved around, take three readings and then develop an average. ■ If clothing is possibly contaminated, place in a plastic bag and insert the instrument probe into the bag to sample the atmosphere. ■ ACGIH TLV/TWA is 0.025 mg/m^3; ACGIH recommends that women of childbearing age should not be exposed to concentrations > 0.010 mg/m^3.

Figure 7-3 (*Continued*)

Monitoring Strategies

The IC and/or the Hazmat Group Supervisor must establish a monitoring strategy. In developing this strategy, the following operational issues must be considered:

- Establish monitoring priorities based on whether the incident is in open air or in an enclosed or confined space environment.
- Always use the appropriate **monitoring instrument(s)** based on dealing with known or unknown materials. The instrument(s) should be able to detect the anticipated hazard(s), measure appropriate concentrations, and operate under the given field conditions.
- Monitoring personnel should have a good idea of what readings to expect. In the event that abnormal or unusual readings are encountered, the possibility of instrument failure should be considered. Try to confirm the initial reading with another instrument.
- The absence of a positive response or reading does not necessarily mean that contaminants are not present. A number of factors can affect contaminant concentrations, including wind, temperature, and moisture.
- Never assume that only one hazard is present.
- Interpret the instrument readings in more than one manner (i.e., always play "devil's advocate").
- Establish action levels based on instrument readings.

Monitoring operations at long-term incidents may be performed by multiple responders using different monitoring instruments at various locations. Reliable monitoring results can only be obtained by using consistent monitoring procedures. Monitoring locations must be identified and described so that subsequent responders will conduct air monitoring at the same location and height. Various systems can be used to identify monitoring and sampling locations, including systematically dividing the incident scene into quadrants (i.e., grid system) or using GPS devices.

Responder Safety Tip

Remember the Rule of Threes—when dealing with unknowns and suspected criminal scenarios involving hazardous materials, use several types of detection technologies to classify or identify the hazard.

Monitoring results should be documented as follows:

- Instrument—the type of monitoring instrument being used
- Location—specific location where the monitoring is conducted
- Time—time at which the monitoring is conducted
- Level—level where the monitoring reading is taken (e.g., foot, waist, head)
- Reading—the actual reading(s) given by the monitoring instrument

Monitoring priorities will be dependent on whether responders have identified the hazmat(s) involved. These priorities should be systematic and continuously evaluated throughout the course of the emergency.

Unknowns will create the greatest challenge for responders. The nature of the incident (e.g., credible threat scenario involving WMD agents), the location of the emergency (e.g., outdoors, indoors, confined space), and the suspected physical state of the unknown (i.e., solid, liquid, or gas) will influence the monitoring strategy. In scenarios involving unknowns, the role of hazmat responders is much like that of a detective. At the conclusion of the testing process, responders may still be unable to specifically identify the material(s) involved; however, they should be able to rule out a number of hazard classes and shorten the list of possibilities.

The following monitoring priority is used by many hazmat responders when dealing with scenarios involving unknown substances in an open-air environment. If a corrosive liquid is suspected, responders should use pH paper wetted with sterilized water to determine if a corrosive atmosphere is present, as these vapors may adversely affect some meters. Initial efforts should be toward determining if IDLH concentrations are present and providing an initial base to confirm or refute the existence of specific hazards.

1. *Radiation*—If there is any doubt, radiation detection should be the first priority. Remember—gamma rays travel the greatest distance and are the primary hazard for external exposure. A positive reading twice above background levels would confirm the existence of a radiation hazard. If initial gamma radiation readings are negative but clues are present indicating the possible presence of radioactive materials, additional testing for beta and alpha sources should be conducted.
2. *Flammability*—Since flammability and oxygen levels are directly related, monitoring for flammability and oxygen is usually implemented simultaneously through combination CGI/O2 meters. Remember that an oxygen-deficient atmosphere will shorten the flammable range, while an oxygen-enriched atmosphere will expand the flammable range.
3. *Oxygen*—Monitoring should evaluate the presence of both oxygen-deficient and oxygen-enriched atmospheres, particularly when dealing with confined spaces. Responders must consider that oxygen levels may also be influenced by the level of contaminants (i.e., an increasing level of contaminants displacing available oxygen).
4. *Toxicity*—The level and sophistication of toxicity monitoring will depend upon available instrumentation. Resources range from the simple to the sophisticated and may include the following:
   - Indicator papers, such as pH paper, and M-8/M-9 tape for chemical agents. Chemical Classifier™ kits can test liquid spills for corrosivity, oxidizers, fluoride, organic solvents/petroleum distillates, and iodine/bromine/chlorine, while the HazMat Smart-Strip™ can test liquids and aerosols for chlorine, pH, sulfides, oxidizer, fluorides, nerve agents, cyanide, and arsenic.
   - Specific or combination air monitors, which detect toxic gases such as hydrogen sulfide or carbon monoxide. Carbon monoxide monitors are useful in fire and post-fire situations. Hydrogen sulfide monitors are useful when dealing with confined spaces and when working around petroleum facilities where "sour" gas is handled.
   - Colorimetric detector tubes can be used for both known and unknown substances. Commercial detector tube kits are available for identifying unknown airborne hazards. By conducting a series of measurements with various pre-established detector tubes, responders can often determine the chemical class (e.g., organic gases, alcohols, acidic gases) present. Detector tube kits are also available for determining the presence of chemical agents. These kits only indicate the presence of several classes of gases and vapors; they provide a gross estimate of concentration and may not differentiate between specific gases within a certain class.
   - Survey instruments, such as flame ionization detectors (FIDs) and photoionization detectors (PIDs).

If the incident involves a confined space scenario, OSHA 1910.126—*Permit-Required Confined Space Standard,* clearly outlines the required monitoring priority as follows:

- Oxygen deficiency and enrichment
- Flammability
- Toxicity

From a practical perspective, the use of multi-sensor instruments (e.g., three to six gas sensors, including a photoionization detector) that allow one instrument to provide monitoring for all three hazards can make this initial monitoring priority discussion a moot point.

Evaluating Monitoring Results—Actions Levels and Guidelines

Initial air monitoring efforts should be directed toward determining if IDLH concentrations are present. Decisions regarding protective clothing recommendations, establishing hazard control zones, and evaluating any related public protective actions should be based on the following parameters:

1. *Radioactivity*—Any positive reading twice above background levels or alpha and/or beta particles that are 200 to 300 counts per minute (cpm) above background would confirm the existence of a radiation hazard and should be used as the basis for initial actions.
2. *Flammability*—The IDLH action level is 10% of the lower explosive limit (LEL).
3. *Oxygen*—An IDLH oxygen-deficient atmosphere is 19.5% oxygen or lower, while an oxygen-enriched atmosphere contains 23.5% oxygen or higher. In evaluating an oxygen-deficient atmosphere, consider that the level of available oxygen may be influenced by contaminants that may be present.
4. *Toxicity*—Unless a published action level or similar guideline (e.g., ERPG-2) is available, the STEL or IDLH values should initially be used. If there is no published IDLH value, responders may consider using an estimated IDLH of 10 times the TLV/TWA. Initial hazard control zones can be established for toxic materials using the following guidelines:
   - Hot Zone—Monitoring readings above STEL or IDLH exposure values.
   - Warm Zone—Monitoring readings equal to or greater than TLV/TWA or PEL exposure values.
   - Cold Zone—Monitoring readings less than TLV/TWA or PEL exposure values.

It is important to note that action levels are guidelines. Many assume that because an action level is achieved, the immediate area must always be evacuated. While there are scenarios where this is true, "actions" may also include ventilating an area to allow responders to continue to work in the environment. Remember—it's all about *risk!*

Identifying or Classifying Unknown Materials

The Hazardous Materials Technician is faced with a number of challenges when encountering unknown substances. The initial challenge is to identify and/or classify the material(s) involved and the type of hazards the material(s) may present. In some instances, there may be little or no information on the substance and few visual clues to help you determine the nature of the released substance.

Using a risk-based response process is a critical element in this assessment. The **risk-based response process** must include (1) analyzing the problem; (2) identifying and assessing the hazards; (3) evaluating the potential consequences; and (4) determining the appropriate response actions based on the facts, science, and circumstances of the incident. In short, this skill requires that you use a logical and methodical approach to identify the problem and evaluate the incident risks.

You should be able to apply and use the basic principles outlined in the General Hazardous Materials Behavior Model for the initial hazard and risk assessment tasks posed by a hazardous materials incident. This task should include predicting potential product and/or container behavior based on known incident facts and information and identifying the likely harm events.

The AHJ will prescribe the specific monitoring and detection equipment that should be used for initial entry and assessment tasks. However, a baseline capability used by most HMRTs will include (1) a radiological detection device; (2) a multi-gas detector to determine flammability and oxygen levels; (3) a photoionization detector (PID) to evaluate potential

Scan Sheet 7-B

Detection and Monitoring for Terrorism Agents

The terrorism problem has created new challenges for the emergency response community. As a result, hazmat trained personnel are now being required to broaden their skills and knowledge-base as it relates to chemical, biological, and radiological hazards.

Detection and monitoring instruments will play a key role in assessing the credibility of a threat. Among the clues in making the transition that an incident may be terrorist or criminal in nature are:

- Is the incident at a target occupancy or event?
- Has there been a history of any previous threats?
- Have there been reports of any unusual odors? Explosions? Hazardous materials?
- Are multiple casualties involved? Are the reason(s) known?
- Are initial emergency responders down?
- Have any secondary events occurred?

These clues will ultimately determine your initial monitoring strategy at a suspected terrorism incident. If a suspected explosive device is involved, the explosives hazard will take precedence over any other threats. If responders have arrived at an incident where an explosion has already occurred, the potential for an improvised radiological device should be initially considered. The radiological threat can be quickly assessed in most instances through the use of a radiation pager.

When conducting detection, monitoring, and sampling at a suspected terrorism incident where hazardous materials or WMD may be involved, consider the following:

- Avoid contact with the suspected material(s).
- When sampling for chemical warfare agents, be aware of the potential for false positives.
- Remember the "Rule of Threes"—use several types of detection technologies to classify, identify, or verify the hazard.
- Grab a sample and do the testing away from the hazard area.
- Remember your product and container specialists for terrorism and WMD agents. Initial sources may include your local FBI WMD Coordinator, Joint Terrorism Task Force (JTTF), or the FBI Hazardous Materials Operations Unit. Additional sources are available through agencies such as the National Guard Civil Support Teams (CST), the U.S. Army Medical Research Institute of Infectious Diseases (USAMRID), and the Centers for Disease Control (CDC).

Responder Safety Tip

The use of secondary devices targeted towards emergency responders is a common tactic used by both terrorists and Bad Guys. Responders must maintain their situational awareness of both the incident scene and surrounding areas for possible secondary threats and risks.

toxicity; and (4) pH paper to evaluate corrosivity hazards. These devices can be used to assess the hazards of both known and unknown liquids, solids, and gases.

Many hazmat response agencies will use a 5-gallon plastic bucket or comparable carrying device for carrying the monitoring, detection, and sampling equipment that may be required for initial identification/classification tasks. The main thing to remember is to take all the tools and equipment you think you might need and to know that you can rely on the findings of each instrument.

To identify or classify unknown materials, follow the steps in **Skill Drill 7-1**:

1. Radiation pagers or radiation survey meters may be used for the initial assessment of radiological hazards. While a radiation pager can detect the presence of beta and gamma radiation, a survey meter is more sensitive and more precise in its measurements. The selection of radiation detection equipment should be based on suspected threats and available equipment. **(photo 1)**
2. A multi-gas detector will assist in identifying or classifying the basic flammability and oxygen deficiency/enrichment hazards that may be present. Knowing how to interpret these initial readings is critical for your safety, since many products do not have any visual clues. Remember, just because you get a "hit" on a single detector does not mean that other hazards are NOT present. Don't stop at the first reading on any device or single technology! **(photo 2)**
3. A photoionization detector (PID) may assist in classifying and identifying many toxic products that emit vapors and gases. **(photo 3)**

Skill Drill 7-1

Identifying or Classifying Unknown Materials

1 Always use radiation survey meters (or other suitable instrumentation) when assessing unknowns.

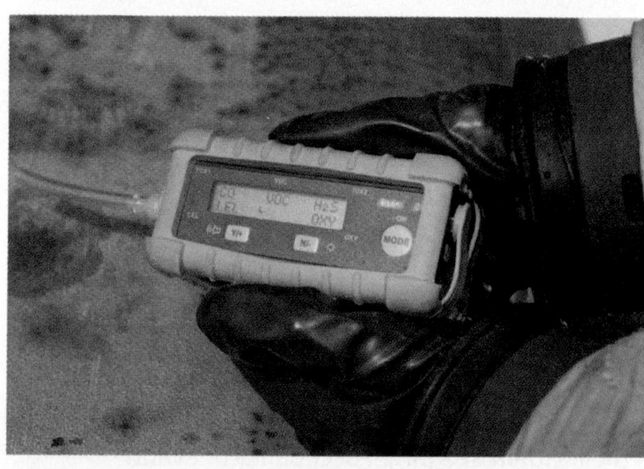

2 Assess flammability, oxygen levels and/or other airborne hazards.

3 Some instruments have a PID function built in. If not, it is recommended to use a separate device.

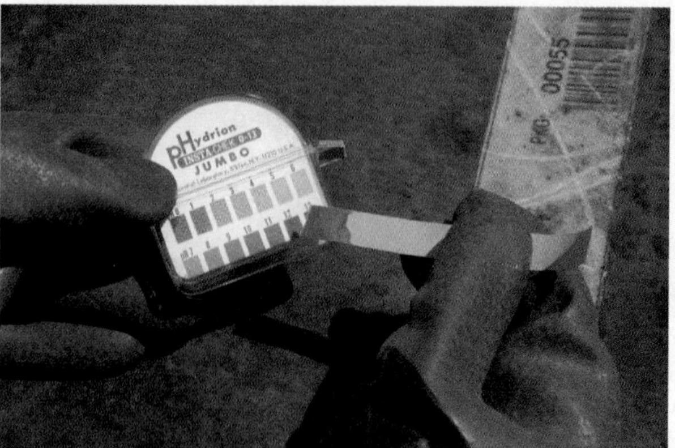

4 Determining pH is an important step in the process. A neutral reading is a valid piece of information when put in context. Solvents, for example, would not register a pH—this can help confirm the identity and/or chemical properties of the substance.

4. pH paper will detect both acids and caustics (bases). By wetting pH paper with an unknown liquid you can determine whether the product is an acid or caustic (base) or a substance that does not register a pH reading. Initially wetting the pH paper with sterilized water and holding it in the vapor space of an unknown liquid or gas can help to determine whether the unknown is producing a corrosive vapor. This, in turn, can lead responders to the classification of the hazard(s) and potentially identification of the product. **(photo 4)**

By using this basic cadre of instruments, you can address the fundamental tasks of (1) product identification and/or classification, and (2) identify potential hazards present. When the findings from these direct-reading instruments are combined with on-scene indicators, you should be capable of making a timely and effective risk assessment of the incident.

Using Equipment Provided By the AHJ

The AHJ or testing authority must determine which detection and monitoring devices and equipment will be available. The selected equipment should reflect a range of monitoring capabilities and hazards, and should not solely focus upon one hazard. The correct operational procedures for using each piece of equipment should be based on the manufacturer operating instructions, where available.

NFPA 472 is focused on the correct use of the detection and monitoring equipment available by the AHJ to the Hazardous Materials Technician. However, the Hazardous Materials Technician must also be skilled and competent in interpreting the results provided through monitoring tasks. This skill drill provides a framework by which the field screening tasks can be combined with the correct application and use of the available equipment and instrumentation.

Classifying and identifying unknown materials, and quantifying the hazards posed by known materials, is a "bread and butter" task for the Hazardous Materials Technician. Field identification of unknowns is part art and part science—it requires the responder to apply the powers of observation, good situational awareness, common sense, and a logical and methodical approach to understanding the results of the detection and monitoring process. While this competency is focused on basic detection and monitoring equipment, the responder may also have access to other tools including wet chemistry testing (e.g., HazCat™) and other technologies.

A thorough field screening is the doorway into the hazard classification process. In general, field screening begins with the selection of the appropriate tools and equipment, including instrumentation, to determine the general characteristics of an unknown substance. This skill drill provides a framework for thinking through the task of field screening—it is not intended to be the only way to approach the task. Any of the following items could be useful in the field screening process. Field screening could be done on an individual container and should be done prior to taking any samples.

To properly use equipment provided by the AHJ in performing field screening, follow the steps in **Skill Drill 7-2 ▼**:

1. A field screen of any scene should include an initial check for explosive devices. This includes not only a thorough visual scan for obvious signs of devices, but also for some of the chemicals that could be used to make explosives. For example, discovering containers of acetone and hydrogen peroxide could be an indicator of the production or attempted production of the explosive tri-acetone tri-peroxide (TATP). In some cases, a Bomb (EOD) Technician could be called to the scene to check for explosive devices. **(photo 1)**
2. Field screening must include a check for radiation. Responders should never assume radioactive sources are not present. Use the instruments provided by the AHJ—it could save your life. **(photo 2)**
3. Flammability risks should be evaluated with a combustible gas indicator. This is commonly performed using a multi-gas instrument with an LEL sensor and oxygen sensor, but it could be any technology that evaluates the atmosphere for the presence of flammable gases or vapors. **(photo 3)**
4. Corrosivity should be checked using either pH paper or a pH meter. Depending on the procedures established by your AHJ, dry or hydrated pH paper may be used. pH paper can be used to swipe across surfaces, wetted and placed gently into the head space of containers after they have been carefully opened, or perhaps inserted into unknown atmosphere at the end of a long pole.

   If you suspect an unknown liquid is a solvent, for example, and a pH check reveals no change on the pH paper—you are one step closer to confirming the presence of a solvent. Conversely, if you suspect an acid, and the result of a pH check confirms your finding, you are closer to confirming your suspicion. In some cases, when a total unknown is being field screened, a pH check can be a quick means to rule out a number of possibilities and help guide you in your effort to identify or classify. Although pH checks are generally only done on liquids, they can be used to evaluate vapors or unknown atmospheres. **(photo 4)**
5. Oxygen levels (deficiency and enrichment) should also be checked, especially when operating in an indoor or confined area. This is commonly accomplished using a multi-gas meter with an oxygen sensor, but single oxygen meters may also be used. **(photo 5)**
6. An unknown container or atmosphere should also be evaluated for the presence of volatile organic compounds (VOCs). This is typically done with a photoionization detector (PID), which may be integrated into a multi-gas instrument. Unknown substances that give off certain readings (e.g., VOC with low or high vapor pressure) can be easily classified and even identified by the reaction time and speed of the PID. **(photo 6)**

Skill Drill 7-2

Using Equipment to Perform Field Screening

1 Visually scan the scene for hazards and or dangerous situations. Pay attention to the details!

2 Radiation cannot be detected by sight, sound, or smell—make sure to evaluate the scene for this potential threat.

3 Evaluate flammability risks with a combustible gas indicator.

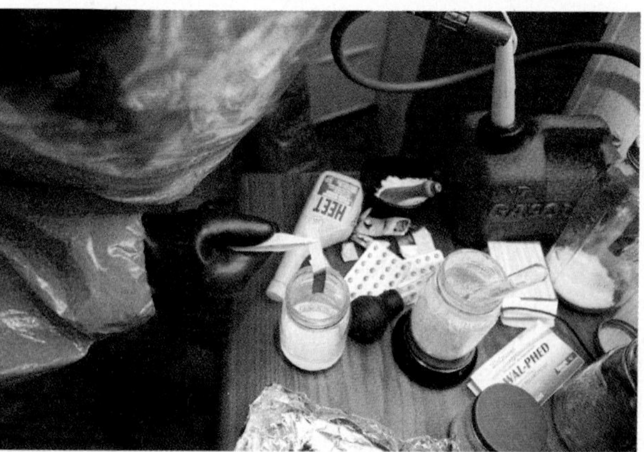

4 If placing pH paper into the head space of a liquid container has not changed the color of the pH paper, draw a small amount of liquid into a pipette and drop onto the end of the pH paper.

5 Check oxygen levels (deficiency and enrichment) using a multi-gas meter with an oxygen sensor.

6 Evaluate an unknown container or atmosphere for the presence of volatile organic compounds (VOCs).

Demonstrating Field Maintenance and Testing Procedures

Hazardous materials responders must be able to rely on the readings obtained from detection and monitoring devices and other tools and equipment used to identify and classify substances. This is accomplished by first understanding how to use and maintain those devices and understanding the limitations inherent in these instruments. There is no single device that can do everything—you must select the right piece of equipment for the job. That level of understanding leads to a certain level of confidence in the results and readings provided. Finally, the responder must be able to interpret that information and form conclusions based on those results to make sound decisions regarding things like PPE selection, evacuation areas, decontamination effectiveness, and other tasks. As you can see, this triangle of "know the instrument—know what it can do—know what it's telling you" is crucial to your own health and safety.

This drill is intended to demonstrate a process and the minimum level of detail you should know about all tools you have available to you. It is not intended to be an exhaustive "how to" guide for every instrument imaginable. It is always recommended that the responder adhere to the use guidelines specified by the manufacturer. To that end, we offer field-level guidance on the start-up procedures, limitations of use, and suggested alarm points for a standard multi gas meter and colorimetric tubes. Similar processes may exist within your agency or AHJ. If not, it is incumbent on you to develop a checklist or process for preparing your instrument for use.

To demonstrate field maintenance and testing procedures using a multi-gas meter with electrochemical sensors or colorimetric tubes, follow the steps in **Skill Drill 7-3 ▼**:

Skill Drill 7-3A: Using a Multi-Gas Meter with Electrochemical Sensors

1. Turn the unit on and let it warm up (usually 5 minutes is sufficient).
2. Ensure the battery has sufficient life for the operational period.
3. Identify the installed sensors (e.g., oxygen, lower explosive limit, hydrogen sulfide, carbon monoxide) and verify they are not expired. (Note: some units have a built in PID function.) Review alarm limits and the audio and visual alarm notifications associated with those limits. Review and understand the types of gases and vapors that could harm or destroy the sensors. Use other methods to check for those substances (e.g., pH paper) to ensure they are not present in the atmosphere to be sampled. Care must be taken to avoid pulling liquids into the device—they are designed to sample air, not liquid! **(photo 1)**
4. Perform a test on the pump by occluding the inlet and ensuring the appropriate alarm sounds. **(photo 2)**
5. Perform a fresh air calibration and 'zero' the unit. **(photo 3)**
6. Perform a bump test—expose the unit to a substance or substances that the unit should detect and react to accordingly. In essence, you are making sure the unit will 'see' what it is supposed to see before it might be called upon to see it for real. **(photo 4)**
7. Allow the device to reset, or return to fresh air calibration state.
8. Review the alarm levels and resetting procedures for addressing sensors that become saturated or are exposed to too much gas or vapor. **(photo 5)**
9. Review other device functions such as screen illumination, data logging if available, low battery alarm, etc. **(photo 6)**
10. Review decontamination procedures.
11. Carry out monitoring and detection. **(photo 7)**

Skill Drill 7-3B: Using Colorimetric Tubes

1. Determine the mission and relevance of using colorimetric tubes for air sampling. **(photo 1)**
2. Select the appropriate tubes for the mission. Ensure the tubes are within their life expectancy—do not use outdated tubes. **(photo 2)**
3. Collect all parts of the tube sampling system including pumps, tubes, extension hoses, etc. **(photo 3)**
4. Review the directions for using each tube, including pump stroke count, tube cross sensitivities, and expected reactions and color changes with the selected tubes. **(photo 4)**
5. Perform a pump test as specified by the manufacturer to ensure there are no leaks. A typical test is to use an unopened tube to hold the negative pressure inside the pump for a given period of time. **(photo 5)**
6. Assemble the tubes and other sampling apparatus prior to entering a contaminated area. **(photo 6)**
7. Perform air sampling in accordance with the directions on the tube, using an unreacted tube as a control to view any color changes. **(photo 7)**
8. Discard used tubes in accordance with manufacturer's guidelines.

Sampling

If air monitoring provides no information on the identity or hazard class of the unknown, responders may collect a sample to conduct field tests of the material, or send the sample to a lab for further analysis. Samples may also be collected as evidence if criminal activity is potentially involved. Although solid and liquid samples are most common, samples of gases and biological materials can also be collected.

Examples of instruments and systems used by responders for analyzing samples include the following:

- Locally developed and commercial chemical identification kits (e.g., HazCat™ Chemical Identification System). These systems rely on a series of field chemical tests that follow a predefined logic sequence to identify unknown liquid and solid materials.
- Fourier-Transform Infrared Spectrometry (FT-IR). This technology records the interaction of the infrared spectra with chemical samples, measuring the frequencies at which a sample absorbs light and the intensities of the absorptions. Since chemical functional groups

Skill Drill 7-3A

Demonstrating Field Maintenance and Testing Procedures Using a Multi-Gas Meter with Electrochemical Sensors

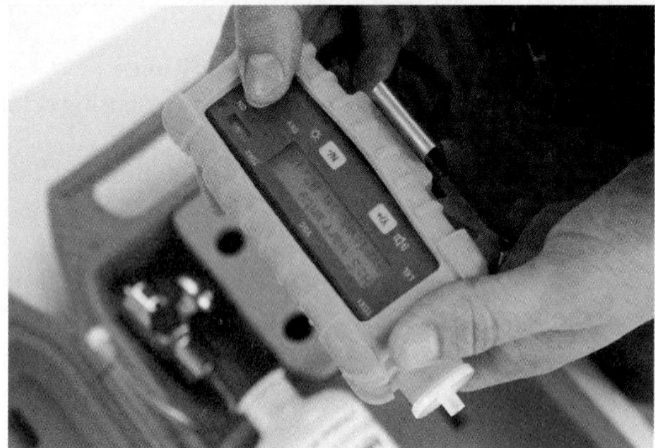

1 Turn the unit on and let it warm up (usually 5 minutes is sufficient).

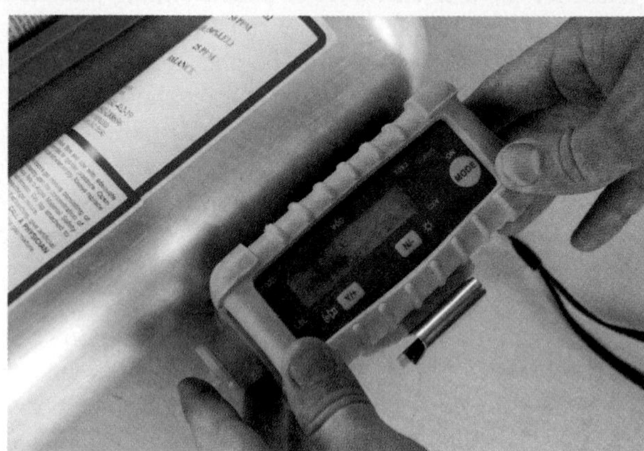

2 Perform a test on the pump.

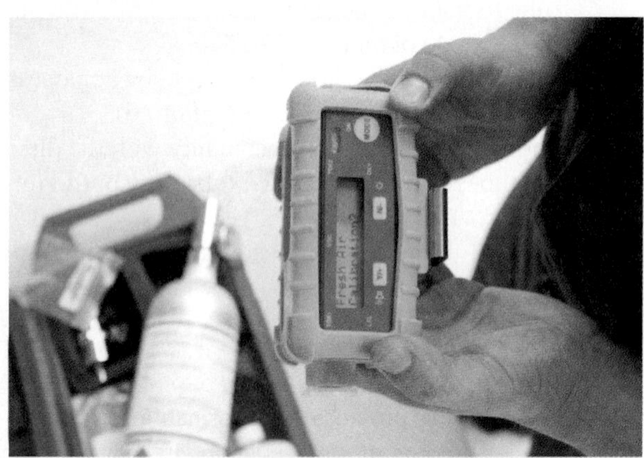

3 Perform a fresh air calibration and 'zero' the unit.

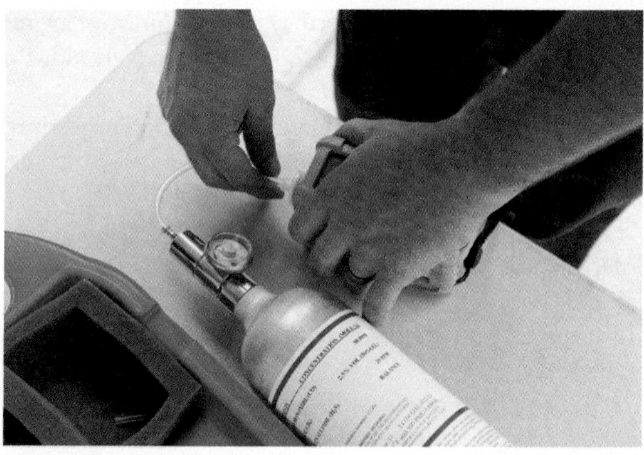

4 Perform a bump test.

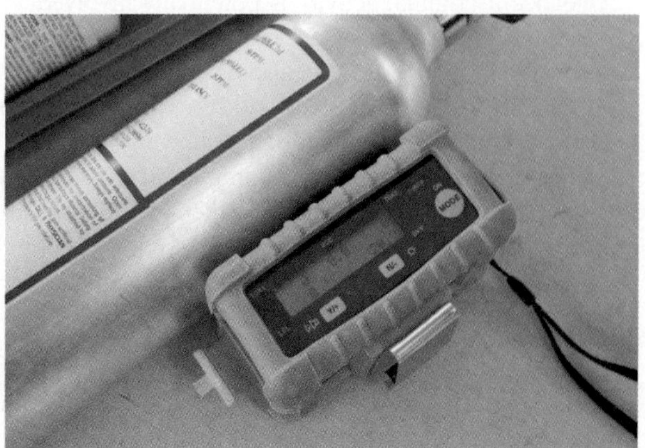

5 Allow the device to reset, or return to fresh air calibration state.

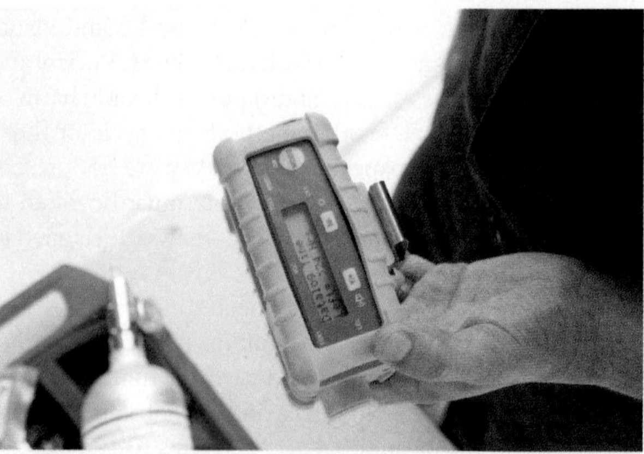

6 Review other device functions such as screen illumination, data logging if available, low battery alarm, etc.

Skill Drill 7-3A

Demonstrating Field Maintenance and Testing Procedures Using a Multi-Gas Meter with Electrochemical Sensors (*Continued*)

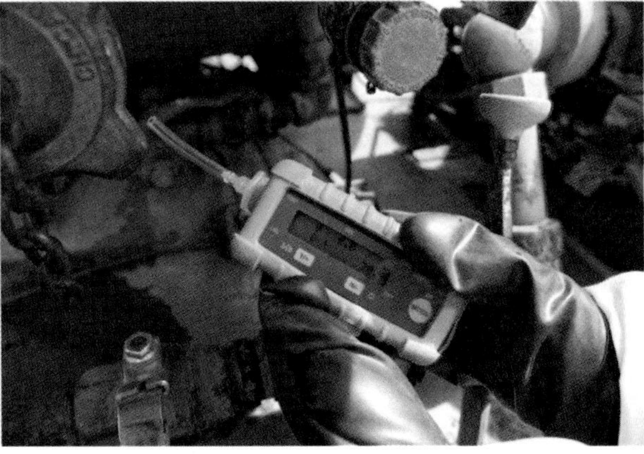

7 Review decontamination procedures. Carry out monitoring and detection.

are known to absorb light at specific frequencies, the FT-IR technology can allow for the specific identification of liquid and solid samples. FT-IR instruments, such as the TravelIR HCI™, HazMatID™, and Avatar ESP, can provide rapid identification of hazardous materials, explosives, precursors, drugs, and other common liquid and solid materials. Although they cannot identify biological materials, they can detect the presence of protein material.

- Biological detection systems currently used in the field rely on responders acquiring a sample and then subjecting the sample to some testing process. These range from hand-held immunoassays (HHA) to a polymerase chain reaction (PCR) test, which compares the DNA of the sample to those in a library. Although the bio testing field is rapidly changing, at the present time field biodetection systems are not a definitive tool to determine the presence of any biological pathogen or toxin and should not be used to make decisions on patient management or prophylaxis. However, they can be useful in assessing credibility of the threat. For definitive results, field tests must be confirmed through lab analysis.

Sampling Considerations

Responders are often required to respond to incidents involving abandoned drums, as well as illicit or clandestine laboratory operations. Containers at these scenes usually have few markings, may be stored in incompatible containers, and may be stressed or damaged. In worst-case scenarios, the container may be undergoing an internal reaction, thereby increasing the risk of a container breach or explosion.

The following are some basic considerations that are applicable at most scenarios where samples may be collected:

- Personal safety and avoiding contamination of samples are key principles in any sampling operation.
- Collect the samples from an upwind position.
- Wide mouth containers should be used when collecting liquid samples, as possible.
- When collecting a sample, record the type of container the sample is being drawn from, any container markings, and any reactions or other relevant information concerning the site.
- Once the sample is properly collected, take it to a safe testing location in the warm zone.
- Any materials and equipment used for evidence collection must be certified "clean," kept sealed, and only used one time to collect each sample. Whenever possible, use precleaned and individually prepackaged containers and sampling tools. Avoid using sampling containers and tools of an unknown origin.
- If a sample may become part of a criminal or regulatory investigation, chain of custody procedures must be followed and documented.

Evidence Considerations

Environmental crimes and criminal events involving hazardous materials will bring additional challenges to the sampling process. Although there are some similarities, there are also distinct differences that must be recognized and understood. Sampling is

Skill Drill 7-3B

Demonstrating Field Maintenance and Testing Procedures Using Colorimetric Tubes

1 Determine the need for using colorimetric tubes.

2 Select the appropriate tubes for the mission.

3 Collect all parts of the tube sampling system, including pumps, tubes, extension hoses, etc.

4 Review the directions for each selected tube. Always take an unopened, unreacted, tube to use as a control for tubes used in air sampling.

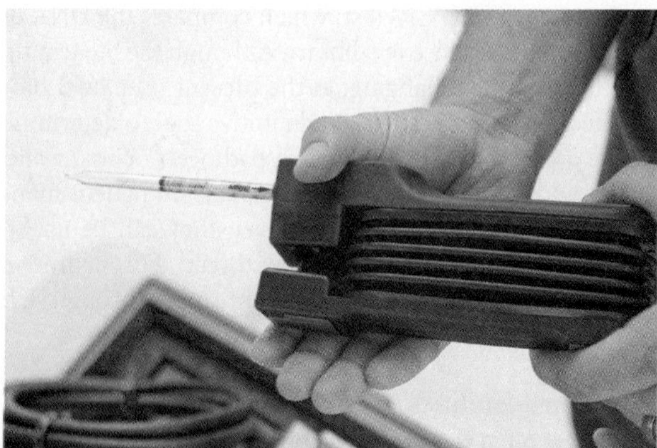

5 Perform a pump test.

Skill Drill 7-3B

Demonstrating Field Maintenance and Testing Procedures Using Colorimetric Tubes (*Continued*)

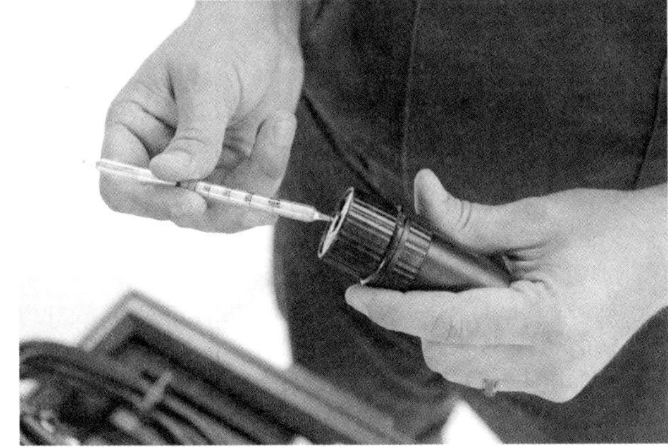

6 Assemble a sampling kit with the needed equipment including the proper tools to open both ends of the selected tubes.

7 Perform air sampling.

used to assess the level of risk, determine material hazards and characteristics, and make emergency response and clean-up decisions. Evidence collection is a process that involves the collection of material that will be used in a legal proceeding. Issues such as scene management, chain of custody, and minimizing any contamination or cross-contamination of samples will become critical.

When collecting evidence samples, additional concerns include the following:

- A sampling plan should be established initially that clearly outlines the type of sampling (e.g., systematic, random), sample team members and their specific tasks, sample points, cross-contamination issues, and other related health and safety issues.
- Sampling tools and gloves must only be used one time for each sample. Sampling tools should not be reused or recycled. Bio samples should be collected in a sterile container, while chemical samples should be collected in certified "clean" containers.

Responder Tip

As a general rule, samples collected for product identification during emergency response operations should not be used for evidentiary purposes. Collect a separate sample for evidence.

- Collected samples should be transported or stored away from unused tools, equipment, and other chemicals to avoid the potential of cross-contamination.
- All sample containers should be clearly labeled with the appropriate identifying information. Labels should include sample identifier, type of material (e.g., powder, liquid), date obtained, and the investigator's name or initials. If known, the nature of the hazard and any special handling procedures may also be included. If possible, it may be easier to pre-label the sample containers in the cold zone.
- Control blanks should be provided as part of the sampling process to later assess cross-contamination and systematic contamination issues. Control blanks are the same container or material that has been used for evidence sample, but that has not been exposed at the scene.
- Sample containers certified as "clean" will have a letter stating that they are cleaned to some specification. This letter should be forwarded with the respective sample and container.
- All sample collection should be incorporated into the overall evidence collection process, following appropriate logging and documentation procedures.
- Protect evidence samples from heat and direct sunlight, and keep as cool as possible. Basic rules of thumb when handling evidence samples are as follows:
  - If the sample is cold, keep it cold but don't freeze it.
  - If the sample is warm, make it cold but don't freeze it.

Scan Sheet 7-C

Joint Hazard Assessment Teams (JHATs)

The threat of an attack on a high-profile individual or large numbers of people gathered for major public assembly events has become a significant problem for public safety agencies. Threats must be evaluated from the perspective of:

1. The credibility of the threat, and
2. The technical capability of the individual or group to do what they are threatening to do.

Threats made against events involving large numbers of people can't be ignored and must be evaluated using a combination of intelligence analysis, real time surveillance and monitoring, and by having trained eyes on the problem. While most threats never materialize, some may involve a suspicious package, an unknown powder on the floor, or a strange odor in the area; all of these need to be evaluated. A balance must be struck in the reaction to a threat—overreacting to a perceived threat could result in disruption of the event or crowd panic; no response to a real threat could result in loss of life.

The JHAT Concept

Joint Hazards Assessment Teams (JHATs) have been used successfully by the Federal Bureau of Investigation (FBI) and U.S. Secret Service (USSS) to manage and respond to threats at major public events. JHATs are a proven concept at the federal level and the same approach has many applications at the state and local level for special events that draw large crowds. Some examples of JHAT implementation at the local level include parades, major sporting events, farm shows and fairs, and rallies or protests.

The JHAT concept is an effective response tool for public safety agencies challenged by major events that draw large numbers of people to a concentrated area where egress, response times, and crowd control could become a concern.

JHATs are defined as organized teams that provide technical support for planning and responding to threats at special events. JHATs are normally comprised of specialized personnel from different agencies and disciplines, such as fire, law enforcement, bomb squad, and public health professionals. JHATs may also include scientists and specialists from private organizations or contractors, depending on the type of event or the possible threats that could occur during the event.

JHAT Mission

The goal of JHATs is to bring together a highly-skilled cohesive response team that provides the IC with technical assessment, operational guidance, and communications support for identified threats that may occur at a major event. The objectives of JHATs are to:

- Provide technical support to the IC
- Provide a rapid technical assessment of suspicious situations or items
- Analyze 911 calls for a WMD nexus
- Assist the IC with the coordination of specialized resources
- Most importantly, provide the IC with the ability to de-escalate an incident when a larger response is unnecessary and may disrupt the event

JHATs function within the predesignated incident command structure that has been established for the special event. They function in a scientific, technical, and operational support capacity to the IC to assist in decision-making concerning suspicious packages, suspicious powders, or other types of problems which may present a hazmat/WMD threat. JHATs can also provide scientific and technical liaison between federal law enforcement national assets and the local emergency response community. Finally, they may also provide support to strategic command centers and support crime scene operations.

The key advantage of a JHAT is described in its name:

- **JOINT**—The team is composed of multiple agencies with different charters and responsibilities. The legal authority and jurisdiction of the fire department, the police department, and the health department are unique and different. When you bring representatives from multiple agencies together in a "joint" operation you bring a lot of authority and expertise to deal with specialized problems.
- **HAZARDS**—Large public events can present a wide range of anticipated hazards like the possibility of a gas leak, a strange odor, a smell of smoke in the area, a suspicious package, or a bomb threat. Evaluating

Scan Sheet 7-C (continued)

these hazards or threats requires various disciplines that usually reside in different agencies.

- **ASSESSMENT**—The strength of a JHAT is that it brings together significant technical assessment capability on the front end of a special event to ensure that the level of response matches the assessed level of the problem and threat. During the event, the JHAT provides rapid on-site assessment of hazards.
- **TEAM**—The JHAT is formed in advance of an event so that members are trained, organized, and function as a team. The different expertise and experience of the individual JHAT members brings strength to the team. JHAT brings together the expertise of fire, hazardous materials, explosives, and public health professionals into one cohesive team.

It is important to remember that JHATs are not "mini-hazardous materials teams"; they are a small group of technical advisers to the IC. The JHAT is designed to move and act quickly to a variety of emergency situations.

JHAT Lessons Learned

Lessons learned from actual JHATs deployments at national-level events include:

- Plan for worst-case scenarios. Special events can produce a variety of challenges, including extraordinary crimes, violence by protesters, a terrorist attack, street closures, security searches of large numbers of people, and challenges to constitutional rights including freedom of speech and assembly.
- Be cautious about fielding unproven technology that has not already been validated in the field. Like major disasters, major events can bring out engineers, scientists, and even politicians touting new gear that may be useful in the future, but is still in the beta test phase or untested in field operations. High risk, high profile missions usually do not make good test platforms.
- Don't allow agencies to do independent hazmat/WMD searches that are not part of the JHAT plan or have an operational role. All hazmat/WMD operations must be coordinated by and through the JHAT.
- There is rarely a requirement to have personnel randomly walking around with detectors. Having serious looking people walking around the crowd in tactical uniforms with impressive detection equipment only draws attention to you and may raise concerns that there is a threat that does not exist.
- Make sure you have a plan when the "alarm" sounds. Based on current intelligence for the day of the event, write your mission statement and goals and post them for everyone to see. Five to seven strategic goals are all that you need for an event.
- Have clearly stated objectives for each operational period. Some events last for only one operational period, while others may be 1 to 2 weeks in length.
- Complete the IAP for each operational period, and develop a written site safety plan. Ask yourself if the plan is safe, legal, and within accepted standards and practices. While your pre-event plan may be volumes, your IAP should be simple and easy to understand across the entire organization.

For additional information, see the following reference:

Hawley, Chris, Gregory G. Noll and Michael S. Hildebrand, "The Need for Joint Hazard Assessment Teams," *Fire Engineering Terrorism Supplement* (September 2009).

    - If the sample is frozen, keep it frozen.
    - If the sample is dry, leave it dry.
    - If the sample is wet, leave it wet.
- Before any evidence is shipped or transported to a lab, ensure that it has been screened for fire, corrosive, toxic, and radioactive hazards.
- Chain-of-custody must be maintained throughout the course of the event. Documentation, such as a chain-of-custody form, should be used to track the movement of the evidence. The only time evidence may be left alone is when stored in a secured location designed for holding evidence (e.g., police department evidence room or locker). Remember—the goal is to prosecute the Bad Guy and not leave any holes for the legal defense team to exploit.

Sampling Equipment

The following general supplies and equipment are commonly used for collecting samples:

- Nonsparking bung wrench.
- Glass tube or disposable polypropylene/PVC bailer.
- Coliwasa waste samplers (composite liquid waste sampler)—used for taking samples at all levels of a liquid container.
- Nonsparking sample pole, extendable to 10 feet.
- Glass and plastic sample cups and bottles. Glass should be used for chemical samples, while plastic may be used with bio samples.
- Plastic bags—positive seal is preferred with evidence tamper-proof bags.
- Bomb sampler or weighted bottle sampler.

In addition, the following tools and equipment are likely to be used when collecting specific forms of materials:

- *Liquid sampling*—Transfer pipettes, syringe, and tubing.
- *Solid sampling*—Stainless steel spoons, scoops, scalpels, and spatulas.
- *Wipe sampling for residues*—Nylon or dacron swabs, transfer swabs, cotton or synthetic gauze, and forceps are used to collect samples. Isopropanol, deionized sterile water, or other appropriate prep solutions may be used. Do not dip the swab into the prep solution; bring the solution to the swab.

Sampling Methods and Procedures

Accepted methods for collecting samples for various scenarios include the following:

- **Powders**—Powders can often be found when encountering illicit laboratories, as well as suspicious powder/package incidents. One of the greatest challenges when dealing with powders is collecting a sample to facilitate field testing, while ensuring that a sufficient amount of the sample is preserved and packaged for laboratory confirmation within the Centers for Disease and Control's Laboratory Response Network (LRN).

  Responders should be familiar with ASTM E2458-10—*Standard Practices for Bulk Sample Collection and Swab Sample Collection of Visible Powders Suspected of Being Biological Agents from Nonporous Surfaces.* These practices should be performed after (1) a risk assessment is conducted; (2) field screening has been conducted as defined by the FBI-DHS-HHS/CDC Coordinated Document for explosive, radiological, and other acute chemical hazards; and (3) a visible powder is deemed a credible biological threat. The ASTM E2458-10 Standard addresses:

    - Sample Collection Method A covers the bulk collection and packaging of suspicious visible powders that are suspected biological agents from solid nonporous surfaces, while Sample Collection B covers swab sampling of such powders.
    - Guidance for the packaging and transport of such powders to ensure compliance with all appropriate federal regulations regarding biosafety and biosecurity.

  The ASTM E2458-10 sampling guidance supports and should be used in conjunction with ASTM E2770-10—*Standard Guide for Operational Guidelines for Initial Response to a Suspected Biothreat Hazard.*
- **Drums**—When opening a drum to collect a sample, use a nonsparking bung wrench. Manual drum opening operations should be performed only with structurally sound drums. If the contents are shock sensitive, reactive, or explosive, the operation may need to be treated as an explosive device incident. Destructive nonsparking drum openers that can be operated remotely may be required.

  When dealing with flammable liquids, bung caps should be unscrewed very slowly, at approximately .25 inches per movement. The bung area should be monitored with a combustible gas indicator both prior to and during opening.

  A hazmat drum thief, coliwasa, or polyethylene bailer is often used to collect drum samples. A poly hazmat thief is a simple, cost-effective, and efficient tool and is the most widely used implement for drum sampling. Drum contents can be sampled by inserting the tube into the drum and then removing it, thereby collecting a cross section of the contents. This is similar to taking soda out of a glass by holding a finger over the end of a straw.

  The coliwasa is designed to collect a sample from drums and other containerized wastes. Some coliwasa kits have coupler sets that allow different coliwasa sets to be combined to increase their length. Drum contents may layer or stratify after sitting for extended periods. The bottom, middle, and top samples within the sampling tube should be placed into separate containers for analysis.
- **Sumps and wells**—Disposable polypropylene bailers were made for monitoring groundwater wells but also work well for retrieving samples from other static water sources. The open top bailer has a ball at the bottom that allows liquids to enter but prevents the liquids from escaping. Bailers can be lowered into the water by a suspension cord.
- **Puddles**—Use a turkey baster or a small plastic/glass cup attached to the end of a pole to collect a sample from the puddle.

Scan Sheet 7-D

Preservation of the Hazmat/WMD Incident Scene

Incident Management Considerations

Criminal events involving the use of hazardous materials and WMD create additional incident management challenges. Once the priorities of life safety and incident stabilization are addressed, evidence of the crime must be collected properly to assist in the prosecution phase.

These incidents will pass through four distinct phases of management: Tactical, Operational, Crime Scene, and Remediation. The establishment of an Incident Command System (ICS) organization early in the incident will assist in the management and facilitation of these phases.

Tactical Phase. Involves the removal of any hostile threats from the hazmat environment. Law enforcement tactical teams and SWAT personnel are typically responsible for the completion of this phase.

Operational Phase. A critical step in this phase is the "render safe" of all explosive devices and anti-personnel devices (aka booby traps). Once the scene is rendered safe, other emergency response operations can be initiated, including rescue, medical treatment and transport, and hazard stabilization.

Although samples may be collected as part of the risk assessment process, it should be stressed that a sufficient amount of any samples collected must be preserved for later collection by law enforcement as part of the crime scene processing phase. Non-law enforcement responders should NOT gather samples with the intent of those samples becoming evidence. Law enforcement officials may place samples collected by emergency responders for the purposes of public safety into evidence at a later time, or may direct responders to collect isolated evidence (e.g., hoax powder letters, envelopes).

All public safety samples collected must be field-screened prior to their admission to a Laboratory Response Network (LRN) laboratory. The field screening process is focused on classification rather than identification and should be performed prior to packaging. The field screening process includes:

- Clearance for explosive devices by an accredited bomb technician
- Radiation survey (alpha, beta, and gamma)
- Flammability survey using a combustible gas indicator (CGI)
- Oxygen level survey using an oxygen meter
- Volatility survey using a photoionization detector (PID)
- Corrosives survey with pH paper (liquids only)

Crime Scene Phase. Although law enforcement coordinates this phase, emergency responders may be requested to support evidence recovery operations with hazmat entry support, lighting, etc. In addition to the hazmat/WMD issues that may be present, investigators will also collect traditional forensic evidence. Law enforcement personnel and agencies follow specific guidelines regarding the acquisition of warrants, affidavits, hazmat evidence, and traditional forensic evidence.

Remediation Phase. Addresses environmental health and safety activities to remediate any remaining hazards after the crime scene is processed. These activities are typically coordinated through the appropriate local, state, or federal environmental agency.

Agencies and individuals seeking additional information on the hazmat/WMD training requirements for those assigned to perform evidence preservation, sampling, and collection should consult NFPA 472, Section 6.5—*Mission Specific Competencies: Evidence Preservation and Sampling.*

- **Slick on top of water**—Use a piece of loosely woven fiberglass cloth attached to a string. Place the fiberglass on top of the water; the floating material will collect on the cloth. Two or three samples may be necessary to obtain a sufficient amount for testing.
- **Heavier than water unknowns (from underwater)**—Lower a bomb sampler or a weighted bottle sampler into the water and open it when the bomb enters the layer to be sampled.
- **Deep holes**—Lower a glass or plastic bottle on a string or fishing line into the hole to collect the sample.
- **Dry piles of solids**—Surface samples from piles can be collected with a plastic or stainless steel scoop or cup attached to the end of a sample pole. This allows for sample collection from a safe distance. A lab spoon, plastic spoon, or shovel will also work, but avoid the use of devices plated with chrome or other materials.

Sample Collection

An important issue to consider when collecting samples is the composition of the equipment used in the sampling process (e.g., vials, jars, containers, etc.). Equipment composed of the same chemicals or material(s) as the samples being collected may influence and taint the sample. Responders should consider using glass, Teflon®, or stainless steel sampling equipment when looking for trace organic compounds, and to using Teflon®, plastic, or glass equipment when sampling for trace metals.

This skill drill assumes that emergency responders have already determined the appropriate level of personal protective clothing and equipment required for entry operations, and have characterized the hazards that are present. Regardless of the physical state of the unknown material (liquid, solid, gas), a field screening for possible hazards of the involved materials should be completed first.

To collect field samples properly, including liquid, solid, and gas samples, follow the steps in **Skill Drill 7-4**:

Skill Drill 7-4A: Collecting a Liquid Sample

1. To collect a liquid sample, assemble sufficient equipment to meet the quantity requirements of the AHJ. **(photo 1)**
2. Prior to completing any sample collection, remember that the appropriate level of PPE should be worn. At a minimum, compatible gloves and respiratory equipment should be worn. **(photo 2)**
3. Using a pipette, syringe, or a syringe with tubing, slowly draw the liquid until the quantity required is met. **(photo 3)**
4. Once quantity is met, properly dispose of pipettes/syringes/tubing. Close the container tightly and place it in a bag. **(photos 4a, b and 5)**

Skill Drill 7-4B: Collecting a Solid Sample

1. To collect a solid sample, assemble sufficient equipment to meet the quantity requirements of the AHJ. **(photo 1)**
2. Prior to completing any sample collection, remember that the appropriate level of PPE should be worn. At a minimum, compatible gloves and respiratory equipment should be worn. **(photo 2)**
3. Using a spatula, scoop, or other tool, remove the solid powder or chunks of material from the area to be collected. **(photo 3)**
4. If the chunks are too big, use a scalpel or spatula to break and then place in the collection container slowly until the quantity required is met. Do not attempt to overfill the sample container; if additional quantities of the sample are needed, use additional collection containers until required quantity is met. **(photo 4a, b)**
5. Properly dispose of all collections tools. Close the sample container tightly so that it may be decontaminated before leaving the hot zone.

Skill Drill 7-4C: Collecting a Gas Sample

1. To collect a gas (air) sample, assemble sufficient equipment to meet the quantity requirements of the AHJ. This can include specialized collection tools such as an electric or manually operated pump, extension hoses, telescoping extension pole, probes, and various sizes of polyvinylidene fluoride (PVDF) or similar gas sample collection bags.
2. Prior to completing any sample collection, remember to wear the appropriate level of PPE. At a minimum, compatible gloves and respiratory equipment should be worn.
3. Employ the use of a pump or vacuum device to draw the air sample into a sample collection bag. Do not attempt to overfill the sample collection bag; use additional collection bags until required quantity is met.
4. Once the quantity is met, the sample collection bags will be submitted to a laboratory for analysis.

Skill Drill 7-4A

Collecting a Liquid Sample

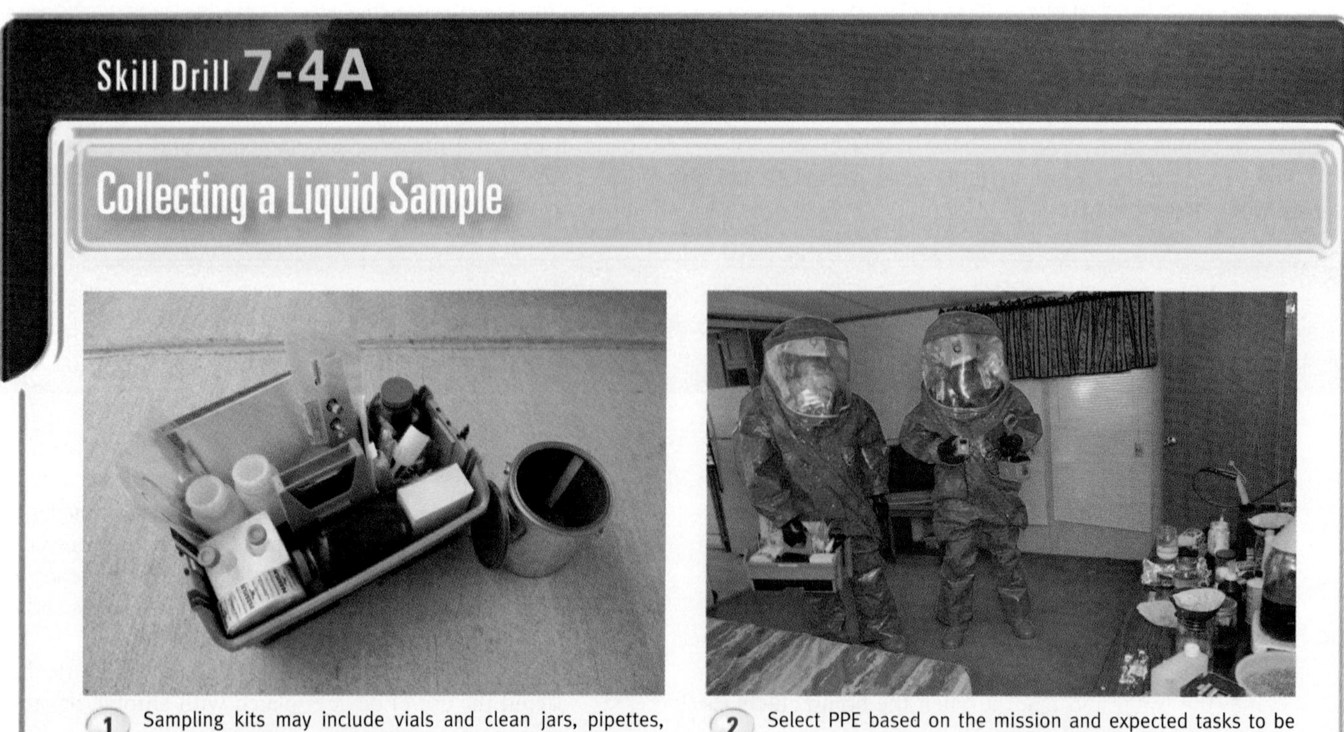

1 Sampling kits may include vials and clean jars, pipettes, syringes, and tubing. These can be either glass or plastic.

2 Select PPE based on the mission and expected tasks to be performed.

Skill Drill 7-4A

Collecting a Liquid Sample (*Continued*)

3 Do not attempt to overfill the sample container. If additional liquid is needed, use additional draws until required quantity is met.

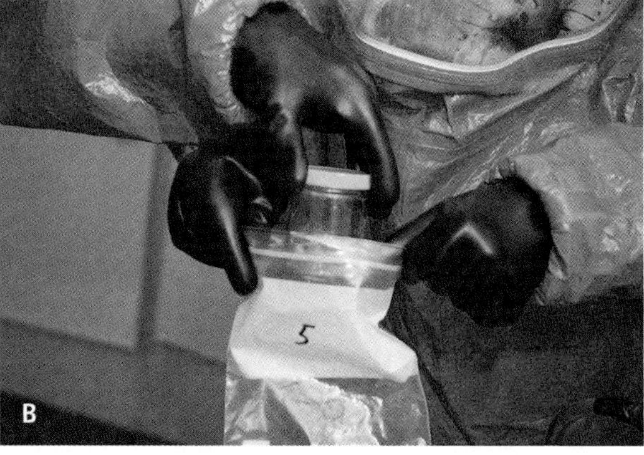

4 Make sure to close the sample container tightly so that it may be decontaminated before leaving the hot zone, then place it in a bag.

Skill Drill 7-4B

Collecting a Solid Sample

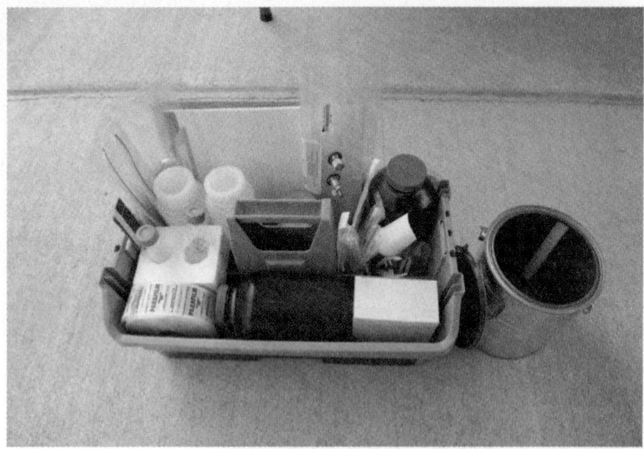

1. Solid sampling kits may include material collection tools like spatulas, scoops, swabs (cotton or synthetic) and forceps. Additional equipment can include glass or plastic vials, jars, and bags.

2. Prior to completing any sample collection, make sure you are wearing the appropriate level of PPE.

3. Using the spatula, scoop, or other tool, remove the solid powder or chunks of material from the area to be collected.

B

4. If the chunks are too big, use a scalpel or the spatula to break and then place in collection container slowly until the quantity required is met.

Managing Hazard Information

In the process of evaluating risks, response personnel will be gathering and updating data and information from various sources. Not only must responders assess the accuracy and validity of the information being collected, but equally important is the management of that information. Responders have sometimes been overwhelmed by the volume of hazard information collected during the course of an incident and have been unable to effectively manage the volume of data.

To minimize these problems and concerns, responders should prioritize their information requirements—what do I need to know right now, in 1 hour, and in 8 hours? For example, accurate hazard information, protective clothing and equipment recommendations, and initial control measures need to be determined as soon as possible. However, equipment disposal and clean-up information are not needed in the initial stages of the incident and can be delayed until later.

Many responders rely on data forms, checklists, and tactical worksheets to ensure all information requirements have been prioritized and addressed. Regardless of their exact design and format, some logical, functional, and user friendly system must be used Figure 7-4.

Evaluating Risks

Risk evaluation is the most critical task performed by emergency responders. Failure to understand and perform the risk evaluation process correctly can lead to scenarios where responders ultimately take great risks in situations where there may be little or no gain. In simple terms, the actions taken by emergency responders do nothing to favorably change or influence the potential outcome.

To understand the risk evaluation process at a hazmat incident, look at it from a systems perspective. In any system, there are input factors that will go through some evaluation process, and result in outputs. The input factors that must be considered at a hazmat incident include:

- Hazardous material(s) involved
- Type of container and its integrity
- Environment and location where the incident occurred
- Resources and capabilities of emergency responders

By understanding the potential behavior of the hazardous material and its container, responders can then estimate likely outcomes and develop an IAP designed to produce a favorable outcome.

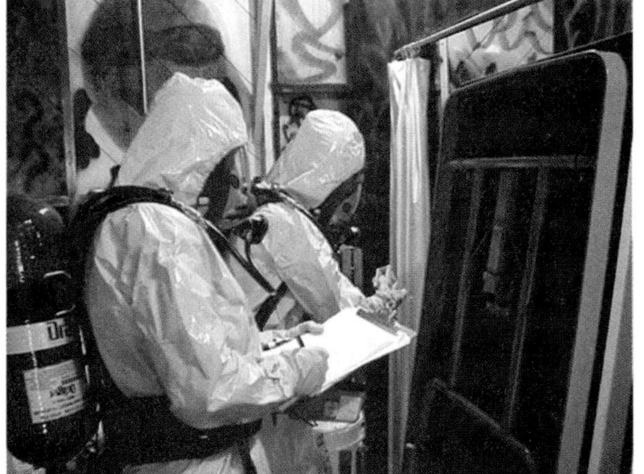

Figure 7-4 Responders use a variety of data forms, checklists, and tactical worksheets to compile and manage incident data and information.

Basic Principles

All emergencies consist of a series of events that occur in some logical sequence. For example, a fire in a confined room or structure will move from the incipient to the free-burning phase and eventually to the smoldering phase, if there is no fire department intervention. We refer to this process as the stages of fire. Similarly, a hazmat incident will follow a logical sequence of events arriving at some stabilized outcome if there is no intervention by emergency responders. Figure 7-5 describes the Hazardous Materials Emergency Model and its components.

The overall objective of emergency responders at any emergency is to favorably change or influence the outcome. Direct outcomes are typically stated as fatalities, injuries, property and environmental damage. Indirect outcomes include systems disruptions (e.g., water, transportation, utility), damaged reputations, and residual fears.

To determine whether to intervene, responders must first estimate the likely harm that will occur without intervention. Simply, what will happen if you do nothing? This analysis is one of the fundamental tenets of a risk-based response. The decision to intervene requires responders to (1) visualize the likely behavior of the hazardous material and/or its container, along with the likely harm associated with that behavior; and (2) describe the outcome of that behavior.

To visualize likely behavior, address five basic questions:

1. **Where** will the hazardous material and/or its container go when released?
2. **How** will the hazardous material and/or its container get there?
3. **Why** are the hazardous material and/or its container likely to go there?
4. **What harm** will the hazardous material and/or its container do when it gets there?
5. **When** will the hazardous material and/or its container get there?

When answering these questions, recognize and understand the factors that will affect hazmat behavior, including the following:

- Inherent properties and quantities of the materials involved (i.e., toxicity, flammability, reactivity, etc.)
- Built-in design and construction features of the container (thermal insulation, pressure relief devices, fixed water spray systems, etc.)
- Natural laws of physics and chemistry, as these will influence dispersion patterns and where the product will travel once it is released from its container

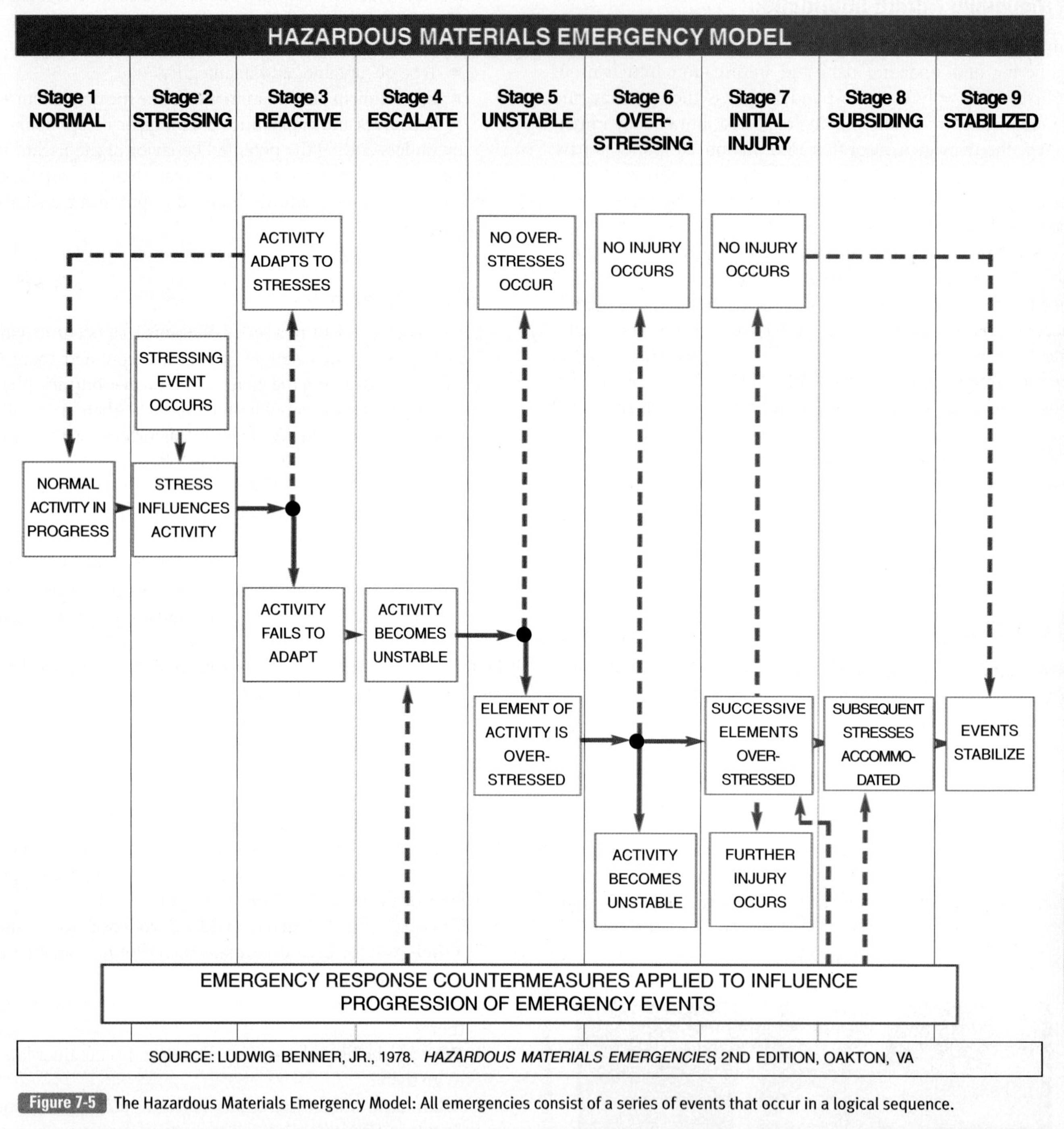

Figure 7-5 The Hazardous Materials Emergency Model: All emergencies consist of a series of events that occur in a logical sequence.

- Pertinent environmental factors (i.e., terrain, weather and atmospheric conditions, wind direction and speed, and the physical surroundings [i.e., rural vs. urban]).

Behavior of Hazmats and Containers

All hazmat releases will follow a logical sequence of events, regardless of the hazard class involved. The concept of events analysis is a useful tool to visualize hazmat behavior and estimate what is likely to occur. Events analysis is defined as the process of breaking down complex actions into smaller, more easily understood parts. It helps responders to (1) understand, track, and predict a given sequence of events; and (2) decide when and how to change that sequence.

An easy way to visualize hazmat behavior is by using the General Hazardous Materials Behavior Model or GHBMO, pronounced "gebmo." Originally developed by Ludwig Benner of the National Transportation Safety Board (NTSB) and published in 1978, the GHBMO is an excellent tool for understanding and predicting the behavior of the container and its contents at a hazmat incident **Table 7-3**. Understanding and applying the GHBMO is the foundation of hazardous materials risk-based response.

Stress Event

Under normal conditions, hazardous materials are controlled within some type of container or containment system. Containment systems can range from nonbulk containers such as bags, bottles, and drums, to bulk containers, including cargo tank trucks, pressurized storage vessels, and chemical reactors. For an emergency to occur, either the container or its contents must first be disturbed or stressed in some fashion.

Stress is defined as an applied force or system of forces that tend to either strain or deform a container (external action) or trigger a change in the condition of the contents (internal action). It is important to recognize that this stress can affect the container and/or its contents.

Three types of stress—thermal, mechanical, and chemical—are common. Less likely, though still possible, are etiological and radiation stresses. These stressors may be present alone or in combination with each other.

- *Thermal Stress*—Generally associated with hot or cold temperatures and their effects upon the container or its contents. Examples include fire, sparks, friction or electricity, and ambient temperature changes. Extreme cold, such as with liquefied gases and cryogenic liquids coming in contact with certain metals, may also act as a stressor. Clues of thermal stress may include the operation of pressure relief devices or the bulging of a container.
- *Mechanical Stress*—The result of a transfer of energy when one object physically contacts or collides with another. Indications would be punctures, gouges, scores, dents, or tears in the container. A container may also be weakened but have no visible signs of potential release. Assessment of damaged pressurized containers will be discussed later in this section.
- *Chemical Stress*—The result of a chemical reaction between two or more materials. Examples include corrosive materials attacking a metal, the pressure or heat generated by the decomposition or polymerization of a substance, or any variety of corrosive actions. For example, a 1988 explosion of a railroad tank car containing methacrylic acid in Houston, Texas was caused by iron contamination and a low inhibitor level.

All three types of stress can also occur in combination with each other. The combination of thermal (i.e., sun and ambient heating) and mechanical stresses (i.e., rail burns) at the Waverly, Tennessee railroad incident in 1978 resulted in 16 deaths and 43 injuries. Fatalities included several firefighters,

> **Responder Tip**
>
> Always look for stressed containers. If the hazmat has escaped before your arrival, look for more than one stressed container.

Table 7-3 General Hazardous Materials Behavior Model©

Event					
Stress	**Breach**	**Release**	**Engulf**	**Impinge**	**Harm**
Event Categories					
Thermal	Disintegration	Detonation	Cloud	Short term	Thermal
Radiation	Runaway cracking	Violent rupture	Plume	Medium term	Radiation reactive
-	Attachments opening	Rapid relief	Cone	Long term	Asphyxiation
Chemical	Punctures	Leak	Stream	-	Chemical
Mechanical	Splits or Tears	Spill	Irregular deposit	-	Etiologic Mechanical
Event Interruption Principles					
Influence Applied Stresses	**Influence Breach Size**	**Influence Quantity Released**	**Influence Size of Danger Zone**	**Influence Exposure Impinged**	**Influence Severity of Injury**
Redirect impingement	Chill contents	Change container position	Initiate controlled ignition	Provide shielding	Rinse off contamination
Shield stressed system	Limit stress level	Minimize pressure differential	Erect dikes or dams	Begin evacuation	Increase distance from source
Move stressed system	Activate venting devices	Cap off breach	Dilute	-	Provide shielding

Source: Ludwig Benner, Jr.

the fire chief, and the police chief. Always look beyond the obvious for damaged or stressed containers.

If the potential types of stressors can be identified prior to an event, a container can be designed and constructed to minimize their impact in the event of an emergency. For example, head shields and shelf couplers are now installed on railroad tank cars transporting hazardous materials to minimize the potential for a mechanical stressor in the event of a train derailment, while jacketing is installed to minimize the impact of a thermal stressor. Likewise, installed water spray systems are used within the petrochemical industry to minimize thermal stress upon exposed bulk pressure vessels and containers.

An obvious question often asked by responders is "how much product and/or pressure is within a container?" Tools and techniques that can be used to answer this question include:

Determining Amount of Product

- Shipping papers and related documents—will show the type (e.g., drum, cylinder, carboy, etc.), number of, and weight of any hazmat containers. Examples include bills of lading, railroad consist or train list, airbills, and dangerous cargo manifest.
- Fixed gauging devices—can be found on both liquid and gas containers. Some examples include fixed level gauges found on bulk liquid and gas storage tanks, rotary gauges commonly used on LPG and anhydrous ammonia cargo tank trucks, and gauging rods located within the protective housing of pressurized railroad tank cars transporting liquefied gases.
- Weigh small nonbulk cylinders—can be undertaken for some liquefied gases in nonbulk cylinders, such as 20 or 30 lb. propane cylinders and 100 or 150 lb. chlorine cylinders.
- Infrared cameras—will detect the temperature difference between the liquid phase and the vapor phase, and clearly show the liquid level within the container **Figure 7-6**.
- Visible frost line on liquefied gas containers (e.g., propane, anhydrous ammonia), especially if the contents have auto-refrigerated.

Determining Container Pressure

- Fixed pressure gauge—permanently attached to some nonbulk and bulk pressurized containers (e.g., MC-331 cargo tank trucks).
- Attach a pressure gauge to sample lines, gauging device, fittings, etc. of a pressurized container.
- Determine the temperature of the product and use a vapor pressure/temperature conversion chart. Vapor pressure/temperature graphs are available from the *Compressed Gas Handbook or the Matheson Gas Data Book*, as well as from the shipper or manufacturer of the material.

Breach Event

If a container is able to adapt to the stress, the incident will be stabilized at that point; however, when the container is stressed beyond its limits of recovery (i.e., its design strength or ability to hold contents), it will open up or **breach**. Different containers breach in different ways—glass bottles shatter, bags tear, pressure cylinders split, and drums tear.

There are five basic types of breach behaviors:

1. *Disintegration*—The total loss of container integrity. Although usually associated with explosives, it can also be easily visualized as the shattering of a glass bottle or carboy.
2. *Runaway cracking*—Occurs in closed containers such as liquid drums or pressure vessels. A small crack in a closed container may suddenly develop into a rapidly growing crack that encircles the container. As a result, the container will typically break apart into two or more pieces in a violent manner. Linear cracking is commonly associated with catastrophic boiling liquid expanding vapor explosion (BLEVE) scenarios, such as the ones in Kingman, Arizona and Livingston, Louisiana.
3. *Failure of container attachments*—Attachments may open up or break off the container, such as pressure relief valves, frangible disks, fusible plugs, discharge valves, or other related attachments.

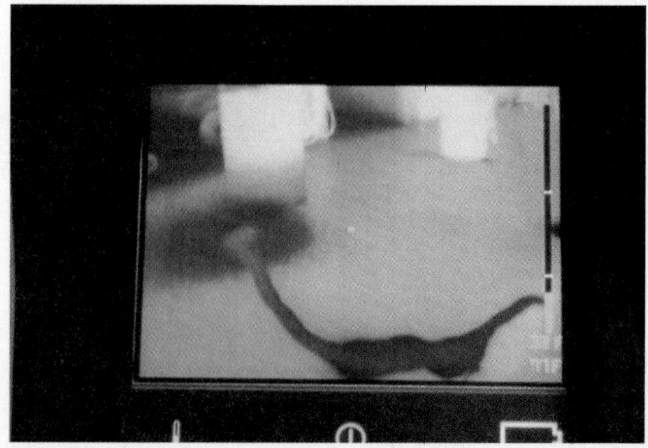

Figure 7-6 Determining product quantity and container pressure can be important factors in predicting container behaviors and outcomes.

4. *Container punctures*—Usually associated with mechanical stressors that result in a breach of the container. Examples include 55-gallon drums being punctured by a forklift truck and coupler punctures of railroad tank cars.
5. *Container splits or tears*—Examples include torn fiber or plastic bags such as those used for certain oxidizers and agricultural chemicals, split 55-gallon drums, and seam or weld failures on both pressurized and nonpressurized containers.

Containers breach differently. The type of breach will depend on (1) the type of container, and (2) the type of stress applied to it. When you are unsure how a container is likely to breach, get technical assistance from product or container specialists.

Responder Tip

Become familiar with how containers breach in an emergency. If you don't know, get help from someone who does.

Release Event

Once a container is breached, the hazardous material is free to escape in the form of energy, matter, or a combination of both. The rate of this **release** is critical since it will directly determine your ability to control it. Generally, the faster the release, the greater likelihood of harm, and the less responders can do to influence the release.

There are four types of release:

1. *Detonation*—An explosive chemical reaction with a release rate of < 0.01 seconds. This gives responders no time to react. Examples include military munitions, commercial explosives such as PETN and Tovex, high-energy oxidizers, and organic peroxides.
2. *Violent rupture*—Associated with chemical reactions having a release rate of 0.01 to 1 second (e.g., deflagration). Again, there is no time to react. This behavior is commonly associated with runaway cracking and overpressures of closed containers.
3. *Rapid relief*—Ranges from several seconds to several minutes, depending on the size of the opening, type of container, and the nature of its contents. Responders may often have time to reach a safe location or develop tactical countermeasures. This behavior is associated with releases from containers under pressure, through pressure relief valve actuations, broken or damaged valves, punctures, splits, tears, or broken piping.
4. *Spills or leaks*—Release rates vary from minutes to hours. These are generally a low-pressure, nonviolent flow through broken or damaged valves and fittings, splits, tears, or punctures. Because of the slower release rate, responders will often have adequate time to develop prolonged countermeasures.

Energy and/or matter *will* be released when a container is breached. Responders must visualize not only how a release will occur but also how quickly. The speed of the release will depend on the nature of the hazmat as well as the stored energy—such as vapor pressure. The greater the stored energy, the faster the release.

Responder Tip

There are only two things jumping out at you—energy and matter. Look for energy and matter to be released when a container is breached.

Engulfing Event

Once the hazardous material and/or energy is released, it is free to travel or disperse, subsequently **engulfing** an area. The farther the contents move outward from their source, the greater the level of problems. How quickly they move and how large of an area they engulf will depend on the type of release, the physical form of the hazmat, the chemical and physical laws of science, and the environment.

To visualize the area the hazmat and/or energy is likely to engulf, consider the following questions:

1. What is jumping out at you? Energy or matter?
2. What form is it in? Solid, liquid, gas, biological organism, pressure wave?
3. What is making it move? Wind, pressure, contaminated individuals?
4. What path will it follow? Linear, random, ground contour, etc.?
5. What dispersion pattern will it create? Cloud, plume, etc.?

These answers will help responders to predict and define where the hazardous material and/or its container will go when released. In turn, responders can then determine the primary danger zone and their exposures.

First responders routinely use the *Emergency Response Guidebook—Table of Initial Isolation and Protective Action Distances* to initially estimate the area potentially impacted by a hazmat release. Computer-based plume dispersion models are another excellent tool for predicting the dispersion patterns of airborne gases and high vapor pressure liquid releases. Similarly, specialized plume dispersion models are available through technical and product specialists for estimating the movement of specific hazardous materials, chemical agents, and radioactive clouds. Common plume dispersion models include ALOHA (part of the CAMEO system) and CHARM® (Complex Hazardous Air Release Model software). A list of additional plume dispersion models that can assist in predicting downwind hazard areas can be referenced from the EPA website at http://www.epa.gov.

Responder Tip

Predict your dispersion patterns.

Table 7-4 Hazardous Materials Behavior—Dispersion Considerations

All hazardous materials will behave in some predictable manner once released from their container. Know dispersion patterns and how they factor into the risk evaluation process.

What is Jumping Out at You?	What is its Form?	What is Making it Move?	What Path Will it Follow?	What Dispersion Pattern Could it Form?
Energy				
	Infra-Red Rays	Thermal Differential	Linear or Radial	Cloud
	Gamma Rays (Nuclear)	Self-Propelled	Linear or radial	Cone or Cloud
	Pressure Waves	Self-Propelled	Linear	Cloud
Matter				
Solids	Dust or Powders	Air Entrainment	With Wind (Linear)	Plume, if Unconfined
		Personal Transport	Random	Irregular Deposits
	Fragments, Shrapnel, or Chunks Organisms	Self-Propelled	Linear	Cloud
		Air Entrainment	With Wind (Linear)	Plume, if Unconfined
		Personal Transport	Random	Irregular Deposits
	Alpha and Beta	Self-Propelled	Linear	Cone or Cloud
Liquids	Pourable Liquids	Gravity	Follow Contour	Stream
Gases	Vapors	Gravity	Follow Contour	Stream
		Air Entrainment	With Wind (Linear)	Plume, if Unconfined
		Vapor Diffusion	Upward From Surface	Cloud Above Liquid
	Gaseous	Gaseous Diffusion	Outward From Surface	Plume, if Unconfined or Shape of Confined Area
Liquefied gases	Vaporizing	Self-Propelled and Boiling	Liquid Follows Contour, Gas Moves With Wind (Linear)	Liquid Forms Stream, Gas Forms Plume

Source: Ludwig Benner, Jr., 1978. Hazardous Materials Emergencies, 2nd Edition, Oakton, VA

Impingement (Contact) Event

As the hazardous material and/or its container engulf an area, they will **impinge** on or come in contact with exposures. Exposures include people (civilian and emergency responders), property (physical and environmental), and systems (critical infrastructure, transportation, community water supply, etc.). They may also impinge on other hazardous materials containers, producing additional problems.

Impinged exposures may or may not suffer any harm. The level of harm is directly dependent on the dose received (remember—dose makes the poison!). Therefore, you must consider factors such as the type of exposure, the duration of the exposure, and the ability of the impinged exposure to minimize the effects of the exposure.

Impingements are categorized based on their duration. Short-term impingements (i.e., a transient vapor cloud) have durations of minutes to hours. Medium-term impingements may extend over a period of days, weeks, and even months. Examples include lingering pesticide residues resulting from fires or spills and asbestos remediation following a process unit fire or explosion.

Long-term impingements extend over years and perhaps even generations. Examples include the contamination of groundwater supplies and radioactive material remediation operations after events at Chernobyl and Fukushima, Japan.

Estimating impingements within an engulfed area must include consideration of all of the following factors:

- Harmful characteristics of the material released (e.g., flammable, toxic, reactive, etc.)
- Concentration of the hazardous material
- Duration of the impingement
- Characteristics of the exposure (i.e., vulnerability)

For example, a toxic vapor cloud released will impinge on all people and objects that fall within its path. However, if the concentration of the toxic chemical is low enough or if the length of impingement is short enough, little harm will be done.

Responder Tip

Identify the exposures likely to be impinged by the release.

Table 7-5 Breach and Release Behaviors for Pressurized Drum Scenarios *(Continued)*

Incidents involving pressurized drums and non-bulk containers can pose significant risks to emergency responders. Scenarios can range from drums storing reactive materials, such as organic peroxides and other materials subject to polymerization, to containers subjected to thermal stressors from fires and ambient heat sources. The failure of a pressurized drum will often result in the rocketing of the drum lid or bottom, ejection of its contents, and a rapid release of pressure or energy.

In 1998 the Los Alamos (New Mexico) National Laboratory conducted a series of tests on various metal and plastic drums commonly used for the transport of hazardous materials. The following information is based upon a summary of that project's findings as published in FIRE ENGINEERING in July 1999.

The objectives of the Los Alamos tests were to:

- Determine the pressure at which 20, 30 and 55-gallon metal and plastic drums and 85-gallon metal overpack drums fail;
- Quantify the amount of deformation 55-gallon metal drums experience at various pressures under separate tests;
- Determine if the data for 55-gallon drums support developing an instrument for determining internal pressures; and
- Conduct a statistical analysis on the mean failure pressures of the collected data for 55-gallon drums.

Drums constructed of different materials (i.e., metal and high-density polyethylene [HDPE]) and different construction (i.e., open head, closed head) were pressurized from 0 psig to failure at 5.0 psig intervals. All drums were new and unused. Tests were conducted at various liquid levels, including empty, 50%, and 75%. A specialized test apparatus was constructed to facilitate pressurization and drum measurements, and to contain any flying debris that may be released from a drum failure.

Although the test sample was relatively small, the test results still provide hazmat responders with background information that can be used in predicting behaviors and evaluating the level of risk at scenarios involving over-pressurized or deformed drums.

Summary

1. The study found significant differences between the failure characteristics of drum types and materials. Among the key findings were the following:

55-Gallon Metal, Open-Head Drums

- Drums appear to vent immediately adjacent to the nut-and-bolt fastener on the ring.
- All drums vented at pressures ≤ 32 psig. Pinging was noticeable between 15–20 psig.
- Drums appeared to bulge only at the top and bottom ends. Body seams experienced no visible distortion or apparent weakening.

55-Gallon Metal, Closed-Head Drums

- Drums failure rate of 95%. Of the catastrophic failures, 68% failed at the bottom end thereby making the entire drum a projectile.
- Pinging was noticeable between 15–20 psig. Rapid and intense pinging occurred immediately before drum failure.
- All drums failed at the top or bottom ends. The chime distorts at approximately 5 psig before failure.
- Drums appeared to bulge only at the top and bottom ends. Body seams experienced no visible distortion or apparent weakening.
- Statistical analysis of test results show a 99% probability that failures will occur at pressures > 48.7 psig.

85-Gallon Metal Overpack Drums

- All overpacks failed ≤ 16 psig and appeared to self-vent immediately adjacent to the placement of the nut-and-bolt fastener on the ring.
- Drums appeared to bulge only at the top and bottom ends. Body seams experienced no visible distortion or apparent weakening.

30-Gallon Metal, Closed-Head Drums

- Extremely high pressures are possible (> 120 psig).
- Violent rupture of the container should be anticipated.
- Drums appeared to bulge only at the top and bottom ends. No apparent failure indicators, such as pinging, were noted.

30-Gallon Metal, Open-Head Drums

- High pressures are possible (> 50 psig).
- Container may self-vent of fail explosively.
- Drums appeared to bulge only at the top and bottom ends. No apparent failure indicators, such as pinging, were noted.

55-Gallon Plastic (HDPE) Drums

- Deformation was observed at the tops, bottoms and sides on the five tests.
- Drums failures occurred through the drum side at no particular location.
- Seamless drum failures for five tests ranged from 30 to 58 psig, with an average of 47 psig.
- A device for estimating internal pressures in 55-gallon drums may not be useful on plastic or plastic-lined drum due to construction differences from metal drums.

20- and 30-Gallon Plastic (HDPE) Drums

- Both seam and seamless drums failed at pressures < 45 psig.
- Deformation was observed at the tops, bottoms and sides.
- Seamless construction drums failures occurred through the drum side at no particular location, thereby making the entire drum a projectile.
- Seam construction drums failed explosively along the bottom seam, thereby mamking the drum a projectile.

Table 7-5 *(Continued)*

2. Tactical considerations for incidents involving pressurized or deformed drums should include the following:
   - Deformation of a drum indicates that the drum has been subjected to internal pressure - not that it is under pressure at the time of inspection. The container's design makes it capable of violent rupture.
   - Consider the physical and chemical properties of the contents. Remember – the violent rupture of drums with nonflammable contents will still result in the release of energy (i.e., pressure wave) and matter (i.e., drum fragments).
   - Establish hazard control zones based upon the violent rupture of the container. Consider what is jumping out at you – energy and matter!
   - If you don't know what to do, get help from someone who does, including product and container specialists, chemists, etc.
   - Tactics to control the situation should be developed in consultation with product and container specialists. Options may include direct cooling through an ice bath, remote venting techniques, or the use of water cannons or disrupter devices.

For additional information see "Pressure Effects on and Deformation of Hazardous Waste Containers" by Michael D. Larrañaga, et. al, FIRE ENGINEERING (July, 1999),

Harm Event

Before responders can favorably influence the outcome of a hazmat incident, they must first understand what **harm** is likely to occur within the engulfed area if they do not intervene. Harm simply refers to the effects of exposure to the hazardous material and/or its container.

Harm can be categorized in the following forms:

- *Thermal*—Harm resulting from exposure to hot and cold temperature extremes. High-temperature harm may result from direct flame contact or radiant heat exposure, while sources of cold temperature harm include exposure to liquefied gases and cryogenic liquids.
- *Toxicity/poisons*—Harm resulting from exposure to poisons and toxic materials. This would include exposure to chemical agents, toxic industrial chemicals, and toxic byproducts that may be released from a fire or chemical reaction.
- *Radiation*—Harm resulting from exposure to radioactive materials.
- *Asphyxiation*—Harm resulting from exposure to simple asphyxiants (e.g., nitrogen, natural gas) or chemical asphyxiants (e.g., carbon monoxide, sarin, methylene chloride).
- *Corrosivity*—Harm resulting from exposure to corrosive materials (acids and bases).
- *Etiologic*—Harm resulting from exposure to biological materials, including bacteria, viruses, and biological toxins.
- *Mechanical*—Harm resulting from contact with fragmentation or debris scattered as a result of a pressure release, explosion, BLEVE, and so on.

Three factors directly influence the level of harm: (1) the timing of the release (speed of escape and travel, length of exposure); (2) the size of the dispersion pattern and the area covered; and (3) the lethality of the chemicals involved (concentration of the chemical or dosage received).

In general, the faster the rate of release, the larger the area covered, and the more lethal the materials involved, the less responders can do to influence the level of harm once the hazmat escapes from its container. Finally, by understanding the type and nature of the potential harm, responders can evaluate the appropriate level of protective clothing and public protective actions required.

Responder Tip

Emergency response personnel must identify the harm that can occur.

Estimating Outcomes

Now that all six events of the GHBMO have been reviewed, how do they relate to "what happens on the street"? As part of the incident size-up process, responders should initially determine exactly where, in the sequence of events, their particular incident is. Knowing where you are in the GHBMO sequence will assist you in estimating outcomes.

In a complex incident, such as a major train derailment or a major fire in a petrochemical process area, multiple containers may be at different stages of the hazmat behavior sequence simultaneously. One or more containers may have already released their contents and pose an immediate threat. However, other containers may also be stressed, waiting only for the proper conditions that will cause them to breach. Responders must visualize and predict the behavior of *all* of the containers involved in the incident in order to accurately reflect potential outcomes.

The GHBMO provides responders with the mental framework to assess incident potential and estimate outcomes within the engulfed areas. To summarize, key factors that should be evaluated to estimate outcomes in the engulfed area will include:

- *The size and dimension of the engulfed area.* This will be influenced by wind speed and direction, as well as topography.
- *The number of exposures within the engulfed area, including people, property, and critical systems.* Related considerations will include special occupancies, time of day, traffic patterns, and so on.
- *The concentration of "bad stuff" within the engulfed area.* Resources for determining concentrations will range from computer plume dispersion models to air monitoring results.
- *The extent of physical, health, and safety hazards within the engulfed area.*

- *Areas of potential harm.* Specifically, is the area safe, unsafe, or dangerous?

Only after responders clearly understand incident potential can they develop and implement a sound IAP to control the incident.

Developing the Incident Action Plan

The IAP is developed based upon the IC's assessment of (1) incident potential (i.e., visualizing hazardous materials behavior and estimating the outcome of that behavior), and (2) the initial operational strategy. Every incident must have an oral or written IAP. The IAP should clearly state the strategic goals, tactical objectives, and assignments that must be implemented to control the problem, as well as required resources and support materials. The IAP provides all command and supervisory personnel with the direction of future actions.

Strategic goals are the broad game plan developed to meet the incident priorities (life safety, incident stabilization, environmental and property conservation). Essentially, strategic goals are "what are you going to do to make the problem go away?" When laid against the GHBMO, the earlier that the events sequence can be interrupted, the more acceptable the loss Table 7-6.

Several strategic goals may be pursued simultaneously during an incident. Examples of common strategic goals at hazmat incidents include the following:

- Rescue
- Public Protective Actions
- Spill Control (Confinement)
- Leak Control (Containment)
- Fire Control
- Recovery

Tactical objectives are specific and measurable processes implemented to achieve the strategic goals. In simple terms, tactical objectives are the "how are you going to do it" side of the equation. Tactical objectives are then eventually tied to specific tasks assigned to particular response units, such as fire, rescue, law enforcement, and HMRTs. For example, tactical objectives for a spill control strategy would include diking, damming, and retention.

Figure 7-7 In a complex incident, multiple containers may be at different stages of the hazmat behavior sequence simultaneously.

Tactical response objectives to control and mitigate the hazmat problem may be implemented in either an offensive, defensive, or nonintervention mode. Criteria for evaluating these options include the level of available resources (e.g., personnel and equipment), the level of training and capabilities of emergency responders, and the potential harm created by the hazmat release.

- **Offensive mode.** These are aggressive leak, spill, and fire control tactics designed to quickly control or mitigate the emergency. Although increasing the risks to responders, **offensive tactics** may be justified if rescue operations can be quickly achieved, if the spill can be rapidly contained or confined, or the fire quickly extinguished. The success of an offensive-mode operation is dependent on having the necessary resources available in a timely manner.
- **Defensive mode.** These are less aggressive spill and fire control tactics where certain areas may be conceded to the emergency, with response efforts directed toward limiting the overall size or spread of the problem. **Defensive tactics** decrease the risk to responders and may be employed as a means of, for instance, reducing the size of a spill or reducing the pressure of an affected pipeline or tank. Examples include isolating a pipeline by closing remote valves, shutting down pumps, constructing dikes, and exposure protection.
- **Nonintervention mode.** Nonintervention is essentially "no action." In brief, the risks of intervening are unacceptable when compared to the risks of allowing the incident to follow a natural outcome, such as scenarios with a high BLEVE or explosion potential. In other cases, environmental impacts may be reduced by allowing a flammable liquids fire to burn itself out.

In some situations, nonintervention tactics may be implemented until sufficient resources arrive on-scene and an offensive attack can be implemented. Defensive tactics are always preferable over offensive tactics if they can accomplish the same objectives in a timely manner.

Most operations will begin from a defensive point of view. The most important question the IC should ask is, "What happens if I do nothing?" There will also be times when an operation is in a marginal or borderline mode. In other words, initial information indicates it is relatively safe to attempt an offensive tactical objective, yet it is also possible that things may turn for the worse during that process. In such a case, it is generally best to take a defensive approach.

The implementation of the IAP, including strategical and tactical options, are discussed in greater detail in the section on "implementing response objectives" found elsewhere in this text.

Evaluating Risks—Special Problems

Three situations that responders commonly deal with are (1) damage assessment of pressurized bulk transport containers, (2) the behavior of chemicals and petroleum products when

Table 7-6 Strategies and Tactics for Hazardous Materials Response Operations

Organizational Levels	Strategic Goals	Tactical Objectives	Operational Tasks
	"What are you going to do?"	"How are you going to do it?"	"Do it."
DEFINITION	STRATEGY: The overall plan to control the incident and meet incident priorities.	TACTICS: The specific and measurable processes implemented to achieve the strategic goals.	TASKS: The specific activities that accomplish a tactical objective.
KEY ELEMENTS	■ Goals ■ Overall game plan ■ Broad in nature ■ Meets incident priorities (life safety, incident stabilization, environmental and property conservation)	■ Objective-oriented ■ Specific and measurable ■ Often builds upon procedures	■ "Hands-on" work to meet the tactical objectives. ■ THE most important organizational level—where the work is actually performed. ■ Most problems "go away" as a result of members performing task-level activities.
DECISION-MAKERS	■ Incident Commander ■ Section Chiefs	■ Operations Section Chief ■ Hazmat Group Supervisor ■ Group / Division Supervisor	■ Individual units and individuals

OPTIONS	■ Rescue
	■ Public Protective Options
	■ Fire Control
	■ Spill Control
	■ Leak Control
	■ Recovery
		■ Rescue
		■ Public Protective Options
		  ■ Evacuation
		  ■ Protection-in-Place
		■ Fire Control
		  ■ Exposure Protection
		  ■ Extinguishment
		■ Leak Control
		  ■ Neutralization
		  ■ Overpacking
		  ■ Patching & Plugging
		  ■ Pressure Isolation & Reduction
		  ■ Solidification
		  ■ Vacuuming
			■ Spill Control (Confinement)
			  ■ Absorption
			  ■ Adsorption
			  ■ Covering
			  ■ Diking, Damming & Diversion
			  ■ Dispersion
			  ■ Retention
			  ■ Vapor Dispersion
			  ■ Vapor Suppression
			■ Recovery

released underground, and (3) the behavior of hazmats in sewer collection systems. Responders often have difficulty in acquiring and interpreting hazard information, as well as evaluating the risks involved in these scenarios. In this section, we build on the basic concepts of the GHBMO and examine these three scenarios more closely.

Damage Assessment of Pressurized Containers

Bulk transport pressurized containers regularly sustain extensive mechanical stress and damage in rollovers and accidents without releasing their contents. Pressurized containers typically have stronger and thicker container shells than those used in liquid or solid service. For example, MC-331 cargo tank trucks must meet the container design requirements of both DOT and the ASME Pressure Vessel Code. However, while not common, there have been catastrophic failures of pressurized containers many hours after the initial event (e.g., the Waverly, Tennessee derailment and BLEVE). Much of the information contained in this section was referenced from materials developed by the Union Pacific Railroad Company and Charles Wright.

Responders can be confronted with a variety of pressurized containers, including cylinders, cargo tank trucks (MC-331), and railroad tank cars (e.g., DOT-105, 112, and 114 tank cars).

Voices of Experience

Cargo tank truck emergencies are often one of the more common response scenarios that an "active" HMRT encounters. Over the years, I've experienced many of them and seen the good, the bad, and the ugly. One common factor these scenarios shared is that they were living laboratories for your understanding and application of hazard and risk evaluation principles.

Here's some of the lessons I've learned:

- **"How long will the road be closed?"** Regardless of the type of cargo tank truck involved, if it is overturned, 8 hours is a good initial estimate for how long the transportation infrastructure will be impacted. Sometimes we do better, sometimes we do worse. But letting law enforcement and local/state/federal transportation officials know that the road will be affected for at least 8 hours on the front end will allow them to develop traffic management plans in a timely manner. In cases like this, remember—minimizing impacts to the transportation infrastructure must be one of your primary response priorities after life safety and property/environmental damage.
- **Know the product and the container.** Having access to knowledgeable, street-smart product and container technical specialists is critical. And if you already know them, that's even better. We've used peers and professional contacts at many incidents, even when the problem didn't involve their product or container. They can often serve as a neutral, unbiased third-party to bounce off questions, issues, and tactical options. We've even called peers while working an incident because we knew they recently experienced a similar incident involving the same product or container.
- **Can you move the problem?** Depending upon the nature of the container breach and release, you might be able to move the problem to an area where there will be fewer exposures or less impact upon the transportation infrastructure. This option worked well when dealing with a MC-312 corrosive cargo tank truck that was leaking due to a bad gasket on the manway. After public safety and industry responders confirmed the exact nature of the problem, the vehicle was moved from an interstate highway onto a state highway maintenance facility where the incident was subsequently mitigated with minimal exposures and impact.
- **Offload vs. Upright.** The question of whether you can upright an overturned cargo tank truck while it is loaded will test your understanding of hazard and risk evaluation principles. There are some "black and white" rules, but there also many "shades of gray." For example, aluminum-shell MC-306/DOT-406 cargo tank trucks commonly used for the transportation of gasoline and fuel oils should NOT be uprighted while loaded. In comparison, MC-331 cargo tank trucks used for the transportation of LPG and anhydrous ammonia are ASME pressure vessels and are built to a much higher standard of care. You can often upright a loaded MC-331, assuming, of course, you have the requisite lifting capability.
- **The Wrecker Industry.** Always verify that the personnel performing the lifting and uprighting operation are trained and qualified to perform the expected tasks. History has taught us that there is a big difference between uprighting a standard overturned eighteen-wheeler box trailer and an MC-312/DOT-412 corrosive tank truck involved in a rollover. While there are many first-rate riggers in the business, there are also plenty of well-meaning tow truck operators and salvage companies who simply do not have the skills to perform this kind of operation.

Gregory G. Noll
Lancaster, PA

Much of the available literature on damage assessment is based on testing and experience with railroad tank cars. While design and construction standards for tank cars are substantially greater than those for cylinders and cargo tank trucks and provide a greater margin of safety, there are some basic principles that have utility for other pressurized containers.

The violent rupture of pressurized containers can be triggered by one of two related conditions: (1) the presence of a crack in the container shell associated with dents and rail burns, or (2) the thinning of the tank shell as a result of scores, gouges, and thermal stress. Factors that can affect the severity of the container damage include the ductility of the container metal and the internal pressure causing stress on the container metal.

Responders should have a basic understanding of ductility. *Ductility* is the relative ability of a metal to bend or stretch without cracking. Ductile materials have a fine grain structure and tend to bend but not crack. Brittle materials have a coarse-grain structure and tend to crack rather than bend. When a ductile steel container cracks, the crack tends to be small; in contrast, a crack in a brittle steel container tends to run linearly and cause the container to fail.

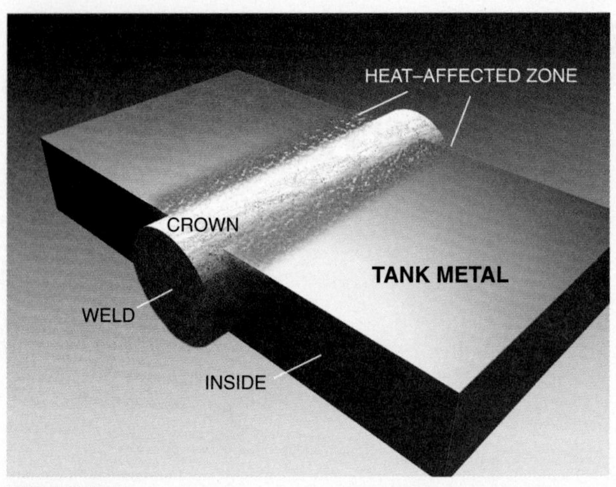

Figure 7-8 The heat-affected zone is the tank metal next to a weld. It may be vulnerable to container stress as cracks are likely to start there.

Key factors that affect tank damage severity are as follows:

1. *Specification of the steel*—Particularly important in evaluating the integrity of damaged railroad tank cars, as different types of steel with different characteristics have been used over time. Product and container specialists should be consulted for further guidance.
2. *Internal pressure*—This is the force against the internal area of the tank that creates stress on the tank. If the stress becomes great enough, cracks associated with dents and rail burns can become unstable, begin to grow, and propagate. As internal pressure increases, so does the risk of container failure. Pressure may increase from ambient temperature increases, solar radiation, radiant heat, chemical reaction, or flame impingement.
3. *Damage affecting the* ***heat-affected zone*** *of the weld*—Area in the tank metal next to the actual weld material. This area is less ductile than either the weld or the steel plate due to the effect of the heat during the welding process and is most vulnerable to damage as cracks are likely to start there (see **Figure 7-8**).
4. *Cold work*—Deformation of steel when it is bent at ambient temperatures or results from an impact or static load. Cold work reduces the ductility of the metal. Almost all container damage in an emergency is cold work.
5. *Rate of application*—This is the speed with which energy is transferred to the stress object (i.e., container). This is typically a qualitative assessment in an emergency response scenario (e.g., high versus low energy transfer).

Guidelines for damage assessment of pressurized containers include the following:

1. Gather information concerning the type of container (e.g., DOT specification number), material of construction (e.g., aluminum, steel), and internal pressure. Methods for determining the internal pressure include:
   - Using pressure gauges attached to sample lines, gauging device, fittings, and so on.
   - Using temperature gauges with vapor pressure/temperature conversion charts. However, tank contents may stratify into layers having different temperatures due to external temperature changes; as a result, the pressure estimated from product temperature readings may be inaccurate (i.e., may be lower than the actual pressure). Laser thermometers and thermal imaging devices may also be used to determine the temperature when a gauge is not available.
   - Using ambient temperature as your reference point, bear in mind that the temperature of the tank's contents may lag ambient temperatures by up to 6 hours.

   The internal pressure in empty railroad tank cars that contain residual vapors may be equal to that in loaded cars (or greater than that in loaded tank cars if some inert gas is used for unloading). Vapor pressure/temperature graphs are available from the *Compressed Gas Handbook,* as well as from the shipper or manufacturer of the material.
2. Determine the amount of material in the container. Sources can include shipping papers and gauging devices.
3. Determine the type of stress applied to the container (e.g., thermal, mechanical, or combination).
4. Evaluate the stability of the container. Take caution when inspecting an unstable container, as it may move or shift during the inspection process. It may be necessary to stabilize the container with blocks, cribbing, or other means.
5. Examine all accessible surfaces of the container, paying attention to the types of damage and the radius (i.e., sharpness) of all dents. Railroad personnel will often use a tank car dent gauge as a "go/no go" device for comparing the radius of curvature of a tank car dent to accepted standards to determine the severity of damage. However, the tank car dent gauge cannot be used for assessing dents on cargo tank trucks due to differences in shell metal and thickness.

Mechanical Stress and Damage

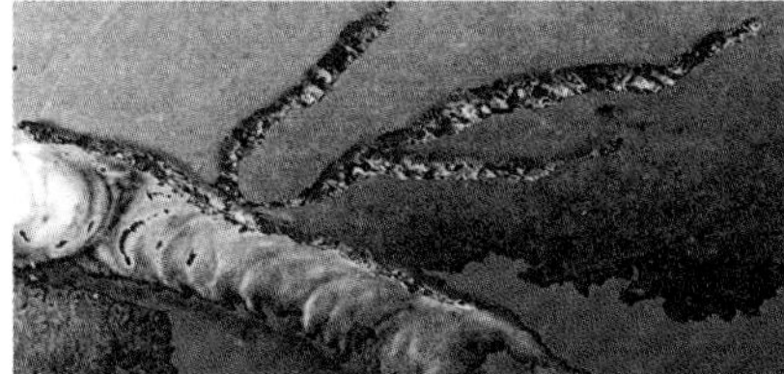

CRACK—narrow split or break in the container metal, which may penetrate through the container metal. It is a major problem that could cause catastrophic failure.

SCORE—reduction in the thickness of the container shell. It is an indentation in the shell made by a relatively blunt object. A score is characterized by the reduction of the container or weld material so that the metal is pushed aside along the track of contact with the blunt object. Scores caused by prolonged contact with a tank car wheel are called "wheel burns."

GOUGE—reduction in the thickness of the tank shell. It is an indentation in the shell made by a sharp, chisel-like object. A gouge is characterized by the cutting and removal of some of the container or weld material along the track of contact.

WHEEL BURN—reduction in the thickness of a railroad tank shell. It is similar to a score but is caused by prolonged contact with a turning railcar wheel.

DENT—a deformation of the tank head or shell. It is caused from impact with a relatively blunt object (e.g., railroad coupler, vehicle, concrete abutment). The sharper the radius of curvature of the dent, the greater the chance of cracking.

RAIL BURN—a deformation in the shell of a railroad tank car. It is actually a long dent with a gouge at the bottom of the inward dent. A rail burn can occur circumferentially or longitudinally in relation to the tank shell. The longitudinal rail burns are the more serious because they have a tendency to cross a weld. A rail burn is generally caused by the tank car passing over a stationary object, such as a wheel flange or rail.

Figure 7-9 Examples of mechanical stress and damage to pressurized containers. *(Continued)*

Mechanical Stress and Damage (*continued*)

STREET BURN—a deformation in the shell of a highway cargo tank. It is actually a long dent that is inherently flat. A street burn is generally caused by a container overturning and sliding some distance along a cement or asphalt road.

Figure 7-9 (*Continued*)

Experience shows that the most dangerous situations will include the following Figure 7-9:

- Cracks in the base metal of a tank or cracks in conjunction with a dent, score, or gouge. Both of these situations justify offloading or reducing container pressure as soon as safely possible.
- Sharply curved dents or abrupt dents in the cylindrical shell section parallel to the long axis of the container. If a dent is considered critical, the container should be first offloaded or pressure reduced before moving it.
- Dents accompanied by scores and gouges.
- Scores and gouges across a container's seam weld or in the heat-affected zone of the weld. If the score crosses a welded seam and removes no more than the weld reinforcement (i.e., that part of the weld which sticks above the base metal), the stress is considered noncritical. However, if the score removes base metal at the welded seam, the stress is considered critical.

Responder Safety Tip

If you are unsure of the container damage or how the container is likely to breach, get assistance from product or container specialists. This may include railroad personnel, gas industry representatives, and cargo tank truck specialists.

Movement and Behavior of Hazmats Underground

When petroleum products or chemicals are released into the ground, their behavior will depend on their physical and chemical properties (e.g., liquid versus gas, hydrocarbon versus polar solvent), the type of soil (e.g., clay versus gravel versus sand), and the underground water conditions (e.g., location and movement of the water table). While such incidents may involve any hazmat class, flammable liquids, gases, and vapors are the most common.

As with hazmat containers and their behavior, responders should have a basic understanding of geology, groundwater, and groundwater movement to evaluate the underground dispersion of hazmat releases and potential exposures and to determine potential outcomes. If responders do not have this background, they must identify local resources that can provide this technical information (e.g., EPA, DEQ). However, the ultimate clean-up and remediation of underground spills and releases is not the responsibility of emergency responders.

Geology and Groundwater

Soil consists of loose, unconsolidated surface materials, such as sand, gravel, silt, and clay. Bedrock is the hard, consolidated material that lies under the soil, such as sandstone, limestone, or shale. Most areas have a soil cover ranging from a few feet to several hundred feet.

Generally, rocks and soils consist of small fragments or sand grains. When compressed together, they may form small voids or pores. Measurement of the total volume of these voids is called the porosity of the rock or soil. If these pores are interconnected, the rock or soil is permeable (i.e., fluid can pass through it). The size of these voids will vary from large (e.g., gravel) to small (e.g., sand and topsoil) to essentially zero (e.g., dense clay). Rock almost never has large voids, but sandstone and limestone have voids similar to a fine sand. Aquifers are permeable sections of soil or rock capable of transmitting water. In contrast, silt and shale have many, but extremely small, pores that are poorly interconnected. Since fluids cannot readily pass through such materials, they are known as impermeable materials.

In most areas, water exists at some depth in the ground. The source of most groundwater is precipitation over land, which percolates into porous soils and rocks at the surface. Rivers and streams that seep water into the subsurface are a second source of groundwater. Groundwater accounts for the majority of the drinking water supply for the United States.

In most areas, groundwater moves extremely slowly. The rate of flow depends on the permeability of the underground aquifer and the slope or "hydraulic gradient" of the water table. Flows can range from 6 feet per year in fine clays to 6 feet per day in gravels. In addition, the location and relative production rates of groundwater wells can significantly disrupt normal flow patterns. Figure 7-10 shows a hypothetical groundwater system.

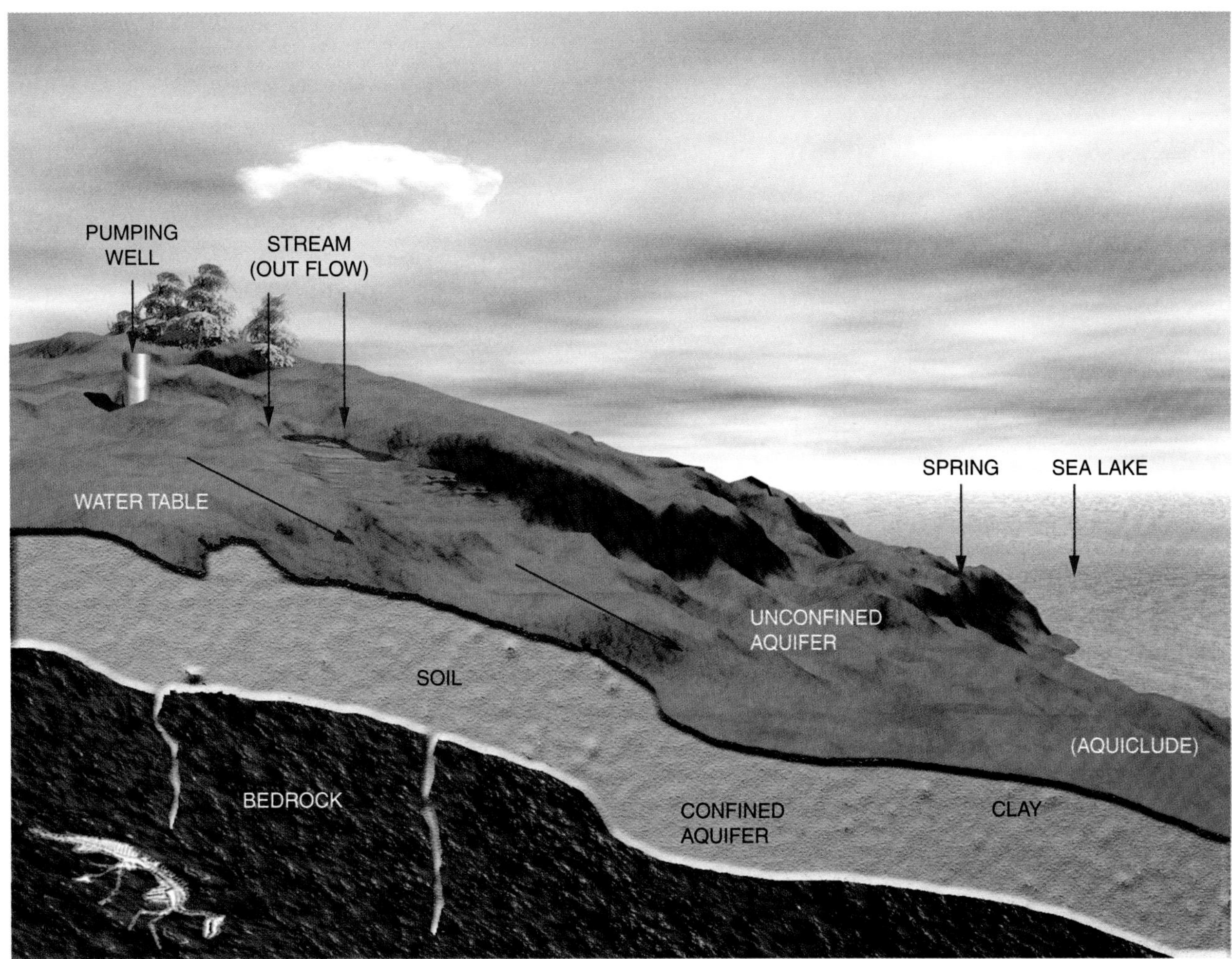

Figure 7-10 Hypothetical groundwater system.

Behavior of Hazmats in Soil and Groundwater

Hazardous materials may be absorbed into the soil through either surface spills or leaks from underground pipelines or storage tanks. Flammable liquids and toxic solvents can create significant problems if they migrate into confined areas or are allowed to flow into waterways.

Flammable and toxic gases, such as natural gas, propane, or hydrogen sulfide, can also accumulate in underground pockets or confined areas. These can occur naturally or may be released from a storage tank or pipeline failure. When dealing with natural gas, recognize that some soils (e.g., clay) can "scrub" or remove the methyl mercaptan odorant commonly used on natural gas distribution lines, thereby removing any physical clues of smell or odor.

The underground movement of hazmats, like water, follows the most permeable, least resistant path. For example, the backfill in trenches carrying utility conduits, sewers, or other piping is often much more permeable than the undisturbed native soil. In urban areas, this can facilitate the rapid and easy movement of liquids and gases to nearby basements, sewers, or other below-grade structures. Identifying these conduits is critical in identifying potential exposures. Utilities and public works agencies, including regional "Miss Utility" and "One Call Systems," can often provide assistance in locating these underground conduits. Responders should also seek technical information specialists, such as geologists, who have either maps or knowledge of the local water tables, soil types and densities, underground rock formations, and so forth.

Liquid hazmats, which are spilled into soil, will tend to flow downward with some lateral spreading Figure 7-11. The rate of hazmat movement in soil will depend upon the viscosity of the liquid, soil properties, and the rate of release. For example, light hydrocarbon products, such as gasoline, will penetrate rapidly while heavier oils, such as #4 fuel oil, will move more slowly. If the soil near the surface has a high clay content and very low permeability, the hazmat may penetrate very little or not at all. However, a porous, sandy soil may quickly absorb the product. Eventually, the downward movement of the hazmat through soil will be interrupted by one of three events: (1) it will be absorbed by the soil; (2) it will encounter an impermeable bed; or (3) it will reach the water table.

Figure 7-11 The movement of hazardous materials through soil will be dependent upon the viscosity of the liquid, properties of the soil, and the rate of the release.

Hazmats absorbed by the soil may move again at some later time as the water table is elevated. For example, responders will often receive a report of hydrocarbon or gasoline vapors in an area, with the source of the odor being unknown. In other situations, recent rains may cause the water table to rise and bring hydrocarbon liquid and vapors to the surface. With these scenarios, the suspected area should be divided into quadrants and monitoring readings "mapped" in a systematic manner to assist emergency responders in identifying the location of the problem and its direction of movement.

Although combustible gas indicators (CGIs) are excellent tools for evaluating flammable atmospheres, they may not be very effective for assessing low-level flammable concentrations such as those found with subsurface and sewer spills. To register a positive reading, many CGIs require a concentration of up to 10,000 ppm (1% in air). While this may not be a flammable concentration, it often represents a significant environmental problem. Photoionization detectors (PIDs) are better suited for these scenarios.

Hazmats that encounter an impermeable layer will spread laterally until becoming immobile or until the hazmat comes to the surface where the impermeable layer outcrops. If the hazmat comes in contact with the water table, there is a high potential for the water supply to become contaminated and for the hazmat to move and accumulate in an underground structure (e.g., basement, sewer system). Remember that groundwater supplies can become contaminated by concentrations as small as 200 parts per billion (ppb). Preventing spills and releases from entering the soil is a critical element in many areas. While responders can do little to influence the underground movement of the material once it has entered the soil, underground aquifers and exposures must still be identified and protected.

Hydrocarbon liquids will not mix with water and will simply float on the surface of the water table. Many oils and refined petroleum products, however, contain certain components that are slightly soluble in water. Gasoline is high in water-soluble components which, when dissolved in water, produce odors and taste that can be detected at levels of only a few parts per million (e.g., ethanol, methanol).

Movement and Behavior of Spills Into Sewer Collection Systems

Sewers, manholes, electrical vaults, french drains, and other similar underground structures and conduits can be critical exposures in the event of a hazmat spill. If the hazmat release penetrates the substructure walls or enters through surface sewers and manholes, significant quantities of liquid can be expected to flow into the sewer collection system. Depending on the hazmat involved, this situation can pose an immediate fire problem, as well as significant environmental concerns regardless of whether ignition occurs or not.

Most sewer emergencies involve flammable and combustible liquids. The probability of an explosion within an underground space will depend on two factors: (1) that a flammable atmosphere exists, and (2) that an ignition source is present. The severity of an explosion and its consequences will depend on the type of sewer collection system, the process and speed at which the hazmat moves through the system, and the ability of emergency responders to confine the release and implement fire and spill control procedures.

Types of Sewer Systems

Sewer systems can be categorized based on their application:

- *Sanitary sewers.* This is a "closed" system that carries liquids and water-carried wastes from residences, commercial buildings, industrial plants and institutions, as well as minor quantities of storm water, surface water, and groundwater that are not admitted intentionally. The collection and pumping system will transport the wastewater to a treatment plant, where various liquid and solid treatment systems are employed to process the wastewater. Sewer pipe diameters from 8 inches to 60 inches are common.
- *Storm sewers.* This is an "open" system that collects storm water, surface water, and street wash and other drainage from throughout a community but excludes domestic wastewater and industrial wastes. A storm sewer system may dump runoff directly into a retention area that is normally dry or into a stream, river, or waterway without treatment. However, large manufacturing facilities that use petroleum or chemicals in their process are required by EPA regulations to collect and treat all on-site surface runoff before it can be discharged. Storm sewers are generally much larger than sanitary sewers, with diameters ranging from 2-foot pipes to greater than 20-foot tunnels. Storm sewers are sometimes used by "illegal dumpers" as a means of hazardous waste disposal.
- *Combined sewers.* Carries domestic and industrial wastewater, as well as storm or surface water. Although separate sanitary sewers are being constructed today, combined sewers may be found in older cities and metro areas. Combined sewers are often very large and can be as much as 20 feet in diameter. Combined sewers may also have regulators or diversion structures that allow any sewer overflow to be discharged directly to rivers or streams during major storm events.

Wastewater System Operations

There are four primary elements of a wastewater system: (1) collection and pumping, (2) filtering systems, (3) liquid treatment systems, and (4) solid treatment systems. The highest risks of a fire or explosion are associated with collection and pumping operations and with the early stages of liquid and solid processing. Similarly, the greatest potential for either environmental damage or shutdown of a wastewater treatment plant operation will take place at the liquid and solid stream treatment processes.

Wastewater, storm water, and surface water initially enter the collection and pumping system through a series of collectors and branch lines that tie together small geographic areas. These collectors and branch lines are eventually tied into a trunk sewer (also known as main sewer), which then carries the wastewater to its final destination for either treatment or disposal.

Where the terrain is flat, the collection system may consist solely of gravity piping. However, in most areas the collection system will require pumping or lift stations. Most pumping stations will have two parts—a wet well and a dry well. The wet well receives and temporarily stores the wastewater. Wet wells often contain electrical equipment such as fans, pumps, motors, and other accessories. In some instances, proper management of the wet well may provide an opportunity for the collection and removal of a flammable liquid that has gotten into the system. The dry well provides isolation and shelter for the controls and equipment associated with pumping the wastewater. They are designed to completely exclude wastewater and wastewater-derived atmospheres, although there may be accidental leakage from pumpshafts or occasional spills.

Depending on the type of sewer system and the specific location, most areas are classified by the National Electrical Code as Class I, Division 2 areas. However, pumping stations should be viewed as potential ignition sources when hydrocarbons are released into the sewer collection system. Pumping stations are sometimes equipped with hydrogen sulfide or fixed combustible gas detectors to detect the presence of flammable vapors and gases.

Primary Hazards and Concerns

There are two basic scenarios involving releases into a sewer collection system. The first scenario is an aboveground release where a spill flows into the sewer collection system through catch basins, manholes, and so on. The second scenario involves underground tank and pipeline leaks where the product migrates through the subsurface structure into the sewer collection system. This type of emergency is usually not obvious from the surface and presents a greater challenge in identifying the source of the problem and controlling the release within the sewer system. Both of these scenarios can also be complicated by the presence of methane and hydrogen sulfide (H2S), which are commonly found in underground structures and response scenarios.

With the subsurface scenario, responders will often receive a report of hydrocarbon or gasoline vapors in an area, with the source of the odor being unknown. In other situations, recent rains may cause the water table to rise and bring hydrocarbon liquid and vapors to the surface. Monitoring readings should be "mapped," as they can assist emergency responders in identifying the location of the problem and its direction of movement.

Some rules of thumb for evaluating monitoring readings are as follows:

1. If readings are high in one area and then drop off or dissipate in a relatively short period of time, the source of the

problem is often a spill or dumping directly into the sewer collection system.

2. If readings are consistent over a period of time, the source of the problem is often a subsurface release, such as an underground storage tank or pipeline.

Spills and releases into the sewer collection system will create both fire and environmental concerns.

Fire Concerns. Flammable liquids, such as gasoline and other low flash point, high vapor pressure liquids, will create the greatest risk of a fire or explosion. The potential for ignition within a sewer collection system will be greatest at points where liquids may enter or where entry is possible. Manholes, storm sewers, catch basins, and pumping station wet wells are likely to present the greatest areas of concern.

If a flammable liquid enters a sanitary or combined sewer, the probability of ignition will be high. If floor traps are not filled with water, flammable liquid in either a sanitary or combined sewer collection system will back up vapors into building basements and other low-lying areas where there are multiple ignition sources, such as pilot lights, hot water heaters, electrical equipment, and so on. Ignition sources should be isolated and controlled and the area ventilated using positive or negative ventilation tactics, as appropriate.

In some instances, vapors have flowed out of the sewer collection system and accumulated in low lying areas, only to be ignited by an ignition source completely outside of the sewer system (e.g., passing vehicle).

In the event of a fire or explosion, secondary and tertiary problems will likely be created by the emergency as the explosion will affect all utilities that occupy the same or nearby utility corridors. Natural gas leaks, electric and telephone utility outages, and a loss of both water and water pressure should be anticipated in areas which suffer a major fire or explosion.

Environmental Concerns. Environmental concerns will be greatest when dealing with poisons and environmentally sensitive materials or when there is no ignition of flammable liquids following their release into the sewer system. Depending on the type of sewer collection and treatment system, environmental impact may range from a shutdown of the wastewater treatment facility and/or destruction of microorganisms necessary for the treatment process, to a spill impacting environmentally sensitive areas (e.g., wetlands, wildlife refuge, etc.) or threatening both potable and aquifer drinking water supplies.

The selection of control agents, such as dispersants and firefighting foams, to control a fire or spill may also have secondary environmental impact. Both sewer department and environmental personnel from the respective on-scene governmental agencies should be consulted on any decision to apply firefighting foams or dispersants either into a sewer collection system or onto a waterway. See **Table 7-7** for Site Safety procedures and considerations.

Table 7-7 Site Safety Procedures for Hydrocarbon Spills into Sewer Collection Systems

- Verify that the public works/sewer department has been notified and enroute. Identifying the direction of flow is critical in identifying exposures and establishing evacuation zones.
- Continuous air monitoring must be established, particularly in low-lying areas. Responders may have difficulty obtaining a sufficient number of monitoring instruments and trained personnel to perform monitoring over a relatively large geographic area.
- Monitoring readings should be mapped as this will assist responders in identifying the location of the spill within the sewer system and its general speed and direction of movement. This information, in turn, will assist in establishing response priorities.
- Control all ignition sources in the area, including vehicles, traffic flares, and smoking materials. Depending on the nature of the emergency, large spills into a sanitary or combined sewer collection system may require the shutdown of gas and electric utilities until the situation is under control.
- If liquids, gases, or vapors are found in tunnels or subways, traffic should be stopped until responders can further investigate and assess the level of risk.
- If the spill is in a sanitary or combined sewer collection system and its speed and direction of movement is known, responders may be able to notify homeowners and facilities ahead of the spill to pour water into their basement floor traps as a quick preventive measure to minimize hydrocarbon vapor migration and build-up.
- Do not allow any personnel to stand on or near manholes and catch basins. In the event of a fire or explosion, manhole lids can be blown into the air and fire can quickly emanate from catch basins and other sewer openings. Manhole lids can be blown into adjoining buildings and vehicles and represent a significant life safety hazard.
- Responders should not enter a sewer collection system unless advised by representatives of the sewer system. In addition to the obvious fire and health hazards, sewer collection systems consist of piping and collection areas of various diameters and depths and pose significant physical hazards.
- There are many confined spaces within a wastewater collection and pumping system. Responders should also consider the presence of oxygen-deficient atmospheres and other toxic and flammable gases, including hydrogen sulfide, methane, and sewer and sludge gases.
- When either flushing or applying control agents into a sewer collection system, the agent must be applied at multiple points along the projected flow path. If control agents are only applied at the source of the release, the agent will never "catch up" with the head of the flow, and there will be a continuous flammable atmosphere within the sewer system.
- The injection of some control agents into a sewer collection system, such as firefighting foams, dispersants, and water, may also introduce air and possibly move the environment into the flammable range.

Coordination with Public Works and the Sewer Department

Preplanning with public works and the sewer department is critical. Responders should have a basic knowledge and understanding of the sewer system and its operations and a good working relationship with sewer department personnel. Responders should identify areas where there is a probability of hazmats entering the sewer collection system and discuss procedures and tactical options for handling such an emergency with the sewer department as part of its pre-incident planning and training activities.

When an emergency occurs, a sewer department representative should be requested to be on-scene as soon as possible. The evaluation and selection of tactical control options should be based upon input from the sewer department and the respective governmental environmental agency. Maps of the local sewer system will be a key element in identifying the direction in which a spill may potentially head and in identifying likely exposures. Effective use of sewer maps will require a sewer department representative who is familiar with the unique aspects of the local system, local construction techniques, and so on. However, sewer maps may not always be accurate and up to date, particularly in identifying all lateral/domestic connections and branch connections in an older, combined sewer system.

Determining Background Radiation Dose

Background radiation is naturally-occurring radiation and it varies depending on location. It is the main source of radiation exposure for most people.

Ionizing radiation is also generated in a range of medical, commercial, and industrial activities. The most familiar and largest of these sources of exposure is medical X-rays. Natural radiation contributes about 88% to the annual background dose to the general population and about 12% to annual medical procedures like X-Rays.

Normal background radiation varies depending on your geographic location. Some geographic areas have low background radiation where other locations are higher. Levels typically range from about 1.5 to 3.5 millisieverts per year but can be more than 50 mSv/yr depending on where one is located.

Each country establishes its own radiation protection standards that are generally in line with the International Commission on Radiological Background Protection (ICRP) to keep exposure as low as achievable. [Source: World Nuclear Association.]

The US EPA has an online application that can be used to calculate your personal radiation dose from any terrestrial location in the USA. The National Council on Radiation Protection Measurements recommends the average radiation dose per person in the US is 620 nrem.

Wrap-Up

Chief Concepts

- Hazards refer to a danger or peril. In hazardous materials response operations, hazards generally refer to the physical and chemical properties of a material. Risks refer to the probability of suffering harm or loss. Although the risks associated with hazmat response will never be completely eliminated, they can be successfully managed. The objective of response operations is to minimize the level of risk to responders, the community, and the environment.
- Hazard and risk assessment is the most critical function in the successful management of a hazardous materials incident. The key tasks in this analytical process are (1) identifying the materials involved; (2) gathering hazard information; (3) visualizing hazmat behavior and predicting outcomes; and (4) based on the evaluation process, establishing response objectives. The system that ties these elements together is the General Hazardous Materials Behavior Model.
- You must know how to use reference materials before the incident in order to use them effectively. Evaluate reference materials before use and make sure your references use the same definitions for hazard terms. A good guidebook should have a well-written "How to Use" section.
- Although reference guidebooks contain data on those chemicals most commonly encountered during hazmat incidents, they are usually not a complete listing of all the chemicals found in your community. There is no replacement for hazard analysis and contingency planning at both the plant and community levels.
- Each information specialist has their own strengths and limitations. It's a good idea to remove the term *expert* from your vocabulary; be wary of self-proclaimed experts without first verifying their background and knowledge.
- Networking and relationships are everything! Local responders and facility personnel must get out into their communities and establish personal contacts and relationships with response partners. These include state, regional, and federal environmental response personnel, law enforcement, clean-up contractors, industry representatives, wrecking and rigging companies, and so on.
- There is no single detection/monitoring device on the market that can do everything. Make sure you understand how an instrument will fit into your standard operating procedures and emergency response strategies. Anyone can use an instrument; the challenge is interpreting what the instrument is (and isn't) telling you and then making risk-based decisions to make the problem go away!
- The nature of the incident and the intent of the monitoring mission will drive the selection of monitoring technologies most appropriate for the incident.
- Emergency responders must understand the operating principles of the detection and monitoring equipment, its application and limitations, and the manner in which the instrument fits into existing response procedures.
- Unknowns will create the greatest challenge for responders. The nature of the incident (e.g., credible threat scenario involving WMD agents), the location of the emergency (e.g., outdoors, indoors, confined space), and the suspected physical state of the unknown (i.e., solid, liquid, or gas) will influence the monitoring strategy. In scenarios involving unknowns, the role of hazmat responders is much like that of a detective. At the conclusion of the testing process, responders may still be unable to specifically identify the material(s) involved; however, they should be able to rule out a number of hazard classes and shorten the list of possibilities.
- Initial air monitoring efforts should be directed toward determining if IDLH concentrations are present. Decisions regarding protective clothing recommendations, establishing hazard control zones, and evaluating any related public protective actions should be based on defined action levels for radioactivity, flammability, oxygen deficiency and oxygen enrichment, and toxicity.
- As a general rule, samples collected for product identification during emergency response operations should not be used for evidentiary purposes—collect a separate sample for evidence.
- An accurate evaluation of the real and potential problems will enable response personnel to develop informed and appropriate strategic response objectives and tactical decisions.
- To visualize likely hazardous materials behavior, five basic questions must be addressed:
  1. Where will the hazardous material and/or its container go when released?
  2. How will the hazardous material and/or its container get there?
  3. Why are the hazardous material and/or its container likely to go there?
  4. What harm will the hazardous material and/or its container do when it gets there?

5. When will the hazardous material and/or its container get there?

- Strategic goals are the broad game plan developed to meet the incident priorities (life safety, incident stabilization, environmental and property conservation). Essentially, strategic goals translate into "what are you going to do to make the problem go away?"
- Tactical objectives are specific and measurable processes implemented to achieve the strategic goals. In simple terms, tactical objectives come down to "how are you going to do it?"
- If you are unsure of the container damage or how the container is likely to breach, get assistance from product or container specialists. This may include railroad personnel, gas industry representatives, and cargo tank truck specialists.
- When petroleum products or chemicals are released into the ground, their behavior will depend on their physical and chemical properties (e.g., liquid versus gas, hydrocarbon versus polar solvent), the type of soil (e.g., clay versus gravel versus sand), and the underground water conditions (e.g., location and movement of the water table).
- Remember—your job is to be a risk evaluator, not a risk taker. Bad risk takers get buried; effective risk evaluators go home.

Hot Terms

Breach The event causing a hazmat container to open up or "breach." It occurs when a container is stressed beyond its limits of recovery (ability to hold contents).

Calibration The process of adjusting a monitoring instrument so that its readings correspond to actual, known concentrations of a given material. If the readings differ, the monitoring instrument can then be adjusted so that readings are the same as the calibrant gas.

Defensive Tactics These are less aggressive spill and fire control tactics where certain areas may be "conceded" to the emergency, with response efforts directed towards limiting the overall size or spread of the problem. Examples include isolating the pipeline by closing remote valves, shutting down pumps, constructing dikes, and exposure protection.

Direct-Reading Instruments These provide information at the time of sampling. They are used to detect and monitor flammable or explosive atmospheres, oxygen deficiency, certain gases and vapors, and ioninzing radiation.

Engulfing Once the hazmat and/or energy is released, it is free to travel or disperse, engulfing an area. The farther the contents move outward from their source, the greater the level of problems. How quickly they move and how large an area they engulf will depend upon the type of release, the nature of the hazmat, the physical and chemical laws of science, and the environment.

Explosion-proof Construction Encases the electrical equipment in a rigidly built container so that (1) it withstands the internal explosion of a flammable mixture, and (2) prevents propagation to the surrounding flammable atmosphere. Used in Class I, Division 1 atmospheres at fixed installations.

General Hazardous Materials Behavior Model A process developed by Ludwig Benner, Jr. in the 1970s used for visualizing hazmat behavior. Applies the concept of events analysis which is simply breaking down the overall incident into smaller, more easily understood parts for purposes of analysis.

Harm Pertains to the harm caused by a hazmat release. Harm events include thermal, mechanical, poisonous, corrosivity, asphyxiation, radiation and etiological.

Hazard Refers to a danger or peril. In hazmat operations, usually refer to the physical or chemical properties of a material.

Heat-affected Zone An area in the undisturbed tank metal next to the actual weld material. This area is less ductile than either the weld or the steel plate due to the effect of the heat of the welding process. This zone is most vulnerable to damage as cracks are likely to start here.

Impinge As the hazmat and/or its container engulf an area, they will impinge, or come in contact with exposures. They may also impinge upon other hazmat containers, producing additional problems.

Instrument Response Time Also known as "lag time," this is the period of time between when the instrument senses a product and when a monitor reading is produced. Depending upon the instrument, lag times can range from several seconds to minutes. Variables will include if the instrument has a pump and the use of sampling tubing.

Intrinsically safe Equipment or wiring that is incapable of releasing sufficient electrical energy under both normal and abnormal conditions to cause the ignition of a flammable mixture. Commonly used in portable direct-reading instruments for operations in Class I, Division 2 hazardous locations.

Lower Detection Limit (LDL) The lowest concentration to which a monitoring instrument will respond. The lower the LDL, the quicker contaminant concentrations can be evaluated.

Monitoring Instruments Monitoring and detection instruments used to detect the presence and/or concentration of contaminants within an environment. They include:

- Combustible Gas Indicator (CGI): Measures the concentration of a combustible gas or vapor in air.
- Oxygen Monitor: Measures the percentage of oxygen in air.
- Colorimetric Indicator Tubes: Measures the concentration of specific gases and vapors in air.
- Specific Chemical Monitors: Designed to detect a large group of chemicals or a specific chemical. Most common examples include carbon monoxide and hydrogen sulfide.
- Flame Ionization Detector (FID): A device used to determine the presence of organic vapors and gases in air. Operates in two modes—survey mode and gas chromatograph.
- Gas Chromatograph: An instrument used for identifying and analyzing specific organics compounds.
- Photoionization Detector (PID): A device used to determine the total concentration of many organic and some inorganic gases and vapors in air.
- Radiation Monitors: An instrument used to measure accumulated radiation exposure. Include both alpha, beta and gamma survey detectors.

Offensive Tactics Aggressive leak, spill and fire control tactics designed to quickly control or mitigate the problem. Although increasing risks to emergency responders, offensive tactics may be justified if rescue operations can be quickly achieved, if the spill can be rapidly confined or contained, or the fire quickly extinguished.

Release Once a container is breached, the hazmat is free to escape (be released) in the form of energy, matter, or a combination of both. Types of release include detonation, violent rupture, rapid relief, and spills or leaks.

Risk The probability of suffering a harm or loss. Risks are variable and change with every incident.

Risk-based Response Process Systematic process by which responders analyze a problem involving hazardous materials/weapons of mass destruction (WMD), assess the hazards, evaluate the potential consequences, and determine appropriate response actions based upon facts, science, and the circumstances of the incident.

Stress An applied force or system of forces that tend to either strain or deform a container (external action) or trigger a change in the condition of the contents (internal action). Types of stress include thermal, mechanical and chemical.

Vapor Pressure The pressure exerted by the vapor within the container against the sides of a container. This pressure is temperature dependent; as the temperature increases, so does the vapor pressure. Consider the following three points:

1) The vapor pressure of a substance at 100°F. is always higher than the vapor pressure at 68°F.
2) Vapor pressures reported in millimeters of mercury (mm Hg) are usually very low pressures. 760 mm Hg. is equivalent to 14.7 psi. or 1 atmosphere. Materials with vapor pressures greater than 760 mm Hg. are usually found as gases.
3) The lower the boiling point of a liquid, the greater vapor pressure at a given temperature.

HazMat Responder *in Action*

You have just arrived at an overturned MC-307/DOT-407 cargo tank truck that has turned over on the interstate that carries an identifying placard number 1294. It is estimated that there is somewhere between 5,000 and 8,000 gallons in the cargo tank truck and there is a moderate leak coming from the dome cover area. You estimate that there are approximately 1,000 gallons on the ground.

Questions

1. What informational resources can be used to identify this product, its chemical and physical properties, and the potential incompatibilities?
   - **A.** DOT ERG/NIOSH/CAMEO
   - **B.** PID/ FID/ERG/NIOSH
   - **C.** NRC/ERG/NOAA/EPA
   - **D.** OSHA/PID/CGI/ERG
2. Which three initial detection devices would you use to establish your hazard control zones?
   - **A.** Radiation monitor/Colorimetric tubes/Passive dosimeters
   - **B.** pH paper/Radiation monitor/Colorimetric tubes
   - **C.** pH paper/Radiation monitor/CGI
   - **D.** pH paper/Radiation monitor/PID
3. What type of stress appears to have impacted the cargo tank truck?
   - **A.** Thermal
   - **B.** Mechanical
   - **C.** Radiation
   - **D.** Chemical
4. According to the General Hazardous Materials Behavior Model, in which event are you?
   - **A.** Stress
   - **B.** Breach
   - **C.** Release
   - **D.** Engulf

References and Suggested Readings

1. Air Force Civil Engineer Support Agency (AFCESA) and PowerTrain, Inc., *Hazardous Materials Incident Commander Emergency Response Training CD,* Tyndall Air Force Base, FL: Headquarters AFCESA (2010).
2. Air Force Civil Engineer Support Agency (AFCESA) and PowerTrain, Inc, *Hazardous Materials Technician Emergency Response Training CD,* Tyndall Air Force Base, FL: Headquarters AFCESA (2010).
3. American Petroleum Institute. API 1628—*Guide to the Assessment and Remediation of Underground Petroleum Releases* (3rd Edition), Washington, DC: American Petroleum Institute (1996).
4. American Society of Testing and Materials. *ASTM E2458-10: Standard Practices for Bulk Sample Collection and Swab Sample Collection of Visible Powders Suspected of Being Biological Agents from Nonporous Surfaces,* West Conshohocken, PA: ASTM International (2010).
5. American Society of Testing and Materials. *ASTM E2770-10: Standard Guide for Operational Guidelines for Initial Response to a Suspected Biothreat Agent,* West Conshohocken, PA: ASTM International (2010).
6. Armed Forces Radiobiology Research Institute—Military Medical Operations, *Medical Management of Radiogical Casualties Handbook* (3rd Edition), Bethesda, MD: AFRRI (June 2010).
7. Benner, Ludwig, Jr., "D.E.C.I.D.E. In Hazardous Materials Emergencies." *Fire Journal* (July 1975), pages 13–18.
8. Benner, Ludwig, Jr., *Hazardous Materials Emergencies* (2nd Edition), Oakton, VA: Lufred Industries, Inc. (1978).
9. Bevelacqua, Armando, *Hazardous Materials Chemistry.* Albany, NY: Delmar–Thomson Learning (2001).
10. Bevelacqua, Armando and Richard Stilp, *Terrorism Handbook for Operational Responders* (2nd edition), Clifton Park, NY: Delmar–Cengage Learning (2006).
11. Brunacini, Alan V., *Fire Command* (2nd Edition), Quincy, MA: National Fire Protection Association (2002).
12. Emergency Film Group, Air Monitoring (two videotape series), Plymouth, MA: Emergency Film Group (2003).
13. Emergency Film Group, Detecting Weapons of Mass Destruction, (videotape), Plymouth, MA: Emergency Film Group (2003).
14. Emergency Film Group, Radiation Monitoring, (videotape), Plymouth, MA: Emergency Film Group (2011).
15. Emery, Rick, "Field Biological Detection Capabilities," *Fire Engineering Terrorism Supplement* (September 2011).
16. Fingas, Merv F. et al., "The Behavior of Dispersed and Nondispersed Fuels in a Sewer System." *American Society of Testing and Materials—Special Technical Publication 1018* (1989).
17. Fingas, Merv F. et al., "Fuels in Sewers: Behavior and Countermeasures." *Journal of Hazardous Materials,* 19 (1988), pages 289–302.
18. Ghormely, David M. "Emergency Response to Polymerizable Materials." Student Handout—Houston, TX: Rohm & Haas Company, Inc. (2003).
19. Hawley, Chris, *Hazardous Materials Air Monitoring and Detection Instruments* (2nd edition). Clifton Park, NY: Delmar/Cengage Learning (2007).
20. Hawley, Chris, "Response Strategies for Unidentified Materials." *Fire Engineering Terrorism Supplement* (November 2005).
21. Hawley, Chris, Gregory G. Noll and Michael S. Hildebrand, "The Need for Joint Hazard Assessment Teams," *Fire Engineering Terrorism Supplement* (September 2009).
22. Johnson, Kevin W., "WMD/Hazardous Materials Evidence Awareness." *Fire Engineering Terrorism Supplement* (September 2008).
23. Ladd, David M. and Cheryl Gauthier, "Bioterrorism Response: Does the Way Forward Lie in Our Past?" *Fire Engineering Terrorism Supplement* (September 2009).
24. Larrañaga, Michael D., David L. Volz and Fred N. Bolton, "Pressure Effects on and Deformation of Hazardous Waste Containers." *Fire Engineering* (July, 1999).
25. Maslansky, Carol J. and Steven P. Maslansky, *Air Monitoring Instrumentation,* New York, NY: Van Nostrand Reinhold (1993).
26. National Fire Protection Association, *Fire Protection Handbook* (20th Edition), Section 7—Managing Response to Hazardous Materials Incidents, Quincy, MA: National Fire Protection Association (2008).
27. National Fire Protection Association, *Hazardous Materials Response Handbook* (6th Edition), Quincy, MA: National Fire Protection Association (2013).
28. National Fire Protection Association, *National Electrical Code Handbook,* Quincy, MA: National Fire Protection Association (2011).

29. National Fire Protection Association, *Recommended Practice for Handling Releases of Flammable and Combustible Liquids and Gases—NFPA 329,* Quincy, MA: National Fire Protection Association (2010).
30. National Fire Protection Association, *Standard for Fire Protection in Wastewater Treatment and Collection Facilities—NFPA 820,* Quincy, MA: National Fire Protection Association (2012).
31. National Transportation Safety Board, "Derailment of a CSX Transportation Freight Train and Fire Involving Butane in Akron, OH." (Report NTSB/HZM-90/2). Washington, DC: National Transportation Safety Board (February 26, 1989).
32. Plog, Barbara A. and Patricia J. Quinlan, *Fundamentals of Industrial Hygiene* (5th Edition), Chicago, IL: National Safety Council (2002).
33. Sidell, Frederick, M.D., William Patrick and Thomas Dashiell, *Jane's Chem-Bio Handbook* (2nd Edition), Alexandria, VA: Jane's Information Group (2003).
34. Union Pacific Railroad Company, "Assessing Tank Car Damage," Participant's Manual—Tank Car Safety Course. Omaha, NE: Union Pacific Railroad Company (April, 2003).
35. U.S. Army Medical Research Institute of Chemical Defense, *Medical Management of Chemical Casualties* (3rd Edition), Aberdeen Proving Ground, MD: USAMIC—Chemical Casualty Care Office (July, 2000).
36. U.S. Army Medical Research Institute of Infectious Diseases (USAMRID), *Medical Management of Biological Casualties* (6th Edition), Fort Detrick, MD: USAMRID (April, 2005).
37. Water Pollution Control Federation, *Emergency Planning for Municipal Wastewater Facilities (MOP SM-8),* Arlington, VA: Water Pollution Control Federation (1989).
38. Wright, Charles, "Predicting Behavior and Estimating Outcomes." Student Handout—Omaha, NE: Union Pacific Railroad Company (2003).

Selecting Personal Protective Clothing and Equipment

CHAPTER 8

Hazardous Materials Technician

Knowledge Objectives

After reading this chapter, you will be able to:

- Describe the types of personal protective equipment that are available for response based on NFPA standards and how these items relate to EPA levels of protection (p. 274).
- Identify and describe personal protective equipment options available for the following hazards (p. 270):
  - Thermal
  - Radiological
  - Asphyxiating
  - Chemical (liquids and vapors)
  - Etiological (biological)
  - Mechanical (explosives)
- Identify the process for selecting personal protective clothing and equipment at Hazmat/WMD incidents (p. 265, 266, 269, 270, 281, 282).
- Describe the following terms and explain their impact and significance on the selection of chemical-protective clothing:
  - Degradation (p. 263)
  - Penetration (p. 264)
  - Permeation (p. 264, 265)
  - Cumulative Permeation (p. 264)
- Describe the differences between limited-use and multi-use chemical protective clothing materials. (p. 266)
- Identify the different designs of chemical vapor-protective and splash-protective clothing and describe the advantages and disadvantages of each type (p. 276–280).
- Describe the advantages, limitations, and proper use of structural firefighting clothing at a Hazmat/WMD incident. (p. 274–276)
- Identify two types of high-temperature protective clothing and describe the advantages and disadvantages of each type. (p. 280, 281)
- Describe safety and emergency procedures for personnel working in chemical protective equipment (p. 281–285).

Skills Objectives

After reading this chapter, you will be able to:

- Given three examples of various hazardous materials, determine the protective clothing construction materials for a given action option using chemical compatibility charts (p. 269, 280).
- Given the personal protective equipment provided by the AHJ, identify the process for inspecting, testing, and maintenance of personal protective equipment (p. 292).
- Demonstrate the ability to don, work in, and doff self-contained breathing apparatus in addition to any other respiratory protection provided by the AHJ (p. 270–273).
- Demonstrate the ability to don, work in, and doff liquid splash–protective, vapor-protective, and chemical-protective clothing in addition to any other specialized protective equipment provided by the AHJ (p. 287–295).

Hazardous Materials Incident Commander

- Identify the four levels of chemical protection (EPA/OSHA) and describe the equipment required for each level and the conditions under which each level is used (p. 274).
- Describe the following terms and explain their impact and significance on the selection of chemical-protective clothing:
  - Degradation (p. 263)
  - Penetration (p. 264)
  - Permeation (p. 264–265)
- Describe three safety considerations for personnel working in vapor-protective, liquid splash–protective, and high temperature–protective clothing (p. 281–283).
- Describe the advantages, limitations, and proper use of structural firefighting clothing at a HazMat/WMD incident. (p. 274–276)

Skills Objectives

There are no Incident Commander skills objectives in this chapter.

You Are the HazMat Responder

Your Hazardous Materials Response Team (HMRT) has been dispatched to a chemical manufacturing plant for a report of a railroad tank car containing dimethylamine (DMA) leaking from the sample line. The tank car is located on an industrial spur within the chemical plant property.

After conducting a hazard assessment and risk evaluation, a joint entry team consisting of the railroad's Hazmat Supervisor and three Hazmat Technicians, all wearing chemical vapor protective clothing, climb on the tank car and are able to install a repair clamp over the leak in the sample line. However, approximately 30 minutes later, the clamp begins to leak and a second entry operation is undertaken. During this second entry, both the railroad supervisor and the Hazmat Technicians begin to experience visibility problems with their CPC face pieces. Problems included melting, clouding, reduced visibility, and lens microcracking, which further reduced visibility Figure 8-1. Leaks also developed in the seams of several of the suits. The face piece in one of the chemical vapor suits shatters when the supervisor drops about 3 ft. while getting off the tank car, subsequently exposing him to the hazardous environment.

Because of their CPC suit failures, the Incident Commander (IC) prohibits further attempts to stop the leak. The incident is ultimately controlled and terminated after a commercial hazardous materials contractor arrives and is able to stop the leak. No difficulties were experienced with the chemical vapor suits used by the contractor; however, the exposure period was very short. All of the CPC used in this emergency were made by the same manufacturer but were not of the same design. Chemical barrier information initially available to emergency responders led them to believe that the CPC suits were adequate for protecting their personnel against the DMA environment.

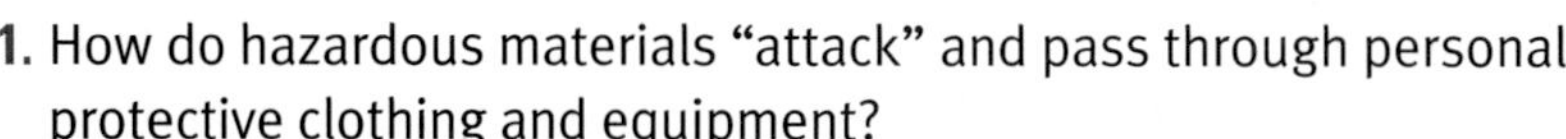

1. How do hazardous materials "attack" and pass through personal protective clothing and equipment?
2. What factors should be considered in selecting personal protective clothing and equipment?
3. What are the strengths and weaknesses of the various types and levels of personal protective clothing and equipment?

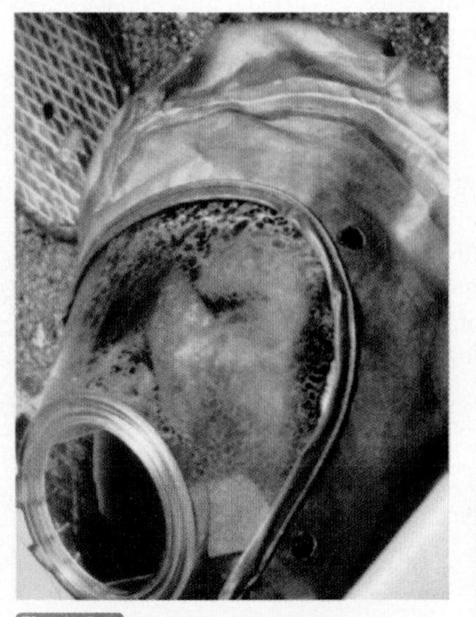

Figure 8-1 Example of a chemical vapor protective clothing failure.

This incident actually occurred in Benicia, California on August 12, 1983, and vividly illustrates some of the personal protective clothing issues and concerns that can be encountered by responders. The willingness of the individuals involved in this emergency to network and "share the lessons learned" eventually led to changes and improvements in both the hazmat emergency response and chemical protective clothing professions. As a result of this incident, there are now NFPA consensus standards that specify minimum documentation, design and performance criteria, and test methods for CPC. In addition, the quality of CPC and the chemical barrier information has significantly improved.

Introduction

This chapter covers the fourth step in the Eight Step Process©—Select Personal Protective Clothing and Equipment. In emergency response applications, the selection of Personal Protective Equipment (PPE) cannot be safely and adequately addressed until tactical response objectives are determined as part of the Hazard and Risk Evaluation process. Likewise, response operations cannot be safely performed if responders are not provided with the proper level of personal protective clothing and are trained in its proper application, limitations, and use.

Protection against hazardous materials can be provided through engineering controls, by the use of safe work practices and administrative controls, and through the use of personal protective clothing and equipment. While engineering controls and safe work practices are preferred for personal protection in controlled workplace environments, emergency responders must typically rely on the safe and effective use of personal protective clothing and equipment.

This chapter will review the application and use of personal protective clothing and equipment at hazmat emergencies. Topics will be presented in the following order:

- Basic principles of **chemical protective clothing (CPC)**, including methods by which hazardous materials attack PPE, protective clothing materials, categories of chemical protective clothing, and chemical barrier and initial selection considerations
- Respiratory protection, including air purification and atmosphere supplying devices
- Levels of protection, including structural firefighting clothing, liquid splash and chemical vapor protective clothing, and high temperature protective clothing.
- Tying the PPE system together, including operational considerations, emergency procedures, donning, doffing, support consideration, and training and inspection/maintenance procedures.

Basic Principles

When evaluating protective clothing for use at a hazmat/WMD (weapons of mass destruction) incident, primary concerns should focus upon chemical resistance of the garment, the integrity of the entire protective clothing ensemble (including the garment, visor, seams, zippers, gloves, boots, overall design, etc.), and the tasks to be performed. Responders must be familiar with the methods by which chemicals may attach and pass through chemical clothing materials Figure 8-2. Key terms and their significance in evaluating protective clothing chemical barriers are as follows:

- **Degradation** is the physical destruction or decomposition of a clothing material due to exposure to chemicals, use, or ambient conditions (e.g., storage in sunlight). Degradation is noted by visible signs such as charring, shrinking, swelling, color change or dissolving, or by testing the clothing material for weight changes, stiffening, loss of fabric tensile strength, and so on. Degradation can occur when chemical barrier data are not properly interpreted or understood by responders, the wrong protective clothing material is used, or exposure recommendations are exceeded.

    Although permeation testing is most common, chemical barrier charts may be found that are based on degradation. These data are normally based upon laboratory tests conducted with pure, undiluted test chemicals on clean, uncontaminated swatches of material over a pre-established time period (often 2 to 6 hours). Virtually all testing is done at room temperature (70°F/21°C).

Responder Safety Tip

Degradation tests do not account for simultaneous exposures to two or more chemicals, exposures at elevated temperatures, or previous exposure to chemicals. Furthermore, degradation testing alone is insufficient to demonstrate the barrier performance of protective clothing materials.

DEGRADATION

PENETRATION

PERMEATION

Figure 8-2 Chemical protective clothing resistance to chemical attack is described in terms of chemical degradation, penetration, and permeation.

- **Penetration** is the flow or movement of a hazardous chemical through closures, seams, porous materials, and pinholes or other imperfections in the material. While liquids are most common, solid materials (e.g., asbestos) can also penetrate through protective clothing materials.

  Penetration testing may be used to demonstrate the liquid barrier properties of protective clothing materials and is important for evaluating the performance of liquid splash CPC. Penetration resistance data are provided as "pass" or "fail" relative to the specific chemical or mixture tested.

  Causes of penetration include clothing material degradation, manufacturing defects, physical damage to the suit (e.g., punctures, abrasions, etc.), normal wear and tear, and PPE defects. CPC can be penetrated at several locations, including the face piece and exhalation valve, suit exhaust valves, and suit fasteners and closures. The potential for penetration generally increases at excessively hot or cold temperatures.
- **Permeation** is the process by which a hazardous chemical moves through a given material on the molecular level. Permeation differs from penetration in that permeation occurs through the clothing material itself rather than through the openings or gaps in the clothing material. Permeation can lead to protective clothing failures when chemical barrier data are not properly interpreted or understood by responders, or if breakthrough times (see below) are exceeded. Because of its significance in evaluating the integrity of PPE, chemical contamination, and decontamination, permeation will be discussed more completely.

Permeation Theory

The process of chemical permeation through an impervious barrier is a three-step process consisting of the following Figure 8-3:

1. Adsorption of the chemical into the outer surfaces of the clothing material; generally not detectable by the wearer
2. Diffusion of the chemical through the clothing material
3. Desorption of the chemical from the inner surface of the clothing material (toward the wearer)

Breakthrough time is defined as the time from the initial chemical attack on the outside of the material until its desorption and detection inside. The units of time are usually expressed in minutes or hours, and a typical test runs up to a maximum of 8 hours. If no measurable breakthrough is detected after 8 hours, the result is often reported as a breakthrough time of > 480 minutes or > 8 hours.

Permeation rate is the rate at which the chemical passes through the CPC material and is generally expressed as micrograms per square centimeter per minute (μg/cm^2/min). For reference purposes, .9 μg/cm^2/min is equal to approximately 1 drop/hour. The higher the rate, the faster the chemical passes through the suit material. Comprehensive chemical barrier charts will contain both the breakthrough time and the permeation rate.

Measured breakthrough times and permeation rates are determined by laboratory permeation testing procedures against a list of chemicals outlined in ASTM F 1001, *Standard Guide for Chemicals to Evaluate Protective Clothing Materials*. The ASTM F 1001 list is also used as the NFPA 1991, *Standard of Vapor Protective Suits for Emergency Response* battery of chemicals for determining CPC material permeation resistance.

Permeation testing is conducted using pure, undiluted test chemicals on clean, uncontaminated swatches over a pre-established period of time (usually 2 to 8 hours). Virtually all testing is conducted at ambient room temperatures (70°F/21°C). Use of the breakthrough times from this testing process then allows responders to estimate the duration of maximum protection under a worst-case scenario of continuous chemical contact.

In evaluating permeation resistance and breakthrough times, several other terms may be found in the CPC manufacturer's barrier information Figure 8-4:

- *Cumulative permeation*—The total mass of chemical that permeates during a specified time from when the material is first contacted by the test chemical (*Source*: ASTM F 1407). NFPA 1994 currently uses a cumulative permeation value of 6.0 μg/cm^2 for one hour for

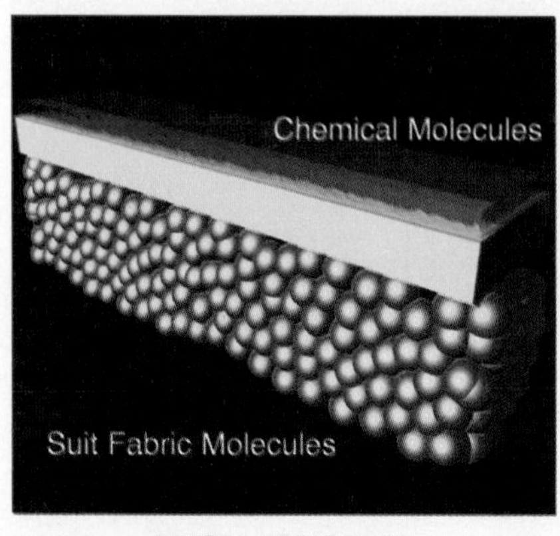

PHASE I: ADSORPTION

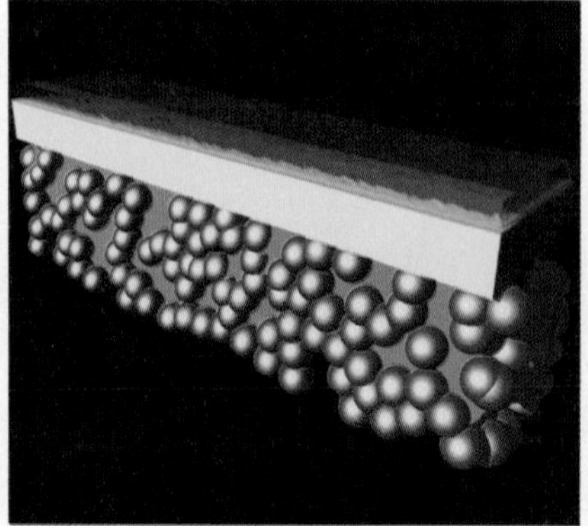
PHASE II: DIFFUSION

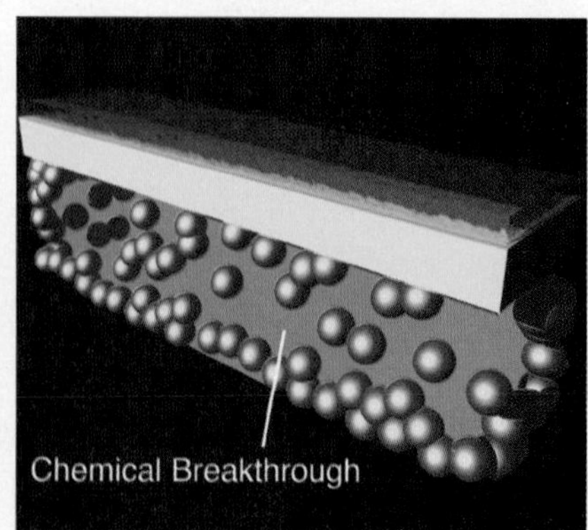

PHASE III: DESORPTION

Figure 8-3 Chemical permeation through protective clothing is a three-step process: absorption, diffusion, and desorption.

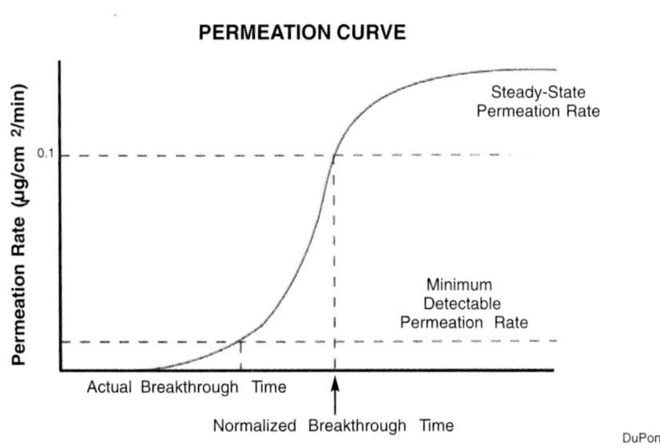

Figure 8-4 Permeation curves can visually illustrate the relationship between permeation rate and time. Breakthrough time is the initial point at which a chemical is detected on the inside of a CPC material.

chemical agents. At the time of publication, cumulative permeation was being considered for use in NFPA 1991 for toxic industrial chemicals (TIC) and toxic industrial materials (TIM).

- *Actual breakthrough time*—Breakthrough time as previously defined.
- *Normalized breakthrough time*—A calculation, using actual permeation results, to determine the time at which the permeation rate reaches 0.1 μg/cm^2/min. Normalized breakthrough times are useful for comparing the performance of several different protective clothing materials. Note that in Europe, breakthrough times are normalized at 1.0 μg/cm^2/min., a full order of magnitude less sensitive.
- *Minimum detectable permeation rate (MDPR)*—The minimum permeation rate that can be detected by the laboratory analytical system being used for the permeation test.
- *System detection limit (SDL)*—The minimum amount of chemical breakthrough that can be detected by the laboratory analytical system being used for the permeation test. Lower SDLs result in lower (or earlier) breakthrough times.

Chemical permeation rates are a function of many factors, including the following:

- *Temperature.* Most chemical barrier tests are conducted at ambient temperatures (68° to 72°F/20° to 22.2°C). However, as temperature increases, permeation rates increase and breakthrough times shorten.
- *Thickness.* Permeation is inversely proportional to the thickness of the clothing material. In other words, doubling its thickness will theoretically cut the permeation rate in half. The breakthrough time will, therefore, become longer.
- *Chemical mixtures and their effects upon chemical resistance are relatively unknown.* It is impossible to test for every possible chemical combination. Tests have also shown that combining data concerning exposures to individual chemicals has limited value. For example, Viton®/chlorobutyl laminate resists hexane for over 3 hours and acetone for 1 hour. However, any mixture of hexane and acetone will permeate Viton®/chlorobutyl laminate in under 10 minutes. Chemical mixtures may result in a stronger attack on protective clothing than with individual exposures.
- *Previous exposures.* Once a chemical has begun the diffusion process, it may continue to diffuse even after the chemical itself has been removed from the outside surface of the material. This is significant when considering the re-use of any protective clothing. *Decontamination does not ensure that permeation has stopped.* Although it is possible to test for permeation, the testing process also destroys the clothing material.

Protective Clothing Materials

Parameters that should be considered when evaluating and choosing CPC materials and barrier fabrics include the following:

1. **Chemical resistance.** This is the most critical factor, as the clothing material must maintain its integrity and protection qualities when it comes in contact with a hazardous material. Chemical resistance (either permeation or penetration) must be evaluated. Permeation resistance test data should be used for vapor CPC, particularly for chemicals that are toxic by skin absorption or present other hazards in vapor form.

   Chemical penetration data should be used for liquid splash CPC. However, penetration data should not be used when chemicals are highly toxic or give off hazardous vapors.
2. **Flammability.** The CPC garment should not contribute to the fire hazard and should maintain its protective capabilities when exposed to elevated temperatures. When burning, the clothing material should not melt or drip.

   Be aware that CPC is not appropriate for firefighting operations or for protection in flammable or explosive environments. While some clothing manufacturers offer "flash" protective overgarments for use with CPC, they offer only limited protection for rapid, short-duration escape and should not be considered for firefighting applications.
3. **Strength and durability.** These characteristics reflect a material's ability to resist cuts, tears, punctures, abrasions, and other physical hazards found at the incident. The breaking strength, seam strength, and closure strength also relate to how well clothing materials withstand repeated use or the stresses of use.
4. **Overall integrity.** CPC should provide complete protection to the wearer. Each vapor protective garment should be tested for integrity with a "pressure" or inflation test. Similarly, liquid splash protective clothing designs should be tested for liquid-tight integrity to verify that the suit design does not allow any liquids onto the wearer's skin.
5. **Flexibility.** This is the ability of the user to move and work in protective clothing and, generally, a factor in how the ensemble is fabricated. Emphasis is on the garment weight, fabrication of seams (i.e., bonded, sewn, glued, sealed, taped), and their resistance to chemical and wear exposures.

Flexibility and dexterity are particularly important with gloves and chemical protective ensembles.

When dealing with laminated fabrics, microcracking may become a concern. Continuous flexing of a laminate material may cause microcracks to develop on the inner laminate layers and lead to a CPC failure with little indication of potential failure.

6. **Temperature characteristics.** Temperature characteristics affect the ability of protective clothing to maintain its protective capacity in temperature extremes. Higher temperatures increase the effects of all chemicals upon polymers. A material suitable for chemical exposures at ambient temperatures may fail at elevated temperatures. Exposure to low temperatures may cause materials to stiffen, crack, flake, or separate.
7. **Shelf life.** Long-term exposure to sunlight or excessive heat (104°F/>40°C) causes many CPC materials to age and deteriorate. Some materials, particularly rubber-like materials, deteriorate over time, much like automobile tires. These changes may also occur without use. Shelf life information should be obtained from the manufacturer for the specific CPC products and materials being used. For example, DuPont suggests that garments older than 5 years old be downgraded for training use only.
8. **Decontamination and disposal.** The ability of the protective clothing material to be cleaned and decontaminated must be evaluated against potential chemical exposures. Limited-use garments often represent a cost-effective option, with the suit being disposed after a single use or hazmat exposure. Although decontamination (decon) may reduce the level of contamination in reusable suits, in many instances the contaminant will not be completely eliminated.

Categories of Chemical Protective Clothing

Protective clothing materials can be classified by use into two broad categories—limited-use garments and reusable garments.

Limited-use materials are protective clothing materials that are used and then discarded. Manufacturers recommend that garments constructed from these materials can be worn until damaged, altered, or contaminated. They are engineered for one or a low number of wearings before disposal. They eliminate many health and safety concerns regarding CPC decon and their return to service.

Most limited-use materials are usually constructed of a nonwoven fiber or a nonwoven fabric with a laminated film (e.g., fabrics such as DuPont Tychem®, Kappler Zytron, Kimberly-Clark Kleenguard® A70 & A80, Lakeland ChemMAX®, W. L. Gore Chempak® fabrics). In general, limited-use garments typically provide a broader range of chemical protection than reusable garments and are lighter in weight than reusable garments with synthetic rubber or vinyl. However, they can be less durable and not as strong as reusable synthetic rubber or vinyl. The chemical resistance of these garments can sometimes be compromised by the physical breakdown of the material which may occur with improper environmental or storage conditions.

Limited-use garments are generally suitable for a single use and should be disposed of in accordance with local, state, and federal environmental regulations. Examples of limited-use garments used in hazmat emergency response include:

- Tychem® QC—polyethylene coated Tyvek®
- Tychem® SL—Dow Saranex® 23P barrier film laminated Tyvek®
- Tychem® F—barrier film laminated to Tyvek®
- Tychem® CPF® 3—barrier film laminates
- Tychem® TK—multiple barrier film laminates
- Tychem® CSM®—multiple barrier film laminates
- Trellchem® TLU—polyamide fabric laminated on each side with a barrier film laminate
- Hazard-Gard™ I—nonwoven fabric laminated with a polyethylene film
- Hazard-Gard™ II—Saranex® 23P film laminated to polypropylene fabric

Advantages of limited-use materials include lower costs, ability to stock a larger and more varied protective clothing inventory, and reduced inspection and maintenance requirements. They are often used for support functions, including decontamination, remedial clean-up of identified chemicals, and training.

Reusable garments are designed and fabricated to allow for decon and reuse. Generally thicker and more durable than limited-use garments, they are used for liquid chemical splash and vapor protective suits, gloves, aprons, boots, and thermal protective clothing. Reusable garment materials are usually made from chlorinated polyethylene (CPE), vinyl (plasticized polyvinyl chloride—PVC), fluorinated polymers (Teflon®), and rubberlike fabrics, such as butyl rubber, neoprene rubber, and Viton®. Some reusable garment materials combine the barrier technology of limited-use garments with the durability of synthetic rubber (e.g., Trellchem® TLU and HPS).

Although these garments are considered reusable, certain chemical exposures may require the disposal of this clothing as well. Disposal must be in accordance with local, state, and federal environmental regulations **Figure 8-5**.

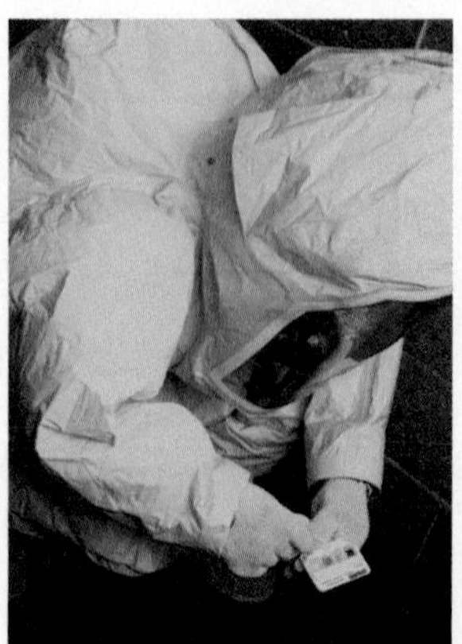

Figure 8-5 Chemical protective clothing must meet national standards in compliance with NFPA 1991, 1992, or 1994.

Scan Sheet 8-A

NFPA Chemical Protective Clothing Standards

The NFPA Technical Committee on Hazardous Materials Protective Clothing and Equipment has developed three consensus standards that specify minimum documentation, design and performance criteria, and test methods for CPC. These standards are often referenced as minimum requirements in purchase specifications and cover the following:

1. NFPA 1991—*Vapor Protective Ensembles for Hazardous Materials Emergencies*
2. NFPA 1992—*Liquid Splash Protective Ensembles for Hazardous Materials Emergencies*
3. NFPA 1994—*Protective Ensemble for Chemical/Biological Terrorism Incidents*

Each standard requires independent, third party certification to ensure the protective clothing meets the respective design, performance, and documentation requirements. Certification agencies, such as Underwriter's Labs (UL) or the Safety Equipment Institute (SEI), certify the garment performance and list the certified products on their websites. NFPA is not involved in the certification process and does not list certified products. Compliant products must carry a product label indicating compliance with the NFPA standard, a technical data package, and user instructions. A chemical protective ensemble can be certified to several NFPA standards.

NFPA standards provide minimum performance requirements for specific situations, which are clearly stated in the Standard's purpose statement. For example, the scope of NFPA 1991 states the following:

- This standard shall specify minimum design, performance, certification, and documentation requirements; and test methods for vapor-protective ensembles and individual elements for chemical vapor protection; and additional optional criteria for chemical flash fire escape protection and liquefied gas protection.
- This standard shall not apply to protective clothing for any fire fighting applications and shall not provide criteria for protection from radiological or cryogenic liquid hazards, or from explosive atmospheres.
- This standard shall not apply to vapor protective ensembles for protection from biological hazards unless the ensemble is certified as compliant with the additional requirements for chemical and biological terrorism incidents.

NFPA 1991 and 1992 performance requirements can be summarized as follows:

PERFORMANCE REQUIREMENT	NFPA 1991 Chemical Vapor Protective Clothing	NFPA 1992 Liquid Splash Protective Clothing
SUIT INTEGRITY & OVERALL BARRIER	*Inflation Test* & *Gas Inward Leakage Test* after exercise	*Shower Test* after exercise
BARRIER	Permeation resistance after flex and abrasion	Penetration resistance after flex and abrasion
CHEMICAL BATTERIES	ASTM F1001 Test Battery for Permeation Resistance (21 chemicals—15 liquids & 6 gases, including 5 dual-use TICs and 2 chemical warfare agents)	ASTM F1001 Test Battery for Penetration Resistance (7 liquid chemicals not carcinogens, skin absorbers and < 5 mm Hg vapor pressure)
PHYSICAL HAZARD RESISTANCE	Puncture/tear, burst resistance, seam strength, closure strength. Specific tests for footwear and gloves	
DURABILITY	Abrasion Resistance, Flex Fatigue, Suit Integrity testing after exercise	
FUNCTIONAL PERFORMANCE	User Mobility, Visor Clarity, Glove Dexterity	
COMPONENT PERFORMANCE	Valve Leakage, Pass Thru Strength	Pass Thru Strength
FLAME RESISTANCE	Primary materials exposed to flame in 2 stages: 3 seconds to determine ease of ignition; and 12 seconds to determine flame extinguishment. Fabric cannot melt or drip.	No flame impingement requirements in base specifications; part of the optional flame fire escape requirements
COMPONENT TESTING	Examples may include visibility through a visor; glove dexterity/gripping ability; footwear slip resistance; exhaust valve operation.	
OTHER SUIT PERFORMANCE TESTS	*Functionality Test*—user demonstrates ability to perform tasks, manipulate tools, etc. *Airflow Test*—assesses ability of CPC to exhaust air if SCBA goes into the bypass mode.	
OPTIONAL PERFORMANCE AREAS	■ Liquefied Gases ■ Flash fire escape protection	■ Flash fire escape protection

NFPA 1994—Protective Ensemble for First Responders to CBRN Terrorism Incidents

NFPA 1994 was originally enacted in 2001 as a result of the growing terrorism problem. It sets minimum protection levels for first responders who may be exposed to terrorism agents or victims during assessment, extrication, rescue, triage, decontamination, treatment, site security, crowd management, and force protection operations. NFPA 1994 defines four classes of ensembles based on the perceived threats at an incident. Ensemble differences are based on (1) the ability of the design to resist inward leakage of chemical and biological contaminants; (2) resistance of the suit to chemical warfare and toxic industrial chemicals; and (3) the strength and durability of these materials. All NFPA 1994 ensembles (i.e., garment, gloves, and footwear) are designed for a single exposure.

Many of the NFPA 1994 testing requirements are similar to those found in both NFPA 1991 and 1992. In addition, the standard permits dual certification.

- **Class 1 ensembles** were specified in the first edition of NFPA 1994. In 2006, the chemical, biological, radiological, and nuclear (CBRN) requirements were added to NFPA 1991 and the Class 1 requirements were removed from NFPA 1994. NFPA 1991 compliant ensembles should be considered as the Class 1 level of performance, and offer the highest level of protection and are intended for use in worst-case circumstances, where the substance creates an immediate threat and is unidentified and of unknown concentrations. Scenarios for use may include an ongoing release with likely gas/vapor exposures, the responder is close to the point of release, and most victims in the area appear to be unconscious or dead from exposure. SCBA and air-supplied units would be used for respiratory protection.
- **Class 2 ensembles** provide limited protection to emergency first responder personnel at terrorism incidents involving vapor or liquid chemical hazards where the concentrations are at or above Immediately Dangerous to Life or Health (IDLH) concentrations. The level of garment barrier performance is tied to the performance of the CBRN-certified SCBAs.
- **Class 3 ensembles** provide limited protection to emergency first responder personnel at terrorism incidents involving low levels of vapor or liquid chemical hazards where the concentrations are below IDLH concentrations, permitting the use of air purifying respirators (APRs) or powered air-purifying respirators (PAPRs).
- **Class 4 ensembles** provide limited protection to emergency first responder personnel at terrorism incidents involving biological hazards or radiological particulate hazards where the concentrations are below IDLH concentrations, permitting the use of APRs or PAPRs.

The NFPA 1994 performance classes can be summarized as shown in Table 8-1.

Test chemicals currently used in NFPA 1994 are as shown in Table 8-2.

Table 8-1 NFPA 1994 Performance Classes

Class	Challenge	Respirator	Vapor Threat	Liquid Threat	Victim's Condition
1 (NPFA 1991)	Vapors Aerosols Pathogens	Open-circuit CBRN SCBA	Unknown or Not Verified	High	Unconscious, not symptomatic and not ambulatory
2	Limited vapors Moderate liquid splash Liquid-borne Pathogens	Open-circuit CBRN SCBA	> IDLH	Moderate	Mostly alive, but not ambulatory
3	Limited vapors Light liquid splash Liquid-borne pathogens	CBRN APR or CBRN PAPR	Below IDLH; O_2 > 19.5%; Chemical Identified & Conc. Is Known; APR Suitable	Low to none	Self-ambulatory
4	Airborne and Liquid-borne Pathogens	CBRN APR or CBRN PAPR	Below IDLH; O_2 > 19.5%; Chemical Identified; Conc. Known; APR Suitable	Low to none	Self-ambulatory

Table 8-2 **NFPA 1994 Test Chemicals**

Type of Material	Chemical
Chemical Agent	■ Distilled Sulfur Mustard (HD) ■ Sarin (GB)
Industrial Chemical (Liquid)	■ Dimethyl Sulfate (DMA) ■ Acrolein ■ Acrylonitrile
Industrial Chemical (Gas)	■ Ammonia ■ Chlorine

NFPA protective clothing standards are reviewed on a 5-year basis. The process is open to public participation. Proposed changes to any NFPA standard can be submitted at any time, and if of an emergency nature, can be implemented within several months. Otherwise, proposed changes are included in the normal revision cycle.

Chemical Barrier and Initial Selection Considerations

No single protective clothing material offers total chemical protection. The initial selection of protective clothing and equipment should be based on a hazard assessment of those chemicals found in the community or the facility. Unfortunately, there may be some chemicals for which there is no adequate protection.

Compliance with any of the CPC standards (e.g., NFPA 1991, 1992, or 1994) does not ensure protection in all situations, as these standards describe *minimum* performance requirements for specific purposes. The user must rely on the manufacturer's representation to determine if the garments are suitable for tasks and situations beyond the scope of the standards. While NFPA 1991 and NFPA 1992 are focused on hazardous material emergencies, NFPA 1994 pertains to CBRN terrorism incidents and is limited to activities during assessment, extrication, rescue, triage, decontamination, treatment, site security, crowd management, and force protection operations. Similarly, the National Institute of Justice (NIJ) standard on *CBRN Protective Ensemble Standard for Law Enforcement* (NIJ Standard 0116.00) reflects the unique operational requirements of law enforcement officers while conducting operations at an incident involving suspected or identified CBRN hazards.

The glove, boot, visors, and garment components of CPC will often be constructed of different material or laminates. Be aware of which of these materials form the basis for the chemical barrier data provided by the manufacturer. Recognize that it may not be possible to determine the specific material(s) used in some limited-use laminate garments due to product proprietary reasons.

The manufacturer should provide technical test data that reflects the chemical barrier performance of both the primary suit material and all secondary components (e.g., gloves, boots, closure assemblies, visors, and exhaust valves). When evaluating chemical vapor protective suits, evaluate all suit components and their construction materials. The performance of CPC is only as strong as its weakest material.

Manufacturers will publish quantitative chemical resistance data for particular chemicals. This data will normally be based on standardized laboratory tests such as those established by the ASTM Protective Clothing (F23) Committee. Standard permeation and penetration tests often incorporate a very large safety factor in predicting failures. The size of this safety factor depends on the established testing criteria.

Chemical resistance data is described in terms of either (1) chemical permeation/breakthrough times and rates; (2) cumulative permeation for a given test duration; or (3) as "pass/fail" chemical penetration testing results. Remember that the longer the breakthrough time or the lower the cumulative permeation value, the better the level of protection. If two CPC materials have comparable breakthrough times, the CPC with the lowest reported permeation rate should normally represent the better option.

Chemical barrier recommendations for boots, gloves, and some garments may also be provided in the form of qualitative chemical resistance ratings or use recommendations for a specific protective clothing material and particular chemicals. These ratings are often in the form of a four- to six-grade scale (e.g., excellent, good, fair, and poor/not recommended) or a color code (e.g., green, yellow, red). They often will not include performance specifications or quantitative data such as

Table 8-3 Personal protective equipment options for the primary hazardous materials hazards using the TRACEM model.

Hazard	Skin Protection	Respiratory Protection
Thermal*	▪ Structural Firefighting Clothing (SFC) ▪ Proximity Firefighting Clothing ▪ Chemical Vapor Protective Clothing w/Liquefied Gas Overcover (Level A Ensemble)	SCBA/SAR
Radiological (Particles)**	▪ Structural Firefighting Clothing (SFC) ▪ Chemical Splash Protective Clothing (Level B/C Ensemble) ▪ Radiological Clothing (per NFPA 1994)	SCBA/SAR APR/PAPR
Asphyxiation	n/a	SCBA/SAR
Chemicals (Liquid)	▪ Chemical Splash Protective Clothing (Level B/C Ensemble)	SCBA/SAR APR/PAPR
Chemicals (Vapor/Gases)	▪ Chemical Vapor Protective Clothing (Level A Ensemble)	SCBA/SAR
Etiological (Biological)	▪ Chemical Splash Protective Clothing (Level B/C Ensemble)	SCBA/SAR APR/PAPR
Mechanical (Explosives)	▪ Specialized Bomb Suit (only used by certified Bomb Technicians)	SCBA/SAR APR/PAPR

*Includes both high (i.e., radiant heat, conduction) and low (e.g., cryogenic, liquefied gases) temperatures

**Protection from Ionizing radiation requires special shielding clothing

breakthrough times. Degradation resistance data may be misleading and should be avoided when selecting CPC materials.

When evaluating chemical barrier recommendations, consider the following guidelines:

- The primary reference source for chemical barrier recommendations should come from documentation published by the suit, glove, or boot vendor. Other credible sources will include CPC reference manuals and computer databases, such as Forsberg and Keith's *Chemical Protective Clothing Permeation/Degradation Database*, and Forsberg and Mansdorf's *Quick Selection Guide to Chemical Protective Clothing*.
- Determine the basis of chemical barrier recommendations. Degradation or immersion testing is not sufficient for barrier assessment. Permeation or penetration test data should be sought since permeation of rubber or plastic fabrics can occur with little or no physical effect on the clothing material. Remember, permeation is an insidious process that can occur with no sign of degradation and in the absence of any holes, voids, or defects.
- Degradation recommendations based upon immersion testing data may be quite old, and they may also be based on subjective evaluations rather than quantitative measurements for swelling, weight, or strength changes. In some cases, the testing criteria and qualitative descriptions for defining *good*, *excellent*, and other key words may not be documented.
- Materials constructed of the same primary fabric or material (e.g., butyl rubber, PVC) are not necessarily equal in performance. Variations in formulations, thickness, and coating and backing materials influence chemical exposure times.
- There may be a conflict in barrier recommendations between sources. Responders should initially rely on the protective clothing manufacturer's chemical resistance recommendations. Always select the most conservative data.

Respiratory Protection

The respiratory system is the most direct and critical exposure route. Inhalation is the most common exposure route and is often the most damaging. Remember that a material does not have to be a gas in order to be inhaled—solid materials may generate fumes or dusts in a dry powdered form, while high-vapor-pressure liquid chemicals can generate vapors, mists, or aerosols that can be inhaled.

The selection of respiratory protection at a hazmat/WMD incident should be based on a number of factors, including the following:

- What is the physical form of the contaminant (i.e., solid, liquid, or gas)?
- Has the contaminant been identified?
- Are concentrations known or unknown?
- What is the purpose of response operations?
- What will be the duration of response operations?
- What is the operating environment and operating conditions (e.g., indoors, outdoors, heat, cold, precipitation, etc.)?
- What type and level of skin protection will be required?

Respiratory protection can be provided by either air purification devices or by atmosphere supplying respiratory equipment.

Air Purification Devices

Air purification devices are respirators that remove particulate matter, gases, or vapors from the ambient air before inhalation. When used for gases or vapors, they are commonly equipped with a sorbent material that absorbs or reacts with the hazardous gas. Particle-removing respirators use a mechanical filter to separate the contaminants from the air. Some cartridges are combined sorbent and mechanical filters. The proper cartridge must be used for the expected contaminants (e.g., acid gas, organic vapor, nuclear/bio/chemical agent, etc.). There is a uniform NIOSH color code system for the identification of cartridges.

The NIOSH Respirator Certification Requirements (42 CFR 84) outline the requirements for particulate respirators. Emergency responders may use particulate filters for suspicious powder scenarios and at structural collapse incidents. Part 84 defines nine classes of filters—three levels of filter efficiency (95%, 97%, and 99.7%), each with three categories of resistance to filter efficiency degradation (N, R, and P). The selection of N-, R- and P-filters depends on the presence or absence of oil particles, where N = Not resistant to oil; R = Resistant to oil, and P = oil Proof.

- If no oil particles are present, use any series (N, R, or P).
- If oil particles are present, use only R or P series.
- If oil particles are present and the filter is to be used for more than one work shift, use only P series.

The selection of filter efficiency (i.e., 95%, 97%, or 99.97%) will depend on how much filter leakage can be accepted. For example, high efficiency particulate air filters (HEPA) are at least 99.97% efficient in removing particles 0.3 micrometers in diameter and larger (e.g., N100, R100, and P100 filters).

Two basic types of air purification devices may be used for emergency response purposes Figure 8-6:

- Air Purification Respirators (APRs) are respirators with an air-purifying filter, cartridge, or canister that removes specific air contaminants by passing ambient air through the air-purifying element. These are negative pressure respirators and can be found with either a full-face or half-face configuration with sorbent, mechanical, or combination cartridges attached. They are commonly used in controlled industrial and workplace environments where the contaminants are known and concentrations measured. For emergency response applications, full-face respirators are typically the respirator of choice. If half-face respirators are used, eye protection must be provided.
- Powered-Air Purification Respirators (PAPRs) are air-purifying respirators that use a blower to force the ambient air through air-purifying elements to a full-face mask. As a result, there is a slight positive pressure in the face piece that results in an increased protection factor. Where an APR has a protection factor of 50:1, a PAPR will have a protection factor of 1000:1 (Note: A protection factor of 1 = no respiratory protection in place. A protection factor of 1,000 means that the concentration of a breathed contaminant is reduced by a factor of one-thousand from the ambient concentration.) PAPRs are being used in a wide range of emergency response and post-emergency response applications, including decon, patient handling in medical facilities, and investigation of hazmat and terrorism crimes.

Operational considerations when using APRs and PAPRs include the following:

- Air purification devices should not be used at hazmat releases unless qualified personnel have first monitored the environment and determined such devices can be safely used (per OSHA 29 CFR 1910.120[q][3][iv] and 1910.134). As a general rule, they should not be used for initial response operations at hazmat incidents and for emergency response operations involving unknown substances.
- Cannot be used in IDLH environments or in oxygen-deficient atmospheres containing less than 19.5% oxygen. When used, both the contaminant and oxygen levels must be constantly monitored.
- Should not be used in the presence or potential presence of unidentified contaminants. Not recommended for areas where contaminant concentrations are unknown or exceed the designated use concentrations. "Designated use concentrations" are based on testing at a given temperature (usually room temperature) over a narrow range of flow rates and relative humidity. Therefore, the level of protection may be compromised in nonstandard conditions.
- Respiratory protection can be downgraded from air-supplied to air-purifying respirators if (1) the contaminants have been identified; (2) the atmosphere is being monitored and contaminant levels are within acceptable limits; (3) oxygen levels are above 19.5%; and (4) the IC approves.
- May present logistical problems for storage and maintenance because of the variety of filters and cartridges required. The shelf life of filters and cartridges will vary depending on the type of cartridge (i.e., sorbent versus mechanical filter), its packaging, and how it's stored. Always consult manufacturer instructions for guidance.
- Have a limited-protection duration. Once opened, sorbent canisters begin to absorb humidity and air contaminants whether in use or not, and their efficiency and service life will decrease dramatically. Where possible, cartridges should have an end-of-service-life indicator (ESLI) that warns the user of the approach of the end of adequate respiratory protection (e.g., the sorbent is approaching saturation or is no longer effective). If the cartridge doesn't have an ESLI indicator, a clear schedule must be established to ensure the cartridges are changed before the end of their service life.
- APRs and PAPRs only protect against specific chemicals and only to specific concentrations. Their effectiveness against two or more chemicals simultaneously is highly questionable. They are well suited for operations involving solids, dusts, powders, and many biopathogens and toxins.

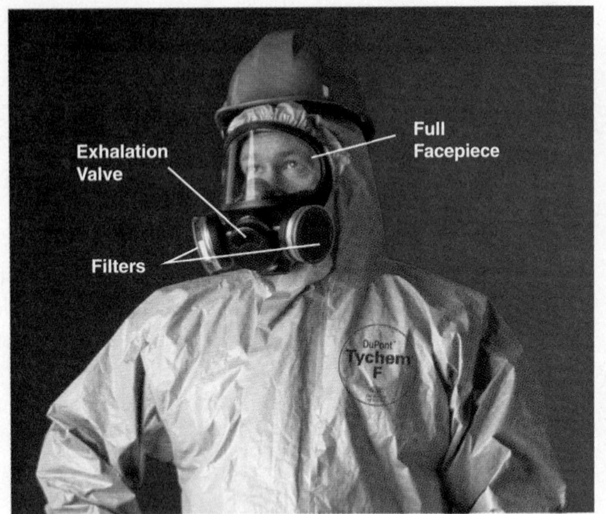

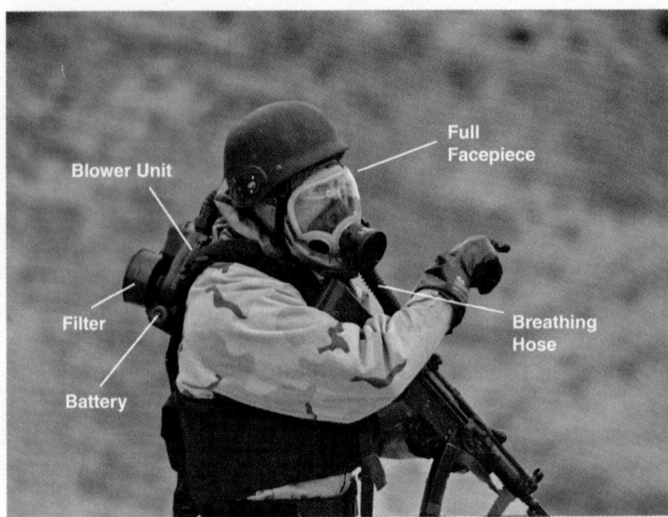

Figure 8-6 Air purifying respirators (APRs) and powered air purifying respirators (PAPRs) are respirators that remove particulate matter, gases, or vapors from the ambient air before inhalation.

- Individuals must meet the fit testing and medical requirements as outlined by OSHA 29 CFR 1910.134—*Respiratory Protection*.

Advantages of APRs and PAPRs include their light weight and lack of physical stress on the user. Limitations include many of the operational considerations in the preceding list, including: (1) the need to identify the contaminants present and to monitor the concentration of each of these contaminants; (2) to verify the respirator is effective against each contaminant at the concentrations present and that these concentrations are below IDLH; and (3) to ensure oxygen levels are above 19.5% and less than 23.5%.

Logistically, based upon incident duration the respirator change-out period must be established and sufficient cartridges must be available to complete the work activities. Nonpowered, negative-pressure respirators also carry a greater risk of leakage than powered, positive-pressure respirators.

Atmosphere Supplying Devices

Respiratory protection devices with an air source are referred to as atmosphere-supplying devices. There are two basic types: self-contained breathing apparatus (SCBA) and supplied air respirator (SAR), which supply air from a source away from the scene connected to the user by an airline hose.

These devices provide the highest available level of protection against airborne contaminants in oxygen-deficient atmospheres. Only positive-pressure devices that maintain positive pressure in the face piece during both inhalation and exhalation should be used for emergency response applications. Positive-pressure respirators will provide a protection factor of 10,000:1, assuming they are properly fitted, maintained, and worn. Atmosphere supplied respiratory protection devices must use compressed breathing air that meets the requirements for Grade D breathing air as described in ANSI/Compressed Gas Association Specification for Air, G-7.1. Additional information on these breathing air requirements can be referenced from OSHA 29 CFR 1910.134(i)(1)(i).

Self-Contained Breathing Apparatus. There are two basic types of SCBA: open-circuit and closed-circuit. *Open-circuit SCBA* use a compressed air cylinder to supply fresh air, which is subsequently exhaled directly into the ambient atmosphere. It is the predominant type of SCBA used in the emergency response community. *Closed-circuit SCBA* are those in which exhaled air is recycled by removing the carbon dioxide with an alkaline scrubber and replenishing the consumed oxygen from a solid, liquid, or gaseous oxygen source. Closed-circuit SCBA are used for specialized response scenarios where long, extended operations may be required, such as tunnel, subway, and mine rescue operations. However, they are not commonly used in conjunction with encapsulating chemical vapor clothing, and can generate heat which may add to the heat stress encountered in totally encapsulating suits.

Open-circuit SCBA used for firefighting applications should meet the requirements of NFPA 1981—*Standard for Open-Circuit SCBA for Fire Fighters*. As a result of the growing terrorism threat, NIOSH has also developed a testing protocol for respiratory protection during CBRN terrorism incidents. By mutual agreement of both NFPA and NIOSH, CBRN-compliant SCBAs must meet the performance requirements of NFPA 1981 and the NIOSH CBRN requirements. The major change initiated by the NIOSH CBRN requirements were improvements in the chemical resistance of the SCBA face piece. NIOSH has also developed a set of CBRN certification requirements for both APRs and PAPRs for which there are no concurrent NFPA performance standards. Additional information on the NIOSH CBRN respirator requirements and those APR/PAPR manufacturers and models that meet the requirements can be referenced at the NIOSH Certified Equipment List Search at *http://www2a.cdc.gov/drds/cel/cel_form_code.asp* or the Responder Knowledge Base at *http://www.rkb.us*.

Advantages of using SCBA include that they are readily available in the emergency response community, most responders are proficient in their use, and they provide the highest level of respiratory protection. Limitations include their size, weight, bulkiness, limited duration of air supply, overall resistance of the SCBA and its components to chemical exposures, and size restrictions when used in confined spaces.

Operational considerations when using SCBA include the following:

- Atmosphere-supplying units are required for initial response operations until the hazards and concentration of air contaminants can be fully assessed.
- Duration of the operation. Although 30-minute air cylinders are commonly used in the fire service, most active hazmat response teams will use 45- to 60-minute air cylinders to provide a sufficient backup air supply for entry, exit, and decon operations.
- Depending on the type of cylinder, certain chemicals may attack the outer shell of an air cylinder. In 1996, a fiberglass-wrapped composite aluminum cylinder that was accidentally exposed to a commercial cleaning fluid containing hydrofluoric acid, phosphoric acid, and sulfuric acid failed explosively approximately 6 days after exposure to the fluid. SCBA components exposed during a hazardous materials incident may have to be discarded.

Supplied Air Respirators. Although SCBA are most common, SARs may be used when extended working times are required for entry, decon, or remedial clean-up operations. Components of a SAR include (1) source of breathing air—usually a cylinder, a cylinder cart or a cascade system; (2) airline hose; (3) positive-pressure respirator; and (4) emergency air supply, such as a small escape cylinder (Figure 8-8).

Operational considerations when using SARs include the following:

- Atmosphere-supplying units are required for initial response operations until the hazards and concentration of air contaminants can be fully assessed.
- NIOSH certification limits the maximum hose length from the source to 300 feet (91.4 meters).
- Use of airlines in IDLH or oxygen-deficient atmosphere requires a secondary emergency air supply, such as an escape pack for immediate backup protection in case of airline failure. In addition, use of a SAR will require personnel to monitor the air supply source.
- Using airline hose will probably impair user mobility and slow the operation. The user must retrace his or her entry path when leaving the work area.
- The airline hose is vulnerable to physical damage, chemical contamination, and degradation. Airline sleeves constructed of disposable materials can provide additional protection. Proving that a decontaminated airline hose is now "clean and safe" will be difficult.

Dual flow SCBAs that have the capability of being supplied by either a SCBA or an airline may provide additional flexibility for both entry and decon operations. The user can operate in either the SCBA or airline hose modes by operating a manual or automatic switch.

Figure 8-7 Supplied air respirators being used with chemical vapor protective clothing.

Advantages of using SAR units include lower profile and weight, increased work durations, and their ability to provide the highest level of respiratory protection. Limitations include a number of the operational considerations listed above.

Air purification devices may be appropriate for operations involving volatile solids and for remedial clean-up and recovery operations where the type and concentration of contaminants is verifiable. However, air-supplied devices such as airline hose units will offer the greatest protection for exposures to gases and vapors **Figure 8-7**.

Levels of Protection

The need for proper protective clothing and equipment in a hostile environment is obvious. Unfortunately, there is no one type of PPE that satisfies our protection needs under all conditions. For example, chemical protection and thermal protection are very difficult to combine into one protective clothing material. The IC and the Hazmat Group Supervisor must be familiar with the various types and levels of protective clothing available.

Three basic types of protective clothing may be used at hazmat incidents:

1. Structural firefighting clothing is designed to protect against extremes of temperature, steam, hot water, hot particles, and the typical hazards of firefighting.
2. CPC is designed to protect skin and eyes from direct chemical contact. There are two basic types of CPC used: chemical splash protective clothing and chemical vapor protective clothing.
3. High temperature protective clothing is designed to protect against short-term exposures to high temperatures, such as proximity and fire entry suits.

The EPA has developed a classification scheme for the various levels of chemical protective ensembles (i.e., clothing and respirators). **Table 8-4** *compares the levels of protection published by EPA as compared to the NFPA CPC Standards.*

Table 8-4 Levels of protection.

NFPA Standard	OSHA/EPA Level	NIOSH-Certified Respirator	NFPA Chemical Barrier Protection Method(s)	Expected Dermal Protection from Suit(s): Chemical Vapor*	Chemical Liquid*	Particulate	Liquid-borne Biological (Aerosol)
1991	A	CBRN SCBA (open circuit)	Protection against permeation and penetration*	X	X	X	X
1992	B	Non-CBRN SCBA (or CBRN SCBA)	Protection against penetration*		X		
	C	Non-CBRN APR or PAPR	Protection against penetration		X		
1994, Class 1	(Note: The NFPA 1994, Class 1 ensemble, was removed in the 2006 edition of the standard because of its redundancy with NFPA 1991.)						
1994, Class 2	B	CBRN SCBA	Protection against permeation	X	X	X	X
1994, Class 3	C	CBRN APR or PAPR	Protection against permeation	X	X	X	X
1994, Class 4	B	CBRN SCBA	Protection against penetration	NA	NA	X	X
	C	CBRN APR or PAPR	Protection against penetration	NA	NA/NT	X	X

*Notes: Vapor protection for NFPA 1994 Class 2 and Class 3 is based on challenge concentrations established for NIOSH certifi cation of CBRN open-circuit SCBA and APR/PAPR respiratory equipment. Class 2 and Class 3 do not require the use of totally encapsulating garments.

For our purposes in this chapter, protective clothing and equipment will be discussed in terms of its use—respiratory protection, structural firefighting clothing, CPC, and high-temperature protective clothing.

Structural Firefighting Clothing

While **structural firefighting clothing (SFC)** is the most common type of PPE used by emergency responders, it has a number of vulnerabilities when worn in hazmat environments. Although SFC may offer sufficient protection to the wearer who is fully aware of the hazards being encountered and the limitations of the protective clothing, it is normally not the first PPE choice for most hazmat response scenarios. An exception to this statement would be flammable gas and liquid fire incidents where SFC and SCBA will provide sufficient protection for most response scenarios.

For our purposes, SFC includes a helmet, positive-pressure SCBA, PASS device, turnout coat and pants, gloves and boots, and a hood made of a fire-resistant material. The ensemble should meet NFPA 1971—*Standard on Protective Ensemble for Structural Firefighting and Proximity Firefighting* requirements and is shown in **Figure 8-8**.

SFC provides limited protection from heat and cold but may not provide adequate protection from hazardous vapors and liquids. SFC may be used when the following conditions are met:

- Contact with splashes of extremely hazardous materials is unlikely.
- Total atmospheric concentrations do not contain high levels of chemicals toxic to the skin. In addition, there are no adverse effects from chemical exposure to small areas of unprotected skin.

The increased presence of plastics and other toxic or carcinogenic synthetic materials found in structural fires has also led to increased concerns with the contamination and decontamination of SFC. Products of combustion include inorganic gases (e.g., hydrogen sulfide, nitrogen oxides), acid gases (e.g., hydrochloric acid, sulfuric acid), hydrocarbons (e.g., benzene), metals, and polynuclear aromatic compounds (PNAs). The inspection, cleaning, and maintenance of SFC should be in accordance with NFPA 1851—*Standard for the Selection, Care and Maintenance of Protective Ensembles for Structural Fire Fighting and Proximity Fire Fighting.*

Hazardous chemicals can both penetrate and permeate firefighting protective fabrics. To be certified to NFPA 1971, structural firefighting moisture barrier materials and seams must resist penetration to surrogate gasoline (50/50 toluene/isooctane), fire-resistant hydraulic fluid, battery acid (37% sulfuric acid), 3% aqueous film-forming foam (AFFF) concentrate, 65% free chlorine hypochlorous acid and a suspension of a viral solution. In addition, garments certified to the CBRN optional NFPA 1971 requirements must resist permeation by sarin, VX, dimethyl sulfate, acrylonitrile, and acrolein applied in a 10-droplet pattern. When evaluating the performance of SFC for hazardous materials, consider the following points:

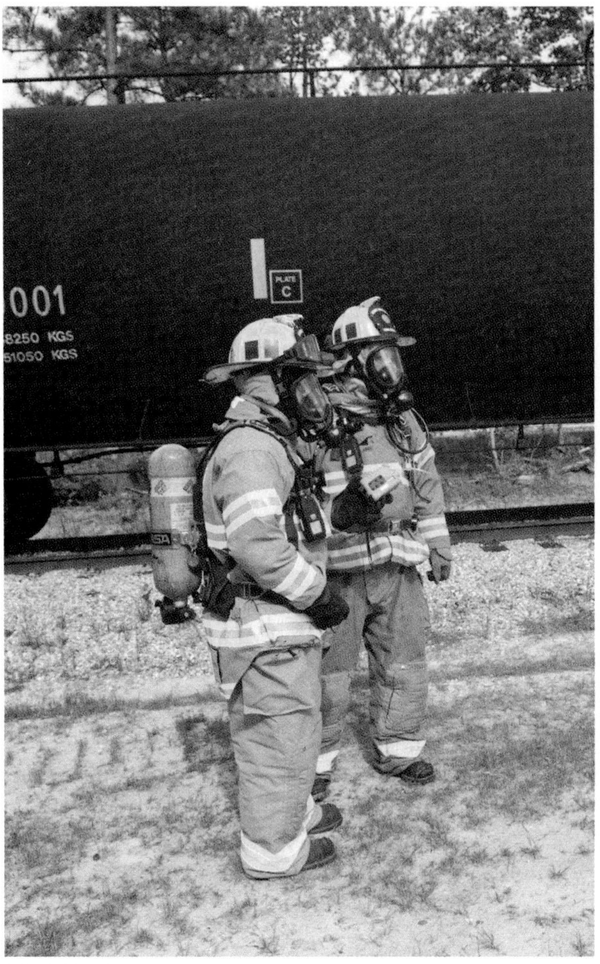

Figure 8-8 Structural firefighting clothing

- The NFPA 1971 tests are performed on an unused garment. The barrier performance of used garments has not been determined and is likely to vary significantly with the amount of garment usage.
- Barrier tests are performed on the garment materials, but the chemical barrier of the wrist, neck, hood to respirator, pant leg to boot interfaces, are not evaluated.
- Clothing materials are tested against a limited list of chemicals and manufacturers do not generally provide any additional barrier data.
- In order to safety reuse the exposed garments, the chemical contamination must be removed or lowered to a safe level.
- Clothing and equipment materials are porous and are easily contaminated by chemical penetration, such as:
  - Turnout clothing outer shells, thermal liners, collars, and wristlets
  - Station/work uniforms
  - Glove shells and liners
  - Fire retardant hoods
  - Boot linings
  - Helmet straps and linings
  - SCBA straps
- Coated or rubberlike materials are more likely to be affected by chemical permeation, such as:
  - Moisture barriers
  - Reflective trim
  - Boot outer layers
  - SCBA masks
  - Hard plastics used in the helmet and SCBA components
- Ash, resins, and other smoke particles can easily become trapped within the protective clothing fibers.
- Infectious bloodborne diseases, including the HIV, hepatitis B, and hepatitis C viruses, can be readily absorbed into the protective clothing fibers. SFC should be certified as protective against bloodborne pathogens. However, common biological decon agents containing active chlorine, such as dilute bleach, cannot be used on SFC fabrics such as Kevlar®.

Body protection. The outer shell that provides thermal protection is constructed of materials such as Kevlar®, PBI™, or Nomex™. Turnout coats and pants should be constructed with a moisture barrier, usually neoprene, Goretex™, or similar materials. Note that the manufacturers of Goretex™ recommend that their fabric not be worn in any type of chemical atmosphere since it will not stop the passage of chemical vapors.

The most serious problem faced when using SFC for hazmat operations is ensuring that all exposed skin surfaces are covered and protected. A hood made of fire-resistant materials such as Nomex® or Kevlar® will provide some protection for the head, ears, neck, and throat. When worn properly, they do not interfere with the SCBA face seal. A disadvantage is that any chemical splashed onto the hood may be absorbed and remain in direct contact with the skin. SFC compliant with the NFPA 1991 CBRN option have sophisticated closures and interfaces to the gloves, respirator, and boot. These closures must not be damaged by wear and be properly closed to provide limited liquid and vapor protection.

In some situations, elastic bands may be applied around the forearms and ankles for additional vapor protection. These materials provide a false sense of security and do not increase the ability of SFC to provide either chemical splash or vapor protection. In addition, bleach materials that may be used for decon operations will also weaken several of the fire retardant materials used in SFC. Remember—SFC is not designed to offer chemical protection!

Gloves must be selected in reference to the tasks to be performed and the specific chemicals they will be exposed to. Because of the likelihood of physical contact, protective gloves should be considered as a critical element in the protective clothing ensemble. Factors to evaluate include chemical resistance, physical resistance, and temperature resistance.

Cotton, synthetic fiber, leather, firefighting, and rescue gloves will absorb liquids, keeping these chemicals in contact with the skin when working around any hazardous liquids. They will also deteriorate when exposed to corrosive liquids. In comparison, synthetic rubber and plastic gloves may melt when exposed to high temperatures associated with firefighting or may deteriorate on contact with certain petrochemical products. They also may not provide adequate protection against

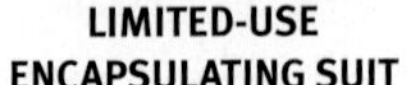

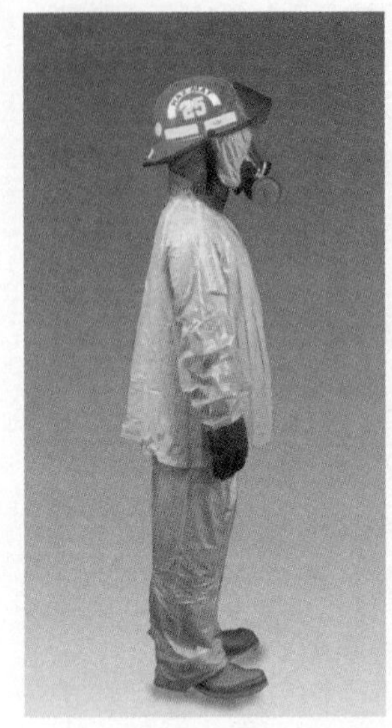

Figure 8-9 Examples of liquid chemical splash protective clothing.

many corrosives and agricultural chemicals. Polyvinyl alcohol (PVA) gloves provide excellent barrier protection against certain petroleum solvents but break down on exposure to water.

Products that penetrate natural rubber and silicone may also create serious exposure problems for gloves, boots, and non-CBRN-compliant SCBA face pieces. Examples include methyl bromide, dichloropropene, and some chemical agents.

Respiratory protection. Since toxic, corrosive, and flammable vapors, along with the products of combustion are present, air-supplied respiratory protection devices are required. Positive-pressure SCBA is the minimum level of respiratory protection. Because of problems with decontamination before refilling, additional SCBA and air supply units are almost always required at hazmat incidents.

It is not uncommon for exposure to a specific chemical or hazmat environment to require the complete discarding of all SFC. Leather and fibrous materials are easily permeated by many chemicals and make decontamination difficult at best. Polycarbonate helmets may be affected by solvents. Pesticides, PCB-related fires, and radioactive materials incidents may make any decon impossible. Disposal should be done in accordance with local, state, and federal environmental regulations.

Liquid Chemical Splash Protective Clothing

There are many hazmat incidents where the hazards and potential harm of the released material may require that specialized clothing be worn. **Liquid chemical splash protective clothing** consists of several pieces of clothing and equipment designed to provide skin and eye protection from chemical splashes. It does not provide total body protection from gases or vapors and should not be used for protection against liquids that give off vapors known to affect or be absorbed through the skin. Depending on the materials involved and the nature of response operations, respiratory protection may be initially provided through SCBA or SAR, or by using APRs or PAPRs once air-monitoring operations have determined that the atmosphere allows for their use (i.e., EPA Level B or Level C configuration).

Liquid chemical splash protective clothing may be used under the following conditions:

- The vapors or gases present are not suspected of containing high concentrations of chemicals that are harmful to, or can be absorbed by, the skin.
- It is highly unlikely that the user will be exposed to high concentrations of vapors, gases, or liquid chemicals that will affect any exposed skin areas.
- Operations will not be conducted in a flammable atmosphere. If operations must be conducted in an environment with the potential for combined thermal and chemical hazards, responders should consider the use of flash protection overgarments available from some CPC manufacturers.
- Wearing a flame-resistant garment underneath flammable chemical protective outer garments does not provide protection against heat and flame exposure. The heat load from the burning and melting of the flammable outer chemical garment will generally overwhelm the thermal protection of normal fire-resistant clothing worn underneath. However, if the chemical outer garment is made from a nonmelting material and is supported by thermal exposure test data, wearing additional underlayers of fire-resistant clothing can provide increased protection.

Skin and body protection. Liquid chemical splash protective clothing is routinely used in hospitals and laboratories where various chemicals, biologicals, and infectious diseases are handled. They are also found in nuclear facilities and installations that handle or process radioactive materials, used in handling mild corrosives and PCBs, in protecting against asbestos fibers and lead dust, and in formulating and applying agricultural chemicals.

In emergency response, liquid chemical splash protective clothing is often used for initial response operations, to protect decon personnel, and for post-emergency response investigation and clean-up operations. Depending upon the application (e.g., emergency response versus law enforcement), CPC fabrics may be available in various colors.

Several common types include the following:

- **Single-piece suits.** Usually coveralls, a splash suit, or an encapsulating suit that is not vapor tight. Hoods and booties may be attached to coveralls and splash suits. Both limited-use and multi-use garments are available. Considering their low cost, absence of decon problems, and the varied needs of responders, single-piece limited-use suits are an excellent alternative used by many hazmat responders.
- **Two-piece suits.** Usually consist of bib overalls or pants worn with a jacket. Some ensembles include an expanded back or "humpback" design that covers a SCBA. Although they may encapsulate the user, they do not provide vapor-tight protection. Accessories such as gloves, boots, and hoods are available.

Closures and interfaces. The overall level of liquid splash protection provided by a garment is directly dependent upon the performance of all seams and the interfaces with other ensemble components. Depending on the use of the CPC, seams may be sewn, bound, thermally welded, taped, or double taped. For a nonencapsulating garment, the critical interfaces are around the respirator face piece, the attachment of the sleeves, and the connection to the boots. The suit closure can also be a source of leakage. NFPA 1991, 1992, and 1994 all incorporate a shower test that evaluates the performance of these closures and interfaces.

Head protection. Hard hat, helmet, or hood. Some form of hard hat protection is recommended when using a hood or encapsulating suit. This gear is designed in various configurations and materials. Some manufacturers offer a respirator-fit hood that will provide a tighter fit around the face piece and completely cover the neck area.

Gloves. Some coveralls and jackets have a sleeve mounted splash guard that prevents wrist exposure. Some manufacturers have also developed a combined O-ring glove and cuff assembly that ensures a leak-proof glove/cuff assembly. These designs allow gloves to be easily interchanged according to the hazard.

For maximum hand protection, *overgloving* and *doublegloving* should be used. Doublegloving involves the use of surgical gloves under a work glove. It permits doffing of the work glove without compromising exposure protection and also provides an additional barrier for hand protection. Doublegloving also reduces the potential for hand contamination when removing protective clothing during decon procedures. Overgloving is the wearing of a second glove over the work glove for additional chemical and abrasion protection during lifting and moving operations.

Gloves should be sized for use by individual responders. In general, size 9 gloves are commonly used for single- or doublegloves, while sizes 11 and 12 are typically used as overgloves for additional strength and chemical protection. Common glove materials include Viton®, neoprene, butyl rubber, Chloropel™, polylaminates such as SilverShield®, and polyvinyl chloride (PVC).

Exposure to liquefied gases and cryogenic liquids (e.g., anhydrous ammonia, chlorine) will embrittle typical rubber gloves and will require the use of a thermal overglove. The thermal overglove may be as simple as a leather glove, providing it insulates the rubber glove from direct contact with the cold liquid.

Footwear and shoe covers. Foot protection may be chemical boots, separate shoe covers, or booties that are part of the CPC ensemble. Boots should provide both chemical and mechanical protection (e.g., cuts, punctures, etc.). Chemical shoe boots are usually commonly used by responders, but work boots over footwear are also found. Common boot materials include PVC, neoprene, PVC/nitrile, and nitrile rubber.

Many liquid chemical splash suits constructed from limited-use materials have integral or connected sock booties. These attached "socks" are designed to be worn inside boots. They are not sufficiently durable or slip resistant to be worn as outer footwear. Specially designed shoe covers and booties are most often used in those areas where radioactive materials, etiological agents, and agricultural chemicals are handled or where cleanliness is a concern. There are also chemical protective suits with attached boots specifically designed for hazmat response operations.

Aprons and body coverings. Aprons, lab coats, sleeve guards, and other body coverings are designed for protection against spills and splashes that occur when physically handling chemicals and other hazardous materials. They are primarily used for routine chemical handling operations rather than emergency response. Aprons and similar clothing may also be worn by responders while handling samples and performing hazardous categorization (HazCat) tests.

Historically, duct tape has been used to ensure that zippers remain closed, to accommodate size differences, secure gloves and boots, and to secure the hood to the SCBA mask on some liquid chemical splash suits.

Responder Safety Tip

Duct tape provides a false sense of security and does not increase the ability of the garment to provide liquid chemical splash protection. Duct tape is not a substitute for a properly designed and fitting garment, and its use should be discouraged.

Chemical Vapor Protective Clothing

Chemical vapor protective clothing (i.e., EPA Level A clothing) provides full-body protection with vapor-tight integrity. When used with air-supplied respiratory devices, it provides a gas-tight envelope around the wearer **Figure 8-10**.

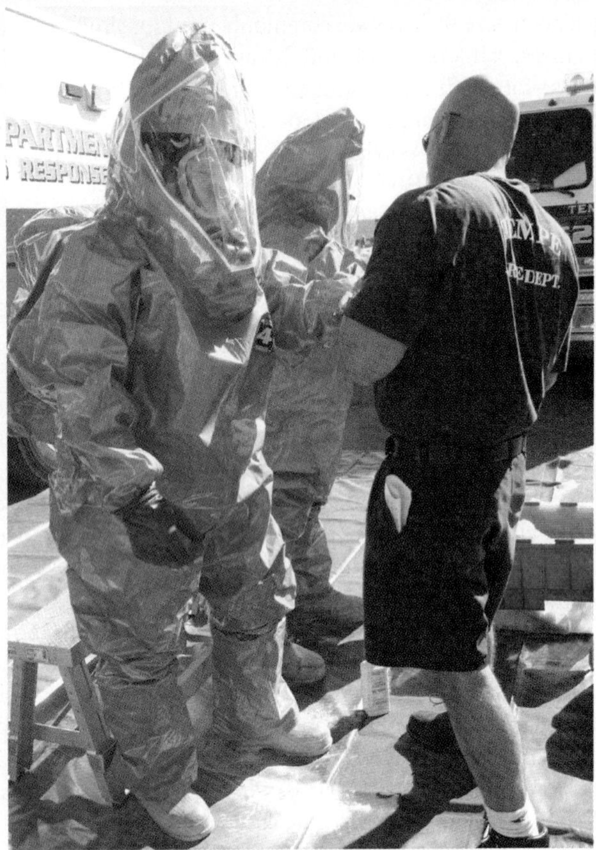

CHEMICAL VAPOR SUITS (SCBA INSIDE SUIT)

Advantages

1. Offers maximum level of protection to the user.
2. Positive internal pressure may help to minimize minor leaks.
3. If SCBA malfunctions, the user may have some time to reach a nonhostile environment.

Limitations

1. Reduced mobility and visibility, with some problems in confined space operations.
2. No easy method of re-supplying air cylinders—must decontaminate before opening suit.
3. Problems implementing SCBA emergency procedures if suit does not have "Batwing Sleeves".
4. Higher weight than Type 2 suits.

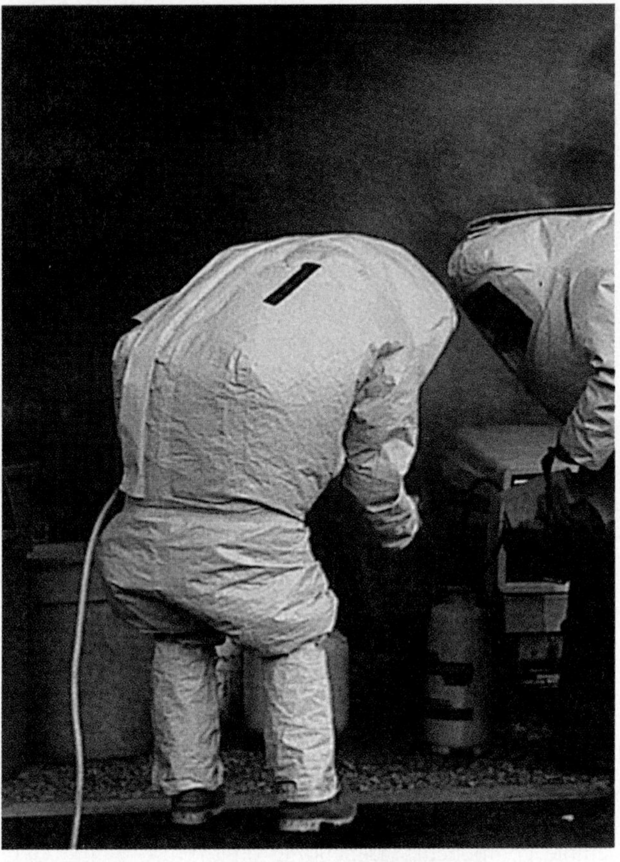

CHEMICAL VAPOR SUIT (SUPPLIED AIR RESPIRATOR)

Advantages

1. Permits extended operations.
2. Positive pressure always maintained in the suit.
3. Airline hose may provide mechanism for minor body cooling.
4. May rely upon SCBA as primary air supply and airline hose or a second SCBA bottle as secondary air supply.

Limitations

1. Limited maneuverability due to hoseline. Creates a tripping hazard and may become tangled.
2. If using airline hose, distance is limited to the length of the airline hose (generally not greater than 300 feet.)
3. Construction of the airline hose in regards to chemical compatibility.

Figure 8-10 Examples of Chemical Vapor Protective Clothing.

Chemical vapor protective clothing should be used when the following conditions exist:

- Extremely hazardous substances are known or suspected to be present, and skin contact is possible (e.g., cyanide compounds, toxic and infectious substances).
- There is potential contact with substances that harm or destroy skin (e.g., corrosives).
- Anticipated operations involve a potential for splash or exposure to vapors, gases, or particulates capable of being absorbed through the skin.
- Anticipated operations involve unknown or unidentified substances and the scenario dictates that vapor-tight skin protection is required.

Skin and body protection. Chemical vapor protective clothing is manufactured in several configurations. The most common is where the SCBA is worn underneath the ensemble, thereby providing total vapor protection by encapsulating the wearer. This configuration is easily identified by its "hump-back" expanded-back design. CPC manufacturers will also incorporate an airline hose bulkhead connection onto the suit if a supplied air respirator will be used. These connections vary by respirator manufacturer and must be specified and installed at the time of garment purchase. This design will require a secondary emergency air source (e.g., SCBA or escape pack).

In European countries, it is very common to find chemical vapor protective clothing where the SCBA is worn outside of the suit. The SCBA face piece is either incorporated into the suit hood or worn over the hood. The gas-tight integrity of the suit is dependent on the quality of the seal between the suit hood and the SCBA face piece. In addition, the face piece serves as the primary barrier for respiratory protection against chemical permeation.

A chemical vapor suit is only as strong as its weakest link. All components of the ensemble, including the face shield, zipper, gloves and boots, and pressure relief valves must provide an equivalent level of protection. Most chemical vapor suits utilize bonded or thermally welded seams that are then taped (inside, outside, or both).

Visibility through the face shield, including peripheral vision, will be critical. Face shields may be constructed of one or more materials that are layered together, such as PVC, Teflon®, and Lexan®. If a SCBA with a mask-mounted regulator is being used, a larger visor may be necessary. If a protective overgarment will be used, consideration must also be given to the impact on operational performance in viewing through multiple face shields.

Vapor Suit Attachments and Accessories

Gloves can be permanently attached to a chemical vapor suit or be detachable. Permanently attached gloves offer integral, vapor-tight wrist protection. However, when the gloves must be replaced, the entire suit must be taken out of service and possibly returned to the manufacturer.

Most manufacturers allow the interchange of gloves. This permits the user to select the glove material that offers the highest level of chemical and/or physical protection yet facilitates maintenance and repair. Methods of attaching the glove to the suit include concentric rings, a ring/clamp system, an interlocking ring/pin system, and connecting rings. This ensures that the glove/suit seal is vapor tight and stops the penetration of vapors and/or liquids. When detachable gloves are changed, the suit should undergo a pressure test to ensure that a vapor-tight suit/glove seal is present.

Gloves must match the chemical resistance of the primary suit material and typically consist of two or more layers of gloves. Common glove materials used with chemical vapor suits include butyl rubber, nitrile rubber, polylaminates, and Viton®. If manual labor tasks will be performed where gloves may become pinched or torn, responders may also use an outer glove constructed of Kevlar®, leather, or a similar durable material. A nonwoven cloth glove may be worn as an inner glove for the moisture absorption.

Boots. Boots can be either an integral part of the suit or a separate item. The "socks" attached to some chemical suits should be worn inside an outer, durable boot, as they do not have sufficient durability or slip resistance to be worn as outer footwear. Some socks have added soles to provide some additional durability and slip resistance but do not provide the level of chemical and physical protection normally found in chemically protective footwear.

Most chemical vapor suits incorporate an integral sock boot or "bootie" design constructed of the same material as the suit. Chemical boots are worn over the booties and a splash guard is then pulled down to prevent liquid product from entering the boot. This feature allows the user to wear footwear that fits, a task more difficult with other boot designs. Chemical boots used for emergency response purposes should have both steel toe and shank protection. Common boot materials include PVC, neoprene, PVC/nitrile, and nitrile.

Suit fit and closure assemblies. Mobility is sacrificed whenever a chemical vapor suit is worn. The degree of restriction depends on the suit type, the primary suit material, and type of respiratory protection used. Most suit manufacturers offer several sizes. Unless suits can be provided for individual users, it is a good idea to order extra-large and double extra-large sizes. Although a smaller person can always modify a larger suit, someone 6′2″ will rarely squeeze into a "medium."

All chemical vapor suits are sealed by a closure assembly. The pressure-sealing zipper is the most common. When closed, the zipper forms a gas-tight seal, and an outer flap then protects the closure from direct splashes. A second type consists of outer extruded sealing lips with an inner restraint zipper. The inner zipper provides closure strength while the sealing lips provide the gas-tight seal. The extruded sealing lip assembly is similar in principle to the Ziploc® closure of plastic bags.

Emergencies may arise when the suit integrity is compromised or when the SCBA malfunctions. Beware of the initial impulse to immediately get out of the suit, as this may endanger the user. Many suits have an expanded sleeve design (i.e., "batwing sleeve"), which allows the user to easily remove one or both arms from the suit to manipulate the SCBA valves. Should the suit integrity remain intact, several minutes of air should remain within the suit, allowing the user to reach a safe haven. Note: This is an emergency measure and should only be used to assist the user in *immediately* evacuating the area.

Overgarments. Overgarments are available from a number of chemical vapor protective clothing manufacturers that offer flash protection or protection against the cold temperatures associated with liquefied gas exposures. While all overgarments may provide additional physical protection and thermal protection, they can also have a negative impact upon visibility (i.e., user may have to look through two or three face shields or visors), mobility, and manual dexterity.

Flash overgarments are not entry or proximity clothing; they lack any insulation and will provide only limited protection (i.e., several seconds) in the event of a flash fire. The flash fire option noted in both NFPA 1991 and NFPA 1992 is for "escape" purposes only. Entry operations into combined chemical and thermal environments are a high-risk operation. Responders should not knowingly enter a flammable or explosive environment in CPC. On September 17, 1984 in Shreveport, Louisiana one firefighter died and a second firefighter received serious burn injuries when an anhydrous ammonia release inside a cold storage warehouse ignited. The entry crew was wearing chemical vapor protective clothing (Level A) while maneuvering a fork-lift truck in an attempt to control the release when the ignition occurred. Post-incident documentation clearly showed that the CPC melted and increased the severity of burn injuries. Remember—CPC is NOT designed for thermal protection applications!

CPC exposures to liquefied gases can cause cold embrittlement and failure of the suit material. Low temperature overgarments offer only limited protection against splashes of liquefied gases, cryogenic liquids, and their associated vapors.

Undergarments. The type of personal clothing worn by hazmat responders underneath CPC can influence heat stress potential, as well as the ability of the user to effectively operate within a hostile environment. Many Hazardous Materials Response Teams (HMRTs) provide fire retardant coveralls that are worn underneath the CPC as an additional layer of both physical and thermal protection. A shirt and gym shorts might seem like an ideal approach but may actually increase potential risks to the wearer. Remember—fire retardant coveralls worn underneath CPC that melts or burns offers little to no added protection.

Head protection. Although some chemical vapor suits may incorporate head protection, most do not. As a result, head protection must be provided through the use of a separate hard hat, bicycle helmet, or comparable protection. When selecting head protection, many responders prefer helmets with a ratchet adjustment (for sizing) and a chin strap to secure the helmet. Losing your head protection when working in a contaminated environment increases the risks of a significant injury.

Cooling and ventilation. Both liquid chemical splash and vapor protective clothing seal the body in a manner that retains body heat and moisture. Heat stress becomes a concern even in moderate ambient temperatures. The heat stress factor, in conjunction with added weight and restrictions in movement, results in a high level of physiological and psychological stress. CPC should only be worn by individuals in good health and physical condition, as profiled by a comprehensive medical surveillance program. Cooling methods are discussed in the section on "health and safety" elsewhere in this text.

Communications. Verbal, person-to-person communications while wearing chemical vapor clothing are virtually impossible. Radio communications are a necessity for entry operations. In addition, other alternatives, such as voice amplifiers, hand signals, and large flash cards, must also be evaluated. The potential failure of radio systems and the need to conduct response operations in high-noise environments makes hand signals a critical safety and operational requirement.

Communications systems for use within protective clothing ensembles include radio headsets, ear, mouth, and bone microphones, and voice amplifiers. Radio communication systems may be activated through three modes: continuous transmit, push to transmit, and voice-activated. Any communication device should be intrinsically safe and approved for use in explosive atmospheres (see discussion on terms in the section on "hazard assessment and risk evaluation"). In addition, entry communications should normally occur on a predesignated tactical radio channel that cannot be "stepped on" or used by other response units.

Visual identification of personnel can be critical, as it enables the Safety Officer and Hazmat Group Supervisor to differentiate between personnel operating in the hot and warm zones. Identification methods can include the use of different-colored suits, large numbers attached to the suit, or color-coded (or numbered) traffic vests. Reflective tape may be helpful when operating at night or in low light environments. Cyalume® lightsticks suspended from the suit can also be used to identify personnel during night operations. Check the manufacturer's recommendations before applying tape, as it may damage the suit material.

High-Temperature Protective Clothing

Hazmat responders may be required to operate in high-temperature environments that exceed the protective capabilities of structural firefighting clothing. These types of hazards are often found during aircraft rescue firefighting (ARFF) operations and in some flammable liquid and gas scenarios. In these circumstances, special aluminized fabric protective clothing may be necessary.

Thermal energy can be encountered in three forms. You must recognize and understand their differences before selecting appropriate thermal protection.

- Ambient heat or the temperature of the surrounding atmosphere in a given scenario. For example, the ambient temperature in a public building is generally 65° to 70°F (18° to 21°C). During a structure fire, it may range from 120° to 200°F (49° to 93°C) at the floor level.
- Conductive heat is the heat generated by direct physical contact with a hot surface. For example, touching a metal valve which has been heated as a result of a flammable liquid spill can expose responders to temperatures over 1,000°F (538°C).
- Radiant heat is the heat generated by a heat source such as a flammable gas or liquid fire and is absorbed by materials that are struck by the radiant heat emitted by the heat source.

High-temperature protective clothing is designed primarily for radiant heat exposures. Its principal application will be in those situations where there is a minimal likelihood of direct contact with chemical vapors or splashes. SCBA should always be considered part of the ensemble.

Types of High-Temperature Clothing

Two types of high-temperature protective clothing may be used in the hazmat response community Figure 8-11:

1. **Proximity suits** are designed for exposures of short duration and close proximity to flame and radiant heat, such as in ARFF operations. The outer shell is a highly reflective, aluminized fabric over an inner shell of a flame-retardant fabric such as Kevlar® or Kevlar®/PBI® blends. These ensembles are not designed to offer any substantial chemical protection. Design and performance criteria for proximity suits is outlined in NFPA 1971—*Standard on Protective Ensembles for Structural Fire Fighting and Proximity Fire Fighting.*

Figure 8-11 The two most common examples of high temperature protective clothing are proximity suits and fire entry suits.

Proximity suits are available as a separate coat and pants ensemble or as coveralls. A hood or a helmet with an aluminized cover and faceshield may be used for head and face protection. The outer surface of the faceshield must be kept clean to ensure maximum reflection of radiant heat. A SCBA is worn for respiratory protection.

The outer shell of aged suits may begin to crack or flake off after several years of regular use. At this point, the protection factor drops significantly and the suit should be replaced.

2. Fire entry suits offer complete, effective protection for short-duration entry into a total flame environment. They are designed to withstand exposures to radiant heat levels up to 2000°F (1093°C). Entry suits consist of a coat, pants, and separate hood assembly. They are constructed of several layers of flame-retardant materials, with the outer layer often aluminized.

Because of the rapid speed with which burn injuries affect the body, entry suits are not effective for rescue operations in total flame environments. Unprotected individuals cannot be safely rescued from a fully involved area. Possible rescues are further impeded by the time necessary to don gear and enter the fire area.

Entry suits may be useful for offensive operations such as valve shutdowns in a flammable gas or liquid facility. However, there is a lack of mobility and flexibility when attempting these manipulations. Fire entry suits are usually a low priority budget item for most fire departments and HMRTs. There are currently no standards specifying the performance of fire entry suits.

Tying the System Together

Operational Considerations

The selection and use of specialized protective clothing at a hazmat/WMD emergency should be approached from a systems perspective. This system begins with an evaluation of four key factors: (1) the hostile environment, (2) the tasks to be performed, (3) the type of protective clothing required, and (4) the capabilities of the user/wearer.

1. **Hostile environment.** In simple terms, what is the challenge? Among the questions that must be considered are the following:
   - What material(s) are involved?
   - What is the physical state of the substance (i.e., solid, liquid, gas)?
   - What are the hazards of the substance (flammable, toxic, corrosive, etc.)?
   - What is the result of contact to the skin?
   - What physical hazards are present?
   - What is the ambient temperature and weather conditions?
2. **Tasks to be performed.** What are the objectives of entry operations? Given the tactical response objectives being implemented, what is the potential for exposure to the substances involved, including the level and duration of exposure? For example, rescue operations to remove symptomatic victims from a known contaminated environment have a high probability of responders becoming contaminated, while entry operations to conduct reconnaissance and air monitoring operations may result in little or no chemical exposure.
3. **Type of protective clothing required.** Strategic goals and tactical response objectives will determine the level of protective clothing required to bring about a more favorable outcome. Remember, this decision must take into account the level of risk associated with the overall response.
4. **Capabilities of the user/wearer.** All responders have personal strengths and weaknesses. Likewise, there are both physical and psychological stressors that will affect responders. A user with unrealistic expectations of what CPC can and cannot do will significantly increase the level of risk to both himself (or herself) and other responders.

Examples of physical stressors include the following:

- Extreme heat or cold operating conditions
- Noise
- Reduced vision from the fogging of CPC or SCBA face pieces
- Operations in low-light or low-visibility environments
- Reduced handling and dexterity due to the need to wear several layers of gloves
- Adverse weather conditions
- Physical hazards and the physical operating environment

Examples of psychological stressors include the following:

- Lack of physical fitness and the physical ability to perform the required tasks while wearing CPC
- Response operations involving injuries, fatalities, or high-risk operations
- Operations within enclosed or confined space environments
- Background and experience levels in both wearing CPC and operating in hostile environments
- Fear of either suit or respiratory protection failure

Consider the following operational issues and safety procedures when using protective clothing at hazmat incidents:

- The selection, maintenance, and use of protective clothing must be an integral part of an overall PPE and safety program. PPE is NOT your first line of defense; it is your last line of defense that comes into play if (1) your selection of tactical objectives and (2) site safety procedures can't keep the bad stuff off you.
- Chemical protection and thermal protection are, for all practical purposes, mutually exclusive. While some ensembles combine flash protection overgarments with CPC, they offer limited thermal protection beyond short-duration exposures.
- In some situations, SFC may be worn in combination with liquid chemical splash protective clothing, such as coveralls or a non-vapor-tight encapsulating suit. There are no hard and fast rules in this area—the response situation will dictate what level of protection will be worn on the inside and outside. However, the combination of the two levels of protective clothing will also impact mobility and increase the potential for heat stress for responders.
- Decon should be established prior to entry operations into the hot zone.
- Always minimize direct contact with any chemicals, regardless of your level of protection. Use common sense—fingers, tongues, and toes are not effective monitoring instruments! Avoid walking into or touching substances whenever possible.

Responder Safety Tip

You are more likely to hurt your responders as a result of heat stress than a chemical exposure. Remember the basics—physical fitness, prehydration, using cooling devices, rotating entry personnel, and providing effective rehab are simple things you can do to take care of your people.

- Depending upon the ambient temperature and humidity, fogging may occur inside the face piece of a chemical vapor suit. Some responders will keep a towel inside of the suit to wipe off the face piece, as needed. Another technique is to duct tape a small towel on top of the user's hard hat and then wipe off the inside of the visor by simply bending forward.
- Ensure entry and backup crews have equivalent levels of protection. When chemical vapor protection is required, at least four suits will be needed to support entry and back-up operations.
- "Two-In/Two-out" may be great if you are doing an entry in an open-air environment, but it won't cut it if your entry team is operating inside a structure and goes down. Think of your backup team as your initial rescue team. However, the reality may be that you will need a Rapid Intervention Team (RIT) capability of 4+ responders for entry operations into structures where responders must climb steps, enter enclosed areas, or have a long entry distance into the hot zone.
- Entry personnel must maintain their situational awareness at all times. Take a periodic 360° look around where you are working and what is occurring. Always have a Plan B for your emergency route of escape in case the incident quickly deteriorates.
- Communications is a critical element in entry operations, especially when using chemical vapor suits. A radio system backed by hand signals is a minimum requirement. Where possible, entry, backup, and safety personnel should operate on a radio channel completely separate from other on-scene units.

Responder Safety Tip

Air supply management will often be the most critical element of entry operations. Work mission duration should be calculated based upon the following factors:

- Entry Time
- Exit Time
- Decon Time (assume minimum of 5 minutes)
- Safety Factor (25% of air supply or expected duration)
- Remaining Time = Work Time

Remember: a "30 minute cylinder" does not provide 30 minutes of air; likewise, a high pressure cylinder will not provide all users with 45 or 60 minutes of breathing air. Air consumption will be based on level of personal fitness, the level of protection being used, ambient environmental conditions, and the nature of the task.

Most response organizations use an initial entry limit of 20 to 30 minutes and primarily rely upon 60-minute SCBAs. The Assistant Safety Officer–Hazardous Materials should consider the impact of ambient weather and humidity conditions upon responders, monitor entry times and air supplies, and notify the entry crew of their time limits at 5-minute intervals.

- A pre-entry safety briefing should be conducted prior to recon or entry operations. This may be provided by either the Assistant Safety Officer–Hazardous Materials or other Hazmat Group personnel (e.g., Entry Leader). All entry and backup personnel must be familiar with the objectives, tasks, and procedures to be followed. Topics should include objectives of the entry operation, a review of all assignments, verification of radio procedures (designated channels) and emergency signals (both hand signals and audible), emergency escape plans and procedures, protective clothing requirements, immediate signs and symptoms of exposure, and the location and layout of the decon area.

- Any plan to allow the same entry personnel to reenter the Hot Zone should be approved by the Hazmat Group Supervisor, Assistant Safety Officer–Hazardous Materials, Medical Supervisor, and, most important, the individuals involved. Do not underestimate the physiological and psychological effects of wearing specialized protective clothing.
- Support personnel are always needed to assist entry and backup crews in donning and doffing protective clothing. As a rule, one support person is necessary for each entry person. Donning times in excess of 10 minutes should be considered as excessive. Pre-entry support checklists can help take the "guess work" out of the donning and doffing process.

Emergency Procedures

Hazmat response organizations should develop procedures to address the following scenarios involving chemical vapor suits as follows:

1. **Loss of air supply.** This constitutes a major, life-threatening emergency. Causes may include simply running out of air to a mechanical failure of the SCBA or SAR unit. With training, responders can develop some basic skills that will allow them to react to this problem in a timely and competent manner. Key elements of any procedure should be as follows:
   - The affected responder must let their partner immediately know that they are having a problem. They should immediately start to exit the hazard area and move toward decon or a less-contaminated environment.
   - Immediately communicate the situation to the Assistant Safety Officer–Hazardous Materials or Entry Leader, as appropriate.
   - Resist any urge to immediately get out of the suit, particularly if still operating within a contaminated environment.
   - If possible, remove the regulator or pull the face piece away from your face and use the air within the chemical vapor suit. Chemical vapor suits with the SCBA inside of the ensemble (i.e., humpback design) typically have a sufficient supply of breathing air inside of the suit that will permit the user to at least exit the hazardous environment. While it won't be "good" air, it will contain some level of oxygen that can buy additional escape time. Note: This is an emergency measure and should only be used to assist the user in *immediately* evacuating the area.

   Remember—this emergency procedure is designed to get the user out of the hostile environment and out of the suit as quickly as possible. Continued use of breathing air from inside of a vapor-tight suit will quickly lead to decreased oxygen levels and a medical emergency.
2. **Loss of suit integrity.** Causes may include physical damage as a result of a puncture or tear, or having a limited-use garment "blow out" as a result of improper sizing. Regardless of the cause, this is a major, life-threatening emergency if it occurs within a contaminated atmosphere. Key elements of any procedure should be as follows:
   - Stay calm! Short-term exposure to contaminants inside of the suit does not necessarily mean you will immediately sustain an injury, providing your respiratory protection remains in place.
   - The affected responder must let their partner immediately know that they are having a problem. They should immediately start to exit the hazard area and move toward decon or a less-contaminated environment.
   - Immediately communicate the situation to the Assistant Safety Officer–Hazardous Materials or Entry Leader, as appropriate
   - Try to determine the size and location of the breach and if the user has been contaminated.
   - If possible, try to cover the breach with a glove or sleeve. The Entry Team partner may be able to assist in this process.
   - Do not remove your face piece or respiratory protection. Opening the SCBA emergency bypass valve may provide a positive-pressure environment and reduce the flow of contaminants into the suit.
   - Decontaminate the wearer and remove them from the suit as soon as possible.
   - Follow up with appropriate personal decon and medical procedures, as appropriate.
3. **Loss of communications.** It's amazing we can send an astronaut to the moon but sometimes cannot communicate via radio with an entry team that is less than 100 yards downrange in a structure! Since verbal communications are impossible when wearing vapor CPC, responders should have an overlapping communication system that incorporates both electronic communications and hand signals.

   Most HMRTs use a hand signal system based upon SCUBA diver hand signals. At a minimum, hand signals should be established for the following:
   - Are you okay? An affirmative response is usually a thumbs-up or okay hand signal. A negative response would be a thumbs-down response.
   - Out of air. Most response organizations have the responder bring their hand up to their throat.
   - In-suit emergency. Most response organizations have the responder raise their hands over their head and cross both hands or wave them.
4. **Buddy down in the hot zone.** The ability to remove entry personnel who are down in the hot zone will be both a labor-intensive and time-sensitive operation. A backup team of only two responders will be inadequate for most scenarios, and the OSHA "two in/two out" rule should be viewed as a minimum requirement. Ultimately, the nature and location of the entry operation should determine the number of personnel who should be assigned to the RIT function.

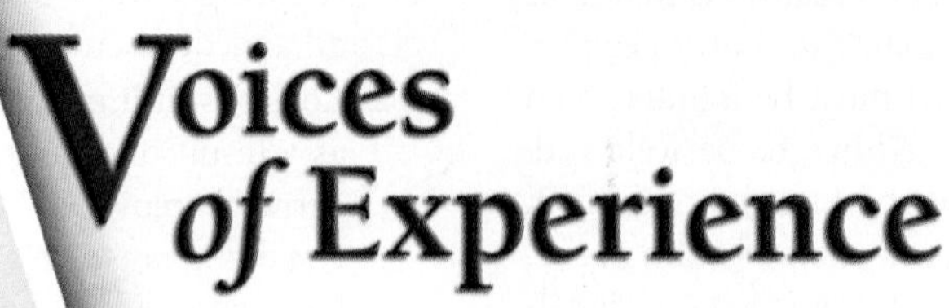

An old Navy Chief Petty Officer once said "It takes about 20 years to get 20 years of experience." Likewise, experience is nothing more than being able to survive your mistakes.

In the early days of hazmat response (around 1982), my partner Greg and I were attending a hazardous materials training class at a national training facility. We wanted to try various plugging and patching tools and techniques while wearing CPC to evaluate functional performance so we could take the lessons learned back to our HMRT.

Because of the time restrictions in the normal training class, we asked for special permission to return to the training field after hours to use chemical vapor protective clothing and SCBA and various tools on the many types of training props that were available. We received approval from the instructor-in-charge and were provided access to the training grounds.

After working through an evolution on a training prop, I noticed that it was getting difficult to inhale air through my face piece. Prior to this sensation I had no unusual signs or symptoms. I was warm but reasonably comfortable inside the suit and had no activation of an SCBA low-pressure warning alarm. Within a few breathing cycles I realized that my SCBA air cylinder was totally empty. This immediately created confusion for me because my training and experience told me that before the air cylinder was empty, the low-pressure alarm should have activated allowing me ample time to leave the area and doff the suit. I quickly realized that the low-pressure alarm had failed to operate and I was out of air and in immediate trouble.

My first instinct was to try to pull my right arm out of the glove and through the wrist gauntlet so I could remove the low-pressure breathing tube from the regulator. This was something I had done many times before while blindfolded in training. It took one try to determine that this was not going to happen. Meanwhile, my face piece remained on my face and was drawing a vacuum. I was suffocating and seconds away from passing out.

My prior training kicked in and I instinctively used my right hand from outside the suit to grab the low-pressure breathing tube and pull the face piece away from my chin. This allowed me to relieve the feeling of suffocation and draw a breath of air from inside the suit. While the air that I had been exhaling inside the suit had reduced oxygen content, it was still breathable. That first breath of stale air I took from inside the suit was wonderful and it kept me from panic.

My next move was to try to unzip my suit. To do this I had to allow the face piece to rest back on my face to free up both hands so I could see out of the hood and locate the zipper. I was quickly able to locate the zipper seam and trace it to the location of the zipper, but the lanyard on the zipper was very short and I could not get a grip on it with my hand, which was double gloved and restricted in range of motion. While attempting this maneuver I had to hold my breath and then alternate between pulling the face piece back from my chin, breathing, and then returning to my attempted self-rescue. After two attempts to unzip my suit I realized that I was consuming energy and precious suit air and that the best course of action was to wait for my partner to see that I was having trouble.

Fortunately, Greg saw that I was experiencing some kind of difficulty. He walked over to me and said, "Are you OK?" I remember saying something to the effect of, "No, get me out of this [deleted] suit!" Greg was able to access the zipper and free me from what could have been a very bad outcome if I were operating solo.

Prior to this incident I had about 12 years of experience in the fire service, including having served as a fire academy breathing apparatus instructor where I spent many hours operating in a smoke house and live training fires, and had the opportunity to observe and respond to numerous breathing apparatus problems. On many occasions my fellow instructors would practice dealing with various types of SCBA emergencies in confined spaces and total darkness. My prior training, and the fact that my partner was alert, allowed us to follow standard operating procedures using the buddy system, and we both remained calm—all of which may have saved my life that day. Trust me, running out of air inside a suit is not fun!

Mike Hildebrand

Operational considerations for conducting rescue operations of downed entry personnel will include the following:

- Immediately communicate the situation to the Assistant Safety Officer–Hazardous Materials or Entry Leader, as appropriate.
- What is the cause of the entry member going down? Suit failure, chemical exposure, heart attack, heat stress, and so on.
- If the responder has gone down as a result of a cardiovascular problem, RIT personnel should initiate life support operations as soon as possible. If possible, emergency decon measures should be taken. However, in most scenarios the basics of life support should not be delayed—it's a much better option than having a fully decontaminated but dead responder.
- For incidents in an open-air environment with no obstructions, the two-in/two-out backup procedure may be effective, providing that RIT personnel can use resources such as a wheeled stokes basket or a Sked® stretcher to facilitate the extrication and removal of the downed responder.
- If a responder falls into a pool or waterway and is wearing chemical vapor protective clothing (i.e., Level A), the inflation of the suit can provide some buoyancy. If the responder cannot self-extricate themselves from the pool or waterway, they should attempt to remain floating on their back in a semi-reclined position until they can be assisted. If necessary, the SCBA emergency bypass valve may be used to provide some additional air and inflation of the suit.
- Any scenario that involves a downed responder inside a structure or which will require egress via a stairwell will probably require a minimum of four or more backup personnel to conduct a timely extrication and removal. The use of a "stair chair" or a table chair may facilitate the removal process.
- Rescue operations where entry personnel are wearing Level A or chemical vapor clothing are generally more difficult than those where nonencapsulating chemical splash clothing is worn Figure 8-12.

5. **Suit over-pressurization.** Suit over-pressurization can occur as a result of a problem with an encapsulating suit's pressure relief valve, including the valve being blocked, malfunctioning, or being defective. The most obvious sign of this problem will be the suit ballooning up with air, thereby resulting in restricted movement or visibility being reduced as the suit visor rises. Responders may also encounter increased pressure in their ears. Key elements of any procedure should be as follows:
   - The affected responder must let their partner immediately know they are having a problem. They should immediately start to exit the hazard area and move toward decon or a less-contaminated environment.
   - Immediately communicate the situation to the Assistant Safety Officer–Hazardous Materials or Entry Leader, as appropriate.
   - If possible, attempt to force air out of the suit by bending or crouching. It will be extremely difficult to check the status/position of the relief valve while wearing the suit in a contaminated environment.
6. **Reduced egress scenarios.** Situations may occur where entry personnel must work in or around areas with limited access and egress. If a responder must exit through a narrow area or "pinch point," a reduced profile technique may be used. The specific process and techniques will be dependent upon the type of Level A suit being worn and the SCBA being used. Key elements of any procedure should be as follows:
   - Immediately communicate the situation to the Assistant Safety Officer–Hazardous Materials or Entry Leader, as appropriate.
   - The responder should attempt to (1) remove their arm from the suit sleeve; (2) loosen the SCBA shoulder strap and remove their arm from the SCBA strap; (3) maneuver

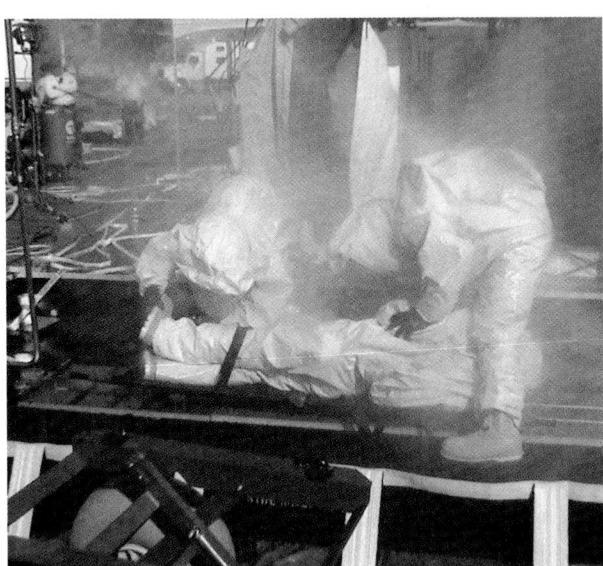

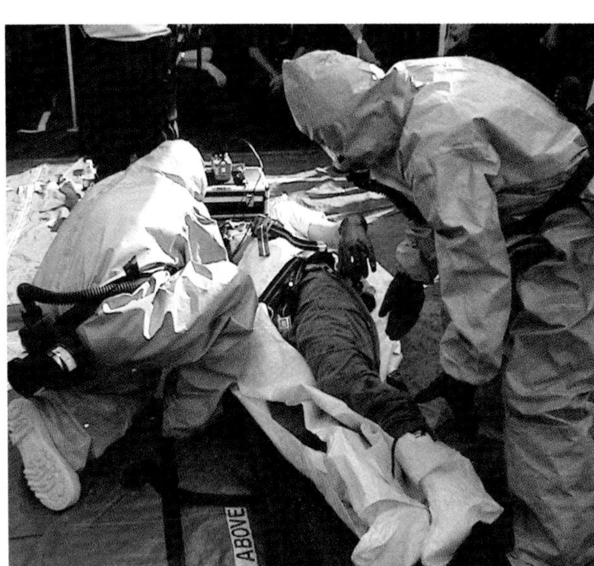

Figure 8-12 Emergency procedures while wearing chemical vapor protective clothing

the SCBA so it aligns with the body; (4) place their arms back into the suit sleeve and proceed through the reduced space; and (5) when clear of the reduced space, reposition the SCBA to its proper position.

Donning, Doffing, and Support Considerations

The donning and doffing of CPC can be a time-consuming process. More than one IC has asked the question, "Why is it taking so long?" Procedures for the donning and doffing of specific CPC ensembles should be based upon the manufacturer's instructions. The following general procedures are presented as a basic guideline.

Donning Procedures

- Coordinate with medical to begin monitoring of entry and backup personnel, as well as prehydration. Ensure that the medical evaluation is maintained and coordinated with the Assistant Safety Officer–Hazardous Materials.
- Determine the appropriate CPC and respiratory protection information and recommendations.
- Remove, tag, and secure all personal items, including wallets, rings, watches, and so on.
- Don any required undergarments that may be worn under the CPC ensemble, such as fire-retardant coveralls.
- Responders who wear glasses should ensure that their corrective lenses are properly installed within their face piece.
- At least one and preferably two support personnel should be assigned to assist each responder in donning their CPC.
- A donning and doffing area should be selected; these two locations should be adjacent to each other. Criteria for area selection should include the following:
  - Located as close as possible to the entry point but still isolated from other response activities.
  - Protected from the weather and elements. HMRTs may use vehicles, portable tents, and structures or fixed facilities as available.
  - Sufficiently large enough to accommodate all donning and support operations.
  - Clearly defined and delineated location, such as through the use of banner tape, cones, color-coded and pre-marked tarps, and so on.
  - Responders who will don PPE should be provided suitable seating. Some HMRTs use modified chairs or benches with the back removed.
- A pre-entry safety briefing should be conducted prior to donning operations. This may be provided by either the Assistant Safety Officer–Hazardous Materials or other Hazmat Group personnel (e.g., Entry Leader). Topics should include objectives of the entry operation, a review of all assignments, verification of radio procedures (designated channel/s) and emergency signals (both hand and audible), emergency escape plans and procedures, protective clothing requirements, immediate signs and symptoms of exposure, and the location and layout of the decon area. At the conclusion of the briefing, entry personnel should repeat their instructions to ensure understanding.
- Entry and backup personnel should dress at approximately the same rate to ensure that entry personnel are not "standing around and waiting." Personnel should not go on air during this process. Items that should be checked and verified during this process include air supply status and communications.
- Once the IC approves the entry operation, entry personnel should be placed on air, complete the donning process, and then led to the access control point. A final check of all PPE and CPC should be made prior to any entry operation, including ensuring that all zippers and suit closures are properly secured, no obvious suit damage is present, and the appropriate gloves and footwear are verified.
- In most response scenarios, backup personnel will remain in a stand-by mode and be "off-air" until needed.

Doffing Procedures

- Ensure that support personnel use the appropriate level of PPE to assist with the doffing process. This will be based on the nature of the contaminants and decon operations. In general, this will typically consist of chemical gloves and eye protection if a splash hazard is present.
- Entry personnel may be tired, extremely hot and sweaty, and anxious to remove their PPE. Vision may also be obscured through fogging of the face piece. Heat stress is a genuine concern.
- PPE should be doffed based on the manufacturer's instructions. General guidelines include the following:
  - Support personnel should only touch the outside of the CPC ensembles.
  - Entry personnel should only touch the inside of the CPC ensembles.
  - Minimize cross-contamination from outside to inside of the CPC ensembles.
  - When dealing with chemical vapor clothing, the top portion of the garment should first be unzipped and removed/pulled down low enough so that the wearer can sit down. Once sitting, support personnel can then remove the arms, legs, and SCBA harness.
  - Entry personnel should be permitted to remove their own inner gloves and face piece.
  - Once removed, CPC should be placed in a bag or container for further decon or disposal.
- Entry personnel should then be hydrated, medically evaluated, and debriefed, as appropriate.

Donning and Doffing Chemical Protective Clothing

As mentioned throughout this chapter, the selection and proper use of PPE is vital to responder safety. To that end, responders must thoroughly understand the use parameters and limitations of all components of chemical protective equipment. The rubber meets the road however, when the time comes to wear all the items you have selected and go to work.

Donning and doffing PPE, while seemingly straightforward, requires a deliberate and methodical approach, executed in a timely fashion. Care must be taken to not damage or compromise the PPE while putting it on; zippers and other closure points must be completely fastened; everything you wear should be the right size—in short, you must pay attention to detail and ensure your ensemble is battle-ready. Working in a contaminated atmosphere is not the time to be fooling around with pieces of your gear that should have been addressed prior to entry. This draws down your work time while on air or otherwise stops the forward momentum of making the problem go away.

To don a Level A ensemble, follow the steps in **Skill Drill 8-1**:

1. Conduct a pre-entry briefing, medical monitoring, and equipment inspection. (**photo 1**)
2. While seated, pull on the suit to waist level and pull on the attached chemical boots. Fold the suit boot covers over the tops of the boots. (**photo 2**)
3. Stand up and don the SCBA frame and SCBA face piece, but do not connect the regulator to the face piece. (**photo 3**)
4. Place the helmet on the head. (**photo 4**)
5. Don the inner gloves. (**photo 5**)
6. Don the outer chemical gloves (if required by the manufacturer's specifications).
7. With assistance, complete donning the suit by placing both arms in the suit, pulling the expanded back piece over the SCBA, and placing the chemical suit over the head. (**photo 6**)
8. Connect the regulator to the SCBA face piece and ensure that the air flow is working correctly. (**photo 7**)
9. Close the chemical suit by closing the zipper and sealing the splash flap. (**photo 8**)
10. Review hand signals and indicate that you are okay. (**photo 9**)

Skill Drill 8-1

Donning a Level A Ensemble

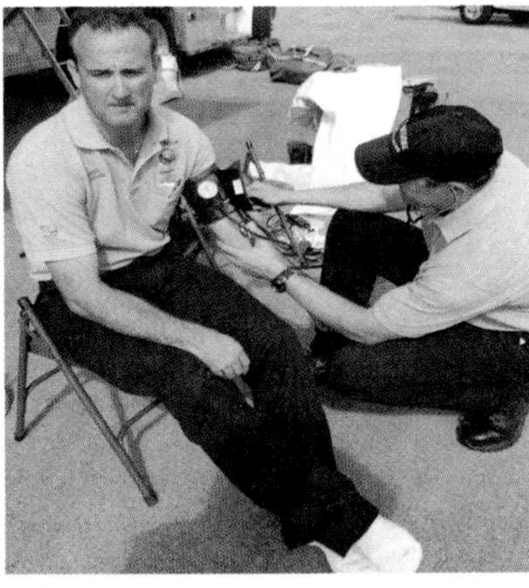

1 Conduct a pre-entry briefing, medical monitoring, and equipment inspection.

2 While seated, pull on the suit to waist level; pull on the chemical boots over the top of the chemical suit. Pull the suit boot covers over the tops of the boots.

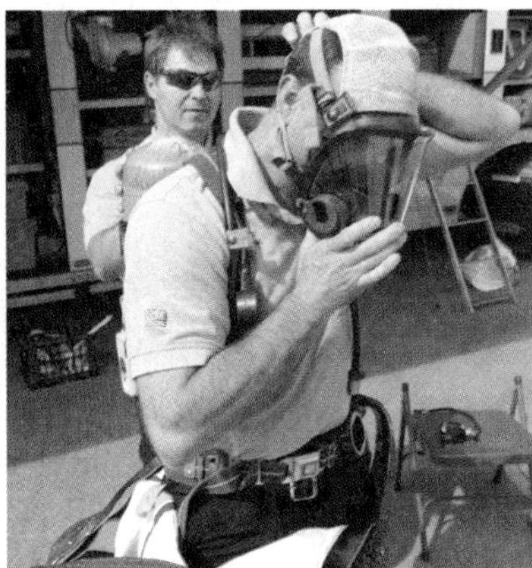

3 Stand up and don the SCBA frame and SCBA face piece, but do not connect the regulator to the face piece.

Skill Drill 8-1

Donning a Level A Ensemble (*Continued*)

4 Place the helmet on the head

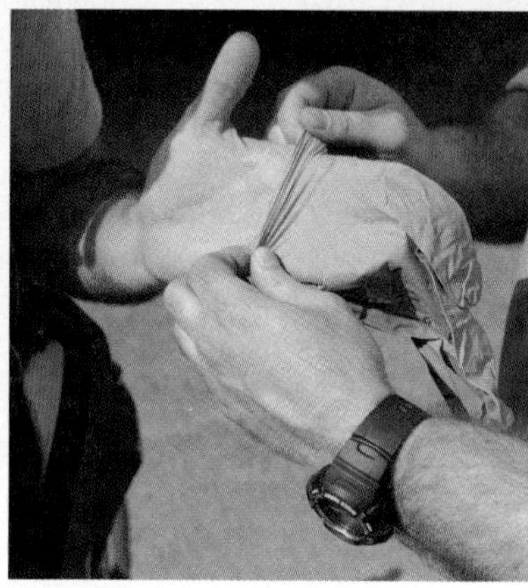
5 Don the inner gloves.

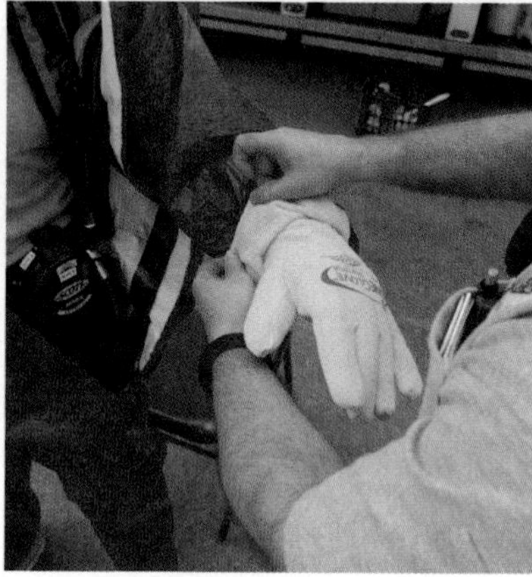
6 Don the outer chemical gloves (if required). With assistance, complete donning the suit by placing both arms in the suit, pulling the expanded back piece over the SCBA, and placing the chemical suit over the head.

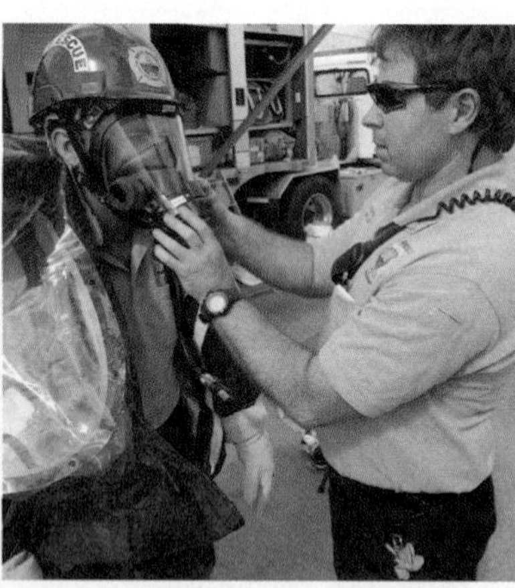
7 Connect the regulator to the SCBA face piece and ensure air flow.

8 Close the chemical suit by closing the zipper and sealing the splash flap.

9 Review hand signals and indicate that you are okay.

To doff a Level A ensemble, follow the steps in **Skill Drill 8-2**:

1. After completing decontamination, proceed to the clean area for suit doffing.
2. Pull the hands out of the outer gloves and arms from the sleeves, and cross the arms in front inside the suit. (**photo 1**)
3. Open the chemical splash flap and suit zipper. (**photo 2**)
4. Begin at the head and roll the suit down and away until the suit is below waist level. (**photo 3**)
5. Complete rolling the suit from the waist to the ankles; step out of the attached chemical boots and suit. (**photo 4**)
6. Doff the SCBA frame. The face piece should be kept in place while the SCBA is doffed. (**photo 5**)
7. Take a deep breath and doff the SCBA face piece; carefully remove the helmet, peel off the inner gloves, and walk away from the clean area.
9. Go to the rehabilitation area for medical monitoring, rehydration, and personal decontamination shower. (**photo 6**)

Skill Drill 8-2

Doffing a Level A Ensemble

1 After completing decontamination, proceed to the clean area for suit doffing. Pull the hands out of the outer gloves and arms from the sleeves, and cross the arms in front inside the suit.

2 Open the chemical splash flap and suit zipper.

3 Begin at the head and roll the suit down and away until the suit is below waist level.

4 Complete rolling the suit from the waist to the ankles; step out of the attached chemical boots and suit.

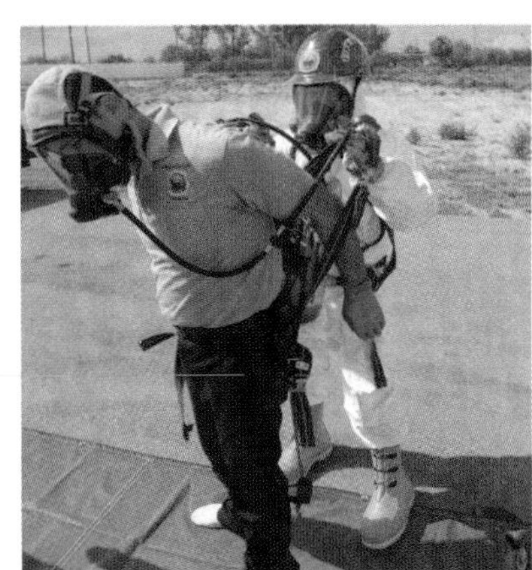

5 Doff the SCBA frame. The face piece should be kept in place while the SCBA frame is doffed.

6 Take a deep breath and doff the SCBA face piece; carefully peel off the inner gloves and walk away from the clean area. Go to the rehabilitation area for medical monitoring, rehydration, and personal decontamination shower.

To don and doff a Level B encapsulated chemical-protective clothing ensemble, follow the same steps found in Skill Drills 8-1 and 8-2. Remember, the difference between the Level A ensemble and Level B encapsulating ensemble is not the procedure—it is the construction and performance of the garment.

To don a Level B nonencapsulated chemical-protective clothing ensemble, follow the steps in **Skill Drill 8-3**:

1. Conduct a pre-entry briefing, medical monitoring, and equipment inspection. (**photo 1**)
2. Sit down, pull on the suit to waist level; pull on the chemical boots over the top of the chemical suit. Pull the suit boot covers over the tops of the boots. (**photo 2**)
3. Don the inner gloves. (**photo 3**)
4. With assistance, complete donning the suit by placing both arms in the suit and pulling the suit over the shoulders.
5. Close the chemical suit by closing the zipper and sealing the splash flap. (**photo 4**)
6. Don the SCBA frame and SCBA face piece, but do not connect the regulator to the face piece (**photo 5**)

Skill Drill 8-3

Donning a Level B Nonencapsulated Chemical-Protective Clothing Ensemble

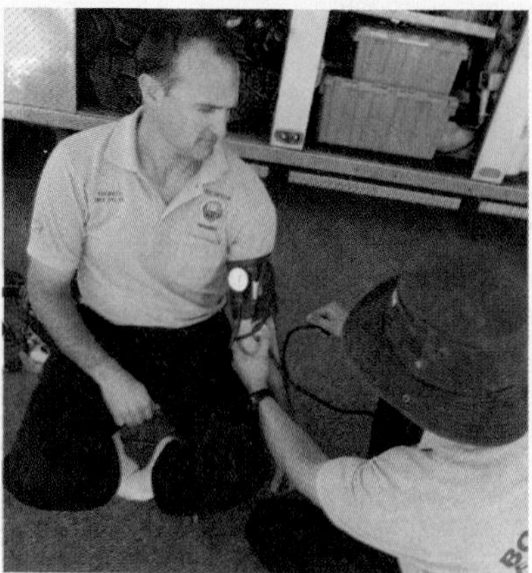

1 Conduct a pre-entry briefing, medical monitoring, and equipment inspection.

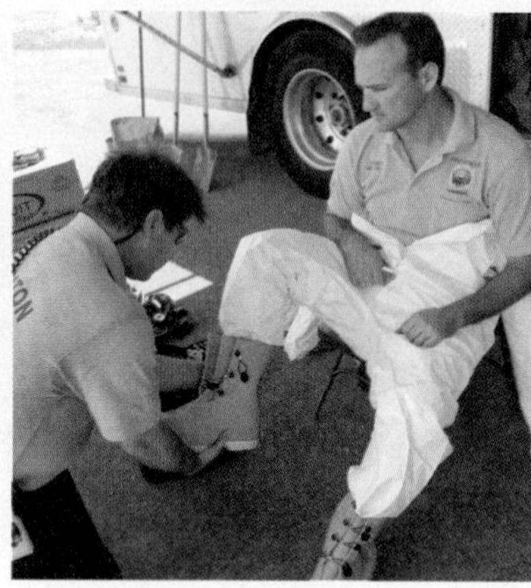

2 While seated, pull on the suit to waist level; pull on the chemical boots over the top of the chemical suit. Pull the suit boot covers over the tops of the boots.

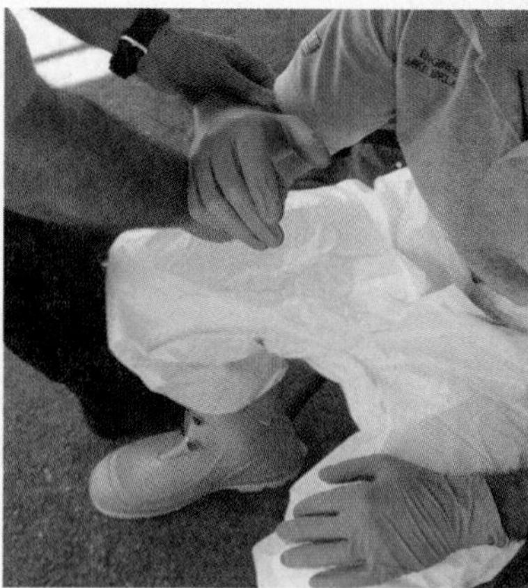

3 Don the inner gloves.

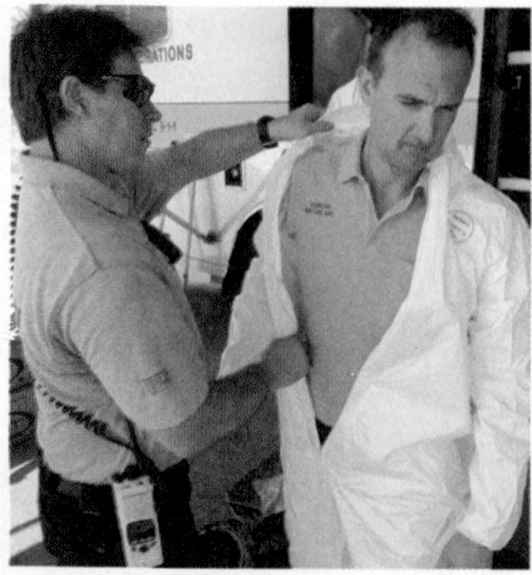

4 With assistance, complete donning the suit by placing both arms in suit and pulling suit over shoulders. Close the chemical suit by closing the zipper and sealing the splash flap.

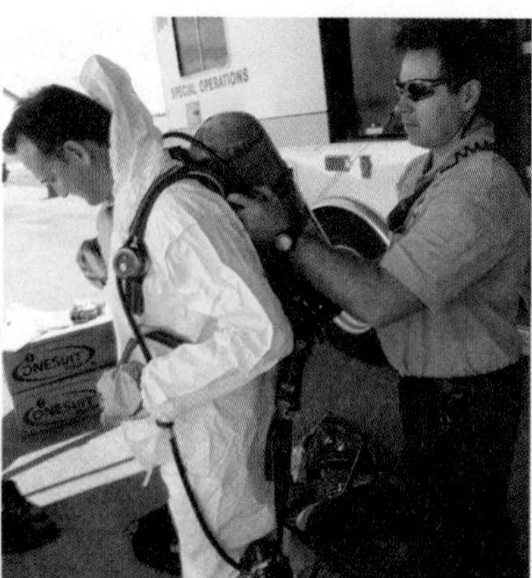

5 Don the SCBA frame and SCBA face piece, but do not connect the regulator to the face piece.

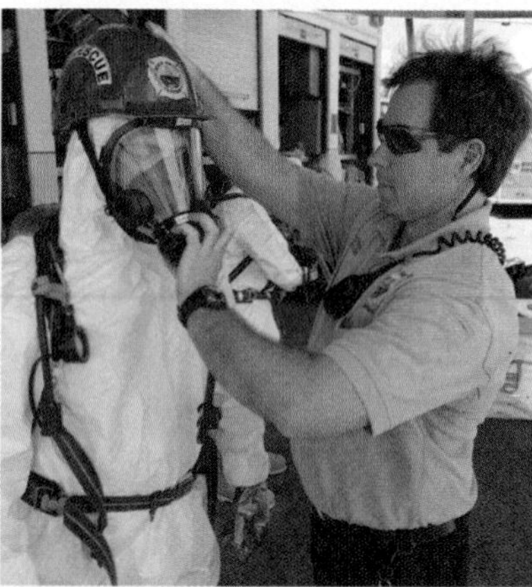

6 With assistance, pull the hood over the head and SCBA face piece. Place the helmet on the head. Put on the outer gloves. Connect the regulator to the SCBA face piece and ensure you have air flow.

7. With assistance, pull the hood over the head and SCBA face piece.
8. Place the helmet on the head.
9. Pull the outer gloves over or under the sleeves, depending on the situation.
10. Instruct the assistant to connect the regulator to the SCBA face piece and ensure that the air flow is working correctly. (**photo 6**)
11. Review hand signals and indicate that you are okay.

To doff a Level B nonencapsulated chemical-protective clothing ensemble, follow the steps in **Skill Drill 8-4**:

1. After completing decontamination, proceed to the clean area for suit doffing.
2. Stand and doff the SCBA frame. Keep the face piece in place while the SCBA frame is placed on the ground. (**photo 1**)
3. Instruct the assistant to open the chemical splash flap and suit zipper. (**photo 2**)

Skill Drill 8-4

Doffing a Level B Nonencapsulated Chemical-Protective Clothing Ensemble

1 After completing decontamination, proceed to the clean area for suit doffing. Stand and doff the SCBA frame. Keep the face piece in place.

2 Open the chemical splash flap and suit zipper.

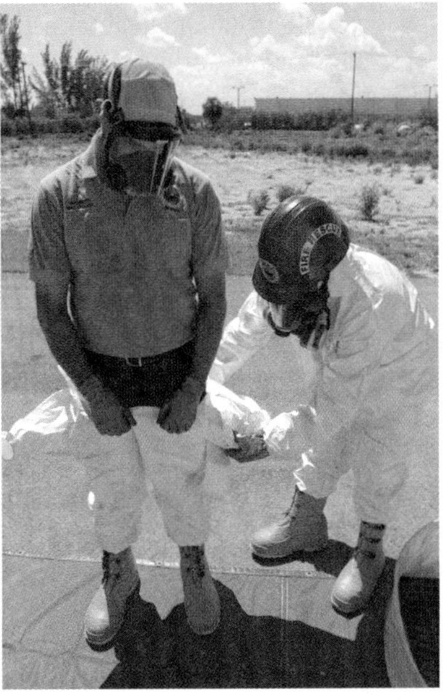

3 Remove your hands from the outer gloves and your arms from the sleeves of the suit. Cross your arms in front inside of the suit. Begin at the head and roll the suit down and away until the suit is below waist level.

4 Complete rolling down the suit to the ankles; step out of attached chemical boots and suit.

5 Doff the SCBA face piece.

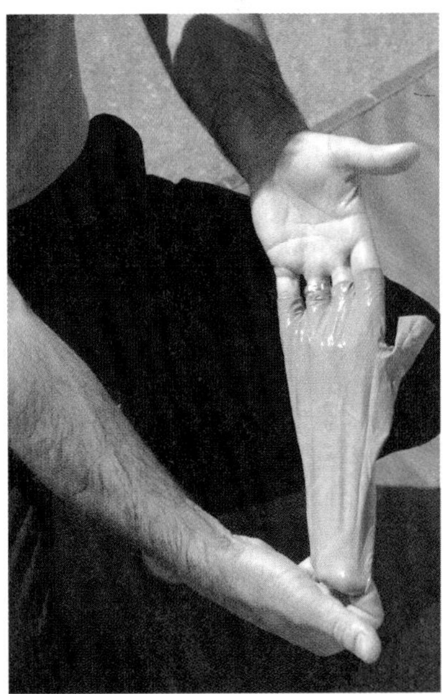

6 Carefully peel off the inner gloves and walk away from the clean area. Go to the rehabilitation area for medical monitoring, rehydration, and personal decontamination shower.

4. Remove your hands from the outer gloves and arms from the sleeves, and cross your arms in front inside the suit.
5. Begin at the head and roll the suit down and away until the suit is below waist level. (**photo 3**)
6. Complete rolling down the suit to the ankles. Step out of the outer boots and suit. (**photo 4**)
7. Stand and doff the SCBA face piece and helmet (**photo 5**).
8. Carefully peel off the inner gloves and go to the rehabilitation area for medical monitoring, rehydration, and personal decontamination shower. (**photo 6**)

To don a Level C chemical-protective clothing ensemble, follow the steps in **Skill Drill 8-5**:

1. Conduct a pre-entry briefing, medical monitoring, and equipment inspection.
2. While seated, pull on the suit to waist level; pull on the chemical boots over the top of the chemical suit. Pull the suit boot covers over the tops of the boots. (**photo 1**)
3. Don the inner gloves. (**photo 2**)
4. With assistance, complete donning the suit by placing both arms in the suit and pulling the suit over the shoulders.
5. Close the chemical suit by closing the zipper and sealing the splash flap. (**photo 3**)
6. Don APR/PAPR face piece.
7. With assistance, pull the hood over the head and the APR/PAPR face piece.
8. Place the helmet on the head.
9. Pull the outer gloves over or under the sleeves, depending on the situation.
10. Review hand signals and indicate that you are okay. (**photo 4**)

To doff a Level C chemical-protective clothing ensemble, follow the steps in **Skill Drill 8-6**:

1. After completing decontamination, proceed to the clean area for suit doffing.
2. As with level B, instruct the assistant to open the chemical splash flap and suit zipper.
3. Remove the hands from the outer gloves and your arms from the sleeves.
4. Begin at the head and roll the suit down and away until the suit is below waist level.
5. Complete rolling down the suit and take the suit and boots away.
6. Remove the inner gloves.
7. Remove the APRPE PAPR. Remove the helmet. (**photo 1**)
8. Go to the rehabilitation area for medical monitoring, rehydration, and personal decontamination shower. (**photo 2**)

To don a Level D chemical-protective clothing ensemble, follow the steps in **Skill Drill 8-7**:

1. Conduct a pre-entry briefing and equipment inspection.
2. Don the Level D suit.
3. Don the boots.
4. Don safety glasses or chemical goggles.
5. Don a hard hat.
6. Don gloves, a face shield, and any other required equipment. (photo 1)

Training Considerations

The use of protective clothing at a hazmat incident should be the final step of a comprehensive system that begins with an analysis of the facility or community's hazards. This system must include organized and documented training with all types of protective clothing ensembles used within the organization, as well as a regular, effective preventive maintenance and testing program.

Both classroom and hands-on training is essential. Aggressive, experienced firefighters and emergency response team (ERT) personnel do not always make effective hazmat responders. As a result, many organizations have developed a protective clothing qualification system that requires both initial and regular qualification in each type of suit and respiratory protection device used by the response organization. Protective clothing training evolutions are most effective when conducted on a "building block" or modular basis and combined with other manipulative skills, and should include practicing emergency procedures for possible scenarios.

Training with protective clothing is essential for the development of effective skills and competencies. To minimize damage to "front-line" liquid chemical splash and vapor suits, response organizations normally purchase training suits, use limited-use garments specifically for training purposes, or use Level A suits that have exceeded their shelf life. When using older suits, ensure that they have been completely decontaminated and do not have a significant chemical exposure history. "Front-line" chemical vapor suits should only be used during actual emergencies.

Inspection and Maintenance Procedures

Preventive maintenance and documentation are integral elements of a comprehensive PPE program. Unfortunately, they are also one of the most neglected.

A records file should be maintained for all chemical vapor clothing and respiratory protection units documenting their respective history. This documentation should indicate date of purchase, manufacturer and vendor, serial number, material of construction, and any other unique or specific information. For chemical vapor protective clothing, a logbook should be established for each suit that records each time the clothing is worn, inspection and maintenance data, unusual conditions or observations, decontamination solutions and procedures, and dates with appropriate signatures. Periodic records review may pinpoint an item with excessive maintenance costs or out-of-service times.

Manufacturer maintenance and testing recommendations should be consulted for maintenance intervals and procedures. At a minimum, protective clothing should be inspected at the following benchmarks:

- Upon receipt from the manufacturer or vendor
- After each use

Skill Drill 8-5

Donning a Level C Chemical-Protective Clothing Ensemble

1 Conduct a pre-entry briefing, medical monitoring, and equipment inspection. While seated, pull on the suit to waist level; pull on the chemical boots over the top of the chemical suit. Pull the suit boot covers over the tops of the boots.

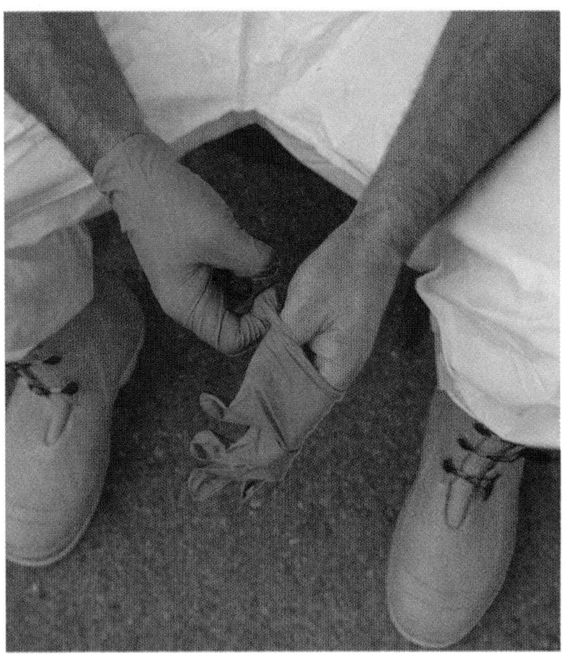

2 Don the inner gloves.

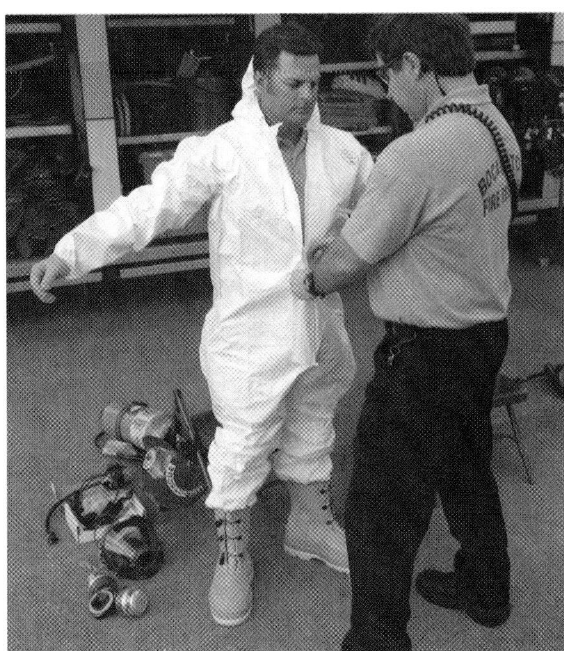

3 With assistance, complete donning the suit by placing both arms in the suit and pulling the suit over the shoulders. Instruct the assistant to close the chemical suit by closing the zipper and sealing the splash flap.

4 Don APR/PAPR. Pull the hood over the head and APR/PAPR. Place the helmet on the head. Pull on the outer gloves. Review hand signals and indicate that you are okay.

Skill Drill 8-6

Doffing a Level C Chemical-Protective Clothing Ensemble

1 After completing decontamination, proceed to the clean area. As with level B, the assistant opens the chemical splash flap and suit zipper. Remove the hands from the outer gloves and arms from the sleeves. Begin at the head and roll the suit down below waist level. Complete rolling down the suit and take the outer boots and suit away. The assistant helps remove inner gloves. Remove APR/PAPR. Remove the helmet.

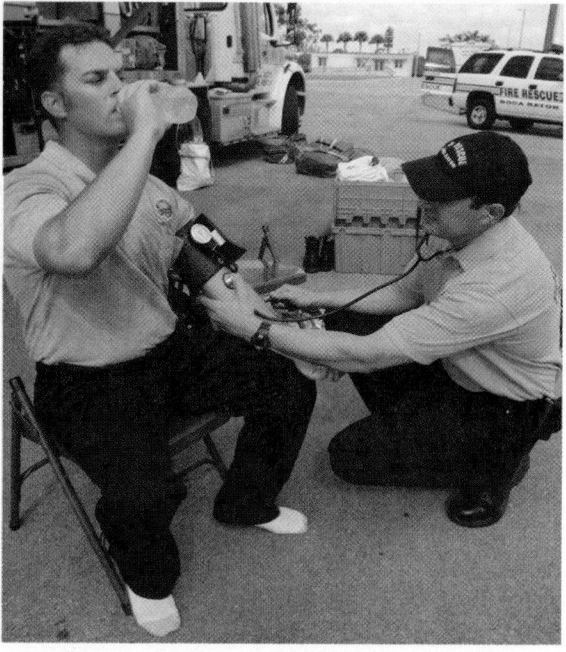

2 Go to rehabilitation area for medical monitoring, rehydration, and personal decontamination shower.

- Periodic inspections (i.e., monthly or quarterly)
- Whenever questions arise regarding selected protective equipment or when problems with similar equipment arise

Each inspection will cover different areas in varying levels of thoroughness depending on the type of protective clothing. Detailed inspection procedures are usually available from the manufacturer.

Documentation and maintenance of all appropriate records is a top priority. Individual inventory or record numbers should be assigned to all reusable pieces of protective clothing, including gloves. This will simplify the process of tracking gloves, liquid chemical splash suits, and chemical vapor suits over a period of time and facilitate monitoring for potential problems.

Chemical vapor clothing should undergo a tightness test at intervals as established by the suit manufacturer and NFPA 1991—*Vapor Protective Ensembles for Hazardous Materials Emergencies*. Tightness tests are normally conducted upon manufacture of the suit and annually thereafter, using protocols established by ASTM F1052—*Standard Test Method for Pressure Testing Vapor Protective Ensembles*. Under the ASTM F1052 test method, the suit is inflated to a specified pressure and then the pressure drop is measured over time (e.g., 4 minutes). Pass/fail criteria is based upon the suit maintaining the minimum allowable pressure. Both chemical vapor and splash protective clothing should also undergo visual inspections for any signs of degradation, stress cracks, or other damage.

Skill Drill 8-7

Donning a Level D Chemical-Protective Clothing Ensemble

1. Conduct pre-entry briefing and equipment inspection. Don the Level D suit. Don the boots. Don safety glasses or chemical goggles. Don a hard hat. Don gloves, a face shield, and any other required equipment.

All protective clothing must be stored properly to prevent damage caused by dust, moisture, sunlight, chemical exposures, temperature extremes, and impact. The manufacturer's storage guidelines should always be followed. Numerous equipment failures during actual use are attributed to improper storage procedures. Many response organizations store all chemical vapor suits in sealed packaging to ensure they are not damaged or tampered.

Wrap-Up

Chief Concepts

- Personal protective clothing and equipment is critical to the success of an organization's hazardous materials response program. They are an integral element of the health and safety program and facilitate the ability of emergency responders to respond and control hazmat releases in a safe, efficient, and effective manner.
- An effective and comprehensive personal protective clothing program should address six fundamental elements: hazard identification, PPE selection and use, medical monitoring, training, inspection, and maintenance.
- Using a risk-based approach is critical when selecting personal protective clothing and equipment. All decisions should be well thought out and realistic, taking into account both the positive and negative effects of the tactical options being pursued.
- Emergency responders should be familiar with the policies and procedures of the Authority Having Jurisdiction (AHJ) so as to ensure a consistent approach for the selection and use of PPE.
- Chemicals may attack and pass through protective clothing materials via three methods: degradation, penetration, and permeation. Barrier compatibility charts are primarily based upon penetration and permeation testing.
- CPC materials are classified as either limited-use (disposable) garments or reusable garments.
- NFPA 1991, 1992, and 1994 provide design and performance criteria for chemical protective clothing. NFPA 1991 addresses chemical vapor protective ensembles, NFPA 1992 addresses liquid splash protective ensembles, and NFPA 1994 addresses protective ensembles for chemical/biological terrorism incidents.
- Chemical resistance data is described in terms of either (1) chemical permeation/breakthrough times and rates; (2) cumulative permeation for a given test duration; or (3) as "pass/fail" chemical penetration testing results. Remember that the longer the breakthrough time or the lower the cumulative permeation value, the better the level of protection.
- The glove, boot, visors, and garment components of CPC will often be constructed of different materials or laminates. Be aware of which of these materials form the basis for the chemical barrier data provided by the manufacturer.
- Air purification devices should not be used at hazmat releases unless qualified personnel have first monitored the environment and determined that such devices can be safely used (per OSHA 29 CFR 1910.120[q][3][iv] and 1910.134). As a general rule, they should not be used for initial response operations at hazmat incidents and for emergency response operations involving unknown substances.
- Three basic types of protective clothing may be used at hazmat incidents:
  - Structural firefighting clothing is designed to protect against extremes of temperature, steam, hot water, hot particles, and the typical hazards of firefighting.
  - CPC is designed to protect skin and eyes from direct chemical contact. There are two basic types of CPC used: chemical splash protective clothing and chemical vapor protective clothing.
  - High temperature protective clothing is designed to protect against short-term exposures to high temperatures, such as proximity and fire entry suits.
- The EPA/OSHA levels of protection (A, B, C, D) reflect the design of the protective clothing ensemble and the respiratory protection provided, but do NOT provide an accurate description of the protection provided.
- Although SFC may offer sufficient protection to the wearer who is fully aware of the hazards being encountered and the limitations of the protective clothing, it is not designed to provide chemical splash or chemical vapor protection.
- Hazmat emergency responders are more likely to be injured as a result of heat stress than a chemical exposure. Many response agencies use some form of cooling technology to reduce the potential for heat stress injuries.
- Manufacturers' guidelines for maintenance, testing, inspection, storage, and documentation should be followed for all PPE provided by the AHJ.

Hot Terms

Air Purifying Respirators (APR) Respirators or filtration devices that remove particulate matter, gases or vapors from the atmosphere. These devices range from full-face piece, dual cartridge masks with eye protection, to half-mask, face piece mounted cartridges.

Breakthrough Time The time from the initial chemical attack on the outside of the material until its desorption and detection inside. The units of time are usually expressed in minutes or hours, and a typical test runs up to a maximum of 8 hours. If no measurable breakthrough is detected after 8 hours, the result is often reported as a breakthrough time of >480 minutes or >8 hours.

Chemical Protective Clothing (CPC) Single or multi-piece garment constructed of chemical protective clothing materials designed and configured to protect the wearer's torso, head, arms, legs, hands, and feet. Can be constructed as a single or multi-piece garment. The garment may completely enclose the wearer either by itself or in combination with the wearer's respiratory protection, attached or detachable hood, gloves and boots.

Chemical Vapor Protective Clothing Chemical protective clothing ensemble that is designed and configured to protect the wearer against chemical vapors or gases. Vapor chemical protective clothing must meet the requirements of NFPA 1991.

Degradation The physical destruction or decomposition of a clothing material due to exposure to chemicals, use, or ambient conditions (e.g., storage in sunlight).

Fire Entry Suits Suits that offer complete, effective protection for short duration entry into a total flame environment; designed to withstand exposures to radiant heat levels up to 2,000°F (1,093°C). Entry suits consist of a coat, pants, and separate hood assembly.

Limited-Use Materials Protective clothing materials that are used and then discarded. Although they may be reused several times (based upon chemical exposures), they are often disposed of after a single use.

Liquid Chemical Splash Protective Clothing The garment portion of a chemical protective clothing ensemble that is designed and configured to protect the wearer against chemical liquid splashes but not against chemical vapors or gases. Liquid splash chemical protective clothing must meet the requirements of NFPA 1992.

Penetration The flow or movement of a hazardous chemical through closures, seams, porous materials, and pinholes or other imperfections in the material.

Permeation The process by which a hazardous chemical moves through a given material on the molecular level.

Permeation Rate The rate at which a chemical passes through a given chemical protective clothing material. Expressed as micrograms per square centimeter per minute (μgm/cm2/min).

Powered-Air Purification Respirators (PAPR) Air-purifying respirators that use a blower to force the ambient air through air-purifying elements to a full-face mask. As a result, there is a slight positive pressure in the facepiece that results in an increased protection factor.

Proximity Suits Designed for exposures of short duration and close proximity to flame and radiant heat, such as in aircraft rescue firefighting (ARFF) operations. The outer shell is a highly reflective, aluminized fabric over an inner shell of a flame-retardant fabric such as Kevlar™ or Kevlar™/PBI™ blends. These ensembles are not designed to offer any substantial chemical protection.

Structural Firefighting Clothing (SFC) Protective clothing normally worn by firefighters during structural fire fighting operations. It includes a helmet, coat, pants, boots, gloves, PASS device, and a hood to cover parts of the head not protected by the helmet. May also be referred to as turnout or bunker clothing.

HazMat Responder *in Action*

You have just arrived at a local chemical warehouse for a report of a chemical spill. You are provided an SDS with the UN identification number of 2047 and the chemical name 1,3-Dichlororopropane. The level of concern is high and HazMat Response Team is approximately three minutes away.

Questions

1. Which protective equipment would you use for a reconnaissance mission?
   - **A.** NFPA Class 4 ensemble with air purifying respirator
   - **B.** EPA Level B ensemble with supplied air
   - **C.** Structural Firefighting Clothing with Self-Contained Breathing Apparatus
   - **D.** NFPA Class 1 ensemble with Self Contained Breathing Apparatus
2. The physical destruction or decomposition of a clothing material due to exposure describes:
   - **A.** Permeation
   - **B.** Degradation
   - **C.** Penetration
   - **D.** Breakthrough
3. The operational considerations for this incident are:
   - **A.** Two in, Two out established/Pre-entry briefing/Tasks to be performed
   - **B.** Type of protective clothing required/Air supply considerations/Entry briefing
   - **C.** Type of hostile environment/Task to be performed/Type of protective clothing/Capabilities of the user
   - **D.** Pre entry briefing/Tasks to be performed/Type of protective clothing required/Type of hostile environment
4. Assume that you have an IDLH atmosphere and no secondary emergency air supply. Which of the following respiratory protection options should you select?
   - **A.** APR
   - **B.** PAPR
   - **C.** SCBA
   - **D.** SAR

References and Suggested Readings

1. Air Force Civil Engineer Support Agency (AFCESA) and PowerTrain, Inc., *Hazardous Materials Incident Commander Emergency Response Training CD*. Tyndall Air Force Base, FL: Headquarters AFCESA (2010).
2. Air Force Civil Engineer Support Agency (AFCESA) and PowerTrain, Inc., *Hazardous Materials Technician Emergency Response Training CD*, Tyndall Air Force Base, FL: Headquarters AFCESA (2010).
3. Carroll, Todd R. "Contamination and Decontamination of Turnout Clothing." Emmitsburg, MD: Federal Emergency Management Agency—U.S. Fire Administration (April 29, 1993).
4. Forsberg, Krister and S. Z. Mansdorf, *Quick Selection Guide to Chemical Protective Clothing* (5th Edition), New York, NY: John Wiley & Sons, Inc. (2007).
5. Forsberg, Krister, Michael Blotzen, Michael and Lawrence H. Keith, *Chemical Protective Clothing Permeation/Degradation Database* (CD), Boca Raton, FL: CRC Press (1992).
6. Jenkins, Tiffany, Josh Smith, Ben Zimmerman and Ron Raab. "Oxygen Depletion in Level A Hazmat Suits." *Fire Engineering* (November, 2010).
7. Kreis, Steve, "Rapid Intervention Isn't Rapid." *Fire Engineering* (December, 2003), pages 56–66.
8. National Fire Protection Association, *Hazardous Materials Response Handbook* (6th Edition), Quincy, MA: National Fire Protection Association (2013).
9. National Fire Protection Association, *NFPA 1500—Fire Department Occupational Safety and Health Program*, Quincy, MA: National Fire Protection Association (2007).
10. National Fire Protection Association, *NFPA 1851—Standard on Selection, Care and Maintenance of Structural Fire Protective Ensembles*. Quincy, MA: National Fire Protection Association (2012).
11. National Fire Protection Association, *NFPA 1991—Standard on Vapor-Protective Ensembles for Hazardous Materials Emergencies*, Quincy, MA: National Fire Protection Association (2012).
12. National Fire Protection Association, *NFPA 1992—Standard on Liquid Splash-Protective Ensembles for Hazardous Materials Emergencies*, Quincy, MA: National Fire Protection Association (2012).
13. National Fire Protection Association, *NFPA 1994—Standard on Protective Ensemble for Chemical/Biological Terrorism Incidents*, Quincy, MA: National Fire Protection Association (2012).
14. National Institute of Justice, *NIJ Standard-0116.00—CBRN Protective Ensemble Standard for Law Enforcement*, Washington, DC: National Institute of Justice (2010).
15. National Institute for Occupational Safety and Health, NIOSH Publication 2009-132: *Recommendations for the Selection and Use of Respirators and Protective Clothing for Protection Against Biological Agents*. Morgantown, WV: National Institute for Occupational Safety and Health (April 2009).
16. National Institute for Occupational Safety and Health, "Safety Advisory on the Potential for Sudden Failures of Fiberglass-wrapped Composite Aluminum Cylinders Following Exposure to Acidic Chemical Cleaning Agents." Morgantown, WV: National Institute for Occupational Safety and Health (July 30, 1996).
17. National Transportation Safety Board. "NTSB Safety Recommendations and Report on the Benicia, California Dimethylamine Railroad Tank Car Incident." Washington, DC: National Transportation Safety Board (April 23, 1984).
18. Schwope, A. D., et al., *Guidelines for Selection of Chemical Protective Clothing* (3rd edition), Cambridge, MA: Arthur D. Little, Inc. (1992).
19. Sendelbach, Timothy E., "SCBA Confidence for Fireground Survival (Part 2)," *Atlantic Firefighter* (April, 2003), page 12.
20. Stull, Jeffrey O., "Dressed for Defense." *Fire Chief* (June, 2003), pages 40–47.
21. U.S. Army Research, Development, and Engineering Command (RDECOM)—Homeland Defense Business Unit. *Risk Assessment of Using Firefighter Protective Ensemble with SCBA for Rescue Operations During a Terrorist Chemical Incident*. Edgewood, MD: RDECOM (June, 2003).
22. U.S. Environmental Protection Agency, *Standard Operating Safety Guidelines*, Washington, DC: EPA (1988).
23. Veasey, Alan., "Coping With Loss of Air Supply in Chemical Protective Suits." *Fire Engineering* (March, 1998), pages 173–178.
24. Veghte, James H., *Physiologic Field Evaluation of Hazardous Materials Protective Ensembles*. Emmitsburg, MD: Federal Emergency Management Agency—U.S. Fire Administration (September, 1991).
25. Zeigler, James P., "NFPA Standards on Hazardous Materials Protective Ensembles." Student Handout—Richmond, VA: DuPont Nonwovens (2010).
26. Zeigler, James P., "Understanding the New Generation of Chemical Protective Clothing." Student Handout—Richmond, VA: DuPont Nonwovens (2010).

Information Management and Resource Coordination

CHAPTER 9

Hazardous Materials Technician

Skills Objectives

There are no Hazardous Materials Technician objectives in this chapter.

Hazardous Materials Incident Commander

Knowledge Objectives

After reading this chapter, you will be able to:

- Define the terms information, data, and facts as related to the function of information management. (p. 303)
- Describe the types of information required to manage a hazmat incident safely and effectively. (p. 304)
- Describe the criteria for evaluating hazardous materials information management systems for field applications. (p. 304)
- Describe the Hazmat Group functions required to manage information at a hazmat incident. (p. 308)
- Define the terms, resources, and support as related to the function of organizing resources within the incident command system. (p. 311)
- Describe the process and procedures for coordinating internal and external resource groups at a hazmat incident. (p. 311, 312)
- List and describe three techniques for improving coordination. (p. 314)

Skills Objectives

There are no Incident Commander skills objectives in this chapter.

You Are the HazMat Responder

You are the acting chief officer on a weekend shift at a regional airport. At 17:45 hours you receive an emergency call for a fuel spill inside of a hangar. You respond with an Aircraft Rescue/Firefighting (ARFF) vehicle and an engine company.

When you arrive on the scene you find an Eclipse-500 private twin-engine jet parked inside the hangar with about 20 gallons of Jet-B fuel pooled on the hangar floor. The door is open on the hangar and there is a tow motor parked inside, which is not running.

You have a Hazardous Materials Technician take a multi-gas meter reading near the hangar door at ground level and determine the atmosphere is approximately 20% of the lower explosive limit (LEL). You have ordered that the area be isolated and ignition sources secured. Your pre-fire plan does not indicate where the overhead heating system master switch is located, and you are concerned about a fire and explosion if the heater activates.

Given that there is no fixed foam system inside the hangar, you make the decision to apply Class B foam on the fuel spill as a way to reduce the vapors and the potential for an explosion inside the hangar. As responders prepare for foaming down the fuel spill, the aircraft's owner shows up. He doesn't want you to make a big mess inside the hangar and requests that fire fighters manually tow the aircraft outside of the hangar. As Incident Commander (IC), you feel this is neither safe nor practical given the size and weight of the aircraft.

1. What type of resources would you need to safely mitigate this incident?
2. What type of information would you need to support your Incident Action Plan?
3. The aircraft's owner may present a problem for you. How would you resolve his concerns? Which techniques would be most effective in helping you avoid a major conflict with the owner?

Introduction

This chapter will describe the fifth step in the Eight Step Process©—Information Management and Resource Coordination. This step focuses on making sure that:

- Information flows to the agencies and people that need to know what is going on in a timely manner.
- Everyone involved in the incident knows what the Incident Action Plan is to make the problem go away.
- Emergency responders have the right resources (people, equipment, and supplies) to get the job done safely and rapidly.

Coordinating the information and resources required to resolve a complex hazmat emergency may seem like an overwhelming problem (and it often is), but it isn't that difficult if you use a system. In this chapter, we will review several structured and systematic methods that responders can use to organize information and resources so they can be used to their fullest advantage.

Before we get into the chapter in detail, let's review several main points we discussed elsewhere in this text that relate to information management and resource coordination:

- Failure to get the right information and resources to the right people at the right time can jeopardize the safety of responders and the overall success of the emergency response effort.

- Information and resources cannot be coordinated effectively if an incident management organization is not in place. The Public Information Officer and the Liaison Officer are the key players for moving information within the command structure and to external agencies and stakeholders. The Logistics Section Chief and the Staging Area Manager are the key players for obtaining and moving resources at the incident.
- Information poorly coordinated among the players at the emergency scene can politically damage the IC's credibility and ultimately undermine the Incident Action Plan. When responders don't understand the plan or perceive that one doesn't exist, they improvise and start to "freelance." Freelancing (also known as running around without a plan and doing your own thing) can get people hurt and waste resources.
- The IC's Incident Action Plan (IAP) must have a solid technical basis (e.g., the actions proposed must be within the limits of science and technology and respect the basic laws of engineering, chemistry, and physics). A heavier-than-air gas will always be heavier than air, and it takes a 50-ton crane to lift 50 tons. If you buy into an action plan that has *Magic Foo-Foo Dust* sprinkled onto the scientific facts you will eventually end up regretting it.

Responder Safety Tip

Failure to get the right information and resources to the right people at the right time can jeopardize the safety of responders and the overall success of the emergency response effort.

Managing Information in the Field

The Basics

Decisions cannot be made without reliable information. But how do we know the information we get is reliable, how much information do we really need to make good decisions, and how often is it required?

The reliability of the information used in decision making depends on the quality of the data and facts used to compile that information (Data + Facts = Information). The IC must make decisions based on reliable data and facts, not assumptions.

Data

Data refers to individual data elements that are gathered and organized for analysis. At a hazmat emergency the data we are most interested in concerns the material's physical (i.e., how it behaves) and chemical (i.e., how it harms) properties, such as specific gravity, flash point, exposure values, and vapor density. The reliability of most published data on the various characteristics and hazards of hazardous materials is pretty high. Some common mistakes made at the emergency scene regarding data include looking up the wrong chemical in the database, not copying the information down correctly, using a Safety Data Sheet with the wrong concentration or mixture, and failure to validate the data using multiple reference sources.

Data is no substitute for thinking! If there is any doubt about the data you are looking at, it's okay to question the source or authenticity. Just because it is written down on paper or shows up on a computer screen does not mean you have to buy into it, subscribe to it, or believe what it says. Be inquisitive and challenge information you may have to rely on in the field.

Facts

Statements or observations about something verified and validated as true are known to be factual. In emergency response work, facts are typically based on objective observations made by trained and experienced personnel. For example, a digital photo taken by your reconnaissance (recon) team of a score across a welded seam on an overturned and damaged MC-331 propane transport is a reliable fact. Comments by the recon team that the metal below the surface of the weld seam might be cracked is only a suspicion or theory; it is not a fact. The recon team's observation that the combustible gas indicator did not detect any flammable vapors near the damaged weld seam is also a fact. When these two facts are combined with the reliable photo, you have credible information that can help you make some decisions, or at least allow you to ask the right questions of product and container specialists.

Information Management Requirements

Imagine you have responded to a freight train derailment at 3:00 am. The train has 80 cars and 20 contain different types of hazardous materials. The railroad presents you with a computer consist profiling how the train is made up, including emergency response information sheets on each hazmat carried in the train. You have many different documents to look at. This doesn't even include other sources of information you may have available through CHEMTREC®, technical reference manuals, computer and smart phone databases, or pre-incident response plans. Sorting, evaluating, interpreting, and communicating this information to people who need it can be a perplexing problem if you don't use a system to manage it.

Information management must begin well before the incident **Figure 9-1**. Important decisions that must be made include the following:

1. What type of information will be needed at the emergency scene? How should the information be compiled?
2. What is the priority of the information needed? What do you need immediately versus an hour into the incident?
3. How will the information be stored for quick recovery at the incident scene—manually or electronically? If electronically, or primarily internet-based, is there a backup plan in the event there is no connectivity or other technical problems?
4. Are the information and retrieval systems suitable for field applications?
5. Who will be responsible for managing and coordinating information at the incident scene? Are they properly trained and equipped for the job? What type of Information Centers may be required to coordinate information?

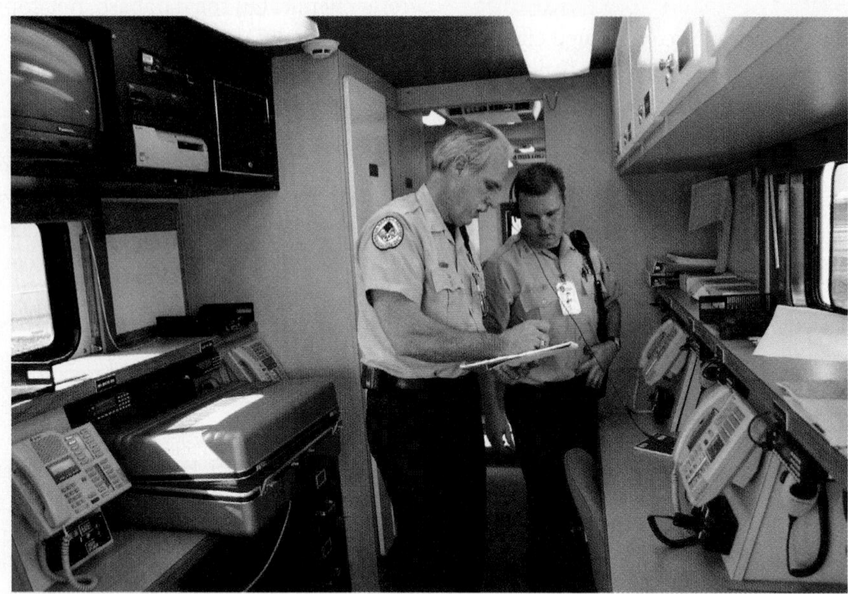

Figure 9-1 Information management must begin well before the incident with some important decisions concerning the type of information needed and how it will be stored and retrieved.

What Do You Really Need to Know?

In the early days of Hazmat response history, responders lacked reliable and accessible hazard data to make decisions. Computers, the Internet, and electronic search engines largely solved that problem. Now we have access to what seems to be unlimited information in a very short period of time. However, what do we really need to know at a hazmat emergency and when do we need to know it?

The information management process starts by identifying what you need to know versus what is nice to know. Example: In a crisis situation, is it more important to know the time or how the watch works and who made it? In the hazmat world your "need to know" information list should read something like this:

1. What are the hazards or how will it hurt me? (e.g., flammability, toxicity, and reactivity)? What is the source of energy and how will it disperse and hurt the things it touches?
2. What are the Personal Protective Equipment (PPE) requirements or what must I do to protect myself?
3. What are the health risks (e.g., exposure values, signs and symptoms of exposure, antidotes, etc.)?
4. What is the container type and condition (integrity, size, orientation, etc.)? Hazmat incidents are a two-part problem—the product and the container. You can have the worst hazardous material in the world involved in an accident, but the situation is usually under control if the container is stable and the product is still in the container. Conversely, even something benign like sugar can present a serious problem if the container is breached and a lot of the product is on the ground in the wrong place at the wrong time. Case in point: The Boston Molasses Disaster of January 15, 1919 killed 21 people and injured 150 when a storage tank containing up to 2 million gallons of molasses failed at the Purity Distilling Company. It took over 87,000 man-hours to clean up the spill. (*Source: Dark Tide: The Great Boston Molasses Flood of 1919,* by Stephen Puleo, Beacon Press, 2004.)
5. What are the initial tactical recommendations (e.g., spill control, leak control, fire control, public protective actions)?
6. What type of decontamination procedures and methods will be required?

Trying to collect and carry a hard copy of every source of hazard and response information in an emergency response vehicle just isn't practical. Responder information needs will be different based on local or facility conditions.

There are four basic groups of information sources that should be immediately accessible from the incident scene:

1. Facility Emergency Response Plans
2. Pre-Incident Tactical Plans
3. Published Emergency Response References
4. Shipping Documents

Facility Emergency Response Plans

Refineries, chemical plants, and facilities that manufacture or store hazardous materials on-site are required by OSHA to have a Facility Emergency Response Plan (29 CFR 1910.)

EPA also requires facilities that store or transfer oil, animal fats, or vegetable oils to have a Facility Response Plan (40 CFR 112.20 and 112.21, including Appendices B through F). Requirements for animal fats and vegetable oils were added in 2000. These plans usually have a hazard analysis section, which identifies special problems that may exist within the plant or community, the potential risks and consequences of a hazmat incident, and the available emergency response resources in the facility.

Although an Emergency Response Plan does not typically have much tactical application in the field, it is a good beginning point to identify target hazards for which a more detailed Pre-Incident Plan should be developed. This process may allow you to determine the types of information to be stored electronically for easy access.

Process safety and risk management documents can also be good sources of facility hazard analysis information. For example, the hazard analysis section may identify that poison gas is used in a particular process at a chemical plant. Having specific toxicity/exposure data and evacuation information would be very useful for immediate retrieval at an actual incident.

The chapter that covers "hazardous materials management systems" describes several types of hazard analysis tools and sources of planning information that are available. Familiarize yourself with these different tools and documents so you know what to ask for when you write or visit a facility for information.

Pre-Incident Tactical Plans

Pre-Incident Tactical Plans (or preplans) are like a football team's playbook. The plan explains exactly who does what and where they are supposed to be to execute the plays. Preplans focus on a specific problem or location, such as a rail yard or a bulk storage facility. In fixed facilities with chemical manufacturing operations, preplans may concentrate on a particular process or tank(s) that present special hazards.

The larger the facility or jurisdiction, the more essential the preplanning process be practical and focused on prioritizing problem areas based on the hazard analysis process. Criteria for developing special preplans in these situations may include the following:

- Type of hazards and risks present—Facilities that present high risks to the community should be preplanned. Facilities that have a low probability of something going wrong, but a high consequence if it does go wrong, are good candidates for a preplan. When we respond to these kinds of facilities we have to get it right the first time. Nuclear power plants, chemical plants, petroleum bulk storage facilities, and refineries are all good examples.
- Critical infrastructure—Facilities that may have a serious impact on national security if they experience a major loss. Examples include facilities that support military, intelligence, or Homeland Security missions.
- Economically sensitive sites—Large businesses that serve as a substantial financial contributor to the local economy in terms of tax base and employment. Losses or reduction in production capacity at these facilities may create political ramifications for the responding agencies. It's a certainty that your community believes you are prepared to handle what would be considered obvious threats such as a fire or chemical release; not having a plan in the event something goes wrong will shake the confidence of your community and probably take years to recover from politically.
- Environmentally sensitive exposures—Facilities in close proximity to waterways or aquifers present high financial and environmental risks that need to be well considered. Example: If the high-risk storage paint warehouse in close proximity to the community's primary aquifer catches on fire, should an offensive fire attack be initiated? If the answer is yes, what objectives and tactics should be implemented to control any runoff and protect the sensitive aquifer and ground water?
- Unusual or poor water supply requirements—Examples include facilities located in an area with water supply problems or installations that may require unusually high fire flows (e.g., large diameter flammable liquid storage tank). You probably know where these areas are. Do you have a fire water supply plan for them?
- Locations that will require large quantities of foam concentrate such as bulk liquid petroleum storage facilities and pipelines. The foam concentrate quantity requirements need to be known as well as application rates, availability of foam supplies, logistical considerations, and so on. In addition, a plan for managing water and foam runoff needs to be considered if water pollution is a concern.
- Restricted or delayed response routes—For example, single approach, access corridors, railroad tracks, or draw bridges that are frequently blocked.
- Poor accessibility—Examples include restricted entrance corridors, secure government installations with strong antiterrorism force protection countermeasures in place, obstacles, and unusual ground slope.

Preplans can provide valuable information if the right type of field survey form is used to record key information during the site visit. While informal site visits may be instructive for the personnel participating in them, they often provide little long-term benefit if there is no mechanism for compiling, maintaining, and disseminating key response information. Bigger is not necessarily better; a simple form or standardized electronic documentation process can reduce the maintenance time required to keep preplans current. Preplan now—you will be thankful later.

NFPA 1620—*Standard for Pre-Incident Planning* provides guidance on the type of information that should be included in pre-incident plans. A well-prepared preplan should include a simple plot plan that shows the basic details of the facility but is not cluttered with extraneous information. The preplan should also indicate the availability of any special plans prepared by the facility that may be referenced during an emergency (e.g., foam calculations, tactical checklist) **Figure 9-2**.

One special note regarding preplans is Operations Security (OPSEC). We must apply good OPSEC in how we store preplan information and control who has access to this information. In the wrong hands, preplanning information could be used by criminals or terrorists to commit a crime or plan an attack against a facility we are trying to protect.

Key transportation areas should be preplanned as well. Obvious areas include rail yards and trucking terminals but should also include high-traffic-density highways, particularly poorly designed interchange ramps with a history of trucks overturning, bridges, tunnels, and toll plazas. These areas should be preplanned for access, topography, environmental exposures, sensitive receptors, high-density populations, and water supply requirements to support firefighting foam operations.

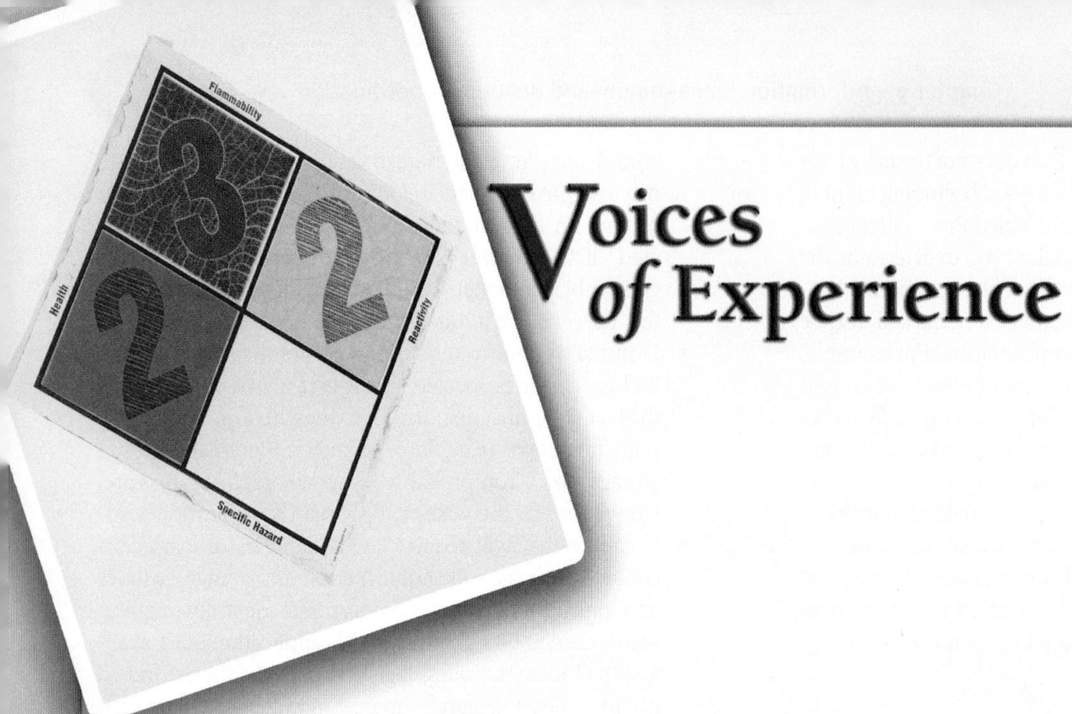

Voices of Experience

Our Hazardous Materials Response Team responded to a train derailment and ensuing fire at approximately 2:30 a.m. The train consisted of 116 ethanol cars and 16 grain cars. All 16 grain cars and around 10 ethanol cars were involved in the derailment. The fire from the ethanol cars was visible from a great distance. Upon arrival at the scene, there were several rapid releases and ensuing fires from the damaged rail cars. There were no immediate exposures in the area, the closest being approximately 600 feet away.

The initial action plan was to secure the area and establish a water supply in a nonhydrated area to protect any exposures from a potential release of burning liquid. It was also decided to uncouple and remove as many ethanol cars from the train to the undamaged side of the derailment. This left three upright ethanol cars on the rails that were still coupled but not on fire. The fire was monitored but no attempt of extinguishment was made. At approximately 9:30 am the fire had appeared to have subsided with light white smoke issuing from several of the grain cars.

An Illinois State Department of Transportation (DOT) helicopter had been requested earlier by the IC to get an aerial reconnaissance of the scene. This helicopter was equipped with an infrared video camera. After the crew made the recon flight, they were able to come back to the incident command post with aerial video of the scene and allow the command staff to view both the normal and infrared images. The infrared video showed that the ethanol cars were still actively burning, which could not be seen on the standard video. This allowed the IC, the hazmat branch director, and the railroad hazmat team to formulate a new action plan based on the intelligence gathered from the infrared video.

Without the available technology of the infrared video and the ability to produce still pictures from the video, the ability to assess the severity of the fire still burning, it would have required putting personnel in close proximity to the fire. Using the still pictures, the railroad hazmat team was able to determine which cars they wanted to target for final extinguishment and to gather temperatures on all the ethanol cars involved without putting response personnel at risk.

Jerry Janick, Team Leader
Illinois MABAS Division 25
Hazardous Materials Response Team

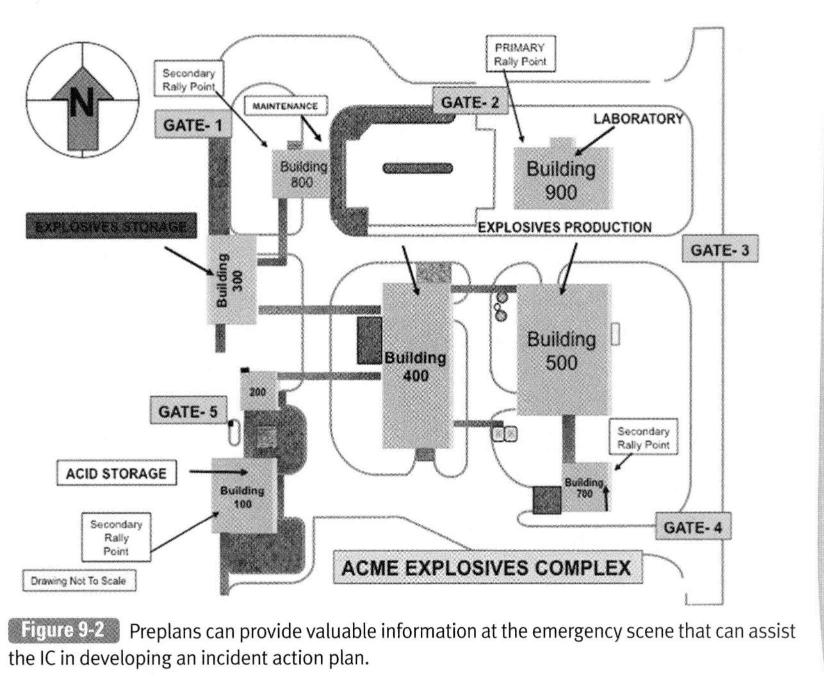

Figure 9-2 Preplans can provide valuable information at the emergency scene that can assist the IC in developing an incident action plan.

Emergency Response References

Most emergency responders have some capability to remotely access hazard information from smart phones, laptops, and onboard computer systems. There are a wide variety of published emergency response references available on the Internet as well as mobile applications that can be consulted on a handheld device.

Electronic and technology-based reference sources have dramatically changed both the timing and efficiency by which emergency responders can access information. While there are many advantages to using these electronic tools, responders must also consider what happens if the electronic tools don't work. In simple terms, what is Method-B and C when the Method-A doesn't work? When buying subscriptions to data bases and references, consideration must be given to keeping them current with the most recent editions.

References are generally divided into the following categories:

- Reference manuals and guidebooks. You need to be selective in what you decide to purchase and carry to the scene of an emergency. It is better to have a few reputable references that you have trained on and are comfortable with than to have a large and unfamiliar reference library.
- Technical information centers accessible by telephone (e.g., CHEMTREC®, CANUTEC).
- Hazardous materials databases accessed through the Internet either as public domain or subscription databases (e.g., WISER, TOXNET, TOMES® System).

The chapter of this text that covers "hazard assessment and risk evaluation" provides a good overview of the different types of widely recognized, published references and databases. Familiarize yourself with these sources of information and know how to use and interpret them before you need them.

Information Storage and Recovery

Most Hazardous Materials Response Teams (HMRTs) have field access to portable computer systems or mobile communications. Having access to terabytes of information on a computer is a big advantage, but don't overlook the simplicity of a few standard hard copy reference books for rapid assessment of hazard data. Books work the first time—every time. A good flashlight and reliable sources like the DOT-ERG and AAR *Emergency Action Guides* can be your best friend at 3:00 a.m. when the electricity is out or a virus has infected your computer's operating system.

No matter how good your onboard computer system is, it does not necessarily guarantee an improvement to your information management system at the incident scene. Don't expect much from your portable computer system and the databases it references if you haven't read the user's manual and had some realistic training that simulates field conditions.

The key to successfully managing and retrieving hazmat information under emergency conditions is good organization and simplicity. The "acid test" for deciding whether one type of information management system is better than another should be, "Will it work on the street?" As previously noted, there will always be some situations where manual information management systems may be better suited than computerized systems.

When evaluating information systems, consider the following:

- **User friendly**—Stressful situations call for simple solutions. Beware of hardware and software that requires a lot of training and experience to operate. While almost every HMRT has an in-house computer wizard, don't build your information system around the expertise of one individual! The same case can be made for every piece of equipment in your inventory.

If you rely on one person to work your gee-wiz toys, you are "people dependent" and not system dependent. The best teams out there are system dependent, not people dependent.

- **Durability**—We don't operate in a nice clean office environment. Special Operations Teams are expected to operate anytime, anywhere, and in any weather conditions. If you rely on a computer or hand-held device in the field, (1) the information should be backed up, and (2) it should be "Mongo," or Hazmat Technician-proof.

Remember—a computer's outstanding attributes of speed, storage capacity, consistency, and the ability to process complex logical instructions are of no value unless they are applied within a good management process and the computer works when you need it.

Coordinating Information Among the Players

Coordinating information in the field becomes particularly important as the IC, Hazmat Group Supervisor, and others evaluate options concerning protective clothing, decontamination (decon) requirements, and public protective actions.

> **Responder Safety Tip**
>
> Coordinating information regarding protective clothing requirements and public protective actions is critical to safety.

Coordinating information is a dynamic process that must adjust its scale over time to provide the correct and clear information to the right people at the right time. The larger and more complex the incident, the larger the command organization needed to manage the incident. The larger the command organization, the more need there is for a formal structure to manage the data and facts that will flow between Command and the various individuals and organizations at the emergency scene. Information must also flow freely to and from the incident scene to off-site support facilities, such as the Emergency Operations Center (EOC), CHEMTREC®, and elected officials.

Hazmat Group Functions

If your organization has some depth in resources, an effective way to organize and manage a working hazardous materials incident is for Command to form a Hazmat Group and then delegate the responsibility for hazmat information to the Hazmat Group Supervisor. The Hazmat Group then sub-divides its functions into specific Units, as described in the "managing the incident" section of this text. Following this general approach, the Hazmat Group Supervisor assigns different functions to the first available and qualified person as the situation requires.

Primary functions and tasks assigned to the Hazmat Group include the following:

- **Safety function**—Primarily the responsibility of the Hazmat Group Safety Officer (or Assistant Safety Officer–Hazmat), he or she is responsible for ensuring that safe and accepted practices and procedures are followed throughout the course of the incident. This officer possesses the authority and responsibility to stop any unsafe actions and correct unsafe practices. This position should be filled by a strong player—someone who is reliable, steady, and has good common sense.
- **Entry/backup function**—Responsible for all entry and backup operations within the hot zone, including recon, monitoring, sampling, and mitigation.
- **Decontamination function**—Responsible for the research and development of the decon plan and set-up and operation of an effective decontamination area capable of handling all potential exposures, including entry personnel, contaminated victims, and equipment. If necessary, will include the coordination of a Safe Refuge Area.
- **Site access control function**—Establish hazard control zones, establish and monitor egress routes at the incident site, and ensure that contaminants are not being spread. Monitor the movement of all personnel and equipment between the hazard control zones. Manage the Safe Refuge Area, if established.
- **Information/research function**—Responsible for gathering, compiling, coordinating, and disseminating all data and information relevant to the incident. This data and information will be used within the Hazmat Group for assessing hazards and evaluating risks, evaluating public protective options, the selection of PPE, and development of the incident action plan.

Secondary support functions and tasks that may be assigned to the Hazmat Group include the following:

- **Medical function**—Responsible for pre- and post-entry medical monitoring and evaluation of all entry personnel, and provides technical medical guidance to the Hazmat Group, as requested. Should also formulate the "what if" plan in the event of a civilian or responder chemical exposure. This includes identifying the appropriate receiving hospital and transmitting pertinent medical and exposure information to the hospital-based caregivers.
- **Resource function**—Responsible for control and tracking of all supplies and equipment used by the Hazmat Group during the course of an emergency, including documenting the use of expendable supplies and materials. Coordinates, as necessary, with the Logistics Section Chief (if activated).

Checklist System

The most simple and reliable method of coordinating information between the various Hazmat Group functions is to use the checklist system. Checklists can be stand-alone worksheets or included as a Job Aid in a Field Operations Guide (FOG). For example, this text has a companion, FOG, which includes critical Hazmat Group position checklists as well as a worksheet for the IC Figure 9-3.

MEDICAL WORKSHEET
TACTICAL GOALS

INCIDENT ALARM #: ______________ DATE: ________ TIME ESTABISHED: ________ Ops Period

INCIDENT ADDRESS:______________ RADIO: **MEDICAL** TIME TERMINATED:________

ENTRY TEAM SUPPORT

TEAM				**TASKS** (See Entry and Medical Status Worksheet)
A	**B**	**C**	**Time**	
				Medical Monitoring of entry and back-up personnel
				All personnel items removed, tagged and secured
				Suit Selection double checked with Information
				Protective Clothing:________________
				Visual Check of Entry Suit
				All zippers and closures properly secured
				No obvious suit damage
				Communication check (Channel:______)
				Respiratory Protection:________________
				Facepiece seal insured
				Air Pressure verified
				Gloves:________________
				Gloving, Overgloving, Doubleglove verified
				Boots:________________
				Footwear verified as appropriate
				Entry Officer notified that teams are ready
Chemical				

	Initial Entry Physical

	Suit Numbers		
	TEAM A	TEAM B	TEAM C

	Respiratory Protection Numbers		
	TEAM A	TEAM B	TEAM C

Figure 9-3 Field Operations Guide Checklist from Hazardous Materials: Managing the Incident Field Operations Guide, 2nd edition

Formal checklists have several distinct advantages as they relate to information management in the field. These include the following:

- Checklists don't panic. We all have bad days now and then, and when you are having a really bad day and are falling apart, a well-thought-out checklist can get you back with the program. If you are confused or lack the field experience, get the checklist out and start working through your assigned tasks.
- Checklists have institutional memory. Pilots routinely use checklists to take off and land their aircraft and to work through in-flight emergencies. Commercial aircraft have these checklists integrated into cockpit information systems when an emergency occurs. Years of good and bad experience go into developing a checklist so that the lessons learned from the past are "institutionalized" and remembered as new people rotate into an assignment. If you routinely critique your incidents and exercises, part of the feedback of the lessons learned should go into improving your checklists.
- Identify the tasks assigned to each Hazmat Group function. A good checklist assigns duties and responsibilities to each member of the team so that work is not duplicated and important information-gathering activities are not overlooked.
- List critical activities and action items required for each function. As elements are addressed, they are formally "checked off" as a method of verifying that the activity has been completed. You might have the checklist in your head and work through it mentally, but an experienced professional eventually gets out the formal checklist and runs through it to make sure that something has not been overlooked.
- Prioritize actions so that important activities are completed early in the incident.
- Provide a framework for development of the IAP.
- Identify which Hazmat Group functions or individuals need to be formally contacted to coordinate information.
- Provide the required documentation of the incident for the post-incident analysis and the critique. See the section found elsewhere in this text on "terminating the incident" for a discussion of these two activities.

For the checklist system to be effective, checklists must be updated on a regular basis. It is important responders take ownership of the checklists and make gradual improvements to the system over time.

Hazmat Information Leader

The person designated as the Hazmat Information Leader (may also be known as Research or Science) will play a key role in the successful mitigation of any hazmat incident. This individual should be chosen because of his or her ability to communicate, comprehend and manage information, work effectively under stress, and coordinate activities with individuals from different backgrounds.

Extended incidents or incidents involving multiple chemicals may require that an Information Unit be formed within the Hazmat group. This allows the workload to be split up into areas such as on-scene technical reference library research, contacting the shipper and chemical manufacturer, and on-scene data collection. CHEMTREC® is a great resource early in the incident, but for a complex incident you will usually be dealing with the product manufacturer or facility personnel after the first 2 hours. The more products involved in an accident, the more product or container specialists you will need to interface with.

In complex incidents of long duration, the Hazmat information function may need to move to nearby offices or houses to complete their assignments. Access to comfortable surroundings with telephones, computer or Internet access, office supplies, bathrooms, and other amenities of life makes the job much more endurable and the work product more reliable. Good lighting, controlled temperature, and an uninterrupted power supply are also essential.

Coordinating Information

Government Agencies and Private Sector Organizations

Major incidents involving hazardous materials or weapons of mass destruction create demands for information across a wide spectrum of local, state, and federal agencies. Effects on public safety and health, disruption of transportation systems, impact on infrastructure, and financial costs are but a few of the many issues of concern to our governing bodies.

Well-planned, established, and applied communications enable accurate and rapid dissemination of information among government agencies being impacted by the emergency. Common communications plans, interoperable communications equipment, and standard operating procedures help build a communications architecture that supports situational awareness among government agencies during disasters.

The Unified Command is responsible for providing regular and factual Situation Status Reports to the EOC of the impacted jurisdiction or facility once it becomes operational. It should be the responsibility of the EOC to assemble a Common Operating Picture (COP) and share it across the matrix of effected jurisdictions, government agencies, and other stakeholders that need to know what is going on.

The purpose of building and maintaining the COP is to provide an overview of the incident by collecting and collating information such as traffic, weather, damage assessment, resource availability, etc., in order to support decision making based on facts. Having a COP across primary and supporting governmental jurisdictions helps policy makers make consistent and timely decisions.

When the size and scope of the incident requires that information flows from the IC to external agencies and organizations outside the command structure, a Liaison Officer should be appointed. The Liaison Officer serves as the "gatekeeper" to the IC, and ensures that incident-specific external information, issues, and concerns flow both into and out of the response organization. If the incident is large or complex, Assistant Liaison Officers may be established.

The Liaison Officer is the point of contact for representatives of government agencies and non-governmental organizations (commonly referred to as LNOs or A-Reps), who provide input on their policies, resource availability, and other incident-related matters. Within NIMS, these external agencies may fall into the following categories:

- Assisting Agency—Agency or organization providing personnel, services, or other resources to the agency with direct responsibility for incident management. An assisting agency provides tactical resources.
- Cooperating Agency—An agency supplying assistance other than direct operational or support functions or resources to the incident management effort.
- Non-Governmental Organizations (NGO)—A legally constituted, non-governmental organization that may work cooperatively with government. NGOs serve a public purpose rather than private purposes, and may serve as either an assisting or cooperating agency depending upon their role at an incident.

Agency representatives are assigned, work for, and report to their "parent" agency. While they are not part of the incident chain-of-command, the Liaison Officer serves as their gateway into the response organization. They should have the authority to speak for their parent agencies after proper consultation with senior leadership. The challenge of being an agency representative is to strike a delicate balance of obtaining and sharing "need to know" versus "nice to know" information to policymakers or executive management. In large-scale incidents, there is an entire flow and subtext of unofficial and/or developing information and stories, which may or may not be useful to help paint the picture up the chain. Keep in mind—not all information is good information.

Public Information

Although governments play a major role in keeping the public informed during a major emergency, the news media is the major communicator. Moving accurate and timely information from the incident scene to news media is a critical task. When media liaison is done well the public stays informed and is safer; when done poorly many problems evolve that can have a lasting impact on the credibility of emergency responders. It is a fact that in some cases, a deliberate and well thought out media strategy is equally important as the strategy used to make the problem go away. Sometimes, the problem goes away with relative ease, but leaves a lingering imprint due to poor or haphazard interactions with the media.

The general public is a consumer of information and they are our customers. It is the IC's responsibility to make sure the product we produce (information) is of good quality and timely. Within the incident command organization, the Public Information Officer (PIO) is responsible for gathering, verifying, coordinating, and disseminating accurate and timely information on the incident. The scope of this information usually includes:

- Incident cause
- Size of the spill or release
- Hazardous materials involved
- Information concerning potential safety, security, and health impact on the public
- Current situation
- Resources committed
- Estimated time to resolve the problem

In a Unified Command setting, the PIO supports command by collaborating with the PIOs from other agencies who have statutory or legislative authority to:

- Identify key information that needs to be communicated to the public.
- Craft messages conveying key information that are clear and easily understood by all, including those with special needs.
- Prioritize messages to ensure timely delivery of information without overwhelming the audience.
- Verify accuracy of information through appropriate channels.
- Disseminate messages using the most effective means available.

As a member of the command staff, assistant PIOs may be appointed based upon the public affairs needs of the incident.

When multi-agency responses require long periods of operation to resolve the incident, a Joint Information Center (JIC) may be established to manage the expanding information requirements. The JIC is typically staffed by information management specialists who are skilled in critical emergency information functions, such as crisis communication and public affairs functions.

Typically, a JIC is established at a single, on-scene location with the appropriate federal, state, and local agencies represented. Incidents of great complexity like the 9-11 attacks or the 2010 Gulf of Mexico Offshore Platform Fire and Oil Spill may require multiple JICs servicing state- or national-level operations centers.

The typical JIC is responsible for:

- Ensuring consistent messages to the public are delivered across agencies participating in the response.
- Releasing information that is accurate and not conflicting.
- Screening for incident-sensitive information. For example, preventing the release of the names of the injured or deceased before notification of the family.
- Ensuring good OPSEC and protecting critical information. For example, law enforcement sensitive information or information useful to our adversaries (i.e., criminals and terrorists).

Resources

Resources are the people, equipment, and supplies required to manage a hazardous materials emergency. As with military operations, efficiently managing and moving resources around the incident scene is a fundamental part of strategy and tactics.

Human Resources

Human resources includes emergency responders, technical specialists, product or container specialists, and support personnel. Effective coordination of human resources is important because people provide the thinking power and manual labor required to bring the situation under control. People also represent the greatest financial, legal, political, and technical exposure for the IC. The chapter on "managing the incident" elsewhere in this text lists some of the different types of human resources (or "players") that may be involved in a hazmat incident.

Equipment Resources

Equipment resources are items that are reusable, such as hand tools, generators, pumps, monitoring instruments, and fire apparatus. Hazmat equipment can represent a substantial cost outlay whether it is rented, leased, or owned. Very few organizations are self-sufficient when it comes to the equipment required to resolve a large-scale hazmat incident. Consequently, good coordination is required between the Planning, Logistics, and Finance Sections to ensure the right equipment resources are available when they are needed, are tracked throughout the operation, their use and costs are monitored, and that they are demobilized in a timely manner. Whenever an equipment resource is assigned to support hazmat operations there should by a parallel service capability to maintain continued operation throughout the incident. For example, if generators and light units are required to support night operations, then these resources need to be supported with diesel fuel and a technician who can service the units. Responsibility for providing the necessary support for equipment falls within the scope of the Support Unit Leader's duties within the Logistics Section.

Supply Resources

Supply resources differ from equipment resources in that they are usually considered expendable. In other words, you use them once or twice, then dispose or recycle them. Examples include foam concentrate, decon solutions, limited-use protective clothing, absorbent pads, calorimetric tubes, and medical supplies. Hazmat supplies require special coordination because they are usually consumed in bulk quantities, may require special hazardous waste disposal, and often require a longer time to replenish inventories. They can also be expensive and may require special approval for procurement.

In major hazmat incidents, the IC will appoint a Logistics Section Chief to manage the resources required. Within the command structure, the Logistics Section is typically organized into two subgroups: a Resources Branch and a Support Branch Figure 9-4.

As we noted in the chapter on "managing the incident," a good Logistics Section Chief doesn't wait until the Operations Section Chief needs a resource; they anticipate what will be needed and then arrange for the resource to go to the staging area or a support base Figure 9-5.

Coordinating Resources

Hazardous materials incidents are unique to the emergency response business because of the specialized resources required to mitigate the problem. There are few special operations in emergency services that involve such a broad spectrum of private and public services. For example, the train derailment previously described may require extensive resources to bring the situation under control. These could include the following:

- Railroad operations and hazmat specialists
- Product and/or container specialists representing a variety of different companies
- Wreck clearing contractors
- Environmental specialists and contractors
- Fire fighters, police officers, EMS, and rescue personnel from multiple jurisdictions
- State and county emergency management officials
- Local, state, and federal environmental officials
- National Transportation Safety Board (NTSB) investigators
- Federal Railroad Administration (FRA) inspectors
- Transportation Security Administration security specialists

This is only the short list! If military munitions or radioactive materials are involved, or if the incident involves terrorism, the list could grow to 20 or 30 different agencies and organizations. These agencies usually bring their own people, equipment, supplies, and agendas to the incident, and thus there is a greater need for the IC to be very well organized and coordinated.

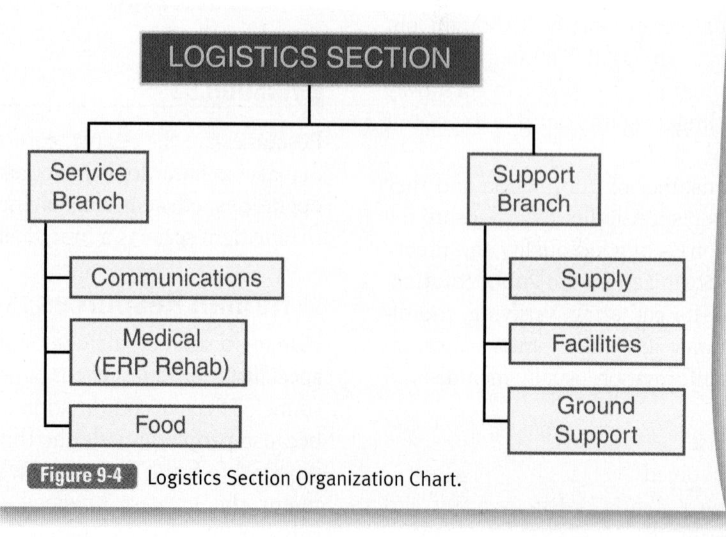

Figure 9-4 Logistics Section Organization Chart.

Figure 9-5 A. Staging areas support the immediate needs of the Operations Section, while a support base (B.) serves the logistical requirements of longer term incidents.

Internal Resource Coordination

Coordinating resource requirements within the internal structure of the emergency response organization can be an easy process if your organization understands and regularly operates within the requirements of the National Incident Management System (NIMS).

As information flows between the IC and Hazmat Group, resource needs and requirements will evolve. It is important the IC implement a system of identifying which resources (people, equipment, and supplies) are needed early in the incident to reduce response time.

Special resource requirements are funneled through the chain of command to the Logistics Section, which coordinates the requirements through the overall command structure and gets the needed resources to the incident scene. Once these resources arrive, they are assigned to a Staging Area and held in reserve until they are required or assigned to the ICS organization or function needing them.

Within the ICS organization, resources are coordinated and tracked by the Resource Unit within the Planning Section. The Resource Unit is responsible for controlling and tracking all equipment and supplies used by all emergency responders during the course of the emergency. This includes all expendable supplies and materials. If incident needs do not require the establishment of a Planning Section, resources would then need to be tracked through the Hazmat Group. Obviously, there must be close coordination between the Planning and Logistics Section Chiefs.

What are some of the attributes of an effective Logistics Section Chief? Important traits and characteristics include the following:

- Self-starter—Looks at the scope and nature of the incident and starts to determine the necessary type and level of resources. Needs little direction.
- "Scrounger"—Knows everyone and has the ability to get resources from different sources.
- Good listener with the ability to quickly establish priorities. Can anticipate what is needed immediately as opposed to 3 hours from now. Can handle multiple tasks at once and keep track of it all. The information tsunami that comes at a Logistics Section Chief can be overwhelming.

As is the case with coordinating information, a formal checklist can provide some structure to coordinating resources.

External Resource Coordination

Outside agencies, whether public or private, will usually be involved in providing resources at a major incident. Each agency contributing significant resources must be coordinated through a single on-scene command organization. If the responding agency has legislative or statutory responsibility for the incident, they will be represented through Unified Command. If it is an assisting agency, they will initially be coordinated through the Liaison Officer. (See the chapter on "managing the incident" for a more detailed discussion on how Unified Command works.)

As resources from external agencies arrive at the scene, a leader or agency representative should be directed to the Incident Command Post (ICP). Important items that should be addressed by the IC or Liaison Officer during this initial check-in process include:

- Making sure all players understand you are running the operation using an ICS structure.
- Determining at what capacity a particular agency or organization is going to participate. If you requested their resources, this may seem obvious. However, resources sometimes just show up. Also, some agency representatives will simply turn out for "fact finding missions" or as observers.
- Physically identifying each agency or organization representative as they arrive. If they are a player they should appear on the incident command organizational chart. If they are an observer they should be accounted for. Request a business card and post it in the ICP so that you are sure of the individual's name, title, rank,

and cell phone number. If the organization has its own communications system, list cellular telephone numbers and radio frequencies on the business card so that you can contact representatives directly.

- Establishing the ground rules for safety and accountability. Don't allow external resources to "freelance" or violate the command structure. Remember that final responsibility for site safety of external resources rests with the IC.

If external agencies have a problem, make sure it is brought to your immediate attention. Remember—bad news doesn't get better with time. If there's a problem, the earlier you know about it, the sooner you can start to fix it.

The best way to guarantee that external resources are properly coordinated and integrated into your command structure is to develop written guidelines or Memorandums of Understanding (MOU) between your respective organizations before the incident occurs.

Resource Coordination Problems

Major emergencies, like train derailments, aircraft crashes, and acts of terrorism, can shift the IC's role from one of a strategist and tactician to a politician and diplomat. These types of incidents bring many agencies to the incident scene with resources ranging in capabilities from the simple to the complex. Managing resources at multi-jurisdiction and complex incidents can create a wide range of incident management problems.

Most resource coordination problems fall into four categories:

1. Failure to understand or work within the incident command structure.
2. Given the type and nature of the incident, failure to anticipate potential problems, as well as "gaps" in information or resources.
3. Inadequate training. When disaster strikes, people seldom "rise to the occasion"—rather they usually default to their level of training. Poor training and preparedness = poor performance on game day.
4. Communications and personality problems between players.

Emergencies don't permit much time for resolving long-standing personality conflicts; they usually intensify under stress. The IC can resolve many communications and personality problems by using a little psychology and some group leadership techniques at the ICP. Some useful techniques that can be effectively applied in stressful situations include the following:

- **Listening:** Pay attention to others as they communicate. Don't just listen to what people are saying, pay attention to their body posture, gestures, mannerisms, and voice inflections. Identify angry players and their issues and objections. Resolve the problem before the situation gets out of hand and eats up your valuable time.
- **Clarifying:** Clarify what the person's issue or concern is. Identify and sort out individual problems. In most cases (but not always) the person has an issue with some kernel of legitimacy. Big problems are more manageable when they are broken into smaller individual issues. If you can identify what the person's concern is, you can address one issue at a time. Issues can also be handed off to subordinates for resolution.
- **Summarizing:** If large groups of people are involved in a Unified Command setting, the decision-making process can get bogged down or fragmented. If several individuals have a problem, summarizing where you are can be helpful. For example, if the discussion is turning into a debate, the IC can interrupt and ask each agency represented in unified command to briefly state how he or she feels about the issue. The IC can then summarize by identifying some common ground and turn the discussion toward acceptable alternatives. Look for issues to agree on, and then build on those as a platform to work through the unresolved issues.
- **Empathizing:** If a special interest emerges and becomes a problem, try empathizing with the individual to reassure him or her that their concern is valid. For example, protecting sea birds from a massive oil spill created by a burning barge at the dock is a high priority to fish and wildlife officials; however, under the circumstances, life safety is a bigger and more immediate concern. Empathize and show respect for the individual's concern, then get a "buy in" to the fact that your options are limited. A little well-placed empathy goes a long way to building support for your plan for managing the resources available.

Develop a good working relationship with supporting agencies, suppliers, contractors, and consultants before the incident and most of the personality issues will dissolve, or at a minimum they won't get in the way of on-scene operations. The Local Emergency Planning Committee (LEPC) is a good place to start laying the foundation.

Wrap-Up

Chief Concepts

- Failure to get the right information to the right people at the right time can jeopardize both the safety of responders and the overall success of the emergency response effort.
- The way other agencies and the public perceive how the incident was handled generally depends on the way information was managed.
- Information and resources must be managed within the framework of the Incident Command System.
- The key to successfully managing and retrieving hazmat information under emergency conditions is good organization and simplicity. If it doesn't work on the street it is useless.
- Information management systems must be user-friendly and durable.
- Coordinating information in the field becomes particularly important as the IC, Hazmat Group Supervisor, and others evaluate options concerning protective clothing, decon requirements, and public protective actions.
- Coordinating information is a dynamic process that must adjust its scale over time to provide the correct and credible information to the right people at the right time. The larger and more complex the incident, the larger the command organization needed to manage the incident. The larger the command organization, the more need there is for a formal structure to manage the data and information that will flow between Command and the various individuals and organizations at the emergency scene. Information must also flow freely to and from the incident scene to off-site support facilities, such as the EOC, CHEMTREC®, and elected officials. In addition, accurate and timely information to the public and the media must be well managed.
- The checklist system is one of the most effective tools for ensuring that information and resources are effectively coordinated both internally and externally.
- Extended incidents or incidents involving multiple chemicals may require that an Information Unit be formed within the Hazmat Group.
- Resources are the people, equipment, and supplies required to manage a hazardous materials emergency. Within the command structure, the Logistics Section is typically organized into two subgroups that include a Resources Branch and a Support Branch.
- A good Logistics Section Chief doesn't wait until the Operations Section Chief needs a resource; they anticipate what will be needed and then arrange for the resource to go to the Staging Area or a Support Base.
- Coordinating resource requirements within the internal structure of the emergency response organization can be an easy process if your organization understands and regularly operates within the requirements of the National Incident Management System (NIMS).

Hot Terms

Assisting Agency An agency or organization providing personnel, services, or other resources to the agency with direct responsibility for incident management. An assisting agency provides tactical resources.

Cooperating Agency An agency supplying assistance other than direct operational or support functions or resources to the incident management effort.

Non-Government Agency (NGO) A legally constituted, non-governmental organization that may work cooperatively with government. NGOs serve a public purpose rather than private purposes, and may serve as either an assisting or cooperating agency depending on their role at an incident purpose rather than private purposes.

HazMat Responder *in Action*

Your regional response organization has just acquired a new HazMat Response Unit, and there is some disagreement with regard to the needs of the Technical Reference Center within the unit. You have been asked to establish some guidelines for this research center. Your task is two-fold. First, what physical resources are needed? Second, what information is required when at the scene of an incident?

Questions

1. Specific gravity, flash point, exposures values, and vapor density are all examples of:
   - **A.** Internet opinion
   - **B.** Facts
   - **C.** Data
   - **D.** Information
2. What are the four basic groups of information sources that should be immediately accessible from the incident?
   - **A.** Facility Response Plans/Safety Data Sheet/Tactical Plans/Emergency Response References
   - **B.** Safety Data Sheet/Tactical Plans/Emergency Response References/Strategic planning
   - **C.** Facility Response Plans/Tactical Plans/Emergency Response References/Shipping Papers
   - **D.** Tactical Plans/Emergency Response References/CAMEO/Computers
3. Supply resources differ from equipment resources in that supply resources are:
   - **A.** Less expensive
   - **B.** Easier to requisition
   - **C.** Expendable
   - **D.** Very expensive
4. Within the HazMat Group, resources are coordinated by the:
   - **A.** Staging Area Manager
   - **B.** Information Officer
   - **C.** Assistant Safety Officer—Hazmat
   - **D.** Hazmat Resource Unit Leader

References and Suggested Readings

1. Department of Homeland Security, National Incident Management System, Washington, D.C.: (December 2008).
2. Drucker, Peter F., *The Effective Executive,* New York, NY, Harper and Row (2002).
3. Engels, Donald W., *Alexander the Great and the Logistics of the Macedonian Army,* Berkeley and Los Angeles, CA: University Press (1978).
4. Hersey, Paul and Ken Blanchard, *Management of Organizational Behavior: Utilizing Human Resources* (9th Edition), Englewood Cliffs, NJ, Prentice-Hall (2007).
5. Hildebrand, JoAnne Fish, "Stress Research: Solutions to the Problem, Part-3." *Fire Command* (July 1984). Pages 23–25.
6. National Fire Protection Association, NFPA 1620—*Standard for Pre-Incident Planning,* Quincy, MA: National Fire Protection Association (2010).
7. Occupational Safety and Health Administration, Principal Emergency Response and Preparedness: Requirements and Guidance, OSHA 3122-06R, Washington, D.C.: (2004).
8. Corey, Marianne Schneider, Corey, Gerald, and Cindy Corey, *Groups: Process and Practice* (8th edition), Monterey, CA: Brooks/Cole Publishing Company (2010).
9. Siu, R. G. H., *The Craft of Power,* New York: John Wiley and Sons, Inc. (1979).

Implementing Response Objectives

CHAPTER 10

Hazardous Materials Technician

Knowledge Objectives

After reading this chapter, you will be able to:

- Describe the following terms and their significance in developing a plan of action for a hazardous materials/ WMD incident (p. 323–326):
  - Strategic goals (NOTE: NFPA 1072/472 uses the term *response objectives*)
  - Tactical objectives (NOTE: NFPA 1072/472 uses the term *action options*)
  - Operational tasks
  - Offensive mode
  - Defensive mode
  - Nonintervention mode
- Describe the purpose of, procedures for, equipment required for, and safety precautions associated with the listed hazardous material/WMD control techniques. (pp. 336–346).
- As part of developing a plan of action, list and describe the safety considerations to be included and the points that should be made as part of the safety briefing prior to working at the scene. (pp. 326, 327).
- List and describe the basic questions that emergency responders must consider when evaluating the risks of conducting a technical rescue operation (p. 335).
- Identify the atmospheric and physical safety hazards associated with hazardous materials/WMD incidents involving a confined space, and the considerations for assessing a leak or spill without entering the confined space. (pp. 333–336).
- Identify the procedures, equipment, and safety precautions for preserving and collecting legal evidence at hazardous materials /WMD incidents (pp. 385–387).
- Identify three safety considerations for product transfer operations (p. 378).
- Identify the methods and precautions used to control a fire involving an MC-306/DOT-406 aluminum shell cargo tank (p. 362).
- Given MC-306/DOT-406, MC-307/DOT-407, and MC-312/ DOT-412 cargo tanks, describe the methods for containing the following types of leaks (pp. 347, 348):
  - Dome cover leak
  - Irregular-shaped hole
  - Puncture
  - Split or tear
- Identify the considerations and risk factors that should be evaluated for the following types of hazmat fire emergencies (p. 351–376):
  - Flammable liquids
  - Flammable gases
  - Liquefied Natural Gas
  - Reactive chemical fires and reactions
- Identify the factors that must be considered in selecting the appropriate Class B firefighting foam concentrate for a given flammable liquid hazard, and determine the Class B foam concentrate and water supply requirements. (p. 356)
- Given MC306/DOT 406, MC-307/DOT-407, MC-312/DOT-412, MC-331, and MC-338 cargo tanks, describe the product removal and transfer consideration and the common transfer methods from each type of cargo tank (pp. 376–384).
- Identify the factors to be considered in uprighting overturned cargo tank trucks transporting hazardous materials. (pp. 382–384)

Skills Objectives

After reading this chapter, you will be able to:

- Demonstrate the methods and application of various control devices for the following types of leaks from the following locations (p. 351, 352):
  - Fusible plug
  - Fusible plug threads
  - Side wall of cylinder
  - Valve blowout
  - Missing valve gland
  - Valve inlet threads
  - Valve seat
  - Valve stem assembly blowout
- Demonstrate the ability to perform the following containment methods on a pressure container (p. 351):
  - Close valves that are open
  - Replace missing plugs
  - Tighten loose plugs

- Contain leaks in a 55 gal (208 L) steel or plastic drum using the tools and materials provided by the AHJ on the following types of leaks (pp. 344–346):
  - Bung leak
  - Chime leak
  - Forklift puncture
  - Nail puncture
- Demonstrate the ability to overpack a 55 gal (208 L) steel or plastic drum using the appropriate overpack drum and equipment as provided by the AHJ using the following methods (p. 342, 343):
  - Rolling slide-in
  - Slide-in
  - Slip-over
- Identify the maintenance and inspection procedures for the tools and equipment provided for the control of hazardous materials releases according to the manufacturer's specifications. (p. 343)
- Given an MC-306/DOT-406 cargo tank and a dome cover clamp, demonstrate the ability to install the clamp on the dome. (p. 347, 348)
- Given scenarios involving hazardous materials/WMD incidents and the incident action plan, evaluate the effectiveness of any control functions identified in the incident action plan. (p. 326–331)
- Identify the procedures, equipment, and safety precautions for preserving and collecting legal evidence at hazardous materials/WMD incidents. (p. 385–387)

Hazardous Materials Incident Commander

Knowledge Objectives

After reading this chapter, you will be able to:

- Describe the following terms and their significance in developing a plan of action for a hazardous materials / WMD incident (pp. 323–326):
  - Strategic goals (NOTE: NFPA 1072/472 uses the term response objectives)
  - Tactical objectives (NOTE: NFPA 1072/472 uses the term action options)
  - Operational tasks
  - Offensive mode
  - Defensive mode
  - Nonintervention mode
- Describe the purpose of the listed hazardous material/WMD control techniques. (p. 324).
- Identify the process and procedures for evaluating whether the response options are effective in accomplishing the objectives (pp. 322–331).
- As part of implementing a plan of action, identify the safe operating practices and procedures that should be followed. (p. 326, 327).
- Identify the safety precautions associated with search and rescue operations at hazardous materials / WMD incidents (p. 333–336).
- Identify the atmospheric and physical safety hazards associated with hazardous materials / WMD incidents involving a confined space (p. 334–336).
- Identify the considerations and risk factors that should be evaluated for the following types of hazmat fire emergencies (pp. 351–376):
  - Flammable liquids
  - Flammable gases
  - Liquefied natural gas
  - Reactive chemical fires and reactions
- Identify the factors that must be considered in selecting the appropriate Class B firefighting foam concentrate for a given flammable liquid hazard, and determine the Class B foam concentrate and water supply requirements. (p. 356)
- Identify the factors to be considered in uprighting overturned cargo tank trucks transporting hazardous materials. (p. 382–384)

Skills Objectives

After reading this chapter, you will be able to:

- Develop an incident action plan (IAP), including site safety and control plan, consistent with the emergency response plan or standard operating procedures of the AHJ. (p. 326, 327)
- Evaluate the completed IAP for progress and/or successful implementation, and determine the effectiveness of the following (p. 327):
- Control, containment, or confinement operations
- Decontamination process
- Established control zones
- Personnel being used
- Personal protective equipment

You Are the HazMat Responder

Personal protective equipment

You are the senior Hazardous Materials Technician and have been assigned as the Hazardous Materials Team Group Supervisor at the scene of an overturned tractor trailer incident on a secondary road. The truck is on its side and the top has been ripped open exposing the cargo. The truck contains a mixed load of hazardous materials in various small packages. A review of the shipping papers and discussions with the driver determined that the containers include:

- Ten (10) 50-lb. bags of fertilizer grade ammonium nitrate
- Ten (10) 20-gallon carboys of 35% concentration hydrogen peroxide
- Twenty (20) 50-lb. bags of low grade potassium permanganate
- Fourteen (14) 20-gallon containers of muriatic acid
- Fifteen (15) 55-gallon drums of industrial grade floor cleaner
- Seventeen (17) 30-lb. fiberboard containers of calcium hypochlorite

An entry team has conducted reconnaissance (recon) and reports back to you that there are two leaking 20-gallon carboys of muriatic acid that will require overpacking. Three bags of ammonium nitrate have been damaged and will also require repackaging.

The current weather forecast calls for light rain to begin falling within the next 2 hours. The sun is beginning to set.

1. What are your priorities at this point? Be specific and explain why the priorities you have selected are important to the overall safety of the operation.
2. Based on your priorities, what are your response objectives?
3. How would you explain your priorities and proposed response objectives to the Incident Commander?

Introduction

This chapter discusses the sixth step in the Eight Step Process©—Implementing Response Objectives. It represents the phase in a hazmat emergency where the Incident Commander (IC) and tactical responders implement the best available strategic goals and tactical objectives which will produce the most favorable outcome. Remember that good and bad outcomes are measured in terms of fatalities, injuries, property and environmental damage, and systems disruption.

In the chapter on "hazard assessment and risk evaluation" found elsewhere in this text, we explain that all hazmat incidents follow a logical sequence of events along a predictable timeline. In other words, every hazmat incident has a beginning and an end. A good IC must quickly size up what events have already occurred, determine what events are happening now, and predict what events will occur. If an event can be predicted, it may be prevented. This is the essence of the risk evaluation process. In summary, the operational strategy for the incident is developed based on the IC's evaluation of the current conditions and forecast of future conditions.

This chapter focuses on the various strategic goals and tactical objectives available to the IC to influence outcomes. Topics include basic principles of decision making, guidelines for determining and implementing strategic goals and tactical objectives, and special tactical problems. We will not use a "how to do" approach that focuses on the specific tasks involved in implementing respective tactical objectives; rather, we focus on the management criteria and guidelines associated with selecting and implementing the appropriate response action.

The authors would like to thank Ludwig Benner, Jr. for permission to reproduce materials from his copyrighted works in the Basic Principles discussion of this chapter. Ludi has contributed greatly to the body of knowledge in hazardous materials emergency response, especially in the areas of hazard and risk assessment, fire fighter safety, and accident investigation. His groundbreaking work at the National Transportation Safety Board (NTSB) and at Montgomery College, Rockville, Maryland in the 1970s under Frank Brannigan's leadership set the standard for hazard and risk assessment that is still used today in emergency services. No doubt many lives have been saved due to Ludi's and Frank's willingness to share their knowledge with the emergency response community.

Basic Principles: Understanding Events

To determine which strategy and tactics are best suited to change the outcome of a particular hazmat incident, responders must be able to understand what has already occurred, what is occurring now, and what will most likely occur in the future. In other words, if emergency responders can visualize what has already happened and understand what is happening now, it may be possible to interrupt the chain of events and favorably change the future.

If you think about every emergency you ever responded to in your career, it could be plotted along a timeline. At some point on that timeline you arrived on the scene, sized up the situation, took some action, and began influencing the outcome in some positive way (e.g., you rescued a victim, extinguished a fire, etc.). See Figure 10-1.

The response options available to change the outcome of an incident has a lot to do with where you are on the risk timeline when you arrive at the scene. If containers have already breached and produced fatalities and injuries, the options available to influence the outcome in a favorable way are fairly limited. On the other hand, if you arrive on the scene while containers are still being stressed, it may be possible to prevent containers from breaching.

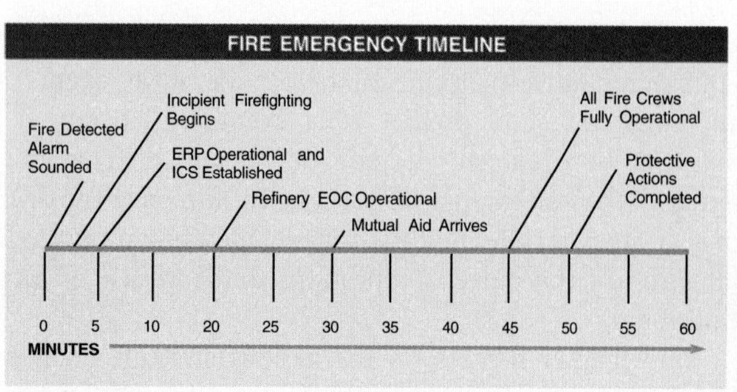

Figure 10-1 Example of an emergency response timeline for a refinery fire. Every emergency can be plotted along a timeline.

The last 50-plus years has been marked by many significant hazmat incidents that produced bad outcomes. Unfortunately, many of these incidents occurred after emergency responders intervened in an incident and tried to change the outcome using offensive tactics. Ironically, by intervening without understanding the risk they were taking, responders actually created a worse outcome by becoming casualties themselves.

In today's risk-based management and decision-making environment it may be hard to imagine that in the 10-year period of 1968 to 1978 there were 51 emergency responder line-of-duty deaths caused by 16 major hazmat transportation-related incidents. These incidents caused over 1,400 injuries to public and emergency responders and destroyed 2,200 homes and commercial buildings. It is your responsibility as emergency responders to make sure we do not go back to the "Dark Days" of hazmat response. Risk-based decision making saves lives!

Due to the willingness of hazmat emergency responders to share their lessons learned, the ability of dedicated instructors to work these lessons learned into their training programs, and the development of competency-based hazmat/weapons of mass destruction (WMD) training standards, there has been a significant reduction in the number of incidents resulting in multiple hazmat-related fatalities and injuries over the last 2 decades. One of the primary reasons for this reduction is that emergency responders are better trained to understand how hazardous materials behave and what responders can realistically achieve with the resources available.

Producing Good Outcomes

Producing good outcomes at hazmat emergencies depends a great deal on the ability of emergency responders to visualize the emergency in a chronological sequence of events. The IC should analyze and visualize events at the emergency scene in three phases: (1) what has already taken place; (2) what is taking place now; and (3) what will most likely be taking place in the immediate future. Stated another way, responders must evaluate the incident in terms of past, present, and future events. This evaluation process is the cornerstone of risk-based decision making at both hazardous materials and WMD incidents.

- **Past Events**—During the initial size-up, the IC should develop a mental picture of what has already occurred before the arrival of responders. Examples include whether containers have already breached, which containers have already been stressed, the dispersion patterns created, and the exposures that have already been contacted and harmed.
- **Present Events**—Understanding what has already happened can help the IC focus on what is occurring now and what may be happening in the next 10 to

15 minutes. Getting the "Big Picture" as quickly as possible is an important step in (1) predicting what will happen in the future, and (2) determining which options are available to influence those events. How quickly the IC develops this "Mental Movie" of what is happening has a lot to do with how quickly command is established and how information is managed to support decision making.

- **FUTURE EVENTS**—If the IC has a good mental picture of what is occurring along a timeline, it may be possible to influence the outcome of the emergency in a variety of different ways. There are five factors, identified by Benner 35 years ago, which can be remembered by the acronym MOTEL:
  - **Magnitude**—Trying to keep the incident as small or limited as possible. For example, rotating a 1-ton chlorine container 180° can change a liquid leak to a vapor leak and reduce the size of the vapor cloud.
  - **Occurrence**—Preventing a future event from occurring by influencing the current events. For example, activating a water spray system or rapidly extinguishing a ground fire to minimize thermal stress upon structural steel and surrounding containers.
  - **Timing**—Trying to change when an event happens and/or how long it lasts. For example, product transfer operations can be initiated to reduce the quantity of product remaining in a fixed facility storage tank while resources are being assembled to implement leak-control tactics.
  - **Effects**—Trying to reduce the size and/or effects of an event. For example, using vapor dispersion tactics (e.g., water monitors and water spray systems) to reduce the size or magnitude of an anhydrous ammonia release.
  - **Location**—Trying to change where the next event occurs. For example, if a tank truck with a minor leak could be safely moved to a less populated and congested area for repairs, the impact of the emergency might be minimized.

When selecting strategic goals to either stop the current event or prevent future events from occurring, remember these two basic principles:

1. You cannot influence events that have already happened or change the outcomes of those events; and
2. The earlier the events sequence can be interrupted, the greater the probability of a producing a favorable outcome.

Responder Safety Tip

The goal of every responder should be to favorably change the outcome of the incident. If you can't change the outcome safely then consider nonintervention.

Strategic Goal

Strategy Defined

Uncertainty is a reality of emergency response. One of the IC's most critical tasks is to minimize uncertainty by using a structured decision-making process to size up the problem and select the safest strategy to make the problem go away. Adopting strategic goals to manage the incident is one of the IC's first priorities before intervening in an incident. Guessing what the best course of action is when dealing with hazardous materials can teach you some hard lessons—acting on instinct without facts and without predicting product and container behavior can get people killed. A systematic thought process is important to keeping yourself and your team members safe.

What are strategy and tactics? Strategy is a plan for managing resources. Strategy becomes the IC's game plan to control the incident. Strategies are very broad and are developed at the Command Level. Strategic goals may be pursued simultaneously during an incident and include:

- **Rescue**—Finding disoriented, trapped, and injured people and moving them to a safer area.
- **Public protective actions**—Protecting-in-place, evacuation, or a combination of both tactics.
- **Spill control (confinement)**—A defensive strategy to keep a released material within a defined or specific area.
- **Leak control (containment)**—An offensive strategy to control a release at its source.
- **Fire control**—Confining, controlling, and sometimes extinguishing the fire.
- **Recovery**—Recovering the hazmat (usually liquids) for recycling or disposal.

Tactics are the specific objectives the IC uses to achieve strategic goals. To be effective, objectives need to be concise, easy to communicate, and achievable in a given timeframe with the resources available.

Tactics are normally decided at the section or group/division levels in the command structure (Table 10-1).

If the IC expects strategic goals to be understood and implemented, they must be packaged and communicated in simple terms. If strategic goals are unclear, tactical objectives will become vague. Hazmat strategic goals and tactical objectives can be implemented from three distinct operational modes as outlined below. Criteria for evaluating and selecting operational modes include the level of available resources (e.g., personnel and equipment), the level of training and capabilities of emergency responders, and the potential harm created by the hazmat release.

- **Offensive Mode**—Offensive-mode operations commits the IC's resources to aggressive leak, spill, and fire control objectives. Offensive strategic goal/tactical objectives are achieved by implementing specific types of offensive operations designed to quickly control or mitigate a problem. Although offensive operations can increase the risk to emergency responders, the risk may

Table 10-1 Strategies and Tactics for Hazardous Materials Response Operations

ORGANIZATIONAL LEVELS	STRATEGIC GOALS "What are you going to do?"	TACTICAL OBJECTIVES "How are you going to do it?"	OPERATIONAL TASKS "Do it."
DEFINITION	STRATEGY: The overall plan to control the incident and meet incident priorities.	TACTICS: The specific and measurable processes implemented to achieve the strategic goals.	TASKS: The specific activities that accomplish a tactical objective.
KEY ELEMENTS	■ Goals ■ Overall game plan ■ Broad in nature ■ Meets incident priorities (life safety, incident stabilization, environmental and property conservation)	■ Objective oriented ■ Specific and measurable ■ Often builds on procedures	■ "Hands-on" work to meet the tactical objectives. ■ The most important organizational level—where the work is actually performed. ■ Most problems "go away" as a result of members performing task-level activities.
DECISION MAKERS	■ Incident Commander ■ Section Chiefs	■ Operations Section Chief ■ Hazmat Group Supervisor ■ Group/Division Supervisors	■ Individual units and individuals
OPTIONS	■ Rescue ■ Public Protective Options ■ Fire Control ■ Spill Control ■ Leak Control ■ Recovery	■ Rescue ■ Public Protective Options • Evacuation • Protection-in-Place ■ Fire Control • Exposure Protection • Extinquishment ■ Leak Control • Neutralization • Overpacking • Patching and Plugging • Pressure Isolation and Reduction • Solidification • Vacuuming	■ Spill Control (Confinement) • Absorption • Adsorption • Covering • Diking, Damming and Diversion • Dispersion • Retention • Vapor Dispersion • Vapor Suppression

be justified if rescue operations can be quickly achieved, if the release can be rapidly confined or contained, or if the fire can be quickly extinguished.

- **Defensive Mode**—Defensive-mode operations commits resources (people, equipment, and supplies) to less aggressive objectives. Defensive strategic goal/tactical objectives are achieved by using specific defensive tactics, such as diverting or diking the hazmat. The IC's defensive plan may require "conceding" certain areas to the emergency while directing response efforts toward limiting the overall size or spread of the problem. As a general rule, defensive operations expose responders to less risk than offensive operations.
- **Nonintervention Mode**—Nonintervention means taking no action other than isolating the area. In short, responders wait out the sequence of events underway until the incident has run its course and the risk of intervening has been reduced to an acceptable level (e.g., waiting for an LPG container to burn off). This strategy usually produces the best outcome when the IC determines that implementing offensive or defensive strategic goals/tactical objectives will place responders at an unacceptable risk. In other words, the potential costs of action far exceed any benefits (e.g., a BLEVE scenario).

In some situations, nonintervention tactics may be implemented until sufficient resources arrive on-scene and an offensive mode can be implemented. Defensive tactics are always preferable over offensive tactics if they can accomplish the same objectives in a timely manner. For example, if a spill diversion tactic (a defensive operation) can achieve the same objective as a containment tactic (an offensive operation), then spill diversion is the better tactic because it can usually be accomplished without coming into close proximity of the spilled material.

Hazmat operations usually begin in the defensive mode. The most important question the IC should ask is, "What happens if I do nothing?" There will also be times when initial information indicates it is relatively safe to attempt an offensive tactical objective, yet it is very possible that things may turn for the worse during that process.

Selecting Strategic Goals

Once the IC has run a "mental movie" for each option being considered, Command must make a decision to do something or to do nothing—both constitute a decision. Selecting the best strategic goal involves weighing what will be gained against the "costs" of what will be lost—a process often easier said than done.

Determining what will be "gained" involves weighing many different variables, including:

- **Potential casualties and fatalities**—Will lives be saved by pursuing aggressive rescue operations? If a civilian is trapped in a confined area contaminated with poison gas, can his or her life be saved by committing personnel to a rescue? Likewise, does the rescue environment present an unreasonable risk to responders?
- **Potential property damage or financial loss**—What will be the financial cost of implementing one option over another? If a small chemical spill occurs in an auto assembly facility, should the entire building be evacuated and production operations be halted? If you don't evacuate and the problem quickly degrades, the potential risk to employees and emergency responders may be unacceptable.
- **Potential environmental damage**—What will be the impact on the environment as a result of your actions? If offensive fire control tactics are implemented at an agricultural chemical warehouse fire, water runoff and pollution will likely result. Allowing the fire to burn out with no application of water may result in widespread air pollution over a much larger area. This may require additional public protective actions downwind. Which option is the most acceptable?
- **Potential disruption of the community**—Will the community be disrupted to an unacceptable level? If a gasoline tank truck has overturned and is burning on a major freeway just before rush hour, is it an acceptable risk to the community to let the tanker burn for 3 hours and consume the product, or will extinguishment be required sooner? Which option will actually open the highway sooner?
- **"Political" considerations**—In this instance, we are *not* referring to partisan politics, but to the interpersonal and inter-organizational dynamics that can surround major Hazmat/WMD incidents. Command officers must recognize that their actions and decisions may be judged in the court of public opinion. Remember this basic point—safety and health decisions based upon a solid technical basis will likely weather any storms that come your way in the days and weeks that follow the incident.

Decision-Making Trade-Offs

Every decision the IC makes involves making trade-offs between two or more factors that are usually in conflict with one another (e.g., life safety vs. property or environmental damage). Decision making at a hazmat incident is not a "black and white" process; it is full of gray areas.

Conflicting information and values can add to the uncertainty of the decision-making process. Consider the following:

- **Conflicting or uncertain information.** The more critical the life safety situation, the less likely accurate and reliable information will be available to make an informed decision. Experience shows that the IC is sometimes pressured into making decisions based on incomplete or inadequate information. For example, fire fighters have been killed and seriously injured pursuing aggressive rescue operations when told that there is someone trapped inside a building, only to discover later that the information was incorrect.

  The most difficult decisions responders must make are those related to life safety. Experience shows that the greater the risks to people, the greater the risks responders are willing to take to save them.

  A corollary to this fact is that the worse the potential outcome, the greater the level of uncertainty the IC should be willing to accept in implementing response goals and objectives. Understanding this point and incorporating it in your decision-making process will save lives!
- **Conflicting or competing values.** Every individual has a set of values they bring to the incident. Regardless of your background, these values influence the decision-making process. Even among similar groups with the same types of values, people have different opinions about the perceived risk involved in carrying out an option as well as the perceived value of what might be gained.

  When lives are at stake, each person can be expected to select a different option as the "best" one in his or her judgment. If you doubt this, conduct an exercise in which you give responders a specific situation and ask them to rate the risks involved in protecting life, property, environment, and in disrupting the community. You may be surprised what you learn about how people think and how much risk they are willing to take.
- There is no single best way to evaluate life safety decisions. Pure risk-taking (the black area) can lead to self-destruction, while pure safety (the white area) can lead to no action at all. Somewhere between these two extremes is the right decision (the gray area).

Figure 10-2 Getting a buy-in from the risk-takers is an important step in getting the IC's game plan implemented.

Getting a buy-in from the risk-takers (i.e., the people in the PPE working in the hot zone or IDLH environment) is an important step in getting the IC's strategic game plan implemented Figure 10-2. The success of obtaining agreement depends on:

1. The IC's ability to understand the differences in how responders perceive risks.
2. The IC's ability to explain the options available to the risk-takers.
3. Getting the other agencies and organizations involved in the incident to understand the big picture concerning what will be lost versus what will be gained.

The more you understand how different organizations and individuals perceive risk and the value of evaluating different options before an incident, the quicker you can implement these decisions at the incident. This is why getting to know the people, personalities, and tendencies of the people you will be working with before the incident is so important.

Tactical Objectives

Once the Command Level has committed to a strategy, subordinate branches, groups and divisions will implement the IC's general game plan by establishing specific tactical objectives. There is no way to make a hazmat emergency go away without tactics!

A well-defined tactic has a stated objective that can be achieved using specific procedures and tasks within a reasonable period of time. The tactical action plan must be easy to understand and laid out with straightforward objectives. A good standard for evaluating a tactical objective is that it should be able to be communicated and implemented without providing too much detail from Incident Command.

An IC who gets involved in making detailed tactical decisions like which type of hand tool to use loses the broad perspective of the incident and develops tunnel vision. This can be a problem with new command officers who have worked their way up through the system and just can't let go of the hands-on details. Using a military analogy, generals are good at strategy, captains are good at tactics, and sergeants are good at tasking privates to make problems go away. Things don't work well when Washington tries to make the process work in reverse. Generals in Washington are not good at tactics, and sergeants in enemy territory do not have the "big picture" required to develop sound long-term strategy.

Developing a Plan of Action

Developing an effective plan of action requires the IC to draw upon experience, training, and situational awareness to make a good decision and implement a well thought-out and realistic plan. Being the person in charge of bringing order to chaos—is best approached with a systematic methodology for decision making and planning.

The IC must use an organized approach to developing a plan of action including:

- Identifying realistic and measureable incident objectives.
- Defining the operational timeframe under which the plan must be executed.
- Determining the organizational structure of the players on the scene.
- Identifying the known or potential hazards at the scene and the mitigation efforts that will make those hazards go away.
- Specifying how the responders will communicate and what actions they will take if something unexpected happens or things go wrong.
- How to communicate all that information to those tasked with any portion of executing the game plan.

This can be a daunting task, especially with complicated or dangerous incidents. Developing a Site Safety Plan can be time-consuming and lead to the "paralysis from analysis" mode, where everything is dissected, second guessed, and besieged by "what if" questions to the point that nothing gets done in a timely fashion. There is a delicate balance between the time and effort expended making the plan, getting it implemented, and the work finally getting done.

A good IC can see several potential outcomes of an incident and constructs a plan that is balanced with what needs to be accomplished with how the work should be done safely. When there is imminent danger or a rescue is required, the site safety plan can be abbreviated and quickly communicated, but the safety plan still needs to be determined and communicated.

The ICS 208 form—Site Safety and Control Plan—is a useful guide and template for an IC to structure mitigation efforts Figure 10-3. The steps listed in Skill Drill 10-1 highlight some of the thinking points an IC should consider when constructing a site safety and control plan. There are several versions of the ICS 208 form, but most contain similar information. Refer to the form used in your jurisdiction as a guide.

To begin the process of assembling a site safety and control plan, the IC (or a combination of players such as the IC and/or assistant safety officer or hazmat group supervisor, depending on your jurisdiction) will need to understand the nature of the problem in order to craft reasonable incident objectives. This may take some time and intelligence gathering, but is critical to developing an appropriate response. If you don't know what the problem is, it's difficult to figure out what you will do about it!

Keep in mind the form is laid out sequentially, but can be filled out in any manner that makes sense. The goal is to come up with a plan that is reasonable and workable. As you become a more experienced IC, you will undoubtedly develop your own style and approach to incident management.

For a suggested approach to completing a site safety and control plan, read the scenario below and follow the steps. (These steps are in a suggested order but may vary depending on your incident). Think through the possible points that should be considered in drawing up your plan.

Imagine this scenario: Your three-person hazmat team arrives on the scene of a vehicle accident involving a small passenger vehicle and a tanker truck carrying 20 tons of anhydrous ammonia. The tanker rolled over on its side as a result of the accident and slid down the highway for approximately 150 feet. There are no injuries to the three victims in the passenger vehicle, but the driver of the tanker truck is pinned inside the cab. You notice a strong smell of ammonia in the air, but see no visible signs of a product release. The regional hazmat team, fully staffed with six technician-level responders, is also responding with a 1 hour estimated time of arrival. There are three engine companies on scene, each with one officer and three fire fighters.

To complete a site safety and control plan, follow the steps in **Skill Drill 10-1**:

1. Understand the nature of the problem. It is often helpful to draw a map of the scene to help you understand what is causing you to take action and what that action might look like. This includes identifying the chemical and physical properties of the released substance or substances, and figuring why the substance is out of its container. This is the "getting to know your problem" piece of incident management. This critical step will inform your strategic decision making.
2. Identify your available and potential resources, including personnel, tools, and equipment. Is your resource pool adequate to carry out your plan? Examples include determining what technologies and/or procedures might be required to perform detection and monitoring; what decontamination solutions might be required; what level of protection might be needed; or deciding if specialized teams or subject matter experts are needed. The IC may need to decide if other disciplines such as law enforcement or an explosive ordnance disposal (EOD) team is needed.
3. Develop incident objectives. These objectives should be based on as much factual information as you can get. Incident objectives should be realistic, measurable, and clearly articulated. It is critical that everyone understands the game plan in simple, straightforward terms. Where incidents with criminal intent are suspected, evidence preservation an/or collection might be a required objective.
4. Develop tactical objectives. This is the main reason we are on the scene—to make the problem go away. IC's should be mindful of the tactical objectives, but as mentioned earlier, should not be micro-managing what detection device is used or how it's being used. IC's should be more interested in information management and seeing the results, or lack of results, from the mitigation efforts. Look at it this way—the IC is the conductor of the orchestra, not the first chair violinist!
5. Identify an operational period. Is this a 1-hour problem, a 1-day problem, or longer? Developing an incident timeline is helpful to keep you moving forward.
6. Identify the players. A simple organizational chart may work or you may choose to list personnel by position (decon team leader; assistant safety officer; hazmat group supervisor, etc.).
7. Identify emergency procedures and/or other safe work practices. This may include requiring electrical bonding or grounding when offloading flammable liquids, or paying particular attention to environmental concerns such as excessive (hot or cold) temperatures, or operating on elevated platforms. You are already operating on the scene of an emergency; this is a plan developed in the event an emergency occurs with the responders, which may include chemical exposures, traumatic injuries, or other unplanned medical emergencies.
8. Conduct the safety and operational briefing (prior to taking action). This part of the plan is more art than science and can be conducted in a number of ways. In short, the briefing should be just as the name implies—brief! There isn't time to get wordy or wander off topic. The idea is to get information out in digestible bites and make sure everyone is on the same page in terms of the strategic and tactical objectives and how business will be conducted on the scene. The assistant safety officer or hazmat group supervisor or IC typically conducts safety briefings. Follow the protocol established in your jurisdiction.

<table>
<tr><td colspan="2">SITE SAFETY AND CONTROL PLAN
ICS 208 HM</td><td colspan="4">1. Incident Name:</td><td colspan="4">2. Date Prepared:</td><td colspan="3">3. Operational Period:
Time:</td></tr>
<tr><td colspan="13">Section I. Site Information</td></tr>
<tr><td colspan="13">4. Incident Location:</td></tr>
<tr><td colspan="13">Section II. Organization</td></tr>
<tr><td colspan="3">5. Incident Commander:</td><td colspan="5">6. HM Group Supervisor:</td><td colspan="5">7. Tech. Specialist - HM Reference:</td></tr>
<tr><td colspan="3">8. Safety Officer:</td><td colspan="5">9. Entry Leader:</td><td colspan="5">10. Site Access Control Leader:</td></tr>
<tr><td colspan="3">11. Asst. Safety Officer - HM:</td><td colspan="5">12. Decontamination Leader:</td><td colspan="5">13. Safe Refuge Area Mgr:</td></tr>
<tr><td colspan="3">14. Environmental Health:</td><td colspan="5">15.</td><td colspan="5">16.</td></tr>
<tr><td colspan="5">17. Entry Team: (Buddy System)
Name: PPE Level</td><td colspan="8">18. Decontamination Element:
Name: PPE Level</td></tr>
<tr><td colspan="2">Entry 1</td><td colspan="3"></td><td colspan="5">Decon 1</td><td colspan="3"></td></tr>
<tr><td colspan="2">Entry 2</td><td colspan="3"></td><td colspan="5">Decon 2</td><td colspan="3"></td></tr>
<tr><td colspan="2">Entry 3</td><td colspan="3"></td><td colspan="5">Decon 3</td><td colspan="3"></td></tr>
<tr><td colspan="2">Entry 4</td><td colspan="3"></td><td colspan="5">Decon 4</td><td colspan="3"></td></tr>
<tr><td colspan="13">Section III. Hazard/Risk Analysis</td></tr>
<tr><td>19. Material:</td><td>Container type</td><td>Qty.</td><td>Phys. State</td><td>pH</td><td>IDLH</td><td>F.P.</td><td>I.T.</td><td>V.P.</td><td>V.D.</td><td>S.G.</td><td>LEL</td><td>UEL</td></tr>
<tr><td></td><td></td><td></td><td></td><td></td><td></td><td></td><td></td><td></td><td></td><td></td><td></td><td></td></tr>
<tr><td></td><td></td><td></td><td></td><td></td><td></td><td></td><td></td><td></td><td></td><td></td><td></td><td></td></tr>
<tr><td></td><td></td><td></td><td></td><td></td><td></td><td></td><td></td><td></td><td></td><td></td><td></td><td></td></tr>
<tr><td></td><td></td><td></td><td></td><td></td><td></td><td></td><td></td><td></td><td></td><td></td><td></td><td></td></tr>
<tr><td colspan="13">Comment:</td></tr>
<tr><td colspan="13">Section IV. Hazard Monitoring</td></tr>
<tr><td colspan="5">20. LEL Instrument(s):</td><td colspan="8">21. O_2 Instrument(s):</td></tr>
<tr><td colspan="5">22. Toxicity/PPM Instrument(s):</td><td colspan="8">23. Radiological Instrument(s):</td></tr>
<tr><td colspan="13">Comment:</td></tr>
<tr><td colspan="13">Section V. Decontamination Procedures</td></tr>
<tr><td colspan="11">24. Standard Decontamination Procedures:</td><td>YES:</td><td>NO:</td></tr>
<tr><td colspan="13">Comment:</td></tr>
<tr><td colspan="13">Section VI. Site Communications</td></tr>
<tr><td colspan="3">25. Command Frequency:</td><td colspan="6">26. Tactical Frequency:</td><td colspan="4">27. Entry Frequency:</td></tr>
<tr><td colspan="13">Section VII. Medical Assistance</td></tr>
<tr><td>28. Medical Monitoring:</td><td>YES:</td><td>NO:</td><td colspan="8">29. Medical Treatment and Transport In-place:</td><td>YES:</td><td>NO:</td></tr>
<tr><td colspan="13">Comment:</td></tr>
</table>

Figure 10-3 Site Safety and Control Plan

Section VIII. Site Map
30. Site Map:
Weather ❑ Command Post ❑ Zones ❑ Assembly Areas ❑ Escape Routes ❑ Other ❑

Section IX. Entry Objectives
31. Entry Objectives:

Section X. SOP S and Safe Work Practices		
32. Modifications to Documented SOP s or Work Practices:	YES:	NO:
Comment:		

Section XI. Emergency Procedures
33. Emergency Procedures:

Section XII. Safety Briefing	
34. Asst. Safety Officer - HM Signature:	Safety Briefing Completed (Time):
35. HM Group Supervisor Signature:	36. Incident Commander Signature:

Figure 10-3 *Continued*

INSTRUCTIONS FOR COMPLETING THE SITE SAFETY AND CONTROL PLAN
ICS 208 HM

A Site Safety and Control Plan must be completed by the Hazardous Materials Group Supervisor and reviewed by all within the Hazardous Materials Group prior to operations commencing within the Exclusion Zone.

Item Number	Item Title	Instructions
1.	Incident Name/Number	Print name and/or incident number.
2.	Date and Time	Enter date and time prepared.
3.	Operational Period	Enter the time interval for which the form applies.
4.	Incident Location	Enter the address and or map coordinates of the incident.
5 - 16.	Organization	Enter names of all individuals assigned to ICS positions. (Entries 5 & 8 mandatory). Use Boxes 15 and 16 for other functions: i.e. Medical Monitoring.
17 - 18.	Entry Team/Decon Element	Enter names and level of PPE of Entry & Decon personnel. (Entries 1 - 4 mandatory buddy system and back-up.)
19.	Material	Enter names and pertinent information of all known chemical products. Enter UNK if material is not known. Include any which apply to chemical properties. (Definitions: ph = Potential for Hydrogen (Corrosivity), IDLH = Immediately Dangerous to Life and Health, F.P. = Flash Point, I.T. = Ignition Temperature, V.P. = Vapor Pressure, V.D. = Vapor Density, S.G. = Specific Gravity, LEL = Lower Explosive Limit, UEL = Upper Explosive Limit)
20 -23.	Hazard Monitoring	List the instruments which will be used to monitor for chemical.
24.	Decontamination Procedures	Check NO if modifications are made to standard decontamination procedures and make appropriate Comments including type of solutions.
25 -27.	Site Communications	Enter the radio frequency(ies) which apply.
28 - 29.	Medical Assistance	Enter comments if NO is checked.
30.	Site Map	Sketch or attach a site map which defines all locations and layouts of operational zones. (Check boxes are mandatory to be identified.)
31.	Entry Objectives	List all objectives to be performed by the Entry Team in the Exclusion Zone and any parameters which will alter or stop entry operations.
32 - 33.	SOP s, Safe Work Practices, and Emergency Procedures	List in Comments if any modifications to SOP s and any emergency procedures which will be affected if an emergency occurs while personnel are within the Exclusion Zone.
34 - 36.	Safety Briefing	Have the appropriate individual place their signature in the box once the Site Safety and Control Plan is reviewed. Note the time in box 34 when the safety briefing has been completed.

Figure 10-3 *Continued*

Tactical Decision Making

Tactical decision making is about whether to use one tactic over another. Most hazmat incidents cannot be resolved using one tactic. Effective tactical decision making requires thinking ahead and planning various tactical options so that the right people, with the right training, and the right equipment, are available at the appropriate time.

When deciding which tactic to use, consideration should be given to how long it will take to accomplish the objective. Conditions can change rapidly as an incident progresses. What seemed like a good tactical objective early in the incident may no longer be an available option as conditions change. Spills can grow larger and leaks can get worse as responders gather their equipment and prepare for entry. Valuable time can be lost if the entry team has to shift from Plan A to Plan B. The challenge is to keep ahead of what the hazmat will be doing when the entry team is ready to go to work.

The length of time required to implement tactical objectives at the task level must be compared to how long the window of opportunity will be open. As the clock ticks, tactical options will often become more limited. All hazmat incidents follow a natural timeline; leaks and spills usually get worse all by themselves before the situation gets better.

Well-defined tactics can be employed to delay events or slow down the clock until entry teams are ready to implement the solution to the problem. Responders may buy time with less effective, but easy to implement defensive tactics until the most effective offensive tactic can be implemented.

Examples of tactical options that can be used to buy time include:

- **Barriers**—Place a physical barrier between the hazmat, its container, and surrounding exposures. Examples: building dikes, retention ponds, or diversion dams well in advance of an oncoming spill can confine the hazmat release to a limited area or slow it down until entry teams are ready to contain the leak.
- **Distance**—Separate people from the hazmat. The farther away you are from the problem, the lower the risk. Increasing the size of hazard control zones or moving the problem to an isolated area can further reduce the life safety risk until the entry team is ready to enter the hot zone to resolve the problem.
- **Time**—Reduce the duration of the release, or trade off or rotate persons exposed to the hazmat. If you can reduce the pressure on a leaking pipe or vessel, the magnitude of the problem can be reduced significantly. This may allow responders to buy some time to work out a more complete solution to the problem.
- **Techniques**—These are specific tasks and procedures performed by responders to stop the leak and control the problem (e.g., up-righting a leaking liquefied gas cylinder so that vapors rather than liquid product will be released).

Tactical decision making sometimes involves trial and error. What sounded like a great idea after recon operations may not be very practical after the entry team begins on-site operations. Not everything works in the field the way you planned and trained. No amount of training and simulation can prepare you for actual field conditions—always have options available.

In the next part of this chapter, we will discuss specific strategic goals and the tactics associated with each option.

Rescue and Protective Actions

Rescue

Saving lives is our number one mission. Life safety is always the IC's highest priority. One of the first concerns after sizing up the incident is search and rescue. Regardless of the nature of the rescue operation, the "First Law of Hot Zone Operations" is that personnel working in the hot zone must:

- Be trained to play
- Be dressed to play
- Use the buddy system
- Have a backup capability
- Have an emergency decon capability, as a minimum
- Have the IC's approval for the entry/rescue operation

Hazmat rescue problems fall into three general categories:

1. **Searching for and relocating people who will be immediately exposed and harmed by the hazmat as the situation gets worse.** Includes everyone inside the Hot Zone not wearing protective clothing and equipment rated for the hazards. This can consist of civilians and employees who have left the immediate hazard area on their own and believe that they are now in a safe location. It may also include the curious, the "I'm authorized to be here," or people with no clue what danger they are in. These people need to be rescued, they just don't know it yet! Normally, all that is needed is a little organization and direction to move these people back and away from the hazard area. This group may also include people who were initially exposed but have not yet shown the signs or symptoms of exposure.
2. **Rescuing victims who have been disoriented or disabled by the hazmat.** Includes individuals or groups of people who have been exposed to the hazmat and are suffering from its harmful effects. Examples include victims who have been burned, poisoned, blinded, and so on. Normally, rescue involves packaging and removing the victim following standard operating procedures. In the section called "decontamination" found elsewhere in this text, there are some specific guidelines on how to handle and treat these contaminated victims.
3. **Planning and executing technical rescue.** Includes rescues of one or more victims who have been exposed to the hazmat and require physical extrication and removal from the hazard. Examples of technical hazmat rescue situations include the following:
   - High-angle rescue situations, including injured or disabled victims found in high areas (e.g., cooling towers and elevated structures in refineries or chemical plants, on top of a large-diameter storage tank, scaffolding, etc.).

Voices of Experience

On a winter evening around 7:30 pm our ladder company was returning from a routine fire when we noticed a very large fire glowing against the night sky. Based on the magnitude of the fire observed in the distance I assumed the incident either involved a large aircraft crash or a tank truck because there were no bulk storage facilities in that general area.

We made our communications operator aware of the situation and headed toward the direction of the fire, working our way onto a three-lane interstate highway. As we progressed down the highway using the shoulder to pass stopped traffic, we came upon an overturned MC-306 gasoline tank truck. The cab was still connected to the tank truck and burning gasoline was flowing from the rear cell which had been breached during the rollover. The burning gasoline was flowing toward the cab of the truck and into a drainage culvert along the road.

By this time many 911 calls had reported the incident and Communications had already struck a Box Alarm, which included a foam unit and a foam tender.

Our first priority while we waited for assistance was to approach the truck to determine if the driver was still in the cab. Fortunately the driver had received only minor injuries and had already escaped.

Meanwhile, the running and burning gasoline pool fire that was flowing into the drainage culvert found its way into a drainpipe and fire popped up on the other side of the highway igniting brush on the hillside. We watched with some amazement as the fire worked its way up the hillside threatening structures at the top of the hill.

As units began arriving on the scene, I was standing across the highway in a "safe area" along a concrete highway barrier trying to get a better look at where the gasoline was entering the drainage pipe in the ditch. I decided I would be safer on the other side of the barrier so I hopped over it and had walked about 20 yards away from the vehicle when one of the truck tires exploded, separating the rim into pieces. Parts of the rim blew across the highway and into the concrete barrier where I had just been standing. If I had not changed locations I would have been severely injured or maybe killed because the metal shrapnel took chucks of concrete out of the wall.

Eventually an adequate water and foam supply was established and the fire threatening the structures was cut off. By this time all of the cells in the tank truck had breached and the tank truck was fully involved. One of the first arriving engine companies had taken up a position that was too close to the burning tanker. As the fire grew in intensity the IC ordered the apparatus moved farther away. The process of disengaging the hoselines took some effort. The fire got pretty hot so the engine and crew had to make a hasty and somewhat embarrassing retreat to safety. Unfortunately, one of the 2½ inch charged lines had not been disconnected, and as the engine drove away the line pulled out of its coupling and whipsawed the hose back, striking a fire fighter, breaking his kneecap.

In the end, this incident was not a pretty picture. I had a near miss of being injured, one fire fighter received a serious injury, and we almost lost a really expensive piece of fire apparatus. The difference of 1 or 2 seconds that night might have resulted in a much worse outcome. Don't underestimate flammable liquids fires, especially when dealing with running and burning pool fires. They can enter storm drains, sewers, and reappear in drainage basins, on waterways, or in the basements of buildings. You have to think well ahead on the incident timeline and visualize what the outcome might be if things go bad.

Mike Hildebrand

- Victims pinned and trapped inside wreckage or debris (e.g., building collapse, train crew pinned in a train locomotive cab, or a tank truck driver inside an overturned vehicle).
- Confined space rescue situations (e.g., inside storage tanks, on lowered floating roof tanks, underground vaults, sewers, etc.).

Technical Rescue

Technical rescue incidents are high-risk operations that often require the use of special operations teams. Technical rescue problems involving hazardous materials have several common elements that make them difficult to plan for and execute. These include:

- **Hazardous atmospheres**—Typically involve multiple combinations of flammability, toxicity, and oxygen deficiency or enrichment that change over time. The longer the victim is exposed to these atmospheres, the less likely the chance for survival.
- **Hazardous work areas**—Slippery or uneven walking surfaces, missing or damaged catwalks, ladders, and handrails, as well as sharp and jagged metal edges. Under normal working conditions, these types of physical hazards require special precautions. Add the restricted vision and motion problems created by wearing SCBA and specialized protective clothing, and these areas become ultra-hazardous.
- **Limited access areas**—Areas with a single access point (only one way out if you are trapped), confined/narrow walkways or ladders, and high angles of egress. If the route to or from the rescue site is blocked or cut off, the rescue team can become trapped.

Historically, the track record of making effective technical rescues has not been very good. Rescues that do occur successfully are typically made by the initial responding Operations-level responders rather than Technician-level responders who arrive later along the incident timeline. Some rescuers have become victims because they either (1) underestimated the hazards and risks, (2) took action without the proper tools or equipment, (3) were not properly trained for the tasks at hand, or (4) failed to understand the emergency response timeline (e.g., if the victim is trapped in an atmosphere above the IDLH, and medical statistics tell you that humans are biologically dead in 5 minutes after breathing has stopped, and it takes you 20 minutes to respond, set up, and start the rescue … you do the math!). Time simply works against the rescuers and the victims in most technical rescues when hazmat are involved **Figure 10-4**.

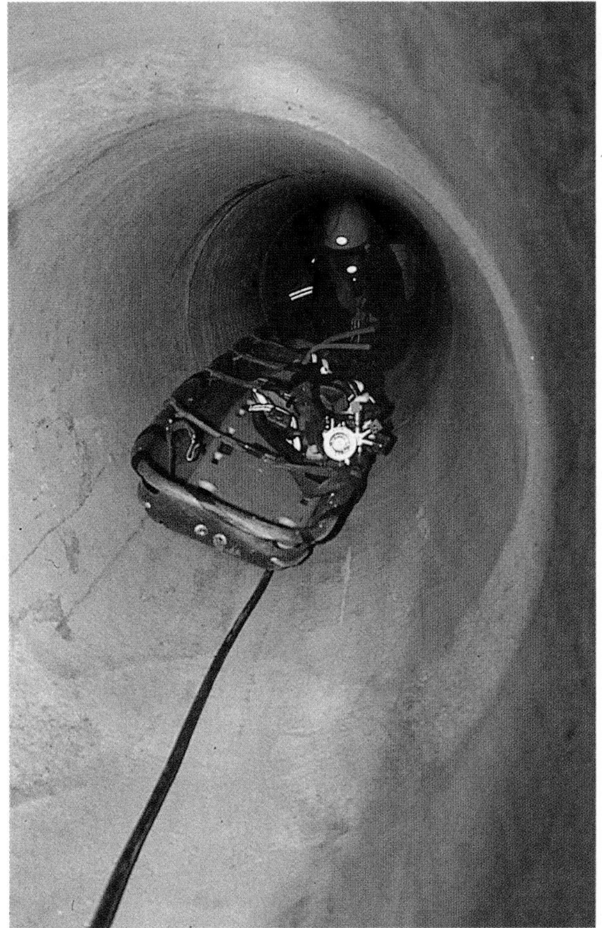

Figure 10-4 Confined spaces include many hazards—entering them is a high-risk operation.

Responder Safety Tip

When entering a confined space to perform a rescue, you are in danger! If there is much to be gained (saving a life) then the risk may be acceptable. If the operation is a body recovery and not a rescue, take more time to plan the operation and look for ways to lower the risk. For example, ventilate the space prior to entry.

Without an oxygen supply to the brain, clinical death occurs in 3 minutes, and biological death occurs in 5 minutes. Exposure to hazardous atmospheres accelerates the timeline. Don't turn rescuers into victims.

Public Protective Actions

For more details on public protective actions, see the chapter called "site management" found elsewhere in this text. However, some of the basics of public protective actions are worth reviewing here as they relate to implementing response objectives.

Public protective actions (i.e., evacuation or protection-in-place) must be continuously monitored during the course of an incident. Incidents are dynamic events—weather conditions may change, the problem may grow, resources may be used up, and so on. Initial protective actions will often be insufficient as we gather more information, get smarter about the incident, and fully assess the level of hazards and risks. Likewise, the IC will often be under tremendous political pressure to allow civilians and employees who have been evacuated to return to their homes and work stations.

Scan Sheet 10-A

Confined Spaces Rescues

Confined spaces present emergency responders with a serious technical rescue challenge. By the nature of their definition, confined spaces are dangerous places. Add in any combination of hazardous materials, and the confined space environment becomes deadly.

According to OSHA, the majority of the fatalities in confined spaces incidents occur from entrapment in (1) toxic atmospheres, (2) asphyxiating atmospheres, and (3) physical hazards inside the confined space.

What is a confined space? According to OSHA (29 CFR 1910.146), a confined space is "any area that has limited or restricted means for entry or exit; is large enough and so configured that an employee can bodily enter and perform assigned work; and is not designed for continuous employee occupancy." Examples of confined spaces and related rescue situations include fixed storage tanks, the interior spaces on mobile tank trucks or rail cars, process vessels, pits, ship and barge compartments, vats, reaction vessels, boilers, ventilation ducts, tunnels, or pipelines. This is not a complete list, but you get the idea—these spaces are confined.

Special Hazards and Risks

Confined spaces have a number of hazardous characteristics, including:

- **Hazardous atmospheres**—Approach a confined space rescue problem with the attitude that the atmosphere inside the space is flammable, toxic, and oxygen-deficient or -enriched. Even if air monitoring indicates that the atmosphere is entirely clean or within acceptable limits, handle the incident as if the atmosphere were contaminated. Never enter a confined space for a rescue operation without using a supplied air respirator (SCBA or airline hose unit). Never remove your facepiece to "revive the victim." If you remove your facepiece in a confined space in a contaminated atmosphere, you will probably not go home at the end of your shift!
- **Limited egress**—Most confined spaces have only one way in and one way out. Egress is further complicated by restricted access points, such as manways that may be poorly designed and placed at unusual angles. Manways create special problems for entry teams using SCBA and specialized protective clothing. Removing the SCBA bottle and harness from the wearer's back while leaving the facepiece in place is a poor option. This is a very difficult task and is dangerous. Many inspectors and maintenance personnel have been overcome by a hazardous atmosphere when they temporarily removed their facepiece to crawl through a manway. Also, accidentally dropping the SCBA harness while crawling through the hole can yank the facepiece off the wearer's head. Supplied air respirators are a better option when working in a confined space.
- **Extended travel distances**—Many confined spaces require extended travel distances to enter and access the confined area to conduct search and rescue operations. The longer the travel distance, the greater the rate of air consumption. This ultimately translates to a short-duration stay inside the confined area. Much like a diver who must calculate air consumption time to allow for a safe ascent to the surface, a confined spaces rescuer must use special care not to overextend the stay and risk running out of air. Even higher pressure, long duration SCBA units will often limit the actual on-site working time to 5 to 10 minutes. The use of an SAR can make entry into the space easier and extend the search and rescue time, but these units are limited to 300 feet of air hose. This may not be very practical for confined spaces such as pipelines and sewers.
- **Unusual physical hazards**—These can include the hazards of uneven or slippery walking and climbing surfaces, being struck by falling objects, or becoming trapped in inwardly converging areas that slope to a tapered cross section (e.g., a grain elevator chute).
- **Darkness**—Almost every confined space is totally dark—you can't see your hand in front of your face. Search and rescue operations must be carried out using portable lights that are brought into the space. The potential for a flammable atmosphere inside most confined areas means that all lighting must be intrinsically safe. In other words, the equipment must be suitable for NEC Class 1, Division 2 locations.
- **Poor communications**—Confined spaces don't like radios. Radio communications are always a problem. Below-grade spaces and steel construction do not promote very good radio reception. Add the protective clothing and SCBA facepiece problem and good radio transmissions become virtually impossible. Like hand lights, radio equipment used in confined spaces must meet NEC Class 1, Division 2 atmosphere requirements.

(*Continued*)

Evaluating Confined Spaces for Rescue Operations

The IC and ICS supervisors should ensure that the following have been evaluated before committing personnel to rescue in a confined space:

1. **Can the confined space be entered safely by emergency responders?**

   Confined spaces may be flammable, toxic, oxygen deficient, or some combination of these.

   - **Flammable atmospheres**—Flammable atmospheres containing 10% or less concentration of flammable vapors are considered within safe limits for conducting rescue operations, but this is not the only factor that requires consideration. Concentrations between 10 and 20% are considered hazardous and should never be entered by rescue teams unless they have the proper PPE and respiratory protection and all electric equipment is rated for Class 1, Division 2 atmospheres. The explosive range of many hydrocarbon vapors range from a 1 to 10% vapor-to-air mixture; however, the explosive range for oxygenated materials like alcohols and glycols is wider. Any mixture of vapor and air between the UEL and LEL will ignite when exposed to an ignition source and should be considered as too dangerous for entry.
   - **Toxic atmospheres**—An atmosphere above the OSHA Permissible Exposure Limit (PEL) or the ACGIH Threshold Limit Value (TLV) does not necessarily prohibit entry into a confined space to perform rescue operations. But realize that the risk to the rescue team increases significantly and there is no margin for error. A rescuer who experiences damaged protective clothing or an air supply problem inside of a confined space with a toxic IDLH atmosphere faces almost certain injury or death. Making an entry under these conditions is a judgment call that must be made on a case-by-case basis by the person taking the risk. Risk looks and feels a lot different when you are the person taking the risks. The people taking the risk always have the final say on Go/No Go.
   - **Oxygen-deficient or -enriched atmospheres**—An oxygen-deficient confined space has less than 19.5% oxygen. An oxygen-enriched atmosphere is 23.5% or greater in oxygen content. The risk of entering an oxygen-deficient atmosphere is similar to entering a toxic atmosphere. Obviously, the less oxygen content, the greater the risk to the rescue team if there is an air supply problem. Oxygen-enriched atmospheres present rescuers with a significant risk because of an increased risk of fire. If a flammable atmosphere is present, the explosive range will become wider.

   All potentially hazardous atmospheres in confined spaces should be confirmed by monitoring instruments.

2. **Are you really rescuing someone or recovering a body?**

   When human life is involved this is a serious and tough question to ask and answer. When the human body is deprived of oxygen death occurs in 3 to 5 minutes so the chances for life decline rapidly. The longer someone goes without oxygen, the less likely he or she can be revived, even under the best of circumstances. The presence of a toxic atmosphere accelerates death's timeline. If a flammable hydrocarbon atmosphere is present, the PEL and the TLV-TWA will usually be exceeded before 10% of the lower flammable atmosphere is reached. Therefore, the atmosphere will almost always be toxic before reaching the flammable concentration.

   As a general guideline, whenever the victim has been subjected to an oxygen-deficient atmosphere of less than 19.5%, or a flammable atmosphere of 10% or greater, or a toxic atmosphere above the PEL/TLV for periods longer than 5 to 15 minutes, the IC should consider the possibility that there is no real chance for a successful rescue. The lower the oxygen content and the higher the toxic and flammable atmospheres inside the confined space, the less likely the victim will survive for periods of exposure exceeding 5 minutes. As is the case with every medical emergency, the condition of the victim, age, preexisting health conditions, etc., affect the chance for survival.

   In addition to these basic medical parameters, the IC must consider the amount of time it will take to set up and safely conduct search and rescue operations. Darkness, limited access, extended travel distances, and difficult working conditions cause delays that work against the victim's chances for survival.

3. **Do you have control of the situation and is there a coordinated incident action plan?**

   Once the frantic pace of a rescue begins, it is very difficult to stop the operation. Emergency responders eagerly do what they do best, save lives. Be sure you are handling the rescue following standard operating procedures and are not "winging it." As the IC or ICS supervisor, the objectives of the rescue and your expectations must be communicated to everyone involved in the operation. Make sure your IAP is clear.

Responders must maintain strict control of the scene throughout the entire incident. The following activities can help maintain site discipline until the problem has been eliminated and the incident safely terminated:

- **Maintain an Incident Safety Officer throughout the incident.** On campaign operations, it may be necessary to rotate both the Incident Safety Officer and Assistant Safety Officer positions at regular intervals to maintain alertness. An Assistant Safety Officer is especially important during hazardous operations such as leak control and technical rescue.
- **Use formal site safety checklists.** Checklists don't make mistakes, tired people do. The longer you work at the hazmat scene, the more likely you will overlook something critical, like making sure your radio is turned on before your suit is zipped up.
- **Enforce isolation perimeter security and the use of hazard control zones.** Make sure the playing field is clearly known and identified to all players. A weak perimeter often leads to loss of site control, which increases the potential for an accident.
- **Establish a crew rotation schedule.** This is especially important during extreme weather conditions where responders sometimes remove their PPE because they are uncomfortable. Don't contribute to the problem by holding personnel in forward positions for an unreasonable time period. Rested people are more alert.

Spill Control and Confinement Operations

Spill-control strategies and confinement tactics are the actions taken to confine a product release to a limited area. These actions usually occur remote from the spill or leak site and are, therefore, defensive in nature. As a general rule, confinement tactics expose personnel to less risk than containment tactics. If responders can accomplish the same objective using defensive confinement tactics such as diking or remotely closing a valve, then they should be implemented before attempting higher risk, offensive-oriented options.

Confinement operations present several advantages over containment options, including the following:

- Avoid direct personnel exposure.
- Can often be performed without special equipment other than some shovels and dirt.
- Can usually be performed by first responders with minimal supervision.

The decision to use confinement tactics is based on the availability of time, personnel, equipment, and supplies. Decisions should be made with a review of the potential harmful effects the leaking material will have on personnel downwind of the spill, where most of the spill-control operations normally take place. For example, a decision to divert a flowing diesel fuel spill from a storm drain to a roadside ditch may be based on the observation that the fuel is flowing too fast and sufficient personnel and equipment are not available to construct a dike. Finally, the fuel will cause substantially less potential damage in the ditch than in the storm system.

Confinement tactics such as diversion can usually begin immediately upon the arrival of first responders trained to the Operations Level. Diking can be started with basic first-responder equipment as more personnel arrive. Retention techniques will then follow as specialty teams and equipment become available.

Don't make the mistake of concentrating all resources on one tactic. It is easy to assign too many responders to the construction of a dike, for example, which may fail and force everyone to move to a safer location to begin again. Recognize that virtually all confinement tactics are "first aid" measures and will usually fail over time.

Confinement Tactics

There are a number of tactical options available to achieve the spill control strategic goal. These include both physical and chemical methods. A summary of the various tactical options is described below.

Absorption is the physical process of absorbing or "picking up" a liquid hazardous material to prevent enlargement of the contaminated area. As the material is picked up, the sorbent will often swell and expand in size. Depending upon the absorbent, it can be used for liquid spills on both land and water.

Operationally, absorbents are effective when dealing with liquids of less than 55 gallons. Larger spills are more difficult to absorb, and often the cost and time exceed the benefits. Materials used as absorbents include clay, sawdust, charcoal, absorbent particulate, socks, pans, pads, and pillows. Absorbent socks and tubes can also be deployed as a circular dike around small spills. When using absorbents, compatibility must be considered (e.g., sawdust used on an oxidizer could start a fire).

Products like PetroGuard™ capture, immobilize, and stabilize hydrocarbon-based chemicals and many reactive compounds. PetroGuard™ has a strong affinity for hydrocarbon-based liquids, particularly the primary aromatic compounds. Examples of organics that can be absorbed and immobilized include titanium tetrachloride, chlorinated hydrocarbons, trichloroethylene, and naptha.

Adsorption is the chemical process in which a sorbate (liquid hazardous material) interacts with a solid sorbent surface. Since the sorbent surface is solid, the sorbate adheres to the surface and is not absorbed, as with absorbents. An example is activated carbon. Characteristics of this chemical interaction include the following:

- The sorbent surface is rigid and there is no increase/swelling in the size of the adsorbent.
- The adsorption process is accompanied by the heat of adsorption, whereas absorption is not. As a result, spontaneous ignition may be a possibility with some liquid chemicals.
- Adsorption can only occur when the sorbent has an activated surface, such as activated carbon. Adsorbents are primarily used for liquid spills on land and should be nonreactive to the spilled material.

Covering is a physical method of confinement. It is typically a temporary measure until more effective control tactics can be implemented Figure 10-5. Depending on the product

Figure 10-5 A tarp is placed over an exposed dome cover as a covering tactic.

Figure 10-6 Damming tactic for diesel fuel spill.

involved, it may be necessary first to consult with a product specialist. Examples of covering include:

- Placing a plastic cover or tarp over a spill of dust or powder
- Placing a cover or barrier over a radioactive source, normally (alpha or beta) to reduce the amount of radiation being emitted.
- Covering a flammable metal or pyrophoric material with the appropriate dry powder agent.
- Covering a spill with a vapor mitigation agent.

Damming is a physical method of confinement by which barriers are constructed to prevent or reduce the quantity of liquid flowing into the environment Figure 10-6. Damming consists of constructing a barrier across a waterway to stop/control the product flow and pick up the liquid or solid contaminants.

There are two types of dams—overflow and underflow.

- **Overflow dam.** Used to trap sinking heavier-than-water materials behind the dam (specific gravity > 1). With the product trapped, uncontaminated water is allowed to flow unobstructed over the top of the dam. Operationally, this is most effective on slow moving and relatively narrow waterways Figure 10-7.
- **Underflow dam**. Used to trap floating lighter-than-water materials behind the dam (specific gravity < 1). Using PVC piping or hard sleeves, the dam is constructed in a manner that allows uncontaminated water to flow unobstructed under the dam while keeping the contaminant behind the dam. Operationally, this is most effective on slow moving and relatively narrow waterways Figure 10-8.

If the pipes are not deep enough on the upstream side of the dam, a whirlpool may be created and pull the hazardous substance through the pipes. This problem can be overcome through the use of a t-siphon on the upstream side. To be effective, several overflow or underflow dams should be placed downstream to catch product that may be missed by the first dam.

Diking is a physical method of confinement by which barriers are constructed on ground used to control the movement of liquids, sludge, solids, or other materials. Dikes prevent the passage of the hazmat to an area where it will produce more harm.

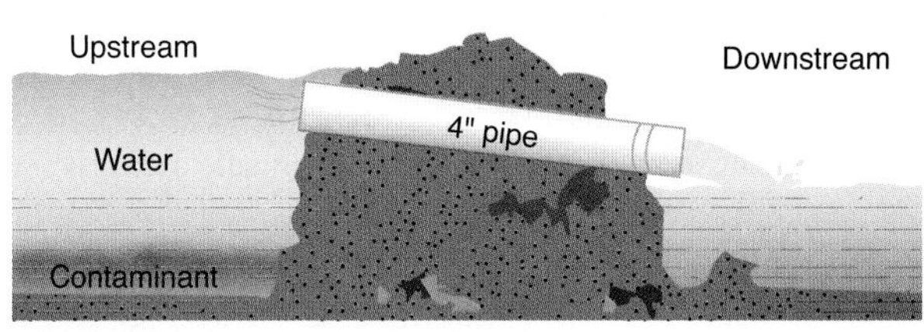

Figure 10-7 An overflow dam is used to contain materials that are heavier than water.

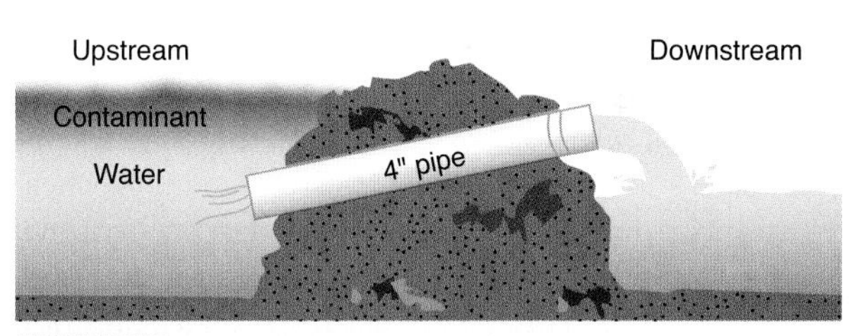

Figure 10-8 An underflow dam is used to contain materials that are lighter than water.

Dikes are most effective when they can be built quickly. Although any available material will do the job, the best quickly acquired supplies are dirt, tree limbs, boards, roof ladders, pike poles, and salvage covers. Bagged materials such as tree bark, sand, and kitty litter can be found at hardware and garden supply stores when more substantial control is required. However, when really large spills occur, dump-truck-sized deliveries will be required.

Dikes can be constructed by first responders using whatever on-scene equipment is available. When considering building a dike, quickly compare your resources to the quantity of the spilled material. Most people overestimate the amount of spill and underestimate the personnel and resources required to complete a dike.

Slow-moving or heavy materials should be confined by use of a circle dike. Faster moving products will require a V-shaped dike located in the best available low-lying area. Always use the land to your advantage.

Dike construction should begin by choosing large, heavier materials for reinforcement, followed by an outer layer of lighter material such as dirt. Operationally, dikes are normally a temporary measure and can begin to leak after a while. Seepage can be minimized by using plastic sheets or tarps at the dike base and within the dike by placing a final layer of dirt along the leading edge between the plastic and the ground. Be aware that plastic sheets may be degraded by certain chemicals.

Factors that can limit dike construction include situations in which:

- The surrounding area is concrete or asphalt with no available soil. Either sacrifice the area for better turf or truck in necessary materials.
- The ground is frozen. Snow may be used in conjunction with materials such as plastic and ladders. Otherwise, truck in necessary materials.
- Essential equipment is unavailable. At least three pointed, long-handled shovels are necessary. When possible, construct dikes upwind in safe areas. Be sure to consider the need for SCBA.

Dilution is a chemical method by which a water-soluble solution, usually a corrosive, is diluted by adding large volumes of water to the spill. There are four important criteria that must be met before dilution is attempted. These include determining in advance that the substance (1) is not water reactive, (2) will not generate a toxic gas upon contact with water, (3) will not form any kind of solid or precipitate, and (4) is totally water-soluble. Before attempting to dilute large spills for mitigation, local or state environmental officials should be consulted, especially if the runoff will end up in waterways.

As a general rule, dilution should only be attempted on liquid and solid substances that are corrosives, and only when all other reasonable methods of mitigation and removal have proven unacceptable. In other words, dilution tactics are a last resort.

Dilution can be effective for small corrosive spills of one quart or less, especially for concentrated corrosives with a pH of 0–2 (acidic) or 12–14 (alkaline). In outdoor situations, local water department or fish and wildlife representatives should be consulted for their approval to use dilution tactics. Federal and most state regulations limit corrosive entries into storm drains and drainage canals to a pH of 6 to 8 as long as no other pollutants are involved that may be harmful to the environment or wildlife.

The major disadvantage to dilution is that it is not well understood by emergency response personnel. It is not a straight, linear one-to-one process. It is important to recognize that dilution is actually a logarithmic process (i.e., on a 1 to 10 scale). For example, a 1-gallon spill of an acid with a pH of zero will require 1 million gallons of water just to bring its pH up to 6! That is a lot of water just to dilute 1 gallon of acid. This same rule applies to the full range of corrosives, from a pH of 0 to 14. The following chart provides some guidelines that can help responders determine whether the dilution is the best tactical option **Table 10-2**.

Table 10-2 Acid Spill of 1 Gallon

Acid Spill of 1 Gallon (pH of 0) Water to add	
10 gallons	1
100 gallons	2
1,000 gallons	3
10,000 gallons	4
100,000 gallons	5
1,000,000 gallons	6

Alkaline Spill of 1 Gallon (pH of 14) Water to add	
10 gallons	13
100 gallons	12
1,000 gallons	11
10,000 gallons	10
100,000 gallons	9
1,000,000 gallons	8

A rule of thumb that can be used in the field to determine the volume of water required to bring the pH to the 6–8 **Table 10-3**.

Diversion is a physical method of confinement by which barriers are constructed on ground or placed in a waterway to intentionally control the movement of a hazardous material into an area where it will pose less harm to the community and the environment.

A flowing, land-based spill can quickly be diverted by placing a barrier (e.g., dirt) ahead of the spill. As when fighting a fast-moving brush fire, the barrier should be placed well ahead of the actual spill. This may require sacrificing some intermediate territory to the hazmat in order to establish complete control at the final diversion site.

Booms can also be placed across streams and waterways to divert the hazardous substance into an area where it can be absorbed or picked up, such as with vacuum trucks.

In constructing a diversion barrier, consider the angle and speed of the oncoming spill. The greater the speed of the flow, the greater the length and angle of the barrier required to slow and divert the flow. For fast-moving spills, barriers constructed at a 45° perpendicular angle will be ineffective; a barrier angle of 60° or more should be used.

Constructing a diversion barrier requires teamwork. When a team with the right equipment works quickly, a large area can be rapidly controlled. A typical four-person crew can build a 20-yard-by-8-inch diversion wall in about 10 minutes if the proper tools and materials are available.

Dispersion is a chemical method of confinement by which certain chemical and biological agents are used to

Table 10-3 Rule of Thumb for Bringing the pH to 6–8

Step 1: Determine the size of the spill to be diluted in gallons (e.g., there are 10 gallons on the ground).

Step 2: Determine the pH of the spilled material using pH paper or a pH meter (e.g., the spill has a pH of 3).

Step 3: Determine the pH that you want to dilute the spill to (e.g., you want to go to a pH of 6 so that the spill can be safely flushed into the storm system).

Step 4: Determine the number of dilution steps between the starting pH and the ending pH. In our example, we started with a pH of 3 and want to end up with a pH of 6. That is three steps.

Step 5: Add three zeros to the beginning gallonage. In our example, we started with 10 gallons, so we add three zeros, which will give us 10,000. This is the number of gallons of water that must be added to the spilled 10 gallons of acid in order to bring the pH up to the desired level of a pH of 6. This rule can be applied to the entire logarithmic scale no matter where you enter it (e.g., you started with a pH of 4 and want to go to 6, that would be two steps, etc.).

Figure 10-9 A Dakota aircraft drops chemical dispersants, as it sweeps over the stern of the stricken tanker Sea Empress off St. Ann's Head on the Welsh coast, Monday Feb. 19, 1996.

disperse or break up the material involved in liquid spills on water. The use of dispersants may result in spreading the hazmat over a larger area **Figure 10-9**.

Dispersants are often applied to hydrocarbon spills, resulting in oil-in-water emulsions and diluting the hazmat to acceptable levels. They do not neutralize or make flammable materials become nonflammable. Experience also shows that some dispersants will separate over time. Use of dispersants may require prior approval of the appropriate environmental agencies.

On April 20, 2010, the BP Oil mobile offshore drilling unit (MODU) *Deepwater Horizon* located approximately 42 miles southeast of Venice, Louisiana in the Gulf of Mexico, experienced an explosion and fire. See **Figure 10-10**. The explosion and fire killed 11 men working on the platform and injured 17 others. The MODU sunk on April 24, scattering debris from the riser pipe across the ocean floor in approximately 5,000 feet of water. It became clear within a few days that the blowout preventer was not functional and oil was leaking into the water from more than one location on the broken riser.

On April 29th, the incident was declared a Spill of National Significance (SONS). At that time, an estimated 12,000 to 19,000 barrels of oil were released into the water every day, making the incident the largest oil spill in U.S. history. The final estimate reported that 53,000 barrels per day (8,400 m^3/d) were escaping from the well just before it was capped on July 15, 2010. More than 990,000 gallons of dispersant were used in the response.

<u>**Retention**</u> is a physical method of confinement by which a liquid is temporarily contained in an area where it can be absorbed, neutralized, or picked up for proper disposal. Retention tactics are intended to be more permanent and may require resources such as portable basins or bladder bags constructed of chemically resistant materials.

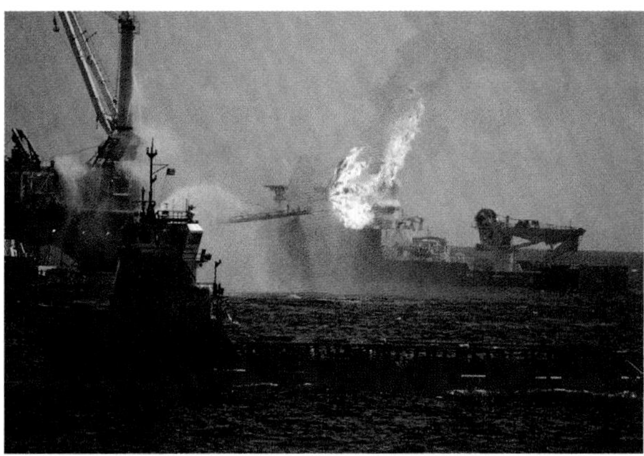
Figure 10-10 The Deepwater Horizon disaster was the largest oil spill in history.

Retention tactics can sometimes be implemented independently and act as a backup to diversion or diking tactics. For example, storm sewer systems can be protected by placing salvage covers or plastic over drains and covering them with dirt. The same procedure can be used for sewer system manways.

When the hazmat is primarily a liquid or slurry, has a specific gravity less than 1.0, and is not water reactive, it may be possible to flood the retention area with water from an engine or hydrant. The hazmat would then float on the water, and any subsequent leakage into the storm system would only be water.

<u>**Vapor dispersion**</u> is a physical method of confinement. Water spray or fans are used to disperse or move vapors away from the surface areas of the product. See **Figure 10-11**. Vapor dispersion is particularly effective on water-soluble materials

Figure 10-11 Vapor suppression tactics may involve the use of water fog, foams, or chemical vapor suppressants.

(e.g., anhydrous ammonia), although the subsequent runoff may involve environmental trade-offs. Fans and positive pressure ventilation may also be used if they are rated for the hazardous atmosphere.

When dealing with flammable materials, such as LP gases, the turbulence created by the water spray may reduce the gas concentration and bring the atmosphere into the flammable range.

Vapor suppression is a physical method of confinement to reduce or eliminate the vapors emanating from a spilled or released material. Operationally, it is an offensive technique used to mitigate the evolution of flammable, corrosive, or toxic vapors and reduce the surface area exposed to the atmosphere. Examples include the use of Class B firefighting foams and chemical vapor suppressants.

While vapor suppression generally does not change the nature of a hazardous material, it can greatly reduce the immediate hazard associated with uncontrolled vapors. This tactic can buy additional time to undertake further measures to control the problem.

Vapor mitigation agents like Ansul's TARGET-7™ can mitigate dangerous vapor releases while simultaneously neutralizing the spilled material without causing additional vapor release. The agent forms a protective cover over the spilled material. The neutralizing agent brings the pH of the spill close to 7. This type of agent can be effective on chlorine dioxide, oleum, chlorsulfonic acids, sulfur trioxide, liquid ammonia, and a variety of fuming acids.

Leak Control and Containment Operations

Leak-control strategies and containment tactics are the actions taken to contain or keep a material within its container. As an offensive operation, containment tactics require personnel to enter the hot zone to control the release at its source and are considered high-risk operations. Examples include uprighting a leaking container, closing and tightening container valves, plugging and patching container shells, and depressurizing vessels by isolating valves or shutting down pumping systems.

Containment tactics may be implemented when defensive options have not produced acceptable results or when people are at great risk from potential chemical exposures. These tactics should only be approved after conducting a thorough hazard and risk evaluation. No emergency situation is worth taking unreasonable risks. Rapid withdrawal from the hot zone is always an option; aggressive/offensive does not mean quick and stupid.

Before initiating containment operations, emergency responders should consider the following:

1. What hazardous material(s) are involved?
2. What is its physical form (i.e., solid, liquid, or gas)? With the possible exception of dusts, solids are usually the easiest materials to contain. High-vapor-pressure liquids and gases present the most difficult challenge to responders.
3. What are its hazards? Whenever working with or near flammable materials, always have charged hoselines and a reliable water supply.
4. What are the risks to both responders and civilians?
5. What are the training levels and physical abilities of the entry team that will perform the operation?
6. Are special tools and equipment needed for the leak-control operation? Are they available (for example, leak-control kits or nonsparking tools)?
7. Are the responders prepared for emergency care and decontamination if an accident happens?

If these questions cannot be answered in an affirmative manner, leak-control operations should be delayed until sufficient information or resources are obtained and the IC feels the operation can be safely conducted.

Although containment operations may pose higher risks, they may be necessary to:

- **Minimize environmental damage**—This is particularly true for hazardous liquids that threaten storm systems or water supplies.
- **Reduce operating response time**—Leaks confined to the area immediately around the container usually limit the spread of the material and minimize the need for evacuation, particularly when faced with a gas or toxic chemical.
- **Reduce clean-up costs**—Contaminants are usually limited to smaller areas or have not entered nearby ground or surface waters.

Situations well-suited for aggressive leak-control offensive strategies include:

1. The hazmat is a vapor or gas and threatens to migrate away from its container.
2. The hazmat is in a solid, powder form and weather conditions threaten to carry it from its original site.
3. Defensive options have been attempted but have not produced the desired results.
4. The situation is getting worse and increasing in risk as time progresses.

Successful offensive operations should be preceded by a thorough reconnaissance. Recon may be as simple as having a team member relay his or her observations or as complex as a Recon Team surveying the entire work site, making sketches, taking photos, and then returning to the incident command post for further analysis.

On large scale incidents like blue water oil spills, tank farm fires, or train derailments the IC may require an aerial reconnaissance to determine the size and scope of the incident or to monitor progress in bringing the incident under control. The U.S. Civil Air Patrol (CAP), an auxiliary of the U.S. Air Force, is well-suited for supporting emergency services with airborne recon. CAP flies more than 85 percent of all federal inland search-and-rescue missions directed from the Air Force Rescue Coordination Center at Langley Air Force Base, Virginia. CAP capabilities include air to ground communications, thermal imaging, and aerial photography.

When planning leak-control tasks, keep the time-tested KISS Principle in mind—Keep It Simple, Stupid—as most leak-control tactics are indeed pretty simple. Consider the following:

- When possible, reduce the rate of release before containing the leak. For example, relieve pressure on the container.
- If piping is involved, isolate the leak by checking the position of upstream and downstream valves.
- Check the integrity of container openings—tighten caps, bungs, lids, and so on.
- Standing up a leaking liquid container may be sufficient to stop the leak.
- Moving the container to place the hole above the release level reduces the hazmat release.
- When dealing with liquefied gases (e.g., chlorine, anhydrous ammonia, LPG), rotate the container to deal with a vapor release rather than a liquid release. If liquid escapes, it will expand the problem and hazard area (e.g., liquid/vapor expansion ratio for chlorine is 460 to 1 and 850 to 1 for ammonia).

When these common-sense techniques aren't effective, try to:

- Plug or patch the opening (i.e., control it at the source).
- Use vapor suppression agents such as foam (i.e., limit its vaporization).
- Neutralize using another chemical (after consulting with a product specialist).
- Control the leak and dispose in-place (e.g., controlled burning, flaring).

Containment Tactics

A number of tactical options are available to achieve leak-control strategies. These include physical and chemical methods. A summary of the various tactical options is given below.

Neutralization is a chemical method of containment by which a hazmat is neutralized by applying a second material to the original spill that will chemically react with it to form a less harmful substance Figure 10-12. The most common example is the application of a base to an acid spill to form a neutral salt.

The major advantage of neutralization is the significant reduction of harmful vapors being given off. In some cases, the hazmat can be rendered harmless and disposed of at much less cost and effort. However, during the initial phases of combining an acid and a base a tremendous amount of energy may be generated, as well as toxic and flammable vapors.

Figure 10-12 Neutralization tactic.

Operationally, many responders recommend that neutralization operations should be limited to spills of less than 55 gallons. It is quite easy to use too much neutralizing agent and end up with a large caustic spill instead of the original acid spill.

Before initiating any neutralization techniques, the following conditions should be satisfied: (1) the hazmat has been identified positively, (2) its physical and chemical properties have been researched properly, and (3) the spill has been controlled and confined to prevent runoff after application of the neutralizing agent. Responders should always contact product specialists before initiating neutralization operations for large spills and releases. Sufficient neutralizing agent should be on hand to complete the process once it is begun. Table 10-4 provides guidance on neutralizing spills.

When a decision has been made to neutralize a spill, consideration should be given to the type of neutralizing agent that will be used. Some materials are more environmentally friendly than others. The key concern is biodegradability. The most widely favored neutralizing agents from an environmental perspective are sodium sesquicarbonate (for acid spills) and acetic acid (for alkali spills). Sodium and calcium hydroxide will not produce a biodegradable end

Table 10-4 Guidelines for Neutralizing Spills

To determine the amount of base necessary for an acid spill, consider the following example: A glass bottle containing 1 gallon of 70% nitric acid falls and breaks open. Responders have an ample supply of neutralizing agent (soda ash) available. How much would be required to bring the pH somewhere close to 7 (neutral)?

1. Determine the weight of 1 gallon of the acid in pounds, using an SDS or the following formula to determine the information:

Quantity of Acid Spilled	×	Specific Gravity	×	8.33 lbs./gal. (weight of water)	×	Percent Acid	=	1 Gallon Weight in Pounds
Example:								
(1 gallon)	×	(1.5)	×	(8.33)	×	(.70)	=	8.75 lbs.

2. Select the appropriate conversion factor:

Neutralizing Agent	Sulfuric Acid	Nitric Acid	Hydrochloric Acid	Phosphoric Acid
Sodium Carbonate (soda ash):	1.082	0.841	1.452	1.622
Calcium Hydroxide (slaked lime)	0.755	0.587	1.014	1.133
Sodium Bicarbonate (baking soda)	1.673	1.302	2.247	2.541

3. Multiply the weight of the spilled acid times the conversion factor for the appropriate neutralizing agent to determine the amount needed to neutralize the acid spill.

   8.75 lbs./gallon × 0.841 = 7.4lbs. of soda ash would be required

Responders can develop a chart by which they can estimate the amount of neutralizing agent available for specific size spills. Using 70% nitric acid as an example:

1-gallon spill = 7.4lbs. of soda ash

10-gallon spill = 74.0lbs. of soda ash

100-gallon spill = 740lbs. of soda ash

1,000-gallon spill = 7,400lbs. of soda ash

product. If the spill is in an environmentally sensitive area, a product specialist should be consulted for advice on which neutralizing agent to use.

Corrosive spills should be neutralized by applying the neutralizing material from the outermost edge inward, thereby protecting responders. Avoid walking through spills, even when wearing proper protective clothing.

Some caustic spills have been neutralized using various types of diluted acids, but this technique should never be attempted without seeking the advice of a product specialist.

Neutralizing agents should be purchased in bulk quantities and stored at key locations. Commercial neutralizing kits are also available for handling small spills; these are normally packaged for smaller laboratory or workshop-type spills of one to five gallons.

Overpacking is a physical method of containment by which a leaking drum, container or cylinder is placed inside a larger undamaged and compatible overpack container. Although commonly used for liquid containers, overpacking can also be used for some compressed gas cylinders like chlorine.

Liquid overpacks are constructed of both steel and polyethylene. They range in size from 5 gallon to 110 gallon capacities (19L to 416L). Lab Packs range in size from 20 gallon to 110 gallon capacity (75L to 416L). When possible, the leak should be temporarily repaired before the container is placed inside of the overpack. Methods of placing a leaking drum into an overpack are as follows:

- *The inverted method* Figure 10-13. Slip the compatible overpack container over the leaking drum. However, if the leaking drum is upright, all bungs and lid openings will now be on the bottom after the overpack operation, thereby making pump-off operations more difficult. One tactic to minimize this problem is to first overturn the leaking drum onto the lid of the overpack drum and then place the overpack drum over the leaking drum, tighten down the retention ring, and invert the overpack drum.

- *The slide-in method can be used if the leaking drum is in a horizontal position* **Figure 10-14**. Place the open end of a horizontal overpack container near the leaking drum, then raise the end of the leaking drum while your entry partner slides the overpack around the leaking drum.

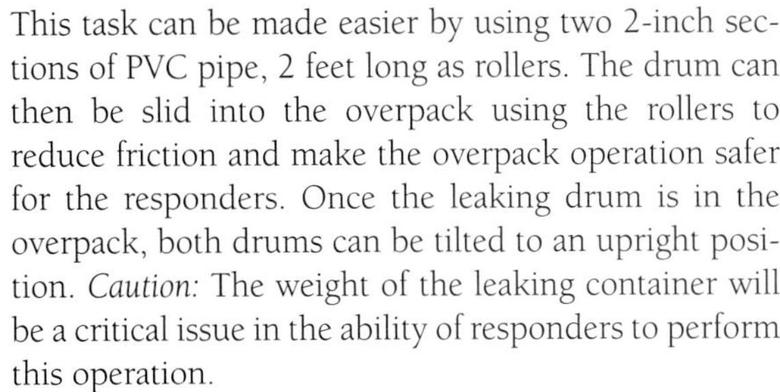

This task can be made easier by using two 2-inch sections of PVC pipe, 2 feet long as rollers. The drum can then be slid into the overpack using the rollers to reduce friction and make the overpack operation safer for the responders. Once the leaking drum is in the overpack, both drums can be tilted to an upright position. *Caution:* The weight of the leaking container will be a critical issue in the ability of responders to perform this operation.

- *The rolling slide-in or V-roll method is a variation of the slide-in method* that can be used if space permits and if the leaking drum is in a horizontal position **Figure 10-15**. Orient the drums so that they form a wide letter V, with the open end of the overpack drum placed under the rim of the leaking drum. Then push the drums from the apex of the "V" so that the rolling motion causes the leaker to roll into the overpack drum. Finally, tilt both drums to an upright position.

Responder Tip

Responders must follow maintenance and inspection procedures for hazmat control tools and equipment to ensure they are operated safely and within their design parameters. Recommendations on maintenance requirements and schedule, inspection and cleaning are provided by the manufacturer, and should be available to the Hazmat Technician.

Figure 10-13 Inverted method of overpacking.

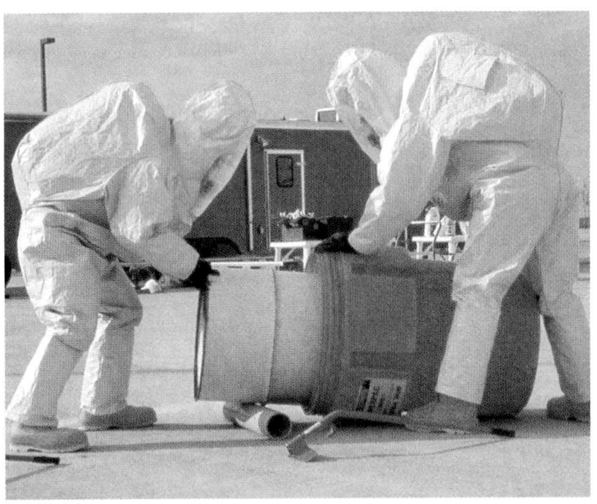

Figure 10-14 Slide-in method of overpacking.

Figure 10-15 V-roll method of overpacking.

Containing Leaks in Drums

Hazmat technicians may be dispatched to a variety of incidents that involve chemical and mechanical stressors that may affect metal, wood, and plastics drums. The mechanical stressors may include events that involve strikes by machinery such as forklifts, pallet jacks, and hand trucks, cold work such as containers rubbing against each other in transit, and drops and strikes on the loading dock. The chemical stressors may include reactions within the container from incompatibles or from temperature and pressurization.

In each of these instances, the "juice" can start leaking from the container. These leaks can come from various locations within the drum. Given a drum leaking a hazardous material, you should be able to develop a plan of action, select the appropriate tools, and control the leak.

To contain various types of leaks, follow the steps in Skill Drill 10-2:

A. Bung Leaks

1. Stand drum or roll drum so that opening is at the top or the bung is in the vapor space. **(photos 1 and 2a and b)**
2. Take standard bung wrench, find appropriate fitting on wrench and tighten bung until leak has stopped or is reduced. **(photos 3 and 4)**
3. Replace the bung with a new, compatible bung plug or reuse the existing. **(photo 5)**

B. Punctures

1. Roll the drum to the point that the hole is higher than the liquid level and make sure to chock the drum to keep it from rolling back to its original position. **(photo 1)**
2. Use manufactured mechanical plug such as T-Bolt with compatible patch that fits size of leak. **(photos 2a and b and 3)**
3. Use a compatible chemical patch that will bond with the container. **(photo 4)**

C. Chime Leak

1. Stand drum or roll drum so that chime is at the top and is in the vapor space.
2. Use a compatible chemical patch that will bond with the container. **(photo 1)**

Depending on container size and weight, mechanical equipment (e.g., forklift or hoist) may be required to raise and lower the leaking container into the overpack. A 55-gallon drum of sulfuric acid can weigh over 600 pounds. In addition, the container may have been weakened as a result of the leak. The overpack container must then be labeled in accordance with DOT hazmat regulations if it will be transported from the scene.

Salvage cylinders are overpack devices used for transporting leaking, damaged, or deteriorated compressed gas cylinders. These cylinders are typically manufactured from carbon steel to meet ASME Lethal Service Standards. Gases inside the cylinder can be withdrawn via a valve and directed to a scrubber or repackaged. Cylinder overpacks do have several drawbacks, including the fact that they aren't readily available, the mobilization process can be very time consuming, and they are extremely expensive to manufacture and purchase Figure 10-16.

Figure 10-16 Plugging tactic.

Patching (plugging) is a physical method of containment that uses chemically compatible patches and plugs to reduce or temporarily stop the flow of materials from small container holes, rips, tears, or gashes. Although commonly used on atmospheric pressure liquid and solid containers, some tactics can also be used on pressurized containers. Figure 10-17.

Patching involves placing a material or device over a breach to keep the hazmat inside of the container. Patches can include both commercial and homemade devices and are used to repair leaks on container shells, piping systems, and valves. Patches must be compatible with the chemicals involved.

Like plugs, patches can also be fabricated on the scene, but you can save a great deal of time by manufacturing a variety of devices before the fact and carrying them on response vehicles. Responders are only limited by their ingenuity and imagination. Common examples include toggle bolt compression patches, gasket patches, glued patches, and epoxy putties.

Container pressure is a critical factor in evaluating the application and use of patching tactics. Leak bandages and

Figure 10-17 Patching and plugging tactics increase risk because they require hands on work in the hot zone. Containers and tanks should be preplanned to determine the types of tools required to perform the tasks.

Skill Drill 10-2

Containing Leaks in Drums

A. Bung Leak

1 Stand drum or roll drum so that opening is at the top or the bung is in the vapor space.

2 Take standard bung wrench, find appropriate fitting on wrench and tighten bung until leak has stopped or is reduced.

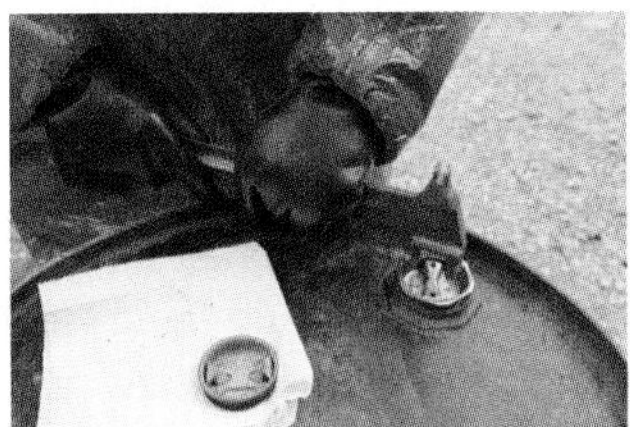

3 Replace the bung with a new, compatible bung plug or re-use the existing plug.

B. Punctures

1 Roll the drum to the point that the hole is higher than the liquid level and make sure to chock the drum to keep it from rolling back to its original position.

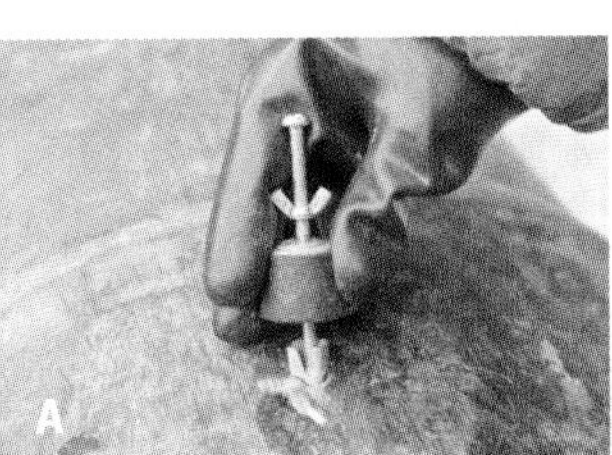

2 Use manufactured mechanical plug such as T-Bolt with compatible patch that fits size of leak.

3 Chemical patches may be used to stop small leaks in drums, or in conjunction with mechanical plugs.

C. Chime Leak

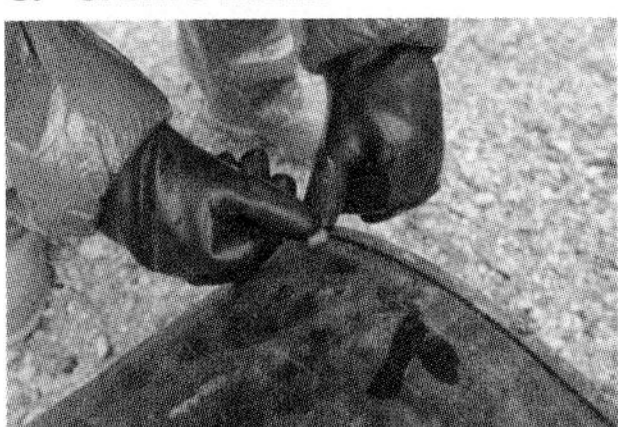

1 Stand drum or roll drum so that chime is at the top and is in the vapor space. Use a compatible chemical patch that will bond with the container.

Note:

- When using mechanical patching devices, it is important to remember that the drum will more than likely be overpacked and if the plug is too long it may impede the overpack drum from being used.
- When using chemical patches, check for compatibility with the container as well as with the product. For example, chemical patches do not work well with plastic drums because of a compatibility issue as well as the type of damage usually associated with plastic drums.

leak sealing kits are effective tools for dealing with liquids with a low head pressure or low-pressure gases (<100 psi or <7 bar). Some inflatable air bag patch kits, like the Vetter Bag™, are effective at pressures less than 25 psi (2 bar). The Chlorine A and B Kits have a side patch kit for container shell leaks on 100- and 150-pound and 1-ton chlorine cylinders.

To properly organize a patching operation, consider the following:

- Select a patching device at least half a size larger than the breach. Smaller devices using nuts, toggle bolts, T-bolts, and so on, can be drawn inside the container as these closure devices are tightened.
- Ensure the patch is compatible with the hazmat involved. This is especially true when dealing with corrosives.
- Patching operations should be planned with air supply operating times in mind. Several entries may be required. Overlap entry crews so that one crew is always working, one crew is always ready to step-in (i.e., backup team), and a third is ready to step into the operations flow.
- If the patching operation is complex and time allows, consider having the entry team walk-through the patching operation in the cold zone. Make sure you have the proper tools and equipment and that all personnel are thoroughly briefed on the entry and patching operation.

When dealing with liquid containers, recognize that plastics and metals behave differently when they are breached and that the methods of repair will vary accordingly.

Plugging involves putting something into a breach or opening to reduce both the size of the hole and the amount of product flow. The plug can be mechanical or chemical and must be compatible with both the chemical and the container. For example, a small hole in an aluminum MC-306/DOT-406 cargo tank truck can sometimes be plugged by driving a wooden wedge into the opening with a rubber mallet. However, a soft pine plug would not be compatible with a strong acid leak.

Plugs can be fabricated on the scene, but you can save a great deal of time by manufacturing a variety of devices before the fact and carrying them on response vehicles. Plugs can be constructed of various materials, including wood, rubber, and metal. Plugs constructed of soft woods, such as yellow pine or Douglas fir, are quite effective for holes whose area is less than 3 square inches. Pneumatic plugs are also available. Preplan potential container leaks and network with other responders to determine what works Figure 10-18.

Plugging techniques are usually used in conjunction with synthetic rubber gaskets, lightweight cloth, or special chemical-resistant putty to ensure a good seal by filling the cracks around the plug. Small holes (less than ½-inch diameter) not under pressure can be filled with putty or epoxy resin compounds. The longevity of these compounds is limited due to material compatibility, the size of the breach, and the head pressure of the container. These should be viewed as only temporary first-aid techniques.

Pressure isolation and reduction is a physical or chemical method of containment by which the internal pressure of a closed container is reduced. The tactical objective is to reduce sufficiently the internal pressure in order to either reduce the flow or minimize the potential of a container failure. Pressure reduction tactics are high-risk operations that require responders to work in close proximity to the container. Examples of tactics include flaring and vent and burn.

Figure 10-18 Isolating valve tactic.

Many hazmat containers are designed to store their contents under pressure. Cylinders, process vessels, MC-331 cargo tank trucks, rail cars, and pipelines are examples. It is also possible for nonpressurized containers to become pressurized because of internal chemical reactions, thermal stress, or accidentally diverted pressure.

Pressurized containers are dangerous because:

- They can rupture under stress and travel great distances as fragments or in one piece. This happens very quickly and allows no reaction time.
- High pressure kills quickly. You cannot usually determine the operating pressure of a given container without close inspection. High pressure can propel valve caps, breach protective clothing, or sever SCBA air lines. Ultra high pressures (5000 to 15,000 psi) can penetrate the skin and cause an air embolism, which will be followed by death within minutes.

Leak Control on Liquid Cargo Tank Trucks

When evaluating leak-control options on liquid cargo tank trucks, remember some basic principles:

- Damage to an insulated or double-shell cargo tank truck may only be to the outer shell and may be difficult to assess.
- Small holes or cracks on lined cargo tanks (e.g., MC-312/DOT-412) may get larger when a plug is inserted, as the internal container shell may be weakened over a larger surface area. In addition, inserting a plug into a small crack on some liquid cargo tanks may actually cause the crack to "run" and increase the problem.
- Valves may not always operate properly when in a different orientation from normal operations. This may also be the cause of a valve leak.
- Some liquid products are shipped under pressure of an inert gas (e.g., nitrogen) to prevent the material from reacting with air or moisture. If the container is still under pressure, the rate of release can often be reduced by simply relieving the pressure, which will create a vacuum. However, do not leave the valve open after the pressure is relieved.
- Always make sure the leak-control materials and equipment are compatible with both the chemical and the container. This is especially critical when dealing with corrosives.
- If you are unsure of anything, always consult product or container specialists.

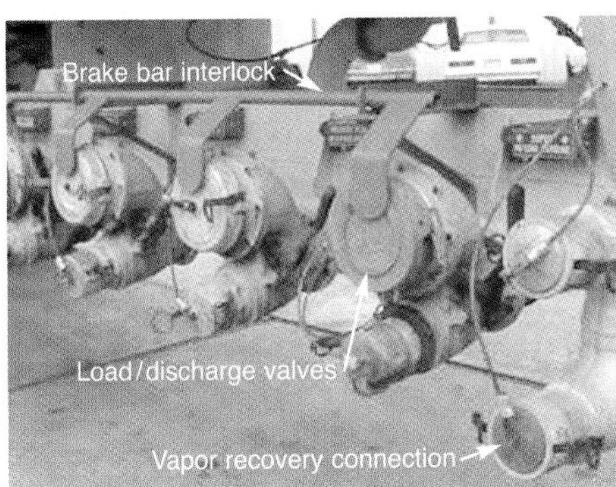

Fire hazards must be secured before leak control tactics are attempted.

Types of Leaks

Piping and Valve Leaks

Piping and valve leaks are among the most common leaks found on cargo tank trucks and may be controlled by either closing valves or tightening the valve packing.

If the piping is leaking, try to isolate any valves both upstream and downstream of the leak. While MC-306/DOT-406 and MC-307/DOT-407 are normally under gravity or atmospheric pressure, MC-312/DOT-412 corrosive cargo tank trucks will often be under pressure.

Splits, Tears, Punctures, and Irregularly Shaped Holes

Most breaches on bulk cargo tank trucks can be controlled in the same manner as those on smaller, nonbulk containers. The big differences will be the size of the container, the quantities of product involved, and the potential for higher pressures. In addition, double-shell or insulated cargo tank trucks can have the product flow into the shell interface, further weakening the structural integrity of the container.

If the breach is in the vapor space, containment is often very easy. Providing the materials are compatible with the product, a patch, plug, or bandage can be applied with minimal product contact. If the breach is below the liquid level, significant product release will likely occur during leak-control operations. In addition, head pressure and the flow of product will make leak-control operations extremely difficult to perform as well as contaminate the entry team.

Pressure-Relief Devices, Vents, and Rupture Disks

Pressure-relief devices include spring-loaded Pressure-relief valves, "Christmas Tree" vents, and rupture disks. All are piped into the vapor space and are designed to allow the release of vapors under normal operating conditions. In addition, they are engineered to minimize a liquid product leak if the cargo tank truck is involved in a rollover. Although a pressure-relief valve may quickly open and lose up to 1 liter (.26 quarts) of liquid product, it should immediately reseat.

If liquid is being released from a pressure-relief device, it has probably failed. Responders should consult with container specialists to assess the level of risk and control options. Remember that blocking off a pressure-relief device may result in additional container stress depending on the scenario, especially in a fire event.

Dome Cover Leaks

Dome covers may become loose and begin to leak as a result of a rollover. However, the stability of the container and the pressure of the product flow may increase the risks of leak-control operations.

For MC-306/DOT-406 cargo tank trucks, dome clamps can typically be used to control the leak. For MC-307/DOT-407 and MC-312/DOT-412 cargo tank trucks, the leak can sometimes be controlled by tightening down the wingnuts found around the manway.

- Cryogenic liquids are stored in pressurized containers at temperatures below −130°F (−90°C). Cryogenics can freeze tissue and damage protective clothing.

Examples of pressure reduction tactics include the following:

- **Isolating valves**—When dealing with valves you should expect the unexpected. Dealing with a "simple valve leak" can turn into a campaign operation while waiting for the right fitting, tool, or specialist to resolve the problem. According to the *Handbook of Compressed Gases,* nearly 200 types of compressed gases are commonly shipped annually. More than 12 different cylinder specifications exist with more than 64 different valve outlets used for these 200 gases.

Pressurized containers often leak in and around valves and fittings. If the leak continues after the valve is closed, try tightening the packing nut on the valve. Most valves can be closed by turning the valve wheel clockwise, unless it is damaged. There are exceptions to the "right-to-tight" rule. Make sure you know what type of cylinder you are handling **Figure 10-18**.

In-service containers may also have piping leaks. These leaks usually stop when the supply valve is isolated and blocked in. Depending upon the situation, it may be necessary to isolate the valve both upstream and downstream.

- **Isolating pumps and pressure/energy sources**—Some containers are pressurized through a separate compressor or pumping system to move and transfer products. In these scenarios, the magnitude of a leak can be significantly reduced by simply lowering the pressure or completely isolating or shutting down the compressor or pumping system. Product and container specialists familiar with the system must be consulted before any shutdown operations are initiated, as a shutdown may over-pressurize other vessels or create related upstream and downstream problems. Lack of pressure may also produce unstable chemical reactions.

  Product specialists and process engineers should also be contacted whenever responders face large, complex process units and related pressure vessels. Shutting down electrical power to pressure systems may ruin chemical batch processing equipment, cause dangerous pressure buildups in other locations, and disable critical safety devices. In some cases, an Emergency Shutdown (ESD) can be executed from a single switch by an operator. ESDs may also activate safety systems, stop ventilation systems, and dump hazardous materials to neutralizing scrubbers and flares. The IC must work closely with plant operators to execute an ESD function.
- **Venting** is the controlled release of a material to reduce and control the pressure, and decrease the probability of a violent container rupture. The method of venting will depend upon the nature of the hazardous material and process involved. For example, nontoxic materials may be vented directly to air (e.g., steam). Venting is typically limited to nonflammable gases.
- **Scrubbing** is usually associated with venting and involves the use of physical or chemical filters or control devices to "scrub" contaminants, particulates or gases during normal or emergency process operations. Emergency scrubbers are found in many chemical plant process operations where a system failure or runaway reaction could cause a discharge of toxic products into the atmosphere causing injuries or fatalities. When activated, an emergency scrubber is normally capable of filtering or neutralizing 100% of the accidental discharge. Scrubbers may also be used in a field-setting as part of product transfer or remediation operations (e.g., chlorine).
- **Flaring** is the controlled burning of a liquid or gas material to reduce or control pressure inside the tank, dispose of residual vapors, and/or to dispose of a product when transfer operations may be impractical. Flares are designed to burn either liquid or vapor product. Flaring is commonly used in the propane industry to safely burn off product in cylinders, trucks, and rail cars when they have been severely damaged or access to the accident site is impractical for offloading to another vehicle. When dealing with bulk containers, liquid flares are often preferred due to the exceptionally long burn-off times of a vapor flare. See Figure 10-19.

  One drawback of flaring is the time required to burn off the product. For example, using a 2-inch-diameter hose of 150 feet in length, it would take approximately 177 hours to flare off 30,000 gallons of propane, 54 hours to burn off 11,500 gallons, and 14 hours to burn off 3000 gallons (assuming the temperature of the propane was 0°F). If the IC decides to go with a flaring option, the time required to accomplish the task must be weighed against other factors such as safety, disruption of transportation systems and businesses, and the safety and speed that other options may present.

  Another factor affecting flaring operations involving flammable liquefied gases is auto-refrigeration. Essentially, as the product is burned off the liquid begins to boil inside the container. Boiling requires energy and this energy is obtained through heat from the air or ground surrounding the tank being flared. When the capacity to make up the loss of heat from the surrounding area is exceeded, the temperature of the product drops as energy is drawn from it. The product actually cools itself, boiling slows down, and the product eventually goes into a state of auto-refrigeration. Of course, this decrease in product temperature also causes a decrease in vapor pressure and a subsequent decrease in the size and intensity of the flaring operation. Auto-refrigeration is discussed in more detail in this chapter under Liquefied Natural Gas (LNG).
- **Hot tapping** is used to gain access to bulk liquid or gas tanks, pipelines, or containers for the purpose of product removal. It involves the welding of a threaded nozzle to the exterior of a tank or pipeline. A valve is

Figure 10-19 Flaring operations involving an overturned and stressed MC-331 cargo tank truck transporting propane.

then attached to the threaded nozzle and a hole is drilled through the container shell with a specially designed machine. The drilling machine is equipped with seals that prevent the loss of product during the drilling operation. Hoses are then attached to the valve outlet and the contents are removed.

Hot tapping is commonly used within the petroleum and petrochemical industries. However, it should only be attempted during an emergency by trained hot tapping specialists. For more information consult American Petroleum Institute Publication 2201, "*Safe Hot-Tapping Practices in the Petroleum and Petrochemical Industries,*" 5th edition (June 2003.)

- **Vent and burn** is a process by which shaped explosive charges are placed by explosives demolition specialists on a flammable container, such as a pressurized tank car, to cut a hole (or holes) in it, thereby allowing the contents to flow into a pit constructed adjacent to the breached car where the product can safely burn off. See Figure 10-20. When dealing with pressurized tank cars, two holes are normally required: Hole #1 is at the high end of the tank car and is designed to relieve the internal pressure so that the tank car will not "rocket," while hole #2 is at a low point and is designed to allow the liquid contents to drain into a containment area. This is a highly specialized technique that should only be attempted by trained explosive specialists under very specific situations.

  Vent and burn is normally an option of last resort and may be used under the following conditions:
  - The tank car has been exposed to fire, resulting in elevated internal pressures and possible tank damage.
  - Conditions do not allow for the safe transfer, venting, or flaring of the tank car.
  - Site conditions prevent rerailing the damaged tank car.
  - There has been damage to leaking valves, and fittings cannot be repaired.
  - The tank car has been damaged to the extent that it cannot be safely offloaded, rerailed, and moved to an unloading point.

  Vent and burn tactics have been used safely and successfully by the railroad industry at a number of major train derailments. In September, 1982, vent and burn was successfully used by explosives specialist Billy Poe on eight tank cars—six vinyl chloride monomer, one styrene, and one toluene diisocyanate—in Livingston, Louisiana. Since that time vent and burn has evolved into an accepted practice, however, it is usually viewed as a last resort to resolving the problem.
- **Solidification** is a chemical method of containment whereby a liquid substance is chemically treated so that a solid material results. The primary advantage of this process is that a small spill can be contained relatively quickly and immediately treated.

  Solidification is often used for both corrosive and hydrocarbon spills. Commercial formulations are available that can be applied to a liquid acid or caustic spill, neutralize the hazard, and form a neutral salt. Commercially available adsorbents can also be used to solidify nonsoluble oily wastes. The spilled hydrocarbon is adsorbed to granules to form a solid, nonflowing mixture. This resulting mixture is actually safer than the original spilled material and can be easily transported and disposed of at a waste treatment facility.
- **Vacuuming** is a physical method of containment by which a hazardous material is placed in a chemically compatible container by simply vacuuming it up. The method of vacuuming will depend upon the hazmats involved. Vacuuming is commonly used for containing releases of certain hydrocarbon liquids, solid particulates, asbestos fibers, and liquid mercury. Vacuuming is usually performed using a High-Efficiency Particulate (HEPA) vacuum. To meet U.S. Department of Energy standards, a HEPA air filter must remove 99.7% of all particles greater than 0.3 microns from the air that passes through Figure 10-21.

  The primary advantage of vacuuming is that there is no increase in the volume of waste materials. In selecting this tactic, care must be taken to ensure

Figure 10-20 Vent and burn tactics are sometimes used when offloading or moving the container presents a high risk.

Figure 10-21 Vacuuming tactics require the use of HEPA filters.

that the vacuum and related equipment is compatible with the hazmats involved and that vacuum exhaust vapors are controlled. The use of vacuum trucks is discussed later in this chapter in relation to product transfer methods.

One last thought about containment options is that if two heads are better than one to come up with a solution, then five heads must be even better! Don't disregard ideas on how to solve a problem until you've seriously considered them. What sounds like a dumb idea in hour one of the response might be looking like a brilliant idea in hour four when nothing else has worked. As they say in the Army, "If it's stupid and works, then it ain't stupid!"

Controlling a Leak in a Hazardous Materials Cylinder

Hazmat technicians may be dispatched to a variety of leaks caused by thermal, chemical, and/or mechanical stressors that affect pressurized cylinders. Thermal stressors may include flame or heat impingement on the cylinder, mechanical stressors may include incidents that involve strikes that damage or sever the valve(s), cold work such as cylinders rubbing against each other in transit or in a system, and drops and strikes during movement of the cylinder. Chemical stressors may include reactions within the container from incompatibles.

Given a hazardous materials container with a leaking valve, plug, or attachment, select the appropriate tools and control the leak by following the steps in **Skill Drill 10-3**:

1. Fusible plug: Apply a clamping device that will hold a compatible material to stop the leak valve assembly. **(photo 1)**
2. Fusible plug threads: Attempt to tighten plug with available tools. If this fails, cut plug off flush with valve body and use a clamp with compatible material over leak. **(photo 2)**
3. Valve stem or assembly blow-out: Drive a compatible drift pin material into opening and secure with load strap. **(photo 3)**
4. Valve seat: Re-open and re-close the valve handle allowing the possible obstruction to clear the seat. This can be done either once hooked back into a process system or in the field to a scrubber depending on the product involved. If this fails, place threaded outlet cap with gasket and tighten. **(photo 4)**
5. Valve packing: Tighten the valve stem. Locate the packing nut on the valve and tighten the nut. **(photo 5)**
6. Valve inlet threads: Using an appropriate size wrench, tighten the entire valve assembly into the cylinder body with slow, constant pressure. **(photo 6)**
7. Cylinder side-wall damage: Before attempting, shift leak into an upright position to maintain in the gas phase of the product and sound the side walls of the container to detect additional container damage. Use a sidewall patch kit with a compatible patch to secure the leak. **(photos 7 a–c)**

In each of these instances, the "juice" can start leaking from the container. These leaks can come from various locations within the cylinder. Please consult with appropriate container and industry technical specialist with regard to various valve configurations.

Fire-Control Operations

There is a wide range of hazardous materials that are flammable in nature. While specific tactics to deal with particular materials are beyond the scope of this text, in this section we will focus on three categories of hazmat that emergency responders may encounter: flammable liquids, flammable gases (including LNG), and reactive chemicals. These hazmat classes have historically produced the most injuries and fatalities to emergency responders and are encountered in virtually every plant and community. The emphasis in this section will be placed on hazards, risks, and tactical response considerations of each.

In the case of LNG, its safety history has been very good, but its use as a fuel source is increasing, and it presents unique hazards and risks that need to better understood by emergency responders.

The following general factors can be applied uniformly to hazmat fire problems. They should be considered early in the incident as part of the hazard and risk evaluation process.

1. **What hazardous material(s) are involved?** Specifically, are we dealing with flammable liquids, gases, a reactive material, or some combination of flammability and other hazards?
2. **What are its hazards?** The physical and chemical properties of a material significantly influence the selection of tactics and fire extinguishing agents. What works well on one type of fire won't necessarily work on another. Critical questions include determining the material's (1) chemical family (i.e., hydrocarbon or polar solvent), (2) water solubility, (3) specific gravity, (4) water reactivity, and (5) control and extinguishing agents. What type of container is involved? This can include storage tanks, pressure vessels, reactors, and pipelines. Responders must also evaluate container features, such as pressure-relief devices, valves, and fixed foam systems. Many reactive materials are shipped in specialized containers and may have unusual features.
3. **What are the risks to responders, employees, and the community?** What is the likelihood of the incident growing and involving other containers? Is there a potential for significant environmental impact?
4. **Are specialized resources required?** What is their availability? This could include personnel, supplies, equipment, extinguishing agents, and related appliances.
5. **What will happen if I do nothing?** Remember—this is the baseline for hazmat decision making and should be the option against which all strategies and tactics are compared.

Skill Drill 10-3

Controlling a Leak in a Hazardous Materials Cylinder

1 Fusible plug: Apply a clamping device that will hold a compatible material to stop the leak valve assembly.

2 Fusible plug threads: Attempt to tighten plug with available tools. If this fails, cut plug off flush with valve body and use a clamp with compatible material over leak.

3 Valve stem or assembly blowout: Drive a compatible drift pin material into opening and secure with load strap.

4 Valve seat: Re-open and reclose the valve handle, allowing the possible obstruction to clear the seat. If this fails, place the threaded outlet cap with gasket and tighten.

5 Valve packing: Tighten the valve stem. Locate the packing nut on the valve and tighten the nut.

6 Valve inlet threads: Using an appropriate size wrench, tighten the entire valve assembly into cylinder body with slow, constant pressure.

A

B

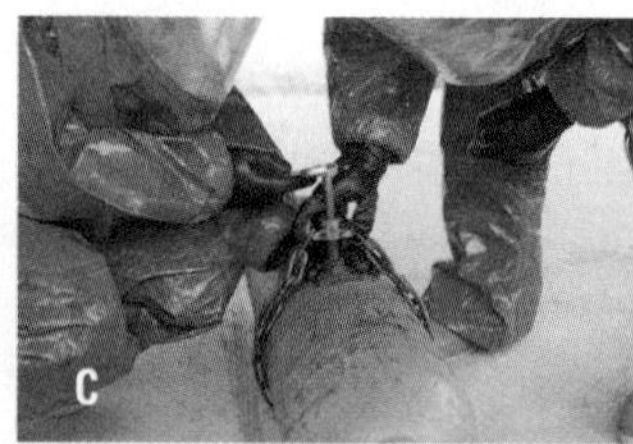

C

7 Cylinder side-wall damage: Before attempting, shift leak into an upright position to maintain in the gas phase of the product and sound the side walls of the container to detect additional container damage. Secure the leak using a sidewall patch kit with a compatible patch.

Flammable Liquid Emergencies

Flammable liquids are the most common hazard class encountered by emergency responders. Most flammable liquid emergencies are relatively small, involve nonbulk containers, and are successfully and safely handled by first responders using Class-B foams **Figure 10-22**.

Dealing with large flammable liquids incidents is another story entirely. The fact is that over 70 fire fighters have been killed in the line of duty in the last 50 years at incidents involving flammable liquid storage tank fires. No other class of hazardous materials or type of container has killed more fire fighters—so pay attention!

When we take an objective look at the history of fighting large flammable liquid fires the cost is very high. For example, according to a 2009 study of losses from 1972 to 2009 conducted by MARSH (a major insurer of petroleum and chemical facilities), some of the largest losses in the history of the hydrocarbon and chemical industries involved petroleum storage tanks. See Scan Sheet 10-C.

Incidents involving bulk flammable liquids pose considerable risks to both fire fighters and emergency responders.

Summary of the Three Largest Petroleum Storage Tank Losses in the United States

Note: All loss figures are shown in their original value. The figure inside the parentheses has been adjusted to 2013 dollars.

November 25, 1990
Denver, Colorado
$32,000,000 ($57,261,392) Loss
A fire at a 16-acre tank farm that supplied jet fuel to the Denver International Airport burned for more than 55 hours, damaging seven storage tanks and consuming more than 1.6 million gallons of jet fuel. The tank farm was comprised of 12 storage tanks.

At approximately 9:20 a.m., the fuel supply company received a "no flow" indication in the pipeline to the tank farm. Shortly after that time, the airport control tower noticed a black column of smoke from the tank farm. An initial fuel leak originating at an operating fuel pump in the valve pit was ignited by the electric motor for the pump resulting in the fire. A cracked fuel supply pipe in the valve pit formed two V shaped streams extending 25 to 30 feet into the air. This provided additional fuel to the pool fire.

As the fire continued to grow, coupling gaskets in the piping deteriorated and more fuel flowed out of the storage tanks, spreading the fire throughout the dike area. The valve controlling the fuel flow to the airport supply line sporadically released fuel into the valve pit. Fire fighters were unable to prevent the back flow of fuel from this line since the nearest manual shutoff valve was 2 miles from the tank farm.

The Denver Airport Fire Department dispatched four aircraft rescue firefighting (ARFF) vehicles and one rapid intervention vehicle to the fire. The second and third alarms provided an additional five engine companies, three truck companies, and one rescue unit from the Denver City Fire Department.

In addition to the foam concentrate on hand at the scene, foam concentrate was received from other local departments as well as from the Seattle, Houston, and Chicago Fire Departments. Unknown to the Denver Fire Department, a pipeline that was reported to have been shut down continued to supply fuel to the fire. After repeated unsuccessful attempts to extinguish the fire by the Denver Fire Department, Williams Fire and Hazard Control was brought in by the owners to assist the fire department with extinguishing the fire. The pipeline was eventually shut down and the fire extinguished.

July 7, 1983
Newark, New Jersey
$35,000,000 ($62,629,648) Loss
Gasoline was being received by pipeline into a 42,000-barrel internal floating roof tank at a products terminal when an overfill occurred, spilling about 1300 barrels into the tank diked area. A slight wind (1 to 5 mph) carried gasoline vapors about 1000 feet to a drum reconditioning plant, where an incinerator provided the ignition source. The resulting explosion caused $10 million damage to the terminal and up to $25 million in over 2000 claims to rail cars and adjacent properties. Although dikes contained the burning spill to the tank that was overfilled, two adjoining internal floating roof tanks and a smaller tank ignited and were eventually destroyed along with 120,000 barrels of product. Since the burning tanks presented little exposure to other facilities, the decision was made to let the fire burn itself out. This incident resulted in tank overfill prevention requirements in NFPA-30—*The Flammable and Combustible Liquids Code.*

September 24, 1977
Romeoville, Illinois
$8,000,000 ($14,315,348) Loss
Lightning struck a 190-foot diameter cone roof tank containing diesel fuel. Roof fragments were thrown 240 feet and struck a 100-foot-diameter covered floating roof gasoline tank. An adjacent 180-foot diameter floating roof gasoline tank 80 feet away was also struck by debris. The entire surfaces of the cone and internal floating roof tanks ignited immediately. The rim fire on the floating roof resulted in the roof sinking after about 4 hours. The two largest tanks were full; the smallest about half full. The two larger tanks and their contents were destroyed. The fire in the internal floating roof tank was extinguished after about 2 hours.

After burning for approximately 46 hours, the fire was extinguished through both top-side and subsurface foam applications. The refinery's five stationary fire pumps supplied up to 10,000 gallons per minute (gpm) of the estimated 14,000 gpm required during the fire. Thirty-five municipal and industrial fire departments, including a 12,000-gpm fire boat, assisted the refinery fire department. About 22,000 gallons of foam concentrate were consumed during the firefighting effort.

Source: Marsh, *The 100 Largest Losses* 1972–2001: Large Property Damage Losses in the Hydrocarbon-Chemical Industries, (20th edition).

HazMat Responder *in Action*

Case Study: The Buncefield Incident, December 11, 2005 at Hemel Hempstead, United Kingdom

Early on Sunday, December 11, 2005, a series of explosions and subsequent fire destroyed large parts of the Buncefield oil storage and transfer depot located at Hemel Hempstead, England.

The main explosion took place at 06:01 hours. It was massive—causing widespread damage to neighboring properties. This was followed by a large fire that eventually engulfed 23 large fuel storage tanks.

The incident injured 43 people and caused significant damage to commercial and residential properties near the Buncefield site. About 2000 people were evacuated from their homes. Sections of the M1 highway were closed. Some schools in Hertfordshire, Buckinghamshire, and Bedfordshire were closed for 2 days following the

explosion. The fire burned for 5 days, destroying most of the site and emitting a large plume of smoke into the atmosphere that dispersed over southern England and beyond.

The incident scenario started late on Saturday, December 10th when a delivery of unleaded gasoline started to arrive by pipeline at Tank 912. The safety systems in place to shut off the supply of gasoline to the tank to prevent overfilling failed to operate, resulting in gasoline flowing down the side of the tank and collecting at first inside the diked area. As the overfill continued, a vapor cloud formed by the mixture of gasoline and air flowed over the dike wall, dispersed, and flowed west offsite towards the Maylands Industrial Complex. A white mist was clearly observable in closed circuit TV footage produced during the investigation. The exact nature of the mist is not known but it is believed to have been a volatile fraction of the fuel (butane) or ice particles formed from the chilled, humid air as a consequence of the evaporation of the escaping fuel.

An estimated 300 tons (approximately 96,000 gallons) of gasoline escaped from the tank, about 10% of which turned to vapor that mixed with the cold air and eventually reached flammable concentrations. The main explosion at Buncefield was unusual because it generated much higher overpressures than would usually have been expected from a vapor cloud explosion.

The exceptional scale of the incident was matched by the scale of the emergency response. Gold Command (the equivalent of a NIMS Type-I Incident Management Team) was established within hours and coordinated by Hertfordshire Police. Unified Command included the Hertfordshire Fire and Rescue Service, Hertfordshire County Council, Dacorum Borough Council, and the Environment Agency with Health and Safety Executive in support.

At the peak of the fire, at noon on Monday December 12th, 25 Hertfordshire fire apparatus were on site with 20 support vehicles and 180 fire fighters. The full-scale operation involved 1000 fire fighters from Hertfordshire and across the country, supported by police officers from throughout the UK. It took 32 hours to extinguish the main blaze, although some of the smaller tanks were still burning on the morning of Tuesday December 13th. The following day a new fire started in a previously undamaged tank, but the fire service let it burn out safely.

Firefighting efforts required 198,129 gallons (750,000 litres) of foam concentrate and 14,529,462 gallons (55 million litres) of water.

Sources: The following sources were referenced for this case study:

Major Incident Investigation Board, "*The Buncefield Incident 11 December 2005, Volume 1*," Health and Safety Executive: Kew, Richmond, Surrey, England (2008).

Major Incident Investigation Board, "*The Buncefield Incident 11 December 2005, Volume 2b: Recommendations on the Emergency Preparedness for, response to and Recovery from Incidents*," Health and Safety Executive: Kew, Richmond, Surrey, England (July 2007).

Figure 10-22 Class-B foam firefighting operations.

History is filled with case studies where responders have been seriously burned, injured, and killed because of their failure to understand how flammable liquids and their containers behave. In light of this fact, the following material will focus on the tactical problems associated with managing larger flammable liquid fires, such as dealing with bulk liquid storage facilities, transportation containers, and storage tank fires.

Hazard and Risk Evaluation

Large flammable liquids fires normally allow enough time to gather the necessary resources before mounting an aggressive, offensive-oriented fire attack.

Keeping track of times and key events at a major flammable liquid fire can be a difficult task. The IC should acquire the following information during the size-up process:

- Time when the incident started. This may not necessarily be the same time the incident was reported.
- Time at which responders arrived on scene.
- Probability that the fire will be confined to its present size.
- Fuel involved (flammable or combustible liquid), including the quantity, surface area involved, and the depth of the spill.
- Hazards involved, including flash point, reactivity, solubility (e.g., hydrocarbon or polar solvent), and specific gravity.
- Estimated pre-burn time. This will help the IC determine factors such as how "hot" the fuel is, identify and prioritize exposures, consider transfer and pump-off options, determine if a heat wave is developing for crude oils, and so on.
- Layout of the incident, including the following specific points:
  - Type of storage tank(s) involved. Common aboveground liquid storage tanks are cone roof, open floating roof, covered floating roof, and dome roof tanks.
  - Size of the dike area(s) involved.
  - Valves and piping systems stressed or destroyed by the fire.
  - All surrounding exposures, including tanks, buildings, process units, utilities, and so on. This should include identifying and prioritizing exposures (e.g., flame impingement, radiant heat exposure). Process unit personnel can help with these decisions.

Tactical Objectives

Hazard and risk evaluation is the cornerstone of decision making. Based on the type and nature of risks involved, the IC will implement the appropriate strategic goals and tactical objectives. Tactical options for flammable liquid emergencies include the following:

- **Nonintervention**—This is a "no win" situation in which responders assume a passive position (i.e., get out the beach chairs, umbrellas, and suntan lotion, and watch the fire burn itself out). This option is sometimes implemented when there are insufficient water supplies, very little product remaining which can be saved, or no exposures in the immediate area.
- **Defensive tactics**—These tactics involve protecting exposures and allowing the fire to burn. In many cases, defensive tactics are a temporary measure until sufficient resources can be assembled to pursue an aggressive, offensive attack.

  The primary concerns during defensive operations are direct flame impingement and radiant heat exposures. Flame impingements must be cooled immediately, while radiant heat exposures should be handled as soon as possible. Exposures should be prioritized in the following manner:
  - **Primary exposures**—Pressure vessels, closed containers, piping systems, or critical support structures exposed to direct flame impingement. Failure of exposed vessels, tanks, and piping systems is likely unless cooling water is quickly applied. If a storage tank is involved, direct flame contact on the tank shell can cause the upper portion of the shell, as well as any associated foam systems, to lose their integrity and fold inward. Water streams should cool all surfaces above the liquid level. Remember, cooling water is a valuable resource; don't waste it.
  - **Secondary exposures**—Pressure vessels, closed containers, piping systems, or critical support structures exposed to radiant heat. Failure of structural components is possible if cooling water is not applied.

    Be careful of applying water onto open floating roofs—sinking the roof with water lines can be somewhat embarrassing as well as hazardous! Also, remember that radiant heat will pass through structures with clear glass and windows. In addition to applying exterior cooling lines, fire fighters should be sent inside to check for any fire extension.
  - **Tertiary Exposures**—Noncritical exposures without life safety concerns.
- **Offensive Tactics**—These tactics are implemented when sufficient water and firefighting foam supplies and related resources are available for a continuous,

Figure 10-23 Offensive tactics are being used here to support the rescue of an injured utility worker.

uninterrupted fire attack. Although the primary focus is on fire extinguishment, it may also be necessary to maintain exposure lines **Figure 10-23**.

Time becomes a critical factor for offensive operations. The IC must determine the duration of any fire attack. For example, NFPA 11—*Technical Standard on Low Expansion Foam and Combination Agents*, recommends an application time of 15 minutes for flammable liquid spill fires and 65 minutes for Class I flammable liquid storage tank fires. The IC should document the time foam operations start, the time at which the fire is controlled, and the time at which the fire is extinguished.

A final note about exposure protection: Flammable liquid spill fires confined to a diked area may accumulate a large quantity of water from both fire attack and exposure streams. If these flows and associated runoff are not closely monitored, dikes may overflow and carry the burning flammable liquid outside of the area. In addition, loss of electrical power to the facility may cause the sewer system pumps to lose power and create additional runoff problems and hazards. This is was a significant factor in the 1975 Gulf Oil refinery fire in Philadelphia, PA when eight fire fighters died. As a general guideline, water streams applied to exposures inside of a diked area (e.g., adjoining storage tank) should be temporarily shut down when they no longer produce steam at the point where water contacts the hot steel surface.

Responder Safety Tip

Avoid walking into a flammable liquid spill even of you have the proper protective clothing and breathing apparatus. Never walk through a flammable liquid spill that has accumulated in depth.

Extinguishing Agents

The availability of water and foam concentrate is a critical factor in evaluating the risks involved in a flammable liquid emergency. The IC must determine whether:

1. An adequate and uninterrupted water supply is available.
2. There is sufficient capability to deliver the required foam solution to control and extinguish the fire.

If an adequate water and foam supply is not available for both protecting exposures and controlling the fire, the IC should consider implementing defensive or nonintervention tactics until sufficient resources are available.

Class B firefighting foam is the workhorse of flammable liquid firefighting. While other agents, such as dry chemicals, are used to extinguish small fires or deliver the knockout punch for three-dimensional fires (e.g., a flange fire), Class B foam is still the "top gun" for large-scale flammable liquid problems. When dealing with three-dimensional fires, be sure the fuel source can be shut off when the fire is extinguished. Class B foam can then be used to secure the fuel surface area.

Selecting Foam Concentrates

Selection of a foam concentrate is an important part of a successful firefighting operation. There are several different types of Class B foam concentrates sold by a variety of reliable manufacturers at different concentrations and for different fire protection applications. The selection of a foam concentrate can be as much a business decision as a technical one.

Firefighting foam concentrates should be selected based on the type of fuel (e.g., hydrocarbon vs. polar solvent), and the type and nature of the hazard to be protected (e.g., spill scenario vs. storage tank). The most common foam concentrates used for flammable liquid storage tank fire protection are as follows:

- **Fluoroprotein foam**—Combination of protein-based foam derived from protein foam concentrates and fluorochemical surfactants. The addition of the fluorochemical surfactants produces a foam that flows easier than regular protein foam. Fluoroprotein foam can also be formulated to be alcohol resistant. Fluoroprotein foam is effective on most spills and tank fires. The fuel-shedding ability of fluoroprotein foam makes it well-suited for subsurface injection into bulk tanks in which hydrocarbons are burning. The foam is injected into the tank above the level of any water that may be present.

Key characteristics of fluoroprotein foam as it relates to tank firefighting include the following:

- Available in 3% and 6% concentrations
- Are oleophobic (they shed oil) and can be used for subsurface injection
- Compatible with simultaneous application of dry chemical extinguishing agents (e.g., Purple K)
- Form a strong blanket for long-term vapor suppression on unignited spills
- Must be delivered through air aspirating equipment
- Has a shelf life of approximately 10 years

- **Aqueous Film-Forming Foam (AFFF)**—Synthetic foam consisting of fluorochemical and hydrocarbon surfactants combined with high boiling point solvents and water. AFFF film formation is dependent on the difference in surface tension between the fuel and the

Scan Sheet 10-D

Gasoline and Ethanol Blends

Nearly all gasoline sold as a motor fuel in the United States contains up to 10% fuel grade ethanol in regular unleaded gasoline. This 10% blend is known as E-10, indicating that a total volume of the gasoline/ethanol blend mixture is 10% alcohol.

With the call by regulation for more blending of hydrocarbon fuels with ethanol, there will be an increase in the type of mixtures found. There is already E10, as mentioned above; however, the market has seen the introduction of E85, which is 15% gasoline and 85% ethanol, and now E15, which is 15% ethanol and 85% gasoline. These blends, as well as others, can be found throughout the United States and elsewhere. The Renewable Fuel Standards has increases that will bring more and more ethanol to market. The automobile industry in turn has introduced more flexible fuel vehicles that will demand more of the fuels at the pump.

The most common ethanol blend being transported today by rail tank car is E95, or denatured ethanol. It may be delivered directly to a petroleum terminal if the facility has the ability to accept rail traffic. If not, it will go to a rail yard or siding that is designed to transload the product from the railcar to a highway cargo vehicle. From this facility, it will then travel to a petroleum terminal where it will be offloaded into separate bulk storage tanks until it is mixed with gasoline, at the appropriate percentage, at the cargo tank truck loading rack. Denatured ethanol may also be transported by barge or ship as well as through pipeline, though pipeline shipments are very limited at this time because of the corrosivity of the ethanol.

Emergency responders are generally familiar with the characteristics of gasoline, its hazards, and how to extinguish flammable liquid fires. There are some similarities between gasoline and ethanol, as both are organic solvents and are highly volatile. The primary differences between the two fuels are that ethanol is a polar solvent (i.e., completely soluble in water) and gasoline is a hydrocarbon (i.e., low solubility in water). As a result, the selection and use of Class B firefighting foams is critical, as regular Class B firefighting foams are not be effective on polar solvents. Both products are toxic by inhalation and present the risk of fire in open unconfined areas. Ethanol has a slightly wider flammable range than gasoline. The chart below provides a comparison of the two fuels.

Hazards of Ethanol vs. Gasoline

Ethanol	IDLH = 3300 ppm	LFL = 3.3%	UFL = 19.0%
Gasoline	IDLH = 500 ppm	LFL = 1.4%	UFL = 7.4%

Based on a series of performance fire tests conducted by the American Petroleum Institute (API) and the Ethanol Emergency Response Coalition (EERC), tactical recommendations for dealing with gasoline–ethanol mixtures can be summarized as follows:

(Continued)

- Fires involving gasoline–ethanol mixtures up to 10% will behave like a hydrocarbon fire, while those from 10 to 15% will assume more of the burning properties of a polar solvent. Regular AFFF and AR-AFFF/ARC will be effective for these mixtures, although an increased application rate will be required when using regular AFFF, especially for those scenarios requiring prolonged burnback resistance.
- Regular Class B firefighting foams (e.g., AFFF, flouro-protein foam) will not be effective for either ethanol or for gasoline–ethanol mixtures greater than 10% ethanol.
- Overall, alcohol-resistant AFFF (AR-AFFF or ARC) is the best agent for dealing with both ethanol and gasoline–ethanol blends.

firefighting foam. The fluorochemical surfactants reduce the surface tension of water to a degree less than the surface tension of the hydrocarbon so that a thin aqueous film can spread across the fuel. AFFF is an effective agent in rapidly knocking down and extinguishing a spill fire. It is also effective in suppressing vapors and reducing the possibility of reignition Figure 10-24.

Key characteristics of AFFF foam as it relates to tank firefighting include the following:

- Available in 1%, 3%, and 6% concentrations for use with either fresh or salt water
- Very effective on spill fires with a good knockdown capability
- Compatible with simultaneous application of dry chemical extinguishing agents (e.g., Purple K)
- Suitable for subsurface injection

Figure 10-24 AFFF foam being applied to a gasoline spill to prevent ignition.

- **Alcohol-Resistant AFFF or (AR-AFFF or ARC)**—Alcohol-resistant film forming foams or ARCs are available at 3% hydrocarbon/3% polar solvent (known as 3 × 3 concentrates), although 3% hydrocarbon/6% polar solvent concentrations (known as 3 × 6 concentrates) may also be found. When applied to a polar solvent fuel, they will often create a polymeric membrane rather than a film over the fuel. This membrane separates the water in the foam blanket from the attack of the polar solvent. Then the blanket acts in much the same manner as a regular AFFF. AR-AFFF foams are effective in dealing with ethanol, alcohol, and methyl ethyl ketone.

Ethanol, also known as ethyl alcohol or grain alcohol, is a flammable liquid and a polar solvent often used to make automobile fuel, usually mixed with 90% gasoline (E-10). Class B foam intended for hydrocarbon fires *will not* be effective on an ethanol fire or for gasoline-ethanol mixtures greater than 10%. Alcohol-resistant foams should be used and applied with Type II techniques rather than direct application. The foam should be streamed onto a vertical surface and allowed to run down onto the fuel. The higher the concentration of ethanol, the quicker the foam blanket will break down.

Key characteristics of alcohol-resistant AFFF foam as it relates to tank firefighting include the following:

- Must be applied gently to polar solvents so that the polymeric membrane can form first
- Should not be plunged into the fuel, but gently sprayed over the top of the fuel
- Very effective on spill fires with a good knockdown capability
- May be used for subsurface injection applications

- **Film-Forming Fluoroprotein Foam (FFFP)**—Based on fluoroprotein foam technology with AFFF capabilities. FFFP combines the quick knockdown capabilities of AFFF along with the heat resistance benefits of fluoroprotein foam.

Key characteristics of FFFP foam as it relates to tank firefighting include the following:

- Available in 3% and 6% concentrations
- Compatible with simultaneous application of dry chemical extinguishing agents (e.g., Purple K)
- Can be used with either fresh or salt water

FFFP is also available in an alcohol-resistant formulation. Alcohol-resistant FFFP has all of the properties of regular FFFP, as well as the following characteristics:

- Can be used on hydrocarbons at 3% and polar solvents at 6%. Newer FFFP concentrates can be used on either type of fuel at 3% concentrations.
- Can be used for subsurface injection.
- Can be plunged into the fuel during application.

A few rules should be observed regarding the compatibility of Class B foams:

1. Similar foam concentrates by different manufacturers are not considered to be compatible in storage applications. The exception to this would be Mil-Spec (i.e., military specification) foam concentrates. The Mil Specs are written so that mixing can be done with no adverse effects.
2. Don't mix different kinds of foam concentrates (e.g., AFFF and fluoroprotein) before or during proportioning.
3. On the emergency scene, concentrates of a similar type (e.g., all AFFFs, all fluoroprotein, etc.) but from different manufacturers may be mixed together immediately before application.
4. Finished foams of a similar type but from different manufacturers (e.g., all AFFFs) are considered compatible.

Determining Foam Concentrate Requirements

The availability of water and foam concentrate are critical factors in evaluating the risks involved in a storage tank emergency. The IC must evaluate the following factors in developing the Incident Action Plan, including:

1. Size of the fire (i.e., area involved, spill fire, tank fire, combination tank and dike fire)
2. Type of fuel (i.e., hydrocarbon or polar solvent)
3. Required foam application rate
4. Amount of foam concentrate required on-scene and the ability to resupply it
5. Ability to deliver the required amount of foam/water onto the fuel surface and sustain the required flow rates

If an adequate water and foam supply is not available for both protecting exposures and controlling the fire, the IC should consider implementing defensive or nonintervention tactics until sufficient resources are available. As a general tactical guideline, foam application operations should not be initiated until sufficient foam concentrate is on-site to extinguish 100% of the exposed flammable liquid surface area.

In evaluating specific types of Class B foam concentrate for the protection and/or extinguishment of specific fire scenarios (e.g., spill, tank fire, hydrocarbon vs. polar solvent, etc.), emergency responders should review the technical data package and the minimum foam application rates published by the respective foam manufacturer.

NFPA 11 recommends minimum foam application rates for specific fuels, foams, and applications, which are listed below:

- 0.10 gpm/ft^2—Fixed system application for hydrocarbon fuels (e.g., cone roof storage tank with foam chambers).
- 0.30 gpm/ft^2—Fixed system application for seal protection on an open-top floating roof tank.
- 0.10 gpm/ft^2—Subsurface application for hydrocarbon fuels in cone roof tanks.
- 0.10 gpm/ft^2 (AFFF, FFFP) to 0.16 gpm/ft^2 (protein, fluoroprotein)—Portable application for hydrocarbon spills (e.g., 1¾-inch handlines with foam nozzles).
- 0.16 gpm/ft^2—Portable application for hydrocarbon storage tanks (e.g., portable foam cannons and master stream devices). A foam application rate of 0.18 to 0.20 gpm/ft^2 has been used to successfully extinguish hydrocarbon fires in large diameter tanks using master stream portable nozzles.
- 0.20 gpm/ft^2—Minimum recommended rate for polar solvents. Higher flow rates may be required depending on the fuel involved and the foam concentrate used.

Determining Water Supply Requirements

As noted in the previously mentioned Buncefield, UK case study, flammable liquid storage tank fires can require tremendous quantities of water for a sustained period of time **Figure 10-25**. Before an effective fire attack can be made, it must be determined whether the water system is capable of:

- Delivering a water flow rate equal to or greater than that required to control the largest potential fire area.
- Delivering the required flow rates at pressures that can be used effectively by water application devices such as fixed systems, portable monitors, and handlines.

Figure 10-25 Large flammable liquid fires require large fire flows up to 20,000 gpm.

How to Calculate Foam Requirements

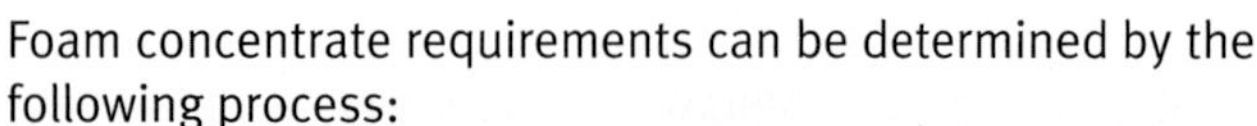

125-ft diameter open top floating roof tank

Foam concentrate requirements can be determined by the following process:

1. Determine the type of fuel involved—hydrocarbon or polar solvent. This will determine the type of foam concentrate to be used.
2. Determine the surface area involved.
   - Calculate storage tank area = (.785)(Diameter²) or (.8)(Diameter²)
     Note: The formula (.785)(Diameter²) is commonly used during the preincident planning process, while the formula (.8)(Diameter²) is commonly used for field applications.
   - Calculate dike or rectangular area around the tank: Area = Length × Width
3. Determine the recommended NFPA 11 foam application rate, as noted above.
4. Determine the duration of foam application per NFPA 11.
   - Flammable liquid spill = 15 minutes
   - Storage tank:
     Flash point 100°–200°F = 50 minutes
     Flash point < 100°F = 65 minutes
     Crude oil = 65 minutes
     Polar solvents = 65 minutes
     Seal application = 20 minutes
5. Determine the quantity of foam concentrate required. This figure will be determined by the percentage of foam concentrate used.

Problem 1: A 125-ft diameter open floating roof tank containing gasoline is fully involved in fire. Determine the amount of foam concentrate required to control and extinguish the fire. The fire department is using a 3% × 3% alcohol-resistant foam concentrate (ARC).

1. *What is burning?* Gasoline–hydrocarbon liquid. The 3% × 3% ARC can be used and will be proportioned at a 3% concentration.
2. *Determine the surface area involved:*
   Area = (.8) (Diameter²)
   Area = (.8) (125 ft)²
   Area = 12,500 ft²
3. *Determine the appropriate foam application rate.* The fire department is using portable application devices—foam cannons. The foam application rate is 0.16 gpm/ft²
   Foam application = area × recommended application rate
   Foam application = (12,500 ft²) (0.16 gpm/ft²)
   Foam application = 2000 gpm
4. *Determine the duration of foam application.*
   Gasoline has a flash point of approximately −45°F. Therefore, the recommended duration is 65 minutes.
   Required foam solution = foam application × duration
   Required foam solution = (2000 gpm) (65 minutes)
   Required foam solution = 130,000 gallons of foam solution
5. *Determine the quantity of foam concentrate required.*
   The fire department is using a 3% foam concentrate.
   Required amount foam concentrate = required foam solution × 3%
   Required amount foam concentrate = (130,000 gallons) (0.03)
   Required amount foam concentrate = 3900 gallons of foam concentrate
   Required amount of water = 126,100 gallons

Problem 2: A 150-ft diameter covered floating roof tank containing gasoline has overflowed and ignited. Both the tank and the dike area (100 ft × 80 ft) are completely involved in fire. Determine the amount of foam concentrate required to control and extinguish the fire. The fire department is using a 3% × 3% alcohol-resistant foam concentrate (ARC).

150-ft diameter covered floating roof tank fully involved with dike fire.

1. *What is burning?* Gasoline–hydrocarbon liquid. The 3% × 3% ARC can be used and will be proportioned at a 3% concentration.
2. *Determine the surface area involved:*

Storage Tank	*Dike Area*
Area = (.8) (Diameter^2)	Area = (Length) (Width)
Area = (.8) (150 ft)^2	Area = (100 ft) (80 ft)
Area = 18,000 ft^2	Area = 8000 ft^2

3. *Determine the appropriate foam application rate.*

   The storage tank is protected with foam chambers designed for full-surface protection and require a foam application rate of 0.10 gpm/ft^2. Portable application devices are required for the dike fire and require a foam application rate of 0.16 gpm/ft^2.

   Storage Tank
   Foam application = area × recommended application rate
   Foam application = (18,000 ft^2) (0.10 gpm/ft^2)
   Foam application = 1800 gpm
   Dike Area
   Foam application = area × recommended application rate
   Foam application = (8000 ft^2) (0.16 gpm/ft^2)
   Foam application = 1280 gpm

4. *Determine the duration of foam application.*

   Gasoline has a flash point of approximately −45°F. Therefore, the recommended duration is 15 minutes for the dike area and 65 minutes for the storage tank.

   Storage Tank
   Required foam solution = foam application × duration
   Required foam solution = (1800 gpm) (65 minutes)
   Required foam solution = 117,000 gallons of foam solution
   Dike Area
   Required foam solution = foam application × duration
   Required foam solution = (1280 gpm) (15 minutes)
   Required foam solution = 19,200 gallons of foam solution

5. *Determine the total quantity of foam concentrate required.*

   The fire department is using a 3% foam concentrate.

   Storage Tank
   Required amount foam concentrate = required foam solution × 3%
   Required amount foam concentrate = (117,000 gallons) (0.03)
   Required amount foam concentrate = 3510 gallons of foam concentrate
   Required amount of water = 113,490 gallons
   Dike Area
   Required amount foam concentrate = required foam solution × 3%
   Required amount foam concentrate = (19,200 gallons) (0.03)
   Required amount foam concentrate = 576 gallons of foam concentrate
   Required amount of water = 125,524 gallons
   Total amount foam concentrate required = 4086 gallons
   Total amount of water required = 239,014 gallons

Problem 3: A MC-306/DOT-406 cargo tank truck containing 8500 gallons of gasoline has overturned on a four-lane interstate highway. All cells are ruptured upon your arrival and the gasoline tank truck is fully involved in fire. The surface area burning on the street is approximately 150 ft × 300 ft (46 m × 91 m). Determine the amount of foam concentrate required to control and extinguish the fire. The fire department is using a 3% × 3% aqueous film-forming foam (AFFF).

Surface fire involving a gasoline tank truck.

1. *What is burning?* Gasoline—primarily hydrocarbon liquid as an E-10 motor fuel. The 3% AFFF can be used for extinguishment using 1 3/4-inch handlines with foam nozzles.
2. *Determine the surface area involved:*

   Surface Spill
   Area = (Length) (Width)
   Area = (150 ft) (300 ft)
   Area = 45,000 ft^2
3. *Determine the appropriate foam application rate.*

   The gasoline surface fire can be extinguished using 0.10 gpm/ft^2 AFFF.

   Surface Area
   Foam application = area × recommended application rate
   Foam application = (45,000 ft^2) (0.10 gpm/ft^2)
   Foam application = 4500 gpm
4. *Determine the duration of foam application.*

   Gasoline has a flash point of approximately −45°F. Therefore, the recommended duration is 15 minutes for the contained surface fire.

   Required foam solution = foam application × duration
   Required foam solution = (4500 gpm) (15 minutes)
   Required foam solution = 67,500 gallons of foam solution
5. *Determine the total quantity of foam concentrate required.*

The fire department is using a 3% AFFF foam concentrate.

Surface Fire
Required amount foam concentrate = required foam solution × 3%
Required amount foam solution = (67,500 gallons) (0.03)
Required amount foam concentrate = 2025 gallons of foam concentrate
Required amount of water = 65,475 gallons

Note: Scenario 3 provides an example of the problems an MC-306/DOT-406 gasoline tank truck fire will present. Although some steel cargo tanks may still be found, the majority of MC-306/DOT-406 cargo tank trucks are constructed of aluminum. Aluminum shell MC-306/DOT-406 cargo tanks will melt down to the liquid level, and the product will burn off in a controlled manner rather than build up internal pressures that would otherwise lead to a catastrophic breach. As a result, these scenarios usually involve a spill fire of some size or magnitude, which may mean the best solution is an open pit fire for any remaining product to burn off.

If the running spill fire can safely be confined using defensive tactics (e.g., diversion and diking), the surface area becomes smaller, the product pooled in the compartments is confined, and less foam concentrate may be required for extinguishment because of the smaller surface area.

Remember in fifth grade when you told your friends you would never really use math? Math is actually very important. Hazmat responders have to get the math right the first time. You have to use the right amount of foam concentrate, at the right application rate, for the right amount of time, or the fire won't go out.

There are several assumptions regarding water supply requirements often misunderstood by emergency responders, especially when relying upon petroleum facility water supply systems. These include:

1. **Assumption:** The facility fire water system is adequate for fire attack because all of the past fires at the facility have been successfully controlled.

   Reality: Most hydrocarbon fires never reach their full potential because they are extinguished in their incipient stages. Therefore, they were controlled well before they reached their full-scale potential and never really taxed the water system.

2. **Assumption:** The water flow rate available in the facility is believed to be adequate because the rated flow of the facility's fixed fire pumps exceeds the maximum foreseeable water flow demand.

   Reality: Fire pumps are not constant flow and constant pressure devices. Therefore, the sum of their rated flows is not equal to the water flow rate available in the facility. When delivering water, fire pumps must provide sufficient pressure to operate water application devices in the area of the fire and to overcome pressure losses in the piping system. If this pressure is higher than the pumps' rated pressure, they will flow less water than the sum of their flows. If the water flow rate demand exceeds the rated flow of the pumps, the pressure available in the fire area will *decrease*, with possible effects on the operation of water application devices.

As a result, the water flow rate available in each area of a facility varies depending on the area's relative elevation and location with respect to the pumps and the pressure requirements of the water application devices to be used. In addition, experience shows that some water systems have been modified or damaged over time. While they may have been properly designed and installed, years of neglect may have rendered them ineffective.

3. **Assumption:** The normal/static system pressure is an adequate indicator of the system's capability (e.g., the system is adequate because it has a pressure of 100 psi when no water is flowing).

   Reality: As soon as water begins to flow, the operating pressure will drop. The greatest pressure drop will be in the area where the water is being used. How far the pressure drops will be dependent on the pressure available at the system's water source(s), the distance between the fire and the system's water sources, the size and arrangement of the piping in the fire water system, and the water flow rate. Remember, the greater the flow rate, the greater the pressure drop. **Figure 10-26** shows an example of how these three problems can work against emergency responders.

During the early stages of a fire, several monitors and handlines may be placed in service to protect exposures. As a result, the available water pressure in the immediate fire area drops slightly. See point A in the graph below.

When it is realized that the initial fire attack cannot be handled with monitors and handlines, additional water streams are ordered into service. As these additional hoselines and devices are placed into service, the water pressure continues to drop even further. At this point one of two outcomes is possible: (1) the flow required to control the fire is obtained and the fire is brought under control with the flow and water pressure remaining fairly constant; or (2) the water pressure in the area begins to fall below that required to maintain full firefighting effectiveness. See point B.

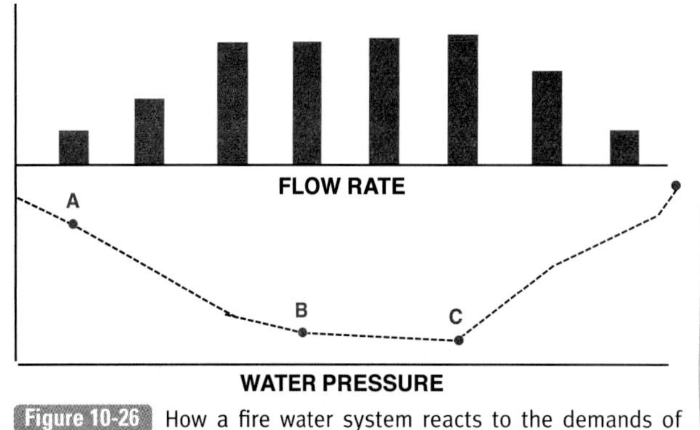

Figure 10-26 How a fire water system reacts to the demands of hydrocarbon firefighting operations.

At this point, the overall water flow rate cannot be significantly increased because placing additional devices in service will further lower available water pressure, causing the flow out of every line or device already in operation to decrease. As a result, the fire continues to grow, possibly out of control, while the water flow rate remains fairly constant. See point C. This is characteristic of an inadequate water system.

Reprinted with permission from Loss Control Associates, Inc.

Determining Cooling Water Requirements for Exposures

Water applications for exposure protection usually start before foam application starts. Cooling water for exposure protection may come from potable or nonpotable sources. As previously noted, exposure lines should be applied when there is direct flame impingement on exposed tanks and/or when radiated heat is sufficient to cause steam at the tank shell when water is applied. Exposure streams should be shut down when steam is no longer produced from the metal surface.

Basic cooling water guidelines for exposed tanks and pressure vessels are as follows:

- Atmospheric storage tanks up to 100-ft diameter require 500 gpm.
- Atmospheric storage tanks from 100-ft diameter to 150-ft diameter require 1000 gpm.
- Atmospheric storage tanks exceeding diameters listed above require 2000 gpm.
- Pressure vessels should have a minimum of 500 gpm applied at the point of fire impingement. This is a widely quoted number that has proven to be a reliable guideline over time. Taking action with less water using offensive and defensive tactics increases risk to personnel significantly. However, lower flow rates may still be effective if they are applied from fixed systems and can be activated without risk to responders.

Scan Sheet 10-F

Fire Safety Features at Bulk Liquid Facilities

A bulk storage facility

Note: Material in Scan Sheet 10-F is intended to meet the requirements of NFPA 7.2.4.2.

Bulk flammable liquids storage facilities that are designed to meet the requirements of NFPA 30—*Flammable and Combustible Liquids Code* or Article 15 of the *Uniform Fire Code.* American Petroleum Institute (API) standards have many fire and safety features engineered into them that can favorably influence the behavior of the products in the event of a spill or fire. These systems and their design features should be factored into the analysis during preincident planning and response operations.

The following types of engineered features and systems may be found at flammable liquid bulk storage facilities:

- **Fire Protection Systems**—The heart of fire protection systems protecting bulk storage facilities is a reliable water supply able to supply the pressure and quantity of the fire demands. Regardless of how many fixed or semi-fixed foam suppression systems are in place at the facility, they are useless without a good water supply to support their operation.

Bulk storage facilities usually have hydrants with or without fixed monitor nozzles. Their number and placement depends on the hazards of the products being stored and exposures. Based on the risks present and the requirements of the Authority Having Jurisdiction (AHJ), fire protection systems may be required and can include:

- Diesel, gas, steam, or electric powered fire water pumps
- Foam concentrate storage
- Detection and fixed fire protection systems at loading racks
- Fixed or semi-fixed foam suppression systems on storage tanks
- Spill containment at loading racks and around transfer pumps
- Water spray or sprinkler systems for exposure protection of critical areas, including fireproofing to protect critical vertical and horizontal support structures

Hazardous Materials Technicians and ICs should be familiar with the basic principles of operation of the fixed and semi-fixed fire protection systems and understand how they may be used to support offensive and defensive firefighting operations.

- **Monitoring and Detection Systems**—Bulk storage facilities may be equipped with a variety of fixed monitoring and detection systems that indicate when a spill or leak has occurred in an unattended area. Normally these systems are tied to an audible alarm that alerts at a specific threshold based on the concentration of hydrocarbon vapors in the atmosphere. In addition, confinement areas around pumps and critical valves may have sensors that activate in the event of a spill. These alarms typically sound in the facility control room to alert operators that there is a problem at a specific location within the facility.

In the past, tank overfills have been the source of several major fires at bulk storage facilities (See Newark, N.J., January 7, 1983 and Buncefield, UK, December 11, 2005 discussed in this chapter.) Bulk storage tanks may be equipped with high-level alarms that alert operators that a tank being filled is approaching an unsafe level and is at risk from overfill. In some jurisdictions, independent high level alarms that can automatically shut down and divert product flow are mandatory for all tanks receiving Class-I liquids by pipeline or from a marine vessel.

Tank overfill alarms are normally equipped with several levels of alarms that are tied to a Standard Operating Procedure that requires specified action based on the type and level of the alarm. For example, a high-alarm may require an operator to go to the site and investigate the occurrence; a high-high-alarm may require operator action to begin a shutdown of filling operations, and so on. Many tank farm areas are also under closed circuit television (CCTV) video surveillance. which allows the control room to observe the area in real time. A layered approach to high-level alarms allows control room operators to intervene and resolve the problem as it is developing before an emergency occurs. Examples of action that can be taken by operators include slowing the delivery of the product into the tank, diverting the liquid to another tank that can accept the additional volume of liquid, or closing down the pipeline. Turning off a pipeline during a bulk storage filling operation is not like turning off the kitchen spigot!

- **Pressure Relief and Vacuum Relief Protection**—Aboveground petroleum storage tanks are designed to operate at atmospheric or low pressures. Cone roof tanks will have a pressure-vacuum valve for normal operations, while open floating and covered floating roof tanks will vent to the atmosphere during normal transfer operations. In the event of an over-pressure, cone roof and covered floating roof tanks designed to API specifications will have a weak roof-to-shell seam that will allow the roof to fail and avoid a catastrophic tank failure.

Excessive filling or pump discharge rates can cause a tank overpressure or vacuum resulting in tank damage. Properly designed venting keeps the tank's internal pressure within the design limitations.

Low-pressure storage tanks are vented for both normal operations and emergency scenarios. Venting for normal bulk storage tanks are sized according to published tables in American Petroleum Institute Standard 2000, *Venting Atmospheric and Low-Pressure Storage Tanks.*

- **Product Spillage and Control** (impoundment and diking)—Impoundment and diking systems are designed to ensure that accidental discharge of flammable liquids will be prevented from endangering important facilities, adjoining property, or entering waterways.

Remote impoundment is designed so that any spilled liquid drains away from the storage tank and exposures by way of grading, swales, or ditches, to an area large enough to contain all of the liquid that can drain into it.

When protection of adjoining property or waterways requires ensurance that no product will escape the area, impoundment by diking is used. The disadvantage of this level of protection is that if the spill inside the dike ignites it exposes the tank's shell to fire and increases the risk of tank failure. Both NFPA 30 and Article 15 require that the capacity of the dike be greater than the amount of liquid that can be released from the single largest tank within the diked area.

Dikes are equipped with drains to remove accumulated rainwater. These valves are normally maintained in the closed position to prevent flammable liquid from flowing out of the dike. During spill control and firefighting operations it is critical that all dike drains be checked to ensure they are in the closed position. When tank exposure to fire is an issue, cooling water applied to the tank's shell can accumulate in depth inside the diked area. If hydrocarbon liquid is floating on top of the spill a serious hazard can develop if the dike fails, overflows, or the facility's storm and sewer system is overwhelmed. This was a critical factor in the death of eight fire fighters at the August 17, 1975 fire at the former Gulf Oil refinery in Philadelphia, Pennsylvannia.

- **Tank Spacing**—Storage tank fires can generate a significant amount of radiant heat and expose nearby tanks to excessive heating, thereby causing damage to unprotected structural steel. To decrease the risk of fire spread from tank-to-tank, adequate spacing between tanks inside a common dike area is required. Tank-to tank and tank-to-property line spacing is based on several factors including: (1) type of tank (i.e., cone, open floater, covered floater), (2) product contained (i.e., Class 1, 2, or 3 liquid), (3) minimum distance from the tank to the property line that can be built upon, and (4) minimum distance from the nearest side of any public right-of-way or the nearest important building on the same property. The key point is that the farther away tanks are from each other and other exposures like buildings, the less the risk of the fire spreading from one tank.

Crude oil tanks can pose unique challenges to emergency responders, including the possibility of a boilover. While boilovers are not very common, a 1982 crude tank boilover at a power plant in Tacoa, Venezuela resulted in over 40 fire fighter deaths. See the glossary for the definition of a boilover.

- **Transfer Operations**—Bulk storage tank operations require moving large volumes of flammable liquids to facilitate storage, pipeline deliveries, and distribution operations. Emergency responders conducting a walk-through of a bulk storage facility should trace the flow of product both into and out of the facility. For example, most terminals are supplied by a liquid transmission pipeline, with product then flowing through a series of internal piping systems and into a storage tank.

Ultimately, product can be transferred to other storage tanks or moved to a cargo tank truck or railroad loading rack through a system of horizontal or vertical pumps and the related internal piping system.

Gasoline—ethanol blends are the most common motor fuel today, with the gasoline and ethanol being blended at the loading rack as it is loaded into a cargo tank truck. Cargo tank trucks being loaded with flammable liquids are normally a bottom-loaded process with an integrated vapor recovery system. Top-loading racks can still be found for combustible liquid loading operations.

Bulk storage facilities include many design features that range from electrical and mechanical to fire protection-engineered systems that can provide emergency responders with many options for spill and fire control. Understanding how these systems function and their limitations could mean the difference between a good or bad outcome.

Source: NFPA 30 and 30A: *Flammable and Combustible Liquids Code Handbook,* by Robert P. Benedetti, National Fire Protection Association, Quincy, MA (2008). This handbook is an excellent source of information on flammable and combustible liquids. The book includes the entire text of the NFPA 30 and 30A codes as well as a very useful commentary by the editor on the history of the code's development and requirements. There are also several historical case studies included.

For more information on bulk storage tanks and fire suppression methods, see *Storage Tank Emergencies,* by Michael S. Hildebrand and Gregory G. Noll, Chester, MD: Red Hat Publishing, Inc. (1997).

The types of bulk storage tanks and the methods of identifying them are discussed in more detail in the chapter on "identifying the problem."

Flammable Gas Emergencies

Flammable gases have claimed many fire fighters' lives. If you lose respect for flammable gases and make a mistake while handling them, they will be very unforgiving. The most commonly encountered flammable gases are natural gas (i.e., methane or CH_4) and the liquefied petroleum gases (propane or C_3H_8 and butane or C_4H_{10}). LNG is an emerging fuel source and will be discussed separately within this chapter.

Since 1993, there have been at least three significant propane incidents in North America involving propane that have claimed the lives of seven fire fighters. These include incidents at Saint Elizabeth de Warwick, Quebec on June 27, 1993; Burnside, Illinois on October 2, 1997; and Albert City, Iowa on April 9, 1998. In each of these incidents, the pressure-relief valves on the propane storage tanks were not functioning due to external fire impingement. The tanks subsequently failed due to a boiling liquid expanding vapor explosion (BLEVE), which resulted in emergency responder injuries and fatalities. BLEVEs will be discussed in more detail later in this section.

Hazard and Risk Evaluation

While most responders are familiar with the potential hazards of large LPG tanks, they often fail to see comparable hazards when dealing with smaller containers with similar hazards. See Figure 10-27. Many fire fighters have been injured from small spray cans or 1-pound containers when they ruptured in dumpsters or structural fires. In addition, flammable gases may also possess multiple hazards. Consider the following examples:

- **Common flammable gases**—Propane, butane, natural gas, compressed natural gas (CNG), liquefied natural gas (LNG), liquid hydrogen, methane.
- **Flammable and toxic gases**—Hydrogen sulfide, ethylene oxide, phosphine, arsine.
- **Flammable and corrosive gases**—Isopropyl chloroasetate, 1 amino-2 propanol.
- **Flammable and reactive gases**—Butadiene, acetylene (without acetone stabilizer), silane.
- **"Fooler gases"**—Anhydrous ammonia.

Figure 10-27 Flammable gas firefighting.

Keeping track of times and key events is critical during such incidents. Flammable gas incidents often start out as vapor releases that eventually ignite once they come in contact with an ignition source. It is important that the IC acquire the following information during the size-up process:

- Time when the incident started. Remember, this may not necessarily be the same time the incident was reported.
- Time at which responders arrived on scene.
- Probability that the fire will be confined to its present size.
- Layout of the incident, including the following specific points:
    - Size and type of pressure vessel(s) involved. Flammable gas containers can range from 20-pound to 250-pound nonbulk propane cylinders, to 30,000- to 60,000-gallon horizontal tanks. Common transportation containers will include the MC-331 cargo tank truck and pressurized railroad tank cars (usually DOT 112 and DOT 114, although DOT 105 specification may also be found). For additional information, see the section on "problem identification" found elsewhere in this text.
    - Valves and piping systems stressed or destroyed by the fire.
    - Presence of any fixed or semifixed fire protection systems, such as water spray systems and lab cylinder cabinets.
    - All surrounding exposures, including other pressure vessels, tanks, buildings, process units, utilities, and so on. This should include identifying and prioritizing exposures (e.g., flame impingement, radiant heat exposure).

Hazard and risk evaluation is essential to the safety of responders and the effectiveness of the response effort. Factors of evaluation will include the following:

- Nature, quantity, and pressure of the gas involved.
- Design and construction of the container/pressure vessel. Factors should include the container size, type of the container breach, types of pressure-relief devices present (e.g., pressure-relief valve, frangible disk, fusible plug), and the ability to isolate the source of the fire/leak.
- Type of stress on the container or piping system. Although thermal stress is the primary concern in a fire situation, mechanical stress (e.g., overturned cargo tank truck, derailed tank car) may be an equal concern.
- Size and type of area affected by the fire as well as the likelihood that the fire will be confined to its present size.
- Identifying and prioritizing exposures. Remember that the highest priority should be given to flame impingement on vessels, piping, and critical support structures.

- Resources available to combat the problem, including the ability to rapidly apply sufficient water to the point of flame impingement.

Tactical Objectives

Based on the type and nature of risks involved, the IC will implement the appropriate strategic goals and tactical objectives. Tactical priorities for managing a flammable gas fire are to:

1. Protect primary and secondary exposures to the fire.
2. Isolate the flammable gas source feeding the fire.
3. Reduce the operating pressure of the line feeding the fire.
4. Permit the fire to self-extinguish and consume residual flammable gas inside the vessel or piping system.
5. Control and extinguish secondary fires.

Tactical options for flammable gas fires include the following:

- **Nonintervention**—This is a "no win" situation in which responders cannot positively change or influence the sequence of events. The best example is an imminent BLEVE situation. BLEVE means different things to different people, including "Blast Levels Everything Very Effectively."
- **Defensive tactics**—These tactics involve protecting exposures and allowing the fire to burn. This may include implementing a controlled burning of any remaining gas or flaring off the product.

As with flammable liquids, both direct flame impingement and radiant heat exposures are critical concerns. Critical exposures must be identified and prioritized in a rapid manner. Flame impingements must be cooled immediately, while radiant heat exposures should be handled as soon as possible. Use fixed systems and unmanned portable monitors whenever possible; they are a "force multiplier" that frees up personnel for other critical tasks. They also reduce risk to responders.

- **Offensive tactics**—These tactics are implemented when sufficient cooling water and related resources are available for a continuous, uninterrupted fire attack. Although the primary focus is toward isolating the fire and the source of the gas, it will be necessary to maintain exposure lines.

Tactical Considerations and Lessons Learned

The nature of pressure vessels and their associated containment systems makes them vulnerable to external heating. A pressure vessel may be thermally stressed and activate one of its safety attachments (e.g., pressure-relief valve, frangible disk, etc.). As the heat is transmitted internally, the container pressure will increase proportionately. Eventually the gas will escape to the outside atmosphere through one of its attachments or it may fail violently Figure 10-28.

Pressure-fed flammable gas fires may produce direct flame impingement on nearby vessels and cause catastrophic tank failure within 5 to 20 minutes of exposure. BLEVEs of bulk containers and process vessels can produce severe fire and fragmentation risks within 3000 feet (914 meters) of the failed container. The IC must evaluate potential BLEVE situations early in the fire and take immediate action to activate fixed water spray systems, monitor nozzles, provide cooling streams, and evacuate the area.

Figure 10-28 The decision to approach a closed container showing direct flame impingement on its vapor space must incorporate the consideration of risks versus benefits.

The IC must determine (1) if an adequate water supply is available to deliver the volume of water required for cooling exposures and (2) if there is adequate pumping capacity to provide the required water pressure at the fire scene. If an adequate water supply is not available for cooling primary exposures, the IC should consider immediate withdrawal of all personnel from the hazard area.

Effectively placed hose streams can lower the pressure in most small containers (e.g., cylinders); however, the risks associated with advancing hoselines or monitors may exceed the possible benefits.

Any decision to approach a closed container showing direct flame impingement on its vapor space must be made on a case-by-case basis after carefully evaluating the hazards and risks. Large LPG containers (8,000 to 30,000+ gallons) have violently ruptured and traveled over 3600 feet (1097 meters) or more (Napanee, Ontario), even with functioning pressure-relief valves. See Scan 10-E.

Flammable Gas Emergencies at Petrochemical Facilities

Incidents at facilities that manufacture, process, store, or use large quantities of flammable gases can create specialized problems for responders. The IC must evaluate the overall fire and hazmat problem, recognizing that products other than flammable gases may be involved.

Boiling Liquid Expanding Vapor Explosions

San Jacinto, Mexico City—The November 19, 1984 San Jacinto, Mexico LPG plant fire and explosion killed over 500 people, injured hundreds, and caused a $29 million loss. At 5:37 am a LPG bullet tank was overfilled from a 12-inch pipeline. At 5:50 am the tank caught fire and subsequently failed due to BLEVE, killing many civilians in a densely populated neighborhood outside the plant. At 5:57 am a second tank BLEVE'd killing the majority of the emergency responders. The explosion produced a fireball estimated at 1200 feet (368 m) in diameter. The radiated heat from the rupturing tank and the missile damage allowed the release of more fuel from other LPG tanks. Eventually, four spheres and 44 bullet tanks failed. Some of the tanks that BLEVE'd weighed 20 tons and were propelled 3937 feet (1200 meters) into a two-story home. LPG tanks can fail within 10 to 20 minutes of direct flame impingement or even days after the container is initially stressed.

Sources: Marsh, *The 100 Largest Losses* 1972–2001; Special Report on the San Juan Ixhuatepec, Mexico LPG Disaster, prepared by B. F. Olson for the American Petroleum Institute, December 28, 1984.

Waverly, Tennessee—The Waverly, Tennessee incident is a classic case study for damage assessment and predicting the behavior of pressurized containers that have been stressed. On February 22, 1978, a Louisville & Nashville railroad train derailed 23 cars in the middle of downtown Waverly. This included two LPG tank cars. At the time of the original derailment, there was no fire or release of LPG. Two days later, one of the propane tank cars failed explosively with no warning or pressure-relief valve activation, as transfer operations were being set up. This mechanical BLEVE resulted in 16 deaths, including several fire fighters, the fire chief, and the police chief. NTSB investigators ultimately concluded that the combination of mechanical (i.e., rail burn) and thermal stressors (i.e., sun and ambient heating) resulted in the container failure.

What is a BLEVE?
BLEVE is an acronym for Boiling Liquid Expanding Vapor Explosion. A BLEVE is defined as a container failure with a release of energy, often rapidly and violently, which is accompanied by a release of gas to the atmosphere and propulsion of the container or container pieces due to an overpressure rupture.

As a liquefied gas, an LPG tank contains both liquid and vapor. Any external fire creating direct flame impingement on the vapor space will heat the tank shell; any fire on the vapor space will heat the tank shell more rapidly than any fire impingement on the liquid area. As temperatures soar, the tank shell's temperature in the vapor space quickly reaches 752°F (400°C). Above 1112°F (600°C) steel weakens significantly. By the time the steel reaches 1800°F (982°C) it has lost 90% of its strength. Thinned and weakened from the fire, the tank will eventually relieve pressure to the outside either through a split in the tank in the form of a jet flame or the container will fail. Not every LPG tank exposed to fire fails by BLEVE. Sometimes the tank just splits open at the weld seam and burns off, and sometimes the pressure-relief valve functions and burns off the LPG. There is no way to accurately predict how and when any type or size of tank will fail. A functioning pressure-relief valve is not an effective way of determining if a tank will fail, but it is a

San Jacinto, Mexico, November 19, 1984.

Crescent City, Illinois (1971).

Scan Sheet 10-G (continued)

good indicator that the internal pressure inside of the tank is higher than normal. The pressure-relief valve on an ASME tank is set for 250 psi (17.2 bar), while a motor fuel tank relief valve would be set between 312.5 to 375 psi (21.5 to 25.8 bar).

When a tank fails due to a BLEVE, projectiles (i.e., pieces of tank, metal parts, and other debris) travel in all directions (360 degrees around the tank) for great distances. If you are standing (or hiding) between where the tank is headed and where it will be landing, you are probably not going to survive the experience!

Distance is your friend. Consult the Emergency Response Guide for protective action and evacuation recommendations based on the size of the container. This isn't rocket science—the larger the container involved, the farther back you need to be. A nonintervention strategy is often the best option if the tank is on fire when you arrive at the scene. Before a defensive or offensive strategy is employed the IC must be satisfied that there is an adequate water supply to support the fire attack, the risks have been adequately evaluated (gain or loss vs. the risk you will be taking), and both the IC and the responders have been properly trained in handling LPG fires (i.e., specifically in evaluating the hazards and risks for "Go or No-Go" decision making).

Tank exposures to high-velocity jet flame (i.e., pressure-fed fire) will require 500 gallons per minute of water (1892 liters per minute) at the point of impingement. Exposures to radiant heat with no direct flame contact will require 0.1 to 0.25 gallons per minute per square foot (0.4 to 0.9 L per minute) to maintain the integrity of the exposure.

Never extinguish a pressure-fed flammable gas fire unless you can control the fuel supply. Isolate the source of the gas and permit the fire to self-extinguish, thereby consuming any residual gas inside the vessel or piping system. Unignited flammable gases and vapors escaping under pressure will rapidly form an unconfined vapor cloud, which will usually be reignited by ignition sources in the area. Explosions of unconfined vapor clouds can cause major structural damage and quickly escalate the size of the emergency beyond responder capabilities.

For more information on LPG fires, consult *Propane Emergencies* (3rd Edition), by Hildebrand and Noll; Washington, DC: Propane Education and Research Council (2007).

Specifically, the IC should determine the following:

- Were there any abnormal operating conditions immediately before the emergency?
- Were there any equipment problems or changes immediately before the emergency (e.g., maintenance operations, changing over pumps, blinding off lines)?
- Are exposures protected with fixed water spray systems or monitors? Are systems operating?
- Are fixed fire protection and chemical mitigation systems available (e.g., scrubbers and neutralizers)? Have they been activated?
- What is the status of the fire pumps? What is the fire water system pressure?
- Is the process unit isolated?
- What is the structural stability and potential failure of the unit? Is fireproofing in place? Experience in the petrochemical industry shows that in a major fire:
  - Instrumentation lines can begin to fail within 5 minutes.
  - Pressure vessels and other closed containers can begin to fail within 10 minutes.
  - Structural steel will begin to fail within 15 minutes.
- What is the status of the process unit? Is the process stable (e.g., temperature, pressure, and reactions)?

Responder Safety Tip

Pressure vessels can fail early in the fire. If you do not have sufficient cooling water to maintain the integrity of the container on all sides, evacuate the area and back away. Distance is your friend. The larger the container, the farther away you should be.

- What types of safety systems are in place? Have they been activated? These would include emergency shutdown systems, pressure-relief devices, flares, scrubbers, and so on.

- What is the status of the utility systems, including electrical, instrument air, steam, fuel gas, and so forth? Isolating utilities without coordinating with facility process personnel may create secondary and tertiary problems greater than the initial event.

Liquefied Natural Gas Emergencies

Liquefied natural gas (LNG) is methane in its liquid state. LNG is a flammable and odorless gas. While it is nontoxic, it can be an asphyxiant when it displaces oxygen in a confined space. LNG is a cryogenic liquid stored and transported at –260°F (–160°C). When cooled to this temperature at atmospheric pressure, natural gas turns into a liquid which provides a practical and economical method for transportation and storage. When released from its containment system as a liquid, it behaves very differently from a natural gas release.

The safety record of LNG has been very good as compared to other fossil fuels. Since the first commercial LNG plant became operational in the United States in 1941, no fire fighter line-of-duty deaths have occurred responding to LNG incidents. In the U.S. LNG industry's 70-year operating history (1941 to 2011), there have been three significant incidents: Cleveland, Ohio (1944), Staten Island, New York (1973), and Cove Point, Maryland (1979).

Emergency responders should become familiar with the hazards and risks of LNG as it is evolving as a primary fuel source to satisfy industrial, commercial, and residential usage requirements in the United States. For example, in 2000 the U.S. produced 19.1 trillion cubic feet of LNG. By 2020, annual production is projected to expand to 29 trillion cubic feet.

Hazard and Risk Evaluation

LNG presents two primary hazards and risks: (1) flammability and 2) liquid or vapor dispersion.

Flammability—When LNG is spilled and its vapors come into contact with an ignition source, the spill will develop into a pool fire and present a thermal radiation hazard. If there is no ignition source, the LNG will vaporize rapidly forming a cold gas cloud that initially is heavier than air, spreads, and is carried downwind until it reaches neutral buoyancy when enough air mixes with it. The vapor is ignitable in the 5 to 15% range. The flammable region of the vapor cloud can be estimated by the visible white cloud that is actually water vapor condensed due to the cold LNG vapor. Once ignited it will burn back to the source.

LNG is primarily composed of 85% to 96% methane, with other light hydrocarbons such as propane, ethane, and butane making up most of the balance. LNG also contains about 1% nitrogen. LNG is flammable in its vapor state between 5% and 15% concentration of gas in air. See **Figure 10-29**. By comparison with other common fuels, propane's flammable range is 2.1% to 9.5%, and gasoline is 1.3% to 7.1%. The ignition temperature of LNG vapor at 1004°F (540°C) is higher than that of other common fuels. For example, LPG = 850°F (454°C), ethanol = 793°F (423°C), diesel = 600°F (315°C), and gasoline = 495°F (257°C).

Figure 10-29 LNG is flammable between 5% and 15% concentrations of gas in air.

Liquid and Vapor Dispersion—LNG has an expansion ratio of 600:1 when vaporized at 1 atmosphere and warmed to room temperature. See **Figure 10-30**. It is usually stored and transported in well-insulated containers at very low pressures, typically less than 5 psig. Heat leaking from the LNG within the container causes the liquid to boil. Removal of the boil-off gas helps maintain the LNG in its liquid state—a phenomenon known as "auto-refrigeration."

Auto-refrigeration is often misunderstood by emergency responders. LNG is stored as a liquid at near atmospheric pressure in a container at –260°F. Heat leaking through the insulation warms the liquid that rises in a boundary layer up the walls of the container to the liquid surface. At temperatures above –260°F (–160°C) the surface liquid will boil off into a vapor until the vapor pressure in the container is equal to its equilibrium vapor pressure. In a closed container this equalization can be achieved by a gradual increase in vapor

Figure 10-30 LNG has an expansion ratio of 1 to 600. If LNG is spilled on the ground it will boil rapidly initially, then boil slowly as the ground cools. In this controlled demonstration at Texas A&M, water is being applied to the spill to show how water will increase vaporization.

Scan Sheet 10-H

Fire Safety Features at Bulk Gas Facilities

Note: Material in Scan Sheet 10-H is intended to meet the requirements of NFPA 7.2.4.3

Modern bulk flammable gas storage facilities have a safe operating history. These facilities primarily handle liquefied petroleum gases, such as propane, butane, and related blends. They are designed and equipped with fire and safety features to help control the behavior of the products during an emergency.

According to NFPA 58—*LP-Gas Code Handbook,* a bulk plant is defined as a facility where the primary function is to store LP-Gas prior to further distribution. Bulk plants can have storage tanks of at least 2000 gallons (7.6 m3) water capacity or more and have container filling or truck loading facilities on the premises. Larger bulk facilities can have multiple storage tanks ranging up to 30,000 gallons and also have a railcar loading rack.

Bulk gas facilities may have the following types of safety and fire protection design features:

- **Fire Protection Systems**—Most LP-Gas bulk plants are equipped with minimal fixed fire protection systems because the risk of fire is low. Fire protection resources commonly found on-site can include municipal or industrial fire hydrants, fixed monitors for protecting tank shell exposures, and portable or wheeled fire extinguishing units. Depending on the authority having jurisdiction (AHJ), fixed water spray systems to protect individual tanks may be required. These systems may be manually or automatically activated.
- **Monitoring and Detection Systems**—Bulk storage and truck and rail car loading rack areas may be equipped with fixed air monitoring equipment designed to detect low concentrations and provide early warning of an LP-Gas leak. These systems may activate local or control room alarms, thereby alerting operators of a potential problem.
- **Excess Flow Valves (EFVs)**—The primary fire protection strategy at bulk LP-Gas storage facilities is built around the requirement to rapidly stop the flow of gas to the surrounding atmosphere and control potential ignition sources. Excess flow valves are designed to close automatically when the flow of liquid or vapor LPG reaches the volume specified in the design of the valve or when flow pressure exceeds a preset value. An EFV is required on the storage tank's liquid internal safety valve, but may also be found on some vapor lines.
- **Pressure-Relief Valve Protection**—Pressure-relief valves (PRVs) are installed in the vapor space of the bulk tank. Bulk LP storage tanks may have as many as four PRVs installed for the protection of the tank. A properly designed and maintained PRV automatically relieves excess pressure that may build up in the tank. To ensure that the vapors, if ignited, are directed away from the tank, the PRV may be equipped with long pipes attached to the valve called stacks or risers. It should be noted that even with a pressure-relief valve operating, the internal tank pressure may exceed the set pressure of 250 psi.
- **Emergency Shutdown System (ESD)**—In the event of an LPG leak or fire, the bulk plant can be shut down by activating the emergency shutdown. Activating the ESD usually requires an operator to initiate the process from a manual station by pulling a switch handle. Once the ESD is activated nitrogen gas in the piping system bleeds down and activates the shutdown of the plant, closing all valves.

Bulk LP-Gas storage facilities should be preplanned. Follow the flow of product from the point where it enters the bulk plant to the various points of distribution. Of particular importance is knowing where the emergency shutdown (ESD) stations are located and how they are activated. While fires at these facilities are rare, establishing an uninterrupted water supply is critical to ensure sufficient cooling water reaches all sides of both the tank on fire and any exposed tanks, especially if there is an activation of the pressure-relief valves. Owners and operators of these facilities should always be present at the Incident Command Post (ICP) during firefighting operations to advise the IC on product control options.

Source: Propane Emergencies (3rd edition), by Michael S. Hildebrand and Gregory G. Noll, Chester, MD: Red Hat Publishing (2007). For a free copy in PDF format go to http://www.propanesafety.com.

For more information see the *LP-Gas Code Handbook,* National Fire Protection Association, Quincy, MA: National Fire Protection Association (2011).

pressure over the liquid. If vapor is drawn out of the container, the pressure in the container will drop. The result is that the surface pressure of the liquid will exceed the vapor pressure in the container and vaporization (boiling) will resume. The boiling will continue until the container again reaches equalization. When vapor is removed from an LNG tank, the LNG actually cools itself and goes into a state known as auto-refrigeration. When the tank is held at a constant pressure by removing vapor, the energy in the vapor equals the heat energy entering the tank through the insulation. This explains why LNG can be maintained as a cryogenic liquid for prolonged periods in transit and storage.

The density of LNG is 3.9 lbs/gallon—about half the weight of water. If LNG is spilled on the ground, it will boil rapidly at first and then boil slowly as the ground cools. If LNG is spilled on water, it will float on top and vaporize very rapidly since even at water temperatures near freezing, the water is significantly warmer than the spilled LNG. The resulting vapor cloud is very cold and quite visible because it condenses water out of the air.

When LNG is released, initially, the vapor cloud is dense and made visible by ice crystals from water vapor in the air. If ignition is delayed, the mixture hugs the ground and spreads laterally. As the cloud becomes warmer than −256°F (−160°C) and mixes with air, the expanding vapor cloud may not be visible. As it continues to disperse, the cloud will eventually become neutrally buoyant −160°F or 107°C). See **Figure 10-31**.

If there is a sufficient ignition source present, the natural gas cloud may ignite, but has not been shown to explode if it is not confined. LNG itself will not burn or explode: it must be vaporized and mixed with air in the right concentrations (5% LFL to 15% UFL) to make combustion possible.

Tactical Scenarios

LNG presents emergency responders with three general tactical problems: (1) fire, (2) vapor cloud release, or (3) cryogenic spill.

Figure 10-31 If an ignition source is encountered, LNG vapors will ignite and burn back to the origin of the spill. In this demonstration at Texas A&M, the LNG is ignited and slowly burns back to the source.

1. **LNG Fires**

LNG vaporizes quickly as it absorbs heat from the surface on which it spills. When LNG vapor concentrations in air are between 5% and 15%, and an ignition source is present, it will burn. At its normal boiling point of −260°F (−160 °C), LNG vapor is 1.5 times denser than air at 77°F (25°C). When LNG vapor is released into the atmosphere it remains negatively buoyant until it warms to approximately −180°F (−117°C); it then rises and disperses below the lower flammable limit. LNG presents three potential fire risk scenarios: pool fire, jet fire, and vapor cloud fire.

- **Pool Fire**—LNG released from a storage tank or transfer pipeline can form a liquid pool. As the spill forms some of the liquid evaporates. If an ignition source is encountered, the vapors will ignite and travel back to the origin of the spill resulting in a pool fire. If the spill occurs inside a properly designed and maintained diked area, the pool fire will remain contained inside and will continue to burn until the fuel is consumed.

  If the spill occurs outside a confined area, the burning pool fire is free to flow based on topography and the geometry of the spill. Spraying water on an LNG pool only increases the vaporization rate and intensifies any fire; spraying a gallon of water will vaporize about two gallons of LNG.

  The preferred extinguishing agent for small LNG fires is a dry chemical agent such as potassium bicarbonate (i.e., Purple K). High expansion foams are not considered to be effective LNG fire extinguishing agents, but they are effective in controlling LNG pool fires in dikes and impoundment areas because the foam blanket reduces the radiant heat generated by the fire. High expansion foams can also prove valuable in vapor control of unignited LNG. When high expansion foam is first applied to the spill, there is some initial warming and an increase in vaporization, but the rate of vaporization eventually stabilizes and slows down the escaping LNG vapor so the flammable region of the spill at ground level is much smaller.
- **Jet Fire**—If there is a release of compressed natural gas or liquefied gases from a storage tank or pipeline, the vapor discharging through the hole in the container will form a gas jet that entrains and mixes with air. If the mixture finds an ignition source while in the flammable range, a jet fire may occur. This type of fire is unlikely for an LNG storage tank since the product is not stored under pressure. However, jet fires could occur in pressurized LNG vaporizers or during unloading or transfer operations when the internal pressure is increased by pumping. A fire occurring under this scenario could cause severe damage, but would be confined to a local area and be limited by safety systems that stop the LNG flow. At base load import terminals, there is little storage of any pressurized liquids, so there is no possibility of a BLEVE.
- **Vapor Cloud Fire**—When LNG is released to the atmosphere a vapor cloud forms and disperses by mixing with air. If the vapor cloud ignites before the vapor

cloud is diluted below the lower flammable limit, a flash fire may occur. Under this scenario, ignition can only occur within the portion of the vapor cloud that has concentrations in the flammable range; i.e., 5% to 15%. The entire cloud does not ignite at once. A flash fire may burn back to the release point producing either a pool fire or a jet fire, but it will not generate damaging overpressures if it is unconfined.

2. **LNG Vapor Cloud Release**
If a LNG vapor cloud with concentrations in the flammable range (5% to 15%) is confined inside a structure and ignited, damaging overpressures may occur. Areas congested with equipment and structures can also help confine LNG vapor and may facilitate an overpressure upon ignition. As indicated earlier, pure methane has not been known to generate damaging overpressures if ignited in an unconfined area. Other vaporized hydrocarbons including propane and butane are more susceptible to vapor cloud explosions.
The LNG production plant explosion that occurred at Skikda, Algeria on January 19, 2004 serves as a good example of the potential of a confined LNG vapor cloud release. At 6:40 pm a steam boiler exploded at the plant after it drew flammable vapors from a hydrocarbon refrigerant leak into its air intake. This triggered a secondary, more massive vapor cloud explosion which destroyed a large portion of the plant. The incident resulted in 27 fatalities and 74 injuries, and created an $800 million loss. The fire and explosion caused material damage outside the plant's boundaries even though none of the LNG storage tanks were damaged. Prior to this tragedy the plant had a good safety record and had operated for over 30 years without a significant incident.

3. **Cryogenic Spills**
Large leaks from containment systems can create pooled LNG cryogenic liquid on the ground or inside diked areas. Contact with cryogenic LNG (–250°F) can cause severe damage to the skin and eyes. Never walk into an LNG spill. Avoid contacting the liquid or any refrigerated pipe, appliance, or tank. Coming into contact with liquid LNG can result in a life-threatening injury.
LNG can make ordinary metals subject to embrittlement and fracture; therefore, cryogenic operations require specialized containers and piping. LNG is stored in containers made of metals such as 9% nickel, steel, or aluminum, and moved through stainless steel pipes capable of handling these low temperatures. Insulation on cryogenic transfer lines protects workers from the potential for contact freeze burns.
LNG carriers are designed with an inner and outer shell or hull that prevents the LNG from coming into contact with the outer shell/hull. International ship design rules require that areas where marine cargo tank leakage might be expected, the metal must be designed for contact with cryogenic LNG. One study conducted in 2001 by international high-risk insurer Lloyd's of London describes 10 LNG spills involving LNG carriers that occurred between 1965 and 1989. Lloyd's reported that 7 of these spills led to brittle fracture of the deck or tank covers on the ship. While the report does not specify the release source, the nature, location, and damage noted suggest that these were all releases from LNG piping.

Note: This section is based on material developed by Hildebrand and Noll for the National Association of State Fire Marshals. For more information on LNG and a more detailed accident history see *Liquefied Natural Gas: An Overview of the LNG Industry for Fire Marshals and Emergency Responders,* National Association of State Fire Marshals: Washington, D.C. (2005).

Reactive Chemicals—Fires and Reactions

"Reactive Chemicals" are a broad family of materials including oxidizers, organic peroxides, certain flammable solids, hypergolics, pyrophorics, and various water reactive substances. Generally speaking, these materials do not like: (1) water, or (2) air, and/or (3) heat.

Fires involving reactive chemicals have resulted in numerous responder injuries and deaths. The following incident summaries spanning more than 65 years make the potential risk of dealing with reactive chemicals very clear:

Texas City, Texas, April 16–17, 1947—The Texas City ammonium nitrate explosion was one of the worst industrial disasters in American history. The explosions of two World War II era Liberty ships loaded with ammonium nitrate on April 16 and 17 killed at least 468 people (25% of the 1947 population of Texas City), injured 3500 people and destroyed more than $700 million in property (in today's monetary value). The blast basically destroyed everything within a 2000-foot radius. The explosions totally destroyed the adjacent Monsanto chemical plant ($20 million loss), burned 1.5 million barrels of petroleum products valued at $500 million, and damaged 50 storage tanks and pipelines within area refineries. One-third of Texas City's 1519 houses were condemned, leaving 2000 people homeless, many of them parentless children. For more information on the historic incident see: *The Texas City Disaster*, 1947 by Hugh W. Stephens, University of Texas Press, Austin, TX (1997).

Roseburg, Oregon, August 7, 1959—Around 1:00 am fire fighters responded to a structural fire at a building supply store. Unknown to fire fighters, a truck containing two tons of dynamite and four-and-a-half tons of the blasting agent nitro-carbo-nitrate had been parked in front of the building. At 1:14 am the truck exploded, killing 14 people and injuring 125. The damage was estimated between $10 to 12 million.

Kansas City, Missouri, November 29, 1988—Six fire fighters were killed instantly when two trailers loaded with 50,000 pounds of ammonium nitrate exploded at a construction site near the 87th street exit of Highway 71. The ammonium nitrate was being stored on-site to support blasting of rock in the construction of Highway 71. Fire fighters were initially dispatched by a report of a pickup truck fire located near the trailers. The responding companies were warned that there were explosives on-site; however, they were unaware that the trailers were essentially magazines filled with explosives. The explosions shattered windows within a 10-mile (16 km) area and could be heard 40 miles (64 km) away.

Newton, Massachusetts, October 25, 1993—Eleven fire fighters were burned, six seriously, one critically, and one extremely critically, in an explosion that occurred while they were attempting to extinguish a sodium fire in a metals processing establishment. A second explosion occurred when fire fighters were attempting to extinguish a residual fire from the initial explosion, subsequently splashing the fire fighters with burning molten sodium.

Los Angeles, California, June 11, 2010 and July 14, 2010—Los Angeles City fire fighters responded to two separate incidents involving two separate titanium storage and recycling facilities. At the July 14th incident, three fire fighters were injured, including two with burn injuries. Fire fighters encountered a number of violent explosions, including large chunks of flaming debris and chunks of titanium flying through the air.

In the section of this text on "hazard assessment and risk evaluation," we note that when a hazmat container breaches, there are only two things jumping out at you—energy and matter. With reactive chemical families, tremendous amounts of energy can be released in an instant. If you are committed to the hot zone when these materials go bad, you probably will not escape without suffering injury.

Chemical reactivity can cover many hazards and properties. These may include the following:

- **Stability**—The resistance of a chemical to decomposition or spontaneous change. Unstable materials would include those which can rapidly and/or vigorously decompose, polymerize, condense, or become self-reactive.
- **Incompatibility**—The chemical reactions and the products of the reactions as a result of the incompatibility will vary with the nature of the chemicals involved. Typical hazards may include the release of toxic materials, the release of flammable materials, the generation of heat, and the destruction of materials.
- **Decomposition Products**—May be produced in dangerous quantities as a result of a chemical reaction or thermal decomposition. This includes toxic products of combustion (e.g., carbon monoxide, hydrogen cyanide, hydrogen chloride) as well as off-gases created by a chemical reaction.
- **Polymerization**—When polymerization occurs spontaneously or without controls, it can give off tremendous levels of heat and pressure. Unplanned polymerization can occur as a result of environmental conditions (excessive heat) or the depletion of inhibitors. The inadvertent introduction of a catalyst can lead to detonation.

Hazard and Risk Evaluation

The hazard and risk evaluation process for reactive chemicals should focus on the following factors:

- **Hazardous nature of the material involved.** Key factors would include the nature of its reactivity (e.g., water, air, heat, other materials, etc.).
- **Quantity of the material involved.** Although risks are often greater when dealing with bulk quantities, small quantities of highly reactive chemicals can pose significant risks. Remember—there is a very fine dividing line between explosives, oxidizers, and organic peroxides. All are capable of releasing tremendous amounts of energy!
- **Design and construction of the container.** The type of container will vary depending upon whether the chemical is a raw material, an intermediate material being used to form another chemical or product, or the finished product.
- **Fixed or engineered safety systems.** Reactive chemical processes and facilities may have a number of engineered safety systems in place, including explosion suppression systems, explosion venting systems (e.g., blowout panels), holding tanks, flares, and scrubbers. Individual containers may also have pressure-relief devices based on the nature of the hazard.
- **Type of stress applied to the hazmat and/or its container.** A number of chemical families will react to the presence of heat and elevated temperatures (not necessarily a fire). Likewise, other chemical families (e.g., oxidizers) can become contaminated with water or dirt, creating heat, and then overpressurize the container.
- **Size and type of area being affected.** The quantity involved will effect the size of the fire or reaction, as well as the likelihood that the problem will be confined to its present size.
- **Identifying and prioritizing exposures.** This includes the proximity of exposures and the rate of release.
- **Level of available resources.** Incidents involving reactive chemicals will typically require the expertise of technical information and product specialists who are inherently familiar with the materials involved.
- **Criminal or terrorist activity.** Reactive chemicals can be used to manufacture devices for arson, assault, or as a weapon of mass destruction. First responders may make the first contact with a terrorist during a traffic stop, traffic accident, or a fire in a building that initially appears to be accidental. See Figure 10-32.

Tactical Objectives

Responders must carefully weigh the hazards and risks of intervention when selecting the strategic goals and tactical objectives. Unlike flammable liquids and gases that leave responders with a lot of options, the tactical options for reactive materials will often be limited unless the fire or reaction can be handled in its initial stages. See the Garfield Heights, Ohio Case Study.

Tactical options include the following:

- **Nonintervention**—This is a "no win" situation in which responders cannot positively change or influence the sequence of events. As a result, responders withdraw to a safe distance and allow the incident to run its natural course. A classic example of nonintervention

Figure 10-32 Reactive chemicals may be used by criminals and terrorists and require assistance by bomb squad and fire and explosion investigators.

would be a vehicle or structure in a remote area containing oxidizers, organic peroxides, or titanium metal powder that is heavily involved in fire. Given the limited benefits, this should be a "no brainer" type of decision.

- **Defensive tactics**—These tactics involve protecting exposures and allowing the fire to burn or the reaction to run its course. In some instances, responders must play a "wait and see" game. Defensive tactics may include implementing a controlled burn, remotely transferring the product to another container, remotely injecting a stabilizer into the reaction, or disposing of the decomposition products by sending them to a flare or scrubber.

If a building containing reactive chemicals is heavily involved in fire, the only option responders may have is to protect exposures. Do not underestimate the rapid speed at which these fires can move. High temperature accelerants, such as flammable metals and pyrotechnics, have been used in a number of arson fires and terrorist incidents in the United States and Canada. A full-scale test fire in a 30,000 ft^2 (2787 m^2) vacant shopping center in Puyallup, Washington resulted in the flashover of the entire structure within 2 minutes!

- **Offensive Tactics**—These tactics are implemented when sufficient resources are available to control and extinguish the fire. However, unless the fire or chemical reaction is observed in its initial stages, there are often limited offensive tactics that can be implemented to change the sequence of events. In addition, offensive tactics will expose responders to a significantly higher level of risk.

Tactical Considerations and Lessons Learned

- Don't wait to ask for help. Unless you work with these chemicals or metals every day—and even if you do—quickly seek out information and assistance from product specialists. The costs of making a mistake at a reactive chemicals incident will often be measured in lives.
- Some reactive chemical incidents may require that both thermal and chemical protective clothing be used simultaneously. For example, a fire of molten sulfur will also result in high levels of sulfur dioxide in the area. When combined with skin moisture and moisture within the respiratory tract, it will also form a mild acidic solution.
- Water reactive materials can react explosively with no warning when they come into contact with even small quantities of water and moisture. When water is applied, results can include steam explosions, burning metal being thrown in all directions, and the production of flammable hydrogen gas, hydroxide compounds, and related toxic gases. Specialized extinguishing agents (i.e., Class D dry powders) will be required for these situations. When dealing with smaller quantities, the best course of action may be to isolate the material outdoors and allow for a controlled burn.
- When dealing with large quantities of strong oxidizers and organic peroxides, responders should consider treating the incident like an explosives fire. (The only difference between some oxidizers and explosives is the speed of the reaction, which is measured in thousandths of seconds). For example, consider the consequences of the May 4, 1988 ammonium perchlorate plant fire and explosion in Henderson, Nevada. The incident killed two employees, injured 300 civilians, and caused $75 million in property damage. This dramatic explosion was caught on video camera taken miles away on a hill looking over the site. When the plant exploded, the viewer can see a powerful shock wave propagating across the ground, pushing air and dust before it, followed by another large explosion. Remember to use distance to your advantage and start emergency evacuations early in the fire.
- Several major fires have occurred involving pool chemicals, such as calcium hypochlorite and chlorinated isocyanurates. Contamination of these materials can lead to the generation of toxic vapors, heat, and oxygen, which can lead to a fire. If large quantities are involved in fire, manufacturers recommend that large, copious amounts of water be used for fire extinguishment. Of course, this may involve a trade-off between air pollution (decomposition products are toxic and will travel a large distance) versus water pollution (runoff is toxic but may not travel as far and can be controlled). The following example demonstrates the potential for these incidents:

Springfield, Massachusetts, June 17, 1988—Rainwater leaked into a storage room where 600 to 800 cardboard drums, each containing 300 pounds of solid swimming pool chemicals were kept. The chemicals exploded, starting a fire which set off the sprinkler system.

The water soaked the remaining drums and set off more explosions, spreading the fire to other rooms in the building. The fire, explosions, and release to air lasted three days. Over 25,000 people were evacuated; 275 people were sent to the hospital with skin burns and respiratory problems.

- Controlled burning may be an appropriate tactical option if extinguishing a fire will result in large uncontained volumes of contaminated runoff, further threaten the safety of both responders and the public, or lead to more extensive clean-up problems. This option is sometimes used at fires involving agricultural chemical facilities or industrial facilities, where runoff may have significant environmental impact upon both surface water and groundwater supplies.
- Chemical process operations at fixed facilities will often have a series of safety features in place in the event of a chemical reaction or decomposition problem. These may include the injection of certain chemicals to reduce the rate of polymerization or "kill" a chemical reaction, the use of emergency tanks in which product can be diverted, gas scrubbers to neutralize toxic or corrosive vapors before being released, and flares to burn off flammable vent gases.

Product Recovery and Transfer Operations

In this section we will discuss the general decision-making criteria that should be made to switch from an emergency response to a recovery operation. We will also discuss basic product recovery and transfer operations and the related safety considerations. Specific product and container recovery and offloading procedures are beyond the scope of this chapter, given the variety of containers, products, and scenarios that might be encountered. Responders should always consult with product and container specialists to develop a safe and efficient plan for recovering product and the container or transport vehicle.

Role of Emergency Responders in Recovery Operations

As we discuss in more detail in the section of this text covering "terminating the incident," determining when the emergency response phase of an incident ends and when the recovery and restoration phase begins can sometimes fall into a gray area. As product and container specialists arrive on the scene to assist, the IC needs to make sure that they understand how and where they fit into the incident command system, and their roles and responsibilities for operating safely at the incident scene. The IC is responsible for making sure that the transition from emergency response phase to the post-emergency response phase is formal, safe, and as seamless as possible.

Public safety emergency response teams have a fairly limited scope of responsibility for clean-up recovery and site restoration. But if the hazmat incident is on public property (e.g., highway), police and fire agencies have a responsibility to ensure that public safety is safeguarded during these operations.

The extent of emergency responder involvement in making the transition to the restoration, recovery, and ultimately post-emergency response operations will vary depending on the type of hazards present, location and extent of the incident, and jurisdiction. Regardless of the situation encountered, the following general activities should be considered during the restoration and recovery phase:

- **PPE requirements**—Personal protective clothing and equipment requirements for restoration should be determined by the IC in consultation with the Incident Safety Officer and other health, safety, or environmental specialists. Requirements should be based upon the results of air monitoring, the potential for reignition, the specific tasks to be accomplished, and other related factors. PPE requirements should take into account the overall safety of restoration personnel, including mobility, comfort, and heat stress.
- **Product/container specialists and contractor safety**—Before transferring site control to nonemergency response personnel, the IC should verify that any environmental spill contractors used for cleanup and recovery operations are trained per the requirements of OSHA 1910.120—*Hazardous Waste Operations and Emergency Response*. Many states now require that hazardous waste contractors be licensed and that they carry credentials listing their level of training and certification. There are many excellent clean-up contractors in the field today; there are also a few bad ones. Don't contribute to an accident by allowing unqualified workers to handle your clean-up. Asking to see credentials at an emergency scene is standard protocol for safety and security reasons. The IC should not hesitate to ask for credentials of clean-up personnel.
- **Sampling**—Residual contamination of the soil, pavement, or surrounding area may require further assessment to determine if they present a threat to public safety. Depending on the nature of the incident, sampling may be used to determine whether the surfaces require decontamination or removal. It may also be necessary to sample soil, surface water, sediments, or groundwater to assess environmental impact.

  Sampling procedures should follow regulatory guidelines, including careful sample technique documentation, chain of custody, and quality assurance/quality control (QA/QC) procedures. If the incident is a crime scene, proper evidence procedures must be followed. If contamination is found that will require specialized clean-up, a written remedial action plan should be developed and coordinated through environmental officials.
- **Disposal concerns**—It may be necessary to dispose of contaminated protective clothing, decontamination solutions, runoff water, or other materials that may be considered as hazardous waste following an emergency. An environmental specialist should be consulted for waste characterization and disposal, as appropriate.

HazMat Responder *in Action*

Case Study: Magnesium Plant Fire, Garfield Heights, Ohio—December 29, 2003

On December 29, 2003 the Garfield Heights Fire Department responded to a magnesium processing plant fire at a metal recycling plant. White-hot flames engulfed the complex and explosions shot sparks into the sky. With prior knowledge of the facility through preplanning, the decision was made by the IC to use a defensive strategy. Evacuation was the first tactical priority, followed by protection of exposures.

Magnesium burns intensely and can explode when it comes into contact with water; consequently the IC decided that no attempt would be made to place water onto the burning metal. Factors that affected decision making included darkness, heavy rain, and the close proximity of a nearby warehouse containing tons of magnesium.

As fire fighters concentrated on protecting a 30,000 ft^2 building (2787 m^2) across the street, the magnesium began to explode. Un-manned master streams were placed to protect several trailers outside the warehouse that were loaded with magnesium. While master streams were being deployed for defensive operations, one of the trailers exploded, knocking fire fighters to the ground, and breaking windows 2000 feet away.

Hundreds of explosions continued throughout the night. A total of five alarms were sounded bringing 18 departments to provide assistance. Preplanning, good site command, and a defensive strategy provided a good outcome to a bad situation. There were no fatalities and no serious injuries to fire fighters.

Source: This case study is based on an article written by Lt. Tom Lisy, Firehouse.com, January 2004. Photos and article used with permission from the Garfield Heights (Ohio) Fire Department.

Site Safety and Control Issues

Product removal operations cannot commence until after the incident site is stabilized. Stabilization means that all fires have been extinguished, ignition sources have been controlled, and all spills and leaks have been controlled, as necessary. Specific site safety considerations that should be addressed during this phase of the incident include the following:

- When flammable or combustible liquids are involved, ensure that backup crews with a minimum of two 1¾-inch foam handlines and at least one 20- to 30-pound dry chemical fire extinguisher are in place to protect all personnel involved in the offloading and uprighting operation.
- Have an escape plan with an alternate escape route. The emergency escape signal must be clearly understood by everyone. An exit pathway out of the work area should be kept clear at all times for personnel working in the immediate hazard area.
- Continuously monitor the hazard area for flammability, toxicity, and oxygen deficiency, as required by the hazards of the materials involved.
- Ensure all personnel remain alert. Both public safety and industry response personnel sometimes become sloppy, less attentive, and may attempt shortcuts as the emergency extends over several hours. Frequent rotation and rehab of personnel can usually minimize this problem.

Product Removal and Transfer Considerations

Specific procedures for product removal and transfer will vary based on the hazmat involved, container design and construction, container stress and actual/potential breach, and the position and location of the container. In addition, shipping documents from the product shipper or manufacturer can also provide guidance and recommendations.

The following are general guidelines and should be used as applicable.

Surveying the Container

The container should initially be surveyed to determine the safest method of offloading. This is particularly true when dealing with bulk transportation containers, such as cargo tank trucks, railroad tank cars, and intermodal tank containers. It should be noted that offloading may reduces stress on a container but is no guarantee that the container won't fail mechanically (e.g., during lifting). Remember the Waverly, Tennessee disaster described earlier. Factors to evaluate include the following:

- The pitch and position of the container. Containers that are upside down can complicate offloading as access to valves may be blocked. If a cargo tank truck or rail car are involved, the pitch and position of the container—front to back and left to right—are particularly important. It's possible that the container may move as product is pumped off and the product load shifts. Even where the unit appears stable, consideration must be given to bracing. Bracing materials may include timber, jacks, or air bags.
- The position and location of the openings or attachments that will be used for product offloading.
- If the container is a liquid cargo tank truck (e.g., MC-306/DOT-406), the position of the baffle holes.
- The product being offloaded (e.g., flammable vs. corrosive vs. poisonous).
- The level of training, resources, and equipment available for product transfer and container uprighting operations.

Bonding and Grounding Considerations

The foundation and justification for electrical bonding and grounding when flammable liquids and gases are involved is well established in NFPA 472—*Standard for Competence of Responders to Hazardous Materials/Weapons of Mass Destruction Incidents*. For more guidance see the *Hazardous Materials/Weapons of Mass Destruction Response Handbook* (2008). Included in the handbook is a special supplement entitled "Bonding and Grounding for Emergency Responders."

Product Transfer Methods

Product transfer and removal operations are normally performed by product and container specialists, as well as environmental contractors. However, public safety responders will often continue to be responsible for site safety and will oversee the implementation of all product transfer and removal operations until the incident is terminated. The information in this section is directed toward these oversight operations. Although much of the information presented here is targeted toward cargo tank trucks, it could also be applied to rail car transfers in the field.

Before discussing the primary methods of product transfer, let's first review the "Golden Rules of Transfers" as taught in the cargo tank truck industry:

1. "If product is being removed from a container, the loss must be replaced by an equal volume of air at a similar rate of flow." *Translation:* If you take something out of a container, you must put something else back in.
2. "If a product is placed in a container, the container must be permitted to displace an equal amount of air/vapor at a similar rate of flow." *Translation:* If you put something into a container, you must take something else out.
3. "Failure to abide by both of these rules may result in a catastrophic failure of the container." *Translation:* No matter how bad your day might be, Rules 1 and 2 do not change!

Liquid Transfers

Liquid transfers would be used on MC-306/DOT-406, MC-307/DOT-407 and MC-312/DOT-412 cargo tank trucks. Product factors that influence the selection of product transfer methods include the weight and viscosity of the liquid, temperature of the liquid, vapor pressure, product hazards, and available resources.

There are three primary methods of liquid product transfer:

1. **Gravity flow**—These are transfer operations relying on the free flow of a liquid product by gravity. It is normally limited to very fluid products, and is often ineffective for thick or

Scan Sheet 10-I

Grounding and Bonding Sequence

What is bonding? Bonding is the process of connecting two or more conductive objects together by means of a conductor; for example, using an approved bonding wire to connect an aircraft being refueled to the fuel truck. Bonding is done to minimize potential differences between conductive objects, thereby minimizing or eliminating the chance of static sparking.

What is grounding? Grounding is the process of connecting one or more conductive objects to the ground through an earthing electrode (i.e., grounding rod)—for example, connecting an aircraft to the ground through an approved grounding wire and connecting the fuel truck to the ground through a separate grounding wire and grounding rod. Grounding is done to minimize potential differences between objects and the ground. An ohm meter is used to measure the electrical resistance and ensure the electrical continuity of bonding and grounding operations.

What is static electricity? Static electricity is an accumulated electrical charge. In order for static electricity to act as an ignition source, four conditions must be fulfilled:

1. *There must be an effective means of static generation.* This can occur when a flammable or combustible liquid is moved from one place to another through pipes, filtering, or by pouring. Some products like gasoline are good static accumulators and can pick up a static charge as they pass through piping during loading operations. Products that easily accumulate static charges must be loaded at slower flow rates to permit downstream relaxation time for the product to lose its charge.
2. *There must be a means of accumulating the static charge buildup.* Not every product lends itself to accumulating a static charge.
3. *There must be a spark discharge of adequate energy to serve as an ignition source (i.e., incendive spark).* We have all experienced a static discharge at one time or another when we walked across a carpet during the winter and touched a metal object or exited a car in winter when the humidity is low, wearing a nylon jacket or wool slacks, and then touched the car door. The spark can be seen jumping between your finger and the car's metal. Not every static spark carries enough energy to cause ignition, and even if it does, there must be an adequate flammable mixture in air present (see item 4 below).
4. *The spark must occur in a flammable mixture.* In order for a fire or explosion to occur in the presence of an adequate ignition source (static spark), the fuel-to-air mixture must fall within the flammable range. By bonding and grounding, you are giving a static charge a pathway in which to travel to earth without creating a spark. The resistance of the grounding field will be affected by weather, type of soil, moisture content of the soil, and the time of year.

Electrical Resistance Guidance

Emergency response organizations should adopt an acceptable electrical resistance level in their Standard Operating Procedures for grounding purposes. For example, the National Electrical Code (NEC) notes that the ground level should be <25 ohms resistance for "residential purposes," a standard that has been adopted by many emergency response agencies.

In 2014, the NFPA 77 *Committee on Recommended Practice on Static Electricity* changed the recommended resistance to 1,000 ohms. Section 7.4.1.3.1.1 of NFPA 77 states: *"In field-based situations such as hazardous materials response operations or flammable/combustible materials spill control and transfer, it might be necessary to establish a temporary or emergency grounding system in a remote location in order to dissipate static charges. In such situations, various types of conductive grounding rods, plates, and wires, which are sometimes used in combination to increase surface area contact with the earth. If the purpose of the temporary grounding system is to dissipate static electricity, a total resistance of up to 1 kohm (1000 ohms) in the ground path to earth is considered adequate. This can be measured using standard ground resistance testing instruments and is realistically and quickly achievable in most types of terrain and weather conditions."* There are many variables that may make achieving 1,000 ohms difficult like location and type of soil.

Grounding and Bonding Sequence for an Overturned Cargo Tank Truck

The figure on next page shows the grounding and bonding sequence for an overturned tank truck. The truck's tractor may be connected to the cargo tank trailer via the fifth wheel or it may have separated from the cargo tank. The cargo tank metal is in contact with the ground. In this configuration, the product must be removed before the cargo tank can be uprighted. This illustration does not depict the proximity or exact spatial layout of the bonding and grounding system, rather, it shows the sequence of the various connections.

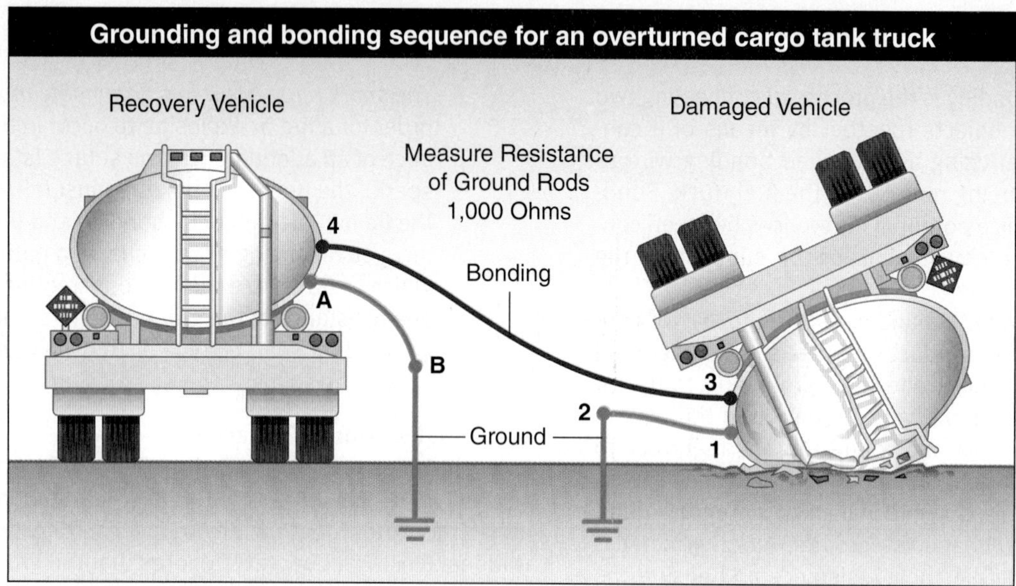

Grounding and bonding sequence for an overturned cargo tank truck.

Note: This illustration does not depict the proximity or exact spatial layout of the bonding and grounding system.

- Make connections in sequence shown to avoid sparks in potentially flammable areas.
- Connections A and B can be made any time prior to pump-off.

viscous liquids, especially if ambient temperatures are cold. Operational considerations include the following:

- Requires that both tanks be vented.
- Allows the use of vapor recovery (i.e., "closed system").
- Gravity flow can be a useful tool when conducting transfer operations when the vehicle is on a hill or large slope area. Gravity transfers are also commonly used for normal transfer operations in fixed facilities, such as the normal offloading of an MC-306/DOT-406 gasoline cargo tank truck at a service station.

2. **Pump transfers**—Transfer operations using pumping systems may be either an "open system" or a "closed system" that allows for vapor recovery or a vapor scrubbing system. Transfer pumps are categorized by their energy source and include gasoline, diesel, power-take-off (PTO), electrical, water, and air-driven pumps. Operational considerations when selecting a pump include the following:
   - Product compatibility. The chemical compatibility of the pump, receiving tank, and all associated hoses and piping is a critical factor. Responders should determine previous products handled by the equipment to ensure there is no residual chemical contamination or reactivity hazards. In addition, product contamination may also be a concern, particularly when dealing with industrial solvents and other high-quality chemicals.
   - Energy source and sparking potential of the pump. Gasoline, diesel, PTO, and electrical pumps can act as an ignition source in a flammable environment. Consider the hazardous classification ratings of any electrical equipment used around flammable liquids or vapors. Class I, Division 2 should be the minimum accepted rating.
   - Hazards of the material being transferred (flammable, corrosive, etc.).
   - Power rating and pressure capacity of the pump, including lift and flow capacities.
3. **Pressure transfers**—Liquid products may be transferred by pressurizing the damaged container with either pressurized air or an inert gas (e.g., nitrogen and CO_2), then moving the product into a receiving container. The regulated pressure creates a positive pressure differential in the damaged

container that pushes the liquid product into the receiving tank. Operational considerations include the following:

- Requires that the tank containing the product to be sealed tight and the receiving tank to be vented. It cannot be used on a "closed system."
- Vapors from the receiving tank may have to be vented to the atmosphere or scrubbed.
- Air transfers cannot be used to transfer flammable or combustible materials, while inert gases permit the transfer of flammables, combustibles, and moisture-sensitive products.
- If an inert gas will be used, ability to acquire sufficient gas in a timely manner to achieve the transfer operation may be an issue. For example, if a bulk source of nitrogen cannot be acquired, consider acquiring 15 to 20 nitrogen cylinders to affect the transfer of a cargo tank truck.

Gas Transfers

Liquefied gases (MC-331 cargo tank trucks) and cryogenic liquids (MC-338 cargo tank trucks) may be transferred through the use of pumps, compressors, and pressure differential. Gas transfers are based on the basic principle that materials will always seek the path of least resistance and will naturally flow from high-pressure to low-pressure areas **Figure 10-33**.

Factors that influence the selection of gas product transfer methods include the nature of stress or damage to the container, position of the container (i.e., upright, on its side, upside down), product vapor pressure, product hazards, and available resources. Gas transfers are always a "closed system" transfer; in essence, there is always a liquid line and a vapor line. Depending upon the position of an MC-331 cargo tank truck, the liquid and vapor eduction lines may be reversed from normal operations.

There are three primary methods of liquid product transfer:

1. **Equalize pressure and pump the liquid**—Under this method, the pressure between the damaged tank and the receiving tank are first equalized. A transfer pump is then used to move the contents into the receiving tank. Operational considerations include the following:

Figure 10-33 Product transfer operation as a result of a MC-331 transport rollover.

   - Using a pump does not increase the internal pressure of the damaged tank. This may be a benefit when dealing with a highly stressed or damaged pressurized container.
   - The operational considerations listed for liquid pumps would also apply here, including pump compatibility, energy source and sparking potential of the pump, and lift and flow capacities. In most instances, a PTO-driven pump on a transport cargo tank truck is used to move the product.
   - Some liquid product will remain in the container, as a pump can only remove liquid product to the bottom of the eduction or dip tube. Industry personnel may refer to this remaining product as a "heel."

2. **Product displacement with compressors**—Similar to a closed system with a pump, the compressor removes vapors from the receiving tank, thereby creating a pressure differential and causing the liquid product to flow from the damaged tank (i.e., high pressure) into the receiving tank (i.e., lower pressure). Compressors can be used to move both liquid and vapor product. Compressor effectiveness and efficiencies are affected by a material's vapor pressure and the ambient temperature. Operational considerations include the following:
   - Compressors are able to move product faster than a pump transfer.
   - Using a compressor will increase the internal pressure of the damaged tank. This may be an issue when dealing with a highly stressed or damaged pressurized container.
   - With a compressor, any remaining liquid and vapors can be pulled off the damaged tank.
   - If there are significant differences in elevation (> 50 feet) or only small liquid lines are available (< 2-inch diameter) to move the liquid product, a booster pump may be required to boost the internal pressure.
   - If there is a concern of building pressure on the damaged container, a smaller container (e.g., 3000-gallon MC-331 bobtail) can be used as the receiving tank. Given the smaller volume of the receiving tank, it is less likely to build up pressure in the damaged tank. The smaller tank can also be used as a nurse tanker, from which a third larger cargo tank (e.g., 10,000 MC-331 transport) would then pull the off the product.

3. **Pressure displacement**—In some instances, an inert gas can be injected into the damaged tank, thereby forcing the liquid to move into the receiving tank. Although not commonly used in the LPG industry, it may be used when dealing with other gas products. Operational considerations include the following:
   - Inert gases permit the transfer of flammables, combustibles, and moisture-sensitive products. If an inert gas will be used, ability to acquire sufficient gas in a timely manner to achieve the transfer operation may be an issue.
   - When dealing with MC-331 cargo tank trucks in LPG service, loading the receiving tank through the spray fill can minimize the pressure buildup in that tank.

Figure 10-34 Vacuum trucks may be used for picking up spilled liquid products, as well as offloading atmospheric pressure and low-vapor-pressure hazardous materials.

Vacuum Trucks

Vacuum trucks may be used for picking up spilled liquid products, as well as offloading atmospheric pressure and low-vapor-pressure hazardous materials. Vacuum trucks rely upon pressure differential to pick up or remove liquid hazardous materials and hazardous waste from an emergency scene. Depending on their rating and design features, they can handle flammable and combustible liquids, corrosives, and some poisons **Figure 10-34**.

A sewer or septic tank pumper is not a vacuum truck rated for hazardous waste. Vacuum trucks are designed for specific operating pressures up to 25 psi (1.72 bar). A vacuum truck generally loads and unloads by reversing its vacuum pump through a four-way valve and manifold, which provides vacuum for loading and pressure for unloading. Some trucks are also equipped with a gear rotary pump for transfer operations.

Vacuum trucks must work close to the damaged hazmat container in order to reach pump-out connections or containment areas. Internal explosions within vacuum truck cargo tanks are rare. However, incidents that have occurred are usually due to pumping incompatible materials inside the tank (e.g., pumping a corrosive into a tank not rated for corrosives) or pumping two incompatible hazardous chemicals into the same tank.

Vacuum trucks can be an ignition source and must be operated at the emergency scene with special precautions. Most vacuum truck fires and explosions are due to either operating the vehicle too close to the spill, pick-up, or discharge point, or failing to vent the vacuum pump discharge to a hazard-free area.

The following safety precautions should be observed when operating vacuum trucks within hazard control zones:

- **Flammable atmosphere test**—Vacuum trucks should not be permitted to enter the hazardous area until flammable vapor monitoring has been conducted. While a reading of zero is preferred, it may not be realistic when open-air spills are being recovered. As a general guideline, concentrations above 10% of the LEL are considered hazardous. Applying foam to a flammable liquid spill before beginning vacuum operations can help maintain the atmosphere within acceptable limits. Monitoring should also continue during the course of the transfer operation.
- **Grounding**—API Publication 2219, *Safe Operation of Vacuum Trucks in Petroleum Service (2005)*, indicates that static electricity does not present an ignition problem with either conductive or nonconductive vacuum truck hoses. However, with nonconductive hoses, any exposed metal, such as a hose flange, can accumulate static electricity and act as an ignition source if the metal touches or comes close to the ground. Since it is often difficult to distinguish between nonconductive and conductive hoses in the field, API recommends that all exposed metal be grounded when any hose is used in other than a closed system with tight connections at both ends of the hose.
- **Venting**—When flammable or toxic liquids are loaded into a vacuum truck, the vacuum pump exhaust should be vented downwind of the truck by attaching a length of hose sufficient to reach an area free from hazards and personnel.
- **Personnel safety**—All unnecessary personnel should leave the area during loading. The vacuum truck driver should leave the truck cab and be in proper PPE. Strict control of ignition sources should be maintained within 100 feet of the truck, the discharge of the vacuum pump, or any other vapor source.

Uprighting Methods

There is no best strategy to handle the offloading or uprighting of hazardous materials containers. Safety should be the number one factor used in making a decision for the best course of action. The decision to offload or upright a rail car, cargo tank truck, or intermodal tank container involving any type of hazmat requires careful consideration and input from a variety of technical specialists. These should include both product and cargo tank specialists and may also include rigging and heavy equipment specialists. All of these groups should be consulted before a plan of action is implemented.

The safety experience in successfully uprighting most cargo tank trucks and rail cars has been very good when (1) the job is performed within the design limitations of the container and the equipment being used for the operations, and (2) safe operating practices are followed. For example, the MC-331 container is an ASME pressure vessel that can sustain heavy mechanical stress and damage without losing its integrity.

Safety Considerations for Approving Lifting

Before any operations are initiated, the IC must verify that the personnel performing the lifting and uprighting operation are trained and qualified to perform the expected tasks. History has taught us some ugly lessons that there is a big difference between uprighting an overturned eighteen-wheeler and an MC-312/DOT-412 corrosive tank truck involved in a rollover. While there are many first-rate riggers in the business, there are also plenty of well-meaning tow truck operators and salvage companies who simply do not know what they are doing. The towing industry is doing a very good job addressing the need for standards and has implemented many training and certification programs to qualify their personnel. The bottom line—make sure the people and equipment on your incident scene are qualified for the job.

Responder Safety Tip

Do not trust your life to people who do not know what they are doing and are using the wrong equipment for the job at hand. Minimize the number of people in the hot zone when lifting operations take place. If you must be close, stay well clear of chains, cables, and straps that are under tension. If they fail they can kill you.

Factors that responders must consider before approving a lifting or uprighting operation include the following:

1. **Conduct a damage assessment of the container before moving or lifting the tank.**

   Containers such as MC-306/DOT-407 aluminum shell cargo tank trucks are not designed to be lifted and uprighted fully loaded. Container specialists, riggers, and crane operators must have access to damage assessment information. Keep in mind that damage assessment is, at best, a qualitative assessment process that requires input from product and container specialists who have experience in evaluating structural damage. For example, is the tank stable versus nonstable, have go/no go safety factors been evaluated, what is the practical field experience with the type of container involved, and so on.

2. **Qualified and experienced rigging specialists and crane operators must be available to supervise and perform the uprighting or lifting operation.**

   Operating hoisting equipment can be a very complex and dangerous process, especially under emergency conditions. A qualified rigger and crane operator should know a wide range of specialized information, including knowledge of:

   - **What can go wrong and how to deal with situations when they go bad.** Riggers and operators must understand the changing conditions of the job and how these conditions will affect the overall safety of the operation. Examples include changes in wind direction, changes in temperature, formation of ice on the ground or lifting surface, and so on.
   - **Is the load properly rigged?** A qualified rigger and operator must be able to read a load chart that shows the lifting capacities for various crane operations.
   - **Cables, chains, ropes, and slings.** Riggers and operators must have a thorough knowledge of the rigging used to lift and stabilize the loads and have the basic training to inspect rigging for problems. Rigging is critical since it forms the interface between the hoisting equipment and the load. The main hazards of rigging are failure due to excessive overloading, deterioration or wear, and improper rigging techniques. Load capacity charts are available for materials handling equipment and rigging.
   - **Safe lifting operations.** An operator must have the ability to evaluate the site for safe lifting operations. For example, for a safe lift with a mobile crane, at least seven items of information are needed: (1) Is the vehicle level? (2) Are outriggers properly extended or retracted? (3) Are extended outriggers supported by stable ground? (4) What is the angle of the boom? (5) What is the boom and jib length? (6) What positions will the boom be in during the lift? and (7) How much does the load weigh?

Operational Considerations

The following are general guidelines that emergency responders should consider while overseeing and approving a cargo tank truck uprighting operation.

1. **Offload or upright?** The decision to either (1) offload the contents and then upright the container or (2) upright the container while still loaded, will be dependent upon a number of variables, including the type of cargo tank truck involved, the nature of container stress and damage, the location of the incident, and resources available to lift the damaged container. Speed, the impact terrain, level of the surface, and weight of the product load will all contribute to the stressing of the container **Figure 10-35**.

   There is only one universal rule—never upright a loaded aluminum-shell MC-306/DOT-406 cargo tank truck. Everything else will be incident specific.

2. **Use enough lift.** Experience shows that in most rollovers the cargo tank truck will be lying on its side. Depending on the type of cargo tank, it may be impossible to remove all of the product load. For example, an MC-331 lying on it side can only have approximately 40 to 50% of the contents removed because the liquid and vapor eduction tubes are at the 3 o'clock/9 o'clock position. While offloading 40% of the load will certainly reduce the weight of the container, it does not change the internal pressure of the contents **Figure 10-36**.

Figure 10-35 The decision to upright or offload a tank truck depends on many variables.

Figure 10-36 Offloading a tank reduces its weight but does not change the internal pressure of the contents.

Figure 10-37 If the vehicle can move or slide during lifting, it should be stabilized.

Make sure you have enough lifting capacity for the situation. An MC-331 transport weighs up to 80,000 pounds when fully loaded. Mobile cranes and heavy wrecker cranes should be used for the uprighting operation, as they typically have boom and lift ratings of 30 tons and higher. Although a 3500-gallon propane bobtail may be uprighted with one wrecker, transports and tractor trailer combinations will require two and sometimes even three wreckers for the uprighting operation.

1. **Stabilize the vehicle.** If the container is in an unstable position where it could shift and injure workers or slide farther off a roadway, shoulder, bridge, and so on, take time to stabilize the entire rig using cables, cribbing, and the like Figure 10-37.

   Tractor and trailer combinations should be lifted as a single unit; separating the tractor from the trailer could contribute to instability. If the tractor becomes separated from the trailer during the rollover, it may be necessary to use a fifth-wheel dolly to support the trailer during the uprighting operation. If the trailer has both a front and rear axle, the axles should be cross-chained together to keep them parallel.

Figure 10-38 If the vehicle is on its side, air bags or mat jacks may be required to gain clearance for the lifting straps.

4. **Position all lifting equipment.** If the container is on its side, wrecker personnel can have difficulty in getting lifting straps or cables under the tank. To facilitate this process, "mat jacks" or airbags can be used to raise the tank a few inches so that the lifting straps can be placed under the tank. Mat jacks are usually rated for over 20 tons of lift and can be inflated from the wrecker's air tanks. *Note:* Do not use hydraulic rescue tools to lift the container, as they may cause further damage to the container shell Figure 10-38.

   Most wreckers use nylon straps 12 to 24 inches wide when lifting cargo tank trucks. These allow the weight to be distributed over a larger surface area as the container is uprighted. While cables may be used, they should be attached to the chassis and should not be placed around the tank shell, as they can cause further damage.

   When dealing with liquid cargo tank trucks, straps should be placed at the strongest points of the container, such as at the fifth wheel, at tank baffles, and at compartment bulkheads. If straps are positioned between compartment bulkheads, the tank may be "crunched" during the uprighting process.
5. **Upright the damaged container.** Most cargo tank truck uprighting operations require one or two wreckers providing the lift, while another wrecker "holds" the cargo tank to prevent it from tipping over. Responders must be aware of the potential for cables or straps to break and whip around, causing serious injury. One individual should be responsible for controlling the overall uprighting operation and to minimize the number of personnel working in the immediate area Figure 10-39.

   Once the overturned container is uprighted, it should be chocked so that the vehicle does not begin to roll away. Even if the tires are flat, a trailer unit can usually be towed a short distance where it can be safely offloaded. However, if it is badly damaged, it may be eventually placed upon a "low boy" or a flatbed trailer and transported from the scene.

For more detailed information on offloading and uprighting damaged MC-331 cargo tank trucks consult *Propane Emergencies* (3rd Edition), Chapter 9 by Hildebrand and Noll.

Figure 10-39 Most cargo tank truck rollovers require one or more heavy duty wreckers to complete the job safely.

Figure 10-40 A wide variety of chemicals available over the counter can be used to improvise weapons by criminals and terrorists.

Evidence Collection in a Hazardous Environment

Hazardous materials have been used by criminals to commit various crimes including murder, arson, extortion, and in the manufacture of illegal explosives and drugs. Terrorists have also improvised weapons of mass destruction using a wide range of common hazardous materials and containers Figure 10-40.

Astute and properly trained first responders have been invaluable in helping law enforcement agencies open and prosecute important criminal and terrorism cases because they: (1) recognized that they were involved in a crime scene, 2) took the correct steps to recognize and preserve potential evidence, 3) rapidly called for assistance from law enforcement agencies, and 4) operated as part of the public safety team to support crime scene operations when requested by the investigating agencies.

When hazardous materials are involved at a crime scene, all evidence preservation and collection operations fall under the health and safety requirements of OSHA 29 CFR 1910.120 (*Hazardous Waste Operations and Emergency Response*) regardless of jurisdiction, agency affiliation, or mission.

Securing and Preserving the Scene

Once an incident is identified as being criminal in nature or potential evidence is observed, the scene must be secured immediately. Securing a crime scene may involve placing hazard tape around the scene's boundaries or assigning personnel to limit access of other nonessential responders and particularly the public.

Early characterization of the scene will also be important. Attempt to identify, at least in broad terms, the type of crime suspected, any major hazards associated with the scene, and the location of significant pieces of evidence. Once suspected evidence is identified, responders should take steps to preserve it. Tactics for protecting evidence may include:

- Physically protect the evidence.
- Removing the evidence from water runoff that occurred during initial firefighting or response operations.
- Placing a larger container over the evidence.
- Place hazard tape, traffic/safety cones, or a larger object near the evidence to prevent it from being stepped on or kicked.
- Any evidence markings or identification must be nondestructive and cannot alter the state of the evidence.

Simultaneously with their efforts to secure the crime scene, responders should notify the law enforcement agency with investigative authority, if members of that agency are not already on-scene. Communication with the investigative authority is also essential whenever an explosive device is suspected to be involved, to ensure that the appropriate explosive ordnance disposal (EOD) personnel are notified.

Responders must maintain an awareness of the potential causation of hazmats and work to detect and protect any evidence that may be discovered during the response efforts. When evidence can be used in the legal process to establish a fact or prove a point, it is often referred to as forensic evidence. To be admissible in court, evidence must be gathered and processed under strict procedures.

To secure, characterize, and preserve the scene, follow the steps in Skill Drill 10-4:

1. Characterize the scene by assessing the number of victims and property damage, if any, and the type (such as a liquid or a solid) and quantity of evidentiary materials on site. **(photo 1)**
2. Secure the scene by placing caution or hazard tape so as to limit access to the scene. **(photo 2)**
3. Preserve any suspected evidence by protecting it from being disturbed. **(photo 3)**

Skill Drill 10-4

Securing, Characterizing, and Preserving the Scene

1. Characterize the scene by assessing the number of victims and property damage, if any, and the type and quantity of evidentiary materials on site.

2. Secure the scene by placing caution or hazard tape so as to limit access to the scene.

3. Preserve any suspected evidence by protecting it from being disturbed.

To collect and process evidence, follow the steps in Skill Drill 10-5:

1. Take photographs of each piece of evidence as it is found and collected. If possible, photograph the item exactly as it was found, before it is moved or disturbed. **(photo 1)**
2. Sketch, mark, and label the location of the evidence. Sketch the scene as near to scale as possible. **(photo 2)**
3. Place evidence in appropriate containers to ensure its safety and prevent contamination. Unused paint cans with lids that automatically seal when closed are the best containers for transporting evidence. Clean, unused glass jars sealed with sturdy sealing tape are appropriate for transporting smaller quantities of materials. Plastic containers and plastic bags should not be used to hold evidence containing petroleum products because these chemicals may lead to deterioration of the plastic. Paper bags can be used for storage of dry clothing or metal articles, matches, or papers. Soak up small quantities of liquids with either a cellulose sponge or cotton batting. Protect partially burned paper and ash by placing them between layers of glass (assuming that small sheets or panes of glass are available at the scene). **(photo 3)**
4. Tag all evidence. Evidence being transported to the laboratory should include a label with the date, time, location, discoverer's name, and witnesses' names. **(photo 4)**
5. Record the time when the evidence was found, the location where it was found, and the name of the person who found it. Keep a record of each person who handled the evidence. **(photo 5)**
6. Keep a constant watch on the evidence until it can be stored in a secure location. Evidence that must be moved temporarily should be put in a secure place that is accessible only to authorized personnel.
7. Preserve the chain of custody in handling all the evidence. A broken chain of custody may result in a court ruling that the evidence is inadmissible.

Skill Drill 10-5

Collecting and Processing Evidence

1. Take photographs of each piece of evidence as it is found and collected.

2. Sketch, mark, and label the location of the evidence.

3. Place evidence in appropriate containers to ensure its safety and prevent contamination. Soak up small quantities of liquids with either a cellulose sponge or cotton batting. Protect partially burned paper and ash by placing them between layers of glass (assuming that small sheets or panes of glass are available at the scene).

4. Tag all evidence. Evidence being transported to the laboratory should include a label with the date, time, location, discoverer's name, and witnesses' names.

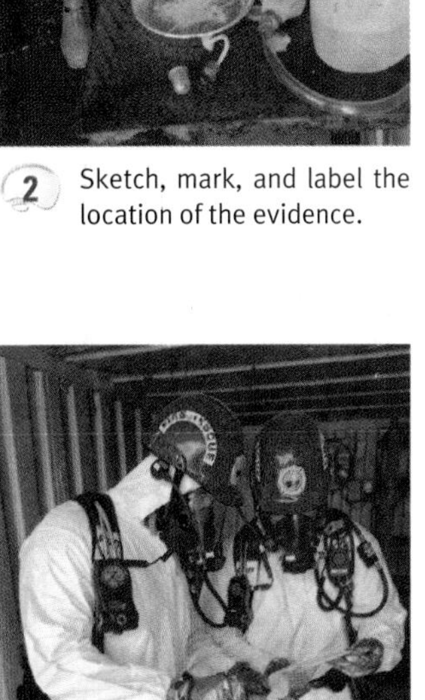

5. Record the time when the evidence was found, the location where it was found, and the name of the person who found it. Keep a record of each person who handled the evidence.

Wrap-Up

Chief Concepts

- The Incident Commander (IC) is responsible for determining the best strategic goals and tactical objectives which will produce the most favorable outcome of the incident.
- The operational strategy for an incident is developed based upon the IC's evaluation of the current conditions and a forecast of future conditions. The effectiveness of this phase of the incident is directly related to how well the hazards were identified and the risks evaluated.
- The IC's hazmat strategic goals include rescue, public protective actions, spill control (confinement), leak control (containment), fire control, and transfer and recovery.
- Tactics are the specific objectives the IC uses to achieve strategic goals. Tactics are normally decided at the section or group/division levels in the command structure.
- Strategy and tactics can be implemented by the IC in the offensive, defensive, or nonintervention mode. Usually, the IC uses a combination of tactics to manage the problem.
- Saving lives is the IC's number one mission! Life safety should always be the IC's highest priority, but remember that in some cases doing nothing and letting the incident run its course is the smartest and safest strategy. As emergency responders we cannot save everyone, and time often works against the responders and the people you are trying to rescue. The IC must weigh the chance for a successful rescue against the hazards and risks.
- Product removal and recovery operations usually begin after the emergency has run its course (e.g., all leaks have been controlled). Product removal and recovery operations should not begin until after the incident site is stabilized and the area has been re-evaluated for hazards and risks. Stabilization means that all fires have been extinguished, ignition sources have been secured, and all product releases have been controlled.
- Product removal and transfer operations involve moving the contents from the damaged or overloaded cargo tank(s) into an undamaged and compatible receiving tank(s), such as a tank car, cargo tank truck, intermodal tank, or fixed tank.
- Product transfer and removal operations are typically performed by either product/container specialists or environmental contractors working on behalf of the carrier or shipper. Public safety responders will often continue to be responsible for site safety and will oversee the implementation of all product transfer and removal.
- Understanding our emergency response culture and history provides perspective. Stated a different way, understanding where we have been in the past helps you better understand where you need to go in the future. Doing it better and safer in the future is our goal!
- As a hazmat professional it is not good enough to simply know what the regulations and standards require; you should understand how and why these standards evolved. Many fire fighters and law enforcement officers gave their lives to help keep their communities safe from hazardous materials. Others have suffered debilitating injuries. Learn from their lessons!

Hot Terms

Absorption 1) The process of absorbing or "picking up" a liquid hazardous material to prevent enlargement of the contaminated area; 2) movement of a toxicant into the circulatory system by oral, dermal, or inhalation exposure.

Adsorption Process of adhering to a surface, a common physical method of spill control and decontamination.

Covering A physical method of confinement; typically a temporary measures.

Damming A physical method of confinement by which barriers are constructed to reduce the quantity of liquid flowing into the environment.

Dilution Chemical method of confinement by which a water-soluble solution, usually a corrosive, is diluted by adding large volumes of water to the spill.

Dispersion Chemical method of confinement by which certain chemical and biological agents are used to disperse or break up the material involved in liquid spills on water.

Diking A physical method of confinement by which barriers are constructed on ground used to control the movement of liquids, sludge, solids, or other materials.

Diversion A physical method of confinement by which barriers are constructed on ground or placed in a waterway to intentionally control the movement of a hazardous material into an area where it will pose less harm to the community and the environment.

Liquefied Natural Gas (LNG) A flammable and odorless gas, LNG is methane in its liquid state. While it is nontoxic, it can be an asphyxiant when it displaces oxygen in a confined space.

Neutralization A chemical method of containment by which a hazmat is neutralized by applying a second material to the original spill which will chemically react with it to form a less harmful substance.

Overpacking Use of a specially constructed drum to overpack damaged or leaking containers of hazardous materials for shipment. Overpack containers should be compatible with the hazards of the materials involved.

Patching/Plugging A physical method of containment, which uses chemically compatible patches and plugs to reduce or temporarily stop the flow of materials from small container holes, rips, tears, or gashes.

Retention A physical method of confinement by which a liquid is temporarily contained in an area where it can be absorbed, neutralized, or picked up for proper disposal.

Solidification A chemical method of containment whereby a liquid substance is chemically treated so that a solid material results.

Overflow Dam Used to trap sinking heavier-than-water materials behind the dam

Pressure Isolation and Reduction A physical or chemical method of containment by which the internal pressure of a closed container is reduced.

Underflow Dam Used to trap floating lighter-than-water materials behind the dam

Vapor Dispersion A physical method of confinement by which water spray or fans is used to disperse or move vapors away from certain areas or materials. It is particularly effective on water-soluble materials (e.g., anhydrous ammonia), although the subsequent runoff may involve environmental trade-offs.

Vapor Suppression A physical method of confinement to reduce or eliminate the vapors emanating from a spilled or released material.

HazMat Responder *in Action*

You are on the scene where a chemical spill is flowing into a river and you need to decide the tactical objectives you would like to accomplish within the next few hours.

Questions

1. The construction of a barrier across a waterway to stop/control the product flow and pick up the liquid or sold contamination is an example of:
   A. Retention
   B. Damming
   C. Diversion
   D. Dispersion
2. Using a chemical dispersant to disperse or break up the material involved in liquid spills in water is called:
   A. Retention
   B. Damming
   C. Diversion
   D. Dispersion
3. Temporarily containing a liquid in an area where it can be absorbed, neutralized, or picked up for proper disposal is an example of:
   A. Retention
   B. Damming
   C. Diversion
   D. Dispersion
4. Barriers constructed on the ground or placed in a waterway to intentionally control the movement of a hazardous material into an area where it will pose less harm to the community and the environment is an example of:
   A. Retention
   B. Damming
   C. Diversion
   D. Dispersion

References and Suggested Readings

1. Air Force Civil Engineer Support Agency (AFCESA) and PowerTrain, Inc., *Hazardous Materials Incident Commander Emergency Response Training*. Tyndall Air Force Base, FL: Headquarters AFCESA (2010).
2. American Petroleum Institute, *Fire Protection Considerations for the Design and Operation of Liquefied Petroleum Gas (LPG) Storage Facilities* (2nd Edition), API 2510-A, Washington, DC: (December 1996).
3. American Petroleum Institute, *Guidelines and Procedures for Entering and Cleaning Petroleum Storage Tanks* (1st Edition), ANSI/API RP 2016, Washington, DC: (Reaffirmed May 2006).
4. American Petroleum Institute, *Guidelines for Work in Inert Confined Spaces in the Petroleum Industry,* (4th Edition), API Publication 2217A, Washington, DC: (July 2009).
5. American Petroleum Institute, *Management of Atmospheric Storage Tank Fires,* (4th Edition), API Publication 2021, Washington, DC: (Reaffirmed May 2006).
6. American Petroleum Institute, *Prevention and Suppression of Fires in Large Aboveground Atmospheric Storage Tanks* (1st Edition), API Publication 2021A, Washington, DC: (1998).
7. American Petroleum Institute, *Safe Access/Egress Involving Floating Roofs of Storage Tanks in Petroleum Service* (2nd Edition), API Publication 2026, Washington, DC: (Reaffirmed June 2006).
8. American Petroleum Institute, *Protection Against Ignitions Arising Out of Static, Lightning, and Stray Currents* (7th Edition), API Publication 2003, Washington, DC: (January 2008).
9. American Petroleum Institute, *Safe Operation of Vacuum Trucks in Petroleum Service* (3rd Edition), API Publication 2219, Washington, DC: (November 2005).
10. Andrews, Jr., Robert C., "The Environmental Impact of Firefighting Foam." *Industrial Fire Safety* (November/December, 1992), pages 26–31.
11. Arnold, David, "Water-Sodium Mix Set off Newton Blast," *The Boston Globe,* Thursday, October 28, 1993, page 22.
12. Association of American Railroads—Transportation Test Center (AAR/TTC), "Grounding and Bonding." Student Handout from the Hazardous Materials Response Curriculum, Pueblo, CO: Association of American Railroads (1999).
13. Bachir, Achour and Ali Hached, "*The Incident at the Skikda Plant: Description and Preliminary Conclusions,*" LNG14, Session 1, 21, Technical Paper presented at the LNG Conference at DOHA-Qatar, (March 21, 2004.)
14. Benner, Ludwig, Jr., "D.E.C.I.D.E. In Hazardous Materials Emergencies." *Fire Journal* (July, 1975), pages 13–18.
15. Benner, Ludwig, Jr., *Hazardous Materials Emergencies* (2nd Edition), Oakton, VA: Lufred Industries, Inc. (1978).
16. Bevelacqua, Armando, *Hazardous Materials Chemistry* (3rd Edition), Albany, NY: Cengage Learning (2012).
17. Bevelacqua, Armando and Richard Stilp, *Hazardous Materials Field Guide* (2nd Edition), Albany, NY: Cengage Learning (2007).
18. Bevelacqua, Armando and Richard Stilp, *Terrorism Handbook for Operational Responders* (3rd Edition), Albany, NY: Cengage Learning (2010).
19. Bradish, Jay, "The Fatal Explosion," A Special Report of the NFPA Investigations and Applied Research Division on the December 27, 1983 Propane Explosion in Buffalo, New York. *Fire Command* (March 1984) pages 28–33.
20. Callan, Michael, *Street Smart Hazmat Response,* Chester, MD: Red Hat Publishing (2002).
21. Callan, Michael, *Responding to Utility Emergencies,* Chester, MD: Red Hat Publishing (2004).
22. Compressed Gas Association, Inc., *Handbook of Compressed Gases* (4th Edition). Norwell, MA: Kluwer Academic Publishers (2003).
23. Crowder, Don, "Emergency Transfers Utilizing Compressors and Pumps." Student Handout from 2003 Propane Emergencies Industry Responders Conference, Washington, DC: Propane Education and Research Council (2003).
24. Davenport, John, A., "Section 7, Chapter 4—Storage and Handling of Chemicals," *Fire Protection Handbook,* (20th edition), Quincy, MA: National Fire Protection Association (2008).
25. Emergency Film Group, *Oil Spill Response* (four-videotape series), Edgartown MA: Emergency Film Group (2003).
26. Emergency Film Group, *Foam* (DVD), Edgartown, MA: Emergency Film Group (2011).
27. Emergency Film Group, *Petroleum Storage Tanks* (DVD), Edgartown, MA: Emergency Film Group (2003).
28. Emergency Film Group, *The Eight Step Process: Step 1 —Site Management and Control* (DVD), Edgartown, MA: Emergency Film Group (2005).

29. Emergency Film Group, *The Eight Step Process: Step 6—Implementing Response Objectives* (DVD), Edgartown, MA: Emergency Film Group (2005).
30. Emergency Film Group, *Intermodal Containers* (DVD), Edgartown, MA: Emergency Film Group (2002).
31. Emergency Film Group, *Cylinders—Container Emergencies* (DVD), Edgartown, MA: Emergency Film Group (2002).
32. Emergency Film Group, *Confined Space Emergency* (DVD), Edgartown, MA: Emergency Film Group (2003).
33. Emergency Film Group, *Terrorism Response,* Nine-part DVD series on Weapons of Mass Destruction, Edgartown, MA: Emergency Film Group:

    Detecting Weapons of Mass Destruction (2003)
    Response to Anthrax Threats (2003)
    Terrorism: 1st Response (2002)
    Terrorism: Biological Weapons (2007)
    Terrorism: Chemical Weapons (2007)
    Terrorism: Explosive & Incendiary Weapons (2003)
    Terrorism: Medical Response (2002)
    Terrorism: Radiological Weapons (2004)
    Terrorism: Roll Call Edition (2005)
34. Federal Emergency Management Agency—U.S. Fire Administration, Report on Tire Fires, Emmitsburg, MD: FEMA (1999).
35. Federal Emergency Management Agency—U.S. Fire Administration, Technical Report on High Temperature Accelerant Arson Fires, Emmitsburg, MD: FEMA (1991).
36. Fire, Frank L., *The Common Sense Approach To Hazardous Materials* (3rd Edition), Tulsa, OK: PennWell Corporation (2009).
37. Goodland, Larie, "Sorbents Selection" *Hazardous Materials Management* (February/March, 2002), page 18.
38. Hawley, Chris. *Hazardous Materials Incidents* (2nd Edition), Albany, NY: Cengage Learning (2007).
39. Hawley, Chris, Gregory G. Noll and Michael S. Hildebrand, *Special Operations for Terrorism and Hazmat Crimes.* Chester, MD: Red Hat Publishing, Inc. (2002).
40. Hawthorne, Edward, *Petroleum Liquids—Fire and Emergency Control.* Englewood Cliffs, NJ: Prentice Hall (1987).
41. Hildebrand, Michael, S., "*The American Petroleum Institute Response to the Winchester, Virginia Oil Spill and Tire Dump Fire.*" A technical paper presented to the API Operating Practices Committee, OPC Paper 9.2-20 (D-120) (1984).
42. Hildebrand and Noll Associates, Inc. "*Liquefied Natural Gas: An Overview of the LNG Industry for Fire Marshals and Emergency Responders,*" National Association of State Fire Marshals: Washington, DC (2005).
43. Hildebrand, Michael S. and Gregory G. Noll, *Propane Emergencies* (3rd Edition), Washington, DC: Propane Education and Research Council (2007).
44. Hildebrand, Michael S. and Gregory G. Noll, *Storage Tank Emergencies,* Chester, MD: Red Hat Publishing, Inc. (1997).
45. Hildebrand, Michael, S., *Hazmat Response Team Leak and Spill Control Guide,* Stillwater, OK: Fire Protection Publications, Oklahoma State University (1992).
46. International Association of Fire Chiefs, et al., "*Guidelines for the Prevention and Management of Scrap Tire Fires.*" Washington, DC: International Association of Fire Chiefs (1993).
47. Marsh Global Energy Practice, "*The 100 Largest Losses 1972–2009: Large Property Damage Losses in the Hydrocarbon Industries,*" (21st Edition), New York, NY (2010).
48. Major Incident Investigation Board, "*The Buncefield Incident 11 December 2005, Volume 1,*" Health and Safety Executive: Kew, Richmond, Surrey, England (2008).
49. Meal, Larie, "*Static Electricity.*" *Fire Engineering* (May, 1989), pages 61–64.
50. Meyer, Eugene, *Chemistry of Hazardous Materials* (5th Edition) Saddle River, NJ: Prentice Hall (2009).
51. National Fire Protection Association, *NFPA 30 and NFPA 30A—Flammable and Combustible Liquids Code Handbook,* Quincy, MA: National Fire Protection Association (2012).
52. National Fire Protection Association, *Hazardous Materials/Weapons of Mass Destruction Response Handbook,* Quincy, MA: National Fire Protection Association (2008).
53. National Fire Protection Association, *LP-Gas Code Handbook,* Quincy, MA: National Fire Protection Association (2011).
54. National Fire Protection Association, "Standard for Low, Medium, and High Expansion Foam," NFPA 11, Quincy, MA: National Fire Protection Association (2010).
55. National Fire Protection Association, "Recommended Practice on Static Electricity," NFPA 77, Quincy, MA: National Fire Protection Association (2007).

56. National Institute for Occupational Safety and Health, "Control of Unconfined Vapor Clouds by Fire Department Water Spray Handlines," (1987).
57. Noll, Gregory G., Michael S. Hildebrand and Michael L. Donahue, *Hazardous Materials Emergencies Involving Intermodal Containers,* Stillwater, OK: Fire Protection Publications—Oklahoma State University (1995).
58. Noll, Gregory G., Michael S. Hildebrand, and Michael L. Donahue, *Gasoline Tank Truck Emergencies: Guidelines and Procedures* (2nd Edition), Stillwater, OK: Fire Protection Publications—Oklahoma State University (1996).
59. Pratt, Thomas, H., *Electrostatic Ignitions of Fires and Explosions,* Hoboken, NJ: John Wiley & Sons, Inc. (1997).
60. Rhoads, Gregory, A., *Emergency Responder's Guide to Railroad Incidents,* Chester, MD: Red Hat Publishing (2007).
61. Sargent, Chase, *Confined Space Rescue,* Saddle Book, NJ: Fire Engineering Books & Videos, (2000).
62. Sachen, John B., "Field Transfers: Drums." *Fire Engineering* (April, 1992), pages 66–67.
63. Safe Transportation Training Specialists, "Low Pressure Tank Truck Construction and Emergency Response." Student Handout, Carmel, IN: Safe Transportation Training Specialists (2003).
64. Santis, Lon D., "Section 6, Chapter 15—Explosives and Blasting Agents," *Fire Protection Handbook* (20th Edition), Quincy, MA: National Fire Protection Association (2008).
65. Schnepp, Rob, *Hazardous Materials Awareness and Operations,* Sudbury, MA: Jones & Bartlett Learning (2010).
66. Shannon, Michael, "The Use of Cranes and Heavy Equipment in Rescue and Hazmat," *Fire Engineering* (March, 1999), pages 145–152.
67. Silensky, Philip. "Plugging and Patching Plastics," *Fire Engineering* (February, 1992), pages 39–43.
68. Skandia International Insurance Corp., *BLEVE! The Tragedy of San Jacinto,* Stockholm, Sweden: Skandia International Insurance Corporation (1985).
69. Slaughter, Rodney A., "Rings of Fire—Handling Tire Fires," Technical paper presented at the Continuing Challenge Hazardous Materials Conference, Sacramento, CA, September 10, 1993.
70. Slye, O. M., Jr., "Section 6, Chapter 12—Flammable and Combustible Liquids," *Fire Protection Handbook,* (20th Edition), Quincy, MA: National Fire Protection Association (2008).
71. Smith, James P., "History As A Teacher: How A Refinery Blaze Killed 8 Firefighters," *Fire House* (December 1997) pages 16–20.
72. Smith, James P., *Strategic and Tactical Considerations on the Fireground* (2nd Edition), Upper Saddle River, NJ: Prentice Hall (2008).
73. Stephens, Hugh W., *The Texas City Disaster*—1947, University of Texas Press: Austin, TX (1997), pp. 1–6.
74. Stuebe, Peter M., "Burning Gasoline Tankers: The Best Action May Be No Action," Fire Engineering (November, 1997), pages 41–46.
75. Tunkel, Steven J., "Fire and Explosion Hazards," A Technical Paper presented at the CONOCO Worldwide Corporate Safety Meeting, Lake Charles, LA, September 17, 1985.
76. U.S. Environmental Protection Agency, Office of Solid Waste and Emergency Response, *Safe Storage and Handling of Swimming Pool Chemicals,* EPA 550-F-01-003 Washington, DC (March, 2001).
77. U.S. Department of Transportation, Federal Railroad Administration, *Field Product Removal Methods for Tank Cars,* Washington, D.C. (February, 1993).
78. U.S. Fire Administration, *Sodium Explosion Burns Firefighters, Newton, Massachusetts,* Washington, DC (1993).
79. U.S. Occupational Safety and Health Administration (OSHA), *Permit Required Confined Spaces for General Industry Standard* (29 CFR 1910.146): Washington, DC.
80. U.S. Occupational Safety and Health Administration, *Selected Occupational Fatalities Related to Toxic and Asphyxiating Atmospheres in Confined Work Spaces as Found in Reports of OSHA Fatality/Catastrophe Investigations,* Washington, DC (July, 1985).
81. Virginia Department of Emergency Management and Technical Hazards Division, "Grounding and Bonding Applications for Emergency Responders." Student handout as part of the Hazardous Materials Technician Training Program (2003).
82. Walls, Wilbur, L., "The BLEVE—Part 1," *Fire Command* (May, 1979), pages 22–24.
83. Walls, Wilbur, L., "The BLEVE—Part 2," *Fire Command* (June, 1979), pages 35–37.
84. Williams, Dwight P., "Over the Top: Techniques and Logistics for Extinguishing Large Tank Fires," *Industrial Fire Safety* (November/December, 1992), pages 21–25.
85. Wray, Thomas K., "Pre-Emergency Planning: Neutralizing Acid Spills," *Hazmat World* (September, 1991), pages 86–87.

Decontamination

CHAPTER 11

Hazardous Materials Technician

Knowledge Objectives

After reading this chapter, you will be able to:

- Describe the following items and their significance in decon and site safety operations:
  - Contaminant
  - Contamination and exposure
  - Surface contamination and permeation contamination
  - Direct contamination and cross contamination
  - Decontamination
  - Disinfection and sterilization
- Given the plan of action, select the appropriate decon procedure, determine the equipment required to implement that procedure, and evaluate the effectiveness of the decontamination procedure (p. 404–406).
- Identify sources of information for determining the applicable decontamination procedure and identify how to access those resources in a hazardous materials/WMD incident (p. 409–411)
- Describe the phases and types of decontamination operations and their application and implementation at a Hazardous Materials/WMD incident (p. 397, 398):
  - Gross decon phase
  - Secondary decon phase
  - Emergency decon
  - Technical decon
  - Mass decon
- Describe the advantages and limitations of each of the following decontamination methods (p. 404–406):
  - Absorption
  - Adsorption
  - Chemical degradation
  - Dilution
  - Disinfecting
  - Evaporation
  - Isolation and disposal
  - Neutralization
  - Solidification
  - Sterilization
  - Vacuuming
  - Washing
- Identify the types of fixed or engineered safety systems that may be used to assist in conducting decon operations within hazardous materials facilities. (p. 421)
- Define the term clean-up and its coordination with decontamination operations. (p. 424, 425)
- Describe the general clean-up concerns when decontaminating equipment. (p. 424, 425)

Skills Objectives

After reading this chapter, you will be able to:

- Demonstrate the ability to set up and implement the following types of decontamination operations:
  - Technical decontamination operations in support of entry operations. (p. 411–415)
  - Technical decontamination operations involving ambulatory and nonambulatory victims. (p. 413–415)
  - Mass decontamination operations involving ambulatory and nonambulatory victims. (p. 415–420)

Hazardous Materials Incident Commander

Knowledge Objectives

After reading this chapter, you will be able to:

- Describe the phases and types of decontamination operations and their application and implementation at a Hazardous Materials / WMD incident (p. 397, 398):
  - Gross decon phase
  - Secondary decon phase
  - Emergency decon
  - Technical decon
  - Mass decon
- Develop an incident action plan, including site safety and control plan, consistent with the emergency response plan or standard operating procedures and within the capability of the available personnel, personal protective equipment, and control equipment (p. 407–409).
- Identify the advantages and limitations of each of the following decontamination methods and describe an example where each decontamination method would be used (p. 404–406):
  - Absorption
  - Adsorption
  - Chemical degradation
  - Dilution
  - Disinfecting
  - Evaporation
  - Isolation and disposal
  - Neutralization
  - Solidification
  - Sterilization
  - Vacuuming
  - Washing

Skills Objectives

There are no Incident Commander skills objectives in this chapter.

You Are the HazMat Responder

You are the Hazardous Materials Group supervisor at the scene of a hazmat incident involving an ammonia release at an ice cream plant. The plant maintenance supervisor advises you that a repair was being made to the refrigeration line and a valve that was supposed to be locked closed was actually left open and is now releasing ammonia vapors inside the plant. Although a mechanic was able to partially close the valve before escaping to safety, there is concern that ammonia vapors will drift downwind and go beyond the plant property, likely impacting a nearby nursing home and medical facility.

A two-person entry team wearing chemical vapor protective clothing (EPA Level A) is sent inside the building to complete the closing of the ammonia valve. Both hazardous materials technicians and the backup crew have an internal suit radio communication capability that you are monitoring.

The entry team locates the correct valve but needs a crowsfoot wrench to get the leverage required to close the valve. Not having the proper wrench and running low on air, the Assistant Safety Officer–Hazardous Materials orders them out of the area. As the entry team approaches the decontamination area, one member indicates by radio that she is experiencing a tingling sensation in her fingers and toes. Her voice is tense and she is clearly excited and breathing heavily. Her partner is not experiencing any difficulties.

1. Does this situation justify an emergency decontamination?
2. How would you handle the decontamination operation when the entry team arrives at the decon station? If you conducted an emergency decon, what risks would there be to the entry team? What risks would there be to the decon team?

Introduction

This chapter will describe the seventh step in the Eight Step Process©, Decontamination or Decon. Proper decon is essential to ensure the safety of emergency responders and the general public. You cannot conduct safe entry operations if you have no way to perform decontamination afterwards. Decon methods and procedures must be considered early in the incident as part of the "hazard assessment and risk evaluation" process as described in the chapter by that name elsewhere in the text.

Twenty-five years ago, when we wrote the first edition of this text, there were many unknowns about the process of decontamination. As a result, emergency responders tended to use a lot of overkill when approaching decon problems in the field. This overcautious approach usually got the job done safely, but the down side was that hazmat operations took a long time to bring to closure because the decon process was too slow. Fortunately, we have learned a great deal about what works and what doesn't work under field conditions. The lessons learned from operations during the 2001 anthrax attacks and the research conducted by government agencies since that time have helped to improve current decon methods.

For the purpose of this text, we have adopted the decontamination terminology referenced in NFPA 472 and 473. Where standard terms or technical guidance did not exist in these standards, we have adopted terms and methods commonly used in the hazmat response community. A list of these additional terms and reference materials can be found in this chapter.

Despite the many improvements to the decon process, the basic principles of decontamination are simple and have not changed much over time. In fact, the simplicity of decontamination has been compared to changing an infant's

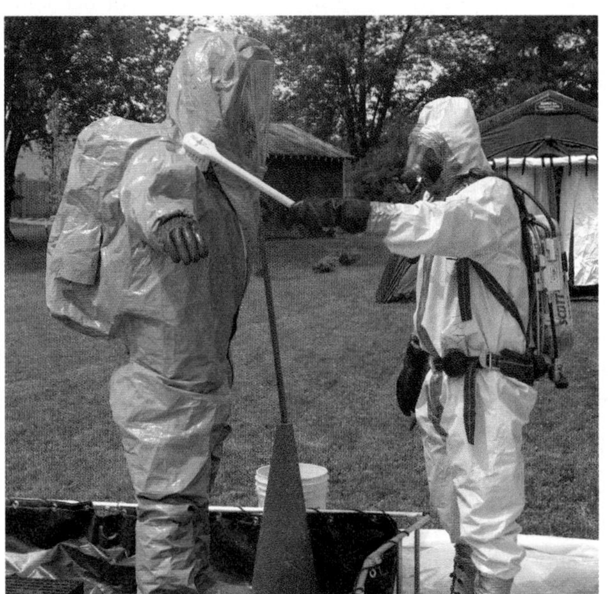

Figure 11-1 Decon needs to be adaptive and flexible.

diaper: (1) remove it from others; (2) keep it off yourself; and (3) don't spread it around! In other words, don't spread the contamination and you won't have a decontamination problem. The basic principle is that the safety and health hazards of the actual contaminants at the incident define how complex decon operations will be.

Decon needs to be an adaptive and flexible process that respects the hazards and behavior of the contaminants, as well as human behavior in stressful and changing events **Figure 11-1**. The hazards and risks presented by the incident will define the scope, nature, and complexity of decon operations. For example, a minimal hazard, such as petroleum oil on turnout boots, can be decontaminated by simply wiping the oil from the boot and then rinsing it with soap and water. In some cases rinsing with water is adequate; nothing more elaborate is needed. In contrast, a major hazard such as a highly poisonous material will require implementing a detailed procedure that includes several intermediate cleaning steps. Exactly how many steps and where they are performed is best determined locally, based on response requirements.

Basic Principles of Decontamination

The process of decontamination basically involves the physical removal of contaminants from personnel and equipment. This procedure is vital to lessen the potential of transferring contaminants beyond the hazard area.

Responder Safety Tip

Proper field decontamination is an essential safety procedure to ensure removal of contaminants from personnel and equipment and to prevent contaminants from leaving the hazard area.

Proper decontamination is especially important in those instances where injured personnel must be transported to medical facilities. Good decontamination of victims in the field speeds up the process of patient care. Most hospitals will not willingly accept nondecontaminated patients and many hospitals are not equipped to perform proper decon.

Responder Safety Tip

Good decontamination of victims in the field ultimately speeds up hospital patient care.

At every incident involving hazardous materials, there is a possibility that personnel, equipment, and members of the general public will become contaminated. The contaminant poses a threat not only to the persons contaminated, but to other personnel who may subsequently have contact with the contaminated individuals or their equipment. The entire process of decontamination should be directed toward confinement of the contaminant within the Hot Zone and removing it within the decon corridor to maintain the safety and health of response personnel and the general public.

The first rule of decontamination is to avoid contamination. If you don't get the stuff on you, you don't get hurt. Methods to minimize contamination and to ensure safe work practices should be part of the standard safe operating procedures at any incident involving potentially hazardous substances.

Terminology and Definitions

To understand the information in this chapter, let's first review some of the basic terminology pertaining to contamination, decontamination, and the establishment of decon operations. Some of these terms should be familiar to you from your First Responder–Operations-level training.

- ALARA—Performing decontamination to a level As Low as Reasonably Achievable.
- Contaminant—A hazardous material that physically remains on or in people, animals, the environment, or equipment, thereby creating a continuing risk of direct injury or a risk of exposure outside of the Hot Zone.
- Contamination—The process of transferring a hazardous material from its source to people, animals, the environment, or equipment, which may act as a carrier.
- Exposure—The process by which people, animals, the environment, and equipment are subjected to or come in contact with a hazardous material. People may be exposed to a hazardous material through any route of entry (e.g., inhalation, skin absorption, ingestion, direct contact, or injection).
- Decontamination Officer—A position within the hazmat branch which has responsibility for identifying the location of the decon corridor, assigning stations, managing all decontamination procedures, and indentifying the types of decontamination necessary.
- Decontamination (a.k.a. decon or contamination reduction)—The physical and/or chemical process of reducing and preventing the spread of contamination

from persons and equipment used at a hazardous materials incident. OSHA 1910.120 defines decontamination as the removal of hazardous substances from employees and their equipment to the extent necessary to preclude foreseeable health effects. *Note:* Some effects may manifest themselves as one or more disease processes many years after exposure and, by definition, may not be foreseeable at the time of the incident.

- **Degradation**—(a) A chemical action involving the molecular breakdown of protective clothing material or equipment due to contact with a chemical. (b) The molecular breakdown of the spilled or released material to render it less hazardous.
- **Disinfection**—The process used to destroy the majority of recognized pathogenic microorganisms.
- **Sterilization**—The process of destroying all microorganisms in or on an object.
- **Safe refuge area**—A temporary holding area for contaminated people until a decontamination corridor is set up.
- Decontamination team—The Decon Team is managed by the Decon Leader and is responsible for determining, implementing, and evaluating the decon procedure.
- **Decontamination corridor**—A distinct area within the Warm Zone that functions as a protective buffer and bridge between the Hot Zone and the Cold Zone, where decontamination stations and personnel are located to conduct decontamination procedures. An incident may have multiple decon corridors, depending upon the scope and nature of the incident.

Basic terminology pertaining to the phases and methods of decon includes the following:

- **Gross decontamination**—The initial phase of the decontamination process during which the amount of surface contaminant is significantly reduced.
- **Secondary decontamination**—The second phase of the decontamination process designed to physically or chemically remove surface contaminants to a safe and acceptable level. Depending on the scope and nature of the incident, multiple secondary decon steps may be implemented.
- **Emergency decontamination**—The physical process of immediately reducing contamination of individuals in potentially life-threatening situations with or without the formal establishment of a decontamination corridor.
- **Technical decontamination**—The planned and systematic process of reducing contamination to a level As Low As Reasonably Achievable (ALARA). Technical decon operations are normally conducted in support of emergency responder recon and entry operations at a hazardous materials incident, as well for contaminated victims requiring medical treatment. Within this chapter injured and ill people are referred to as "victims" until they have undergone decontamination in the field. Once decontaminated they are referred to as "patients" who can then be safely transported to a medical facility for treatment. (Hospitals want to receive decontaminated patients, not contaminated victims.)
- **Mass decontamination**—The process of decontaminating large numbers of people in the fastest possible time to reduce surface contamination to a safe level. It is typically a gross decon process utilizing simply water or soap and water solutions to reduce the level of contamination.

Understanding the Basics of Contamination

Before responders can understand the basics of decontamination, it is first necessary to establish a foundation by learning some important concepts about how contamination occurs. In simple terms, the level of decon will always be based on the level of contamination. Four basic concepts of contamination that will be discussed are:

1. How to prevent contamination
2. Surface versus permeation contamination
3. Direct versus cross-contamination
4. Types of contaminants

Preventing Contamination

Emergency responders sometimes get the terms "contamination" and "exposure" confused. They actually have very different meanings. Knowing these differences can help a great deal in understanding the importance of preventing contamination. For instance, when dealing with ionizing radiation, radioactive alpha particles sticking to the outside of your PPE would be an example of contamination of the garment. Gamma rays passing through the PPE and impacting your body would be an example of an exposure. Poor decontamination practices can ultimately result in a contamination problem becoming a life-threatening exposure problem.

Contamination is any form of hazardous material (solid, liquid, or gas) that physically remains on people, animals, or objects. In the emergency response business, contamination generally means any contaminant on the outside of PPE or equipment while it is still being worn or after it has been taken off. In simple terms, it means that something is "dirty."

Exposure means that a person has been subjected to a toxic substance or harmful physical agent through any route of entry into the body (e.g., inhalation, ingestion, injection, or by direct contact [skin absorption]). In other words, they have the contaminant ("the Bad Stuff") on the outside or on the inside of their bodies.

A responder who has been "contaminated" when wearing PPE has not necessarily been "exposed." For example, an entry team can be contaminated with pesticide dust on the outside of their PPE without having their respiratory system or skin exposed to the contaminant. In order for exposure to occur, the contaminant must come in direct contact with the person, such as through a breach in the protective clothing or respiratory protection Figure 11-2.

Even if the contaminant makes direct contact with the responder inside the PPE, it still does not necessarily mean that the person will be harmed by the contaminant, however, they would be "exposed" and need to take further decon precaution and evaluate the risk of their exposure. Remember the basic

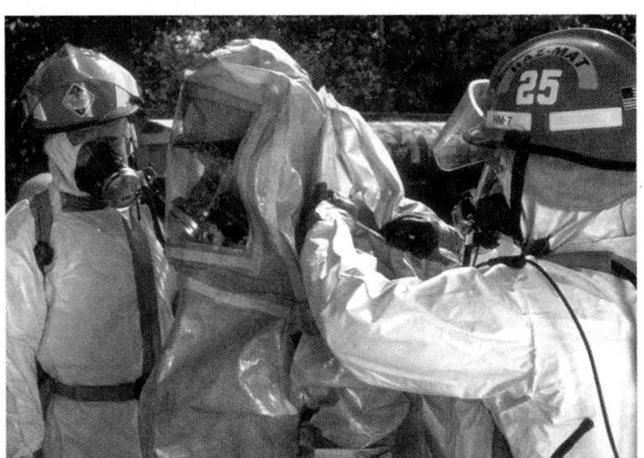

Figure 11-2 Secondary contamination may result from poor decontamination and improper doffing techniques.

principles discussed in the section on "health and safety"—harm depends on the dose, the route of exposure, and the hazards and properties of the contaminant.

The most common cause of responder and response equipment contamination comes from poor organization and discipline exercised during decontamination and clean-up operations. Decon is a critical safety benchmark, and this is the reason why so much emphasis is placed on developing and implementing a well-planned, structured, and disciplined approach to decon operations.

If you are good at decon, you will probably be good at preventing exposures. If contact with the contaminant can be controlled, the risk of exposure is reduced, and the need for decon can be minimized. Consider the following basic principles to prevent contamination:

- Stress work practices that minimize contact with hazardous substances. Stay out of areas that potentially contain hazardous substances. Don't walk through areas of obvious contamination. Special care must be taken to avoid slips, trips, and falls into the contaminants. Remember, if you don't get the "Bad Stuff" on you, you won't get hurt!
- If contact is made with a contaminant, move contaminated personnel to a safe refuge area within the Hot Zone until they can be decontaminated. Remove the contaminant as soon as possible.
- Keep your respiratory protection on as long as possible during the decon process.
- Once you have unzipped your chemical protective clothing, be aware of where your hands go and what they touch. Don't touch the outside surfaces of your PPE, and keep your hands away from your face and mouth until you have washed your hands.
- Use of limited-use/disposable protective clothing or overgarments can significantly "lighten" your decon requirements.
- Use a systematic approach to decon.
- When performing decontamination duties, "Do unto others as you would like them to do unto you." Be methodical and don't skip any steps for expediency.

Responder Safety Tip

Decon is sometimes viewed by responders as a "necessary evil." Treat your fellow responders with care and respect. Make sure the decon process is complete and thorough.

Surface vs Permeation Contamination

Contaminants can present problems in any physical state (i.e., solid, liquid, or gas). There are two general types of contamination—surface and permeation.

1. **Surface contaminants** are found on the outer surface of a material but have not been absorbed into the material. Surface contaminants are normally easy to detect and remove to a reasonably achievable and safe level using standard field decon methods. Examples include dusts, powders, fibers, and so on Figure 11-3A.
2. **Permeation contaminants** are generally viewed as being absorbed into a material at the molecular level but they may also permeate into the microscopic physical spaces between fibers and fabrics. Permeated contaminants are often difficult or impossible to detect and remove. If the contaminants are not removed, they may continue to permeate through the material. Permeation through chemical protective clothing can cause an "exposure" inside of the suit Figure 11-3B.

If the material is a tool or piece of equipment, it could lead to the failure of the item (e.g., an airline hose on a supplied air respirator). Remember that permeation can occur with any porous material, not just PPE.

Factors that influence permeation include the following:

- Contact time—The longer a contaminant is in contact with an object, the greater the probability and extent of permeation.
- Concentration—Molecules of the contaminant will flow from areas of high concentration to areas of low concentration. All things being equal, the greater the concentration of the contaminant, the greater the potential for permeation to occur.
- Temperature—Increased temperatures generally increase the rate of permeation. Conversely, lower temperatures will generally slow down the rate of permeation.
- Physical state—As a rule, gases, vapors, and low-viscosity liquids tend to permeate more readily than high-viscosity liquids or solids. Contact with high pressure releases can affect the level of contamination.

A single contaminant can present both a surface and permeation threat. This is especially the case when corrosive liquids are involved.

Direct vs Cross-Contamination

Direct contamination occurs when a person comes in direct physical contact with a contaminant or when a person comes into contact with any object that has the contaminant on it (e.g., contaminated clothing or equipment). Direct contamination usually occurs while working in the Hot Zone performing leak and spill control tasks but can also occur during the decon

Figure 11-3A Surface contamination—contaminant remains on the surface of the fabric.

Figure 11-3B Permeation contamination—the contaminant has permeated the fabric.

process. Gloves and boots are the most common areas that get contaminated Figure 11-4.

Secondary contamination occurs when a person who is already contaminated makes contact with a person or object that was not previously contaminated. Secondary contamination is typically the result of poor site management and control, inadequate decon and site safety procedures, or failure to follow safety procedures. Secondary contamination is a greater problem when dealing with liquids and solids.

Types of Contaminants

The more you know about the contaminant, the faster and more focused your decontamination operation can be. The types of contaminants can be divided into nine different

Figure 11-4 Secondary contamination can occur when responders come into contact with contaminated victims.

categories based on their primary hazards. These include the following:

1. High acute toxicity contaminants
2. Moderate to highly chronic toxicity contaminants
3. Embryotoxic contaminants
4. Allergenic contaminants
5. Flammable contaminants
6. Highly reactive or explosive contaminants
7. Water reactive contaminants
8. Etiologic contaminants
9. Radioactive contaminants

In the following sections, we will examine each of these nine categories individually; however, as discussed in the chapter called "hazard assessment and risk evaluation" found elsewhere in the text, a material may have more than one hazardous property (e.g., a flammable liquid may also be poisonous). Beware of contaminants that have multiple hazardous properties.

1. **High acute toxicity contaminants**—Can cause damage to the human body as a result of a single or short-duration exposure. These can be found in solid, liquid, or gas forms and will present risks to responders from any route of exposure. Examples include red fuming nitric acid, sulfur trioxide, nitrogen tetroxide, dimethyl hydrazine, chlorine, potassium cyanide, and nerve agents. *Note:* Some acute toxicity contaminants can have delayed system onset (e.g., some nerve and blister agents, as well as many toxic industrial chemicals).
2. **Chronic toxicity contaminants**—Repeated exposure over time (even decades) to these substances can cause damage to target internal organs or the onset of serious debilitating injuries. These contaminants include, but are not limited to, certain heavy metals, their derivatives, and potent carcinogens. They can be found in solid, liquid, or gas forms and will present risks to responders from any route of exposure. Examples include mercury, ethylene dibromide (EDB), and benzene.
3. **Embryotoxic/Teratogenic contaminants**—These are substances that can act during pregnancy to cause adverse effects on the fetus, including death, malformation, retarded growth, and postnatal functional deficits. These substances may also be called teratogens. The teratogenic period of greatest susceptibility to embryotoxins is the first eight to

twelve weeks of pregnancy. This includes a period when a woman may not know she is pregnant; therefore, special precautions must be taken at all times. However, as an additional precaution, pregnant members of hazardous materials teams should avoid working in contaminated areas involving known embryotoxins. Examples include lead compounds, ethylene oxide, and formamide.

4. **Allergenic contaminants**—Allergens are substances that produce skin and respiratory hypersensitivity. Responders are at risk from allergens from both inhalation and direct skin contact. Two people exposed to the same allergen at the same level of exposure may react differently. A sudden outbreak of a mystery rash from one individual, while no other group members show effects, is not cause to dismiss the individual's signs or symptoms. Examples include diazomethane, chromium, dichromates, formaldehyde, and isocyanates.
5. **Flammable contaminants**—These are substances that readily ignite and burn in air and are persistent (they hang around and can stick to PPE). Although most are liquids, some flammable solids and gases may also present problems, especially when indoors. While flammable and combustible contaminants demand respect because of their obvious fire hazard, they also include a wide range of secondary and tertiary hazards. For example, responders contaminated with toluene may experience respiratory distress if they remove their SCBA face piece before adequate decontamination. Exposure to toluene vapors can potentially lead to pulmonary edema and eventually pneumonia. Always expect flammables and combustibles to present more than one contamination problem. Examples include gasoline, acetone, benzene, and ethanol (liquids).
6. **Highly reactive or explosive contaminants**—These contaminants include ethers; olefins (compounds with one double bond) with hydrogen, chlorine, and flourine atoms attached; dienes and vinyl acetylenes; and vinyl monomers. Some highly reactive or explosive contaminants can react with the oxygen present in the atmosphere. Others factors include heat, shock, and friction. Among the more dangerous are the peroxide-forming compounds—chemicals that are members of some specific classes of hazardous materials that are well known to form peroxides due to age, exposure to heat, or outside contaminants. These classes include the aldehydes (some), ethers (some), the alkenes (most), and vinyl and vinyldiene compounds. Specific examples of highly reactive or explosive contaminants include cyclohexane, diethyl ether, diisopropyl ether, and tetrahydrofuran (THF).

The concentration of these contaminants plays an important role in determining the risk involved in decontamination. For example, hydrogen peroxide below 30% concentration presents no serious fire or explosion hazard. However, hydrogen peroxide hazards increase at concentrations above 52%. It is also important to recognize that the evaporation and distillation processes of these materials can create high-risk response scenarios.

Metal tools, such as spatulas and shovels, should not be used to clean up peroxide contaminants because metal contamination can lead to explosive decomposition. Obviously, responders should avoid friction, grinding, and other forms of impact.

7. **Water reactive contaminants**—These contaminants react on contact with water or moisture and therefore water should not be used for decon. Examples include perchloric acid in powder form, fumigants such as aluminum phosphide and magnesium phosphide, and most metal hydrides.
8. **Etiologic contaminants**—Etiologic or biological contaminants are microorganisms such as viruses, fungi, and bacteria or their toxins that can cause illness, disease, or death. Etiological contaminates can enter the body by ingestion, direct contact, or through the respiratory system. They are not always labeled or clearly identifiable, and field detection techniques and equipment may be limited. Without proper identification labels and packaging it is difficult for responders to determine whether etiologic contaminants are present. Specific examples of these materials include bacillus anthracis (anthrax), clostridium botulinum (botulism), and hepatitis A, B, C, D, and E.

The fact that an area is contaminated with an etiological contaminant does not necessarily mean that a person has been exposed or is susceptible to the effects of an exposure. Basic infectious disease control precautions can often significantly reduce the risk to responders. Good pre-entry planning using safe work practices can reduce risk even further. For example, if an etiologic has been spilled in a laboratory setting, there will likely be an aerosol hazard. Waiting 30 to 60 minutes for the aerosol to settle before entry can significantly reduce the risk of exposure. A release inside a vented lab hood will pose even lower risks for responders. Four factors influence an etiologic or biological material's ability to invade and alter the human body:

- *Virulence*—The ability of the biological material to cause disease. It is important to realize that there are a wide variety of etiologic organisms. Their ability to survive, reproduce, and cause harm will depend on the type of organism. If the organism is not strong enough to survive in its environment long enough to enter the human body, or if it is too weak or lacking in its ability to cause illness, then the potential harm to an exposed individual is significantly reduced. Some etiologics cannot cause disease unless the body's defense system has been compromised. For example, if a person's skin is intact and the respiratory system has not been compromised, many organisms cannot penetrate the body. However, cuts, abraded or chapped skin, or exposed mucous membranes increase the opportunity for the organism to enter the body.
- *Dose*—Refers to the number of organisms that have been ingested, absorbed, or inhaled during an exposure period. The size, composition, and population of an etiological agent will determine its ability to affect an exposed person. If an organism is not compatible with the host, or if there are not enough of the organisms to alter the natural balance within the human body, then the etiologic agent cannot survive, regardless of the dose.
- *Physical environment*—The physical environment can also determine the ability of the etiologic to enter the

body. Factors such as heat, cold, skin, and membrane acidity and alkalinity can affect the invading organism's potential to survive and cause harm.

- *Personal health status*—If the exposed person is in good health, with a fully functioning immune system, then the etiologic will have more difficulty surviving. All things being equal, the very young and the very old are most susceptible to etiologic exposures.

Fortunately, most etiologic contaminants are relatively easy to kill using a wide variety of commercially available decontamination solutions. These are divided into disinfectants and antiseptics and are discussed in more detail later in this chapter.

9. **Radioactive Contaminants**—Include some isotopes and radioactive nuclides in the form of a very fine dust or powder or as a gas. These contaminants can emit alpha or beta particles or gamma rays. Examples include cesium 137, cobalt 60, radon 222, uranium hexafluoride, and plutonium. If alpha-emitting contaminants are airborne (e.g., carried in dust or smoke), they can become surface contaminants on protective clothing or tools carried into the Hot Zone. These solid contaminants can enter the body through inhalation or ingestion.

Beta emitters present both the possibility of becoming airborne and creating high radiation levels near the surface of the contaminated area. As is the case of alpha emitters, beta contaminants may enter the body through any route of exposure, especially through the respiratory system.

If the contaminant is a gamma emitter, there is an additional hazard of gamma rays coming off tiny bits of radioactive materials that may be spread all around the area. Of course, the possibility exists that all three types of radioactive emitters may be present.

The single most important factor in eliminating radioactive material contamination problems is to avoid contact with the material. It is very easy to spread radioactive contaminants over a wide area by cross-contamination. If the radioactive contaminant has a short half-life, it is sometimes simpler to isolate contaminated equipment until the radioactivity has declined by the natural process of passing through its half-life. For example, if a reading of 100 milliroentgens (mR) were taken from a gamma emitter with a 24-hour half-life, in 24 hours the reading would be down to 50 mR; in 48 hours it would be down to 25 mR, and in 72 hours it would be down to 12.5 mR. Waiting several days to clean equipment significantly reduces the risk to responders.

A good rule of thumb for making an on-scene decision to clean contaminated equipment now or wait out the half-life is the fact that the passage of 7 half-lives will bring a radiation level down to 1% of what it is at the time you take the first reading. In 10 half-lives, the level will be down to 0.1% **Figure 11-5**.

Remember the basic principles of emergency response to radioactive materials—time, distance, and shielding between you and the radioactive material will lower the risk to responders. The goal is to reduce contaminants to a level As Low As Reasonably Achievable (ALARA).

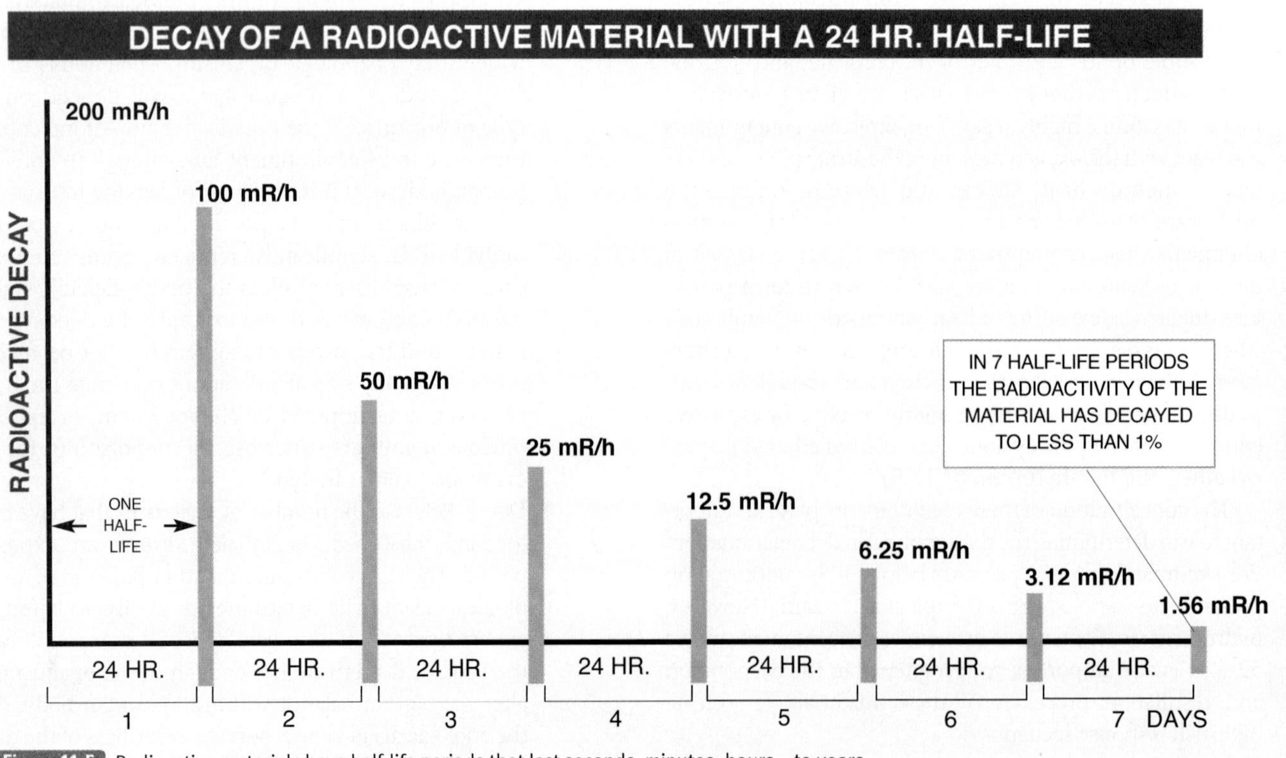

Figure 11-5 Radioactive materials have half-life periods that last seconds, minutes, hours—to years.

Case Study 1

Case Study: The Fukushima Nuclear Disaster—March 11, 2011

Damage to the reactor buildings at the Daiichi nuclear plant.

The Disaster

On March 11, 2011 a magnitude 9 earthquake struck Japan, causing a series of tsunamis that devastated the east coast of Japan. The highest wave was 127 feet high (38.9 meters) at Aneyoshi, Miyako. The earthquake and tsunami waves caused widespread devastation, taking the lives of an estimated 15,800 people, injuring 6,000, with 3,200 people missing.

Tokyo Electric Power Co.'s Fukushima Daiichi atomic plant was damaged by the earthquake and resulting tsunami and began leaking radiation when disabled cooling systems led to the meltdown of uranium fuel rods in three reactors. Cracks in the containment vessels for the melted fuel caused radiation leaks that spread radiation downwind and forced 80,000 people to leave their homes after the government banned entry within a 12-mile radius (20 km) of the plant. The Japanese government estimates that some areas may be uninhabitable for two decades or more.

Radiation Contamination

An estimated 960 square miles (2,400 square km) of land will have to be decontaminated to achieve an annual radiation exposure limit of 5 millisieverts (0.5 rem). A radiation level of 4 microsieverts per hour (0.5 millirem/hr) exceeds the Japanese government benchmark for designated evacuation zones set following the disaster.

As an example of the difficulty in identifying contamination—what is safe for humans and determining what needs to be decontaminated—consider the following example:

One radiation reading taken in October 2011 recorded at Higashi-Fuchie Primary School under the drainpipe attached to a gutter of a machinery room found radiation levels of 3.99 microsieverts per hour (0.399 millirem/hr). At the same location just 2 inches (5 cm) above the ground the reading was 0.41 microsieverts per hour (0.041 millirem/hr) At a height of 20 inches (50 cm) the reading was 0.24 microsieverts per hour (0.24 millirem/hr).

A radiation reading of 3.99 microsieverts per hour equates to a cumulative dose of about 21 millisieverts a year (2.1 millrem/yr), surpassing the 20-millisieverts-a-year standard the Japanese government used to designate expanded evacuation zones after the crisis erupted at the Fukushima No. 1 nuclear power plant.

Conversion factors:

1 Sv = 100 rem

1 microsievert = 0.1 millirem

1 rem = 10 mSv

Decontamination Challenges

Decontamination is a prerequisite for people to safely return to their homelands in Fukushima Prefecture. There are many challenges ahead in the years to come to restore the region for healthy occupation. Among the most significant challenges are:

- Removing and disposing an estimated 37 million cubic yards (29 million cubic meters) of soil from a sprawling area in Fukushima, which is located 150 miles (240 km) northeast of Tokyo, and four nearby prefectures. This daunting task will cost more than $13 billion dollars.
- Lack of available disposal sites for radioactive waste.
- Establishing realistic goals and adoption of decontamination methods easy to implement.
- 70 percent of Fukushima Prefecture is mountainous; recontamination may occur over time affecting a broad area. Every rainfall brings new radioactive substances (like contaminated leaves and soil) washing down from the hills. Areas in and around mountains must be repeatedly decontaminated. Guidelines warn that conifer (pine tree) needles also accumulate radioactive cesium over time and can normally be expected to fall after 3 to 4 years. This will require a constant and long-term effort to keep clearing fallen needles.

Taking the Long View

Whether the mountains and forests will be completely decontaminated remains to be seen. Decontamination of leaf soil (soil made up of decaying leaves) within 246 feet (75 meters) of local properties has been planned pending the consent of landowners. The decontamination process will continue for a long time. Decontamination of all the forest in the contaminated region will likely continue over the next 20 years, but villages need the forests to guarantee their source of fresh water.

Sources: "Fukushima Daiichi—A Disaster in the Wings?" by Dorit Zimerman, ***HazMat Responder World,*** Autumn 2011.
Fukushima Clean-Up Attracts Bids for $14 Billion in Projects by Tsuyoshi Inajima and Yuji Okada. Bloomberg, October 13, 2011.
"High Radiation Detected at Tokyo School," ***The Yomiuri Shimbun,*** October 19, 2011.
"IAEA Urges Japan to Set Realistic Decontamination Goals," ***Wall Street Journal,*** Dow Jones, October 14, 2011.
"Japan to Spend at Least $13 billion for Decontamination," Reuters, October 20, 2011.
"Nuclear Agency Urges Japan to Fix Cleanup Plan," ***The Wall Street Journal,*** Dow Jones, October 15, 2011.
"Residents Near Fukushima Mountains Face Nuclear Recontamination Every Rainfall," ***The Mainichi Daily News,*** October 21, 2011.

Decontamination Methods

There is no universal decon method that will work for every hazmat incident. Although there is no "silver bullet" that works on everything, there is a wide range of options in the decon arsenal from which to choose. Most of the techniques described in this chapter have been revised and improved through trial and error over the years.

Decon and cleaning of equipment and fixed facilities can be conducted to a point where only trace contaminants at ppb levels are detectable. In contrast, field-based decon in support of emergency response operations are geared toward reduction of contaminant levels to an acceptable level so that responders are safe from the short- and long-term health effects of the contaminant (e.g., below the TLV/TWA or PEL).

Respect the hazmat—the more you understand the hazards and behavior of the hazmats involved, the easier it is to select the right combination of decon methods. For example, a chemical plant that manufactures hydrofluoric (HF) acid has probably perfected their HF decon techniques; years of good and bad experience handling the product have taught them what gets results and what doesn't. This is one reason why product specialists are often the best source of decontamination information for public safety organizations.

Responder Safety Tip

Respect the hazmat—the more you understand the hazards and behavior of the hazmats involved, the easier it is to select the right combination of decon methods.

Unlike fixed industrial facilities, emergency responders must be prepared to handle a wide range of hazmat decon problems. The following discussion has been written with the special challenges of these emergency responders in mind.

Decontamination methods can be divided into two basic categories: physical and chemical. Physical methods generally involve physically removing the contaminant from the contaminated person or object. While these methods are often easier to perform and may dilute the contaminant's concentration (i.e., reducing its harmful effects), it generally remains chemically unchanged. Examples of physical methods include:

- Absorption
- Adsorption
- Brushing and scraping
- Dilution
- Heating and freezing
- Isolation and disposal (i.e., "dry decon")
- Pressurized air
- Vacuuming
- Washing
- Evaporation

Chemical methods generally involve removing the contaminant through some type of chemical process. In other words, the contaminant is undergoing some type of chemical change that facilitates its removal. Some chemical methods of decon may introduce other hazards into the process (e.g., neutralization of corrosive materials). Examples of chemical methods include:

- Chemical degradation
- Neutralization
- Solidification
- Disinfection
- Sterilization

The following sections briefly review these various types of physical and chemical decon methods.

Physical Methods of Decontamination

Absorption is the process of "soaking up" a liquid hazardous material to prevent enlargement of the contaminated area. It is primarily used in decon for wiping down equipment and property. Beyond wiping off protective clothing and equipment with towels or rags, it has limited application for decontaminating personnel.

Contaminants in absorbents remain chemically unchanged. In other words, a gallon of oil has the same properties once it has been soaked up into absorbent spill pads.

With some exceptions, use of absorbents is limited to flat surfaces (e.g., soaking up liquids on the ground or contaminants floating on water). The most readily available absorbents are soil, diatomaceous earth, and vermiculite. Other acceptable materials include anhydrous fillers, sand, and commercially available products (e.g., pads, pillows). Absorbent materials should be inert (i.e., have no active properties).

Adsorption is the process of a contaminant adhering to the surface of another material. The adhesion takes place in an extremely thin layer of molecules between the contaminant and the adsorbent. It is primarily used for the clean-up of equipment or a specific area. Examples of adsorption include activated charcoal, silica, and fuller's clay. Commercial spill pads that adsorb petroleum spills while excluding water are another example. In some instances, the adsorption process can produce heat and can cause spontaneous combustion.

An easy way to remember the difference between these two methods is that ABsorbtion works like a sponge by "soaking up" the contaminant while an ADsorbant adds itself or "sticks to" the contaminant like the way Velcro™ works.

Brushing or scraping basically involves using "elbow grease" to remove the contaminant. It can be used for the decon of personnel, PPE, and equipment in both dry decon or wet decon operations using liquid decon solutions. The object is to remove as much of the "big chunks" as possible before progressing on in the decon process. For example, contaminated dirt and mud on boots and gloves should be scraped off or washed off before stepping into decon showers.

Dilution/washing is the use of water or soap and water solutions to flush the hazmat from protective clothing and equipment. The use of detergent or soap takes advantage of the surfactant properties and works well on oils, greases, polar solvents, dirt, grime, and powders. Dilution and washing using detergents are the most commonly used methods for decontaminating personnel, since large amounts of water are almost always available. Common water sources include safety showers and fire hydrants. Engine company or tanker water might be suitable if the water does not include tank saver additives that might create contamination/reactions.

Tactical considerations when using dilution methods include the following:

- Some studies indicate that the dilution of the contaminant is not as important as the physical removal due to water pressure.
- Dilution is especially effective for use on water-soluble materials, such as anhydrous ammonia.
- The application of water only reduces the contaminant's concentration; it does not change the chemical structure.
- While dilution may be an effective method for removing non-water-soluble dusts and fibers, care must be used to prevent the spread of these contaminants to larger areas (e.g., asbestos fibers suspended in water).
- Before using water, consideration should be given to whether the contaminant will react with water, be soluble in water, or whether water will spread the contaminants to a larger area.
- Dilution and washing usually integrate brushing and scrubbing within the decon process.
- The more water consumed in decon operations, the greater the cost and potential disposal problems. Always consult with environmental officials before disposing of any contaminated runoff.

Freezing has limited use in the field by emergency responders, but this method has been used by clean-up contractors to solidify runny and sticky liquids into a solid so that they can be chipped, scraped, or flaked off and handled for a limited time in a solid state. Freezing can be accomplished by using ice, dry ice, carbon dioxide, or, if the outside ambient air temperature is below the hazmat's freezing point, allowing it to cool off and solidify on the ground (e.g., hot wax or tar).

Heating usually involves the use of high-temperature steam in conjunction with high-pressure water jets to heat up and blast away the contaminant. It is primarily used for the decon of vehicles, structures, and equipment. When detergent or solvents are added, this technique can be very effective on petroleum-based materials, such as used motor oil or high-viscosity, water-soluble materials. Heating may also be used to simply evaporate the contaminant. Heating techniques should not be used to decontaminate chemical protective clothing or people.

Isolation is a two-step process that does not involve the use of any water or liquid decon solutions. First, contaminated articles are removed and isolated in a designated area. When enough contaminated items are collected (e.g., disposable clothing), the materials are bagged and tagged. The final step involves packaging the contaminated material in a container suitable for transportation to an approved hazardous waste facility, where they may be incinerated or buried in an approved landfill.

One limitation with dry decon is the increased potential for secondary contamination of the decon crew engaged in the isolation process.

Pressurized air may be used to blow dusts and liquids out of hard-to-get places (e.g., cracks and crevices) from equipment and structures. However, it should never be used for personnel decon because pressurized air on human skin may result in fatal embolisms. One complication of using pressurized air is the aerosolization of the contaminant into the surrounding atmosphere, where it can create secondary contamination problems. For example, following the 2001 anthrax attacks, pressurized air hoses were initially used to clean conveyors and mail handling equipment at the Brentwood, Maryland Post Office. This was responsible for the spread of anthrax spores throughout the building.

Vacuuming involves the use of electric or pneumatic vacuums to collect a contaminant. This method is primarily used to decon structures and equipment, and it can be used on a wide range of contaminants, including flammable and combustible liquids, mercury, lead, asbestos, and other hazardous dusts and fine powders. The vacuum must be rated for its service and application (e.g., explosion proof and dust ignition proof).

High Efficiency Particulate Air (HEPA) vacuums are used when the contaminant is a hazardous dust, powder, or fiber. HEPA filters physically capture the contaminant by allowing air to pass through the filter while capturing the larger particulates floating in the air, and can capture particles as small as 0.1 microns. To be effective, filters must be replaced frequently.

Evaporation is simply allowing a contaminant to evaporate or "off-gas," particularly if its vapors do not present a hazard. It can be used to decon equipment and vehicles and structures when the contaminant is a high-vapor-pressure liquid or gas. Its effectiveness can be limited when dealing with porous surfaces and large quantities of materials.

Chemical Methods of Decontamination

Chemical degradation is the process of altering the chemical structure of the contaminant through the use of a second chemical or material. Commonly used degradation agents include calcium hypochlorite bleach, sodium hypochlorite bleach, sodium hydroxide as a saturated solution (household drain cleaner), sodium carbonate slurry (washing soda), calcium oxide slurry (lime), liquid household detergents, and isopropyl alcohol. Chemical degradation may be used to decon outside surfaces on buildings, walking surfaces, roads, motor vehicles, and heavy equipment. It should not be used to decon chemical protective clothing, people, or animals. Degradation chemicals should never be applied directly to the skin!

Technical advice for chemical degradation procedures should be obtained from product specialists to ensure the solution used is not reactive with the contaminant. When preparing solutions for decontamination follow the directions to avoid damage to protective clothing and equipment from degradation chemicals. (Beware of the Firefighter's Rule: If 0.5% is good, then 25% will be even better! More is not necessarily better.)

The physical and chemical compatibility of the decon solutions must be determined before they are used. Any decon method that permeates, degrades, damages, or otherwise impairs the safe function of PPE should not be used unless there are plans to ultimately isolate and dispose of the equipment. *Note:* The use of sodium hypochlorite or bleach solutions can have adverse effects on any firefighting protective clothing or equipment using Kevlar® or Kevlar® blends. Research by protective clothing manufacturers has shown that the level of degradation will be dependent upon the duration of the exposure and temperature, but any degradation processes will shorten the life of the garment or material.

Neutralization is the process used on corrosives to bring the pH of the final solution to somewhere within the range of pH 5 to pH 9. The neutralization process uses an acid substance to neutralize alkaline or an alkali substance to neutralize an acid. Preferably, the less harmful byproduct produced is a neutral or biodegradable salt.

According to the U.S. EPA, the ideal substances to use for the neutralization of alkaline corrosives in emergencies are citric acid (a powder in 25-pound bags) and sodium sesquicarbonate (a powder in 50-pound bags) when neutralizing acids. Either one forms neutral salts, and depending on the substance being neutralized, sometimes forms a biodegradable salt. Neutralization is primarily used to decon equipment, vehicles, and structures contaminated with a corrosive material. The objective of neutralization is to bring the spilled material's pH towards a pH of 7.

Solidification is a process by which a contaminant physically or chemically bonds to another object or is encapsulated by it. This method is primarily used to decon equipment and vehicles. Commercially available solidification products can be used for the clean-up of spills.

In some situations, large pieces of contaminated equipment have been covered with a cement-like material so that the contaminant is permanently bonded to the object. The contaminated object can then be buried in a hazardous waste landfill. After the Chernobyl disaster, large contaminated objects were covered with cement and entombed in-place.

Disinfection is becoming increasingly important due to the threats posed by biological warfare agents. Disinfection is the process used to inactivate (kill) recognized pathogenic microorganisms. Proper disinfection results in a reduction in the number of viable organisms to some acceptable level. It does not cause complete destruction of the microorganism you are trying to remove. Consequently, it is important that emergency responders obtain technical advice about disinfection techniques prior to their use. Likewise, some disinfectants work better on certain etiologics than others. Commercial disinfectants usually include detailed information outlining the capabilities and limitations of the product. If you have research labs, hospitals, and universities in your response area, you should perform a hazard assessment and familiarize yourself with the specific types of biological hazards present and the best disinfectant(s) for the type of hazard you may encounter.

There are two major categories of disinfectants:

- **Chemical disinfectants** are the most practical for field use. The most common types of chemical disinfectants are commercially available, including phenolic compounds, quaternary ammonium compounds, chlorine compounds, iodine, alcohols, glutataldehydes, and iodophors.
- **Antiseptic disinfectants** are designed primarily for direct application to the skin. These include alcohol, iodine, hexachlorophene, and quaternary ammonium compounds. Some of these compounds are also classified as disinfectants, but alterations in concentration allow them to be classified as antiseptics.

The terms disinfection and sterilization are sometimes used interchangeably. It is important to recognize that sterilization is not the same as disinfection. A decontamination recommendation from an etiologic specialist to sterilize a piece of equipment must not be misunderstood to mean disinfect it.

Sterilization is the process of destroying all microorganisms in or on an object. The most common method of sterilization is by using steam, concentrated chemical agents, or ultraviolet light radiation. Because of the size of the equipment involved in the sterilization process, it has limited field application and cannot be used to decontaminate personnel, but it does play an important role in decontaminating medical equipment. Contaminated medical equipment is sometimes initially disinfected at the site, then transported as contaminated equipment to a special facility, where it is then sterilized or discarded. Contaminated emergency response equipment may be sterilized through autoclaving, but the ability for the item to withstand this process has to be confirmed by the manufacturer.

Evaluating the Effectiveness of Decon Operations

Decon methods vary in their effectiveness for removing different substances. The effectiveness of any decon method should be assessed at the beginning of the decon operation and periodically throughout the operation. If contaminated materials are not being removed or are permeating through protective clothing, the decon operation must be revised.

Five simple criteria can be used for evaluating decon effectiveness:

1. No personnel are exposed to concentrations above the TLV/TWA.
2. Personnel are not exposed to skin contact with materials presenting a skin hazard.
3. Contamination levels are reduced as personnel move through the decon corridor.
4. Contamination is confined to the hot zone and decon corridor.
5. Contamination is reduced to a level of ALARA.

Methods for assessing the effectiveness of decontamination include the following:

- **Visual observation**—Stains, discolorations, corrosive effects, and so on.
- **Monitoring devices**—Devices such as photoionization detectors (PIDs), detector tubes, radiation detection instruments, and survey meters can show that contamination levels are at least below the device's detection limit. For example, placing CPC in a closed bag after decon will allow any residual organic vapors to accumulate; monitoring devices can then be used later to detect any vapors present.
- **Wipe sampling**—Wipe sampling can be used to assess decon effectiveness on CPC, equipment, and skin, as well as vehicles and structures. The sample provides after-the-fact information on the effectiveness of the decon. In some cases, wipes can be taken with pH paper or other types of indicator papers based on the identified product. Some wipe sampling papers for

chemical agents like the M8 and M9 chemical agent detector paper can indicate the type of chemical agent in the field by color changes. It should be noted that there is currently no practical way to determine the effectiveness of decontamination in the field for most etiologic hazards by wipe sampling. However, many new bio sampling products are coming on the market every year, getting responders closer to legitimate on-scene sampling results.

Decon Site Selection and Management

The success of decontamination is directly related to how well the Incident Commander and the Decon Leader control on-scene personnel and their operations. Before initiating decontamination, the Hazmat Group Supervisor and the Decon Leader must decide (1) how much and what decon method is required; and (2) to what extent decon will be accomplished in the field.

These decisions should be based on the answers to the following questions:

- Can decon be conducted safely? Dilution may be impractical due to cold weather or because it presents an unacceptable risk to emergency personnel.
- Are existing resources immediately available to decon personnel and equipment? If not, where can they be obtained, and how long will it take to get them?
- Can the equipment used be decontaminated? The toxicity of some materials may render certain equipment unsafe for further use. In these cases, disposal may be the only safe alternative.

The Decontamination Team

At a working hazmat incident, a decontamination team should be established to manage and coordinate all decon operations. The decon team's functions include research and development of the decon plan and set-up and operation of an effective decontamination area capable of handling all potential exposures, including entry personnel, contaminated victims, and equipment. If necessary, the decon team will also coordinate the establishment of a safe refuge area.

Responder Safety Tip

Having an operational decon area is a prerequisite to Hot Zone entry. You must have a decon capability before making entry.

The Decon Team is managed by the Decon Leader, who reports to the Hazmat Group Supervisor. The Decon Leader and the Decon Team should be trained to the Hazardous Materials Technician (as described in NFPA 472) level (or equivalent). Personnel providing support to the Decon Team and not involved in direct decon operations should be trained to at least the First Responder Operations level.

The Decon Leader performs the following activities:

- Determine the appropriate level of decontamination to be provided.
- Ensure proper decon procedures are used by the Decon Team, including decon area set-up, decon methods and procedures, staffing, and protective clothing requirements.
- Coordinate decon operations with the Entry Leader and other personnel within the Hazmat Group. This will ensure that decon is set up before entry operations begin.
- Coordinate the transfer of decontaminated patients requiring medical treatment and transportation with the Hazmat Medical Unit.
- Ensure the Decon Area is established before any entry personnel are allowed to enter the Hot Zone, whenever possible.
- Monitor the effectiveness of decon operations. Adjust procedures and operations as necessary.
- Appoint or act as an accountability officer to limit access to personnel entering and operating within the decon area.

The Decon Leader should use a formal checklist to ensure that important items are not overlooked.

Decon Site Selection

When a hazmat incident occurs outdoors, the decon site should be accessible from a hard-surfaced road. Water supply, access to safety showers, runoff potential, and proximity to any environmentally sensitive areas, such as streams or ponds, should be considered **Figure 11-6**.

The ideal outdoor decon site is upwind and uphill from the incident and remote from drains, manholes, and waterways, but close enough to the scene to limit the spread of contaminants. Unfortunately, it is often not possible to actually choose such an ideal site. Shifting winds and dispersing vapors can further complicate this selection process. Such real-life problems may force the movement of the decon area once it has been in operation if initial site planning has been hasty.

If decon will be conducted indoors, consideration should be given to quick access such as hallways, type and slope of floor, floor drains, and ventilation airflows in the area.

Complete decon may be impractical to achieve at a single location, so a combination of on-site and off-site contingencies may be necessary. Any time decon is conducted offsite, the entire operation will become more complicated due to the logistics of moving people offsite. If you intend to use an off-site location for decon, special preparations will need to be made to prevent the spread of contaminants.

Decontamination Corridor

Once a decon site has been selected, a decontamination corridor should be marked. The decon corridor is simply a pathway from the Hot Zone into the decon area, with the exit point near the Warm Zone/Cold Zone interface. The decon corridor and the boundaries of the decon area should be clearly identified. Methods for identification include using fence post stakes and colored fire line tape, traffic cones, and

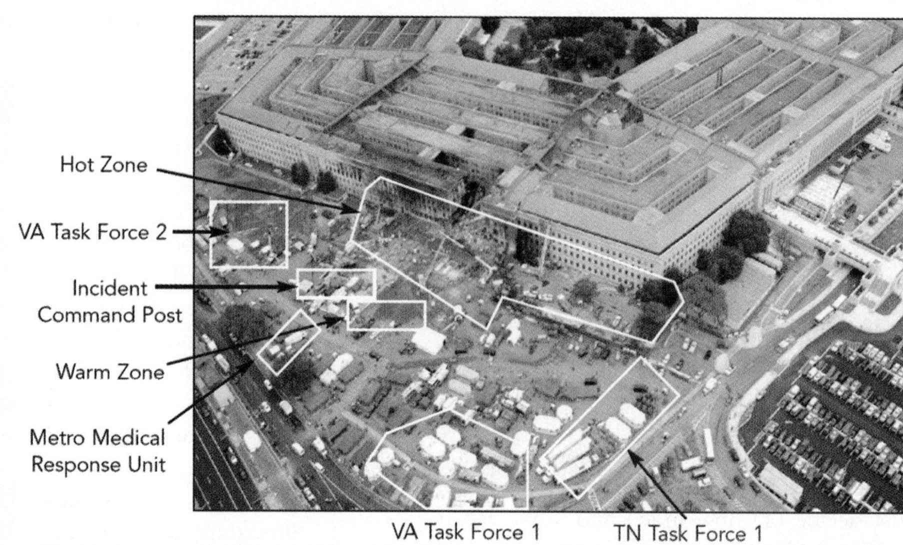

Figure 11-6 Decontamination operations may need to be conducted in congested areas. This photo shows multiple task force operations close to the Hot Zone, due to necessity following the 9-11 attack on the Pentagon.

so on. Signs may also be used to indicate the entry and exit points to the decon area. Pre-sized tarps (minimum of 20 ft by 40 ft) with various stations marked on them can also help organize the decon area and provide secondary containment for decon runoff **Figure 11-7**.

Some scenarios will require multiple decon corridors. For example, a mass casualty scenario involving hazardous materials or weapons of mass destruction (WMD) could have a decon corridor for both mass decon operations and a technical decon operations. The mass decon corridor would likely be established by first-due fire units. Upon the arrival of a hazmat team, a separate decon corridor for technical decon operations would then be established. Similarly, an incident covering a large area (e.g., the World Trade Center or the Pentagon) will have multiple decon corridors, simply because of the size and complexity of the incident site.

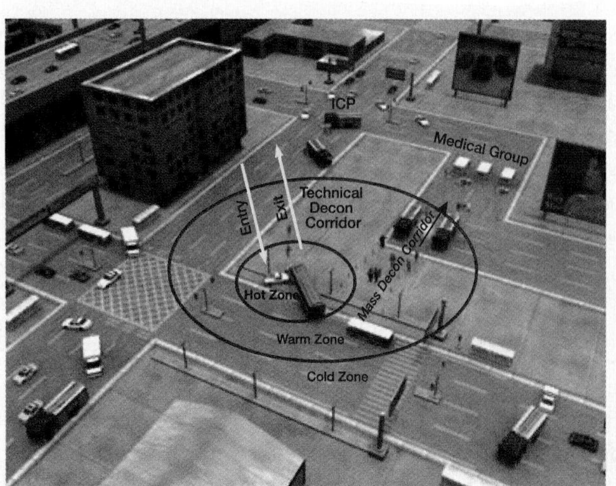

Figure 11-7 The decon area should be organized with clear entry and exit areas for the technical decon corridor.

Response scenarios in which the decon area is physically separated from the "problem" by a large distance can also pose additional challenges for responders. These scenarios will often require some form of transportation from the incident scene to the entry control point, such as small all-terrain utility vehicles or pick-up trucks. Regardless of the type of vehicle, it must have the capability of carrying multiple personnel, PPE, and any required control equipment. Continuous air monitoring may be required when dealing with flammable substances to ensure that these transport vehicles do not enter an unsafe atmosphere. PPE must also be provided for the vehicle operator, including SCBA, which can be accomplished by placing the SCBA beside the driver on the seat. Also, placing a plastic tarp in the vehicle bed can help minimize the spread of contaminants. (Nobody said this was going to be easy!)

Staffing and Personal Protection of the Decon Team

When setting up the decon area, consideration must be given to the staffing of the decon operation and the safety of the Decon Team. Decon is a labor-intensive operation. Although there are tools and equipment that can expedite the decon process, it still requires time and personnel to set up and operate the decon corridor.

The Decon Leader should be trained to the Hazardous Materials Technician level (or equivalent). The Decon Leader is a supervisor and should not be involved in the actual decon process. However, they may still be "dressed out" in the event of an emergency along the decon corridor that requires additional personnel. At a minimum, each staffed decon station should have one decon member assigned. Those stations that require entry personnel to be scrubbed or to remove their PPE should be staffed with two decon members, if possible.

The level of skin and respiratory protection required by Decon Team members will be dependent upon (1) the type of contaminants involved, (2) the level of contamination encountered by entry personnel, and (3) where individuals are working along the decon line.

Tactical and safety considerations will include the following:

- The Decon Team should be dressed in chemical protective clothing and equipment based on the hazards and risks of the contaminants they will be decontaminating. Experience in the field over the last 20 years indicates that the Decon Team can usually operate safely using chemical protective clothing (i.e., skin protection) one level down from the PPE being used by the Entry Team. Example: If the Entry Team is using chemical vapor protective clothing (Level A), then the Decon Team may use chemical splash protection (Level B). This is just a guideline, there will always be exceptions, but exceptions should be based on a hazard and risk assessment and made in conjunction with the Incident Safety Officer and the Assistant Safety Officer–Hazardous Materials. The risk of heat exhaustion should be factored into decision making when selecting the right PPE for decon.
- The most common level of protection for the Decon Team is chemical splash protective clothing (i.e., Level B) and self-contained breathing apparatus (SCBA). Most field-based decon operations involve wet decon methods with a splash hazard. In addition, Level B disposable or limited-use garments and SCBA are readily available in most response organizations.
- Extended entry operations will require that the Decon Team be provided with an uninterrupted air supply. The use of airline hose units (see the section on "selecting personal protective clothing and equipment") can provide an extended air supply but can also decrease Decon Team mobility.
- Respiratory protection for Decon Team members may be downgraded to PAPRs or APRs providing that air monitoring is continually conducted and filter cartridges are compatible with the concentration of contaminants present.
- All Decon Team personnel must be decontaminated before leaving the decon area. The extent of this decon process will be determined by the types of contaminants involved and an individual's work station along the decon corridor. The use of disposable or limited-use garments by the Decon Team can simplify this decon process.

Types of Field Decontamination: Basic Concepts

This section will review the different types of decontamination that may be implemented as part of hazmat response operations. Let's first review some basic concepts and operational procedures that are "horizontal" in nature and influence all decon operations.

Decontamination can be carried out safely and smoothly if responders follow a standard operating procedure that is practical and suitable for use in the field. As is the case with any emergency response procedure, successful implementation requires regular training and practice. Many good "model" procedures are available to serve as a basis for developing and improving your own system. While there are a variety of different approaches to decon, the better standard operating procedures are built around several basic principles:

- Contaminated people and equipment generally flow from the dirty end (area of highest contamination) to the clean end (area of least contamination). The best analogy we have come across to explain this concept is an automated car wash. You drive your dirty car in one end and it comes out clean at the other end.
- Just like the car wash, decontamination requires a multiple-step process to reduce contaminants to an acceptable level. There are two main reasons for adopting a multistep cleaning process:
  1. Conducting all of the cleaning process at one station concentrates all of the contamination in one area. The more things you clean in the same spot, the greater the contamination level becomes. Would you really want to step out of your protective clothing into a highly contaminated puddle of water?
  2. Multiple decontamination stations make you cleaner. The further along the decon line you progress, the cleaner you should become. (Think about it: Does a dirty shirt get cleaner the more times you launder it?)

Multistep decon operations can be broken into two broad phases: gross decon and secondary decon. Gross decon is the initial rinse and secondary decon is the follow-up process that removes the contaminants to a safe and acceptable level. These phases can be defined as follows:

- Gross decontamination—The initial phase of the decontamination process during which the amount of surface contaminant is significantly reduced. In simple terms, it is designed to remove most of the "big chunks."

At minor incidents, gross decon may be the only field decon performed. At incidents where greater hazards or higher levels of contamination are suspected, gross decon is the initial station in the decon process. Gross decon may be provided through a variety of means, including fixed or portable showers, tent units, hoselines, or wands. Emergency decon and mass decon are typically only gross decon operations.

- Secondary decontamination—The second phase of the decontamination process is designed to physically or chemically remove surface contaminants to a safe and acceptable level. Depending on the scope and nature of the incident, multiple secondary decon steps may be implemented. Technical decon incorporates both gross and secondary phases.

Once responders who perform decontamination have incorporated these basic concepts into their procedures, it doesn't matter whether there are two or eight different decon stations or what they are called (e.g., stations, steps, phases, etc.).

Decon Information Sources

The decision to implement all or part of a decon procedure should be based on a field characterization of the hazards and risks involved. Sources for determining the appropriate decon methods and related information can include the following:

- Product specialists representing the chemical manufacturer.

Scan Sheet 11-B

The Decon Team: Tasks and Functions

Serving as a member of the Decon Team is much like working in a service industry. The customers are those individuals who must go through the decon corridor. Your job is to make sure your customers are decontaminated in a safe, timely, and effective manner.

Here are a few pointers to keep in mind:

- Regardless of your level of protective clothing and equipment, keep the "Bad Stuff" off of you!
- At night and in low light conditions, provide sufficient lighting so that those going through the decon corridor can see clearly. It is not uncommon for frost to form on the inside of a SCBA face piece or chemical vapor protective clothing during extremely cold weather conditions. This frost will make vision extremely difficult in artificial light.
- Provide clear verbal directions and instructions to your customers. Remember, their ability to clearly see and hear may be compromised by the type of PPE being worn, exhaustion and stress, as well as background noise and conditions within the decon corridor.
- Except for those at the doffing station, there is normally no need for you to physically contact your customer. Utilize decon equipment that will allow you to physically clean, wash, and rinse without coming in contact with potentially contaminated personnel or equipment.
- During night-time incidents, the location of the decon corridor entrance in the Hot Zone can be marked with a colored light, to help entry team members with fogged-up face pieces or visors find where they need to be heading. This is especially useful where terrain, berms, or other obstacles obscure the actual decon area.

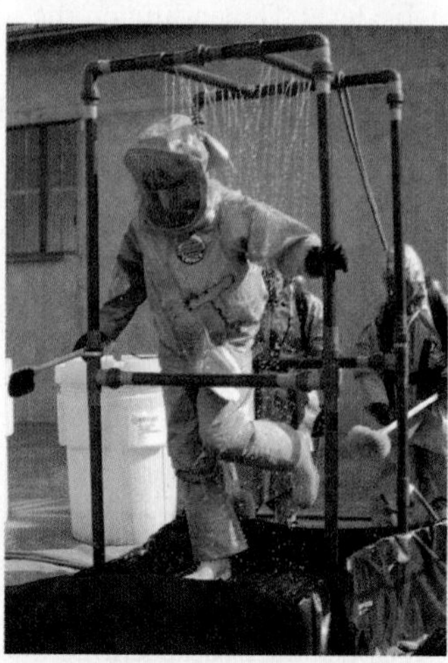

- Technical information specialists with knowledge and understanding of the behavior and hazards of the contaminants.
- Technical information centers, including CHEMTREC™, CANUTEC, SETIQ, Centers for Disease Control, and regional poison control centers.
- Safety data sheets (SDSs). Decon information is usually found in the first-aid section and is often vague and nonspecific.
- Emergency response references. These will generally provide information on the physical and chemical properties of the hazmat you're dealing with, which will be important in assessing the risks and decon requirements. However, they rarely provide specific guidance on how to decon personnel or equipment.
- Some online computer and electronic databases now incorporate decon information.

The bottom line: There is no one single best resource for acquiring field decontamination information. Although technical data and advice from product specialists may be difficult to obtain early in the incident, make every effort to get it **Table 11-1**.

Types of Field Decontamination

Different types of decontamination may be required based on the nature of the problem, the urgency of the situation, the number of people contaminated, the scope of the contamination, and the hazards and risks of the contaminants. This section will provide an overview of the types of decontamination, their application, and implementation. These types include emergency decon, technical decon, and mass decon.

Emergency Decontamination

Emergency decontamination is the physical process of immediately reducing contamination of individuals in potentially life-threatening situations with or without the formal establishment of a decontamination corridor. It is used when emergency responders encounter a life-threatening situation that requires some type of immediate decontamination; e.g. they are exhibiting signs and symptoms of a chemical exposure. The victim(s) may be either emergency responders or civilians. This problem is normally associated with routine events, when hazardous materials are not suspected or anticipated. It may also be required if a responder's PPE is breached and there is immediate skin contact with the contaminant. The urgency of the situation does not allow for the formal establishment of a technical decon operation **Figure 11-8**.

Remember, this is not a controlled situation. Typically, little or no chemical protective clothing is being used, and there is usually no specialized equipment or expertise on scene to assist. Here are some key points to remember:

- Establish an area of refuge as soon as possible. Minimize the potential for cross-contamination.
- Establish a gross decon area using some form of water. Hoselines are best, but a garden hose or a bucket of soap and water can work. Depending on the scenario and location, emergency decon may also be provided through safety showers, pump sprayers, or even a building sprinkler system. The key to limiting harm is to remove the contaminant as rapidly as possible.
- Remove any contaminated clothing. Studies by the U.S. Army Soldier and Biological Chemical Command (SBCCOM) of nerve agents using harmless simulants to track contaminants have shown that approximately 80% of contaminants can be removed by simply undressing.

Figure 11-8 Emergency decon may be required for civilians, responders, or the injured before a formal decon corridor is established.

Table 11-1 Types of Decontamination

TYPES OF DECONTAMINATION	EMERGENCY DECON	TECHNICAL DECON		MASS DECON	
	Field Response	Field Response	Medical Facility	Fog Lines Corridor	Tent Corridor
Gross Decontamination Phase	X[1]	X	X	X[1]	X[1]
Secondary Decontamination Phase		X	X		

X[1]—Any patients who will require medical treatment must receive secondary decontamination before being treated.

- *If time allows*, construct a basin to collect runoff, or use a place that will hold the runoff, such as a depression in the ground. Don't delay life-saving decon to collect runoff.
- Emergency decon operations are graded on speed, not neatness. The sooner you decontaminate, the better.
- Provide emergency medical care for the victims within the scope of your medical training and the available resources on hand.

Emergency decon can be innovative, but the most important concept is to clean the contaminated person as soon as possible. Remember the basics: FLUSH—STRIP—FLUSH. Soap and water is a universal solution and should be applied in large quantities. When acids or bases have contacted bare skin, the minimum amount of time for a water flush is at least 20 minutes.

In summary, emergency decon is a gross decon operation. It should include the removal of any contaminated clothing as soon as possible and as thorough a washing as possible. Victims requiring follow-up medical treatment or evaluation should still undergo secondary decon either on scene or at a medical facility.

Technical Decontamination

Technical decontamination is the planned and systematic process of reducing contamination to a level of ALARA. Technical decon operations are normally conducted in support of emergency responder recon and entry operations at a hazardous materials incident and by medical facilities handling contaminated patients. The key variable with technical decon operations is the time and resources required to become operational.

Technical decon is a multistep process in which contaminated individuals are cleansed with the assistance of trained personnel. While the process has changed over the years, it still follows some basic operational concepts. Technical decon is similar to a car wash:

1. There is an entry point and an exit point (i.e., the decon corridor).
2. In between the entry point (i.e., Hot Zone) and the exit point (i.e., Warm Zone/Cold Zone interface), several progressive cleaning steps takes place to remove the dirt (same as the contaminants).
3. Most of the dirt is removed in the initial stage of the car wash (i.e., gross decon), and the car becomes cleaner as it moves through the various cleaning stages (i.e., secondary decon).

Now, replace the words car wash with decon corridor and you have a technical decon operation. While the car wash facility is permanent, our technical decon process and decon corridor must typically be set up in the field Figure 11-9.

The Technical Decon Process

There are nine basic steps in technical decon:

1. **Tool and equipment drop**—Tools that may be reused at the job site are placed here. When the job is completed, they will receive further cleaning. Monitoring instruments should be separated from other equipment to protect them from damage. If the incident involves law enforcement officers, it may also be necessary to incorporate a drop area for evidence and weapons. A weapons safety officer should be established, as appropriate.
2. **Overglove/overboot drop**—Overgloves and overboots are used to minimize the amount of contamination on PPE. The majority of personal contamination occurs to the hands and feet. Efficient removal of the overgloves and overboots can minimize the potential for secondary contamination.
3. **Gross decon**—The initial decon step in which the entry crew is rinsed off. Special attention is given to the hands and feet. A soap and water wash may also be used, followed by a rinse.
4. **Secondary decon**—Additional washing and rinsing steps designed to further reduce the level of contamination. Special attention is given to the hands and feet. Based on the level of contamination and the hazards of the contaminant, this phase may not be necessary.

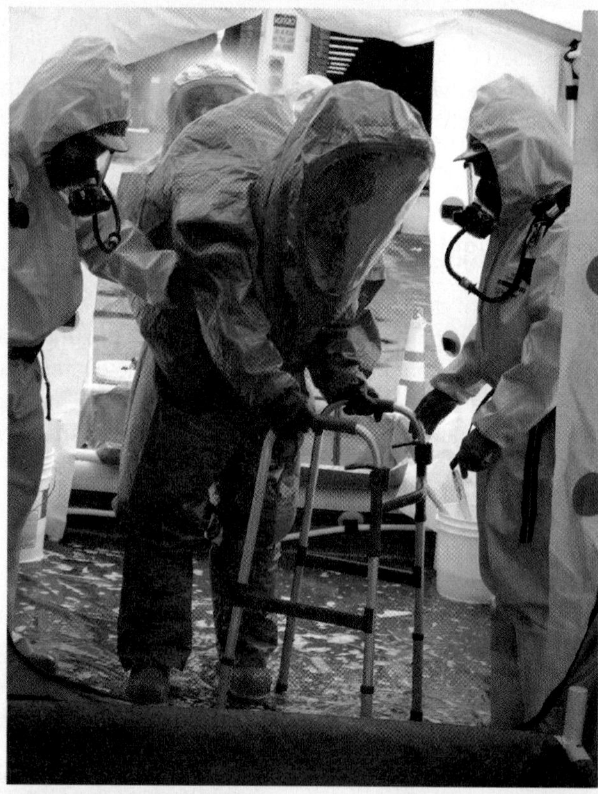

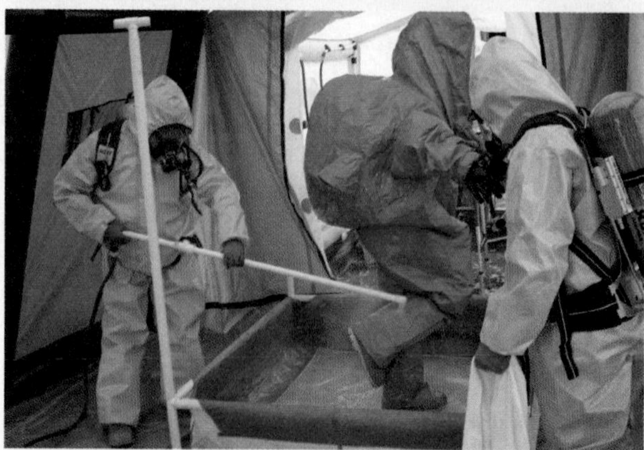

Figure 11-9 Technical decon is intended to render responder PPE safe to remove.

Scan Sheet 11-C

Decontamination of Injured Personnel

Contaminated nonambulatory victims pose unique challenges for emergency responders. At incidents where there are both ambulatory (walking) and nonambulatory (nonwalking) contaminated individuals, separate decon corridors should be established to facilitate victim flow.

Decon of injured and/or contaminated individuals should be accomplished in the field before transport to a medical facility—failure to decon these people before transport will only lead to bad outcomes, including cross-contamination to EMS personnel, ambulances, and emergency room facilities.

In some instances, hospitals and medical facilities may encounter "surprise packages" where contaminated victims simply show up at the front door. Approximately 80% of those who sought medical treatment after the 1995 Tokyo subway incident involving sarin sought treatment outside of the emergency medical system by self-evacuating to the nearest hospital and overwhelming the resources.

Regardless of whether emergency, technical, or mass decon is being provided, there are some basic principles that need to be recognized when dealing with the injured. These include the following:

- Remove all clothing, jewelry, and shoes as soon as possible. If possible, remove clothing from head to foot to limit the risk of inhalation.
- Protect the victim's airway. Blot away any obvious liquids using a soft sponge or washcloth, and/or brush away any obvious solid or dust materials.
- Hazardous materials will tend to enter the body more readily through wounds, the eyes, or mucous membranes than through intact skin. Therefore, these areas should be decontaminated first.
- Begin to rinse the victim around the face and head area; ensure that all fluids flow away from the eyes and respiratory system. Then move to any open wounds, followed by a head-to-toe rinse in a systematic fashion. Pay close attention to skin folds, armpits, genital areas, fingernails, and the feet. These areas of the body are important for chemicals that like moisture to hydrolyze (e.g., blister agents).
- Water is the universal decon agent. However, if the adherent solids or liquids are not water soluble, a mild liquid detergent can be used to facilitate skin washing. Do not use hot water, as this will cause pores in the skin to open. It is better to use slightly cooler than body temperature water, ideally 30°C (86°F).
- If the eyes are symptomatic, irrigate the eyes continuously.
- Attempt to isolate contaminated areas on the victim if the whole victim is not contaminated. For example, cover uncontaminated areas on the victim with a waterproof material and avoid washing a contaminated area into an open wound. This may require placing goggles, earplugs, or an oxygen mask on the victim.
- Provide initial medical treatment based on the signs and symptoms of the suspected material. Remember the basic ABCs!
- Bag and tag all clothing and possessions. Depending on the scenario, these may need to be disposed of as hazardous waste or they may be treated as evidence if a criminal event is suspected.

5. **PPE removal**—PPE should be removed in a manner that minimizes the potential for the Decon Crew to contact those being decontaminated. Large trash bags make handling and disposal easier. If the materials will require an administrative chain of custody (e.g., samples, weapons, evidence), clear plastic bags should be used.
6. **Respiratory protection removal**—Should always be the last item removed. If individuals being deconned are wearing coveralls, special attention should be given so that the user will continue to wear the face piece until PPE is removed.
7. **Clothing removal**—If necessary, change out of the undergarments. In this case, it will also be necessary to provide an additional change of clothing for personnel.
8. **Personal hygiene**—Normal hygiene procedures are usually adequate following removal of protective clothing. There is seldom a situation where it is necessary for someone to totally undress and shower on scene. If a whole body wash is required immediately it should be done inside a temporary portable decon shelter where modesty and comfort are respected. If a breach in a suit has occurred and the contaminants are believed to be inside the suit, dangerous, and an

exposure has occurred, then the whole body wash, or at least the potentially contaminated area, might be justified. This activity may be done on-site using temporary inflatable shelters or off-site at a locker room facility. If a full body shower is required, it should always be done prior to going home.

9. **Medical evaluation**—Responders should always be medically evaluated after an entry. Vital signs are taken 10–15 minutes after rest and oral rehydration to ensure adequate recovery from the stress of entry. Responders who are not recovering appropriately should go to rehabilitation (Rehab) for further evaluation, hydration, and further treatment. For guidance on Rehab, consult NFPA 1584—*Standard on the Rehabilitation Process for Members During Emergency Operations and Training Exercises.*

Portable tents, rapidly inflatable temporary structures, and specially designed vehicles or trailers may also be integrated into the decon corridor and the technical decon process. In addition, some hospitals and high-hazard facilities have constructed specialized rooms or areas where technical decon can be completed. There is not necessarily a right way or a wrong way—there are many different options. Use what works for you.

Performing Technical Decontamination

To begin the technical decontamination process, anyone leaving the hot zone should place any belongings, oversuits, or tools in a drop area near the entrance of the decontamination corridor (these items can be cleaned later, after the contaminated responders are taken care of). This drop area can consist of a container, a recovery drum, a special tarp or other collection device. If another trip into the hot zone is required, subsequent teams may use the same tools.

The responder, still wearing full PPE, proceeds or is moved into the decontamination corridor for gross decontamination (if required). The gross decontamination step is optional, depending on the amount and nature of the contaminant. A portable shower using a low-pressure, high-volume water flow may complete this step (the shower contains the water). Technical decontamination typically involves one to three wash-and-rinse stations, again depending on the nature of the expected contamination. Only one contaminated responder is allowed in a wash and rinse station at a time. The technical decontamination team is responsible not only for containing the water used, but also for scrubbing the PPE worn by personnel. The decontamination team member who is scrubbing should pay special attention to the gloves, crevices in the PPE, and boot bottoms, as these are areas in which hazardous materials are likely to collect.

After the chemical-protective equipment is thoroughly scrubbed and rinsed, it can be safely removed from the responder. The SCBA face piece, air-purifying respirators, or fan powered air-purifying respirators, should remain in place. The members of the decontamination team who are responsible for assisting responders with doffing the PPE should fold or roll the PPE back so that the contaminated side of the garment contacts only itself. If the procedure is done properly, the contaminated side of the garment will not touch the interior of the suit or the person wearing it.

If the responder is wearing outer gloves, they can be removed now; inner gloves will be removed later. The responder then proceeds through the decontamination corridor to an area where helmets, respiratory protection, and any other ancillary equipment are removed. Deposit respiratory protection in a plastic bag or on a tarp. Highly contaminated respiratory protective equipment should be removed and isolated until it can undergo complete decontamination.

Remove inner gloves, and sort them into individual containers for clean-up or disposal. Plastic bags can be used for this purpose because they provide sufficient temporary protection from most materials. They should be sealed with tape and transported elsewhere for disposal. Place the bags in a properly marked recovery drum when disposing of them.

Most chemical-protective equipment in use today is considered to be disposable. In most cases, the decontamination process is done to safely remove the person from the garment; the garment is then discarded.

Decontaminated personnel can now don clean clothes. Disposable cotton coveralls, hospital gowns, hospital booties, slippers, and flip-flops are inexpensive and easy-to-use options for this purpose. They can be prepackaged according to size and stored for easy access. After personnel are thoroughly decontaminated and have showered and donned clean clothes, they should proceed to a medical station for evaluation. Respiratory protection should be left in place as long as possible to protect the lungs and eyes from potential injury.

Technical decontamination can be performed in a number of ways, and your AHJ may have established a specific procedure for it. Check your policies and procedures for instructions on the preferred way to carry out technical decontamination.

To perform technical decontamination on a responder, follow the steps in Skill Drill 11-1 ▼:

1. Drop any tools and equipment into a tool drum or onto a designated tarp. (**photo 1**)
2. Perform gross decontamination, if necessary. (**photo 2**)
3. Perform technical decontamination. Wash and rinse the responder one to three times. (**photo 3**)
4. Remove outer hazardous materials–protective clothing. (**photo 4**)
5. Remove personal clothing.
6. Proceed to the rehabilitation area for medical monitoring, rehydration, and personal decontamination shower. (**photo 5**)

Responder Tip

The concept of removing PPE should follow the same principle as applies when extricating a victim pinned inside an automobile: Remove the car from the person; don't take the person from the car. When it comes to decontamination and PPE removal, remove the PPE from the person; don't take the person out of the PPE.

Skill Drill 11-1

Performing Technical Decontamination

1 Drop any tools and equipment into a tool drum or onto a designated tarp.

2 Perform gross decontamination, if necessary.

3 Perform technical decontamination. Wash and rinse the responder one to three times.

4 Remove outer hazardous materials–protective clothing.

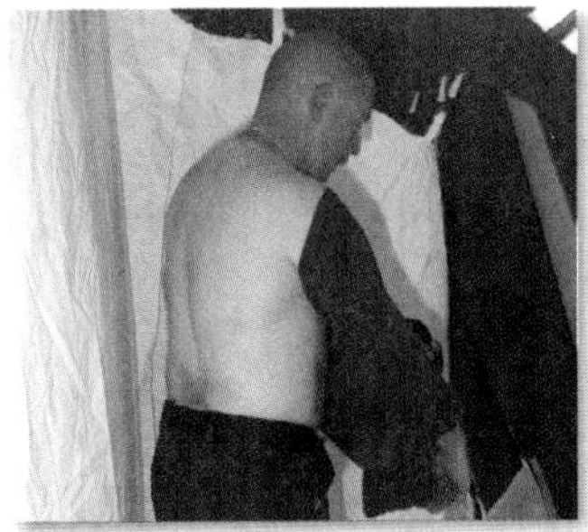
5 Remove personal clothing. Proceed to the rehabilitation area for medical monitoring, rehydration, and personal decontamination shower.

Mass Decontamination

Mass decontamination is established when large numbers of people (i.e., civilians or responders) need to be decontaminated at the scene of a hazmat emergency. The general goal is to provide the greatest good to the greatest number of people in the shortest amount of time.

What is considered a "large" number of people is really based on your local resources. From a training and exercise perspective, begin mass decon practice evolutions with a school bus-sized group (approximately 20 to 40 people) and then develop expanded capabilities from there. A major difference between technical and mass decon is the wide range of victims likely to be involved in a mass decon scenario, including ambulatory and nonambulatory, children, the elderly, and people with disabilities.

Mass decon is a gross decontamination process that relies on the use of water or soap and water solutions to flush the majority of the contaminant from individuals. Mass decon operations are based on the following basic principles:

1. Removing clothing is a form of decon that can remove the majority of the contaminants. SBCCOM tests with simulated chemical agents show that this can account for the removal of up to 80% of the contaminant. If victims are unwilling to remove their clothing, they should be showered while clothed. Responders should not spend any significant amount of time trying to convince people to disrobe since

this impedes the goal of rapidly decontaminating as many people as possible.

2. Once clothing has been removed, flushing a victim with a low pressure/high-volume shower of water will remove additional amounts of the contaminant. Tests conducted by SBCCOM and the Virginia Department of Emergency Services have shown that high volume, low-pressure streams are most effective. Optimal operating pressures for both automatic and standard fire department nozzles are approximately 30 to 50 psi, with average flows of 150 gpm per nozzle.
3. Water flushing can be delivered in a variety of methods, including the following:
   - Establishment of mass decon corridors using either handlines or master fog streams from fire engines or aerial devices. *Note:* Caution should be used if tank saver additives are mixed with fresh water. As more units arrive on scene, they can be added into the process.
   - Use of portable shower set-ups (commercial and available systems).
   - Use of indoor or outdoor swimming pools (victims wade into the water).
   - The use of sprinkler heads in a building with a fire protection sprinkler system.
   - Multiple bathroom showers, such as in a school or health club locker room.

There are many options and methods to provide mass decon. It is our intent to focus on the key concepts, operational issues, and lessons learned that should be incorporated into your mass decon procedure. Again—use the method that works for you.

It's All about People

The greatest challenge in a mass decon operation will be the management and control of people. Among the lessons learned from research and exercises are the following:

- Timing is everything! In order to create a more favorable outcome, mass decon must be established by the first-arriving responders in a rapid and efficient manner. Speedy operations are critical to success.
- Law enforcement will play a key role in assisting with mass decon. Depending on the scenario, their training and expertise in controlling and containing large groups of people will be a major factor in a successful operation. More than a few people will want to leave the scene and seek medical treatment on their own, which defeats the goal of rapid decon.
- The ability to establish a safe refuge area and maintain effective site management and control will be critical until the mass decon corridor is established.
- Psychology will play a major factor in a mass decon scenario; a majority of the victims may not be contaminated and are "worried well." For example, approximately 80% of the 5,510 people who sought medical attention as a result of the 1995 Tokyo subway sarin attack were not exposed to any significant amount of the chemical agent. These individuals needed to be decontaminated for their own peace of mind as well as for their health care providers. The "worried well" may also complain of symptoms similar to that of truly contaminated individuals but which cannot be differentiated in the short time that responders have to decide who warrants emergency decon and treatment.
- When dealing with large numbers of victims, quick air monitoring and detection is crucial. Early identification of the presence of contaminants followed by risk characterization is critical to survival.

The Mass Decon Process

The following tasks and factors should be incorporated into the mass decon procedure:

1. **Protective clothing**—Responders should use either structural firefighting clothing or Level B chemical splash protective clothing with self-contained breathing apparatus. Butyl gloves should be used for hand protection.
2. **Decon prioritization**—Whether you call it triage or decon prioritization process, responders must have a system for initially sorting and prioritizing those individuals who are potentially contaminated or injured. Decon prioritization works in a similar way. In the initial response window, most responders will have a limited amount of personnel and equipment. The general goal of mass decon is to ensure the survival of the greatest number of people.

   When people show symptoms and it is evident that they have been contaminated, they will generally willingly submit to some form of decontamination. Don't slow down the overall process to argue with the victims. Studies involving chemical warfare agents show that victims should be washed for at least 3 minutes but no longer than 5 minutes.

   People who may have been exposed to the problem but who were not physically contaminated (e.g., most gas and vapor exposures) do not need decon, although they may be "psychologically decontaminated" at a later time. People who have material on them or are severely symptomatic must be decontaminated in a timely manner. Others who were just in the building or in the area in most cases do not need to be decontaminated.
3. **Ambulatory versus nonambulatory victims**—Injured victims should be sorted into ambulatory and nonambulatory groups. Medical triage procedures (e.g., S.T.A.R.T.—Simple Triage and Rapid Treatment Process) can facilitate this process. However, responders should recognize that most triage protocols, including S.T.A.R.T., were designed to be used for victims of acute traumatic injuries. These protocols may not accurately identify and classify ill patients from a hazmat or WMD scenario or predict which patients will clinically deteriorate.
   - **Ambulatory**—Victims are able to understand directions, talk, and walk unassisted. Classified as Priority 3 or minor (walking wounded).
   - **Nonambulatory**—Victims who are unconscious, unresponsive, or unable to move unassisted. Classified as Priority 1—immediate (life threatening injury) or

Priority 2—delayed (serious non-life-threatening injury). Types of injuries may include Figure 11-10:

- Serious medical symptoms (e.g., shortness of breath, chest tightness)
- Evidence of liquid on clothing or skin
- Conventional injuries
- Casualties reporting exposure to vapor or aerosol
- Casualties closest to the point of release

4. **Cold weather operations**—Victims can be washed off in cold weather and survive hypothermia if they are then moved to a warm building or vehicle as soon as possible. The rule of thumb is simple: If the victims may die from the contamination, then wash them off, regardless of the theoretical risk of hypothermia. You should then move them to a covered environment with an ambient temperature of at least 70°F (preferably higher) as soon as possible. There is no evidence that cold weather decon will cause permanent injury or harm Figure 11-11.

   Nothing in these preceding statements should imply that these victims will be happy; they will be extremely cold and may be shivering violently. Hypothermia develops when an individual is continuously exposed to a cold stimulus for a period of time, but it is unlikely that responders will cause a significant core temperature drop by performing emergency decon. A person who has stopped shivering or who does not respond to verbal or physical stimuli may be developing hypothermia. Victims who are young, elderly, or have chronic health problems will succumb to hypothermia faster. These people should be moved to a warm area as soon as possible since hypothermia in this group may begin in approximately 10 minutes.

5. **Mass Decon Operations Using Fire Apparatus**—Numerous options using fire apparatus have been developed and tested throughout the fire service. All of these options are typically based upon using the items immediately available to emergency responders—hoselines, water, and the firefighting apparatus itself. SBCCOM's report entitled *Guidelines for Mass Casualty Decontamination During a Terrorist Chemical Agent Incident* is an excellent reference and recommended reading for anyone considering implementation of this mass decon method.

Mass decon operations can be implemented using a single engine company, multiple engine companies, a truck company, or multiple engine/truck companies in combination with each other. Regardless of the resources available, all involve the establishment of a decon corridor using a combination of overhead and/or side water sprays from fire apparatus. The following information is referenced from the Virginia Department of Emergency Services mass decon training program.

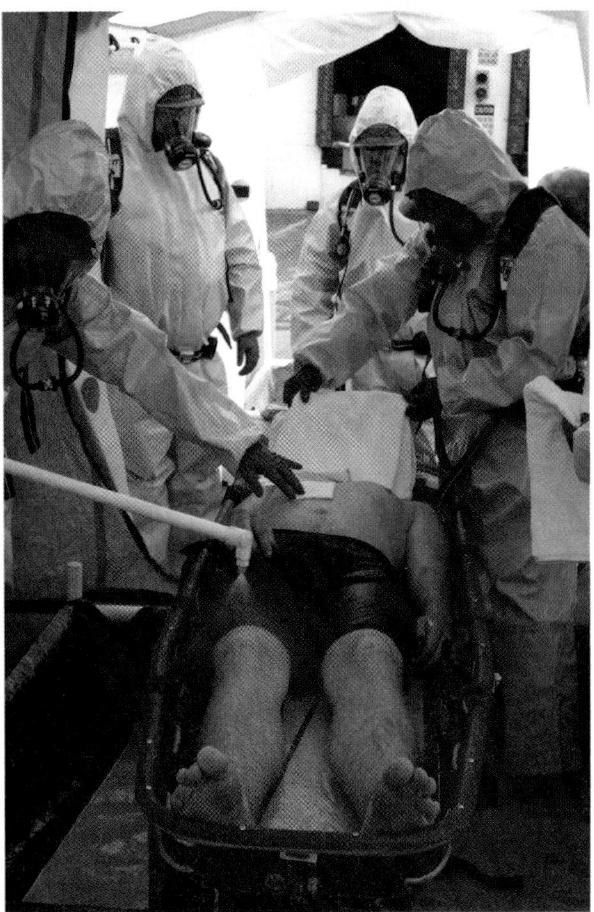

Figure 11-10 Nonambulatory victims are unconscious, unresponsive, or unable to move unassisted.

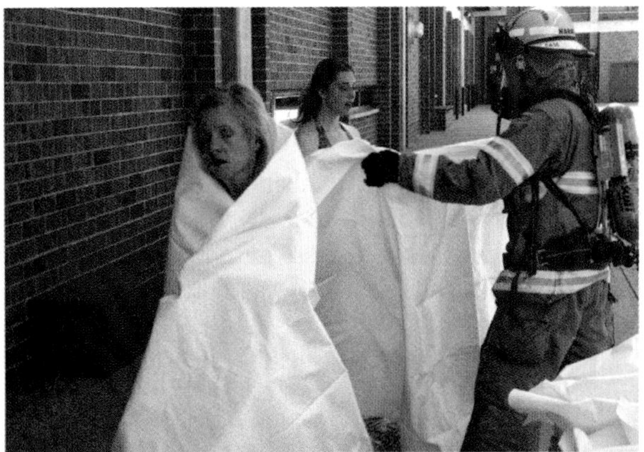

Figure 11-11 The risk of hypothermia must be considered for victims who are decontaminated.

Single Engine Company Response Procedure

Position the engine to create a herding lane (think cattle) between the apparatus and a building wall or other structure. The lane should be 12 to 16 feet wide. By positioning the engine in this manner, victims' modesty can be somewhat protected.

Use the apparatus PA system to direct the victims to the safe refuge area and the Mass Decontamination Corridor. Instructions must be simple, clear, and authoritative (e.g., where to put clothing, what to do in the shower area, where to go after exiting).

Engine company personnel should attach 2 to 3 nozzles to the opposite side of the pump panel and set them for a wide fog pattern. The operator should engage the pump and maintain a pressure of between 30–50 psi at the panel if using automatic or regular nozzles.

Upon exiting the decon corridor a cover should be provided to all victims. Black trash bags, disposable gowns, blankets, etc. are acceptable. Tactics to protect personal modesty (e.g., tarps and poles) may be required.

If emergency tape is available, set up lanes to the entrance and from the exit of the decon corridor. At the exit point personnel should be stationed to direct victims to the EMS area to be triaged. If EMS has not yet set up or has not arrived, place victims in a secure area (e.g., building lobbies, schools, large warehouses, etc.). The goal is to get victims out of the weather and into a contained area suitable to deliver medical treatment as soon as possible.

Multiple Engine Company Response Procedure

If using engine companies with a side-mount pump, the pump panels must face to the street curbside. The engine companies should be positioned approximately 12 to 16 feet apart. Personnel will attach nozzles to all discharges on the side of the pumper facing the decon corridor. If there is only a single discharge, another nozzle should be attached to a section of hose and tied off to either the front bumper or rear of the pumper (e.g., trash line). This corridor will also establish a modesty corridor for the victims. Emergency tape and personnel should be positioned to direct victims into the lane (herding).

Multiple Engine/Ladder Apparatus Response Procedure

If using engine companies with a side-mount pump, the pump panels must face to the street curbside. The engine companies should be positioned approximately 12 to 16 feet apart. Personnel will attach nozzles to all discharges on the side of the pumper facing the decon corridor. If there is only a single discharge, another nozzle should be attached to a section of hose and tied off to either the front bumper or rear of the pumper (e.g., trash line).

The truck company should approach from behind the engine company so that the operator will be able to work upwind and uphill. The aerial apparatus should position the ladder pipe to form a water shower. The water shower should be directed just ahead of the pumpers to give a final rinse to the victims exiting the primary shower set up by the engine companies.

Mass Decontamination Methods

Over the last several years, a number of mass decontamination methodologies have evolved. In today's marketplace, there is no shortage of prepackaged mass decontamination showers for ambulatory victims (able to walk) and nonambulatory victims (unable to walk without assistance). Several versions of pre-plumbed, rapid-deploy shelters are available that contain intricate showerheads and spray wands, along with segregated areas for gender-specific showering. Self-contained decontamination trailers with pop-out sides and overhead tents are also available, as are a wide array of portable showers. Some of these units come equipped with portable water heaters, space heaters, water collection bladders, and sections of rollered platforms for sliding nonambulatory victims (supine, on rigid backboards, or in Stokes baskets) through a series of wash–rinse stations.

Decontaminating nonambulatory victims is a much slower process than performing mass decontamination on ambulatory victims. Handling casualties who cannot walk, or who are unconscious or otherwise unresponsive, requires a significant number of emergency response personnel to complete the decontamination process. It is physically more taxing to carry unresponsive victims, and work times for emergency responders will be limited. Some manufacturers offer stretchers with wheels or other types of carts or sleds to carry those victims who cannot walk. Nevertheless, using these devices may not significantly speed up the mass decontamination process; but it does ease the workload on the responders.

Many jurisdictions set up two separate areas for mass decontamination: one for nonambulatory victims and one for ambulatory victims. It is up to your AHJ to determine the specific procedures for handling both types of victims.

From first responders' perspective, it is important to choose a system that fits the responding agency's needs, based on staffing levels, anticipated numbers of casualties, topography, and proximity to other mass decontamination units in the region. There is no one perfect setup for all occasions. The AHJ must evaluate all operational facets of its own operations and choose a process that best suits the anticipated need.

To set up and use a mass decontamination system on ambulatory victims, follow the steps in Skill Drill 11-2 ▼:

1. Ensure you have the appropriate PPE to protect against the chemical threat.
2. Stay clear of the product, and do not make physical contact with it.
3. Direct victims out of the hazard zone and into a suitable location for decontamination. **(photo 1)**
4. Set up the appropriate type of mass decontamination system based on the type of apparatus, equipment, and/or system available. **(photo 2)**
5. Instruct victims to remove their contaminated clothing and walk through the decontamination corridor.
6. Flush the contaminated victims with water. (A water temperature of 70°F (21°C) is ideal but may not be possible. Try to avoid using water that is uncomfortably hot or cold.) **(photo 3)**
7. Direct the contaminated victims to the triage area for medical evaluation, which may include on-scene treatment and/or transport to an appropriate receiving hospital. (Many agencies provide modesty/comfort packages after decontamination that include gowns, booties, towels, and other pertinent items). **(photo 4)**

To set up and use a mass decontamination system on nonambulatory victims, follow the steps in Skill Drill 11-3 ▼:

1. Set up the appropriate type of mass decontamination system based on the type of apparatus, equipment, and/or system available. **(photo 1)**
2. Ensure you have the appropriate PPE to protect against the chemical threat.
3. Remove the appropriate amount of the victim's clothing. Do not leave any clothing underneath the victim; these items wick the contamination to the victim's back and hold it there, potentially worsening the exposure. (Medical trauma scissors are a helpful and rapid way to accomplish this step.) **(photo 2)**

4. Flush the contaminated victims with water. (A water temperature of 70°F (22°C) is ideal but may not be possible. Try to avoid using water that is uncomfortably hot or cold.) Make sure to rinse well under and around the straps that may be holding the victim to a backboard or other extrication device. *Take care to avoid compromising the victim's airway with water during the process.* **(photo 3)**
5. Move the victims through the decontamination corridor and into the triage area for medical evaluation, which may include on-scene treatment and/or transport to an appropriate receiving hospital. In most cases, significant medical treatment should be provided after decontamination, in a designated medical treatment area. **(photo 4)**

Training is a key component in ensuring an effective mass decontamination operation. Without regular and frequent training, it is impossible to remain proficient in this infrequently used, but potentially high-impact technique. Contaminated victims will not wait around for responders to establish a formal decontamination area. Instead, they will hurry to the hospital by self-transport while still contaminated, which will ultimately affect the ability of the hospital-based providers to render proper medical care.

Because water is a good general-purpose solvent, washing off as much of the contaminant as possible with a massive water spray is the best and quickest way to decontaminate a large group of people.

Regardless of the mass decontamination methodology employed by the responding agency, the theory behind the work focuses on one of three ways to reduce or eliminate contamination: dilution, isolation, and washing.

Skill Drill 11-2

Performing Mass Decontamination on Ambulatory Victims

* This skill drill begins after the victim(s) have been extricated from the contaminated environment and transported in some manner to the mass decontamination corridor.

1 Ensure that you have the appropriate PPE to protect from the chemical threat. Stay clear of the product and do not make physical contact with it. Make an effort to contain runoff by directing victims out of the hazard zone and into a suitable location.

2 Set up the appropriate type of mass decontamination system based on the type of apparatus, equipment, and/or system available.

3 Instruct victims to remove all contaminated clothing and walk through the decontamination corridor. Flush the contaminated victims with water.

4 Direct the contaminated victims to the triage area.

Skill Drill 11-3

Performing Mass Decontamination on Nonambulatory Victims

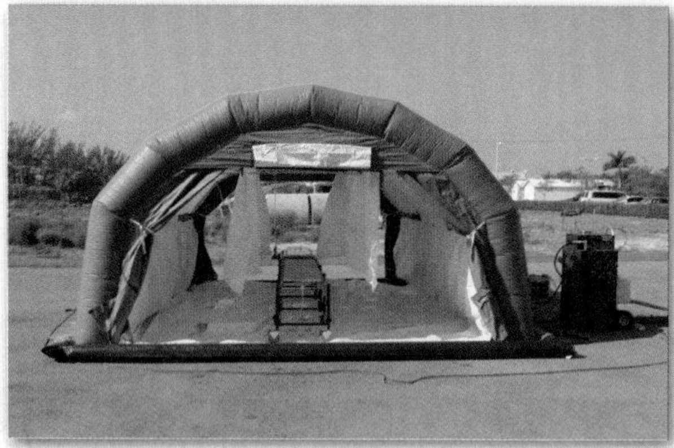

1 Set up the appropriate type of mass decontamination system based on the type of equipment available.

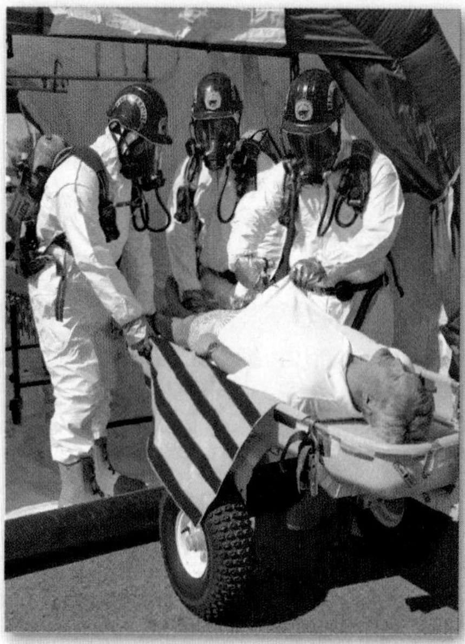

2 Ensure that you have the appropriate PPE to protect against the chemical threat. Remove the victim's clothing. Do not leave any clothing underneath the victim; these items may wick the contamination to the victim's back and hold it there, potentially worsening the exposure.

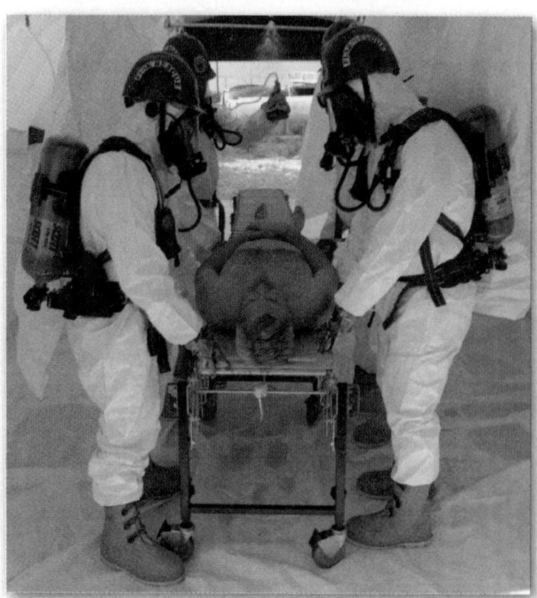

3 Flush the contaminated victim with water.

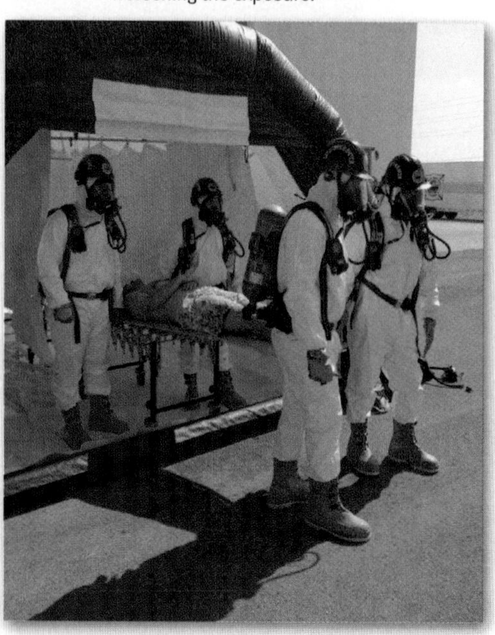

4 Move the victim to a designated triage area for medical evaluation.

Decon Operations Inside Special Buildings

High hazard research and manufacturing facilities are often designed and engineered for specific emergency response scenarios. An increasing number of hospital emergency departments are now being designed to provide immediate decon capability. For guidance on designing decon capability within a health care facility, one should consult NFPA 99—*Health Care Facilities*. Many of these facilities have features that can facilitate the delivery of timely and effective decon operations.

The following factors should be considered when evaluating fixed facilities for decon operations:

- **Decon rooms**—The size of the room should be based upon the number of patients the facility expects to handle. The room's floor, walls, and ceiling should be coated with an inert material that allows the area to be decontaminated. Electrical fixtures should be rated based upon the expected hazards. If flammable vapors are expected, the room or area should meet NFPA 70—*National Electric Code*™ requirements for Class 1, Division 2, Group D classifications.
- **Ventilation systems**—Most facilities engaged in hazardous materials research or manufacturing have specially engineered ventilation systems to reduce employee exposure to hazardous materials. Removal of air contaminants at the source is the most effective method of preventing employee chemical exposure. These systems can be an asset when conducting decon and for determining entry points into a spill area.

The exhaust system should be filtered to control the materials encountered in the facility (e.g., infectious materials, chemicals, radioactives, etc.). Responders must not shut down the power supply that maintains these special systems.

In some facilities, fixed monitoring and detection systems (e.g., oxygen, toxicity, flammability, etc.) may also be installed in the ventilation system to provide early warning of a release. These fixed systems may be monitored in a process control room, security center, or by an off-site alarm company.

- **Positive and negative pressure atmospheres**—These systems maintain a negative air pressure within the hazard area and a positive air pressure outside of the hazard area. If an incident occurs, the flow of air is from the outside toward the inside, thereby minimizing the spread of contaminants (dust, vapors, and gases) from the involved area.
- **Safety showers**—Safety showers are usually located throughout high risk and medical facilities and they are restricted for emergency use in the event of accidental chemical contamination. If a hazmat incident occurs indoors, consider using these safety showers for gross decon. Safety showers have special "deluge" heads that deliver 30 to 50 gallons per minute, far more than a bathroom shower head. This large flow is essential in the initial stage to sweep away, rather than just to dilute, the strong contaminant. Safety showers are a good gross decon option if they have drains and provisions to confine runoff.
- **Emergency eyewash fountains**—Emergency eyewash fountains or hoses are located throughout most industrial and laboratory facilities. Their location should be identified for responders during the pre-entry briefing.
- **Fixed air supply systems**—Some facilities may provide air outlets for use with airline hose respirators. These may be supplied by a bank of air cylinders or from a dedicated breathing air compressor.
- **Personal protective clothing and equipment**—Some facilities will maintain an inventory of protective clothing and equipment to be used for emergency response activities.

The systems and capabilities outlined above are designed based on the physical, chemical, and toxicological characteristics of the materials being handled or anticipated. Responders should consider the use of these specialized systems as a method of reducing the potential contaminants within the Hot Zone. Activation of fixed ventilation systems may also significantly reduce the contamination level before entry.

Always consult with the facility building engineer and safety personnel before using fixed ventilation options. Activation of a ventilation system not designed and rated for the hazards present under emergency conditions could worsen the situation (e.g., cause an explosion or spread the contaminants to a larger area).

Decontamination of Pets and Animals

Animal decontamination is not the same as people decontamination, although the goals and objectives are the same. Most fire departments, Hazardous Materials Response Teams, and law enforcement agencies are not trained and prepared to deal with contaminated animals. In many areas of North America, county and state-based Animal Response Teams (ART) have been developed to address animal handling and response issues. Specialized training courses on animal response and safety have been developed and are usually coordinated through the state agriculture or emergency management agencies. While emergency services may not take the lead role on this issue, they may be asked to play a support role by the lead animal safety and health agencies.

Animals may become contaminated with hazardous materials during an emergency at a fixed facility or from a transportation accident. This could include biological agents, hazardous chemicals, or radioactive materials. Higher probability animal contamination scenarios include agricultural pesticide storage building fires, train derailments near farms and ranches, and petroleum spills on a farm. Lower probability, but credible scenarios include a radiological release effecting a large area or biological attack targeting our animal-related food supply **Figure 11-12**.

Pets and farm animals can become contaminated due to exposure to standing floodwaters caused by severe inland flooding, tropical storms, and hurricanes. Contaminants from floodwater exposure can range from a matrix of hazardous substances in the water to biological materials, sewage and

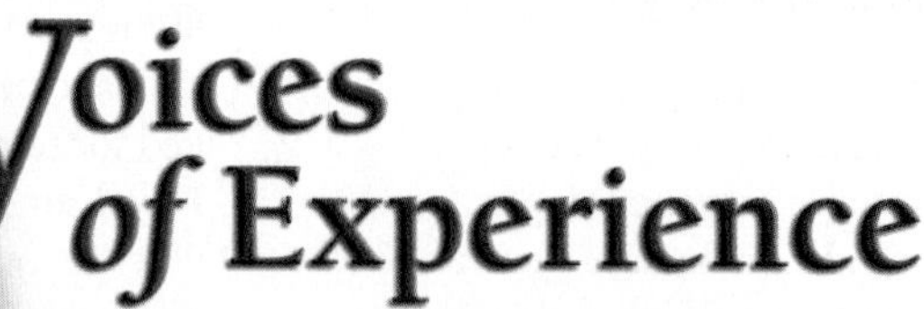

Voices of Experience

In an effort to maintain control of a Hazardous Materials or CBRNE emergency, First Responders must be prepared to continue through the Eight Step Process© and properly decontaminate their victims, fellow emergency response personnel, equipment, and the environment by the systematic isolation, neutralization, or removal of contamination.

Effective decontamination guidance can be found by consulting regulations, consensus standards, and best practices. Some examples include 29 CFR 1910.120, National Fire Protection Association (NFPA) Standard 472, and the *OSHA Best Practices for Hospital-Based First Receivers of Victims from Mass Casualty Incidents Involving the Release of Hazardous Substances.*

For generations prefire planning has helped fire fighters to mitigate potential problem areas in their response zones and to make faster, safer, and more efficient responses to a citizen call for help. My personal experience in Hazardous Materials planning and response, as well as in Emergency Management for the healthcare industry, has shown that preplanning is equally important to the Hazardous Materials Responder. By taking the time to preplan their community, a response team can perform a hazard vulnerability analysis, determining where their focus on training and response protocols need to lie. In addition, preplanning will help to ensure that appropriate types and quantities of decontamination agents are readily available and their application strategies are understood through appropriate training.

In the early morning hours of January 6, 2005, a Norfolk Southern train derailed in the small rural South Carolina town of Graniteville, resulting in the release of over 60 tons of liquefied chlorine gas into the night air, killing nine people and displacing over 5000 from their homes. In all, 554 people were treated at area hospitals, 75 were admitted, and nine would eventually die from the poison. The expected response of local, state and federal agencies ensued and a Unified Command was established to manage the scene.

While many lessons were learned and opportunities for improvement presented themselves in the aftermath (as is true for most disasters), one particularly effective practice involved the use and coordination of a variety of decon resources from area agencies, which resulted in effective and timely decontamination of first responders and victims alike. Local and state agencies performed various functions throughout the incident using the Eight Step Process©, and federal resources were employed in a joint decon effort, which included the use of a heated, two-lane decontamination trailer from the Fort Gordon Fire & Emergency Services organization. Through preplanning of local resources, inter-operative exercises across county and state lines, and use of Unified Command, a tragic situation was kept from becoming even worse.

Preplanning of the hazards in the community has also become a critical piece of the Emergency Management puzzle for hospitals and other healthcare organizations. As part of ongoing accreditation standards a myriad of EM functions must be in place, among them the ability to decontaminate victims of a chemical spill or release prior to their entry into the facility for medical treatment. Eliminating the potential for cross-contamination is critical to a healthcare system's operational capabilities and the continuous availability of a standard of care that the local community expects.

During the Tokyo subway sarin attacks in 1995, these principles were not in place. As self-selected ambulatory victims arrived at area hospitals in need of care—the fact they were still contaminated and there was no plan in place to decontaminate them—cross-contamination became a major concern. In the years since, the international healthcare community has taken to heart an invaluable lesson from this experience. Stateside, OSHA published a best practice document on how to develop a decon team at a healthcare facility, and an entire new subset of first response was born in *First Receivers*. I have personally visited healthcare organizations in upwards of 25 states as well as internationally and have witnessed their evolving adaptation to this new mission. Overall, I have been very impressed with the dedicated professionals I have met who are learning about the industrial dangers in their local communities, as well as investigating the local commodity flow patterns on their highways and railways in an effort to ensure their decon teams are well versed in the dangers that exist and well prepared to respond when duty calls.

Decon does not need to be the most difficult of the Eight Steps, but it also should not be disregarded as simple or something that doesn't require preplanning. Taking the time to research the resources available in your district to assist in a complex event, learning of the specific hazards that may be present during an emergency, and ensuring that the appropriate decon agents are on hand—is truly time well spent.

John Francis Ryan
Fort Gordon Fire & Emergency Services
Fort Gordon, Georgia

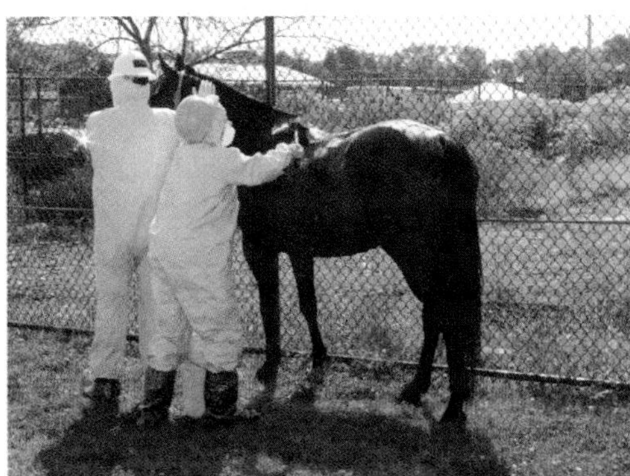

Figure 11-12 Animal decon is not the same as people decon, although the goals and objectives are similar.

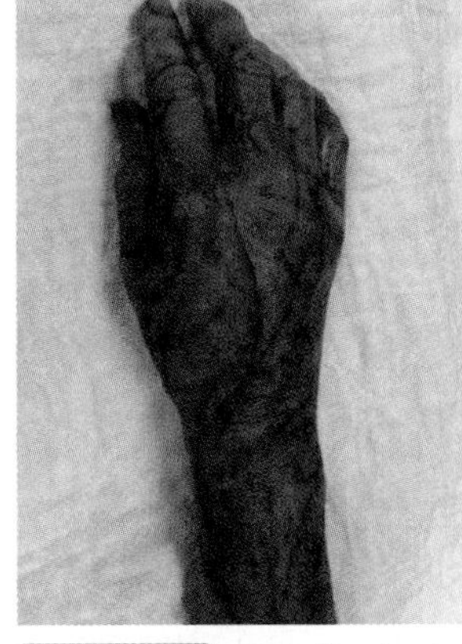
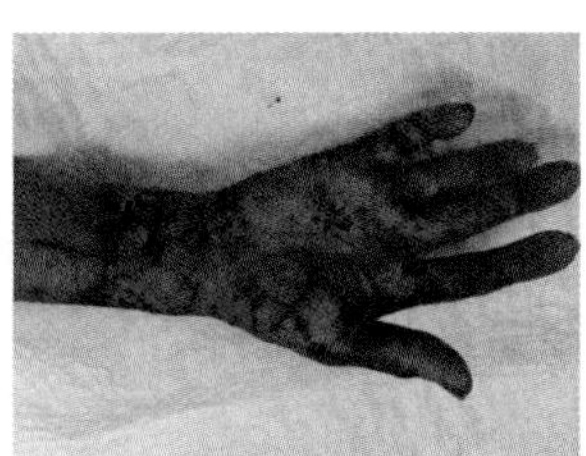

Figure 11-13 A small animal does not mean a lack of danger. A common house cat attacked her handler's hand, causing 52 lacerations and puncture wounds in under a minute. Emergency medical treatment required five hours in the emergency room and 3 weeks to recover.

manure, mud, and sludge. Loose pets and farms animals, animal contamination, animal euthanizing, and dead animal carcass disposal were problematic issues created by Hurricanes Katrina (Louisiana) and Ike (Texas). The September 13, 2008 hurricane Ike disaster caused widespread coastal flooding on the Texas Gulf Coast, including the Bolivar Peninsula which lost an estimated 4,000 head of cattle due to drowning. Some of these animals washed up along shorelines as dead and decaying carcasses, some were washed out to sea and never seen again, and a few came ashore alive on Galveston Island and other locations.

The following issues should be anticipated when dealing with contaminated animals:

- **Animal 101**—Before you can address the issue of how to decon animals, you must first understand the fundamental issues of animal movement and handling. This should include some hints on how to "read" animal behavior. Unlike people, animals can't help you decontaminate themselves nor tell you where it hurts. To humans, "ouch" means "it hurts!" To a human, "moo" simply means, "moo." It probably means something else to fellow cattle, like "I can't believe they want me to raise my left hoof! I weigh 1,500 pounds and need four legs to stand!" Responders should identify veterinarian and large animal specialists, such as farmers and agriculture extension representatives, who can provide assistance on animal behavior.
- **Human safety**—Animals that have been under stress or are out of their natural environment can present handlers with the potential for being injured by bites, scratches, kicks, and crushes. Consider for example that an average horse weighs between 900 and 1,100 pounds. Cattle are even heavier. If a horse steps on your foot, you will remember the experience for the rest of your life. On a smaller scale, a common house cat can do some serious damage when it is stressed out and in the fright and flight mode Figure 11-13.
- **Animal safety**—Preventing escape of the animal using proper and safe restraints to ensure the safety of the animal requires expert knowledge and experience. The method of containment and restraint may be different from one animal to the next. Like people, some animals do not always get along together, especially when sharing the same confined space.
- **Working dogs**—K-9s assigned to search and rescue, arson and bomb, and law enforcement operations may require single animal decon on completion of the mission. When these animals are separated from their handler they can become unsettled. If possible keep the handler with the K-9 and ask the handler for advice on restraints and the best way to manage the animal during the decon process.
- **Outer covering**—With worst-case human decontamination you are primarily dealing with clothing removal and a shower. With animals you are dealing with hair, fur, feathers, or scales. They don't like it when you try to remove them!
- **Type of animal**—If you live and work in the city, your experience with animals is probably limited to household pets and the zoo. The wildlife and farm animal population is very diverse. You need to identify what type of animal you are decontaminating and what its needs are and how it might behave around people.
- **Animal routines**—Animals like dairy cattle follow a daily routine. Daily feeding and milking need to continue regardless if their products enter the food supply or not.
- **Animal first-aid and euthanasia**—If animals have been injured from exposure to hazardous materials a veterinarian should be brought to the scene to assist. If the animal requires first-aid or is unconscious you should have a restraint plan in place. If an animal's injuries are obviously fatal and needs to put down, the animal's owner (if known) must be consulted as well as the authority having jurisdiction. The authority to sedate or euthanize an animal may fall within various government agencies depending on the type of animal, its condition, and the interest of public safety and

health. Knowing the law and having a plan for this scenario will help reduce the animal's suffering.

Large-scale contamination of animals can generate economic, environmental, cultural, and emotional impact within the community. Single and isolated cases involving injured or contaminated animals can draw media attention. If you do not deal with contaminated animal scenarios well you can expect some negative feedback from a broad section of the community you protect.

It is important to have an Animal Decontamination Plan in place that has been coordinated with law enforcement, health, animal control, agricultural, and wildlife agency officials. The plan should also be coordinated with volunteer organizations that specialize in animal care and rescue and veterinary specialists. When large-scale decontamination of animals is required it should be conducted using trained teams with a written Incident Action Plan.

Clean-Up Operations

What Is Clean-Up?

The term *clean-up* means different things to different emergency response personnel. For our purposes, clean-up activities consist of any work performed at the emergency scene by emergency responders, which is directed toward removing contamination from protective clothing, tools, dirt, water, and so on. Clean-up may also involve responder decontamination of some debris, damaged containers, and so on.

From a decon perspective, not all work related to restoring the contaminated site to its previous (nonpolluted) state is considered "clean-up." The section on "implementing response objectives" found elsewhere in the text provides a detailed review of the short- and long-term site restoration and recovery activities associated with hazmat emergency response.

General Clean-Up Options

The Incident Commander has two clean-up options:

1. Conduct a limited-scale clean-up of key emergency response equipment such as fire apparatus. The objective of this option is to place essential equipment back in service as soon and as safely as possible. Responders may also get involved in the more technical aspects of clean-up by working directly under the supervision of an outside agency or contractor.
2. Conduct clean-up using a qualified and authorized environmental contractor. This option is usually exercised when large pieces of heavy equipment have been contaminated and are not part of the emergency response organization's fleet. Examples of such equipment are bulldozers and end loaders used by responders, for instance, to construct dikes inside the Hot Zone.

Before jumping into clean-up activities, make sure you have a sufficiently detailed and coordinated decon plan that is documented within the overall Site Safety Plan. Once responders are decontaminated, the rules of the game shift from one of being an "Emergency Responder" to "Hazardous Waste Generator." In addition, there are clear differences between emergency response and post-emergency response operations (PERO) under OSHA 1910.120. Make sure your clean-up activities are conducted within regulatory guidelines.

Agencies that should be consulted in the development of equipment decon plans can include the following:

- **Water/sewage treatment facilities**—Prior arrangements will be necessary before large quantities of waste water can be flushed into storm/sewer systems via drains connected to the street or from facilities such as fire stations, gyms, etc. As a general rule, all waste should be contained until permission is received for disposal.
- **Pollution control (State Environmental Quality, U.S. Environmental Protection Agency, or U.S. Coast Guard)**—The Incident Commander's authority to create a "runoff" situation during an emergency involving life-threatening materials is well established. Creating additional runoff from equipment decontamination is questionable in most cases. Failure to isolate the runoff could result in a citation from regulatory agencies, generate bad publicity, and increase your potential for civil liability.
- **Product specialists**—May be able to provide clean-up recommendations based on their practical knowledge. Some chemical manufacturers are trained and experienced in equipment decon and can be invaluable when their information is correct, but very damaging when wrong. Product specialists should be interviewed to establish credibility, as discussed in the section on "hazard assessment and risk evaluation" found elsewhere in the text. They do not have legal authority and cannot assume the responsibility for approving on- or off-site disposal.

Equipment Clean-Up General Guidelines

Decon of equipment and apparatus can be difficult and very expensive. Liquids can soak into wood and flow into metal cracks and seams or under bolts. Consult product specialists before initiating decon and clean-up operations. A responder with authority should supervise this phase to ensure that proper planning and coordination between all parties takes place.

While decontaminating, avoid direct contact with contaminated equipment. Brooms and sponge mops can be used to apply cleaning agents to equipment. Protective clothing and respiratory protection must be worn unless proven to be unnecessary by technical specialists who have conducted an appropriate analysis of the contaminants.

Small and Portable Equipment Clean-Up

All small to medium-sized equipment, such as monitoring instruments and hand tools, should be decontaminated before leaving the site. The following issues and concerns should be considered:

- **Hand tools**—May be cleaned for reuse or disposed. Cleaning methods include hand cleaning or, more commonly, pressure washing or steam cleaning. You must weigh the cost of the item against the cost of decontamination and the probability that it can be completely cleaned. Wooden and plastic handles on tools should be evaluated to determine if they can be completely decontaminated. The scientific literature referenced at the end of this chapter provides overwhelming evidence that many pesticide-contaminated wooden parts, like shovel handles, cannot be adequately decontaminated. Consult with product specialists for the best advice.

- **Monitoring instruments**—Follow the manufacturer's recommendations with respect to decon. If the instrument becomes damaged or disabled during the emergency, most instrument manufacturers will not accept the device for repair unless it has been properly decontaminated. If instruments were covered with protective plastic, remove and discard the plastic covering, and tape properly.
- **Fire hose**—Should be cleaned following the manufacturer's recommendations. For most materials, detergents will perform adequately. However, strong detergents and cleaning agents may damage the fire hose fibers. The fire hose should be thoroughly rinsed to prevent any fiber weakening. The hose should then be marked and pressure tested before being placed back into service. Severe exposure to some chemicals such as chlorine will result in damage that may require taking the hose out of service (e.g., when the cleaning water turns to hydrochloric acid, $Cl_2 + H_2O = HCL$).

Motor Vehicle and Heavy Equipment Clean-Up

If a large number of vehicles need to be decontaminated, consider implementing the following recommendations:

- Establish a decon pad as a primary wash station. The pad may be a concrete slab or a pool liner covered with gravel. Each of these should be bermed or diked with a sump or some form of water recovery system to collect the resulting rinse. Get some engineering help and do it right the first time.
- Completely wash and rinse vehicles several times with an appropriate detergent. Pay particular attention to wheel wells and the chassis. Depending on the nature of the contaminant, it may be necessary to collect all runoff water from the initial gross rinse, particularly if there is contaminated mud and dirt on the underside of the chassis.
- Engines exposed to toxic dusts or vapors should have their air filters replaced. Mechanics sometimes blow dust out of air filters during routine maintenance, exposing themselves unnecessarily to this hazardous dust. Contaminated air filters should be properly disposed.

If vehicles have been exposed to minimal contaminants such as smoke and vapors, they may be decontaminated on site and then driven to an off-site car wash for a second, more thorough washing. Car washes may be suitable if the drainage area is fully contained and all runoff drains into a holding tank. Car washes are not recommended if they drain into the sanitary sewer. Car washes used for decon should be inspected and approved in advance.

When a vehicle is exposed to corrosive atmospheres, it should be inspected by a mechanic for possible motor damage. Equipment sprayed with acids should be flushed or washed as soon as possible with a neutralizing agent such as baking soda and then flushed again with rinse water.

Post-Incident Decon Concerns

Debriefing

A debriefing should be held for those involved in decontamination and clean-up as soon as practical. Responders and contractors involved in the operation should be provided with as much information as possible about any delayed health effects of the hazmats.

If necessary, follow-up examinations should be scheduled with medical personnel and exposure records maintained for future reference by the individual's personal physician.

Site Security and Custody

In many instances, contaminated materials must remain at the incident scene until they can be removed for off-site cleaning or disposal. In this case, special precautions should be taken:

- Take appropriate security measures. Potential problems that validate the need for security include the potential for vandalism, curious children who may be injured, or folks who simply see an opportunity to dump their waste. Always ensure that security is provided for hazardous waste and that the proper chain of custody is maintained. Additional lighting may be necessary if the materials must remain overnight.
- Make sure appropriate warning signs are posted and labels are attached to containers.
- Make sure containers are properly sealed.

Hazardous Waste Handling and Disposal

Regulatory Compliance

Local, state, and federal regulations require that hazardous wastes be disposed in a specific manner. All personnel involved in the disposal of hazardous waste must be trained in the provisions of the Federal Resource Conservation and Recovery Act (RCRA) and any related state or local regulations for waste disposal.

Hazardous Waste Containers

All hazardous waste containers should be visibly identified with the proper markings and labels per DOT 215K (Global Harmonization) and EPA regulations.

Only approved chemically compatible containers of sufficient strength should be used for hazardous waste. The containers should be kept covered at all times and arranged so that easy access exists. Care should be taken during all handling to maintain the integrity of the container. Any container stored outdoors must be waterproof.

Wrap-Up

Chief Concepts

- Decontamination is the process of making people, equipment, and the environment safe from hazardous materials contaminants. The more you know about how contamination occurs and spreads, the more effective decontamination will be.
- The basic concepts of decontamination are relatively simple. If contact with the contaminant can be controlled and minimized, the need for decontamination can be reduced.
- Decon needs to be an adaptive and flexible process that respects the hazards and the behavior of the contaminants, as well as human behavior under stress. The hazards and risks presented by the incident will define the scope, nature, and complexity of decon operations.
- Contamination is any form of hazardous material (solid, liquid, or gas) that physically remains on people, animals, or objects.
- Direct contamination occurs when a person comes in direct physical contact with a contaminant or when a person comes into contact with any object that has the contaminant on it (e.g., contaminated clothing or equipment).
- Cross-contamination occurs when a person who is already contaminated makes contact with a person or object not contaminated. Cross-contamination is typically the result of poor site management and control, inadequate decon and site safety procedures, or a failure to follow safety procedures.
- Exposure means that a person has been subjected to a toxic substance or harmful physical agent through any route of entry into the body (e.g., inhalation, ingestion, injection, or by direct contact [skin absorption]).
- The safety and health hazards of the contaminants at any incident will define how complex decon operations will be.
- The best field decontamination procedures emphasize the need to confine contaminants to a limited area. Establishing a designated decontamination corridor and decontamination area are the first steps in limiting the spread of contaminants.
- There is no universal decon method that will work for every hazmat incident or release. Regardless of the number of decontamination steps required, decontamination is most effective when it is carried out by a trained Decontamination Team using multiple cleaning stations.
- The pathway is from the Hot Zone into the decon area, with the exit point near the Warm Zone/Cold Zone interface. The decontamination corridor should be clearly marked.
- Decon is a labor-intensive operation. When setting up the decon area, consideration must be given to the staffing of the decon operation and the safety of the Decon Team.
- Multistep decon operations can be broken into two broad phases: gross decon and secondary decon. Gross decon is the initial rinse and secondary decon is the follow-up process that removes the contaminants to a safe and acceptable level.
- Emergency decontamination is the physical process of immediately reducing contamination of individuals in potentially life-threatening situations with or without the formal establishment of a decontamination corridor.
- Emergency decon can be innovative, but the most important concept is to clean the contaminated person as soon as possible. Remember the basics—FLUSH—STRIP—FLUSH. Soap and water is a near-universal solution and should be applied in large quantities. When acids or bases have contacted bare skin, the minimum amount of time for a water flush is at least 20 minutes.
- Technical decon is a multistep process in which contaminated individuals are cleansed with the assistance of trained personnel.
- Mass decontamination is established when large numbers of people (i.e., civilians or responders) need to be decontaminated at the scene of a hazmat emergency. The general goal is to provide the greatest good to the greatest number of people in the shortest amount of time.
- A debriefing should be held for those involved in decontamination and clean-up as soon as practical. Responders and contractors involved in the operation should be provided with as much information as possible about any delayed health effects of the hazmats.

Hot Terms

ALARA Performing decontamination to a level As Low As Reasonably Achievable.

Chemical Degradation The process of altering the chemical structure of a contaminant through the use of a second chemical or material.

Contaminant A hazardous material that physically remains on or in people, animals, the environment, or equipment, thereby creating a continuing risk of direct injury or a risk of exposure outside of the Hot Zone.

Contamination The process of transferring a hazardous material from its source to people, animals, the environment, or equipment, which may act as a carrier.

Decontamination The physical and/or chemical process of reducing and preventing the spread of contamination from persons and equipment used at a hazardous materials incident. OSHA 1910.120 defines decontamination as the removal of hazardous substances from employees and their equipment to the extent necessary to preclude foreseeable health effects.

Decontamination Corridor A distinct area within the Warm Zone that functions as a protective buffer and bridge between the Hot Zone and the Cold Zone, where decontamination stations and personnel are located to conduct decontamination procedures.

Decontamination Officer A position within the hazmat branch which has responsibility for identifying the location of the decon corridor, assigning stations, managing all decontamination procedures, and identifying the types of decontamination necessary.

Degradation (a) A chemical action involving the molecular breakdown of a protective clothing material or equipment due to contact with a chemical. (b) The molecular breakdown of the spilled or released material to render it less hazardous.

Disinfection The process used to destroy the majority of recognized pathogenic microorganisms.

Direct Contamination Occurs when a person comes in direct physical contact with a contaminant or when a person comes into contact with any object that has the contaminant on it (e.g., contaminated clothing or equipment).

Emergency Decontamination The physical process of immediately reducing contamination of individuals in potentially life-threatening situations with or without the formal establishment of a decontamination corridor.

Exposure The process by which people, animals, the environment, and equipment are subjected to or come in contact with a hazardous material.

Gross Decontamination The initial phase of the decontamination process during which the amount of surface contaminant is significantly reduced.

Mass Decontamination The process of decontaminating large numbers of people in the fastest possible time to reduce surface contamination to a safe level.

Sterilization The process of destroying all microorganisms in or on an object.

Safe Refuge Area A temporary holding area for contaminated people until a decontamination corridor is set up.

Secondary Decontamination The second phase of the decontamination process designed to physically or chemically remove surface contaminants to a safe and acceptable level.

Technical Decontamination The planned and systematic process of reducing contamination to a level As Low As Reasonably Achievable (ALARA).

HazMat Responder *in Action*

As an Emergency Medical Technician, you and your Paramedic partner are in charge of the EMS unit that has been dispatched to the Hazardous Materials Incident within your community. This is your first real incident since finishing EMT school. You are preparing a report and are looking for terminology that addresses what you have provided on scene within your Patient Report.

Questions

1. A hazardous material that physically remains on or in people or equipment and that thereby creates a continued risk of direct injury or a risk of exposure outside of the hot zone is called:

A. Contamination
B. Contaminant
C. Disinfection
D. Exposure

2. The process of transferring a hazardous material from its source to people or equipment, which may count as a carrier.

A. Contamination
B. Contaminant
C. Disinfection
D. Exposure

3. What is the process by which people and equipment are subjected to or come in contact with a hazardous material?

A. Contamination
B. Contaminant
C. Disinfection
D. Exposure

4. What is the process used to destroy the majority of recognized pathogenic microorganisms?

A. Contamination
B. Contaminant
C. Disinfection
D. Exposure

References and Suggested Readings

1. Air Force Civil Engineer Support Agency (AFCESA) and PowerTrain, Inc., *Hazardous Materials Incident Commander Emergency Response Training,* Tyndall Air Force Base, FL: Headquarters AFCESA (2010).
2. Air Force Civil Engineer Support Agency (AFCESA) and PowerTrain, Inc, *Hazardous Materials Technician Emergency Response Training CD,* Tyndall Air Force Base, FL: Headquarters AFCESA (2010).
3. Armed Forces Radiobiology Research Institute—Military Medical Operations, *Medical Management of Radiogical Casualties Handbook* (3rd Edition), Bethesda, MD: AFRRI (June 2010).
4. Amlot, R., et al., *Comparative Analysis of Showering Protocols for Mass-Casualty Decontamination,* Prehospital and Disaster Medicine, 25(5), 435–439 (2010).
5. Black, R. H., "Protecting and Cleaning Hands Contaminated by Synthetic Fallout Under Field Conditions." *Industrial Hygiene Journal* (April, 1960), pages 162–168.
6. Bledsoe, Bryan E., D. O., Robert S. Porter, and Richard Cherry, *Essentials of Paramedic Care* (2nd Edition), Upper Saddle River, NJ: Pearson Prentice Hall, Inc. (2007).
7. Carroll, Todd, R., "Contamination and Decontamination of Turnout Clothing," Washington, DC: Federal Emergency Management Agency (April 1993).
8. Cleaning Pesticide Contaminated Clothing: A Special Report on Safety. *Pest Control Technology Magazine* (November, 1984) pages 42–44.
9. Dawson, Gaynor, W., and B. W. Mercer, *Hazardous Waste Management,* New York, NY: John Wiley and Sons (1986).
10. DOL/HHS Joint Advisory Notice, "Protection Against Occupational Exposure to HBV and HIV," U.S. Department of Labor (DOL) and U.S. Department of Health and Human Services (HHS) Publications (October, 1987).
11. Easley, J. M., R. E. Laughlin, and K. Schmidt, "Detergents and Water Temperature as Factors in Methyl Parathion Removal from Denim Fabrics." *Bulletin of Environmental Contamination and Toxicology* (1982), pages 241–244.
12. Emergency Film Group, *AIDS, Hepatitis & the Emergency Responder* (DVD), Edgartown, MA: Emergency Film Group (2000).
13. Emergency Film Group. *Decon Team* (DVD), Edgartown, MA: Emergency Film Group, Edgartown, MA: Emergency Film Group (2003).
14. Emergency Film Group, *Mass Decontamination* (DVD), Edgartown, MA: Emergency Film Group (2004).
15. Emergency Film Group, *Medical Operations at Hazmat Incidents* (DVD), Edgartown, MA: Emergency Film Group (2000).
16. Emergency Film Group, *Patient Decontamination* (DVD), Edgartown, MA: Emergency Film Group (2007).

    Emergency Film Group, *Terrorism: Medical Response* (DVD), Edgartown, MA: Emergency Film Group (2002).
17. Emergency Film Group, *The Eight Step Process: Step 7—Decontamination* (videotape), Plymouth, MA: Emergency Film Group (2005).
18. Federal Emergency Management Agency, Hazardous Materials Workshop for Hospital Staff, Washington, DC, (July, 1992).
19. Finley, E. L., G. I. Metcalfe, and F. G. McDermott, "Efficacy of Home Laundering in Removal of DDT, Methyl Parathion and Toxaphene Residues From Contaminated Fabrics." *Bulletin of Environmental Contamination and Toxicology* (1974), pages 268–274.
20. Finley, E. L., and R. B. Rogillio, "DDT and Methyl Parathion Residues Found in Cotton and Cotton-Polyester Fabrics Worn in Cotton Fields." *Bulletin of Environmental Contamination and Toxicology* (1962), pages 343–351.
21. Friedman, William, J., "Decontamination of Synthetic Radioactive Fallout from Intact Human Skin," *Industrial Hygiene Journal* (February, 1958).
22. Ganelin, Robert, M. D., Gene Allen Mail, and L. Cueto, Jr., "Hazards of Equipment Contaminated with Parathion." *Archives of Industrial Health* (June, 1961) pages 326–328.
23. Gold, Avram, William A. Burgess, and Edward V. Clough, "Exposure of Firefighters to Toxic Air Contaminants." *American Industrial Hygiene Association Journal* (July, 1978).
24. Hawley, Chris, Gregory G. Noll, and Michael S. Hildebrand, *Special Operations for Terrorism and Hazmat Crimes.* Chester, MD: Red Hat Publishing, Inc. (2002).
25. Hildebrand, Michael, S., "Complete Decontamination Procedures for Hazardous Materials: The Nine Step

Process, Part-1." *Fire Command* (January, 1985), pages 18–21.

26. Hildebrand, Michael, S., "Complete Decontamination Procedures for Hazardous Materials: The Nine Step Process, Part-2." *Fire Command* (February, 1985), pages 38–41.
27. Hildebrand, Michael, S., *Hazmat Response Team Leak and Spill Control Guide,* Oklahoma State University, Stillwater, OK (1984).
28. Hughes, Stephen M., David W. Berry, and Edward D. Hartin, "What Does a Car Wash and a Baby Have to Do with Hazardous Materials Decon?" An independent technical paper prepared by HazMat-TISI, Columbia, MD, (1992).
29. Hsu V. P., et al., "Opening a Bacillus Anthracis-Containing Envelope," Capitol Hill, Washington, D.C.: The Public Health Response. *Emerging Infectious Diseases,* Volume 8, No. 10., U.S. Centers for Disease Control and Prevention, Atlanta, GA. (October 2002).
30. Kampmier, Craig, "Decon Design," *NFPA Journal* (March/April, 2000), pages 59–61.
31. LeMaster, Frank, "Why Protective Clothing Must Be Cleaned." *The Voice,* (August/ September 1993), pages 19–20.
32. Lillie, T. H., et al., "Effectiveness of Detergent and Detergent Plus Bleach for Decontaminating Pesticide Applicator Clothing." *Bulletin of Environmental Contamination and Toxicology* (1982), pages 89–94.
33. Macintyre, Anthony G., MD, et al., "Weapons of Mass Destruction Events With Contaminated Casualties—Effective Planning for Health Care Facilities." *Journal of the Amercian Medical Association* (January 12, 2000), pages 242–249.
34. McGary, Roger, A., "Disinfection of SCBA." *The Voice* (October, 1993), pages 26–28.
35. Molino, Louis N., Sr. "The Big One: Proven Methods for the Management and Mass Decontamination of a Crowd." *Homeland First Response* (July/August, 2003), pages 14–21.
36. National Alliance of State Animal and Agricultural Emergency Programs, *Animal Decontamination: Current Issues and Challenges,* Report by the Animal Decontamination Best Practices Working Group (December 2010).
37. National Institute for Occupational Safety and Health, "Report to Congress on Workers' Home Contamination Study Conducted Under The Workers' Family Protection Act (29 U.S.C. 671a)." Publication No. 95-123, Department of Health and Human Services, Washington, D.C. (September, 1995).
38. National Fire Protection Association, *NFPA 1581—Standard on Fire Department Infection Control Program,* Quincy, MA: National Fire Protection Association (2010).
39. National Fire Protection Association, *NFPA 1851—Standard on Selection, Care, and Maintenance of Protective Ensembles for Structural Fire Fighting and Proximity Fire Fighting,* Quincy, MA: National Fire Protection Association (2008).
40. National Fire Protection Association, *Hazardous Materials Response Handbook* (6th edition), Quincy, MA: National Fire Protection Association (2013).
41. National Fire Protection Association, *NFPA 473—Standard for Competencies for EMS Personnel Responding to Hazardous Materials/Weapons of Mass Destruction Incidents,* Quincy, MA: National Fire Protection Association (2013).
42. Olson, Kent R., M. D. and Ilene B. Anderson, *Poisoning and Drug Overdose* (5th Edition), New York: Lange Medical Books/McGraw-Hill, (2007).
43. OSHA Instruction CPL 2–2.44B, "Enforcement Procedures for Occupational Exposure to HBV and HIV," U.S. Department of Labor (DOL) and U.S. Department of Health and Human Services (HHS) Publications (February 27, 1990).
44. Perkins, John, J., *Principles and Methods of Sterilization in Health Sciences* (2nd Edition), Chicago, IL: Charles Thomas Publishing Co. (1980).
45. Ronk, Richard, and Mary Kay White, "Hydrogen Sulfide and the Probabilities of Inhalation Through a Tympanic Membrane Defect." *Journal of Occupational Medicine* (May, 1985), pages 337–340.
46. Rudner, Glen D, "First Responder Considerations for Decontamination at Mass Casualty Incidents," Student handout material developed for Virginia Department of Emergency Services (January, 2003).
47. Teller, Robert, "Developing Proper Decontamination Procedures for Emergency Response." *Econ Magazine* (March 1993), pages 32–33.
48. United Kingdom Home Office, *The Decontamination of People Exposed to Chemical, Biological, Radiological or Nuclear (CBRN) Substances or Material—Strategic National Guidance* (1st Edition), London, England: Home Office (February, 2003).
49. U.S. Army Medical Research Institute of Chemical Defense, *Medical Management of Chemical Casualties*

(3rd Edition), Aberdeen Proving Ground, MD: USAMIC—Chemical Casualty Care Office (July, 2000).

50. U.S. Army Medical Research Institute of Infectious Diseases (USAMRID), *Medical Management of Biological Casualties* (6th Edition), Fort Detrick, MD: USAMRID (April, 2005).
51. U.S. Army Soldier and Biological Chemical Command (SBCCOM), *Guidelines for Mass Casualty Decontamination During a Terrorist Chemical Agent Attack,* Aberdeen Proving Ground, MD (January, 2001).
52. U.S. Centers for Disease Control (CDC) Publications, *A Curriculum Guide for Public Safety Emergency Response Workers: Prevention of Human Immunodeficiency Virus and Hepatitis B Virus* (February, 1989).
53. U.S. Centers for Disease Control (CDC) Publications, *Guideline for Handwashing and Hospital Environmental Control* (1985).
54. U.S. Centers for Disease Control (CDC) Publications, *Guideline for Prevention of HIV and HBV Exposure to Health Care and Public Safety Workers* (February, 1989).
55. U.S. Environmental Protection Agency, *Guide for Infectious Waste Management* (1986).
56. Walter, Frank G., Raymond Klein and Richard G. Thomas, *Advanced Hazmat Life Support (AHLS) Course—*Provider Manual (3rd Edition), Tucson, AZ: University of Arizona Emergency Medicine Research Center (2003).

Terminating the Incident

CHAPTER

12

Hazardous Materials Technician

Knowledge Objectives

After reading this chapter, you will be able to:

- Assist in the incident debriefing process. (p. 437)
- Assist in the incident critique process, including the three types of critique formats. (p. 439)
- Complete incident reporting and documentation requirements. (p. 439)
- Describe the types of legal liability and operations security issues that should be considered when documenting lessons learned and After Action Reports. (p. 440–441)

Skills Objectives

There are no Hazardous Materials Technician skills objectives in this chapter.

Hazardous Materials Incident Commander

Knowledge Objectives

After reading this chapter, you will be able to:

- Identify five activities for terminating a HM/WMD incident and the procedures for conducting an incident debriefing. (p. 435, 437)
- Conduct an incident critique, including the three types of critique formats. (p. 439)
- Complete incident reporting and documentation requirements. (p. 439)
- Describe the types of legal liability and operations security issues that should be considered when documenting lessons learned and After Action Reports. (p. 440–441)

Skills Objectives

After reading this chapter, you will be able to:

- Conduct an incident debriefing. (p. 437)
- Conduct an incident critique. (p. 439)

You Are the HazMat Responder

In a hazardous materials incident, two contractors died while attempting to clean sludge out of the bottom of an ethylene dibromide (EDB) tank. The first worker entered the tank without respiratory protection and was immediately overcome. The second worker crawled through the manway without respiratory protection to rescue his partner. He was also overcome by the EDB. When rescue personnel arrived on the scene 5 minutes later, they determined that the workers were discovered missing at lunch. A coworker was sent to investigate and found both men trapped in the tank.

When it became obvious that the workers had been trapped for several hours before the 911 call was placed, the Incident Commander (IC) stopped rescue operations and treated the incident as a body recovery operation. A confined space rescue team conducted the body recovery operation without any unusual problems.

The IC has asked you to coordinate the termination phase of the incident and conduct an on-site debriefing of fire and rescue personnel. He reminds you that ethylene dibromide is a bad actor and is a known carcinogen. He wants to make sure the incident is terminated correctly with the proper documentation.

1. What post-incident safety and health concerns do you have that need to be worked into the debriefing?
2. How long do you think it will take to conduct a debriefing that will adequately cover all of the points you will need to make?
3. What documentation would be required to support termination activities?

Introduction

Termination is the final step in the Eight Step Incident Management Process©. It represents the transition between the termination of the emergency phase and the initiation of clean-up and restoration and recovery operations. It also is the phase where responders document incident response operations, including the problem, agencies involved, hazards and risks encountered, safety procedures, site operations, and lessons learned.

Terminating the incident usually consists of five distinct activities Figure 12-1:

1. Termination of the emergency phase of the incident
2. Transfer of on-scene command from the IC of the emergency phase to the individual responsible for managing and coordinating Post-Emergency Response Operations (PERO)
3. Incident Debriefing
4. Post-Incident Analysis
5. Critique

Declaring the Incident Terminated

Terminating a hazardous materials incident is a formal process. Unlike fire emergencies where it is usually obvious that the fire is out, hazardous materials incidents sometimes slowly creep from the emergency phase to the restoration and recovery phase. This type of "mission creep" can lead to an unsafe incident scene. Emergency responders and support personnel sometimes get mixed signals concerning whether there are actually hazards present at this phase of the

Responder Safety Tip

A Hazard and Risk Assessment process should be used to determine if the incident scene is safe, unsafe, or dangerous before terminating the incident.

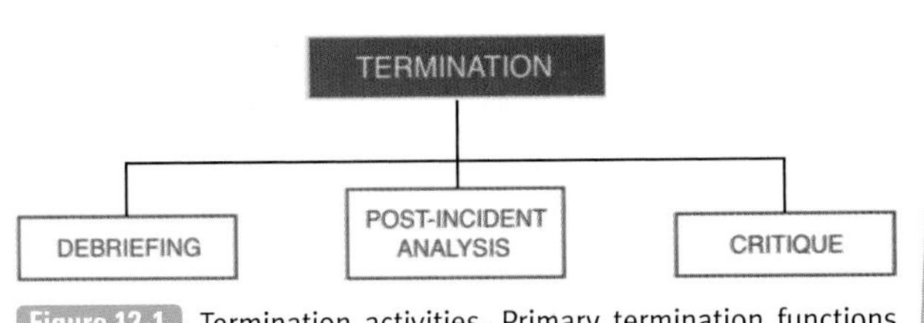

Figure 12-1 Termination activities. Primary termination functions include a debriefing, post incident analysis, and a critique.

operation. If spills have been controlled and personnel are standing around waiting for product transfer operations to begin, boredom and complacency sometimes set in. Control of the perimeter becomes relaxed, protective clothing comes off, and the incident scene can become unsafe or dangerous again.

As a Hazardous Materials technician, you should be able to recognize when hazards are still present at the incident scene and determine if and when the incident scene is safe.

As the IC, you should be satisfied with answers to the following questions before the incident is declared terminated **Figure 12-2**:

- Is the incident scene dangerous? (See the chapter on "hazard and risk assessment" elsewhere in this text.) If the incident scene is still dangerous, this is a strong indicator you should still be in the emergency response mode and that the emergency response agency should still be in command.
- Is the incident scene unsafe? In some instances, responders may terminate the emergency response phase and transfer command of PERO to an environmental agency or contractor (i.e., the PERO Incident Commander). Although the incident may be "stabilized," some hazards may still remain at the incident scene. If the contractor still needs emergency response resources at the incident scene to deal with these hazards, then you should probably still be in the emergency response mode. The lead emergency response agency should retain incident command/control until the hazards are mitigated to the point that emergency response resources are not needed.
- Is the incident scene safe? Safe means totally safe. Use "The Mom Test"—if you would feel completely comfortable having your mother sitting on a lawn chair having a nice picnic where the emergency used to be, then you probably have a safe incident. But guessing is not a good indicator of what is safe. Follow proper standard operating procedures (SOPs) and use instrumentation to evaluate the hazards. The Safety Officer and Hazmat Group Supervisor should help make this call. Returning to the scene of an incident you just left is a pretty bad feeling, especially if someone has been injured.

Remember that the requirements of the OSHA-HAZWOPER Regulation (29 CFR 1910 .120) clearly delineate between "emergency phase" and "post-emergency response operations (PERO)."

Responder Safety Tip

When determining that the incident scene is safe, the authority having jurisdiction makes the final determination.

Transferring Responsibility of the Incident Scene

When the decision has been made to terminate the emergency response phase and additional work is still required at the scene for restoration and recovery, the IC should meet with the senior representatives from the responsible party or contractors taking over to formally hand off the incident scene **Figure 12-3**. It is unprofessional to pick up and leave an incident scene without briefing the personnel who are responsible for carrying out the next phase of the operation. There may also be liability issues for you and your organization if you do not inform these people of the hazards and associated risks remaining at the incident scene. Just because emergency response agencies depart the scene doesn't mean that the need for a post-emergency response incident command organization doesn't exist. Although the players may be different, statutory or regulatory authorities will clearly define the key player(s).

The IC should make it clear that the emergency response phase is being terminated and then formally transfer command to the PERO Incident Commander in a face-to-face meeting. The transfer briefing should cover:

- The nature of the emergency
- Actions taken to stabilize and resolve the emergency
- Name(s) amounts of hazardous material(s) involved
- Hazards and risks that were mitigated and those that still exist
- Safety procedures
- Relevant documentation and points of contact
- Parties responsible for the spill
- Law enforcement agencies responsible for traffic control
- State, municipal, or other regulatory authority having jurisdiction

If the incident has legal or criminal implications that would require documentation or evidence, it is critical that chain-of-custody procedures be followed. The IC should ensure that response operations are fully coordinated with law enforcement or investigation agencies involved.

Finally, before leaving the scene, the IC should document the time of departure, names, companies, and contact information for the personnel assuming control of the scene. This information should be placed in the official log. Be sure to leave your contact information with the group now taking over responsibility for the event.

Incident Debriefing

The purpose of the **incident debriefing** in the field is to provide accurate information concerning the hazards and risks involved directly to the people who may have been exposed,

TERMINATION WORKSHEET
STRATEGIC GOALS

HAZARDOUS MATERIALS EVENT

CONFINED SPACE OPERATIONS

Combined Special Operations (Clan Lab, EOD, SWAT)

Response Objectives in Progress

☐ ☐ ☐ **Evaluate Isolation Area**
- ☐ Maintain control of perimeter
- ☐ Evaluate human resource to hold such perimeter
- ☐ Evaluate isolation area through Hazard/Risk assess.

☐ ☐ ☐ **Evaluate behavior event to current status**
- ☐ Influence breach size
  - ☐ Contents chilled
  - ☐ Stress level limited
  - ☐ Venting activated
- ☐ Influence quantity released
  - ☐ Position of container changed
  - ☐ Pressure differential controlled
  - ☐ Breach capped off
- ☐ Influence of engulfment
  - ☐ Ignition sources controlled
  - ☐ Dikes or damming erected
  - ☐ Dilution or neutralization
- ☐ Influence of impingement
  - ☐ Shielding provided
  - ☐ Evacuation - relocation

☐ ☐ ☐ **Perform Hazard/Risk Assessment**
- ☐ Containment system evaluation
- ☐ Confinement system evaluation
- ☐ Level of hazard through active analysis
- ☐ Establish degree of hazard comparison

☐ ☐ ☐ **Identify current PPE requirements**
- ☐ Hazard comparison establishes:
  - ☐ Respiratory protection still required
  - ☐ Skin protection still required
- ☐ Tasks to be performed
  - ☐ Entry / Back-up mission
  - ☐ Current Hot zone objectives
    - ☐ Behavior not controlled
    - ☐ Waiting evaluation
  - ☐ Correlated towards objective status
- ☐ Decontamination level
  - ☐ Warm zone mission objectives
  - ☐ Correlated towards objective status
- ☐ Evaluation of personnel - medical status
  - ☐ Heat stress
  - ☐ Cold stress
  - ☐ Rehabilitation

☐ ☐ ☐ **Establish resources required and anticipated**
- ☐ Internal Resources
  - ☐ Human resource status
    - ☐ Units working
    - ☐ Personnel required maintaining current objectives
    - ☐ Establish return to service plan
  - ☐ Equipment
    - ☐ Equipment required to maintain current objectives
    - ☐ Decontamination of equipment
    - ☐ Re-supply of equipment used or damaged
      - ☐ Equipment identified damaged or consumed
    - ☐ Establish return to service plan
- ☐ External resources
  - ☐ Determine level of activity required for each external resource on scene
  - ☐ Identify when each resource can be released
  - ☐ Briefing with resource assigned to section, branch, or sector

☐ ☐ ☐ **Evaluate Decontamination Status**
- ☐ Containment of run-off
- ☐ Objectives not completed identified
  - ☐ Decontamination of personnel
  - ☐ Decontamination of equipment
- ☐ Medical surveillance completed

☐ ☐ ☐ **Identify Goals for Transfer**
- ☐ Active analysis –Hazard/Risk Assessment
- ☐ Evaluation of incident by safety officer/HazMat group leader
- ☐ Responsibility Transfer procedures identified
- ☐ Incident debriefing
- ☐ Information gathering responsibilities identified
- ☐ Reports from each section, branch, or sector received for report

Termination Activities in Progress

Scene Termination Activities
- ☐ Active Analysis
  - ☐ Monitoring and Detection results
    - ☐ Hot Zone
    - ☐ Decontamination corridor
    - ☐ Equipment
  - ☐ Reconnaissance analysis
    - ☐ Behavior Event status
    - ☐ Containment system
    - ☐ Confinement System
- ☐ Scene stabilization
  - ☐ Needs assessment for each branch
    - ☐ Information
    - ☐ Recon/Entry
    - ☐ Resource
      - ☐ Internal
      - ☐ External
    - ☐ HazMat Medical
    - ☐ Decon
  - ☐ Level of service assessment
    - ☐ Equipment/Supplies identified
      - ☐ Disposal
      - ☐ Decontamination
      - ☐ Re-supply
    - ☐ Functional re-deployment status
      - ☐ Adequate supplies
      - ☐ Adequate manning

Protect in Place
- ☐ Contact via Media
  - ☐ Announce termination process
  - ☐ Time frame in which termination is complete
- ☐ Public Service Announcement

Evacuation
- ☐ Contact Relocation facilities
  - ☐ Transportation schedule
  - ☐ Transportation plan
- ☐ Contact via Media
  - ☐ Announce termination process
  - ☐ Time frame in which termination is complete
- ☐ Public Service Announcement

Responsibility Transfer
- ☐ Notification and evaluation of resources
  - ☐ Internal evaluation and status
  - ☐ External evaluation and status
- ☐ Identify agency/authority-assuming control
- ☐ Owner/Contractor Brief
  - ☐ Initial nature of the incident
  - ☐ Actions taken during incident
  - ☐ Concerns during the incident
  - ☐ Considerations of the incident
  - ☐ Chemicals involved
  - ☐ Hazard/Risk Assessment
    - ☐ Initial Assessment
    - ☐ Current Assessment

Incident Debriefing
- ☐ Identify Hazards found
- ☐ Signs and symptoms of chemical exposure
- ☐ Damaged or expended equipment/supplies
- ☐ Conditions at time of transfer
- ☐ Assign information-gathering responsibilities

Figure 12-2 The incident termination process is included as a checklist in the *Hazardous Materials: Managing the Incident Field Operations Guide* 2nd edition.

Figure 12-3 Contractors should be briefed by the incident commander before responders turn over the scene.

Figure 12-4 HazMat Team debriefing should be conducted at the incident scene.

contaminated, or in some way affected by the response. The debriefing *is not* a critique of the incident.

An effective debriefing should:

- Inform responders exactly what hazmats they were (potentially) exposed to, the associated signs and symptoms, and how long after exposure signs of exposure could be expected to occur. Any information concerning precautions to protect family members should be discussed.
- Identify damaged equipment requiring servicing, replacement, or repair.
- Identify equipment or expended supplies that will require specialized decontamination or disposal.
- Identify unsafe site conditions that will impact the clean-up and recovery phase. Owners and contractors should be formally briefed on these problems before responsibility for the site is transferred.
- Assign information gathering responsibilities for a post-incident analysis and critique.
- Assess the need for a Critical Incident Stress Debriefing.
- Assign a point of contact for incident-related issues (e.g., concern for delayed symptoms).

Debriefings should begin as soon as the emergency phase of the operation is completed. Ideally, this should be before any responders leave the scene, and it should include the Hazardous Materials Response Team (HMRT), emergency response officers, and other key players, such as Information Officers and agency representatives, who the IC determines have a need to know Figure 12-4. On larger incidents, these representatives will return to their personnel and pass on the essential information, including (when applicable) who to contact for more information.

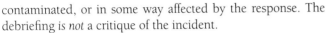

Conducting Debriefings

Debriefings should be conducted in areas free from distractions. In poor environmental conditions, such as extremely cold or hot weather or environments with loud ambient noise, the debriefing should be conducted in a nearby building or vehicle.

The debriefing should be conducted by one person acting as the leader. The IC may not be the best facilitator for the debriefing because special knowledge of the hazards may be required. The IC should at least be present to summarize his or her evaluation of the entire incident and to reinforce its positive aspects.

Debriefings longer than 15–20 minutes are probably too long. A complex incident may require more time. The intent is to briefly review the incident and send everyone home, not analyze every action of every player. If more interaction is needed on a specific subject or operation, continue it after the debriefing or schedule it for another day.

Debriefings should cover certain subjects, in the following order:

1. **Health information**—Describe what personnel may have been exposed and the signs and symptoms of exposure. Some substances may not reveal signs and symptoms of exposure for 24 to 48 hours. Provide guidance to personnel on what they should do if they experience any of these signs and symptoms subsequently while off duty. When appropriate, cover responsibilities that warrant follow-up evaluations and identify the health exposure forms that should be completed in follow-up evaluations.
2. **Equipment and apparatus exposure review**—Ensure that equipment and apparatus unfit for service is clearly "red tagged" for repair and plans are made for special cleaning or equipment disposal. Contaminated firefighting gear and personal clothing should be bagged to prevent the spread of contaminants and be properly cleaned and/or laundered upon return to the station. Someone must be delegated the responsibility for ensuring that contaminated garments are properly laundered or disposed.
3. **A follow-up contact person**—Ensure that anyone involved after the release of responders from the scene, such as clean-up contractors and investigators, have access to a single information source that can share the needed data. This contact person should also be responsible for collecting and maintaining all incident documents until they are delivered to the appropriate investigator or critique leader.

4. **Problems requiring immediate action**—Equipment failures, safety, major personnel problems, sample or evidence chain-of-custody procedures, or potential legal issues should be quickly reviewed on scene. If it is not crucial, save it for the critique.
5. **Thank you**—Most hazmat incidents are hard work and often test personal endurance and everyone's spirits. Reinforcement of the things that went right, a commitment to work on the problems uncovered through the critique process, and a genuine "thank you" from the boss go a long way. Never end on a sour note.

Post-Incident Analysis

The Post-Incident Analysis (PIA) is the reconstruction of the incident to establish a clear picture of the events that took place during the emergency. It is conducted to:

- Ensure the incident has been properly documented and reported to the right regulatory agencies.
- Determine the level of financial responsibility (i.e., who pays?).
- Establish a clear picture of the emergency response for further study. Focus on the General Hazardous Materials Behavior Model as a framework for this discussion (see the section on "hazard assessment and risk evaluation" elsewhere in this text).
- Provide a foundation for the development of formal investigations, which are usually conducted to establish the probable cause of the accident for administrative, civil, or criminal proceedings.

There are many agencies and individuals who have a legitimate need for information about significant hazmat incidents. They may include manufacturing, shipping and carrier representatives, insurance companies, government agencies, and even citizens groups. A formal PIA is one method for coordinating the release of factual information to those who have a need to know.

The Post-Incident Analysis Process

The PIA begins with the designation of one person (or office) to collect information about the response. This person is usually appointed during the on-scene debriefing. The PIA Coordinator should have the authority to determine who will have access to information. This method guarantees that sensitive or unverified information (e.g., injured personnel) is not released to the wrong organization or in an untimely manner.

The PIA should focus on six key topics:

1. **Command and control.** Was the Incident Command System established and was the emergency response organized according to the existing Emergency Response Plan and/or SOPs? Did information flow within the command structure to those who needed it and through appropriate channels? Were response objectives clearly communicated to field personnel at the task level? Were all arriving responders and organizations documented while in staging?
2. **Tactical Operations.** Were tactical operations completed in a safe and effective manner? What worked? What did not? Were tactical operations conducted in a timely and coordinated fashion? Do revisions need to be made to tactical procedures or worksheets? Did personnel involved in tactical operations have the proper level of training to perform their tasks safely?
3. **Resources.** Were resources adequate to conduct the response effort? Was the management of resources adequate? Are improvements needed to equipment or facilities? Were mutual aid agreements implemented effectively?
4. **Support Services.** Were quantity and type of support services adequate and provided in a timely manner? What is needed to increase the provision of support to the necessary level?
5. **Plans and Procedures.** Were the Emergency Response Plan and associated Tactical Procedures current? Did they adequately cover notification, assessment, response, recovery, and termination? Were roles and assignments clearly defined? How will plans and procedures be upgraded to reflect the "lessons learned"?
6. **Training.** Did this event highlight the need for additional basic or advanced training? Multi-agency/jurisdictional training? Were personnel trained adequately for their assignments?

The PIA should attempt to gather factual information concerning the response as soon as possible. The longer the delay in gathering information, the less likely it will be accurate and available. Suggested sources of information include the following:

- Incident reporting forms.
- Activity logs, entry logs, and personnel exposure logs.
- Notes and audio recordings from the Incident Command Post (ICP).
- Photographs, videos, maps, diagrams, and sketches. If photographs or video were taken, copies should also be obtained for the incident file. If future litigation is a concern, a photo/video log should be made recording the following information:
  - Time, date, location, direction, and current and ongoing weather conditions during the incident.
  - Description or identification of subject and relevance of photographs or video.
  - Camera type and serial number.
  - Name, telephone number, and identification number (e.g., agency ID number, Social Security number) of the photographer.
- Results of air monitoring and sampling, including types of instruments used, calibration information, and specific location(s) from where readings or samples were taken.
- Incident command organizational charts, notes, and completed checklists.
- Business cards or notes from agency, organization, or company representatives.
- Tape recordings from the 911/communications center(s) involved.
- Videotape recordings made by the media. Obtain the unshown, unedited video taken by the media within the first 24 hours after the incident. Only the video used in the broadcast is archived.

- Photographs, film, and videotape taken by responders or bystanders or relevant citizen video postings or comments on various social media.
- Interviews of witnesses conducted by investigators that may help establish where responders were located at the incident scene.
- Responder interviews.
- Verification of shipping documents or Safety Data Sheet (SDS).
- Owner/operator information.
- Chemical hazard information from checklists, computer printouts, and so on.
- Lists of apparatus, personnel, and equipment on scene.
- Time and date the incident was turned over to clean-up contractors or other outside agencies.

As soon as practical, construct a brief chronological review of who did what, when, and where during the incident. A simple timeline placing the key players at specific locations at different times is a good start. Cooperation between the PIA Coordinator and other official investigators will save time and combine resources to reconstruct the incident completely.

Once all available data have been assembled and a rough draft report developed, the entire package should be reviewed by key responders to verify that the available facts and events are arranged properly and actually took place.

Incident Reporting

Each emergency response organization has its own unique requirements for recording and reporting hazmat incidents. These requirements may be self-imposed as administrative and management controls or may be mandatory under federal or state laws.

For private shippers, carriers, and manufacturers, the regulatory reporting requirements for leaks, spills, and other releases of specified chemicals into the environment are significant. These include the following:

- Section 304 of the Superfund Amendments and Reauthorization Act (SARA, Title III)
- Section 103 of the Comprehensive Environmental Response, Compensation and Liability Act (CERCLA)
- 40 CFR Part 110—Discharge of Oil
- 40 CFR Part 112—Oil Pollution Prevention
- 40 CFR Part 302—Reportable Quantities
- Any additional local, state, or regional reporting requirements

Under CERCLA, the responsible party must report to the National Response Center (NRC) any spill or release of a specified hazardous substance in an amount equal to or greater than the reportable quantity (RQ) specified by EPA. In addition, SARA, Title III requires that releases be reported immediately to the NRC and the Local Emergency Planning Committee (LEPC).

Many industrial organizations have developed initial incident reporting forms as a way to ensure that key corporate and regulatory reporting requirements are correctly documented. The section on "the hazardous materials management systems" elsewhere in this text provides an overview of some of the more important federal laws that have reporting provisions.

Most major public fire departments participate in the National Fire Incident Reporting System (NFIRS). NFIRS is a product of the National Fire Information Council and is sponsored by the U.S. Fire Administration. This system includes a special category for recording hazardous materials incidents.

After Action Review or Critique

The **After Action Review (AAR)** is a structured and a participatory process of the senior leadership, commanders, and key responding agencies and other participants involved in the response. The goal of the AAR is to identify strengths and opportunities for improvement based on the lessons learned from the response.

In the past, this self-improvement process was often known as the **critique.** Just hearing the word "critique" often invokes a bad feeling in some of us. Words that you might associate with critique could include, criticism, finding blame, etc. If you can personally identify with this feeling it is probably because you had a past bad experience with the way the critique was conducted.

The origin of the word critique comes from the Greek word *kritikē tekhnē*, which when translated means "critical art." When conducted using a systematic process and with a little bit of "art," the critique process can leave everyone feeling positive about the experience. A well-done AAR makes you and the team better. What separates world-class performers from everybody else is regular practice (training and exercises) combined with the ability to self-improve over time.

Many injuries and fatalities have been prevented as a result of lessons learned through the AAR process. An effective critique program must be supported by top management and is the single most important way for an organization to self-improve over time. OSHA (1910.120[l][2][xi]) requires that a critique be conducted of every hazardous materials emergency response.

The primary purpose of a critique is to develop recommendations for improving the emergency response system and responder safety. A critique is not a grading process; rather, it should be a learning process that considers the strengths, vulnerabilities, and opportunities for improvement of the organization and its team members.

In the military, critiques may sometimes be referred to as a "Hot Wash," however, the intent and process are the same.

The critique process should never be used to find fault with the performance of individuals. If individual performance during a response is an issue, other internal agency procedures outside of the critique process should be used to address any deficiencies. These may include one-on-one employee performance reviews with a supervisor, the need for refresher training, skills assessment, etc.

A good critique promotes:

- Objectivity. To be objective, the input from the participants must be honest and based upon incident performance. The critique leader should be aware that differences in personalities among the critique participants can sometimes distort an opinion about what

took place at an incident, especially when considerable time has elapsed.
- A willingness to cooperate through teamwork.
- Improvement of safe operating procedures.
- Sharing information among emergency response organizations.

The critique leader is the crucial player in making the critique session a positive learning experience. A critique leader can be anyone who is comfortable and effective working in front of a group. The critique leader need not necessarily be part of the emergency response team. For example, an organization may select one or two respected and credible individuals to act as neutral parties to critique the larger, more sensitive incidents. Examples of outsiders who may be effective critique leaders include Fire Science or Law Enforcement faculty from a local community college, Regional Fire Training Coordinators, or professionals in your community who are experienced with people skills.

The primary role of the critique leader is to act as a facilitator for the critique process. The leader provides an opportunity for the participants to introduce themselves, establishes rules for the process of conducting the critique, facilitates the discussion, keeps the critique moving along, and ends it on time. Critiques lasting longer than 60 to 90 minutes quickly lose their effectiveness, and the quality of the discussion degrades. An agenda published in advance that is tied to a stated beginning and ending time helps set the goal of ending on time.

When you know going into a critique that a major confrontation between the players is anticipated, set up a meeting before the critique with only the affected parties attending. These types of smaller meetings can be useful to identify points of contention and diffuse the problem.

At the end of the critique the leader should sum up some of the opportunities for improvement identified from the critique and thank everyone for their response to the event and for their involvement in making the next response even more successful.

There are many different ways to conduct a critique. The following is a suggested format for conducting a hazardous materials incident critique:

- **Participant-level critique.** After explaining the rules for the critique, the critique leader calls on each key player to make an individual statement relevant to his or her on-scene activities and what he or she feels are the major issues that need to be discussed with the group. Depending on time, more detail may be added. There should be no interruptions from other participants during this phase. For obvious time reasons, the leader should limit this phase of the process to two or three minutes per person.
- **Operations-level critique.** After determining a feel for the group, the leader moves on to a structured review of emergency operations. Through a spokesperson, each Section Chief or subordinate supervisor presents an activity summary of challenges encountered, unanticipated events, and lessons learned. Each presentation should not exceed five minutes.
- **Group-level critique.** At the end of the operations-level critique, the leader moves the meeting into a wider and more open forum. The facilitator encourages discussion, reinforces constructive comments, and records important points.

As the critique draws to a close, the leader should summarize the more important observations and conclusions revealed by the participants. For large groups, the critique leader should have one or two assistants who act as recording secretaries throughout the session. These notes become the beginning point for writing a post-critique report. Critique reports should be short and to the point. Simply describe what happened in one or two pages and move on to the lessons learned. If recommendations for improvement are appropriate, they should be listed at the end of the report.

When larger incidents are involved or injuries have occurred, formal critique reports should be circulated so that everyone in the response system can share the lessons learned. Other forums that may be appropriate to share lessons learned include trade magazines and technical conferences.

For major incidents where the hazardous materials response was part of a larger incident involving other hazards, such as a flood or hurricane, the findings from the Critique Report may be incorporated into a broader AAR. When multiple agencies or jurisdictions were involved in mitigating the incident, the AAR development and writing process may require a multi-agency working group. There may be many individual incident critiques conducted in order to develop the big picture of what occurred across many different disciplines. For example, there may also be a Hazardous Materials Response Team, law enforcement, and an emergency medical services (EMS) critique conducted independently, with these respective critique findings then distilled into a final AAR.

It is important that any lessons learned that have been identified through the critique process be incorporated into a formal Improvement Plan (IP) for the emergency response system. Set aside a quarterly review date to make sure all action items have been addressed. Management should assign someone in the organization to track the implementation of recommendations, otherwise they tend not to get implemented and the opportunity to improve the system is lost. Failure to change after a bad incident also sends the wrong message to the people in the field (management doesn't care and doesn't want to change).

Operations Security (OPSEC) Issues

AARs may contain critical information useful for criminals and terrorists. Sensitive but unclassified information generated through written critiques and AARs can be used to defeat emergency response operations. The good news about the critique process is that, if done properly, it improves the emergency response system. The bad news is that in our American democratic society, if we share everything that we learned with everyone in the world, this information can actually teach the bad guys (criminals and terrorists) our vulnerabilities and weaknesses. Think about it. If an adversary understands our weaknesses by reading our AARs, does this give them a tactical advantage to use this information to hurt us and the community we are sworn to protect? Of course it does! So where do we draw the line on how much information should be released?

Emergency responders have an obligation to share lessons learned from major hazmat and WMD incidents. But we don't need to share everything we know with everyone. Does it make sense to place the complete report of the lessons we learned from a major WMD incident on the Internet so that anyone in the world can access it and download it for evaluation? At the other end of the spectrum is sharing nothing we know with anyone.

A reasonable approach to this problem is to limit information concerning vulnerabilities and weaknesses learned from critiquing the incident to the people who really need to know it (e.g., public safety agencies, investigators, accident review boards). Complete reports can be circulated through secure email, your agency's secure Intranet (restricted to authorized personnel), or distribution of hard copies using controlled and numbered copies.

Any sensitive information that can be used by criminals and terrorists to hurt the public or target first responders should be edited from critique reports intended for the general public. These edited reports can then be made available to the general public through the department's official website.

To learn more on Operations Security see *Special Operations for Terrorism and HazMat Crimes* (Red Hat Publishing, 2002) by Hawley, Noll, and Hildebrand (Chapter 3), or go to the Interagency Operations Security Support Staff website at http://www.ioss.gov for more information on OPSEC.

Liability Issues

Many managers express concerns that the critique process can expose weaknesses that can be exploited to build a liability case against emergency responders. There is no question that:

- Civil suits and regulatory citations against government and industry for emergency response operations are a problem.
- Chances of losing the suit or citation increase if you do not meet the accepted Standard of Care. See the section on "the hazardous materials management systems" elsewhere in this text for a discussion of standard of care.
- Fines and award sizes have increased.

Nevertheless, organizations must balance the potential negatives against the benefits gained through the critique process. Remember—the reason for doing the critique in the first place is to improve your operations and to keep your personnel safe. An organization that does not improve, doesn't meet the standard of care, and performs poorly makes itself a target for both regulatory citations and lawsuits.

There are five primary reasons for liability problems in emergency response work. They are worth considering as a case for actually building a strong critique program.

1. **Problems with planning.** Plans and procedures are poorly written, out-of-date, and unrealistic. In addition, what is written in the SOP is not followed in the field.
2. **Problems with training.** No training is conducted, training evolutions reinforce unsafe practices, and the training is undocumented.
3. **Problems with identification of hazards.** Hazards were not identified, were misidentified, were not prioritized, or were ignored even though they were known to exist.
4. **Problems with duty to warn.** Warnings concerning safety hazards and design limitations of equipment were not given or were improper.
5. **Problems with negligent operations.** Equipment was not employed properly, plans and procedures were not followed, and equipment was not maintained to an acceptable standard.

A little common sense goes a long way when developing a critique policy. If the critique and follow-up report are properly written in the first place, there will be little useful information that attorneys may use against emergency responders. While the critique report is a critical item, recognize that the legal process of discovery can also reveal organizational shortcomings in other ways. For example, official investigations conducted by OSHA, the fire marshal, or an insurance company can be used to build a case against you or your organization.

Do not let attorneys make management decisions for your organization. Don't let the person that doesn't have to risk his or her life in the field talk you out of using the critique process to improve the emergency response system. A quality emergency response system is your best liability defense. If you run an operation that meets national standards, you have gone a long way toward reducing your liability exposure.

Wrap-Up

Chief Concepts

- Termination is the final step in the Eight Step Process©. It is important that every hazmat incident be formally terminated following a formal procedure.
- Terminating the incident usually consists of five distinct activities: (1) declaring that the incident is "Terminated" either by radio or in a face-to-face meeting, (2) officially transferring responsibility of the incident scene to another agency or contractor, (3) incident debriefing, (4) post-incident analysis, and (5) critique.
- The incident debriefing is done at the incident scene, lasts less than 15 minutes, and focuses on safety and health exposure issues.
- When the decision has been made to terminate the emergency response phase and additional work is still required at the scene for restoration and recovery, the IC should meet with the senior representatives from the agencies or contractors taking over to formally hand off the incident scene.
- The post-incident analysis is conducted after the incident is over. It is a focused effort to gather information concerning what actually happened, why it happened, and who the responsible parties are. It also provides a record of resources and events, which may affect the public health, financial resources, and political well-being of a community.
- The critique is usually conducted within several weeks after the incident is terminated. It is designed to emphasize successful, as well as unsuccessful, operations and to improve the emergency response system. To be successful, management must support the critique process and action items must be tracked to ensure they are implemented.
- The critique process can reveal critical information about our weaknesses and vulnerabilities that can be exploited by criminals and terrorists. A strong critique program designed to improve the emergency response system reduces potential liability by helping to ensure that the organization is committed to constant improvement and meeting the highest standards in the industry.

Hot Terms

After Action Report A written report summarizing the lessons learned and recommendations for improvement following a major incident involving a multiple agency response.

Critique An element of incident termination that examines the overall effectiveness of the emergency response effort and develops recommendations for improving the organization's emergency response system.

Incident Debriefing The purpose of the incident debriefing is to provide accurate information concerning the hazards and risks involved directly to the people who may have been exposed, contaminated, or in some way affected by the response.

Post-Incident Analysis (PIA) The reconstruction of the incident to establish a clear picture of the events that took place during the emergency.

HazMat Responder *in Action*

You have been given the responsibility to conduct an incident debriefing for a working hazmat incident. You are reviewing the components of a proper termination procedure in your Standard Operating Procedures.

Questions

1. Debriefings should take place:
   - **A.** After the critique
   - **B.** After the incident has been properly documented and reported to the proper regulatory agencies
   - **C.** As soon as the emergency phase of the operation is completed
   - **D.** As soon as the post-emergency response activities are concluded
2. Which of the following topics should be addressed first during the debriefing?
   - **A.** Health information
   - **B.** Potential legal issues
   - **C.** Assignment of a follow-up contact person
   - **D.** Assignment of a critique leader
3. Your responsibility during this meeting as the post incident analysis coordinator is:
   - **A.** Appointed during the on-scene debriefing
   - **B.** As the authority who determines who will have access to the information
   - **C.** As the authority who cooperates with other official investigations to reconstruct the incident completely
   - **D.** All the above
4. The need for a Critical Incident Stress Debriefing is accessed during the:
   - **A.** Debriefing
   - **B.** Incident notification
   - **C.** Post incident Analysis
   - **D.** Critique

References and Suggested Readings

1. Benner, Ludwig, Jr., and Michael S. Hildebrand, *Hazardous Materials Management Systems: The M.A.P.S Method,* Prentice-Hall, Upper Saddle River, NJ (1981).
2. Colvin, Geoff, *Talent Is Overrated,* Penguin Group, New York, NY (2008).
3. Emergency Management Institute, *Liability Issues In Emergency Management,* Federal Emergency Management Agency, Washington, D.C. (April, 1992).
4. Hawley, Chris, Gregory Noll, and Michael S. Hildebrand, *Special Operations For Terrorism And Hazmat Crimes,* Red Hat Publishing, Chester, MD (2002).
5. Hildebrand, Michael, S., "An Effective Critique Program." *Fire Chief,* Volume 27, No. 4 (April, 1983).
6. Murley, Thomas, E., "Developing a Safety Culture." A technical paper presented at the Nuclear Regulatory Commission, Regulatory Information Conference, Washington, D.C. (April, 1989).
7. National Fire Protection Association, NFPA 1500—Fire Department Occupational Safety and Health Program Handbook. Quincy, MA: National Fire Protection Association (2002).

NFPA 1072 and 472 Correlation Guide

NFPA 1072 – HazMat Technician – 2017 Edition

NFPA 1072 *Standard for Hazardous Materials/ Weapons of Mass Destruction Emergency Response Personnel Professional Qualifications, 2017 Edition—* Chapter 7 *Hazardous Materials Technician*	Hazardous Materials: Managing the Incident, Revised 4th Edition
7.2.1 (A)	Chapter 2 p 32–35, 45–46, 55, Chapter 6 p 143, 253, 176–179, Chapter 7 p 194–199, 209–217, 218–222, 223–227, 230–231, 277, Glossary p 401, 461, 465, 478, 480, 485
7.2.2 (A)	Chapter 2 p 34, 35, 44, Chapter 6 p 153, 167–171, Chapter 7 p 194–203, 218, 235, 238, Chapter 8 p 280, Chapter 9 p 302–309, Chapter 10 p 370–373, Chapter 11 p 402
7.2.3 (A)	Chapter 6 p 149¬–174, Chapter 7 p 236–241, 244–248, Chapter 10 p 346–372, 383
7.2.4 (A)	Chapter 2 p 42, Chapter 7 p 196–242, Chapter 10 p 364–371, 374
7.2.5 (A)	Chapter 2 p 36–46, Chapter 5 p 127–129, Chapter 7 p 197–199, 207–210, 218, 221–250, Chapter 8 p 303
7.3.1 (A)	Chapter 4 p 103, 104, Chapter 10 p 324–331
7.3.2 (A)	Chapter 2 p 48, Chapter 4 p 102, Chapter 7 p 201, Chapter 8 p 268–278, 280, 281–286
7.3.3. (A)	Chapter 7 p 200–202, Chapter 11 p 404–407, 409–411
7.3.4 (A)	Chapter 3 p 79, Chapter 7 p 230, 243, Chapter 8 p 281–287, Chapter 10 323–352, 385, 386
7.4.1 (A)	Chapter 3 p 77, 79–84, Chapter 9 p 308, 311, 312
7.4.2 (A)	Chapter 1 p 10, Chapter 2 p 55, 58, Chapter 3 p 79, Chapter 8 p 267, 274–288, 292–295, Chapter 9 p 243, Chapter 11 p 411–416, Chapter 12 p 439
7.4.3.1 (A)	Chapter 2 p 55, Chapter 6 p 155–160, Chapter 7 p 243, Chapter 8 p 266, 270, Chapter 10 p 324–351
7.4.3.2 (A)	Chapter 2 p 55, Chapter 6 p 148–154, 159, 161, 166, 167, 173, 174, Chapter 8 p 276–281, Chapter 10 p 339–340, 342–347, 348, 371
7.4.3.3 (A)	Chapter 2 p 35–45, 55, Chapter 4 p 102, Chapter 6 p 149, 166, 167, 179, Chapter 8 p 270, Chapter 10 p 342–345, Chapter 11 p 402, 425
7.4.3.4 (A)	Chapter 2 p 55, 58, Chapter 6 p 159, 160, Chapter 8 p 276–281, Chapter 10 p 358, 359, 378–382
7.4.4.1 (A)	Chapter 3 p 83, Chapter 5 p 115, Chapter 8 p 268–275, Chapter 11 p 409, 411–425, Chapter 12 p 438
7.4.4.2 (A)	Chapter 2 p 82, 83, Chapter 5 p 117–119, Chapter 11 p 396–416, Chapter 12 p 439
7.5 (A)	Chapter 3 p 79, Chapter 4 p 104, Chapter 7 p 218, Chapter 10 p 322–326
7.6 (A)	Chapter 1 p 10, Chapter 12 p 435–440

NFPA 472 - HazMat Technician - 2018 Edition

SECTION	Hazardous Materials: Managing the Incident, Revised 4th Edition
7.1.1.2	Hazardous Materials: Awareness and Operations by Rob Schnepp, ISBN 978-0-7637-3872-3
7.2.1	Chapter 6
7.2.1.1	Chapter 6 (pp 162–166), Figures 6-23, 6-24
7.2.1.1.1	Chapter 6 (pp 162–166), Figures 6-23, 6-24
7.2.1.1.2	Chapter 6 (pp 155–157), Figures 6-15 to 6-20
7.2.1.1.3	Chapter 6 (pp 158–161), Figures 6-21, 6-22
7.2.1.1.4	Chapter 6 (pp 167–171), Figures 6-26 to 6-30
7.2.1.1.5	Chapter 6 (pp 148–152), Figures 6-3 to 6-9
7.2.1.1.6	Chapter 6 (pp 166, 167), Figure 6-25
7.2.1.1.7	Chapter 6 (pp 152–154), Figures 6-10 to 6-14
7.2.1.2	Chapter 6 (pp 155–171), Figures 6-15 to 6-30
7.2.1.2.1	Chapter 6 (pp 155–166), Figures 6-15 to 6-24
7.2.1.2.2	Chapter 6 (pp 167–171), Figures 6-26 to 6-30
7.2.1.3	Practical/Performance Skill; Chapter 7 (pp 218–221)
7.2.1.3.1	Chapter 7 (pp 217–222)
7.2.1.3.2	Chapter 7 (p 217–221)
7.2.1.3.3	Chapter 7 (p 210), Scan Sheet 7-B
7.2.1.3.4	Chapter 7 (pp 210–217), Scan Sheet 7-B, Figure 7-3
7.2.1.3.5	Practical/Performance Skill; Chapter 7 (pp 221–222), Skill Drill 7-2
7.2.1.3.6	Practical Performance Skill; Chapter 7 (pp 224–227), Skill Drill 7-3A and B
7.2.1.4	Chapter 6 (pp 178,179)
7.2.1.5	Practical Performance Skill; Chapter 7 (pp 232–234), Skill Drill 7-4A and B
7.2.2	Chapter 7
7.2.2.1	Chapter 7 (pp199–218)
7.2.2.2	Chapter 7 (pp 194–199)
7.2.2.3	Chapter 7 (pp 195–197)
7.2.2.4	Practical Performance Skill; Chapter 7, Table 7-2 (pp 199–203)
7.2.2.5	Chapter 7 (p 238)
7.2.2.6	Chapter 7 (p 238)
7.2.3	Chapter 6
7.2.3.1	Chapter 6 (pp 148–171)
7.2.3.1.1(1)	Chapter 6 (pp 158–161), Figures 6-21, 6-22
7.2.3.1.1(2)	Chapter 6 (pp 167–171), Figures 6-26 to 6-30

SECTION	Hazardous Materials: Managing the Incident, Revised 4th Edition
7.2.3.1.1(3)	Chapter 6 (pp 153, 154), Figures 6-11 to 6-14
7.2.3.1.1(4)	Chapter 6 (pp 155–157), Figures 6-15 to 6-20
7.2.3.1.1(5)	Chapter 6 (p 152), Figure 6-10
7.2.3.1.1(6)	Chapter 6 (p 173–175)
7.2.3.1.1(7)	Chapter 6 (pp 162–166), Figures 6-23, 6-24
7.2.3.1.2	Chapter 6 (pp 148–152), Figures 6-3 to 6-9
7.2.3.1.3	Chapter 6 (pp 166, 167), Figure 6-25
7.2.3.2	Chapter 6 (pp 173)
7.2.3.3	Chapter 6 (pp 173–175)
7.2.3.4	Chapter 7 (pp 236–242)
7.2.3.5	Practical/Performance Skill; Chapter 7, Skill Drill 7-1 (pp 218–221)
7.2.4	Chapter 7
7.2.4.1	Chapter 7 (pp 200–202), Table 7-2
7.2.4.2	Chapter 10 (pp 364–365), Scan Sheet 10-F
7.2.4.3	Chapter 10 (p 371), Scan Sheet 10-H
7.2.5	Chapter 7
7.2.5.1	Practical/Performance Skill; Local Policy; Chapter 7 (pp 204, 205)
7.2.5.2	Chapter 7 (pp 236–242)
7.2.5.2.1	Chapter 2 (pp 36–39, pp 44–46), Chapter 7 (pp 198–199)
7.2.5.2.2	Chapter 7 (pp 239, 240)
7.2.5.3	Chapter 7 (pp 236–242)
7.2.5.4	Practical/Performance Skill; Chapter 7 (pp 217–219)
7.3.1.1	Practical/Performance Skill; Chapter 7 (pp 243, 244), Chapter 10
7.3.1.2	Chapter 7 (pp 243, 244), Chapter 10 (pp 322–331)
7.3.2.1	Chapter 7 (pp 243, 244), Chapter 10 (pp 331–384)
7.3.2.2	Chapter 10 (p 324)
7.3.3	Chapter 8
7.3.3.1	Chapter 8 (p 274), Figure 8-4
7.3.3.2	Chapter 8 (p 270), Figure 8-3
7.3.3.3	Chapter 8 (pp 270–273, pp 281–283)
7.3.3.4	Chapter 8 (pp 265, 266, pp 281–283)
7.3.3.4.1	Chapter 8 (pp 263–265)
7.3.3.4.2	Chapter 8 (p 263)
7.3.3.4.3	Chapter 8 (pp 276–280)
7.3.3.4.4	Chapter 2 (p 48)

(*continued*)

SECTION	Hazardous Materials: Managing the Incident, Revised 4th Edition
7.3.3.4.5	Chapter 8 (p 281–283)
7.3.3.4.6	Chapter 8 (pp 269, 270)
7.3.3.4.7	Chapter 2 (pp 46–50)
7.3.3.4.8	Practical/Performance Skill; Chapter 8 (pp 292–295); Based on Local PPE Protocol
7.3.4	Chapter 11 (pp 404–406)
7.3.4(1)	Chapter 11 (pp 404–406)
7.3.4(2)	Chapter 11 (pp 409–411)
7.3.5	Chapter 10
7.3.5.1	Chapter 10 (pp 336–351), Specific fire scenarios (pp 352–376)
7.3.5.2	Practical/Performance Skill; Chapter 10 (pp 328–330), Skill Drill 10-1
7.3.5.2.1	Chapter 2 (pp 56–60), Chapter 8 (pp 281–283) Chapter 10 (pp 326, 327)
7.3.5.2.2	Chapter 2 (pp 56–60), Chapter 8 (pp 281–283)Chapter 10 (pp 326, 327)
7.3.5.3	Chapter 10 (pp 333–335), Scan Sheet 10-A
7.3.5.4	Chapter 2 (pp 57, 58), Chapter 8 (pp 281–283)
7.3.5.5	Chapter 7 (pp 225–230), Scan Sheet 7-E, Chapter 10 (pp 386–387), Skill Drills 10-4 and 5
7.4.1	Chapter 3
7.4.1.1	Chapter 3 (pp 80–84)
7.4.1.2	Chapter 1 (p 4)
7.4.2	Chapters 2, 8
7.4.2(1)	Chapter 2 (pp 56–60), Chapter 8 (pp 281–283)
7.4.2(2)	Chapter 8 (pp 283–285)
7.4.2(3)	Practical/Performance Skill, Chapter 8 (pp 270–273)
7.4.2(4)	Practical/Performance Skill; Chapter 8 (pp 287–295), Skill Drills 8-1 through 8-7
7.4.3	Chapter 10
7.4.3(1)	Practical/Performance Skill; Chapter 10 (p 352), Skill Drill 10-3
7.4.3(2)	Practical/Performance Skill; Chapter 10 (p 352), Skill Drill 10-3
7.4.3(3)	Practical/Performance Skill; Chapter 10 (p 345), Skill Drill 10-2
7.4.3(4)	Practical/Performance Skill; Chapter 10 (pp 342, 343)
7.4.3(5)	Practical/Performance Skill; Local Procedures, Chapter 10
7.4.3(6)	Chapter 10 (pp 334–335), Scan Sheet 10-A
7.4.3(7)	Chapter 10 (pp 378–382)

SECTION	Hazardous Materials: Managing the Incident, Revised 4th Edition
7.4.3(8)	Practical/Performance Skill; Chapter 10 (pp 347, 348), Scan Sheet 10-B
7.4.3(9)	Chapter 10 (pp 352–362), Scan Sheet 10-E
7.4.3(10)	Chapter 10 (pp 347, 348), Scan Sheet 10-B
7.4.3(11)	Chapter 10 (pp 376–384)
7.4.4	Chapter 10 (pp 376– 384)
7.4.5	Practical/Performance Skill; Chapter 11
7.4.5(1)	Chapter 11, Skill Drill 11-1 (pp 411–415)
7.4.5(2)	Chapter 11, Scan Sheet 11-B (pp 413)
7.4.5(3)	Chapter 11, Skill Drills 11-2 and 3 (pp 415–420)
7.5.1	Practical/Performance Skill; Chapter 10 (pp 326–331)
7.5.2	Chapter 11 (pp 406, 407)
7.6.1	Chapter 12
7.6.1(1)	Chapter 12 (pp 435–438)
7.6.1(2)	Chapter 12 (pp 435–438)
7.6.1(3)	Chapter 12 (pp 435–438)
7.6.1(4)	Chapter 12 (pp 435–438)
7.6.2	Chapter 12
7.6.2(1)	Chapter 12 (pp 438–440)
7.6.2(2)	Chapter 12 (pp 438–440)
7.6.2(3)	Chapter 12 (pp 438–440)
7.6.2(4)	Chapter 12 (pp 438–440)
7.6.3	Chapters 2, 12; Local Procedures
7.6.3(1)	Chapter 12 (pp 438–441)
7.6.3(2)	Chapter 12 (pp 438–441)
7.6.3(3)	Chapter 2 (p 55)
7.6.3(4)	Chapter 12 (pp 438–441)
7.6.3(5)	Chapter 12 (pp 438–441)
7.6.3(6)	Chapter 2 (p 55)
7.6.3(7)	Chapter 12 (pp 438–441)
7.6.3(8)	Chapter 2 (p 55)
7.6.3(9)	Chapter 2 (pp 55, 56)
7.6.3(10)	Chapter 12 (pp 438–441)

NFPA 1072 – Hazardous Materials Incident Commander – 2017 Edition

NFPA 472 *Standard for Competence of Responders to Hazardous Materials/Weapons of Mass Destruction Incidents, 2018 Edition*-Chapter 8 *Incident Commander*	Hazardous Materials: Managing the Incident, Revised 4th Edition
8.1.5	Chapter 3 p 68–84
8.2 (A)	Chapter 3 p 70, 72, 78, 131, Chapter 6 p 144, Chapter 7 p 203–217, 236–242, 246–248, Chapter 10 p 322–323
8.3. (A)	Chapter 2 p 55, 56, Chapter 3 p 68–81, Chapter 4 p 101, 102, Chapter 5 p 120–134, Chapter 7 p 243, Chapter 8 p 263–281, Chapter 10 p 327–340
8.4 (A)	Chapter 3 p 68–83, Chapter 9 p 310, 311
8.5 (A)	Chapter 2 p 39–42, Chapter 3 85–87, Chapter 10 p 322, 323
8.6 (A)	Chapter 1 p 7–11, Chapter 7 p 206, Chapter 12 p 433–440

NFPA 472 - Incident Commander - 2018 Edition

SECTION	Hazardous Materials: Managing the Incident, Revised 4th Edition
8.1.1.2	Multiple chapters
8.1.2.2	Multiple chapters
8.1.2.2(1)	Chapter 7
8.1.2.2(2)	Chapter 1
8.1.2.2(3)	Chapter 9, 10
8.1.2.2(4)	Chapter 1
8.1.2.2(5)	Chapter 12
8.1.2.2(5)(a)	Chapter 10
8.1.2.2(5)(b)	Chapter 11, 12
8.1.2.2(5)©	Chapter 12
8.1.2.2(5)(d)	Chapter 2
8.2.1.1	Chapter 7 (pp 199–218, 228, 229), Scan Sheet 7-D
8.2.1.2	Chapter 7 (pp 199–218)
8.2.2	Chapter 7
8.2.2(1)	Chapter 7 (pp 236–242)
8.2.2(2)	Chapter 2 (pp 36–39, pp 44–46, Chapter 7 (pp 198,199)
8.2.2(3)	Chapter 7 (pp 239–240)
8.2.2(4)	Practical/Performance Skill; Local Policy
8.2.2(5)	Chapter 2 (pp 32–35)
8.2.2(6)	Chapter 7 (pp 198,199)

SECTION	Hazardous Materials: Managing the Incident, Revised 4th Edition
8.3.1	Chapter 7 (pp 243, 244), Chapter 10 (pp 322–331)
8.3.2	Chapter 7
8.3.2(1)	Chapter 7 (pp 243, 244), Chapter 10 (pp 331–384)
8.3.2(2)	Chapter 10 (pp 336–351), Specific fire scenarios (pp 352–376)
8.3.3	Chapter 8
8.3.3(1)	Chapter 8 (p 274), Figure 8-4
8.3.3(2)	Chapter 8 (pp 263–265)
8.3.3(3)	Chapter 2 (pp 46–48, 56, 57), Chapter 8 (pp 281–286)
8.3.3(4)	Chapter 2 (pp 46–50)
8.3.4	Chapters 5, 7, 10, 11
8.3.4.1	Chapter 5 (pp 114–119), Chapter 7 (pp 243, 244), Chapter 10 (pp 322–331); ICS-200
8.3.4.2	Chapter 5 (pp 119–125), Chapter 10 (pp 333–334)
8.3.4.3	Practical/Performance Skill; Local Policy
8.3.4.4	Chapter 10 (pp 322–331)
8.3.4.5	Chapter 2 (pp 56–60), Chapter 8 (pp 281–283) Chapter 10 (pp 326, 327)
8.3.4.5.1	Chapter 1 (pp 14–17), Chapter 9 (pp 303–305)
8.3.4.5.2	Chapter 2 (pp 56–60), Chapter 8 (pp 281–284)
8.3.4.5.3	Chapter 10 (pp 331, 333–335), Scan Sheet 10-A
8.3.4.5.4	Chapter 11 (pp 404–406)
8.3.4.5.5	Chapter 10 (pp 334–336), Scan Sheet 10-A
8.4.1	Chapters 1, 3, and 9
8.4.1(1)	Chapter 3 (pp 68, 69, 74), Local Policy, ICS-200
8.4.1(2)	Chapter 3 (pp 68, 79); ICS-200
8.4.1(3)	Chapter 3 (pp 80–83)
8.4.1(4)	Chapter 1 (pp 6–9,14–17); Local Policy
8.4.1(5)	Practical/Performance Skill; Local Policy
8.4.1(6)	Chapter 3 (pp 74–80), Chapter 9; ICS-200

(*continued*)

SECTION	Hazardous Materials: Managing the Incident, Revised 4th Edition
8.4.1(7)	Chapter 1 (pp 6–11)
8.4.1(8)	Chapter 3 (pp 68–72); Local Policy
8.4.2	Practical/Performance Skill; Chapter 9 (pp 313–316); ICS-200
8.4.3	Chapters 3, 9
8.4.3(1)	Practical/Performance Skill; Local Policy
8.4.3(2)	Chapter 3 (pp 76, 85–88), Chapter 9 (pp 310, 311); ICS-200
8.4.3(3)	Chapter 3 (p 76), Chapter 9 (p 313); ICS-200
8.5.1	Chapter 7, 10
8.5.1(1)	Chapter 7 (pp 235–244); ICS-200 Chapter 10
8.5.1(2)	Chapter 7 (pp 235–243)
8.5.1(3)	Practical/Performance Skill, Chapter 10
8.5.1(4)	Practical/Performance Skill
8.5.2	Chapter 12 (p 435); ICS-200
8.6.1	Chapter 12
8.6.1(1)	Chapter 12 (pp 434, 435)
8.6.1(2)	Chapter 12 (pp 437, 438)
8.6.2	Chapter 12
8.6.2(1)	Chapter 12 (p 435–438)
8.6.2(2)	Chapter 12 (p 435–438)
8.6.2(3)	Chapter 12 (p 435–438)
8.6.2(4)	Chapter 12 (p 435–438)
8.6.2(5)	Chapter 12 (p 435–438)
8.6.3	Chapter 12
8.6.3(1)	Chapter 12 (p 438–440)
8.6.3(2)	Chapter 12 (p 438–440)
8.6.3(3)	Chapter 12 (p 438–440)
8.6.3(4)	Chapter 12 (p 438–440)
8.6.3(5)	Practial/Performance Skill; Local Policy
8.6.4	Chapters 1, 2, 12

SECTION	Hazardous Materials: Managing the Incident, Revised 4th Edition
8.6.4(1)	Chapter 1 (p 20), Chapter 12 (p 439)
8.6.4(2)	Chapter 2 (p 55), Chapter 12 (p 439–441)
8.6.4(3)	Chapter 2 (p 55), Chapter 12 (pp 438, 439)
8.6.4(4)	Practical/Performance Skilll; Local Policy
8.6.4(5)	Practical/Performance Skill; Local Policy
8.6.4(6)	Practical/Performance Skill; Local Policy

Correlation Matrix for National Fire Academy's Fire and Emergency Services Higher Education (FESHE) Bachelor's (Non-Core) Managerial Issues in Hazardous Materials Course (C0274)

Managerial Issues in Hazardous Materials (C0274) Course Description–This course presents current issues in management of a department-wide hazardous materials program. It includes issues that are pertinent to officers and managers in public safety departments, including regulations and requirements for hazardous materials (hazmat) preparedness, response, storage, transportation, handling and use, and the emergency response to terrorism threat/incident. Subjects covered include State, local and Federal emergency response planning, personnel and training, and operational considerations, such as determining strategic goals and tactical objectives.

Modules and Course Objectives	**Hazardous Materials: Managing the Incident, Revised Fourth Edition by Gregory G. Noll, Michael S. Hildebrand Contributions by: Glen Rudner and Rob Schnepp**
Module 1: Introduction to Hazardous Materials	
Explain the correlation between trends in chemical use and emergency-release incidents.	Chapter 1
Recognize and define common terms used in hazmat response and regulation.	Chapters 1 and 2
Summarize the intent of major pieces of legislation and standards that affect hazmat planning and emergency response.	Chapter 1
Explain the purpose of the State and local emergency-response commissions and their role in managing hazmat situations in the community.	Chapter 1
Identify the Federal agencies that are responsible for enacting and enforcing hazmat regulations, and explain each agency's specific area of concern.	Chapter 1
Module 2: Community-Centered Managerial Issues	
List and explain the basic components of emergency planning for hazmat response and management.	Chapter 1
Recognize the difference between protection-in-place and evacuation strategies.	Chapter 5 (pages121–123)
Explain the legal basis for the requirement of using Incident Command.	Chapters 1 and 3
Differentiate between public information and public education.	Chapter 1 (pages 20, 28), Chapter 9 (page 311)
Explain the legal requirements governing public access to information.	Chapter 9 (page 305), Chapter 12 (pages 436–437)
Describe the benefits of community education programs.	Chapter 1 (page 20), Chapter 5 (page 123)
Identify at least one automated community information program currently in use.	Chapter 5 (pages 127–132)
Module 3: Department-Centered Managerial Issues	
Compare the similarities and critical differences between a "normal" fire emergency and a hazmat emergency.	Chapter 1
Describe the capabilities and limitations of first responders with regard to equipment, protective clothing, training, and experience.	Chapter 8
Explain the training and emergency response requirements mandated in regulation 29 CFR 1910.120(q), and compare them to standard NFPA 1072.	Chapter 1, Appendix-A
Explain the certification-of-competency requirement and recordkeeping requirements specified in the regulations.	Chapter 1
Describe regulated occupancies and activities related to hazardous materials.	Chapter 6
Demonstrate methods of ascertaining code compliance for storage, handling, and use of hazmat.	Chapter 6, Chapter 10
Locate applicable codes and regulations pertaining to storage, handling, and use of hazmat.	Chapter 1, Chapter 10

APPENDIX B

Modules and Course Objectives	Hazardous Materials: Managing the Incident, Revised Fourth Edition by Gregory G. Noll, Michael S. Hildebrand Contributions by: Glen Rudner and Rob Schnepp
Module 4: Incident-Response Managerial Issues	
Assess the strategic goals and tactical options for managing a hazmat incident.	Chapter 10
List and describe the steps involved in the management process at a hazmat incident.	Chapters 3, 4, and 10
Explain additional risk and response considerations for a hazmat incident that is also a terrorist incident.	Chapter 2 (page 54), Chapter 7 (pages 219, 223–225, Chapter 10 (page 384)
State the differences between a Command Post and an Emergency Operations Center.	Chapter 3 (pages 76–78)
Name the different interest groups in the *Command* Post, and explain their goals and concerns.	Chapter 3 (Pages 68–72)
Define the terms *recovery* and *termination.*	Chapter 10 (pages 323), Chapter 10 (pages 364–384, 379–380)
Discuss the necessary documentation to be produced in conjunction with incident management.	Chapter 6 (171–181, 435–436, 292, 230
Explain debriefing, post-incident analysis, and after-action reports.	Chapter 12 (421, 433–434–435–436–438
Explain the Federal precedents for cost-recovery legislation.	Chapter 1 (page 240)
Describe the four phases of termination.	Chapter 12
Make response decisions based on risk analysis.	Chapter 4 (pages 101–102), Chapter 7 (pages 193–194, 219, 235–236, 256)

Glossary

Aboveground Bulk Storage Tank A horizontal or vertical tank that is listed and intended for fixed installation, without backfill, above or below grade, and is used within the scope of its approval or listing. (NFPA 30A).

Absorbent Material A material designed to pick up and hold liquid hazardous material to prevent contamination spread (e.g., sawdust, clays, charcoal, and polyolefin-type fibers).

Absorption 1) The process of absorbing or "picking up" a liquid hazardous material to prevent enlargement of the contaminated area; 2) movement of a toxicant into the circulatory system by oral, dermal, or inhalation exposure.

ACGIH See American Conference of Governmental Industrial Hygienists.

Acids Compound that form hydrogen ions in water. These compounds have a pH < 7, and acidic aqueous solutions will turn litmus paper red. Materials with a pH < 2.0 are considered a strong acid. Examples of the most common acids involved in emergencies include nitric acid, sulfuric acid, hydrochloric acid, and phosphoric acid.

Activity The number of radioactive atoms that will decay and emit radiation in 1 second of time. Measured in curies (1 curie = 37 billion disintegrations per second), although it is usually expressed in either millicuries or microcuries. Activity indicates how much radioactivity is present and not how much material is present.

Acute Effects Results from a single dose or exposure to a material. Signs and symptoms may be immediate or may not be evident for 24 to 72 hours after the exposure.

Acute Emergency Exposure Guidelines (AEGL) Developed by the National Research Council's Committee on Toxicology to provide uniform exposure guidelines for the general public. The Committee's objective is to define AEGLs for the 300+ EHS materials listed in SARA, Title III. AEGLs represent an exposure value specifically developed for a single short-term exposure, such as those encountered in the emergency response community. Three tiers of AEGLs have been developed covering five exposure periods: 10 minutes, 30 minutes, 1 hour, 4 hours, and 8 hours.

Acute Exposures An immediate exposure such as a single dose that might occur during an emergency response.

Administration/Finance Section Responsible for all costs and financial actions at the incident. Includes the Time Unit, Procurement Unit, Compensation/Claims Unit, and the Cost Unit.

Adsorption Process of adhering to a surface, a common physical method of spill control and decontamination.

Aerosols Liquid droplets or solid particles dispersed in air, that are of fine enough particle size (0.01 to 100 microns) to remain dispersed for a period of time.

Agency for Toxic Substances and Disease Registry (ATSDR) An organization within the Centers for Disease Control, it is the lead federal public health agency for hazmat incidents.

Air Monitoring To measure, record, and/or detect contaminants in ambient air.

Air Purifying Respirator (APR) Respirator or filtration device that removes particulate matter, gases, or vapors from the atmosphere. These devices range from full-face piece, dual cartridge masks with eye protection, to half-mask, face piece mounted cartridges. They are intended for use only in atmospheres where the chemical hazards and concentrations are known.

ALS Advanced life support emergency medical personnel, such as paramedics.

Alcohol Resistant Concentrate (ARC) Alcohol resistant concentrate–aqueous film-forming foams (AFFF) are Class B firefighting foams that can be applied to both hydrocarbons and polar solvents. They can be applied at various concentrations, including 3% hydrocarbon / 3% polar solvent (known as 3 × 3 concentrates), and 3% hydrocarbon / 6% polar solvent concentrations (known as 3 × 6 concentrates). When applied to a polar solvent fuel, they will often create a polymeric membrane rather than a film over the fuel. This membrane separates the water in the foam blanket from the attack of the polar solvent. Then, the blanket acts in much the same manner as a regular AFFF.

Alpha Particles A type of ionizing radiation. Largest of the common radioactive particles, alpha particles have extremely limited penetrating power. They travel only 3 to 4 inches in air and can be stopped by a sheet of paper or a layer of human skin.

American Chemistry Council The parent organization that operates CHEMTREC.

American Conference of Governmental Industrial Hygienists (ACGIH) A professional society of individuals responsible for full-time industrial hygiene programs who are employed by official governmental units.

American National Standards Institute (ANSI) A 501(c)3 private, not-for-profit organization that oversees the creation, promulgation, and use of thousands of norms and guidelines that directly impact businesses in nearly every sector: from acoustical devices to construction equipment, from dairy and livestock production to energy distribution, and many more. ANSI is also actively engaged in accrediting programs that assess conformance to standards, including globally-recognized cross-sector programs such as the ISO 9000 (quality) and ISO 14000 (environmental) management systems.

American Petroleum Institute (API) National trade association that represents all aspects of America's oil and natural gas industry. Corporate members include the largest major oil company to the smallest of independents and come from all segments of the industry. They are producers, refiners, suppliers, pipeline operators, and marine transporters, as well as service and supply companies that support all segments of the industry.

Anhydrous Free from water, dry. For example, anhydrous ammonia and anhydrous hydrogen chloride.

API Uniform Marking System American Petroleum Institute's marking system used at many petroleum storage and marketing facilities to identify hydrocarbon pipelines and transfer points. Classifies hydrocarbon fuels and blends into leaded and unleaded gasoline (regular, premium, super).

APR See Air Purifying Respirator.

Aqueous Film Forming Foam (AFFF) Synthetic Class B firefighting foam consisting of fluorochemical and hydrocarbon surfactants combined with high boiling point solvents and water. AFFF film formation is dependent upon the difference in surface tension between the fuel and the firefighting foam. The fluorochemical surfactants reduce the surface tension of water to a degree less than the surface tension of the hydrocarbon so that a thin aqueous film can spread across the fuel. AFFF is not an effective extinguishing agent for natural gas since it is a lighter than air gas.

Area of Refuge Area within the Hot Zone where exposed or contaminated personnel are protected from further contact and/or exposure. This is a "holding area" where personnel are controlled until they can be safely decontaminated, treated, or removed.

Aromatic Hydrocarbons A hydrocarbon containing the benzene "ring" which is formed by six carbon atoms and contains resonant bonds. Examples include benzene (C_6H_6) and toluene (C_7H_8).

Asphyxiation Harm Events Those events related to oxygen deprivation and/or asphyxiation within the body. Asphyxiants can be classified as simple or chemical.

Association of American Railroads (AAR) Professional trade association that coordinates technical information and research within the United States railroad industry.

Atmospheric Tank A storage tank designed to operate at pressures from atmospheric through 0.5 psig (760 mm Hg through 786 mm Hg) measured at the top of the tank.

Atmosphere-supplying devices Respiratory protection devices coupled to an air source. The two types are self-contained breathing apparatus (SCBA) and supplied air respirator (SAR).

Authority Having Jurisdiction (AHJ) An organization, office, or individual responsible for enforcing the requirements of a code or standard, or for approving equipment, materials, an installation, or a procedure.

B-End The end of a railroad car where the handbrake is located. Is typically used as the initial reference point when communicating railroad tank car damage.

BLS Basic life support emergency medical personnel such as emergency medical technicians.

Beta Particles A type of ionizing radiation. Particle which is the same size as an electron and can penetrate materials much further than large alpha particles. Depending on the source, beta particles can travel several yards in air and penetrate paper and human skin but cannot penetrate internal organs.

Biological Agents and Toxins Biological threat agents consist of pathogens and toxins. Pathogens are disease-producing organisms and include bacteria (e.g., anthrax, cholera, plague, e coli), and viruses (e.g., small pox, viral hemorrhagic fever). Toxins are produced by a biological source and include ricin, botulinum toxins, and T2 mycotoxins.

Blister Agents See Vesicants.

Blood Agents Chemical agents that consist of a cyanide compound such as hydrogen cyanide (hydrocyanic acid) and cyanogen chloride. These agents are identical to their civilian counterparts used in industry.

Boiling Liquid Expanding Vapor Explosion (BLEVE) A container failure with a release of energy, often rapidly and violently, which is accompanied by a release of gas to the atmosphere and propulsion of the container or container pieces due to an overpressure rupture.

Boilover Violent ejection of flammable liquid from its container caused by the vaporization of water beneath the body of liquid. It may occur after a lengthy burning period of products such as crude oil when the heat wave has passed down through the liquid and reaches the water bottom in a storage tank. It will not occur to any significant extent with water-soluble liquids or light hydrocarbon products such as gasoline.

Boiling Point The temperature at which a liquid changes to a vapor or gas. The temperature where the vapor pressure of the liquid equals atmospheric pressure. A liquid with a low flash point will also have a low boiling point, which translates into a large amount of vapors being released.

Bonding A method of controlling ignition hazards from static electricity. It is the process of connecting two or more conductive objects together by means of a conductor; for example, using an approved bonding wire to connect an aircraft being refueled to the fuel truck.

Boom A floating physical barrier serving as a continuous obstruction to the spread of a contaminant.

Branch That organizational level within the Incident Command System having functional/geographic responsibility for major segments of incident operations (e.g., Hazmat Branch). The Branch level is organizationally between Section and Division/Group.

Breach Event The event causing a hazmat container to open up or "breach." It occurs when a container is stressed beyond its limits of recovery (ability to hold contents).

Breakthrough Time The elapsed time between initial contact of a hazardous chemical with the outside surface of a barrier, such as protective clothing material, and the time at which the chemical can be detected at the inside surface of the material.

Buddy System A system of organizing employees into work groups in such a manner that each employee of the work group is designated to be observed by at least one other employee in the work group (per OSHA 1910.120 [a][3]).

Bulk Packaging Bulk packaging has an internal volume greater than 119 gallons (450 liters) for liquids, a capacity greater than 882 pounds (400 kg) for solids, or a water capacity greater than 1000 pounds (453.6 kg) for gases.

Bulk Plant or Terminal The portion of a property where liquids are received by tank vessel, pipelines, tank car, or tank vehicle, and are stored or blended in bulk for the purpose of distributing the liquids by some mode of transportation.

Bung A threaded plug used to close a barrel or drum bung hole.

CAA See the Clean Air Act.

Calibration The process of adjusting a monitoring instrument so that its readings correspond to actual, known concentrations of a given material. If the readings differ, the monitoring instrument can then be adjusted so that readings are the same as the calibrant gas.

Canadian Transport Emergency Center (CANUTEC) Operated by Transport Canada, it is a 24 hour, government-sponsored hotline for chemical emergencies. (The Canadian version of CHEMTREC.)

Carboy Glass or plastic bottles used for the transportation of liquids. They range in capacity to over 20 gallons, and may be encased in an outer packaging such as polystyrene boxes, wooden crates, or plywood drums. Often used for the shipment of corrosives.

Carcinogen A material that can cause cancer in an organism. May also be referred to as "cancer suspect" or "known carcinogen."

Cargo Tanks Tanks permanently mounted on a tank truck or tank trailer which is used for the transportation of liquefied and compressed gases, liquids, and molten materials. Examples include MC-306/DOT-406, MC-307/DOT-407, MC-312/DOT-412, MC-331 and MC-338.

Caustics (Base, Alkaline) Compounds that forms hydroxide ions in water. These compounds have a pH > 7, and caustic solutions will turn litmus paper blue. Materials with a pH >12 are considered a strong base. Also known as alkali, alkaline, or base.

CAS Number See Chemical Abstract Service Number.

Catalyst Used to control the rate of a chemical reaction by either speeding it up or slowing it down. If used improperly, catalysts can speed up a reaction and cause a container failure due to pressure or heat build-up.

Centers for Disease Control and Prevention (CDC) The federally funded research organization tasked with disease control and research.

CERCLA See Comprehensive Environmental Response, Compensation, and Liability Act.

Chemical Abstract Service (CAS) Number Often used by state and local Right-To-Know regulations for tracking chemicals in the workplace and the community. Sequentially assigned CAS numbers identify specific chemicals and have no chemical significance.

Chemical Agents Chemical agents are classified in military terms based on their effects on the enemy. Categories of chemical agents are: nerve (neurotoxins), choking (respiratory irritants), blood (chemical asphyxiants), vesicants or blister (skin irritants), and antipersonnel (riot control agents).

Chemical Degradation The process of altering the chemical structure of a contaminant through the use of a second chemical or material. Commonly used degradation agents include calcium hypochlorite bleach, sodium hypochlorite bleach, sodium carbonate slurry (washing soda), and liquid household detergents.

Chemical Interactions Reaction caused by mixing two or more chemicals together. Chemical interaction of materials within a container may result in a build-up of heat and pressure, leading to container failure.

Chemical Protective Clothing (CPC) Single or multi-piece garment constructed of chemical protective clothing materials designed to protect the wearer's torso, head, arms, legs, hands, and feet. Can be constructed as a single or multi-piece garment.

Chemical Protective Clothing Material Any material or combination of materials used in an item of clothing for the purpose of isolating parts of the wearer's body from contact with a hazardous chemical.

Chemical Reactivity A process involving the bonding, unbonding, and rebonding of atoms, that can chemically change substances into other substances.

Chemical Resistance The ability to resist chemical attack. The attack is dependent on the method of test and its severity is measured by determining the changes in physical properties.

Chemical Resistant Materials Materials specifically designed to inhibit or resist the passage of chemicals into and through the material by the processes of penetration, permeation, or degradation.

Chemical Stress The result of a chemical reaction of two or more materials. Examples include corrosive materials attacking a metal, the pressure or heat generated by the decomposition or polymerization of a substance, or any variety of corrosive actions.

Chemical Transportation Emergency Center (CHEMTREC™) The Chemical Transportation Center, operated by the American Chemistry Council (ACC), provides information and technical assistance to emergency responders.

Chemical Vapor Protective Clothing Chemical protective clothing ensemble designed to protect the wearer against chemical vapors or gases. Chemical vapor protective clothing must meet the requirements of NFPA 1991. This type of protective clothing is a component of EPA Level A chemical protection.

Chlorine Emergency Plan (CHLOREP) Chlorine industry emergency response system operated by the Chlorine Institute and activated through CHEMTREC.

Chlorine Kits Standardized leak control kits used for the control of leaks in chlorine cylinders (Chlorine A kit), one-ton containers (Chlorine B kit), and tank cars, tank trucks, and barges (Chlorine C kit).

Choking Agents Chemical agent that can damage the membranes of the human lung. Examples include phosgene and chlorine.

Chronic Effects Result from a single exposure or from repeated doses or exposures over a relatively long period of time.

Chronic Exposures Low exposures repeated over time.

Clandestine Laboratory (Also referenced in NFPA 472 as an illicit laboratory.) An operation consisting of a sufficient combination of apparatus and chemicals that either have been or could be used in the illegal manufacture/synthesis of controlled substances.

Classes (Electrical) As used in NFPA 70—*The National Electric Code* to describe the type of flammable materials that produce hazardous atmospheres. There are three classes:

- Class I Locations—Flammable gases or vapors may be present in quantities sufficient to produce explosive or ignitable mixtures.
- Class II Locations—Concentrations of combustible dusts may be present (e.g., coal or grain dust).
- Class III Locations—Areas concerned with the presence of easily ignitable fibers or flyings (e.g., cotton milling).

Classes (PPE Ensembles) As used in NFPA 1994—*Protective Ensemble for Chemical/Biological Terrorism Incidents* to describe the types of protective clothing available for terrorism response. There are three classes:

- Class 1 Ensembles offer the highest level of protection and are intended for use in worst-case circumstances, where the substance creates an immediate threat, is unidentified, and is of unknown concentrations.
- Class 2 Ensembles offer an intermediate level of protection and are intended for circumstances where the agent or threat may be identified, when the actual release has subsided, or in an area where live victims may be rescued.
- Class 3 Ensembles offer the lowest level of protection and are intended for use long after the initial release has occurred, at relatively large distances from the point of release, or for response activities such as decontamination, patient care, crowd control, traffic control, and clean-up operations.

Clean Air Act (CAA) Federal legislation that resulted in EPA regulations and standards governing airborne emissions, ambient air quality, and risk management programs.

Clean Water Act (CWA) Federal legislation that resulted in EPA and state regulations and standards governing drinking water quality, pollution control, and enforcement. The Oil Pollution Act (OPA) amended the CWA and authorized regulations pertaining to oil spill preparedness, planning, response, and clean-up.

Clean-up Incident scene activities directed toward removing hazardous materials, contamination, debris, damaged containers, tools, dirt, water, and road surfaces in accordance with proper and legal standards, and returning the site to as near a normal state as existed prior to the incident.

Code of Federal Regulations (CFR) A collection of regulations established by federal law. Contact with the agency that issues the regulation is recommended for both details and interpretation.

COFC See Container-On-Flat-Car.

Cold Zone The hazard control zone of a hazmat incident that contains the incident command post and other support functions as are deemed necessary to control the incident. This zone may also be referred to as the clean zone or the support zone.

Coliwasa Tube (Composite Liquid Waste Sampler) A glass or plastic waste sampling kit commonly used for collecting samples from drums and other containerized wastes.

Colorimetric Tubes Glass tubes containing a chemically treated substrate that reacts with specific airborne chemicals to produce a distinctive color. The tubes are calibrated to indicate approximate concentrations in air.

Combination Package Packaging consisting of one or more inner packagings and a nonbulk outer packaging. There are many different types of combination packagings.

Combined Liquid Waste Sampler See Coliwasa Tube.

Combined Sewers Carries domestic wastewater as well as storm water and industrial wastewater. It is quite common in older cities to have an extensive amount of these systems. Combined sewers may also have regulators or diversion structures that allow overflow directly to rivers or streams during major storm events.

Combustible Liquid A liquid that has a closed-cup flash point at or above

- 37.8°C (100°F) (NFPA-30). Under the NFPA 30—*Flammable and Combustible Liquids Code*, there are three classes of combustible liquids:

- **Combustible Liquid Class II** Any liquid that has a flash point at or above 37.8°C (100°F) and below 60°C (140°F)
- **Combustible Liquid Class IIIA** Any liquid that has a flash point at or above 60°C (140°F), but below 93°C (200°F)
- **Combustible Liquid Class IIIB** Any liquid that has a flash point at or above 93°C (200°F)

Command The act of directing, ordering, and/or controlling resources by virtue of explicit legal, agency, or delegated authority.

Command Staff The command staff consists of the Public Information Officer, the Safety Officer and the Liaison Officer, who report directly to the Incident Commander.

Community Awareness and Emergency Response (CAER) A program developed by the American Chemistry Council for chemical plant managers to assist them in developing integrated hazardous materials emergency response plans between the plant and the community.

Compatibility 1) The matching of protective chemical clothing to the hazardous material involved to provide the best protection for the worker; 2) the matching of patching and plugging materials to the leaking product type (pressure/nonpressure) and/or leaking materials.

Compatibility Charts Permeation and penetration data supplied by manufacturers of chemical protective clothing to indicate chemical resistance and breakthrough time of various garment materials as tested against a battery of chemicals. This test data should be in accordance with ASTM and NFPA standards.

Composite Packaging Packaging consisting of an inner receptacle, usually made of glass, ceramic, or plastic, and an outer protection (e.g., sheet metal, fiberboard, etc.), constructed so the receptacle and the outer package form an integral packaging for transport purposes. Once assembled, it remains an integral, single unit.

Compound Chemical combination of two or more elements, either the same elements or different ones, that is electrically neutral. Compounds have a tendency to break down into their component parts, sometimes explosively.

Comprehensive Environmental Response, Compensation and Liability Act (CERCLA) Known as CERCLA or SUPERFUND, it addresses hazardous substance releases into the environment and the clean-up of inactive hazardous waste sites. It also requires those who release hazardous substances, as defined by the Environmental Protection Agency (EPA), above certain levels (known as "reportable quantities") to notify the National Response Center.

Compressed Gas Any material or mixture having an absolute pressure exceeding 40 psi in the container at 70°F (21°C), having an absolute pressure exceeding 104 psi at 130°F (54°C), or any liquid flammable material having a vapor pressure exceeding 40 psi at 100°F (37.7°C), as determined by testing. Also includes cryogenic liquids with boiling points lower than 130°F (54°C), at 1 atmosphere.

Computer Aided Management of Emergency Operations (CAMEO) A computer database storage-retrieval system of preplanning and emergency data for on-scene use at hazardous materials incidents. Developed and maintained by the U.S. EPA.

Computerized Telephone Notification System (CT/NS) A computerized autodial telephone system used for notifying a potentially large number of people in a short period of time. CT/NS systems are often used around high hazard facilities to ensure the timely notification of nearby citizens.

Concentration The percentage of an acid or base dissolved in water. Concentration is not the same as strength.

Confined Space A space that (1) is large enough and so configured that an employee can bodily enter and perform assigned work; (2) has limited or restricted means for entry or exit (e.g., tanks, vessels, silos, storage bins, hoppers, vaults, and pits); and (3) is not designed for continuous employee occupancy.

Confined Space (Permit Required) Has one or more of the following characteristics:

1) Contains or has the potential to contain a hazardous atmosphere. A hazardous atmosphere would be created by any of the following: (a) vapors exceed 10% of the lower explosive limit (LEL); (b) airborne combustible dust exceeds its LEL; (c) atmospheric oxygen concentrations below 19.5% or above 23.5%; (d) atmospheric concentration of any substance for which a dose or PEL is published and which could result in employee exposure in excess of these values; (e) any other atmospheric condition immediately dangerous to life or health (IDLH).
2) Contains a material that has the potential for engulfing an entrant.
3) Has an internal configuration such that a person could be trapped or asphyxiated by inwardly converging walls or by a floor, which slopes downward and tapers to a smaller cross section; or
4) Contains any other recognized serious safety or health hazard.

Confinement Procedures taken to keep a material in a defined or localized area once released.

Consignee Person or company to which a material is being shipped.

Consist A railroad shipping document that lists the order of cars in a train.

Contact Being exposed to an undesirable or unknown substance that may pose a threat to health and safety.

Container Any vessel or receptacle that holds a material, including storage vessels, pipelines, and packaging. Includes both bulk and nonbulk packaging and fixed containers.

Container-On-Flat-Car (COFC) Intermodal container shipped on a railroad flat car.

Containment Actions necessary to keep a material in its container (e.g., stop a release of the material or reduce the amount being released).

Contaminant A hazardous material that physically remains on or in people, animals, the environment, or equipment, thereby creating a continuing risk of direct injury or a risk of exposure outside of the Hot Zone.

Contamination The process of transferring a hazardous material from its source to people, animals, the environment, or equipment, which may act as a carrier.

Contingency (Emergency) Planning A comprehensive and coordinated response to a hazmat problem. This planning process builds upon the hazards analysis and recognizes that no single public or private sector agency is capable of managing the hazmat problem by itself.

Control The offensive or defensive procedures, techniques, and methods used in the mitigation of a hazardous materials incident, including containment, extinguishment, and confinement.

Controlled Burn Defensive or nonintervention tactical objective by which a fire is allowed to burn with no effort to extinguish it.

Corrosive A material that causes visible destruction of, or irreversible alterations to, living tissue by chemical action at the point of contact.

Corrosivity Harm Events Those events related to severe chemical burns and/or tissue damage from corrosive exposures.

Covalent Bonding The force holding together atoms that share electrons between atoms.

Crack Narrow split or break in the container metal which may penetrate through the container metal; may also be caused by metal fatigue.

Crew Resource Management (CRM) Originally defined in 1977 by aviation psychologist Dr. John Lauber as "... using all available resources—information, equipment, and people—to achieve safe and efficient flight operations." Key components of CRM include command, leadership, and resource management.

Crisis An unplanned event that can exceed the level of available resources and has the potential to significantly impact an organization's operability, credibility, and reputation, or pose a significant environmental, economic, or legal liability.

Critical Temperature and Pressure Critical temperature is the minimum temperature at which a gas can be liquefied no matter how much pressure is applied. Critical pressure is the pressure that must be applied to bring a gas to its liquid state. Both terms relate to the process of liquefying gases.

Critique An element of incident termination that examines the overall effectiveness of the emergency response effort and develops recommendations for improvement.

Cross-Contamination (aka secondary contamination) Occurs when a person who is already contaminated makes contact with a person or object that is not contaminated.

Crude Oil or Crude Petroleum Hydrocarbon mixtures that have a flash point below 150 degrees F (65 degrees C) and which have not yet been processed in a refinery.

Cryogenic Liquids Gases that have been transformed into extremely cold liquids and stored at temperatures below -130°F (-90°C). Cryogenic liquid spills will vaporize rapidly when exposed to the higher ambient temperatures outside of the container. Expansion ratios for common cryogenics range from 694 (nitrogen) to 1445 (neon) to 1.

Cryogenic Tank Containers Containers that transport refrigerated liquefied gases such as argon, helium, oxygen, nitrogen, and ethylene. They include the UN Portable Tank T75 and IMO Type 7 specification containers.

Dam A physical method of confinement by which barriers are constructed to prevent or reduce the quantity of liquid flowing into the environment.

Damage Assessment The process of gathering and evaluating container damage as a result of a hazmat incident.

Dangerous Cargo Manifest A list of the hazardous materials carried as cargo on board a vessel. Includes the location of the hazmat on the vessel.

Dangerous Goods In international transportation hazardous materials are commonly referred to as "dangerous goods."

Debriefing An element of incident termination that focuses on the following: (1) informing responders exactly what hazmat they were (possibly) exposed to and the signs and symptoms of exposure; (2) identifying damaged equipment requiring replacement or repair; (3) identifying equipment or supplies requiring specialized decontamination or disposal; (4) identifying unsafe work conditions; (5) assigning information-gathering responsibilities for a post-incident analysis.

Decon Popular abbreviation referring to the process of decontamination.

Decontamination The physical and/or chemical process of reducing and preventing the spread of contamination from persons and equipment at a hazardous materials incident. OSHA 1910.120 defines decontamination as the removal of hazardous substances from employees and their equipment to the extent necessary to preclude foreseeable health effects.

Decontamination Corridor A distinct area within the "Warm Zone" that functions as a protective buffer and bridge between the "Hot Zone" and the "Cold Zone," where decontamination stations and personnel are located to conduct decontamination procedures.

Decontamination Unit Leader A position within the Hazardous Materials Group who has responsibility for identifying the location of the decontamination corridor, assigning stations, managing all decontamination procedures, and identifying the types of decontamination necessary.

Decontamination Unit A group of personnel and resources operating within a decontamination corridor.

Defensive Tactics These are less aggressive spill and fire control tactics where certain areas may be "conceded" to the emergency, with response efforts directed towards limiting the overall size or spread of the problem.

Degradation (1) A chemical action involving the molecular breakdown of a protective clothing material or equipment due to contact with a chemical. (2) The molecular breakdown of the spilled or released material to render it less hazardous during control operations.

Degree of Solubility An indication of the solubility and/or miscibility of a material: negligible—less than 0.1%; slight—0.1 to 1.0%; moderate—1 to 10%; appreciable—greater than 10%; complete—soluble at all proportions.

Dent Deformation of the tank head or shell. It is caused from impact with a relatively blunt object (e.g., railroad coupler, vehicle). If the dent has a sharp radius, there is the possibility of cracking.

Dermatotoxins Toxins of the skin which may act as irritants or cause ulcers, chloracne, or skin pigmentation disorders (e.g., halogenated hydrocarbons, coal tar compounds).

Detonation An explosive chemical reaction with a release rate less than 1/100th of a second. Examples include military munitions, dynamite, and organic peroxides.

Dike A defensive confinement procedure consisting of an embankment or ridge on the ground used to control the movement of liquids, sludges, solids, or other materials. Barrier which prevents passage of a hazmat to an area where it will produce more harm.

Lethal Concentration, 50 Percent Kill (LC50) Concentration of a material expressed as parts per million (PPM) per volume, which kills half of the lab animals in a given length of time. Refers to an inhalation exposure, the LC50 may also be expressed as mg/liter or mg/cubic meter. Significant in evaluating the toxicity of a material; the lower the value, the more toxic the substance.

Lethal Concentration Low (LCLOW) The lowest concentration of a substance in air reported to have caused death in humans or animals. The reported concentrations may be entered for periods of exposure that are less than 24 hours (acute) or greater than 24 hours (subacute and chronic).

Lethal Dose, 50 Percent Kill (LD50) The amount of a dose which, when administered to lab animals, kills 50 percent of them. Refers to an oral or dermal exposure and is expressed in terms of mg/kg.

Lethal Dose Low (LDLOW) The lowest amount of a substance introduced by any route, other than inhalation, reported to have caused death to animals or humans.

Level I Staging The initial location for emergency response units at a multiple unit response to a hazmat incident. Initial arriving emergency response units go directly to the incident scene taking standard positions (e.g., upwind, uphill as appropriate), assume command, and begin site management operations. The remaining units stage at a safe distance away from the scene until ordered into action by the Incident Commander.

Level II Staging Location where arriving units are initially sent when an incident escalates past the capability of the initial response. It is a tool usually reserved for large, complex, or lengthy hazmat operations. Units assigned to Staging are under the control of a Staging Officer or Staging Area Manager.

Liaison Officer Serves as a coordination point between the Incident Commander and any assisting or coordinating agencies who have responded to the incident, but who are not part of unified command or are not represented at the Incident Command Post.

Limited-Use Materials Protective clothing materials that are used and then discarded. Although they may be reused several times (based upon chemical exposures), they are often disposed of after a single use. Examples include Tyvek™ QC, Tyvek™/Saranex™ 23-P, Hazard-Gard™, I Hazard-Gard™ II, and the Tychem Responder™.

Liquid Chemical Splash Protective Clothing The garment portion of a chemical protective clothing ensemble designed and configured to protect the wearer against chemical liquid splashes but not against chemical vapors or gases. Liquid splash chemical protective clothing must meet the requirements of NFPA 1992. This type of protective clothing is a component of EPA Level B chemical protection.

Local Effect The health effects of a hazardous materials exposure at the point of contact.

Local Emergency Planning Committee (LEPC) A committee appointed by a state emergency response commission, as required by SARA, Title III, to formulate a comprehensive emergency plan for its corresponding local government or mutual aid region.

Lower Detection Limit (LDL) The lowest concentration to which a monitoring instrument will respond. The lower the LDL, the quicker contaminant concentrations can be evaluated.

Low-Pressure Tank A storage tank designed to withstand an internal pressure above 0.5 psig (3.5 kPa) but not more than 15 psig (103.4 kPa) measured at the top of the tank.

Manifest A shipping document that lists the commodities being transported on a vessel.

Markings The required names, instructions, cautions, specifications, or combinations thereof found on containers of hazardous materials and hazardous wastes.

Mass Decontamination The process of decontaminating large numbers of people in the fastest possible time to reduce surface contamination to a safe level. It is typically a gross decon process utilizing water or soap and water solutions to quickly reduce the level of contamination.

Maximum Safe Storage Temperature (MSST) The maximum storage temperature that an organic peroxide may be maintained, above which a reaction and explosion may occur.

Mechanical Harm Events Those harm events resulting from direct contact with fragments scattered because of a container failure, explosion, or shock wave.

Mechanical Stress The result of a transfer of energy when one object physically contacts or collides with another. Indicators include punctures, gouges, breaks, or tears in a container.

Medical Monitoring An ongoing, systematic evaluation of individuals at risk of suffering adverse effects of exposure to heat, stress, or hazardous materials as a result of working at a hazmat emergency.

Medical Surveillance Comprehensive medical program for tracking the overall health of its participants (e.g., HMRT personnel, public safety responders, etc.). Medical surveillance programs consist of pre-employment screening, periodic medical examinations, emergency treatment provisions, nonemergency treatment, and recordkeeping and review.

Melting Point The temperature at which a solid changes to a liquid. This temperature is also the freezing point depending on the direction of the change. For mixtures, a melting point range may be given.

Minimum Detectable Permeation Rate (MDPR) The minimum permeation rate that can be detected by a laboratory's analytical system being used for the permeation test.

Miscible Refers to the tendency or ability of two or more liquids to form a uniform blend or to dissolve in each other. Liquids may be totally miscible, partially miscible, or non-miscible.

Mitigation Any offensive or defensive action to contain, control, reduce, or eliminate the harmful effects of a hazardous materials release.

Mixture Substance made up of two or more compounds, physically mixed together. A mixture may also contain elements and compounds mixed together.

Monitoring The act of systematically checking to determine contaminant levels and atmospheric conditions.

Monitoring Instruments Monitoring and detection instruments used to detect the presence and/or concentration of contaminants within an environment. They include:

- Combustible Gas Indicator (CGI): Measures the concentration of a combustible gas or vapor in air.
- Oxygen Monitor: Measures the percentage of oxygen in air.
- Colorimetric Indicator Tubes: Measures the concentration of specific gases and vapors in air.
- Specific Chemical Monitors: Designed to detect a large group of chemicals or a specific chemical. Most common examples include carbon monoxide and hydrogen sulfide.
- Flame Ionization Detector (FID): A device used to determine the presence of organic vapors and gases in air. Operates in two modes—survey mode and gas chromatograph.
- Gas Chromatograph: An instrument used for identifying and analyzing specific organic compounds.
- Photoionization Detector (PID): A device used to determine the total concentration of many organic and some inorganic gases and vapors in air.
- Radiation Monitors: An instrument used to measure accumulated radiation exposure.
- Radiation Dosimeter Detector: An instrument which measures the amount of radiation to which a person has been exposed.
- Corrosivity (pH) Detector: A meter, paper, or strip that indicates the relative acidity or alkalinity of a substance, generally using an international scale of 0 (acid) through 14 (alkali-caustic). (See pH.)
- Indicator Papers: Special chemical indicating papers which test for the presence of specific hazards, such as oxidizers, organic peroxides, and hydrogen sulfide. Are usually part of a hazmat identification system.

Multiple Element Gas Container (MEGC) Assemblies of UN cylinders, tubes, or bundles of cylinders interconnected by a manifold and assembled within a rigid frame with corner castings for the transport of gases. Also known as a tube module.

Multi-Use Materials Based upon the chemical exposure, multi-use materials are designed and fabricated to allow for decontamination and reuse. Generally thicker and more durable than limited-use garments, they are used for chemical splash and vapor protective suits, gloves, aprons, boots, and thermal protective clothing. The most common materials include butyl rubber, Viton, polyvinyl chloride (PVC), neoprene rubber, and Teflon™.

Mutagen A material that creates a change in gene structure, which is potentially capable of being transmitted to offspring.

National Animal Poison Control Center (NAPCC) Operated by the University of Illinois at Urbana-Champaign, their 24-hour hotline provides consultation in the diagnosis and treatment of suspected or actual animal poisonings or chemical contamination.

National Contingency Plan (NCP) Outlines the policies and procedures of the federal agency members of the National Oil and Hazardous Materials Response Team (also known as the National Response Team or NRT).

National Fire Protection Association (NFPA) An international voluntary membership organization that promotes improved fire protection and prevention, establishes safeguards against loss of life and property by fire, and develops and publishes national voluntary consensus standards (e.g., NFPA 472—*Professional Competence of Responders to Hazardous Materials Incidents*).

National Institute for Occupational Safety and Health (NIOSH) A Federal agency which, among other activities, tests and certifies respiratory protective devices, air sampling detector tubes, and recommends occupational exposure limits for various substances.

National Incident Management System (NIMS) A standardized systems approach to incident management consisting of five major sub-divisions collectively providing a total systems approach to all risk incident management.

National Pesticide Information Center (NPIC) Operated by Oregon State University in cooperation with the EPA, NPIC provides information on pesticide-related health/toxicity questions, properties, and minor clean-up.

National Response Center (NRC) Communications center operated by the U.S. Coast Guard in Washington, DC. It provides information on suggested technical emergency actions, and is the federal spill notification point.

National Response Team (NRT) The National Oil and Hazardous Materials Response Team consists of 14 federal government agencies which carry out the provisions of the National Contingency Plan at the federal level.

National Transportation Safety Board (NTSB) Independent federal agency charged with responsibility for investigating serious accidents and emergencies involving the various modes of transportation (e.g., highway, pipeline, air), as well as hazardous materials. Issues investigation reports and non-binding recommendations for action.

Nonpressure Intermodal Tank Containers Standardized 20-foot (6.058 m) bulk liquid stainless steel vessels supported and protected by a steel frame that can easily be moved from ship to highway or rail transport vehicles and can be stacked for storage or transit.

NCP See National Contingency Plan.

Nephrotoxins Toxins that attack the kidneys (e.g., mercury, halogenated hydrocarbons).

Neurotoxins Toxins that attack the central nervous system (e.g., organophosphate pesticides).

Neutralization A chemical method of containment by which a hazmat is neutralized by applying a second material to the original spill which will chemically react with it to form a less harmful substance.

Neutron particles A form of high-speed particle radiation that consists of a "neutron" emitted at a high speed from the nucleus of a radioactive atom. Neutrons are considered a whole body hazard.

NFPA See National Fire Protection Association.

NIMS See National Incident Management System.

Nonbulk Packaging Any packaging having a capacity meeting one of the following criteria: liquid—internal volume of 119 gallons (450 L.) or less; solid—capacity of 882 lbs (400 kg) or less; and compressed gas—water capacity of 1001 lb. (454 kg) or less. Nonbulk packaging may consist of single packaging (e.g., drum, carboy, cylinder) or combination packaging—one or more inner packages inside of an outer packaging.

Nonintervention Tactics Essentially "no action." It is useful at certain emergencies where the potential cost of action far exceeds any benefit (e.g., BLEVE scenario).

Nonionizing Radiation Waves of energy, such as radiant heat, radio waves, and visible light. The amount of energy in these waves is small as compared to ionizing radiation. Examples include infrared waves, microwaves, and lasers.

Non-Persistence Refers to the length of time a chemical agent remains as a liquid. A chemical agent is said to be "non-persistent" if it evaporates within 24 hours.

Normal Physical State The physical state or form (solid, liquid, gas) of a material at normal temperatures (68°F or 20°C to 77°F or 25°C). Determining the physical state of a material can allow responders to assess potential harm.

Normalized Breakthrough Time A calculation, using actual permeation results, to determine the time at which the permeation rate reaches 0.1 µg/cm2/min. Normalized breakthrough times are useful for comparing the performance of several different protective clothing materials.

Not Otherwise Specified (NOS) A shipping paper notation which indicates that the material meets the DOT definition for a hazardous material but is not listed by a generic name within the DOT Regulations.

NRT See National Response Team.

Occupational Safety and Health Administration (OSHA) Component of the U.S. Department of Labor; an agency with safety and health regulatory and enforcement authority for most U.S. industries, businesses, and states.

Odor Threshold (TLVODOR) The lowest concentration of a material's vapor in air that is detectable by odor. If the TLVodor is below the TLV/TWA, odor may provide a warning as to the presence of a material.

Offensive Tactics Aggressive leak, spill, and fire control tactics designed to quickly control or mitigate a problem.

Oil Pollution Act (OPA) Amended the Federal Water Pollution Act in 1990, OPA's scope covers both facilities and carriers of oil and related liquid products, including deepwater marine terminals, marine vessels, pipelines, and railcars. Requirements include the development of emergency response plans, training and exercises, and verification of spill resources and contractor capabilities. The law also requires the establishment of Area Committees and the development of

Area Contingency Plans (ACPs) to address oil and hazardous substance spill response in coastal zone areas.

On-Scene Coordinator (OSC) The federal official predesignated by EPA or the USCG to coordinate and direct federal responses and removals under the National Contingency Plan.

On-Scene Incident Commander See Incident Commander.

OPA See Oil Pollution Act.

Operations Section Responsible for all tactical operations at the incident. The Hazmat Branch falls within the Operations Section.

Operations Security (OPSEC) OPSEC is a systematic process of identifying the critical information that needs to be safeguarded from criminals and terrorists. The process consists of: (1)identifying the critical information that needs to be protected; (2) identifying the threat of potential adversaries gaining access to the critical information and evaluating their intelligence collection capabilities; (3) analysis of the identified vulnerabilities; (4) assessing the risks associated with the identified threat and vulnerabilities, and (5) implementing countermeasures to lower risks in order to protect critical information and mission.

Organic Materials Materials that contain carbon. Organic materials are derived from materials that are living or were once living, such as plants or decayed products. Most organic materials are flammable.

Organic Peroxide Strong oxidizers, often chemically unstable, containing the -o-o- chemical structure. May react explosively to temperature and pressure changes.

OSC See On-Scene Coordinator.

Other Regulated Materials D (ORM D) A material, such as a consumer commodity, which presents a limited hazard during transportation due to its form, quantity, or packaging.

Overflow Dam Spill control tactic used to trap sinking heavier-than-water materials behind the dam (specific gravity > 1). With the product trapped, uncontaminated water is allowed to flow unobstructed over the top of the dam.

Overgarments Protective clothing ensembles worn over chemical vapor protective clothing to provide either additional flash protection or low temperature protection.

Overgloving The wearing of a second glove over the work glove for additional chemical and/or abrasion protection during operations.

Overpack (1) Packaging used to contain one or more packages for convenience of handling and/or protection of the packages. (2) A term used to describe the placement of damaged or leaking packages in an overpack or recovery drum. (3) The outer packaging for radioactive materials.

Overpacking Use of a specially constructed drum to overpack damaged or leaking containers of hazardous materials for shipment. Overpack containers should be compatible with the hazards of the materials involved.

Oxidation Ability The ability of a material to (1) either give up its oxygen molecule to stimulate the oxidation of organic materials (e.g., chlorate, permanganate and nitrate compounds), or (2) receive electrons being transferred from the substance undergoing oxidation (e.g., chlorine and fluorine). Result of either activity is the release of energy.

Oxidizer A chemical, other than a blasting agent or an explosive, that initiates or promotes combustion in other materials.

Oxygen Deficient Atmosphere An atmosphere that contains an oxygen content less than 19.5% by volume at sea level.

Packaging One or more receptacles and any other components or materials necessary for the receptacles to perform their containment and other safety functions.

Packing Group Classification of hazardous materials based on the degree of danger represented by the material. There are three groups: Packing Group I indicates great danger, Packing Group II indicates medium danger, and Packing Group III indicates minor danger.

PAPR See Powered-Air Purifying Respirator.

Patching (Plugging) A physical method of containment, which uses chemically compatible patches and plugs to reduce or temporarily stop the flow of materials from small container holes, rips, tears, or gashes.

PCB Contaminated Any equipment, including transformers, that contains 50 to 500 ppm of PCBs.

Penetration The flow or movement of a hazardous chemical through closures, seams, porous materials, and pinholes, or other imperfections in the material.

Permeation The process by which a hazardous chemical moves through a given material on the molecular level. Permeation differs from penetration in that permeation occurs through the clothing material itself rather than through the openings in the clothing material.

Permeation Rate The rate at which a chemical passes through a given chemical protective clothing material.

Expressed as micrograms per square centimeter per minute (μgm/cm2/min).

Permissible Exposure Limit (PEL) The maximum time-weighted concentration at which 95% of exposed, healthy adults suffer no adverse effects over a 40-hour work week and are comparable to ACGIH's TLV/TWA. PELs are used by OSHA and are based on an 8-hour, time-weighted average concentration.

Persistence Refers to the length of time a chemical agent remains a liquid. A chemical agent is said to be "persistent" if it remains as a liquid for longer than 24 hours and non-persistent if it evaporates within that time. Among the most persistent chemical agents are VX, tabun, mustard, and lewisite.

Personal Protective Equipment (PPE) Equipment provided to shield or isolate a person from the chemical, physical, and thermal hazards that may be encountered at a hazardous materials incident. Adequate personal protective equipment should protect the respiratory system, skin, eyes, face, hands, feet, head, body, and hearing. It includes: personal protective clothing, self-contained positive-pressure breathing apparatus, and air purifying respirators.

pH (Power of Hydrogen) Acidic or basic corrosives are measured to one another by their ability to dissociate in solution. Those that form the greatest number of hydrogen ions are the strongest acids, while those that form the hydroxide ion are the strongest bases. The measurement of the hydrogen ion concentration in solution is called the pH (power of hydrogen) of the compound in solution.

Pipeline System All parts of a pipeline facility through which a gas or hazardous liquid or carbon dioxide moves in transportation, including, but not limited to, line pipe, valves, and other appurtenances connected to line pipe, pumping units, fabricated assemblies associated with pumping units, metering and delivery stations and fabricated assemblies therein, and breakout tanks. See 49 CFR Part 195.2.

Physical State The physical state or form (solid, liquid, gas) of a material at normal ambient temperatures (68°F to 77°F).

Placards Approximately 10.75 inch (273 mm) square markings required under DOT regulations and applied to both ends and each side of freight containers, cargo tanks, and portable tank containers.

Planning Section Responsible for the collection, evaluation, dissemination, and use of information about the development of the incident and the status of resources. Includes the Situation Status, Resource Status, Documentation, and Demobilization Units as well as Technical Specialists.

Plume A vapor, liquid, dust, or gaseous cloud formation, which has shape and buoyancy.

Pneumatic Hopper Trailer Covered hopper trailers that are pneumatically unloaded and used for transporting solids. Have a capacity up to 1500 cubic feet.

Polymerization A reaction during which a monomer is induced to polymerize by the addition of a catalyst or other unintentional influences such as excessive heat, friction, contamination, etc. If the reaction is not controlled, it is possible to have an excessive amount of energy released.

Portable Bin Portable tanks used to transport bulk solids. Are approximately 4 feet square and 6 feet high, with weights up to 7700 pounds. Normally loaded through the top and unloaded from the side or bottom.

Portable Tank Any packaging (except a cylinder having 1000 lbs or less water capacity) over 110 gallons capacity and designed primarily to be loaded into, on, or temporarily attached to a transport vehicle or ship., and equipped with skids, mountings or accessories to facilitate handling of the tank by mechanical means.

Post-Emergency Response Operations That portion of an emergency response performed after the immediate threat of a release has been stabilized or eliminated, and the clean-up of the site has begun.

Post-Incident Analysis An element of incident termination that includes completion of the required incident reporting forms, determining the level of financial responsibility, and assembling documentation for conducting a critique.

Powered-Air Purification Respirators (PAPR) Air- purifying respirators that use a blower to force the ambient air through air-purifying elements to a full-face mask. As a result, there is a slight positive pressure in the face piece that results in an increased protection factor. Where an APR has a protection factor of 50:1, a PAPR will have a protection factor of 1,000:1.

Pressure Isolation and Reduction A physical or chemical method of containment by which the internal pressure of a closed container is reduced. The tactical objective is to sufficiently reduce the internal pressure in order to either reduce the flow or minimize the potential of a container failure.

Private Sector Specialist Employee A Those persons specially trained to handle incidents involving chemicals and/or containers for chemicals used in their organization's area of specialization. Consistent with the organization's response plan and standard operating procedures, the Specialist Employee A shall have the ability to analyze an incident involving chemicals within the organization's area of specialization, plan a response to that incident, implement the planned response within the capabilities of the resources available, and evaluate the progress of the planned response.

Private Sector Specialist Employee B Those persons who, in the course of their regular job duties, work with or are trained in the hazards of specific chemicals and/or containers for chemicals used in their individual area of specialization. Because of their education, training or work experience, the Specialist Employee B may be called upon to gather and record information, provide technical advice, and provide technical assistance (including work within the hot zone) at an incident involving chemicals consistent with their organization's emergency response plan and standard operating procedures and the local emergency response plan.

Private Sector Specialist Employee C The Specialist C should be able to provide information on a specific chemical or container and have the organizational contacts needed to acquire additional technical assistance. This individual need not have the skills or training necessary to conduct control operations and is generally found at the command post providing the IC or their designee with technical assistance.

Process Safety Management (PSM) The application of management principles, methods, and practices to prevent and control releases of hazardous chemicals or energy. Focus of both OSHA 1910.119—*Process Safety Management of Highly Hazardous Chemicals, Explosives and Blasting Agents* and EPA Part 68—*Risk Management Programs for Chemical Accidental Release Prevention*.

Product Name Brand or trade name printed on the front panel of a hazmat container. If the product name includes the term "technical," as in "Parathion Technical," it generally indicates a highly concentrated pesticide with 70% to 99% active ingredients.

Proper Shipping Name The DOT designated name for a commodity or material. Will appear on shipping papers and on some containers. May also be referred to as shipping name.

Protection In-Place Directing fixed facility personnel and the general public to go inside of a building or structure and remain there until the danger from a hazardous materials release has passed. It may also be referred to as in-place protection, sheltering-in-place, sheltering, and taking refuge.

Protective Clothing Equipment designed to protect the wearer from heat and/or hazardous materials contacting the skin or eyes. Protective clothing is divided into four types:

- Structural firefighting protective clothing
- Liquid splash chemical protective clothing
- Vapor chemical protective clothing
- High temperature protective clothing

Proximity Suits Designed for exposures of short duration and close proximity to flame and radiant heat, such as in aircraft rescue firefighting (ARFF) operations. The outer shell is a highly reflective, aluminized fabric over an inner shell of a flame-retardant fabric such as Kevlar™ or Kevlar™/PBI™ blends.

PSM See Process Safety Management.

Public Information Officer Point of contact for the media or other organizations seeking information directly from the incident or event. Member of the Command Staff.

Public Protective Actions The strategy used by the Incident Commander to protect the general population from the hazardous material by implementing a strategy of either (1) Protection in-Place, (2) Evacuation, or (3) a combination of Protection In-Place and Evacuation.

Purging Totally enclosed electrical equipment is protected with an inert gas under a slight positive pressure from a reliable source. The inert gas provides positive pressure within the enclosure and minimizes the development of a flammable atmosphere. Used in Class I, Division 1 atmospheres at fixed installations.

Pyrophoric Materials Materials that ignite spontaneously in air without an ignition source.

Radiation Harm Events Those harm events related to the emission of radioactive energy. There are two types of radiation—ionizing and nonionizing.

Radiation Dispersal Device (RDD) A device designed to spread radioactive material through a detonation of conventional explosives or other (non-nuclear) means. RDD is used interchangeably with the term "dirty bomb."

Radiation Exposure Device (RED) RED consists of radioactive material, either as a sealed source or as material within some type of container, or a radiation-generating device, such as an X-ray device, that directly exposes people to ionizing radiation.

Radioactivity The ability of a material to emit any form of radioactive energy.

Rail Burn Deformation in the shell of a railroad tank car. It is actually a long dent with a gouge at the bottom of the inward dent. A rail burn can be oriented circumferentially or longitudinally in relation to the tank shell. The longitudinal rail burns are the more serious because they have a tendency to cross a weld.

RCRA See Resource Conservation and Recovery Act.

Reactivity/Instability The ability of a material to undergo a chemical reaction with the release of energy. It could be initiated by mixing or reacting with other materials, application of heat, physical shock, etc.

Recommended Exposure Levels (REL) The maximum time-weighted concentration at which 95% of exposed, healthy adults suffer no adverse effects over a 40-hour work week and are comparable to ACGIH's TLV/TWA. RELs are used by NIOSH and are based on a 10-hour, time-weighted average concentration.

Regional Response Team (RRT) Established within each federal region, the RRT follows the policy and program direction established by the NRT to ensure planning and coordination of both emergency preparedness and response activities. Members include EPA, USCG, state and local government, and Indian tribal governments.

Rehabilitation (Rehab) Process of providing for EMS support, treatment and monitoring, food and fluid replenishment, and mental rest and relief from extreme environmental conditions associated with a hazmat incident. May function as a group within the Incident Management System.

Release Event Once a container is breached, the hazmat is free to escape (be released) in the form of energy, matter, or a combination of both. Types of release include detonation, violent rupture, rapid relief, and spills or leaks.

Reportable Quantity (RQ) Indicates a material is a hazardous substance by the EPA. The letters "RQ" (reportable quantity) must be shown either before or after the basic shipping description entries. This designation indicates that any leakage of the substance above its RQ value must be reported to the proper agencies (e.g., National Response Center).

Reporting Marks and Number The set of initials and a number stenciled on both sides and both ends of railroad cars; can be used to obtain information on the contents of a car from either the railroad or the shipper.

Residue The material remaining in a package after its contents have been emptied and before the packaging is refilled, or cleaned, or purged of vapor to remove any potential hazard.

Resource Conservation and Recovery Act (RCRA) Law that establishes the regulatory framework for the proper management and disposal of all hazardous wastes, including treatment, storage, and disposal facilities. It also establishes installation, leak prevention and notification requirements for underground storage tanks.

Respiratory Protection Equipment designed to protect the wearer from the inhalation of contaminants; includes positive-pressure self-contained breathing apparatus, positive-pressure airline respirators, powered-air purifying respirators, and air purifying respirators.

Respiratory Toxins Toxins that attack the respiratory system (e.g., asbestos, hydrogen sulfide).

Response That portion of incident management in which personnel are involved in controlling (offensively or defensively) a hazmat incident. The activities in the response portion of a hazmat incident include analyzing the incident, planning the response, implementing the planned response, and evaluating progress.

Responsible Party (RP) A legally recognized entity (e.g., person, corporation, business or partnership, etc.) that has a legally recognized status of financial accountability and liability for actions necessary to abate and mitigate adverse environmental and human health and safety impacts resulting from a non-permitted release or discharge of a hazardous material. The person or agency found legally accountable for the clean-up of an incident.

Retention A physical method of confinement by which a liquid is temporarily contained in an area where it can be absorbed, neutralized, or picked up for proper disposal.

Reusable Garments Chemical protective clothing garments designed to allow for decontamination and re-use. Generally thicker and more durable than limited-use garments, they are used for liquid chemical splash and vapor protective suits, gloves, aprons, boots, and thermal protective clothing. Reusable garment materials are usually made from chlorinated polyethylene (CPE), vinyl (plasticized polyvinyl chloride—PVC), fluorinated polymers (Teflon®), and rubber-like fabrics, such as butyl rubber, neoprene rubber, and Viton™.

Riot Control Agents Usually solid materials that are dispersed in a liquid spray, and which and cause pain or burning on exposed mucous membranes and skin. Common examples include Mace™ (CN) and pepper spray (i.e., capsaicin).

Risks The probability of suffering a harm or loss. Risks are variable and change with every incident.

Risk Analysis A process to analyze the probability that harm may occur to life, property, and the environment and to note the risks to be taken in relation to the incident objectives.

Risk-Based Response Process Systematic process by which responders analyze a problem involving hazardous materials/weapons of mass destruction (WMD), assess the hazards, evaluate the potential consequences, and determine appropriate response actions based on facts, science, and the circumstances of the incident.

Risk Management Program Required under EPA's proposed 40 CFR Part 68, a risk management program consists of three elements: (1) hazard assessment of the facility; (2) prevention program; and (3) emergency response considerations.

RMP See Risk Management Program.

Roentgen A measure of the charge produced in air created by ioninzing radiation, usually in reference to gamma radiation.

Roentgen Equivalent Man (REM) The unit of dose equivalent; takes into account the effectiveness of different types of radiation.

Runaway Cracking Cracking occurring in closed containers under pressure, such as liquid drums or pressure vessels. A small crack in a closed container suddenly develops into a rapidly growing crack, which circles the container. As a result, the container will generally break into two or more pieces.

Safety Data Sheet A form with data regarding the properties of a particular substance. It is intended to provide workers and emergency personnel with procedures for handling or working with that substance in a safe manner, and includes information such as physical data (melting point, boiling point, flash point, etc.), toxicity, health effects, first aid, reactivity, storage, disposal, protective equipment, and spill-handling procedures.

Safety Officer Responsible for the safety of all personnel, including monitoring and assessing safety hazards, unsafe situations, and developing measures for ensuring personnel safety. The Incident Safety Officer (ISO) has the authority to terminate any unsafe actions or operations, and is a required function based upon the requirements of OSHA 1910.120 (q).

Safe Refuge Area A temporary holding area within the hot zone for contaminated people until a decontamination corridor is set up.

Sampling The process of collecting a representative amount of a gas, liquid, or solid for evidence or analytical purposes.

Sampling Kit Kits assembled for the purpose of providing adequate tools and equipment for taking samples and documenting unknowns to create a "chain of evidence."

Sanitary Sewer A "closed" sewer system which carries wastewater from individual homes, together with minor quantities of storm water, surface water, and ground water that are not admitted intentionally. May also collect wastewater from industrial and commercial businesses. The collection and pumping system will transport the wastewater to a treatment plant, where the wastewater is processed.

SAR See Supplied Air Respirator.

SARA See Superfund Amendments & Reauthorization Act.

Saturated Hydrocarbons A hydrocarbon possessing only single covalent bonds. All of the carbon atoms are saturated with hydrogen. Examples include methane (CH4), propane (C3H8), and butane (C4H10).

SCBA See Self-Contained Breathing Apparatus.

Scene The location impacted or potentially impacted by a hazard.

Score Reduction in the thickness of the container shell. It is an indentation in the shell made by a relatively blunt object. A score is characterized by the reduction of the container or weld material so that the metal is pushed aside along the track of contact with the blunt object.

Scrubber System A diverse group of pollution control devices used to filter particulates of contaminants and gases from industrial exhaust streams. Emergency scrubbers are found in many chemical plant process operations where a system failure or runaway reaction could cause a discharge of toxic products into the atmosphere causing injuries or fatalities.

SDS See Safety Data Sheet.

Secondary Contamination The process by which a contaminant is carried out of the hot zone and contaminates people, animals, the environment, or equipment outside of the hot zone.

Secondary Decontamination The second phase of the decontamination process designed to physically or chemically remove surface contaminants to a safe and acceptable level.

Section That organization level within the Incident Command System having functional responsibility for primary segments of incident operations, such as

Operations, Planning, Logistics, and Administration/Finance. The Section level is organizationally between Branch and the Incident Commander.

Self-Accelerating Decomposition Temperature (SADT) The temperature at which an organic peroxide or synthetic compound will react to heat, light, or other chemicals, and release oxygen, energy, and fuel in the form of an explosion or rapid oxidation. When this temperature is reached by some portion of the mass of an organic peroxide, irreversible decomposition will begin.

Self-Contained Breathing Apparatus (SCBA) A positive-pressure, self-contained breathing apparatus (SCBA) or combination SCBA/supplied air breathing apparatus certified by NIOSH and the Mine Safety and Health Administration (MSHA), or the appropriate approval agency for use in atmospheres immediately dangerous to life or health (IDLH).

Sensitizer A chemical that causes a substantial proportion of exposed people or animals to develop an allergic reaction in normal tissue after repeated exposure to the chemical.

SERC See State Emergency Response Commission.

Shipper A person, company, or agency offering material for transportation.

Shipping Documents/Papers Generic term used for documents that must accompany all shipments of goods for transportation. These include Hazardous Waste Manifest, Bill of Lading, and Consists, etc.

Shipping Name The proper shipping name or other common name for the material; also any synonyms for the material.

Single Trip Container (STC) Container that may not be refilled or reshipped with a DOT regulated material except under certain conditions.

Site Management and Control The management and control of the physical site of a hazmat incident. Includes initially establishing command, approach and positioning, staging, establishing initial perimeter and hazard control zones, and public protection actions.

Size-Up The rapid yet deliberate consideration of all critical scene factors.

Skilled Support Personnel Personnel skilled in the operation of certain equipment, such as cranes and hoisting equipment, and who are needed temporarily to perform immediate emergency support work that cannot reasonably be performed in a timely fashion by emergency response personnel.

Skin Absorption The introduction of a chemical or agent into the body through the skin. Skin absorption can occur with no sensation to the skin itself.

Slopover Can result when a water stream is applied to the hot surface of a burning oil, providing the oil is viscous and its temperature exceeds the boiling point of water. It can also occur when the heat wave contacts a small amount of stratified water within a crude oil. As with a boilover, when the heat wave contacts the water, the water converts to steam and causes the product to "slopover" the top of the tank. Slopovers can range from a quiet-like boiling of the product over the tank, to a large explosion of burning slop.

Slurry Pourable mixture of a solid and a liquid.

Sludge Solid, semi-solid, or liquid waste generated from a municipal, commercial, or industrial waste treatment plant or air pollution control facility, exclusive of treated effluent from a waste water treatment plant.

Solidification Process by which a contaminant physically or chemically bonds to another object or is encapsulated by it. May be used as a chemical method of confinement or decontamination.

Solubility The ability of a solid, liquid, gas, or vapor to dissolve in water or other specified medium. The ability of one material to blend uniformly with another, as in a solid in liquid, liquid in liquid, gas in liquid, or gas in gas.

Solution Mixture in which all ingredients are completely dissolved. Solutions are composed of a solvent (water or another liquid) and a dissolved substance (known as the solute).

SOP See Standard Operating Procedure.

Specialist Employee Employees who, in the course of their regular job duties, work with and are trained in the hazards of specific hazardous substances; they may be called upon to provide technical advice or assistance to the Incident Commander at a hazmat incident.

Specific Gravity The weight of a material as compared with the weight of an equal volume of water. If the specific gravity is less than one, the material is lighter than water and will float. If the specific gravity is greater than one, the material is heavier than water and will sink.

Specification Marking Found in various locations on railroad tank cars, intermodal portable tanks, and cargo tank trucks, it indicates the standards to which the container was built.

Spill The release of a liquid, powder, or solid hazardous material in a manner that poses a threat to air, water, ground, and the environment.

Stabilization The point in an incident at which the adverse behavior of the hazardous material/s is controlled.

Staging The management of committed and uncommitted emergency response resources (personnel and apparatus) to provide orderly deployment. The safe area established for temporary location of available resources closer to the incident site to reduce response time. See Level I Staging and Level II Staging.

Staging Area The designated location where emergency response equipment and personnel are assigned on an immediately-available basis until they are needed.

Standard of Care The minimum accepted level of hazmat service to be provided as may be set forth by law, current regulations, consensus standards, local protocols and practice, and what has been accepted in the past (precedent).

Standard Transportation Commodity Code (STCC) A number found on all shipping documents accompanying rail shipments of hazmat. A seven-digit number assigned to a specific material or group of materials and used in determination of rates. For a hazardous material, the STCC number will begin with the digits "49." Hazardous wastes may also be found with the first two digits being "48." It will also be found when intermodal containers are changed from rail to highway movement.

State Emergency Response Commission (SERC) Formed under SARA, Title III, the SERC is responsible for developing and maintaining the statewide hazmat emergency response plan.

Statement of Practical Treatment Located near the signal word on the front panel of an agricultural chemical or poison label, it is also referred to as the "First Aid Statement" or "Note to Physician." It may have precautionary information as well as emergency procedures. Antidote and treatment information may also be added.

Static Electricity An accumulated electrical charge. In order for static electricity to act as an ignition source, four conditions must be fulfilled:

- There must be an effective means of static generation.
- There must be a means of accumulating the static charge build-up.
- There must be a spark discharge of adequate energy to serve as an ignition source (i.e., incendive spark).
- The spark must occur in a flammable mixture.

Sterilization The process of destroying all microorganisms in or on an object. Common methods include steam, concentrated chemical agents, or ultraviolet light radiation.

Storm Sewer An "open" system that collects storm water, surface water, and ground water from throughout an area, but excludes domestic wastewater and industrial wastes. A storm sewer may dump runoff directly into a retention area which is normally dry or into a stream, river or waterway without treatment.

Strategic Goals The general plan used to control an incident. Strategic goals are broad in nature and are achieved by the completion of tactical objectives. Examples include rescue, spill control, leak control, and recovery.

Street Burn Deformation in the shell of a highway cargo tank. It is actually a long dent that is inherently flat. A street burn is generally caused by a container overturning and sliding some distance along a cement or asphalt road.

Strength The degree to which a corrosive ionizes in water. Those that form the greatest number of hydrogen ions are the strongest acids (e.g., pH < 2), while those that form the greatest number of hydroxide ions are the strongest bases (pH > 12).

Stress Event An applied force or system of forces that tend to either strain or deform a container (external action) or trigger a change in the condition of the contents (internal action). Types of stress include thermal, mechanical, and chemical.

Structural Firefighting Protective Clothing Protective clothing normally worn by fire fighters during structural firefighting operations. It includes a helmet, coat, pants, boots, gloves, PASS device, and a hood to cover parts of the head not protected by the helmet. Structural firefighting clothing provides limited protection from heat, but may not provide adequate protection from harmful liquids, gases, vapors, or dusts encountered during hazmat incidents.

Sublimation The ability of a substance to change from the solid to the vapor phase without passing through the liquid phase. An increase in temperature can increase the rate of sublimation. Significant in evaluating the flammability or toxicity of any released materials which sublime. The opposite of sublimation is deposition (changes from vapor to solid).

Subsidiary Hazard Class Indicates a hazard of a material other than the primary hazard assigned.

Superfund Amendments & Reauthorization Act (SARA) Created for the purpose of establishing Federal statutes for right-to-know standards, emergency response to hazardous materials incidents, re-authorized the Federal Superfund program, and mandated states to implement equivalent regulations/requirements.

Supplied Air Respirator (SAR) Positive-pressure respirator supplied by either an airline hose or breathing air cylinders connected to the respirator by a short airline (or pigtail). When used in IDLH atmospheres, it requires a secondary source of air supply.

Synergistic Effect The combined effect of two or more chemicals, which is greater than the sum of the effect of each agent alone.

System Detection Limit (SDL) The minimum amount of chemical breakthrough that can be detected by a laboratory's analytical system being used for the permeation test. Lower SDLs result in lower (or earlier) breakthrough times.

Systemic Pertaining to the internal organs and structures of the human body.

Systemic Effect The health effects of a hazardous materials exposure when the material enters the bloodstream and attacks target organs and internal areas of the human body.

Tactical Objectives The specific operations that must be accomplished to achieve strategic goals. Tactical objectives must be both specific and measurable.

Tank Container An intermodal container for transporting liquids, solids, and gases in bulk with a capacity of 118.9 gallons (450 L) or more and includes Multiple Element Gas Containers (MEGC), also called tube modules, which are the high-pressure equivalent of tank containers.

Technical Decontamination The planned and systematic process of reducing contamination to a level As Low As Reasonably Achievable (ALARA). Technical decon operations are normally conducted in support of emergency responder recon and entry operations at a hazardous materials incident, as well for handling contaminated patients at medical facilities.

Technical Information Centers Private and public sector hazardous materials emergency "hotlines" that (1) provide immediate chemical hazard information; (2) access secondary forms of expertise for additional action and information; and (3) act as a clearinghouse for spill notifications. Includes both public (e.g., CHEMTREC) and subscription-based systems.

Technical Information Specialists Individuals who provide specific expertise to the Incident Commander or the HMRT either in person, by telephone, or through other electronic means. They may represent the shipper, manufacturer, or be otherwise familiar with the hazmat or problems involved.

Temporary Emergency Exposure Limit (TEEL) TEELs are temporary levels of concern designed to be used as toxic exposure limits for chemicals for which AEGLs or ERPGs have not yet been defined. Like AEGLs and ERPGs, they are designed to represent the predicted response of members of the general public to different concentrations of a chemical during an incident. TEELs do not incorporate safety margins. There are four TEEL Levels.

Teratogen A material that affects the offspring when the embryo or fetus is exposed to that material.

Termination That portion of incident management where personnel are involved in documenting safety procedures, site operations, hazards faced, and lessons learned from the incident. Termination is divided into three phases: debriefing, post-incident analysis, and critique.

Thermal Harm Events Those harm events related to exposure to temperature extremes.

Thermal Stress Hazmat container stress generally indicated by temperature extremes, both hot and cold. Examples include fire, sparks, friction or electricity, and ambient temperature changes. Extreme or intense cold, such as those found with cryogenic materials, may also act as a stressor.

Threshold The point where a physiological or toxicological effect begins to be produced by the smallest degree of stimulation.

Threshold Limit Value/Ceiling (TLV/C) The maximum concentration that should not be exceeded, even instantaneously. The lower the value, the more toxic the substance.

Threshold Limit Value/Short Term Exposure Limit (TLV/STEL) The 15-minute, time-weighted average exposure which should not be exceeded at any time, nor repeated more than four times daily with a 60-minute rest period required between each STEL exposure. The lower the value, the more toxic the substance.

Threshold Limit Value/Skin (TLV/Skin) Indicates a possible and significant contribution to overall exposure to a material by absorption through the skin, mucous membranes, and eyes by direct or airborne contact.

Threshold Limit Value/Time Weighted Average (TLV/TWA) The airborne concentration of a material to which an average, healthy person may be exposed repeatedly for 8 hours each day, 40 hours per week, without suffering adverse effects. TLVs are based upon current available information and are adjusted on an annual basis by organizations such as the American Conference of Governmental Industrial Hygienists (ACGIH).

Threshold Planning Quantity (TPQ) The quantity designated for each extremely hazardous substance (EHS) that triggers a required notification from a facility to the State Emergency Response Commission (SERC) and the Local Emergency Planning Committee (LEPC) that the facility is subject to reporting under SARA, Title III.

TOFC See Trailer-on-Flat-Car.

Toxic Products of Combustion The toxic byproducts of the combustion process. Depending upon the materials burning, higher levels of personal protective clothing and equipment may be required.

Toxicity The ability of a substance to cause injury to a biologic tissue. Refers to the ability of a chemical to harm the body once contact has occurred.

Toxicity Harm Events Those harm events related to exposure to toxins. Examples include neurotoxins, nephrotoxins, and hepatotoxins.

Toxicity Signal Words The signal word found on product labels of poisons and agricultural chemicals that indicates the relative degree of acute toxicity. Located in the center of the front label panel, it is one of the most important label markings. The three toxicity signal words and categories are DANGER (high), WARNING (medium), and CAUTION (low).

Toxicology The study of chemical or physical agents that produce adverse responses in the biologic systems with which they interact.

Trailer-on-Flat-Car (TOFC) Truck trailers shipped on a railroad flat car.

Transfer The process of physically moving a liquid, gas, or some forms of solids either manually, by pump, or by pressure transfer from a leaking or damaged container.

Transportation Index (TI) The number found on radioactive labels which indicates the maximum radiation level (measured in milli-roentgens/hour or mR/hr) at 1 meter from an undamaged package. For example, a TI of 3 would indicate the radiation intensity that can be measured is no more than 3 mR/hr at 1 meter from the labeled package.

Type-A Packaging Packaging used to transport small quantities of radioactive material with higher concentrations of radioactivity than those shipped in Industrial Packaging. Designed to ensure the package retains its containment integrity and shielding under normal transport conditions. However, they are NOT designed to withstand the forces of an accident.

Type-B Packaging Packaging used to transport radioactive material with the highest levels of radioactivity, including potentially life-endangering amounts that could pose a significant risk if released during an accident. Must meet all of the Type A requirements, as well as a series of tests which simulate severe or "worst case" accident conditions. Accident conditions are simulated by performance testing and engineering analysis.

Type C Packaging Packaging used to transport by aircraft high-activity radioactive materials that have not been certified as "low dispersible radioactive material" (including plutonium). They are designed to withstand severe accident conditions associated with air transport without loss of containment or significant increase in external radiation levels. The Type C packaging performance requirements are significantly more stringent than those for Type B packaging and is not authorized for domestic use but can be authorized for international shipments of high-activity radioactive material consignments.

Type I Discharge Outlet A devise that conducts and delivers foam onto the burning surface of a liquid without submerging the foam or agitating the surface; for example, a foam trough.

Type II Discharge Outlet A device that delivers foam onto the burning liquid, partially submerges the foam, and produces restricted agitation of the surface; for example, a foam chamber.

Type III Discharge Outlet A device that delivers foam so that it falls directly onto the surface of the burning liquid in a manner that causes general agitation; for example, lobbing with a foam nozzle.

Unified Command The process of determining overall incident strategies and tactical objectives by having all agencies, organizations or individuals who have jurisdictional responsibility, and in some cases those who have functional responsibility at the incident, participate in the decision-making process.

Unified Commanders (UC) Command-level representatives from each of the primary responding agencies who present their agency's interests as a member of a unified command team. Depending upon the scenario and incident timeline, they may be the "lead" Incident Commander or play a supporting role within the command function.

Underflow Dam Spill control tactic used to trap floating, lighter-than-water materials behind the dam (specific gravity < 1). Using PVC piping or hard sleeves, the dam is constructed in a manner that allows uncontaminated water to flow unobstructed under the dam while keeping the contaminant behind the dam. Operationally, this is most effective on slow moving and relatively narrow waterways.

UN/NA Identification Number The four-digit identification number assigned to a hazardous material by the Department of Transportation; on shipping documents may be found with the prefix "UN" (United Nations) or "NA" (North American). The ID numbers are not unique and more than one material may have the same ID number.

Unsaturated Hydrocarbons A hydrocarbon with at least one multiple bond between two carbon atoms somewhere in the molecule. Generally, unsaturated hydrocarbons are more active chemically than saturated hydrocarbons, and are considered more hazardous. May also be referred to as the alkenes and alkynes. Examples include ethylene (C2H4), butadiene (C4H6), and acetylene (C2H2).

Vacuuming A physical method of confinement by which a hazardous material is placed in a chemically compatible container by simply vacuuming it up. The method of vacuuming will depend on the hazmat involved. Vacuuming is also a physical method of decontamination.

Vapor An air dispersion of molecules in a substance that is normally a liquid or solid at standard temperature and pressure.

Vapor Density The weight of a pure vapor or gas compared with the weight of an equal volume of dry air at the same temperature and pressure. The molecular weight of air is 29. If the vapor density of a gas is less than one, the material is lighter than air and may rise. If the vapor density is greater than one, the material is heavier than air and will collect in low or enclosed areas. Significant physical property for evaluating exposures and where hazmat gas and/or vapor will travel.

Vapor Dispersion A physical method of confinement by which water spray or fans is used to disperse or move vapors away from certain areas or materials. It is particularly effective on water-soluble materials (e.g., anhydrous ammonia), although the subsequent runoff may involve environmental trade-offs.

Vapor Pressure The pressure exerted by the vapor within the container against the sides of a container. This pressure is temperature dependent; as the temperature increases, so does the vapor pressure.

Vapor Suppression A physical method of confinement to reduce or eliminate the vapors emanating from a spilled or released material. Operationally, it is an offensive technique used to mitigate the evolution of flammable, corrosive, or toxic vapors and reduce the surface area exposed to the atmosphere.

Vent and Burn The use of shaped explosive charges to vent the high pressure at the top of a pressurized container and then, with additional explosive charges, release and burn the remaining liquid in the container in a controlled fashion.

Venting The controlled release of a liquid or compressed gas to reduce the pressure and diminish the probability of an explosion. The method of venting will depend on the nature of the hazmat.

Vesicants (Blister Agents) Chemical agents that pose both a liquid and vapor threat to all exposed skin and mucous membranes. These are exceptionally strong irritants capable of causing extreme pain and large blisters upon contact. Examples include mustard, lewisite, and phosgene oxime.

Violent Rupture Associated with chemical reactions having a release rate of less than one second (i.e., deflagration). There is no time to react in this scenario. This behavior is commonly associated with runaway cracking and over-pressure of closed containers.

Viscosity Measurement of the thickness of a liquid and its ability to flow. High viscosity liquids, such as heavy oils, must first be heated to increase their fluidity. Low viscosity liquids spread more easily and increase the size of the hazard area.

Volatility The ease with which a liquid or solid can pass into the vapor state. The higher a material's volatility, the greater its rate of evaporation.

Warm Zone The area where personnel and equipment decontamination and hot zone support takes place. It includes control points for the access corridor and thus assists in reducing the spread of contamination. This is also referred to as the "decontamination," "contamination reduction," "yellow zone," "support zone," or "limited access zone."

Water Reactivity Ability of a material to react with water and release a flammable gas or present a health hazard.

Weapon of Mass Destruction (WMD) (1) Any destructive device, such as any explosive, incendiary, or poison gas bomb, grenade, or rocket having a propellant charge of more than four ounces, missile having an explosive or incendiary charge of more than one quarter ounce (7 grams), mine, or device similar to the above. (2) Any weapon involving toxic or poisonous chemicals. (3) Any weapon involving a disease organism. (4) Any weapon designed to release radiation or radioactivity at a level dangerous to human life. See 18 USC 2332a.

Waybill A railroad shipping document describing the materials being transported. Indicates the shipper consignee, routing, and weights. Used by the railroad for internal records and control, especially when the shipment is in transit.

Wheel Burn Reduction in the thickness of a railroad tank shell. It is similar to a score, but is caused by prolonged contact with a turning railcar wheel.

Abbreviations and Acronyms

The following list is alphabetized by abbreviation. For more information, cross reference the terminology following the abbreviation/acronym with the glossary or use the index to reference back to the discussion in the respective chapters.

AAR **After Action Report**

ACC **American Chemistry Council**

ACGIH **American Conference of Governmental Industrial Hygienists**

ACP **Area Contingency Plan**

AEGL **Acute Exposure Guideline Level**

AFFF **Aqueous Film Forming Foam**

AHJ **Authority Having Jurisdiction**

AIHA **American Industrial Hygiene Association**

ALARA **As Low as Reasonably Achievable**

ALOHA **Aerial Locations of Hazardous Atmospheres**

ALS **Advanced Life Support**

ANSI **American National Standards Institute**

API **American Petroleum Institute**

APR **Air Purifying Respirator**

AR-AFFF **Alcohol-Resistant Aqueous Film Forming Foam**

ARC **Alcohol Resistant Concentrate**

ARFF **Aircraft Rescue and Firefighting**

ASME **American Society of Mechanical Engineers**

ASO **Assistant Safety Officer**

ASTM **American Society for Testing and Materials**

AT/FP **Anti-Terrorism/Force Protection**

ATSDR **Agency for Toxic Substances and Disease Registry**

BLEVE **Boiling Liquid Expanding Vapor Explosion**

BLS **Basic Life Support**

BP **Blood Pressure**

Bq **Becquerel**

C **Centigrade**

CAA **Clean Air Act**

CAMEO **Computer Assisted Management of Emergency Operations**

CANUTEC **Canadian Transport Emergency Centre**

CAP **Civil Air Patrol**

CBRNR **Chemical, Biological, Radiological, Nuclear, and Explosive Materials**

CCTV **Closed Circuit Television**

CDC **Centers for Disease Control**

CEPPO **Chemical Emergency Preparedness and Prevention Office**

CERCLA **Comprehensive Environmental Response, Compensation, and Liability Act**

CFAT **Chemical Facility Anti-Terrorism Standards**

CFR **Code of Federal Regulations**

CGA **Compressed Gas Association**

cGy **centi-Gray**

CHEMNET **Chemical Industry Emergency Mutual Aid Network**

CHEMTREC **Chemical Transportation Emergency Center**

Ci **Curie**

COMMS **Communications**

COP **Common Operating Picture**

CNG **Compressed Natural Gas**

CNS **Central Nervous System**

CPC **Chemical Protective Clothing**

CPG **Comprehensive Preparedness Guide**

cpm **Counts per Minute**

CRM **Crew Resource Management**

CSB **U.S. Chemical Safety Board**

CTN **Critical Transportation Needs**

dBA **Decibels on the A-Weighted Scale**

DECON **Decontamination**

DEQ **Department of Environmental Quality**

DHS **Department of Homeland Security**

DOD **Department of Defense**

DOT **Department of Transportation**

dps **Disintegrations per Second**

EAS **Emergency Alerting System**

EBA **Escape Breathing Apparatus**

EFV **Excess Flow Valve**

EHS **Extremely Hazardous Substance**

EH&S **Environmental Health & Safety**

EKG **Electrocardiogram**

EMS **Emergency Medical Services**

EMT-B **Emergency Medical Technician—Basic**

EMT-I **Emergency Medical Technician—Intermediate**

EMT-P Emergency Medical Technician—Paramedic

EOC Emergency Operations Center

EOD Explosive Ordinance Disposal

EOP Emergency Operations Plan

EPA Environmental Protection Agency

EPCRA Emergency Planning and Community Right-to-Know Act

ERG Emergency Response Guidebook

ERP Emergency Response Plan

ERPG Emergency Response Planning Guideline

ERT Emergency Response Team

ERT Evidence Recovery Team

ESD Emergency Shutdown

ESLI End-of-Service Life Indicator

eV Electron Volts

F Fahrenheit

FAA Federal Aviation Administration

FBI Federal Bureau of Investigation

FEMA Federal Emergency Management Agency

FFFP Film Forming Fluoroprotein Foam

FMECA Failure Modes, Effects, and Criticality Analysis

FOG Field Operations Guide

FOSC Federal On-Scene Coordinator

FRA Federal Railroad Administration

FRA First Responder Awareness

FRO First Responder Operations

FTIR Fourier Transform Infrared Spectrometry

GC Gas Chromatograph

GEBMO General Hazardous Materials Behavior Model

GHS Global Harmonization System (of Classification and Labeling of Chemicals)

Gy Gray

HAZCOM Hazard Communication

HAZMAT Hazardous Materials

HAZOP Hazard and Operability Study (or Analysis)

HAZWOPER Hazardous Waste Operations and Emergency Response

HEPA High-Efficiency Particulate Air

HMT Hazardous Materials Technician

HMRT Hazardous Materials Response Team

HSPD Homeland Security Presidential Directive

IAP Incident Action Plan

IAEM International Association of Emergency Managers

IARC International Agency for Research on Cancer

IC Incident Commander

ICP Incident Command Post

ICP Integrated Contingency Plan

ICS Incident Command System

IDLH Immediately Dangerous to Life or Health

IED Improvised Explosive Device

IMS Incident Management System

IMT Incident Management Team

IND Improvised Nuclear Device

IP Improvement Plan

IP Ionization Potential

IST Incident Support Team

JIC Joint Information Center

kg Kilogram

L Liter

Lbs Pounds

LC Lethal Concentration

LD Lethal Dose

LDH Large Diameter Hose

LEPC Local Emergency Planning Committee

LNG Liquefied Natural Gas

LNO Liaison Officer

LOC Level of Concern

LPG Liquefied Petroleum Gas

μg Microgram

m^3 Cubic Meter

MARPLOT Mapping Applications for Response, Planning, and Local Operational Tasks

MDPR Minimum Detectable Permeation Rate

MEG Multi Element Gas Container

mg Milligram

mm Hg Millimeter of Mercury

MOU Memorandum of Understanding

MOTEL Magnitude, Occurrence, Timing, Effects, and Location

mR Milliroentgens

Mrem Millirem
MS Mass Spectrometer
MSST Maximum Safe Storage Temperature
mSv Millisievert

NAPCC National Animal Poison Control Center
NCP National Contingency Plan
NEC National Electrical Code
NFIRS National Fire Incident Reporting System
NFPA National Fire Protection Association
NIJ National Institute of Justice
NIMS National Incident Management System
NIOSH National Institute of Occupational Safety and Health
NOAA National Oceanic and Aeronautical Administration
NOP Next Operational Period
NPIC National Pesticide Information Center
NRC National Response Center
NRP National Response Plan
NRT National Response Team
NTP National Toxicology Program
NTSB National Transportation Safety Board

OHME Office of Hazardous Materials Enforcement (DOT/PHMSA)
OHMS Office of Hazardous Materials Safety (DOT/PHMSA)
OIM Off-Shore Installation Manager
OPA Oil Pollution Act (of 1990)
OPS Office of Pipeline Safety
OPSEC Operations Security
OSC On-Scene Coordinator
OSHA Occupational Safety and Health Administration

PACE Primary Plan, Alternate Plan, Contingency Plan, Emergency Plan
PAPR Powered Air-Purifying Respirator
PAR Personal Accountability Review
PASS Personal Alert Safety System
PCB Polychlorinated Biphenyls
PEL Permissible Exposure Limit
PERO Post-Emergency Response Operations
pH Power of Hydrogen
PHMSA Pipeline and Hazardous Materials Safety Administration (DOT)
PIA Post Incident Analysis
PID Photo Ionization Detectors
PIO Public Information Officer
PNA Polynuclear Aromatic Compounds
PPA Public Protective Actions
ppb Parts per Billion
PPE Personal Protective Equipment
ppm Parts per Million
PPV Positive Pressure Ventilation
PRV Pressure Relief Valve
PSAP Public Safety Answering Point
PSM Process Safety Management
PVA Polyvinyl Alcohol
PVC Polyvinyl Chloride

QAQC Quality Assurance and Quality Control

R Roentgen
RAD Radiation Absorbed Dose
RCRA Resource Conservation and Recovery Act
RDD Radiation Dispersal Device
RDECOM Research, Development, and Engineering Command (U.S. Army)
RED Radiation Exposure Device
REHAB Rehabilitation (Area)
REL Recommended Exposure Levels
REM Roentgen Equivalent Man
RIT Rapid Intervention Team
RMP Risk Management Program
ROE Rules of Engagement
RP Responsible Party
RPM Remedial Project Manager
RPP Respiratory Protection Program
RQ Reportable Quantity
RRT Regional Response Team

SADT Self-Accelerating Decomposition Temperature
SAR Supplied Air Respirator

SARA Superfund Amendments and Reauthorization Act (of 1986)

SBCCOM Soldier and Biological Chemical Command (U.S. Army)

SBS Sick Building Syndrome

SCBA Self-Contained Breathing Apparatus

SDL System Detection Limit

SDS Safety Data Sheet

SEI Safety Equipment Institute

SERC State Emergency Response Commission

SETIQ Emergency Transportation System for the Chemical Industry (Mexico)

SFC Structural Firefighting Clothing

SI International System (of Units)

SOP Standard Operating Procedure

SONS Spill of National Significance

SOSC State On-Scene Coordinator

STAM Staging Area Manager

START Simple Triage and Rapid Treatment

STEL Short-Term Exposure Limit

Sv Sievert (Sv)

SWAT Special Weapons and Tactics (Team)

TLV Threshold Limit Value

TLV/C Threshold Limit Value/Ceiling

TLV/STEL Threshold Limit Value/Short-Term Exposure Limit

TLV/TWA Threshold Limit Value/Time-Weighted Average

TOXNET Toxicology Data Network

TRACEM Thermal, Radiological, Asphyxiation, Chemical, Ethologic, Mechanical

TSA Transportation Safety Administration (DHS)

UC Unified Command

USCG United States Coast Guard

US&R Urban Search and Rescue

UV Ultraviolet

UL Underwriters Laboratories

VOC Volatile Organic Compounds

WISER Wireless Information System for Emergency Responders

WMD Weapon of Mass Destruction

Index

D

E

F

G

O

S

Photo Credits

Chapter 1

Figure 1-1 Courtesy of Don Sellers; **1-2** FEMA CPG 101 (Version 2.0).

Chapter 2

Figure 2-1 Courtesy of DuPont CPC; **2-4** © HENRIK G. DE GYOR/AP Photos.

Chapter 3

Figure 3-5 Courtesy of Eric Bachman.

Chapter 4

Opener © TFoxFoto/ShutterStock, Inc.

Chapter 5

Opener © FIORENTINI MASSIMO/ShutterStock, Inc.; **5-4** Courtesy of Rebecca Crawford, Dragonfi Photography; **5-5** © Ric Feld/AP Photos; **5-8** Courtesy of National Transportation Safety Board; **Case Study 1** Courtesy of National Transportation Safety Board, HAR-77/01, NTIS Number: PB-268251.; **5-9** © Kyodo/AP Photos; **5-10** © AP/World Wide; **5-11** © Aiken County Sheriff's Dept/AP Photos; **5-13** Courtesy of William McEnroe; **5-14** Courtesy of Steve Danzig.

Chapter 6

Opener Courtesy of Bill Hand.

Chapter 7

Figure 7-3 Courtesy of Alexeter Technologies, Courtesy of BioFire Defense, Courtesy of Thermo Fisher Scientific; **Figure 7-5, Table 7-3, Table 7-4** Ludwig Benner, Jr., 1978. Hazardous Materials Emergencies, 2nd edition, Oakton, VA; **Figure 7-6** Richard Emery; **7-7** John Spaulding.

Chapter 8

Opener Courtesy of Tempe Fire Department; **8-1** Courtesy of Jerry Grey; **8-6A** DuPont® Personal Protection; **8-6B** Courtesy of Chris Hawley; **8-9A&B** DuPont.

Chapter 10

Opener Courtesy of the University of Nevada, Reno Fire Science Academy; **10-3** Courtesy of OSHA; **10-10** Courtesy of NOAA; **Scan Sheet 10-B** *both* Safe Transportation Safety Specialists, LLC; **10-19** Ray Haring; **10-20** © AP/Wide World; **10-21** Courtesy of Bill Hand; **Case Study** *center, bottom* © PA Wire/AP Photos; **Scan Sheet 10-D** Quick Reference Guide: 1) "Fuel Grade Ethanol Spills (Including E-85)" by the Ethanol Emergency Response Coalition and 2) "Evaluating the Use of Firefighting Foam," Fire Engineering (February 1986), pages 44-49; **10-26** Reprinted with permission from Loss Control Associates, Inc; **Scan Sheet 10-G** *left* Marsh, The 100 Largest Losses 1972–2001 .2) Special Report On The San Juan Ixhuatepec Mexico LPG Disaster, prepared by B. F. Olson for the American Petroleum Institute, December 28, 1984; *right* Courtesy of the National Transportation Safety Board; **Scan Sheet 10-H** Propane Emergencies (3rd edition), by Michael S. Hildebrand and Gregory G. Noll, Chester, MD: Red Hat Publishing (2007). For a free copy in PDF format go to www.propanesafety.com; **10-35** Courtesy of Bill Hand; 10-36 Courtesy of Ray Haring; **10-37** Courtesy of Bill Hand; **10-38** Courtesy of Phil Baker.

Chapter 11

11-4 Courtesy of Chris Hawley; **page 401** © Kyodo/AP Photos; **11-6** Courtesy of Glen Rudner; **Case Study 1** *bottom right* Courtesy of Photographer's Mate 1st Class William R. Goodwin/U.S. Navy; *bottom left* Courtesy of Airman 1st Class Scherrie Gates/U.S. Air Force.

Chapter 12

© TFoxFoto/ShutterStock, Inc.

Unless otherwise indicated, all photographs and illustrations are under copyright of Jones & Bartlett Learning, courtesy of Maryland Institute for Emergency Medical Services Systems, the National Fire Protection Association, or have been provided by the authors.